2018 IEEE Asian Solid-State Circuits Conference (A-SSCC 2018)

Tainan, Taiwan
5-7 November 2018

IEEE Catalog Number: CFP18SSC-POD
ISBN: 978-1-5386-6414-8

Copyright © 2018 by the Institute of Electrical and Electronics Engineers, Inc.
All Rights Reserved

Copyright and Reprint Permissions: Abstracting is permitted with credit to the source. Libraries are permitted to photocopy beyond the limit of U.S. copyright law for private use of patrons those articles in this volume that carry a code at the bottom of the first page, provided the per-copy fee indicated in the code is paid through Copyright Clearance Center, 222 Rosewood Drive, Danvers, MA 01923.

For other copying, reprint or republication permission, write to IEEE Copyrights Manager, IEEE Service Center, 445 Hoes Lane, Piscataway, NJ 08854. All rights reserved.

*** *This is a print representation of what appears in the IEEE Digital Library. Some format issues inherent in the e-media version may also appear in this print version.*

IEEE Catalog Number: CFP18SSC-POD
ISBN (Print-On-Demand):   978-1-5386-6414-8
ISBN (Online):   978-1-5386-6413-1

Additional Copies of This Publication Are Available From:

Curran Associates, Inc
57 Morehouse Lane
Red Hook, NY 12571 USA
Phone: (845) 758-0400
Fax: (845) 758-2633
E-mail: curran@proceedings.com
Web: www.proceedings.com

Table of Contents

P 1

Plenary Session 1

1-1 (8233) Circuit Design in Nano-Scale CMOS Technologies
Kevin Zhang
Taiwan Semiconductor Manufacturing Company, Taiwan V

1-2 (8234) Open the New World of 5G
Seizo Onoe
NTT DOCOMO, INC. and DOCOMO Technology, Inc., Japan IX

Industry 2

Advanced Techniques for Industrial Applications

2-1 (8063) A 12.4TOPS/W, 20% Less Gate Count Bidirectional Phase Domain MAC Circuit for DNN Inference Applications
Yosuke Toyama, Kentaro Yoshioka, Koichiro Ban, Akihide Sai, and Kohei Onizuka
Toshiba Corporation, Japan 1

2-2 (8095) A Power and Area Efficient 2.5-16 Gbps Gen4 PCIe PHY in 10nm FinFET CMOS
Shenggao Li, Fulvio Spagna, Ji Chen, Xiaoqing Wang, Luke Tong, Sujatha Gowder, Wenyan Jia, Roan Nicholson, Rui Song, Lily Li, Meng-hung Chen, Sita Iyer, Amanda Tran, Michael De Vita, Deepar Govindrajan, Marcus Pasquarella, David Bradley, Frank Verdico, Matt Duwe, Eric Lee, and Michelle Wigton
Intel Corporation, United States 5

2-3 (8028) 40-nm 64-kbit Buffer/Backup SRAM with 330 nW Standby Power at 65°C Using 3.3 V IO MOSs for PMIC Less MCU in IoT Applications
Yoshisato Yokoyama, Tomohiro Miura, Yukari Ouchi, Daisuke Nakamura, Jiro Ishikawa, Shunya Nagata, Makoto Yabuuchi, Yuichiro Ishii, and Koji Nii
Renesas Electronics Corporation, Japan 9

2-4 (8069) Logic Process Compatible 40nm 256Kx144 Embedded RRAM with Low Voltage Current Limiter and Ambient Compensation Scheme to Improve the Read Window
Chien-An Lai, Chung-Cheng Chou, Chi-Hsiang Weng, Zheng-Jun Lin, Pei-Ling Tseng, Chien-Fan Wang, Chih-Chen Wang, Chin-I Su, Wei-Chi Chen, Yu-Cheng Lin, Tong-Chern Ong, Chi Chang, Yu-Der Chih, and Tsung-Yung Chang
Taiwan Semiconductor Manufacturing Company, Taiwan 13

ETA 3

Intelligent Sensor and Imager Systems

3-1 (8161) **A 10-Bit 1026-Channel Column Driver IC with Partially Segmented Piecewise Linear Digital-to-Analog Converters for Ultra-High-Definition TFT-LCDs with One Billion Color Display**
Chih-Wen Lu[1], You-Gang Chang[1], Xing-Wei Huang[1], Jhih-Siou Cheng[2], Po-Yu Tseng[2], and Chih-Hsien Chou[2]
[1]National Tsing Hua University, Taiwan
[2]Novatek Microelectronics Corp., Taiwan 17

3-2 (8080) **A WDR CMOS Image Sensor Employing In-Pixel Capacitive Variation Using a Re-configurable Source Follower for Low Light Applications**
Neha Priyadarshini, Chandani Anand, and Mukul Sarkar
Indian Institute of Technology, Delhi, India 21

3-3 (8147) **A CMOS Imager for Reflective Pulse Oximeter with Motion Artifact and Ambient Interference Rejections**
Hsiang-Lin Chen, Sung-En Hsieh, Tzu-Hsiang Hsu, and Chih-Cheng Hsieh
National Tsing Hua University, Taiwan 25

3-4 (8205) **A 4-Channel 5.04 µW 0.325 mm² Orthogonal Sampling-Based Parallel Neural Recording System**
Reza Ranjandish and Alexandre Schmid
Swiss Federal Institute of Technology, Switzerland 27

ACS 4

Power Converters and Sensors

4-1 (8144) **An Integrated DC-DC Converter with Segmented Frequency Modulation and Multiphase Co-Work Control for Fast Transient Recovery**
U-Fat Chio[1], Kuo-Chih Wen[1], Sai-Weng Sin[1], Chi-Seng Lam[1], Yan Lu[1], Franco Maloberti[2], and Rui Paulo Martins[1]
[1]University of Macau, Macau
[2]University of Pavia, Italy 31

4-2 (8096) **An 88% Efficiency 2.4µW to 15.6µW Triboelectric Nanogenerator Energy Harvesting System Based on a Single-Comparator Control Algorithm**
Karim Rawy[1], Ruchi Sharma[1], Hong-Joon Yoon[2], Usman Khan[2], Sang-Woo Kim[2], and Tony Tae-Hyoung Kim[1]
[1]Nanyang Technological University , Singapore
[2]Sungkyunkwan University, Korea 33

4-3 (8141) **A Wide-Range Capacitive DC-DC Converter with 2D-MPPT for Soil/Solar Energy Extraction**
I-Che Ou[3], Jia-Ping Yang[3], Chia-Hung Liu[3], Kai-Jie Huang[3], Kun-Ju Tsai[1], Yu Lee[1], Yuan-Hua Chu[2], and Yu-Te Liao[3]
[1]Industrial Research and Technology Institute, Taiwan
[2]Industrial Technology Research Institute, Taiwan
[3]National Chiao Tung University, Taiwan 37

4-4 (8036) **A 13.56 MHz 88.7%-PCE Voltage Doubling Rectifier Using Adaptive Delay Time and Pulse-Width Control**
Ye-Sing Luo[2], Hsing-Hung Lin[1], and Shen-Iuan Liu[1]
[1]National Taiwan University, Taiwan
[2]Richtek Technology, Taiwan 39

4-5 (8091) **A 6800-µm² Resistor-Based Temperature Sensor in 180-nm CMOS**
Jan Angevare and Kofi Makinwa
Delft University of Technology, Netherlands 43

FPGA 5

FPGA-based AI Computing

5-1 (8128)   **FPGA-based CNN Processor with Filter-Wise-Optimized Bit Precision**
Asuka Maki, Daisuke Miyashita, Kengo Nakata, Fumihiko Tachibana, Tomoya Suzuki, and Jun Deguchi
Toshiba Memory, Japan ···47

5-2 (8064)   **An Asynchronous Energy-Efficient CNN Accelerator with Reconfigurable Architecture**
Weijia Chen[2], Hui Wu[2], Shaojun Wei[2], Anping He[1], and Hong Chen[2]
[1]Lanzhou University, China
[2]Tsinghua University, China ···51

5-3 (8213)   **Hardware Architecture for Fast General Object Detection Using Aggregated Channel Features**
Koichi Mitsunari, Jaehoon Yu, and Masanori Hashimoto
Osaka University, Japan ···55

5-4 (8025)   **A Neural Network Accelerator with Integrated Feature Extraction Processor for a Freezing of Gait Detection System**
Val Mikos[2], Chun-Huat Heng[2], Arthur Tay[2], Shih-Cheng Yen[2], Nicole Shuang Yu Chia[1], Karen Mui Ling Koh[1], Dawn May Leng Tan[3], and Wing Lok Au[1]
[1]National Neuroscience Institute, Singapore
[2]National University of Singapore, Singapore
[3]Singapore General Hospital, Singapore ·······································59

WLN 6

Optical Link and CDR

6-1 (8019)   **A 0.25-27Gb/s Wideband PAM4/NRZ Transceiver with Adaptive Power CDR for 8K System**
Yoshihide Komatsu[2], Akinori Shinmyo[1], Masami Funabashi[1], Shuji Kato[1], Kazuya Hatooka[1], Kenji Tanaka[2], Mayuko Fujita[1], and Kouichi Fukuda[1]
[1]Panasonic, Japan
[2]Panasonic Corporation, Japan ···63

6-2 (8221)   **A Fully-Integrated 25Gb/s Low-Noise TIA+CDR Optical Receiver Designed in 40nm-CMOS**
Juncheng Wang[2], Xuefeng Chen[3], Shang Hu[3], Yaxin Cai[3], Rui Bai[3], Xin Wang[3], Yuanxi Zhang[3], Shenglong Zhuo[3], Liu Chang[2], Bozhi Yin[2], Jianxu Ma[3], Hao Yan[3], Jiangao Xuan[3], Milton Lu[3], Tao Xia[3], Qi Nan[1], and Patrickyin Chiang[2]
[1]Chinese Academy of Sciences, China
[2]Fudan University, China
[3]PhotonIC Technology, China ···67

6-3 (8130)   **A Low Input Referred Noise and Low Crosstalk Noise 25 Gb/s Transimpedance Amplifier with Inductor-Less Bandwidth Compensation**
Akitaka Hiratsuka[1], Akira Tsuchiya[3], Kenji Tanaka[2], Hiroyuki Fukuyama[2], Naoki Miura[2], Hideyuki Nosaka[2], and Hidetoshi Onodera[1]
[1]Kyoto University, Japan
[2]NTT Device Technology Laboratories, NTT Corporation, Japan
[3]University of Shiga Prefecture, Japan ···69

6-4 (8012)   **A 10-Gb/s, 0.03-mm², 1.28-pJ/bit Half-Rate All-Digital Injection-Locked Clock and Data Recovery with Maximum Timing-Margin Tracking Loop**
Min-Seong Choo, Han-Gon Ko, Sung-Yong Cho, Kwangho Lee, and Deog-Kyoon Jeong
Seoul National University, Korea ···73

RF 7

Millimeter-Wave Transceivers and Terahertz Sensors

7-1 (8072) **A 28.16-Gb/s Area-Efficient 60Ghz CMOS Bi-Directional Transceiver for IEEE 802.11ay**

Jian Pang, Korkut Tokgoz, Shotaro Maki, Zheng Li, Xueting Luo, Ibrahim Abdo, Seitarou Kawai, Hanli Liu, Bangan Liu, Makihiko Katsuragi, Kento Kimura, Atsushi Shirane, and Kenichi Okada

Tokyo Institute of Technology, Japan ···77

7-2 (8157) **A 77-GHz Mixed-Mode FMCW Generator Based on a Vernier TDC with Dual Rising-Edge Fractional-Phase Detector**

Jianxi Wu[3], Zipeng Chen[3], Wei Zheng[2], Yibo Liu[3], Shufu Wang[2], Nan Qi[1], and Baoyong Chi[3]

[1]*Institute of Semiconductors, Chinese Academy of Sciences, China*
[2]*Radarchip Technology Co., Ltd., China*
[3]*Tsinghua University, China* ···79

7-3 (8184) **A CMOS 76-81 GHz 2TX 3RX FMCW Radar Transceiver Based on Mixed-Mode PLL Chirp Generator**

Taikun Ma[3], Zipeng Chen[3], Jianxi Wu[3], Wei Zheng[2], Shufu Wang[2], Nan Qi[1], and Baoyong Chi[3]

[1]*Institute of Semiconductors, Chinese Academy of Sciences, China*
[2]*Radarchip Technology Co., Ltd., China*
[3]*Tsinghua University, China* ···83

7-4 (8139) **A 25 fps 32×24 Digital CMOS Terahertz Image Sensor**

Tong Fang[2], Run-Jiang Dou[1], Li-Yuan Liu[1], Jian Liu[1], and Nan-Jian Wu[1]

[1]*Institute of Semiconductors, Chinese Academy of Sciences, China*
[2]*Institute of Semiconductors, Chinese Academy of Sciences / University of Chinese Academy of Sciences, China* ········87

PD 8

Panel Discussion

The Circuits and Systems for Mobile AI

P 9

Plenary Session 2

9-1 (8235) **Practical Challenges in Supporting Functions in Memory**

Nam Sung Kim

Samsung, Korea ·· XIII

9-2 (8236) **AI Drive Domain Specific Processors**

Yi Kang

UNISOC Technologies, China ·· XVII

ACS+DC 10

Analog and Data Converter Techniques

10-1 (8105) A 3.9μW, 81.3dB SNDR, DC-Coupled, Time-based Neural Recording IC with Degeneration R-DAC for Bidirectional Neural Interface in 180nm CMOS

Hyuntak Jeon[2], Jun-Suk Bang[3], Yoontae Jung[2], Taeju Lee[2], Yeseul Jeon[2], Seok-Tae Koh[2], Jaesuk Choi[2], Doojin Jang[2], Soonyoung Hong[1], and Minkyu Je[2]

[1]Daegu Gyeongbuk institute of Science and Technology, Korea

[2]Korea Advanced Institute of Science and Technology, Korea

[3]Samsung Electronics, Korea 91

10-2 (8074) A Second-Order Purely VCO-Based CT ΔΣ ADC Using a Modified DPLL in 40-nm CMOS

Yi Zhong[4], Shaolan Li[3], Arindam Sanyal[2], Xiyuan Tang[3], Linxiao Shen[3], Siliang Wu[1], and Nan Sun[3]

[1]Beijing Institute of Technology, China

[2]State University of New York at Buffalo, United States

[3]University of Texas, Austin, United States

[4]University of Texas, Austin / Beijing Institute of Technology, United Statesa 93

10-3 (8010) An 11b 1GS/s Time-Interleaved ADC with Linearity Enhanced T/H

Yan Zhu[1], Chi-Hang Chan[1], and Rui Paulo Martins[1,2]

[1]University of Macau, China

[2]Universidade de Lisboa, Portugal 95

10-4 (8162) A 1-V 3.1-ppm/°C 0.8-μW Bandgap Reference with Piecewise Exponential Curvature Compensation

Hongrui Luo, Quan Sun, Ruizhi Zhang, and Hong Zhang

Xi'an Jiaotong University, China 97

10-5 (8117) A 40nW, Sub-1V Truly 'Digital' Reverse Bandgap Reference Using Bulk-Diodes in 16nm FinFET

Matthias Eberlein[2], Georgios Panagopoulos[2], and Harald Pretl[1]

[1]Intel, Austria

[2]Intel, Germany 99

10-6 (8031) A 0.6V 1.63fJ/c.-s. Detective Open-Loop Dynamic System Buffer for SAR ADC in Zero-Capacitor TDDI System

Yao-Sheng Hu, Li-Yu Huang, and Hsin-Shu Chen

National Taiwan University, Taiwan 103

ETA 11

Technology and Circuit Techniques for IoT

11-1 (8189) Design of a 2.45-GHz RF Energy Harvester for SWIPT IoT Smart Sensors
Pengcheng Xu, Denis Flandre, and David Bol
Université catholique de Louvain, Belgium ·······107

11-2 (8136) A 6.78–200 MHz Offset-Compensated Active Rectifier with Dynamic Logic Comparator for mm-Size Wirelessly Powered Implants
Jianming Zhao and Yuan Gao
Agency for Science, Technology and Research, Singapore ·······111

11-3 (8060) Photovoltaic-Assisted Self-Vth-Cancellation CMOS RF Rectifier for Wide Power Range Operation
Ren Usami, Takao Komiyama, Yasunori Chonan, Hiroyuki Yamaguchi, and Koji Kotani
Akita Prefectural Univesity, Japan ·······115

11-4 (8182) Stable, Self-Biased and High-Gain Organic Amplifiers with Reduced Parameter Variation Effect
Masoud Seifaei[1], Daniel De Dorigo[1], David Ingvar Fleig[1], Matthias Kuhl[1], Ute Zschieschang[2], Hagen Klauk[2], and Yiannos Manoli[1]
[1]*Albert Ludwigs University of Freiburg, Germany*
[2]*Max Plank Institute for Solid State Research, Germany* ·······119

11-5 (8043) An Encryption-Authentication Unified A/D Conversion Scheme for IoT Sensor Nodes
Vinod Gadde, Hiromitsu Awano, and Makoto Ikeda
University of Tokyo, Japan ·······123

MEM 12

Intelligent Memory System

12-1 (8115) A 28nm 320Kb TCAM Macro with Sub-0.8ns Search Time and 3.5+x Improvement in Delay-Area-Energy Product Using Split-Controlled Single-Load 14T Cell
Cheng-Xin Xue[2], Wei-Cheng Zhao[2], Tzu-Hsien Yang[2], Yi-Ju Chen[2], Hiroyuki Yamauchi[1], and Meng-Fan Chang[2]
[1]*Fukuoka Institute of Technology, Japan*
[2]*National Tsing Hua University, Taiwan* ·······127

12-2 (8148) A 6.8TOPS/W Energy Efficiency, 1.5μW Power Consumption, Pulse Width Modulation Neuromorphic Circuits for Near-Data Computing with SSD
Kota Tsurumi, Kenta Suzuki, and Ken Takeuchi
Chuo University-Takeuchi Laboratory, Japan ·······129

12-3 (8228) A 28nm FD-SOI 4KB Radiation-hardened 12T SRAM Macro with 0.6 ~ 1V Wide Dynamic Voltage Scaling for Space Applications
Le Dinh Trang Dang, Dongkyu Seo, Jin-Woo Han, Jinsang Kim, and Ik-Joon Chang
Kyunghee University, Korea ·······133

12-4 (8049) Nonvolatile Crossbar 2D2R TCAM with Cell Size of 16.3 F^2 and K-means Clustering for Power Reduction

Keji Zhou[1], Xiaoyong Xue[1], Jianguo Yang[1], Xiaoxin Xu[2], Hangbing Lv[2], Mingyu Wang[1], Ming'e Jing[1], Wenjun Liu[1], Xiaoyang Zeng[1], Steve S. Chung[3], Jing Li[4], and Ming Liu[2]

[1]Fudan University, China

[2]Institute of Microelectronics of the Chinese Academy of Sciences, China

[3]National Chiao Tung University, Taiwan

[4]University of Wisconsin-Madison, United States 135

12-5 (8035) An Enhanced Built-Off-Test Transceiver with Wide-range, Self-calibration Engine for 3.2 Gb/s/pin DDR4 SDRAM

Joung-Wook Moon, Hye-Sung Yoo, Hundai Choi, Il-Won Park, Seok-Yong Kang, Jun-Bae Kim, Haeyoung Chung, Kiho Kim, Dong-Hun Lee, Ki-Jae Song, Seok-Hun Hyun, Indal Song, Young-Soo Sohn, Yong-Ho Cho, Jung-Hwan Choi, Kwang-Il Park, and Seong-Jin Jang

Samsung Electronics, Korea 139

12-6 (8166) 8T SRAM with Vertical Read Word Line and Data Aware Write Assist for Low Power Application

Lu Lu, Taegeun Yoo, Le Van Loi, and Tony Tae-Hyoung Kim

Nanyang Technological University, Singapore 143

DCS 13

Circuit Technologies for Security Enhancement

13-1 (8041) A 0.46V-1.1V Transition-Detector with In-Situ Timing-Error Detection and Correction Based on Pulsed-Latch Design in AES Accelerator

Xinchao Shang[1], Weiwei Shan[1], Jiaming Xu[1], Minyi Lu[1], Yiming Xiang[2], Longxing Shi[1], and Jun Yang[1]

[1]Southeast University, China

[2]Spreadtrum Communications, China 145

13-2 (8055) Ultra-Lightweight 548 – 1080 Gate 166Gbps/W – 12.6Tbps/W SIMON 32/64 Cipher Accelerators for IoT in 14nm Tri-Gate CMOS

Himanshu Kaul, Mark Anders, Sanu Mathew, Vikram Suresh, Sudhir Satpathy, Amit Agarwal, Steven Hsu, and Ram Krishnamurthy

Intel Corporation, United States 149

13-3 (8149) 31.3 µs/Signature-Generation 256-bit Fp ECDSA Cryptoprocessor

Shotaro Sugiyama, Hiromitsu Awano, and Makoto Ikeda

University of Tokyo, Japan 153

13-4 (8103) A Physically Unclonable Function with 0% BER Using Soft Oxide Breakdown in 40nm CMOS

Kai-Hsin Chuang[2], Erik Bury[1,] Robin Degraeve[1,] Ben Kaczer[1,] Dimitri Linten[1,] and Ingrid Verbauwhede[2]

[1]imec, Belgium

[2]KU Leuven, Belgium 157

13-5 (8011) A 373 F^2 2D Power-Gated EE SRAM Physically Unclonable Function with Dark-Bit Detection Technique

Kunyang Liu, Yue Min, Xuan Yang, Hanfeng Sun, and Hirofumi Shinohara

Waseda University, Japan 161

ACS 14

Inductive DC-DC Converters

14-1 (8065) An 82.1%-Power-Efficiency Single-Inductor Triple-Source Quad-Mode Energy Harvesting Interface with Automatic Source Selection and Reversely Polarized Energy Recycling
Chih-Lun Lo, Hao-Chung Cheng, Pei-Chun Liao, Yi-Lun Chen, and Po-Hung Chen
National Chiao Tung University, Taiwan ··165

14-2 (8138) A Transient-Enhanced Constant on-Time Buck Converter with Light-Load Efficiency Optimization
Mao-Ling Chiu, Tzu-Hsuan Yang, and Tsung-Hsien Lin
National Taiwan University, Taiwan ··169

14-3 (8112) A 99.2% Tracking Accuracy Single-Inductor Quadruple-Input-Quadruple-Output Buck-Boost Converter Topology with Periodical Interval Perturbation and Observation MPPT
Chao-Jen Huang[4], Yao-Sheng Ma[3], Wen-Hau Yang[3], Yen-Ting Lin[3], Chun-Chieh Kuo[3], Ke-Horng Chen[3], Hsiao-Jung Liu[1], Pei-Shan Yu[1], Fang-Chih Chu[1], Ching-Ju Lin[1], Hong-Wen Huang[1], Kuo-Chih Hung[1], Yuan-Hua Chu[1], Ying-Hsi Lin[5], Suhwan Kim[2]2, and Krishnan Ravichandran[2]
[1]*Industrial Technology Research Institute, Taiwan*
[2]*Intel Corporation, United States*
[3]*National Chiao Tung University, Taiwan*
[4]*National Chiao Tung University / Industrial Technology Research Institute, Taiwan*
[5]*Realtek Semiconductor Corp, Taiwan* ··171

14-4 (8186) A Digital Multiphase Converter with Sensor-Less Current and Thermal Balance Mechanism
Kai-Yu Hu, Yu-Sin Chen, and Chien-Hung Tsai
National Cheng Kung University , Taiwan ··175

14-5 (8008) A Fully-integrated LC-Oscillator Based Buck Regulator with Autonomous Resonant Switching for Low-Power Applications
Tianyu Jia and Jie Gu
Northwestern University, United States ··179

DCS 15

Energy-Efficient Circuits and Architectures

15-1 (8073) A Bulk 65nm Cortex-M0+ SoC with All-Digital Forward Body Bias for 4.3X Subthreshold Speedup
Pranay Prabhat2, Graham Knight[2], Supreet Jeloka[1], Sheng Yang[2], and James Myers[2]
[1]*Arm Inc., United Kingdom*
[2]*Arm Ltd., United Kingdom* ··183

15-2 (8053) A 2.1 pJ/bit, 8 Gb/s Ultra-Low Power In-Package Serial Link Featuring a Time-based Front-end and a Digital Equalizer
Po-Wei Chiu, Muqing Liu, Qianying Tang, and Chris H. Kim
University of Minnesota, United States ··187

15-3 (8102) A 2.69 Mbps/mW 1.09 Mbps/kGE Conjugate Gradient-based MMSE Detector for 64-QAM 128 × 8 Massive MIMO Systems

Guiqiang Peng, Leibo Liu, Qiushi Wei, Yao Wang, Shouyi Yin, and Shaojun Wei

Tsinghua University, China ·······191

15-4 (8040) A Fully Standard-Cell Based On-Chip BTI and HCI Monitor with 6.2x BTI Sensitivity and 3.6x HCI Sensitivity at 7 nm Fin-FET Process

Mitsuhiko Igarashi, Yuuki Uchida, Yoshio Takazawa, Yasumasa Tsukamoto, Koji Shibutani, and Koji Nii

Renesas Electronics Corporation, Japan ·······195

15-5 (8006) A 140 nW, 32.768 kHz, 1.9 ppm/°C Leakage-Based Digitally Relocked Clock Reference with 0.1 ppm Long-Term Stability in 28nm FD-SOI

Guénolé Lallement[3], Fady Abouzeid[2], Thierry Di Gilio[2], Philippe Roche[2], and Jean-Luc Autran[1]

[1]*Aix-Marseille Université, France*

[2]*ST Microelectronics, France*

[3]*ST Microelectronics / Aix-Marseille Université, France* ·······197

WLN 16

Advanced Wireline Equalization

16-1 (8027) A 2× Blind Oversampling FSE Receiver with Combined Adaptive Equalization and Infinite-Range Timing Recovery

Seuk Son[2], Hwanseok Yeo[1], Sigang Ryu[2], and Jaeha Kim[2]

[1]*Samsung Electronics, Korea*

[2]*Seoul National University, Korea* ·······201

16-2 (8192) A Bimodal (NRZ/PAM-4) ISI Tolerant Timing Recovery with Adaptive DDJ Equalization

Masum Hossain[2], Aurangozeb Aurangozeb[2], and Nhat Nguyen[1]

[1]*Rambus Inc, Sunnyvale, United States*

[2]*University of Alberta, Canada* ·······205

16-3 (8022) A 12-Gb/s AC-Coupled FFE TX with Adaptive Relaxed Impedance Matching Achieving Adaptation Range of 35-75Ω Z_0 and 30-550Ω R_{RX}

Minsoo Choi, Myungguk Lee, and Byungsub Kim

Pohang University of Science and Technology, Korea ·······209

16-4 (8114) A 40 Gb/s PAM-4 Receiver with 2-Tap DFE Based on Automatically Non-Even Level Tracking

Chia-Tse Hung, Yu-Ping Huang, and Wei-Zen Chen

National Chiao Tung University, Taiwan ·······213

RF 17

Oscillators and Synthesizers

17-1 (8015) A 1.6-GHz 3.3-mW 1.5-MHz Wide Bandwidth $\Delta\Sigma$ Fractional-N PLL with a Single Path FIR Phase Noise Filtering

Jingcheng Tao and Chun-Huat Heng

National University of Singapore, Singapore 215

17-2 (8203) A 37-GHz-Input Divide-by-36 Injection-Locked Frequency Divider with 1.6-GHz Lock Range

Sangyeop Lee, Kyoya Takano, Ruibing Dong, Shuhei Amakawa, Takashi Yoshida, and Minoru Fujishima

Hiroshima University, Japan 219

17-3 (8177) A 37.5-45.1GHz Superharmonic-Coupled QVCO with Tunable Phase Accuracy in 28nm Bulk CMOS

Luya Zhang[1], Ali Ameri[1], Yi-An Li[1], Nai-Chung Kuo[1], Mekhail Anwar[2], and Ali Niknejad[1]

[1]University of California, Berkeley, United States

[2]University of California, San Francisco, United States 223

17-4 (8133) A Fast Auto-Frequency Calibration Technique for Wideband PLL with Wide Reference Frequency Range

Zhao Zhang, Jincheng Yang, Liyuan Liu, Nan Qi, Peng Feng, Jian Liu, and Nan-Jian Wu

Institute of Semiconductors, Chinese Academy of Sciences, China 227

17-5 (8020) A Sub-Picosecond Hybrid DLL for Large-Scale Phased Array Synchronization

Matan Gal-Katziri and Ali Hajimiri

California Institute of Technology, United States 231

DC 18

ADCs and Calibration Techniques

18-1 (8200) A 7b 2 GS/s Time-Interleaved SAR ADC with Time Skew Calibration Based on Current Integrating Sampler

Wenning Jiang[2], Yan Zhu[2], Chi-Hang Chan[2], Boris Murmann[1], Seng-Pan U[2], and Rui Paulo Martins[2]

[1]Stanford University, United States

[2]University of Macau, Macau 235

18-2 (8126) A 15.1-mW 6-GS/s 6-bit Flash ADC with Selectively Activated 8x Time-Domain Interpolation

Il-Min Yi, Naoki Miura, Hiroyuki Fukuyama, and Hideyuki Nosaka

NTT Device Technology Laboratories, NTT Corporation, Japan 239

18-3 (8106) A 38-mW 7-bit 5-GS/s Time-Interleaved SAR ADC with Background Skew Calibration

Yung-Hui Chung, Chia-Yi Hu, and Che-We Chang

National Taiwan University of Science and Technology, Taiwan 243

18-4 (8220) **A 0.6-to-1V 10K-to-100kHz BW 11.7b-ENOB Noise-Shaping SAR ADC for IoT Sensor Applications in 28-nm CMOS**

Young-Ha Hwang, Yoonho Song, Jun-Eun Park, and Deog-Kyoon Jeong

Seoul National University, Korea ···247

18-5 (8067) **A Calibration-Free 0.7-V 13-bit 10-MS/s Full-Analog SAR ADC with Continuous-Time Feedforward Cascaded (CTFC) Op-Amps**

Kwuang-Han Chang and Chih-Cheng Hsieh

National Tsing Hua University, Taiwan ···249

18-6 (8030) **An 89.55dB-SFDR 179.6dB-FoM$_\text{s}$ 12-bit 1MS/s SAR-Assisted SAR ADC with Weight-Split Compensation Calibration**

Yao-Sheng Hu, Jhao-Huei Lin, Ding-Guo Lin, Kai-Yue Lin, and Hsin-Shu Chen

National Taiwan University, Taiwan ··253

DCS+FPGA 19

Multimedia and Signal Processing Hardware

19-1 (8100) **An Image Recognition Processor with Time-domain Accelerators Using Efficient Time Encoding and Non-linear Logic Operation**

Zhengyu Chen and Jie Gu

Northwestern University, United States ··257

19-2 (8180) **A 95pJ/label Wide-Range Depth-Estimation Processor for Full-HD Light-Field Applications on FPGA**

Li-De Chen, Yu-Ta Lu, Yu-Ling Hiao, Bo-Hsiang Yang, Wei-Chih Chen, and Chao-Tsung Huang

National Tsing Hua University, Taiwan ··261

19-3 (8024) **A 280mV 3.1pJ/code Huffman Decoder for DEFLATE Decompression Featuring Opportunistic Code Skip and 3-way Symbol Generation in 14nm Tri-gate CMOS**

Sudhir Satpathy, Sanu Mathew, Vikram Suresh, Vinodh Gopal, James Guilford, Mark Anders, Himanshu Kaul, Amit Agarwal, Steven Hsu, and Ram Krishnamurthy

Intel Corporation, United States ··263

19-4 (8110) **A Wearable Auto-Patient Adaptive ECG Processor for Shockable Cardiac Arrhythmia**

Syed Muhammad Abubakar, Muhammad Rizwan Khan, Wala Saadeh, and Muhammad Awais Bin Altaf

Lahore University of Management Sciences, Pakistan ·····················267

SOC 20

Intelligent Low-Power SoCs

20-1 (8032) **A Capacitance-to-Digital Converter Integrated in a 32bit Microcontroller for 3D Gesture Sensing**
Mitsuru Hiraki, Sugako Otani, Masao Ito, Takuya Mizokami, Masahiro Araki, and Hiroyuki Kondo
Renesas Electronics Corporation, Japan ··269

20-2 (8013) **A 104.8TOPS/W One-Shot Time-Based Neuromorphic Chip Employing Dynamic Threshold Error Correction in 65nm**
Luke Everson, Muqing Liu, Nakul Pande, and Chris H. Kim
University of Minnesota, United States ··273

20-3 (8026) **A 137-µW Area-Efficient Real-Time Gesture Recognition System for Smart Wearable Devices**
Taegeun Yoo[2], Van Loi Le[2], Ju Eon Kim[1], Ngoc Le Ba[2], Kwang-Hyun Baek[1], and Tony Tae-Hyoung Kim[2]
[1]Chung-Ang University, Korea
[2]Nanyang Technological University, Singapore ···277

20-4 (8066) **A 655Mbps Successive-Cancellation Decoder for a 1024-bit Polar Code in 180nm CMOS**
Hye-Yeon Yoon, Seung-Jun Hwang, and Tae-Hwan Kim
Korea Aerospace University, Korea ···281

20-5 (8201) **A Generated Multirate Signal Analysis RISC-V SoC in 16nm FinFET**
Stevo Bailey[3], Jaeduk Han[3], Paul Rigge[3], Richard Lin[3], Eric Chang[3], Howard Mao[3], Zhongkai Wang[3], Chick Markley[3], Adam Izraelevitz[3], Angie Wang[3], Nathan Narevsky[3], Woorham Bae[3], Steve Shauck[2], Sergio Montano[2], Justin Norsworthy[2], Munir Razzaque[2], Wen Hau Ma[2], Akalu Lentiro[2], Matthew Doerflein[2], Darin Heckendorn[1], Jim McGrath[1], Franco DeSeta[1], Ronen Shoham[1], Mike Stellfox[1], Mark Snowden[1], Joseph Cole[1], Dan Fuhrman[1], Brian Richards[3], Jonathan Bachrach[3], Elad Alon[3], and Borivoje Nikolić[3]
[1]Cadence Design Systems, Inc., United States
[2]Northrop Grumman Corporation, United States
[3]University of California, Berkeley, United States ··285

RF 21

Low-Power RF Transmitters and Receivers

21-1 (8021) **A Compact High Efficiency and High Power Front-End Module for GSM/EDGE/TD-SCDMA/TD-LTE Applications in 0.13um CMOS**
Shihai He, Fengxiong Peng, Linjian Xu, Hao Meng, and Yongxue Qian
Beijing Huntersun Electronic Co., Ltd., China ···289

21-2 (8093) **A 2.4-GHz Single-Pin Antenna Interface RF Front-End with a Function-Reuse Single-MOS VCO-PA and a Push-Pull LNA**
Kai Xu[1], Jun Yin[2], Pui-In Mak[2], Robert Bogdan Staszewski[1], and Rui Paulo Martins[2]
[1]University College Dublin, Ireland
[2]University of Macau, China ··293

21-3 (8195) **A 6-8Ghz 200MHz Bandwidth 9-Channel VWB Transceiver with 8 Frequency-Hopping Subbands**
Haixin Song, Dang Liu, Woogeun Rhee, and Zhihua Wang
Tsinghua University, China ···295

21-4 (8087) **A 0.46-2.1GHz Spurious and Oscillator-Pulling Free LO Generator for Cellular NB-IoT Transmitter with 23 dBm Integrated PAs in 28nm CMOS**
Jaewon Choi, Nam-Seog Kim, Juyoung Han, and Thomas B. Cho
Samsung Electronics, Korea ··299

21-5 (8210) **A 152µW -99dBm BPSK/16-QAM OFDM Receiver for LPWAN Applications**
Avish Kosari, Milad Moosavifar, and David Wentzloff
University of Michigan, United States ··303

Author Index ··307

Panel Discussion ··318

Committees ··322

Welcome Message

On behalf of the organizing committee, I would like to welcome all of you to A-SSCC 2018 on November 5 to 7, 2018. A-SSCC is a premiere solid-state circuits conference, sponsored by IEEE SSCS society and Region-10 SSCS chapters, while Asia plays a significant and pivotal role in the global semiconductor and IC industry, being number one in foundry, memory, OSAT, and second place in fabless design companies.

The theme of A-SSCC 2018 is "Silicon enabling mobile intelligence". It has a full three-day program consisting of 4 useful tutorials, 4 outstanding plenary keynotes, one panel discussion, and 86 excellent contributed papers in 16 technical sessions covering circuits and technologies for mobile intelligence, 5G, smart sensors, communication, analog, signal processing, ... , in addition to exhibition, demos, and excellent reception, social hour, and banquet events.

2018 is the 60th anniversary of the invention of integrated circuits. The relentless semiconductor and IC innovation have begun to unleash the full potential of machine learning and AI. The miniaturized silicon technology enabled big success in the realization of machine learning, big data, virtual and augmented realty with applications in image and speech recognition, medical diagnosis, and autonomous driving, etc.. Recently we have witnessed powerful GPU-based AI supercomputers for cloud/data centers and low-power neuro-engines on smartphone AP, implemented in 3Dx3D system superchips. Besides continuing to push energy-efficient cloud- and data center-based AI, mobile intelligence with low-power, fast, secure, and safe on-device AI, on-sensor AI, and embedded AI, etc., represents tremendous R&D and business opportunities down the road for academia, device makers, EDA, fabless design houses, and system companies. Algorithm/Architecture, Circuit, and Technology Co-innovation (ACT-COIN) will be key to our collective success. We encourage you to engage, interact, debate, and brainstorm at this A-SSCC, and hopefully come up with new innovations to collaborate on mobile intelligence and more.

On the side, we invite you to experience and enjoy the hospitality, local delicacy, historical sites, and the rich culture of Tainan City which is the first and oldest port city of Taiwan, the "Formosa". Hope you find other inspiration with fond memory besides technical programs.

Many dedicated volunteers made this A-SSCC possible. I would like to express my sincere gratitude and appreciation to the members of the organizing committee chaired by Prof. L.H. Lu, the steering committee chaired by Prof. Tadahiro Kuroda, and the technical program committee chaired by Hong-June Park and assisted by Prof. Mototsugu Hamada (Co-Chair), Jae-Yoon Sim (Vice-Chair), Jun Deguichi (Vice-Co-Chair), as well as all speakers, authors, and sponsors.

Again, thank you all for coming and attending the conference. Please enjoy it and share your positive experience with colleagues and friends. You are the most important to the continuing success of A-SSCC into the future.

Jack Y.-C. Sun, PhD.
Conference Chair, A-SSCC 2018
IEEE Fellow
Senior Consultant, TSMC
Former VP of R&D, and CTO, TSMC

Foreword

Welcome to the IEEE Asian Solid-State Circuits Conference (A-SSCC) 2018! The conference is held in Tainan, Taiwan from Nov. 5th to 7th, 2018. Being one of the four conference fully sponsored by the IEEE Solid-State Circuits Society, A-SSCC has grown to be a leading conference in the fields of integrated circuits and systems design.

The conference theme for this year is "Silicon Enabling Mobile Intelligence." The miniaturized silicon technology enabled big success in the realization of software solutions such as machine learning, big data, virtual and augmented realty in the image and speech recognition, the medical diagnosis and the autonomous driving automobiles. The current software solutions, however, consume huge power by employing cloud computers along with many graphic processing units and a large amount of memory. Nowadays, the integrated circuit design community tries to develop efficient low-power mobile intelligence solutions by taking challenges in the design of digital and analog circuits, processor architecture, and system for compact IoT devices.

This year, we received 213 submissions from 23 countries around the world. Either four-page or two-page manuscripts were received, while the two-page manuscript could include additional two-page supplements with figures and figure captions. Among all submissions, 95% of papers presented measurement results with silicon chips. After rigorous review process, including the on-site meeting on July 27, the Technical Program Committee (TPC) selected 86 high-quality papers from 17 countries. The acceptance rate is 40.4%.

The conference starts with four tutorials on Nov.5, 2018. Prof. Nan Sun of University of Texas at Austin presents "When SAR Meets Delta-Sigma - A Tale of Two ADC Architectures", Prof. Shuenn-Yuh Lee of National Cheng Kung University presents "Wireless ECG Acquisition and Cardiac Stimulation SOCs for Body Sensor Networks", Prof. Minoru Fujishima of Hiroshima University presents "Terahertz CMOS Technology for Beyond 5G", and Dr. Kyomin Sohn of Samsung Electronics presents "Memory System for Next Generation AI".

Four plenary speeches are presented by distinguished industry leaders. Dr. Kevin Zhang VP Business Development TSMC Taiwan presents "Circuit Design in Nano-Scale CMOS Technologies", Mr. Seizo ONOE CTA NTT DOCOMO & President DOCOMO Tech. Japan presents "Open the New World of 5G", Dr. Nam Sung Kim SVP Memory Division Samsung Korea presents "Practical Challenges in Supporting Functions in Memory", and Dr. Yi Kang Chief Scientist & SVP UNISOC China (Tsinghua Unigroup) presents "AI Drive Domain Specific Processor".

A panel discussion is held on Tuesday afternoon with the topic of "Circuits and Systems for Mobile AI", moderated by Prof. Marvin Frank Chang, National Tsinghua University, Taiwan. The panel invites experts from National Tsung Hsing University, POSTECH, Toshiba, Chuyo University, and Tsinghua University.

The industry session, held on Tuesday, highlights advances in DNN inference engine, PCIe PHY and memory. Four outstanding industry papers are presented by speakers from Toshiba, Intel, Renesas and TSMC. The Student Design Contest(SDC) provides live demos from the top 14 student-authored papers. Three winners are selected and recognized at the conference banquet on Tuesday evening. This year, A-SSCC started the FPGA Session, which accepts papers describing FPGA implementation with novel algorithm and/or architecture. The FPGA demos will be in progress at the same time as the SDC demos.

A-SSCC 2018 TPC consists of 100 members divided into 10 technical subcommitees. The members come from both industry and academia around the world. This year, TPC members gathered in Yilan Taiwan in late July to select excellent papers. Their contributions to maintain a high-quality A-SSCC are highly appreciated. Furthermore, I would like to acknowledge the leadership of the technical subcommittee chairs: Prof. Hidetoshi Onodera of Kyoto University (Analog Circuits and Systems), Prof. Seng-Pan(Ben) U of University of Macau (Data Converters), Prof. Robert Chen-Hao Chang of National Chung Hsing University/National Chi Nan University (Digital Circuits and Systems), Prof. Kazutami Arimoto of Okayama Prefectural University (SoC and Signal Processing), Dr. Satoshi Tanaka of Murata Manufacturing (RF), Prof. Chulwoo Kim of Korea University (Wireline and Mixed-Signal Circuits), Prof. Woogeun Rhee of Tsinghua University (Emerging Technology and Applications), Dr. Junghwan Choi of Samsung Electronics(Memory), Dr. Stefan Rusu of TSMC (Industry Program), and Dr. Shigeki Tomishima of Intel (FPGA ad-hoc). I would also like to acknowledge Prof. Baoyong Chi of Tsinghua University and Prof. Jung-Hoon Chun of Sungkyunkwan University for organizing the Student Design Contest, Prof. Hoi-Jun Yoo of KAIST for preparing the plenary and panel programs, Prof. Hoi-Jun Yoo of KAIST and Prof. Byeong-Gyu Nam of Chungnam National University for the tutorial planning.

I would also like to extend my sincere appreciation to all authors and speakers, conference organizers, committee members, moderators, panelists, and, last but not least all the participants. I hope you will enjoy the technical program of A-SSCC 2018, take this opportunity to network with experts around the world, and bring back good memories with you!

Hong-June Park
Technical Program Committee Chair of A-SSCC2018
POSTECH(Pohang University of Science and Technology), Korea

Circuit Design in Nano-Scale CMOS Technologies

Kevin Zhang
Vice President of Business Development
Taiwan Semiconductor Manufacturing Company
Hsinchu, Taiwan
kevin_zhang@tsmc.com

Abstract—As CMOS technology scaling drives down the feature size of transistor into a nano-scale regime, there are many new challenges facing today's circuit design. In this paper, the landscape of technology scaling is first examined. The paper then focuses on how to overcome some of the key design challenges through innovative circuit approaches. Several real design examples in various areas, including SRAMs and critical analog building blocks, are used to illustrate these advanced circuit design techniques that will help further technology scaling in the future.

Keywords—circuit design; SRAMs; analog circuits; device scaling.

I. INTRODUCTION

The semiconductor industry has experienced tremendous growth over the past 60 years and has now grown into a business valued over US$400 billion. It has also essentially changed different aspects of human life in every imaginable way. The growth drivers for the semiconductor industry have evolved over the years, from mainframe computing to personal computing and from mobile computing to today's ubiquitous computing with embedded artificial intelligence (AI), as illustrated in Fig. 1. With the arrival of AI and 5G high-speed network, it is fitting to project that semiconductors will continue to serve as the foundation for new innovations that will constantly change society as a whole.

Underlying this tremendous growth is the relentless pursuit of Moore's law, which largely comes down to pursuing faster and smaller transistor. To enable continuous transistor scaling, technology has evolved in many different ways, ranging from materials to architecture. The simple poly-gate and planar MOSFET has evolved into a sophisticated 3D FinFET that employs many different materials, including high-k dielectrics, metal gates, and strained silicon channels [1]. With these continuous innovations, the most recent 7nm transistor delivers a more than 40% performance gain or a 65% power reduction, when compared to a 16nm FinFET [2]. Many devices, some based on new device physics well beyond the classical electron-transport theory, are on the horizon. These new devices should help the industry continue on its path toward higher performance with lower power consumption.

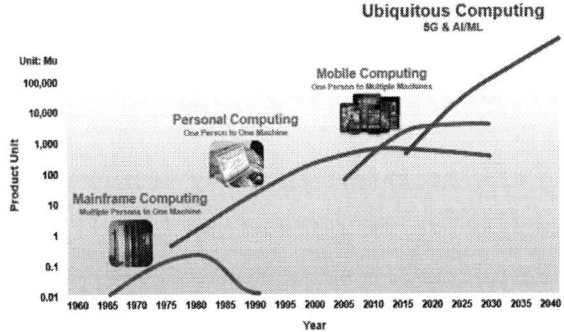

Fig. 1. The evolution of semiconductor growth engines.

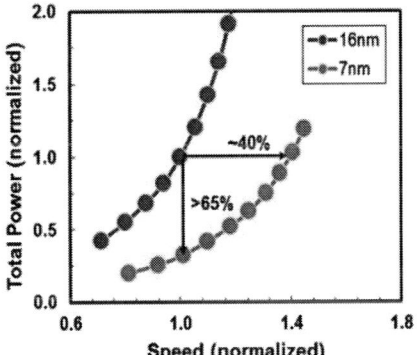

Fig. 2. Transistor performance and power scaling at 7nm [2].

II. SRAM SCALING

On-die SRAM has always been a crucial component of microprocessor performance because it directly provides the data bandwidth needed by compute engines. With the growing gap between fast-growing on-die compute engines and external DRAM bandwidth, shown in Fig. 3, SRAMs are becoming even more crucial for further scaling to enable future high-performance computing.

With continuous geometry reduction of transistors, variation has become a key factor for further SRAM density scaling since that has become the limiting factor to achieve minimum operating voltage, Vmin, to address future product needs. Fig. 4 illustrates the key design challenge in today's SRAM cell design where a balanced read and write margin needs to be optimized simultaneously.

A dynamic voltage control has been introduced to manage both Vmin scaling and power reduction needs [3]. Fig. 5 plots the SRAM cell scaling trend, along with innovations that have enabled the SRAM cell reduction over past generations [4]. In addition to device improvement, e.g., lightly doped Fin substrate, voltage based circuit design techniques have played an increasingly important role in SRAM scaling over the last decade or so.

A recent design implementation is shown in Fig. 6 [5]. To minimize read disturbance and improve read Vmin, it is often required to lower the wordline voltage at the cost of performance degradation. In this particular design, the wordline voltage level is dynamically restored once the bitline voltage is reduced and the cell disturbance is lowered. During the initial signal sensing phase, this would allow the SRAM cell to minimize the strong disturbance when the bitlines are fully charged, while preserving efficient charge transference between the bitline and cells as the wordline voltage is quickly restored to a higher level. A similar timing control mechanism is also implemented in the "negative" bitline control in order to ease the write operation, while avoiding excessive power consumption.

With these innovative design techniques, along with continuous improvements in transistor technologies, the minimum operation voltage of SRAMs has been scaled down further at each technology node as shown in Fig. 7 [4].

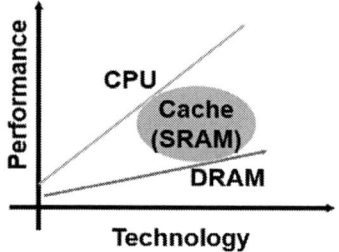

Fig. 3. SRAM plays a key role in providing the high-speed bandwidth to processors.

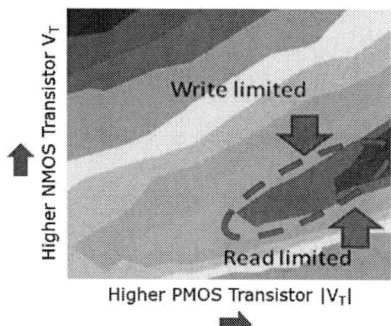

Fig. 4. SRAM cell read and write margin reduction due to transistor Vt variations.

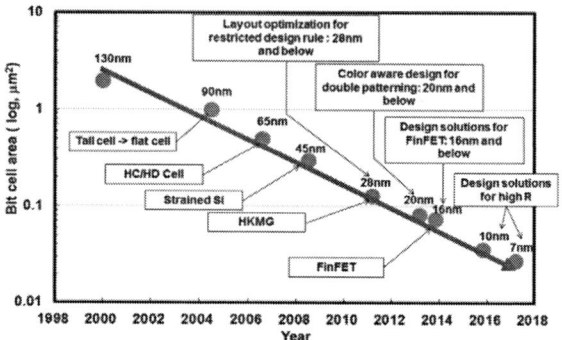

Fig. 5. Key process and design innovations that have enabled the SRAM scaling [4].

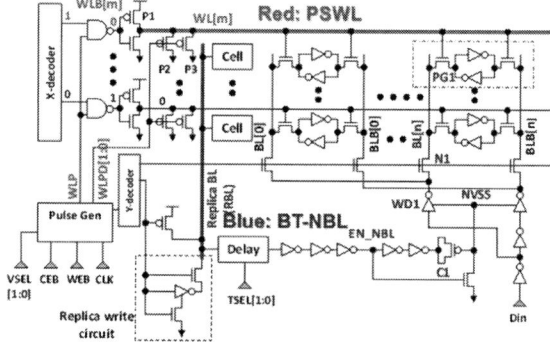

Fig. 6. A design implemention with dynamic Read and Write assist to achieve lower operation voltage (Vmin) [5].

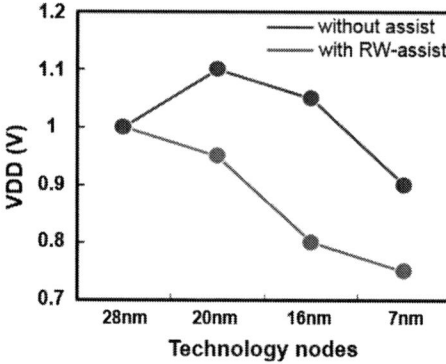

Fig. 7. SRAM Vmin scaling trend with technologies [4].

III. ANALOG CIRCUIT DESIGN

The latest 3D FinFET transistors have proven to provide significant improvements in transistor gain with much higher transconductance (Gm) and larger output resistance (Rout). But layout digitization of the transistor with a FIN structure, coupled with the limitation in the use of long gate length (Lg) devices, have taken conventional design freedom away from analog designers. At the device level, it has become imperative to apply transistor stacking to achieve the results of effective longer Lg devices, as shown in Fig. 8. An intelligent device ratio can further help improve the area efficiency of the design by maximizing the stacking benefits at circuit level.

In order to achieve area scaling of conventional analog and mixed-signal circuit design, it also has become apparent that digital-based architecture implementations are needed. Two essential analog building blocks, thermal sensor and phase-lock-loop (PLL), are discussed here to illustrate the "digital implementation" of "analog functions."

Fig. 9 describes a conventional analog-based thermal sensor design that employs many current sources and operational amplifiers (Ops Amp) to realize the conversion from the differential current of sensing bipolar devices to temperature [6]. While the design is robust for volume production, its area and power scaling are severely constrained, due to the difficulties inherent in analog circuit requirements. Fig. 10 provides a high-level view of an alternative digital-based implementation [7]. It still uses the conventional parasitic bipolar device to provide temperature sensing, but the analog signal coming out the bipolar devices is immediately converted into digital signals using a single-bit ADC counter. It then uses its digital circuits to convert the information into a final digital read out. The new implementation allows this "analog function" to follow the digital scaling trend since the underlying technology is also further scaled.

The PLL has always been a critical functional block of all modern very-large-scale integration (VLSI) systems and has been deeply rooted for generations in analog circuits. The increased number of PLLs used in a modern digital system and the lack of power and area scaling in a PLL have stimulated a tremendous amount of research and progress in digital PLL design. Fig. 11 describes some key differences between analog and digital PLLs. Among these differences, a conventional RC-loop filter is replaced by digital filter, which removes a big area scaling bottleneck in the conventional RC-filter based PLL design. A voltage controlled oscillator (VCO) is converted into a Digital-Controlled-Oscillator (DCO). Fig. 12 provides an implementation example of a DCO [8]. By applying the multi-level digitally controlled ring-oscillator arrays, a large frequency tuning range can easily be achieved at fine granularity. These new digital implementations have keyed overall power and area scaling for PLLs implemented in CMOS technologies.

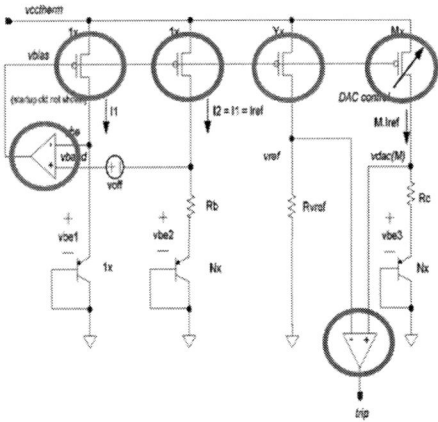

Fig. 9. Conventional thermal sensor design with analog circuit topologies [6].

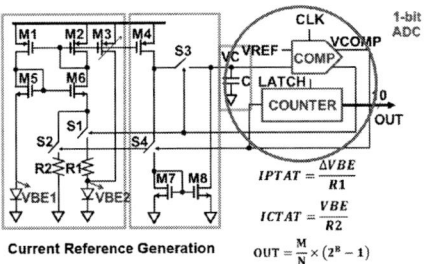

Fig. 10. Digital based thermal sensor design with analog-to-digital converter (ADC) [7].

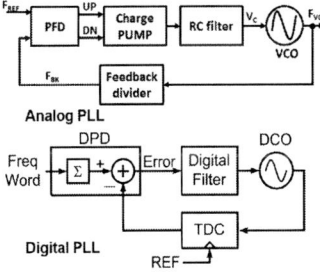

Fig. 11. Analog and digital phase-lock-loop (PLL) design.

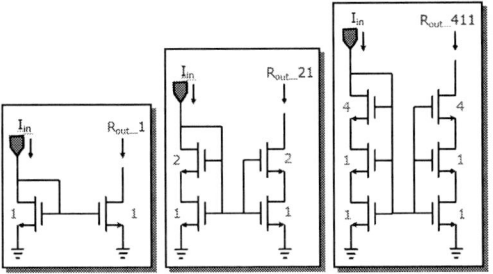

Fig. 8. Leverage transistor stacking effects to achieve long-Le results.

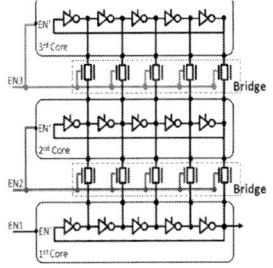

Fig. 12. Digital-Controlled-Oscillator based on capacitor arrays [8].

vii

IV. ADAPTIVE DESIGN

Variation control has always been critical to achieving high-yield volume manufacturing. With continuous geometry reduction of the circuitry and growth of fabrication complexities, e.g., mask counts and etching steps, it has become more and more challenging to preserve the product design margins needed to overcome the effects of variations. Die-to-die or wafer-to-wafer differences, which are often referred to as systematic variation, are key contributors to variations in large SRAMs [9]. To mitigate the impact of systematic variation in SRAMs, an adaptive voltage control on the wordline is implemented as shown in Fig. 13 [10]. By leveraging a simple on-die "sensor," which is used to detect transistor N/P skews, a feedback loop is used to dynamically control the wordline voltage level. This reduces the read disturbance to achieve lower Vmin. With the new adaptive control, Vmin improvement can be achieved across the entire population of materials, whereas a "fixed" wordline control failed to achieve broad benefits. This adaptive design concept has become increasingly useful in combating systematic variations and helping to further the scaling of transistor technologies.

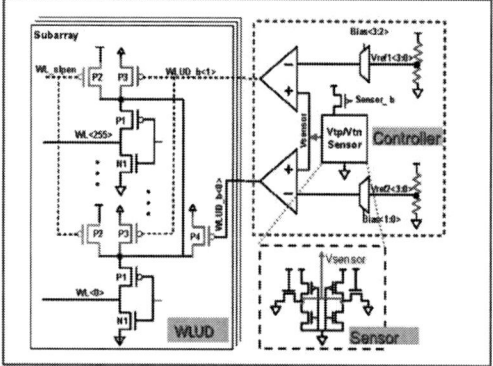

Fig. 13. On-die sensor to control wordline voltage for optimal SRAM Vmin [10].

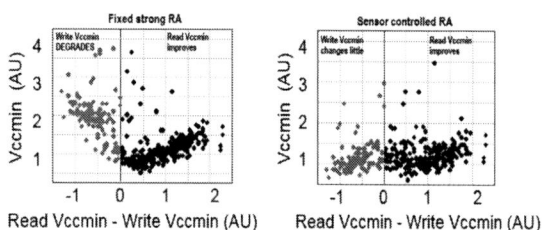

Fig. 14. Vmin distributions with and without adaptive wordline voltage control, respectively [10].

V. SUMMARY

The success of the semiconductor industry is built upon countless innovations over the past 60 years. The industry has become a foundation of modern society and has improved nearly all aspects of our lives today. With strong collaborative efforts across different technical fields -- process, device, circuit, and architecture -- new waves of applications will continue to drive the industry forward in the decades to come.

ACKNOWLEDGMENT

The author is truly grateful to have had the opportunities to be associated with many talented and dedicated engineers across the globe who have contributed to the technical contents of this paper. A special thank goes to Ted Chiang, who helped to prepare the manuscript of this paper.

REFERENCES

[1] S-Y Wu et al., "Demonstration of a sub-0.03 µm^2 high density 6-T SRAM with scaled bulk FinFETs for mobile SOC applications beyond 10nm node," VLSI Dig. Tech. papers, Jun. 2016

[2] S-Y Wu et al., "A 7nm CMOS Platform Technology Featuring 4th Generation FinFET Transistors with a 0.027 µm^2 High Density 6-T SRAM cell for Mobile SoC Applications," IEDM Dig. Tech. papers, pp. 2.6.1 - 2.6.4, Dec. 2016

[3] K. Zhang et al., "A 3-GHz 70Mb SRAM in 65nm CMOS Technology with Integrated Column-Based Dynamic Power Supply," ISSCC Dig. Tech. papers Vol. 1, pp. 474-611, Feb. 2005

[4] J. Chang et al., "Embedded Memories for Mobile, IoT, Automotive and High Performance Computing," VLSI Dig. Tech. papers, pp. T26 - T27, Jun. 2017

[5] J. Chang et al., "A 20nm 112Mb SRAM in High-κ Metal-Gate with Assist Circuitry for Low-Leakage and Low-VMIN Applications," ISSCC Dig. Tech. papers, pp. 316-317, Feb. 2013

[6] D. Durate et al., "Temperature Sensor Design in a High Volume Manufacturing 65nm CMOS Digital Process," CICC Dig. Tech. papers, pp. 221-224, Sep. 2007

[7] YC Hsu et al., "An 18.75µW Dynamic-Distributing-Bias Temperature Sensor with 0.87°C(3σ) Untrimmed Inaccuracy and 0.00946mm^2 Area," ISSCC Dig. Tech. papers, pp. 102-103, Feb. 2017

[8] TH Tsai et al., "A 1.22ps Integrated-Jitter 0.25-to-4GHz Fractional-N ADPLL in 16nm FinFET CMOS," ISSCC Dig. Tech. papers, pp. 346-347, Feb. 2015

[9] Y. Wang et al., "A 4.0 GHz 291 Mb voltage-scalable SRAM design in a 32 nm high-k + metal-gate CMOS technology with integrated power management," in IEEE ISSCC Dig. Tech. Papers, Feb. 2009, pp. 376–377

[10] H. Nho et al., "A 32nm High-κ Metal Gate SRAM with Adaptive Dynamic Stability Enhancement for Low-Voltage Operation," ISSCC Dig. Tech. papers, pp. 346-347, Feb. 2010

Open the New World of 5G

Seizo Onoe

NTT DOCOMO, INC. and DOCOMO Technology, Inc.
Tokyo, Japan
Email: onoe@nttdocomo.com

Abstract—5G is expected to open a new world by meeting a wide range of requirements such as further enhanced mobile broadband, massive machine type communications and ultra-reliable, low-latency communications. Furthermore, it is highly expected to invent new business models and ecosystems across the industries. Some concerns about the economics of 5G deployment are seen due to limited coverage caused by higher spectrum bands. History shows that those challenges have been resolved by technologies driven by semiconductor technology evolution. Laws of mobile communication generations are also derived from history: the law of previous generations' boom just before the next, the law of great success only in even-numbered generations and the law of next generation service advent after network launch. It is expected some laws may be broken for 5G and 5G will spread in a different way from previous generations.

Keywords—*5G, LTE*

I. INTRODUCTION

5G is stimulating people's imagination and expectations for a new world that it may bring about by the year 2020. With visions for the new world, 5G is aimed at meeting a wide range of requirements such as enhanced mobile broadband, massive machine type communications and ultra-reliable, low-latency communications. Furthermore, it is highly expected to invent new business models and ecosystems across the industries.

On the other hand, some concerns about the economics of 5G deployment are raised in various occasions because people tend to believe that 5G would need a huge number of small cells due to its limited coverage caused by higher spectrum bands. In the past, more than 30 years ago, no one believed the realization of a 2GHz cellular system. History shows that mobile communication technologies have achieved many things that were once thought impossible, driven by semiconductor technology evolution. In the future, cost-effective 5G systems with millimeter waves must be realized by further evolved mobile and semiconductor technologies.

Predictions introduced in this paper are based on the laws regarding the generations of mobile communications being derived from the history by the author. The first one is the law of previous generations' boom just before the next and the second is the law of great success only in even-numbered generations. The third is the law of next generation service advent after next generation network launch. It is expected that some laws may be broken for 5G and 5G will spread in a

different way from previous generations. Another aspect of the third law is the law of next generation service birth in the previous generations. This may be a hint for next generation service creation.

II. 5G FEATURES

5G is characterized by three features concerning capabilities and use cases: enhanced mobile broadband; massive machine type communications; and ultra-reliable, low-latency communications. In addition, business aspect is also important. 5G has been a kind of boom started in recent years and is attracting interest from various vertical industries. The business opportunities will be expanded and new business models and ecosystem are expected to be created through cross-industry collaboration. As such, the business aspect will be another key feature for 5G as shown in Figure 1.

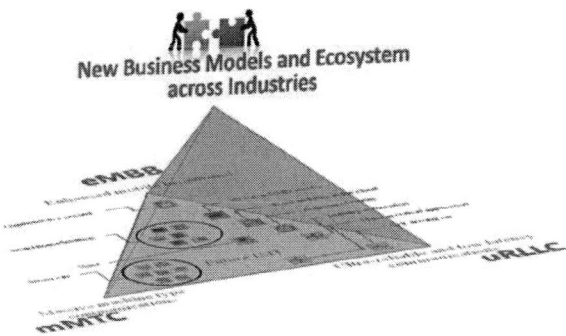

Figure 1. 5G key features with a business aspect.

III. 5G ECONOMICS

People tend to believe that the launch of new generation technology will lead to huge investment. However, Figure 2 shows otherwise. According to this figure, DOCOMO's CAPEX over the past 20 years has shown little sign of increasing trends. This points out that the launch of new generation technologies does not necessarily lead to a sharp ramp-up in investment.

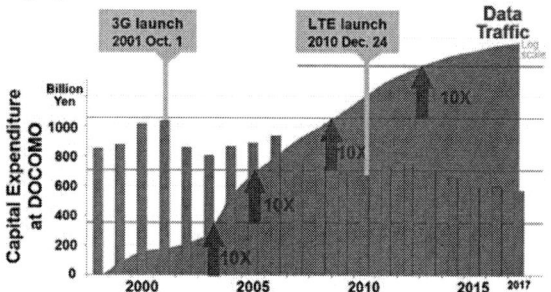

Figure 2. CAPEX and Data Traffic at DOCOMO.

Another fact that defies people's conventional beliefs is the growth of data traffic volume as also shown in Figure 2. Incredible capacity increase has been achieved with no increasing trend in CAPEX thanks to new generation technologies.

Most concerns about 5G investment come from the use of higher frequency bands such as millimetre waves, which usually results in a short coverage distance and a huge number of small cells. This concern does not necessarily turn out to be correct, either. 5G can be deployed with combinations of high and low frequency bands as shown in Figure 3.

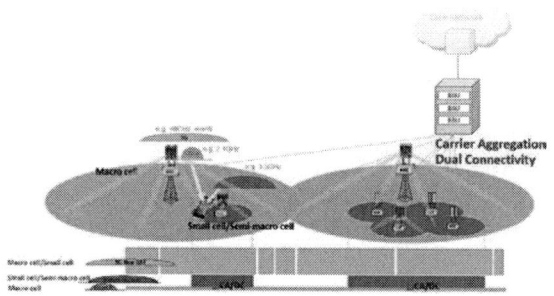

Figure 3. Massive MIMO and Advanced C-RAN.

Massive Multiple-Input Multiple-Output (Massive MIMO) will extend the coverage distance. It is challenging however to implement extreme high-order massive MIMO to realize macro-cell based network.

Semiconductor technology evolution will be indispensable as it has enabled a series of breakthroughs, making what people thought impossible in the past generations such as 2GHz cellular systems and compact size mobile phones as shown in Figure 4.

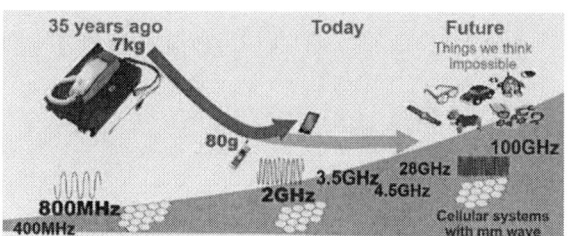

Figure 4. Massive MIMO and Advanced C-RAN.

In the future, what people think impossible today will be achieved. The industry should tackle challenges to achieve favorable economics with a macro-cell based network even with millimetre waves.

IV. LAWS OF MOBILE COMMUNICATION GENERATIONS

This section discusses predictions in the form of 3 laws defined by the author as a predictor based on the observations of previous generations. The first and second laws were presented at Brooklyn 5G Summit 2017[1] in April, 2017. IEEE Spectrum reported the whole story of the event [2]. The laws were elaborated at IEEE 5G Summit Tokyo [3] in September, 2017. The third law is newly discussed although the phenomena have been known by many people in the previous generations.

A. Law of Previous Generations' Boom Just Before the Next

The first law is the law of previous generations' boom just before the next generations. The phenomena described by this law always happened in the previous generations: EDGE before 3G W-CDMA and HSPA+ before 4G LTE. Applying this law to 5G, 4G LTE-Advanced Pro will regain popularity just before 5G launch. Even if 4G regains popularity, the evolved 4G is likely to be called "5G" as a marketing gimmick.

The above was an observation presented by the author in 2017. However, the situation was originally tricky because 5G includes both of 5G NR (5G New Radio) and eLTE (enhanced LTE). Furthermore, some recent developments seem to be raising a possibility that the first law would likely be broken; because there will be no time for 4G boom before 5G since 5G launch would be moved forward before 2020. 5G boom will continue until 5G launch and beyond.

B. Law of Great Success Only in Even-Numbered Generations

The second law is the law of great success only in even-numbered generations as shown in Figure 5.

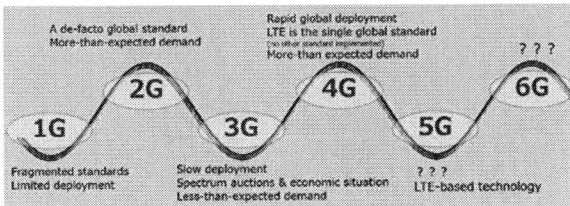

Figure 5. Law of Great Success Only in Even-Numbered Generations.

2G GSM was very successful as the first de-facto global standard. 4G LTE global deployment was remarkably quick while 3G deployment was relatively slow. As such, even-numbered generations, 2G and 4G, had great success. Applying this law to 5G, the industry will need to wait until 6G to see all of its expectations for 5G fulfilled. In other words, 6G will be a complete form of 5G, or 5G will be the final generation that will keep evolving ever after 2030.

The presentation of this prediction in 2017 did not intend to send any negative message about 5G. At the Brooklyn 5G Summit 2018[4] in April, 2018, the author emphasized that there must be something that would lift up 5G to a great success. Cross-industry collaboration will be one that brings 5G to a great success in spite of the downward trend predicted by the law as shown in Figure 6.

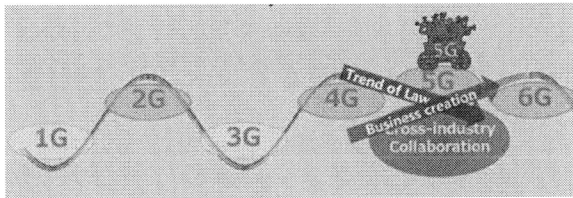

Figure 6. Key to a Great Success for 5G against Trend of Law.

C. Law of Next Generation Service Advent after Next Generation Network Launch

The third is the law of next generation service advent after next generation network launch. There were always doubts as to the necessity of next generation systems in early stages of standardization. The frequently asked questions were "What would you use 2Mbps for?" for 3G and "Do you really need 100Mbps?" for 4G. Historically, however, it has been always the case that unexpected services of next generation became popular after the launch of next generation networks. Services that emerge and become popular are often different from what people imagined or expected beforehand.

The representative services and products for each generation may vary from market to market. Figure 7 shows representative services and products for each generation in the case of Japan.

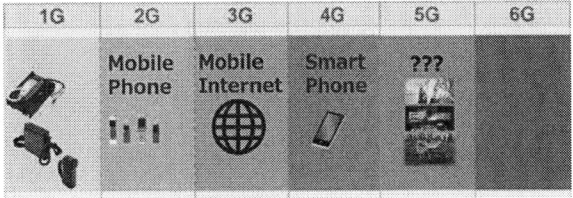

Figure 7 Representative Services and Products in Japan for each Generation.

Another aspect of the third law is the law of next generation service birth in their previous generations. Figure 8 resembles the previous figure but illustrates what this law intends to convey more precisely. As shown in this figure, things that represent each of the generations were in fact born in their previous generations. What can be extrapolated from this law is that people must be seeing 5G products and services today although it is difficult to be aware of them. This law may serve as a hint in creating next generation services. It is important to find them and develop them into successful business opportunities.

Figure 8 Next Generation Services and Products Birth in Previous Generations.

5G trials through cross-industry collaboration are underway worldwide. One of the aims of these trials is to create 5G services before 5G launch and to provide 5G services from DAY ONE of 5G by breaking the third law.

V. CONCLUSION

The 5G features have been described and it has been pointed out that business aspect is also one of the key features in addition to enhanced mobile broadband, massive machine type communications and ultra-reliable, low-latency communications. It is highly expected to invent new business models and ecosystems through cross-industry collaboration.

The economics of 5G have been discussed. The author believes that 5G will bring in favorable economics while there are still many concerns among operators. Combinations of technologies and combinations of spectrum bands are key to realize macro-cell based networks even with millimeter waves. The continuous evolution of semiconductor technology is indispensable in its implementation.

The laws of mobile communication generations have been introduced from the observation of previous generations: the law of previous generations' boom just before the next, the law of great success only in even-numbered generations and the law of next generation service advent after next generation network launch. All laws are a kind of warning that can be heeded to make a great success of 5G. The first law has already been expected unlikely to happen. It has been pointed out that cross-industry collaboration may break the second law. The efforts to break the third law are ongoing. When all of these are taken into consideration, these laws may not be applicable to 5G. 5G may spread in a different way from previous generations. 5G will be a great success and open the new world through cross-industry collaboration.

REFERENCES

[1] Seizo Onoe, "5G Getting Close to Reality," Brooklyn 5G Summit 2017 [Online]. Available: https://ieeetv.ieee.org/ieeetv-specials/5g-getting-close-to-reality-seizo-onoe-brooklyn-5g-summit-2017?rf=events|78&

[2] Amy Nordrum, "5G Progress, Realities Set in at Brooklyn 5G Summit," IEEE Spectrum, 21 April 2017. [Online]. Available: https://spectrum.ieee.org/tech-talk/telecom/wireless/5g-progress-realities-set-in-at-brooklyn-5g-summit

[3] Seizo Onoe, "5G: Past and Future," Journal of ICT Standardization ISSN:2246-0853(Online Version) Vol:5 Issue:3 Published In December 2017 [Online]. Available: https://www.riverpublishers.com/journal_read_html_article.php?j=JICT S/5/3/4

[4] Seizo Onoe, "5G Ready to Launch; Its Past and Future," Brooklyn 5G Summit 2018 [Online]. Available: https://ieeetv.ieee.org/conference-highlights/5g-ready-to-launch-its-past-and-future-seizo-onoe-brooklyn-5g-summit-2018?rf=events|149&

Practical Challenges in Supporting Function in Memory

Nam Sung Kim

Memory Division
Samsung Electronics
Hwaseong, Republic of Korea

Abstract—**The performance of computer systems is often limited by the bandwidth of their memory channels, but further increasing the bandwidth is challenging under the stringent pin and power constraints of packages. To further increase performance under these constraints, various processing-in-memory (or function-in-memory) architectures, which tightly integrate processing functions with DRAM devices using 3D/2.5D-stacking technology, have been proposed. However, they have not been successfully commercialized by the industry yet because of various technical and practical challenges. In this article for the plenary talk, I will briefly discuss what challenges have been overlooked by researchers to shed light on successful commercialization and wide adoption of processing-in-memory architecture.**

Keywords—DRAM; Processing in Memory; Function in Memory; 3D Stacking

I. INTRODUCTION

As the trends driven by Moore's Law come to an end, increased heterogeneity at all levels of computing is required to deliver computing performance needed for emerging applications such as artificial intelligence (AI), leading to the proliferation of various application- or domain-specific accelerators. This in turn demands more memory bandwidth, as the accelerators consume data at a much higher rate than traditional computing, limiting the computing performance.

Under the tight pin and power constraints of packages, however, increasing the bandwidth has been stagnant because a higher data transfer rate per pin comes at the cost of worsened signaling integrity and super-linearly increased power consumption. Furthermore, interconnect capacitance has scaled at a much slower rate than logic capacitance with technology scaling [1], significantly increasing the fraction of data transfer energy in the total system energy [2, 3]. These aforementioned challenges motivate researchers to reconsider the past processing-in-memory (PIM) architectures (e.g., [4, 5, 6, 7, 8, 9, 10]) that aimed to improve performance by integrating processing functions and DRAM on the same die. Such PIM architectures, however, suffered from high manufacturing complexity (i.e., low yield), poor processor logic performance, large DRAM area per bit [10, 11], and design/verification cost associated with custom DRAM architectures.

Recently, 3D-stacking technology [12, 13, 14, 15, 16] has emerged as an alternative integration technology. It can solve some of the critical problems faced by the previous PIM architectures or recent function-in-memory (FIM) architectures because it integrates logic and DRAM layers, each of which is manufactured with dedicated and separate process technology, with high-bandwidth and low-energy TSVs. Leveraging 3D-stacking technology, researchers have proposed near-DRAM acceleration (NDA) architectures that integrate accelerator logic and custom DRAM devices to reap the performance and energy-efficiency benefit of both accelerators and near-memory processing and demonstrated promising results [17, 18, 19, 20].

In this article, regardless of the potential of PIM or FIM with emerging technology such as 3D integration, I will bring up some practical challenges to support PIM or FIM for commercial DRAM. This is not to discourage researchers from pursuing PIM or FIM research but to point the researchers to the right direction for the first successful commercialization of this amazing computing paradigm for the future computing.

II. PRACTICAL CHALLENGES

A. Will 3D-integration cost-effectively offer higher bandwidth and lower latency for the logic layer?

Many researchers have assumed that TSVs are small and cheap. However, I believe this is one of the most common misconceptions, which have led PIM or FIM research to the wrong direction and make it impractical. In fact, the 3D integration itself cannot automatically offer significantly higher bandwidth for 3D-stacked accelerators than standard DRAM devices. This is because the bandwidth of each DRAM bank is often designed to match that of off-chip I/O (e.g., DDR4). To providing notably higher bandwidth for the accelerators, considerable customization of DRAM is required and it significantly increases the cost. For example, analyzing the die photos of high-bandwidth memory (HBM) from Hynix [21] and Samsung [22], we discover that the through silicon vias (TSVs) for 1024-bit I/O consume nearly 20% of each DRAM layer, let alone a separate logic layer for PHYs. This is because the pitch between TSVs is often significantly larger than the diameter of each TSV. In such a case, simply exposing 2× (4×) higher bandwidth to the logic layer will consume 20% (40%) more space of each DRAM layer.

Moreover, the 3D-integration technology cannot give notably lower latency for 3D-stacked accelerators than standard DRAM devices either, because the latency is dominated not by off-chip interconnects but by DRAM core [23]. Therefore, the use of 3D-integration technology and the significant customization of DRAM for PIM or FIM are too expensive to be widely used for commodity computing segments although it may be promising for niche computing segments.

B.   Will popular PIM/FIM substrate such as HBM and HMC be able to handle power consumption by the processor or accelerators in DRAM?

Many researchers have proposed to place processors/accelerators on the logic or DRAM layer to exploit the high internal bandwidth (that can be very expensive to obtain according to my concern expressed in Section II.A). To supply the data for such high internal bandwidth, many more DRAM banks need to be activated than normal DRAM operations, while it has been reported that both HBM and HMC consume a significant amount of power even for the normal DRAM operations to support their high off-chip bandwidth. Then, it is not that difficult to imagine how much more power HBM or HMC will consume to simply supply the data for PIM/FIM operation. Moreover, if the power consumption of processors/accelerators, which need to be very powerful and/or many to fully consume the given high internal bandwidth, is considered, the power delivery architecture and thermal solution of HBM and HMC should be significantly enhanced. Meanwhile, the cost of the power delivery architecture and thermal solution for HBM and HMC just for normal operations is very high due to significantly higher off-chip bandwidth than commodity DRAM.

C.   Will the popular PIM/FIM substrate give enough capacity for traditional applications?

I believe it is important to make DRAM supporting PIM/FIM serve both PIM/FIM and traditional applications so that computers deploying such DRAM are general-purpose (or generic) and able to run any given applications efficiently. This requirement is primarily from an industry point of view, when what both manufacturers and customers desire is considered. That stated, many traditional applications demand larger memory capacity. However, many popular PIM/FIM substrates such as HBM and hybrid memory cube (HMC) suffer from limited capacity because they employ point-to-point interconnections to memory controllers for high-speed data transfers. That is, increase the memory capacity requires more memory controllers, channels, and I/O circuits in contrast to traditional DDR memory modules. This in turn requires (1) use of separate traditional DDR DRAM modules and (2) data transfers between HBM/HMC and DDR DRAM modules. (1) significantly increase the cost of the memory system, and (2) may limit the overall PIM/FIM performance. Lastly, HMC with an abstracted memory interface and a network of HMC modules can increase the memory capacity, but it significantly increases memory access latency. In contrast, the scalability of memory capacity and low latency of memory accesses to large memory systems are critical for servers running popular big-data applications such as in-memory database.

D.   Will it be easy to handle the conflics between a host memory controller and a PIM memory controller that share the DRAM?

Many researchers have proposed to place a small memory controller in the logic layer. Such a memory controller allows the processor/accelerator on the logic layer to access its local DRAM. This demands the handover of the DRAM control ownership between the host memory controller and the PIM memory controller. Since the current memory interfaces lack a mechanism to handle such a handover, many tricks are necessary to support such a handover. Besides, the host memory controller expects deterministic DRAM operations because it has to keep track of many in-flight operations for memory scheduling and refreshing. Placing a local PIM memory controller will make the host memory controller's job even more complicated, and it is unlikely that the host processor manufacturers will take that responsibility. It is often known that one of the reasons that HMC did not get much traction and support from the host processor industry because HMC has its own local memory controller on its logic layer. In other words, one of the keys to make PIM or FIM commercially successful is to make PIM/FIM DRAM operations deterministic without interfering with the host memory controller even when the PIM/FIM DRAM operates in PIM/FIM mode. Otherwise, the host processor industry is not likely to support PIM/FIM DRAM.

E.   Will it be cheap to handle cache coherence?

Another thorny issue stemming from adopting PIM is cache coherence. The easiest option is to flush the on-chip cache in the host processor before processors/accelerators in DRAM start their operations. However, flushing the cache itself is a very slow process and the host processor cannot perform any useful operation while it flushes its on-chip cache. Furthermore, when the processors/accelerators in DRAM complete their operations, the host processor need to bring all these data back to its on-chip cache over the traditional memory interconnect, which will be very slow as well. If we consider the overall performance, it may not make sense to offload computations to the processors/accelerators in DRAM.

An alternative approach would be to implement a cache coherence protocol between the host processor and processors/accelerators in DRAM. For such an approach, first, the host processor industry needs to be convinced. Second, the current coherence protocol for on-chip cache consumes a significant amount of the bandwidth of on-chip interconnects. Consider the fact that the coherence traffic needs to occur over the traditional off-chip DRAM interfaces, which offer much lower bandwidth than the on-chip interconnects, for PIM/FIM. It will hurt the performance of traditional applications.

F.   Will PIM or FIM not hurt memory-level parallelism?

Modern memory systems rely on memory-level parallelism to maximize the utilization of the memory interfaces under long latency of accessing DRAM. Such memory-level parallelism comprises channel-, rank-, and bank-level parallelism. That is, data in contiguous memory address space will in fact spread across different channels, ranks in a given channel and banks in a rank. Furthermore, a rank in a traditional DRAM module often consists of 4~16 DRAM chips where each chip will store 4 to 16 bits of a 64-bit data word. This requires communications among banks in a DRAM chip, DRAM chips in a rank, ranks, and/or channels for processors/accelerators in a DRAM chip to process data in contiguous memory address space. This in turn requires significant changes in how we design DRAM chip, modules, and/or channels. Albeit such changes can be accepted by the industry, most of the PIM/FIM benefits would be negated by the cost of exchanging necessary data pieces among processors/accelerators across DRAM chips. To avoid the aforementioned cost, one may consider remap the data such that each processor/accelerator in a DRAM chip can access a complete 64-bit data word in the DRAM chip. However, it will prevent the host processor to efficiently access such remapped data or hurt the performance of traditional applications by losing the memory level parallelism offered by the traditional memory system.

G. Will programmers/users embrace new programming models for PIM or FIM?

Most PIM proposals cited in this article requires changes in the instruction set architecture (ISA) of the host processors and/or require a new programming model. However, the history of the computing industry has repeatedly shown the reluctance to accepting a new ISA or programming model. For some cases like GPUs, the industry has been amazingly patient until GPUs are used for general-purpose computing with new programming model. This was possible because GPUs can be sold for their original purpose, i.e., graphics and the (hardware) cost of providing general-purpose computing capability was negligible. On the other hand, the cost of supporting PIM or FIM in DRAM sounds significant based on what I have described in this article. That is, the DRAM industry may quickly discontinue its effort to develop DRAM supporting PIM or FIM unless it is an almost immediate success in the market.

III. CONCLUSION

In this article for the plenary talk, I discussed many challenges that can be substantial hurdles for a wide adoption of PIM or FIM. When researchers propose PIM or FIM architectures, they have often neglected or downplayed these described challenges. However, for the commercial success of PIM or FIM, the complex interplay between the host processor industry, the DRAM industry, and the programmer/user community must be taken into account. Otherwise, any PIM or FIM architecture proposal would end up with an academic exercise with no commercial impact.

DISCLAIMER

Some of the practical challenges described and sentences used in this article are from a paper that I co-authored before I joined Samsung [24]. The rest of the described challenges are my personal view that I formed before I joined Samsung. In other words, the described challenges are not a Samsung's official view.

REFERENCES

[1] R. Ho, K. Mai, and M. Horowitz, "The Future of Wires," Proceedings of the IEEE, vol. 89, no. 4, pp. 490–504, Apr 2001.

[2] S. Borkar, "Role of Interconnects in the Future of Computing," IEEE Journal of Lightwave Technology, vol. 31, no. 24, Dec 2013.

[3] S. W. Keckler, W. J. Dally, B. Khailany, M. Garland, and D. Glasco, "GPUs and the Future of Parallel Computing," IEEE Micro, vol. 31, no. 5, pp. 7–17, Sep 2011.

[4] M. F. Deering, S. A. Schlapp, and M. G. Lavelle, "FBRAM: a New Form of Memory Optimized for 3D Graphics," in ACM Conference on Computer Graphics and Interactive Techniques (SIGGRAPH), Jul 1994, pp. 167–174.

[5] J. Draper, J. Chame, M. Hall, C. Steele, T. Barrett, J. LaCoss, J. Granacki, J. Shin, C. Chen, C. W. Kang, I. Kim, and G. Daglikoca, "The Architecture of the DIVA Processing-in-memory Chip," in ACM International Conference on Supercomputing (ICS), Jun 2002, pp. 14–25.

[6] D. G. Elliott, W. M. Snelgrove, and M. Stumm, "Computational RAM: A Memory-SIMD Hybrid and its Application to DSP," in IEEE Custom Integrated Circuits Conference (CICC), May 1992, pp. 30.6.1–30.6.4.

[7] Y. Kang, W. Huang, S.-M. Yoo, D. Keen, Z. Ge, V. Lam, P. Pattnaik, and J. Torrellas, "FlexRAM: Toward an Advanced Intelligent Memory System," in IEEE International Conference on Computer Design (ICCD), Oct 1999, pp. 192–201.

[8] K. Mai, T. Paaske, N. Jayasena, R. Ho, W. J. Dally, and M. Horowitz, "Smart Memories: A Modular Reconfigurable Architecture," in IEEE/ACM International Symposium on Computer Architecture (ISCA), Jun 2000, pp. 161–171.

[9] M. Oskin, F. Chong, and T. Sherwood, "Active Pages: A Computation Model for Intelligent Memory," in IEEE/ACM International Symposium on Computer Architecture (ISCA), Jun 1998, pp. 192–203.

[10] D. Patterson, T. Anderson, N. Cardwell, R. Fromm, K. Keeton, C. Kozyrakis, R. Thomas, and K. Yelick, "A Case for Intelligent RAM," IEEE Micro, vol. 17, no. 2, pp. 34–44, Mar 1997.

[11] M. Chu, N. Jayasena, D. P. Zhang, and M. Ignatowski, "High-level programming model abstractions for processing in memory," in Workshop on Near-Data Processing, 2013.

[12] R. G. Dreslinski, D. Fick, B. Giridhar, G. Kim, S. Seo, M. Fojtik, S. Satpathy, Y. Lee, D. Kim, N. Liu, M. Wieckowski, G. Chen, D. Sylvester, D. Blaauw, and T. Mudge, "Centip3De: A 64-Core, 3D Stacked Near-Threshold System," IEEE Micro, vol. 33, no. 2, pp. 8–16, Mar. 2013.

[13] D. H. Kim, S. Mukhopadhyay, and S. K. Lim, "Through-silicon-via aware interconnect prediction and optimization for 3D stacked ICs," in Intl. workshop on System level interconnect prediction, 2009, pp. 85–92.

[14] B. Black, "Die stacking is happening (keynote)," in Intl. Symp. on Microarchitecture, 2013.

[15] D. H. Kim, K. Athikulwongse, M. Healy, M. Hossain, M. Jung, I. Khorosh, G. Kumar, Y.-J. Lee, D. Lewis, T.-W. Lin, C. Liu, S. Panth, M. Pathak, M. Ren, G. Shen, T. Song, D. H. Woo, X. Zhao, J. Kim, H. Choi, G. Loh, H.-H. Lee, and S. K. Lim, "3D-MAPS: 3D massively parallel processor with stacked memory," in IEEE Intl. Solid-State Circuits Conference, 2012, pp. 188–190.

[16] D. H. Woo, H.-H. S. Lee, J. B. Fryman, A. D. Knies, and M. Eng, "POD: A 3D-Integrated Broad-Purpose Acceleration Layer," IEEE Micro, vol. 28, no. 4, pp. 28–40, Jul. 2008.

[17] Q. Zhu, B. Akin, H. E. Sumbul, F. Sadi, J. C. Hoe, L. Pileggi, and F. Franchetti, "A 3D-stacked logic-in-memory accelerator for application-specific data intensive computing," in IEEE Intl. 3D Systems Integration Conf., 2013, pp. 1–7.

[18] D. P. Zhang, N. Jayasena, A. Lyashevsky, J. Greathouse, L. Xu, and M. Ignatowski, "TOP-PIM: throughput-oriented programmable processing in memory," in Intl. Symp. on High Performance Parallel and Distributed Computing, 2014.

[19] Q. Zhu, T. Graf, H. E. Sumbul, L. Pileggi, and F. Franchetti, "Accelerating sparse matrix-matrix multiplication with 3D-stacked logic-in-

memory hardware," in IEEE High Performance Extreme Computing Conf., 2013, pp. 1–6.

[20]   R. Sampson, M. Yang, S. Wei, C. Chakrabarti, and T. F. Wenisch, "Sonic Millip3De: A massively parallel 3D-stacked accelerator for 3D ultrasound," in 2013 IEEE 19th International Symposium on High Performance Computer Architecture (HPCA), 2013, pp. 318–329.

[21]   D. Lee, et al., "25.2A 1.2V 8Gb 8-channel 128 GB/s high-bandwidth memory (HBM) stacked DRAM with effective microbump I/O test methods using 29nm process and TSV," in 2014 IEEE International Solid-State Circuits Conference (ISSCC), 2014, pp.432–433.

[22]   K. Sohn, et al., "18.2A 1.2V 20nm 307GB/s HBM DRAM with at-speed wafer-level I/O test scheme and adaptive refresh considering temperature distribution," in 2016 IEEE International Solid-State Circuits Conference (ISSCC), 2016, pp.316–317.

[23]   D. Chang, G. Byun, H. Kim, M. Ahn, S. Ryu, N.S. Kim, and M. Schulte, "Reevaluating the latency claims of 3D stacked Memories," in IEEE Asia and South Pacific Design Automation Conference (ASP-DAC), 2013, pp. 657–662.

[24]   H. Asghari-Moghaddam, Y. H. Son, J. H. Ahn and N. S. Kim, "Chameleon: Versatile and practical near-DRAM acceleration architecture for large memory systems," 2016 49th Annual IEEE/ACM International Symposium on Microarchitecture (MICRO), 2016, pp. 1-13.

AI Drives Domain Specific Processors

Yi Kang, Ph.D.
UNISOC Technologies Inc *
Beijing, China

Abstract—in this paper we first list some of basic requirements for domain specific processors. Then we discuss several commonly used architectures for artificial intelligence domain applications. Their pros and cons are also compared. A new architecture defined as In-Cluster Coprocessor is presented which can best utilize existing memory hierarchy in a general processor and has an easy programming model. CBC has potential advantages of power saving and low cost. Further investigation on CBC is underway.

Keywords—Domain-specific; GPU; TPU; SIMD; Vector; Cache Coherence; Cluster; Coprocessors

I. INTRODUCTION

Modern microprocessors have gone to a stage that most of performance gains are from semiconductor process technology and the gain from architecture improvement is minimal [1]. After instruction level parallelism is almost fully exploited. Parallelism at coarse granule level is driven to its extreme by multicore. However, multi-core approaches are near to its limit because of end of Moore's Law and limitation of heat dissipation issue in a single chip.

AI (Artificial Intelligence) is a recent technology trend that is believed to shape future world. Deep Learning draws most of attention in recent years with reports [2] that it can outperform human beings in some object recognition cases. CNN (Convolution Neural Network) is a popular deep learning technique that has been widely used in recently years for various AI applications. In fact the widely used CNN is benefit from enough computation power accumulated by Moore's law that a single chip processor can provide computation power of hundreds of GFLOPS. The infamous AlphaGo beats the best human player in Go game [7]. In AlphaGo Tensor Processing Unit (TPU) is used, each TPU processor chip has TFLOPS capability of copulation power for inference. For server cloud computing providers where training services are provide for Deep Learning it is believed that 100T computation capability falls into a comfort zone for an application [2].

The gap between performance requirement for AI applications and the performance ceiling of general microprocessor that is limited by ending of Moore's law is huge. Recent microprocessors with most advanced semiconductor process technology and with multi-core built onto a single chip can only provide tens of GFLOPS from a single chip. Hundreds even thousands of processor chips are then required to build a system that can meet AI requirement for computation power. Actually the trend of increasing requirement in computation power in AI in recent year is faster than Moore's law. This can be illustrated from the performance increase between each generation of Cuda chip from Nvidea that is used largely for deep learning applications. As Moore's law goes to its end, there is no choice but finding a better domain-specific architecture and its implementation for AI in order to meet the challenges of computation performance, low power and low cost.

II. REQUIREMENT

A. Computation Power

A typical training task may require 100TFlops if a server can finish the task in a reasonable time. Moreover, future AI technologies may enable on-site real-time learning that mimics human leaning behaviors. To be able to meet such an astonishing case, computation power requirement shall exceed far from any general purpose microprocessor that is currently deployed. More computation units need to be added to processors in order to reach such a high level of computation power. In Google's TPU [3] 65536 Multiply Accumulation Unit (MAC) are used in a single TPU unit with peak performance of 92TOPS.

For deep learning we found that 80% or more computations are associated with some convolution loops. Therefore introducing a large number of MAC to processors can effectively speed up computations, but it is not good enough if higher speedup is needed. Amdahl's law states that the theoretical speedup from parallelism is limited by the sequential part that cannot be speedup. For domain processor in AI, 20% of operations such as division, comparison and shift which cannot be speed up by MAC array should also be given enough attention because of Amdahl's law.

B. Bandwidth

Memory bandwidth is a critical issue for AI domain processors. This can be reasoned from a simple fact that number of computation are huge, each computation needs source data and output destination data, thus data movement shall be huge. A processor with 1TFlops computation capability needs multiple of 1T data to keep the processor running efficiently. In this case, an on-chip memory/register file of a processor should provide multiple of 1TB bandwidth to keep computation units busy. And the path from lower hierarchy of the memory system to on chip cache should provide data bandwidth at range of 100GB/s or more even with data reuse being considered. Standard system bus like HBMx or PCIEx are used for this purpose and system bus are being pushed to its limit in high performance implementation of AI domain processor.

C. Precision

Instead of using 32bit or 64 bit floating point, 16bit floating point in IEEE 754 format can be used for training in deep learning [2]. For inference there are many publications which show 8bit integer can be used in CNN with reasonable error rate increase from using 8bit. Actually 1bit or 2bit can be used in certain applications. However there are little publications on using integer for training [2]. Therefore 16bit floating point should be supported in training applications and 8 bit integer should be supported in inference applications for AI processors

D. Flexibility

AI is in a fast development stage nowadays. Algorithms are changing fast as more and more resources and investment in research are devoted. Deep learning has gone from classical CNN to RNN in just a couple of years. It is expected that better algorithms will always emerge. On the other side as AI gets deep into various applications and is used more common as a tool, it is more closely coupled with other processing elements such as database, thus AI processor shall be integrated with the general microprocessor in order to achieve easy programming. It is an ideal case that AI processor and general CPU should share resources and work in sync. That would require that AI processor have enough flexibility to accommodate the dynamic of AI development. Its programming model should be general enough to easily work with CPU.

III. ARCHITECTURE

A direct implementation to processor that optimized for computation, power and cost is the ASIC approach that optimizes for the common used algorithms in an application. However ASIC approach is a fit for a specific application as its name indicates, and it is not good for being used in general purpose thus it is out of domain specific processor scope that is being discussed in this article.

There are many processor architectures for AI applications. Although some have much more market share than others, it is too early to call a clear winner for AI processors. The following architectures are competing for survival in AI domain

A. DSP

DSP (Digital Signal Processor) has a long history of being a domain specific processor used for heavy data computing in signal processing such as speech, image and video. The most significant feature of DSP is its capability of doing multiplications and accumulations. The multiplication and accumulation are just right for deep learning, perfect for convolution loops in CNN. The disadvantage of DSP is that it has quite different architecture with mainstream microprocessor. For example it has separate code and data space, over-simplified memory hierarchy and lower clock frequency due to its long data path. DSP is a good option for

*UNISOC is the sole sponsor for this publication.

AI domain processing in edge computing or at a terminal device and good for inference applications, however it is not a good choice for training applications.

B. GPU

GPU is the most popular used processor today, represented by Nvidia's huge success in market. Processor core in GPU are connected in a mesh structure in order to achieve a high data bandwidth. The good architecture example is Pascal from Nvidia [4]. The host interface of GPU can be on a coherent or non-coherent system bus. In both case there are data move from CPU main memory to GPU local memory. GPU are good for batch processing due to its mesh structure. For single thread the utilization of processor array are low.

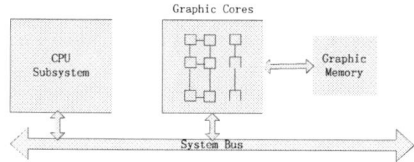

Fig. 1. GPU System Diagram

C. TPU

The well-known AI processor in the world Google's TPU is a coprocessor on PCIe bus so that it can be easily plugged into a server system [3]. See Fig 2 for its system configuration. In TPU systolic data setup unit is well designed so that its matrix array can work in a systolic way to generate results in parallel. TPU has a separate local memory so data move from server memory to its local main memory is a must

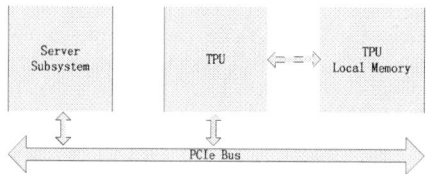

Fig. 2. TPU System Diagram

D. Enhencement of General CPU

This architecture is based on a general multiprocessor architecture with some modifications on top of it.

- **SIMD** The most used microprocessors in server and PC market and in mobile application processor market are now all equipped with SIMD instructions in their ISA. The convolution loops are natural fit to these SIMD instructions for carrying out repeated multiplications and accumulations. However, the computation power for these SIMD extensions are limited by number and length of general purpose register in these processors, and limited by memory bandwidth to feed general purpose SIMD registers.

- **Vector** To resolve the register problem with SIMD structure, general vector instructions can be added as the extension in current superscalar CPU core. The vector instruction extension not only can increase computation power by evoking vector computation units but also efficiently perform operations needed in computing sparse matrix. In deep learning algorithms, particularly in middle layers of neural network there are many zero or near-zero values that might change to zero under quantization. The scatter and gather instructions in vector extension can attack sparse matrix problem well.

- **Coprocessor** The ordinary coprocessor architecture is not good for AI applications as the data path through processor to coprocessor becomes bottleneck because of high data bandwidth requirement for AI applications. A new architecture of coprocessor is proposed in this paper which places coprocessor inside a CPU cluster in parallel with main processors.

IV. IN-CLUSTER COPROCESSOR

In this section we discuss a novel architecture for machine learning. It is called In-Cluster Coprocessor (ICC). It is proposed for being used in mobile computing environment. See Fig 3 for illustration. ICC has the following aspects which is different from other machine learning engines. It is in cache coherence domain while others are mostly in non-cache coherent, IO domain. It is running the same clock speed as the CPU core, which others are running lower clock speed as IO devices. It utilizes main memory including caches as its memory storage, which others have to have large dedicated memory to facilitate the acceleration computation demand. For ICC, virtual address can be used while others are mostly using physical address.

In mobile computing, reducing the SOC area real estate is the key to reduce overall power and minimize the cost. Naturally, allowing resource sharing between Application Processor and AI engine, would allow the design to save resource on dedicated memory and buffers;

Enabling an AI engine to run at CPU core speed can effectively reduce the communication overhead in between the application and acceleration routines running on AI data path.

Programming by virtual address is vital to accommodate AI processing in a complicated application while AI is only part of the application. Virtual address usually involves multiple level of MMU and caches for page tables [5]. If a device uses physical addressing only, data have to be moved between different tasks. In a system with ICC, large amount of data moving is substantially reduced, that brings huge benefits of performance increase and power saving.

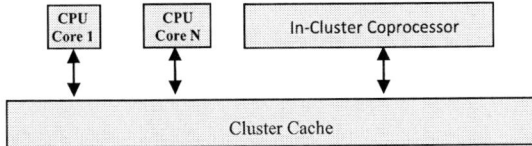

A. ICC Design

1) General Description

Figure 1 shows ICC block diagram and its connection to the L2 cache in a CPU cluster. ICC engine is only targeting on the inference applications here. Acceleration for computations is implemented in the form of 2D array of MACs (multiply-accumulate), which works well in matrix multiplication cases.

- Control Sequencer: This block contains the control for ICC. It facilitates the load requests to L2 cache filling the Fill Buffer and the output of the output buffer. It also has the uTLB to maintain the cache coherency in the system. In addition, it also sequences the data flow among all the rest of blocks.
- Fill Buffer stores load cache lines returned from the L2 cache.
- Staging Arrays converts data into staggered format for MAC engine grid.
- Input Buffer stores input data in staggered format.
- MAC Grid performs matrix multiplication using a 2 dimension systolic array.
- Output Buffer stores output tiles in between passes. The output data is drained to L2 cache through uTLB lookup.

Multiple ICC units can be implemented in a system to improve the overall throughput.

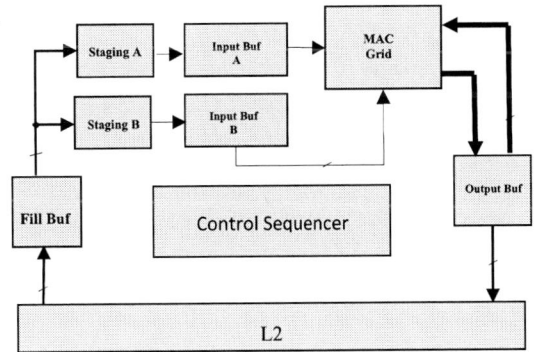

^{a.} Sample of a Table footnote. *(Table f)*

Fig. 4. ICC Design

2) Task Initilization

Task descriptor is formatted in 128 byte memory block. It is 128 byte aligned. During task initialization, the 128 byte data that contain the task descriptor is read and stored in flops.

The task descriptor fields are checked for errors. The ill-formed descriptor triggers an error. The status field in the descriptor is updated with the error code. The task is aborted. ICC signals done to the CPU core with an interruption.

If there are no errors, the next step is for ICC to compute the execution state based on the number of Blocks and Tiles that have already been completed. For a task that had been interrupted, this step allows the task to be restarted at the point where it was interrupted.

3) Task Execution

Execution of the matrix multiplication is through loops on Tiles in a Block. Partial Sums on each Tile are written into Output Buffer. At the last Tile in the Block the result is written back into memory

4) Fill Buffer

The Fill Buffer can be a multiport buffer with a right structure that is used to store cache lines returned from the L2. The buffer serves two purposes.

- It is used to buffer prefetched cache lines that are not ready to be sent to the Staging Arrays because the Staging Arrays are backed up.
- Some half lines may need to be read more than once. For example, a half line that contains the end of one row of a matrix and the beginning of the next row of the matrix is requested from the L2 once but read out of the Fill Buffer twice.

The read port and the write port of the Fill Buffer are half or full cache line wide. When a cache line is read out of the buffer, it goes through a rotator. The rotator shifts the bytes so that they are row-aligned. The row-aligned bytes are then sent to the Staging Arrays.

5) Staging Arrays

The Fill Buffer produces data in row-aligned format. But the Input Buffers store data in the staggered format that is needed for a systolic array. The Staging Arrays hold the data that have been read in row-aligned form until they can be written to the Input Buffers in staggered format.

The output of the Staging Arrays may go through an Input Transform Pipeline before being written to the Input Buffers.

6) MAC Array

The multiply-accumulate grid contains an MxN interconnected array of MAC units. Each MAC unit receives two sources for the multiplication. The sources are passed on to the neighbor to the right and the neighbor below.

Each MAC unit also accepts the accumulated partial sum from the neighbor to the left. Its own partial sum is passed to the neighbor to the right. For the last pass

through the block, this partial sum is also the final computed value of an element in matrix C.

B. Task set up

A task descriptor data structure is memory mapped. Each compute task describes a matrix multiplication and optional pre-processing and post-processing operations.

Software initializes the Read Only Fields with the description of the matrix multiplication task that is to be performed. The Status Fields and Result Field must be initialized to zero before starting the task.

C. Memory Model

Software is responsible for doing the necessary synchronization to make sure that the source matrices are not modified while they are being used.

D. Quatization

Quantization is the process by which floating-point values are represented as fix-point values. Each matrix is quantized separately. I.e. all elements of a matrix are quantized using the same scaling factor, which is different from the scaling factor used for another matrix. ICC supports the signed quantization model. Unsigned quantization is not supported.

ACKNOWLEDGMENT

Part of this paper is from the work done by Liuxi Yang, Will Lin and Andrew Huang while working with UNISOC. Due to ASSCC policy to papers for plenary speech, their names are not in author list. The special acknowledgement for them is presented here for their contribution to this publication.

REFERENCES

[1] David. Patterson, " 50 Years of Computer Archiecture From the Mainframe CPU to the Domain-Specific TPU and the Open RISC-V Instruction Set", p24-p31, International Solid-State Circuits Conference (ISSCC), 2018

[2] Jeff Dean, David. Patterson and Cliff Young, "A New Golden Age in Computer Architecture: Empowering the Machine-Learning Revolution" , p21-p29, IEEE Micro, March/April 2018

[3] Norman P. Jouppi, Cliff Young, and et al, "In-Datacenter Performance Analysis of a Tensor Processing Unit", p1-p12, International Symposium on Computer Architecture (ISCA), June 2017

[4] Manish Arora, Siddhartha Nath, and et al, "Redefinging the role of the CPU in the Era of CPU-GPU Integration", p4-p16, IEEE Micro, November/December, 2012

[5] Denis Foley, John Danskin, "Ultra-Performance Pascal GPU and NVLink Interconnect", p7-17, IEEE Micro March/April 2017

[6] Abhishek Bhattacharjee, "Preserving Virtual Memory by Mitigrating the Address Translation Wall", p6-p10, IEEE Micro, September/October 2017

[7] D Silver. A. Huang and et al "2016, Mastering the game of Go with Deep Neural networks and Tree Search", p529(7587) Nature,

2018 IEEE Asian Solid-State Circuits Conference (A-SSCC)
PROCEEDINGS OF TECHNICAL PAPERS

PROCEEDINGS

November 5 – 7, 2018

Shangri-La's Far Eastern Plaza Hotel,
Tainan, Taiwan

Sponsored by
IEEE Solid-State Circuits Society (IEEE SSCS)

`PROCEEDINGS`

2018 IEEE Asian Solid-State Circuits Conference (A-SSCC)
PROCEEDINGS OF TECHNICAL PAPERS

November 5 – 7, 2018
Shangri-La's Far Eastern Plaza Hotel, Tainan, Taiwan

A 12.4TOPS/W, 20% Less Gate Count Bidirectional Phase Domain MAC Circuit for DNN Inference Applications

Yosuke Toyama, Kentaro Yoshioka, Koichiro Ban, Akihide Sai, Kohei Onizuka

Corporate Research & Development Center, Toshiba Corporation
Kawasaki, Japan
Email: yosuke1.toyama@toshiba.co.jp

Abstract — A small gate count 8 bit bidirectional phase domain MAC (PMAC) circuit is proposed for DNN inference applications targeting IoT edge. PMAC consumes significantly smaller power than standard fully digital MACs, owing to its efficient analog accumulation nature based on Gated-Ring-Oscillator (GRO). Compared with the previous first PoC of PMAC, the bidirectional architecture proposed in this paper achieves 20% less gate count, which is comparable with fully digital MACs, and relaxes system design constraints by eliminating phase error originating in leakage current. Asynchronous readout technique and 2-step DTC for the better system throughput and compact implementation, respectively, are presented for the first time. The PMAC achieves peak efficiency of 12.4 TOPS/W in 28 nm CMOS.

Keywords—DNN accelerator; Phase Domain MAC; GRO

I. INTRODUCTION

Deep learning accelerator (DLA) is a key enabler of edge DNN inference opportunities ranging from typical image recognition to anomaly detection and applications in the rapidly expanding robotics field. The key feature of DLA is the high computational efficiency, or low power consumption, that enables compact housing eliminating a cooling mechanism and realizing long-lasting battery operation of IoT edge devices. Batching is a powerful scheme to minimize the data transfer between cache and processing element and save DLA power consumption. Assuming a batch size of 64 in a fully digital DLA, computation power is three times as large as the memory power [1]. Thus several analog techniques have been explored to save the multiply and accumulation circuit (MAC) power, which dominates the DNN computation energy. One approach is the switched capacitor MAC [2]. But it occupies too much area for a practical design. Another approach is the time domain [3]. Even though it can reduce signal line and shows excellent power efficiency, it is specialized for 1 bit operation. Furthermore, accumulation is not feasible due to the lack of a sampling mechanism for time information.

We recently proposed phase domain MAC (PMAC) [1], which is >20x area efficient than conventional analog MACs[2] and 5x power-efficient than standard fully digital MACs (DMACs) as shown in Fig.1. 8 bit resolution is available in PMAC, which is sufficient, and thus it potentially executes complex tasks such as speech and sensor time series data analysis requires 6-8 bits today[4].

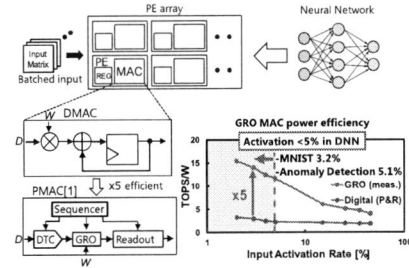

Fig.1 Concept of DLA efficiency improvement by PMAC.

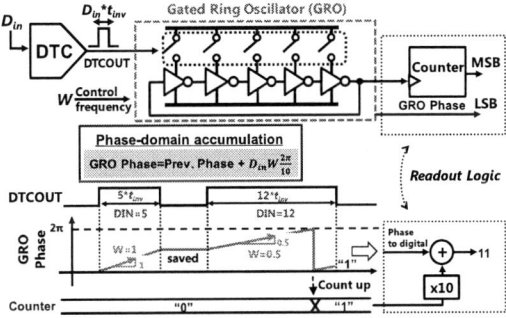

Fig.2 PMAC operation principle.

Fig.2 explains the concept and operation principle of PMAC. GRO composed of a ring oscillator and power gating switches is the core block for multiply and accumulation. GRO oscillation is conducted only when the switch is shorted, and when the switch opens, the phase information is saved [5]. A DTC generates a pulse (DTCOUT) proportional to the input signal (Din), which is applied to every GRO gating switch, and weight signal (W) controls the GRO frequency to enable the "multiply" operation. The phase and the cycles of the GRO are sampled by the flipflops(FFs) and the counter, and the values represent MAC results. In the previous design [1], negative accumulation was performed by the readout logic, which results in larger gate count than that of fully digital MACs.

In this paper, the bidirectional PMAC technique is proposed to relax the readout function and reduces the total gate count by 20%. The architecture is advantageous in terms of leakage tolerance as well.

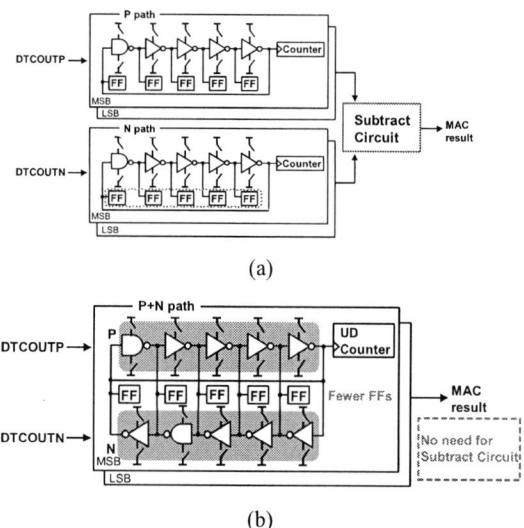

(a)

(b)

Fig.3 (a) Conventional unidirectional implementation and (b) proposed bidirectional implementation.

II. BIDIRECTIONAL PHASE DOMAIN MAC

The original PMAC shown in Fig.3 (a) has a pair of unidirectional GROs for both the positive and negative accumulations, namely p-path and n-path GROs that are mandatory for executing general DNN inference applications. This doubles the numbers of FFs and counters to read the phase and oscillation cycles. In addition, the final subtraction is performed in the digital readout logic which requires almost 3000 gates. To compress the logic scale, we propose the bidirectional GRO architecture shown in Fig.3 (b). In this scheme, phase path is shared by inverter chains in opposite directions, and hence both positive and negative phase accumulations are available inside the GRO core. This eliminates a subtraction block in the readout logic, leading to a 52% gate count reduction of the logic part. The number of FFs is simply halved as well. Since the output sampling counter should be modified so that it can count both up and down, the two independent counters in the original architecture are merged in a single up/down counter. The resulting number of counter logics is reduced by 10%. Fig.4 compares the total gate counts to implement 8bit MAC circuits in fully digital MAC, the original PMAC, and the proposed bidirectional PMAC. The bidirectional architecture requires roughly the same gate counts as the standard fully digital design, whereas the previous unidirectional PMAC requires in excess of 20% more gates.

In addition, the proposed bidirectional architecture is advantageous for phase retention, which is important for accurate MAC operations. In the case of the original unidirectional implementation, while DTC pulse is input to p-path GROs for instance, the n-path GRO is paused and must keep the original phase for more than few tens to hundreds of nanoseconds depending on the continuity of series positive inputs. But in a deep submicron process, the leakage current of the inverter stages pull up or down to the common mode voltage even if the gating transistors are turned-off, and

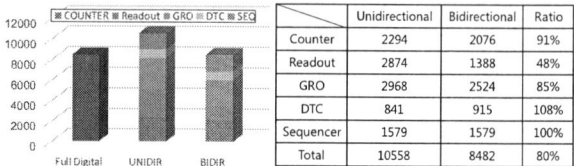

	Unidirectional	Bidirectional	Ratio
Counter	2294	2076	91%
Readout	2874	1388	48%
GRO	2968	2524	85%
DTC	841	915	108%
Sequencer	1579	1579	100%
Total	10558	8482	80%

Fig.4 Gate count comparison with full-digital and unidirectional PMAC.

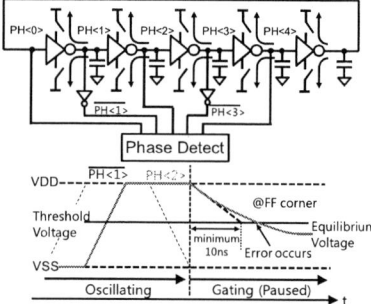

Fig.5 Leakage current and error mechanism of the paused GRO.

accumulates phase error as shown in Fig.5. At the worst FF corner in 28nm CMOS, the paused core can retain the phase error within ±1LSB for only up to around 10ns, and hence active retention is mandatory to guarantee the phase accuracy for arbitrary series of input data.

In the proposed bidirectional architecture on the other hand, all the GRO output nodes are shared by p-path and n-path, and remain updated regardless of the sign of input data. The GRO pauses only when sequencer loads next data, and the time is approximately 1 ns, which is much shorter than the minimum retainable time of 10ns. This drastically minimizes the phase error caused by the leakage current and does not require any special care in the system design.

III. 8 BIT BIDIRECTIONAL PMAC IMPLEMENTATION

A. System Architecture with Asynchronous Readout Sequence

The PMAC circuit shown in Fig.6 is designed to demonstrate the concept of bidirectional operation. To realize an 8 b multiplication function with minimum area, a dual-GRO architecture is employed in which each GRO consists of the bidirectional core. The gate count saving discussed in the previous section simply stands even in this architecture. 7 b weight and frequency resolutions are realized by assigning two individual GROs to 3b MSB and 4b LSB, respectively. The readout circuit generates the MAC output by summing the bit-shifted MSB GRO output to the LSB GRO output. Each GRO has a segmented configuration that is described in the next section. The oscillation frequency proportional to the number of inverters turned-on is linearly controlled depending on W and achieves phase domain multiplication with 7 b resolution.

The asynchronous sequencer is a key block in terms of the better throughput in the GRO-based architecture. Since the GRO operation time is proportional to the value of Din, setting

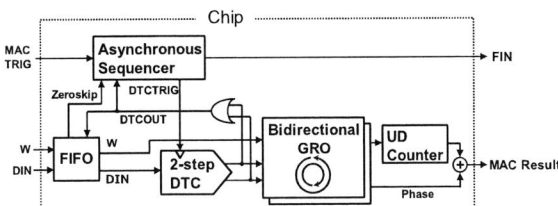

Fig.6 PMAC architecture based on bidirectional GRO.

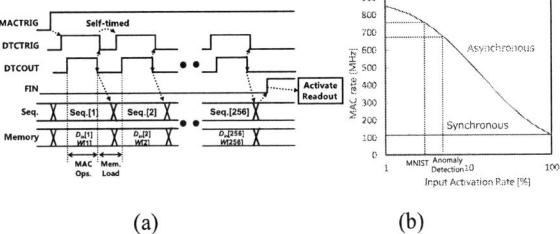

(a)   (b)

Fig.7 (a) Asynchronous operation timing chart and (b) improvement of average MAC rate.

the clock frequency to assure operation at the longest condition greatly worsens the MAC throughput. Furthermore, while high-speed MACs require GHz clock to operate, the clock generation and distribution consume a significant amount of power. Asynchronous operation shown in Fig.7(a) is proposed to achieve high MAC throughput without high-speed clocks to improve the DLA system power efficiency; only a trigger signal (MACTRIG) is required for the entire MAC operation. When the MACTRIG rises, DTC generates a pulse (DTCOUT) proportional to Din to oscillate the GRO. Then, reflecting the DTCOUT falling edge, the MAC operation moves to the next sequence and the SRAM-based memory circuit switches the output. Moreover, if either Din or W is 0, the memory circuit will output a Zeroskip signal to skip the next sequence, since accumulation is not required. The asynchronous operation enhances the GRO MAC throughput significantly as estimated in Fig.7(b), and thus the operation time is minimalized against Din activation rate. The upper DLA system can easily take synchronization by monitoring the FIN signal, which rises after all 256 MAC operations have been completed.

B. Bidirectional GRO Core

Bidirectional phase manipulation is implemented by connecting each node of the forward-direction and reverse-direction GROs as shown in Fig.8. Both the inverters and gating switches are segmented in this configuration, and thus both Din and W can be applied through the gating switches. The DTC generates control pulses DTCOUTP/DTCOUTN depending on whether the product of D and W is positive or negative. By taking AND of the DTC pulse and the W, one of the two directions of GROs is selected to operate. De-glitch circuit to prevent incorrect count described in [5] is modified as shown in Fig.8 in order to adjust bidirectional phase shift.

C. Area Efficient 2-step DTC

The DTC has an impact on gate counts. It would increase exponentially to the bit if DTC were composed by inverter

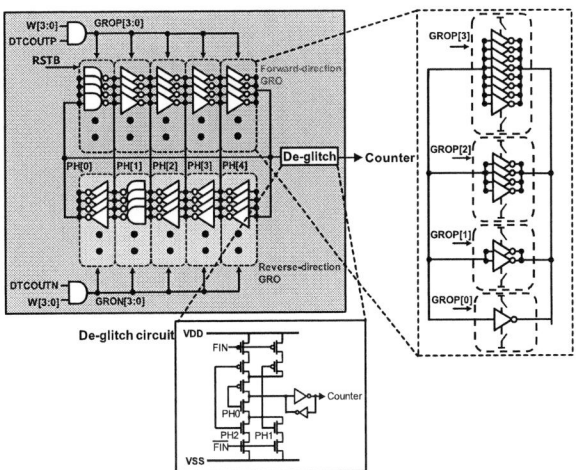

Fig.8 Bidirectional GRO Core.

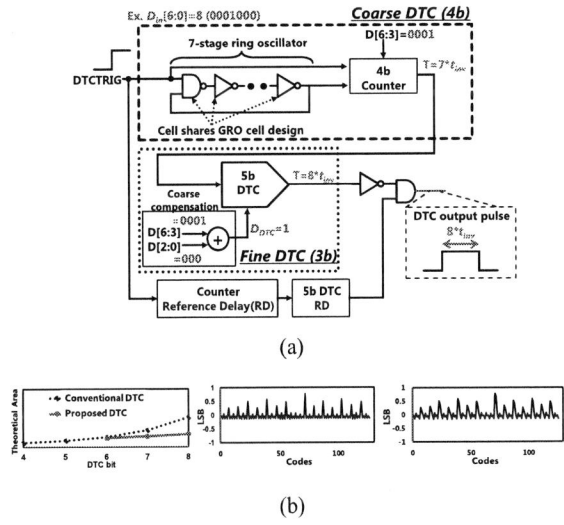

(a)

(b)

Fig.9 (a) Area Efficient 2-step DTC and (b) Area comparison with conventional DTC, DNL and INL.

chain. A 7 bit 2-step DTC circuit is proposed as Fig.9 (a) to minimize the high-resolution DTC area: Whereas conventional DTC circuit area increases exponentially with resolution, the proposed 2-step DTC utilizes a 4 bit counter-based DTC as the coarse DTC; the area increase is linear upon resolution in our proposed circuit, since only the counter impacts the area, and at 7b and 8b resolution, the area savings exceed 40% and 60%, respectively. Simulated DNL and INL of the DTC with LPE are shown in Fig.9 (b) and the non-linearity error is within 1 LSB. The gain mismatch between DTC and GRO due to variations affects the MAC results, but it can be calibrated in peripheral logic to be implemented on the accelerator SoC.

IV. CHARACTERIZATION RESULTS

The bidirectional PMAC circuit shown in Fig.10 is fabricated in 28nm CMOS and its outline is 90 um x 45 um.

978-1-5386-6414-8/18 $31.00 © 2018 IEEE   3

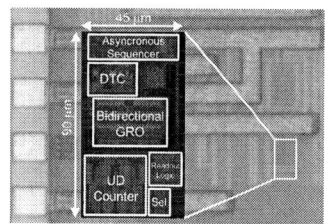

Fig.10 Chip photo and Layout

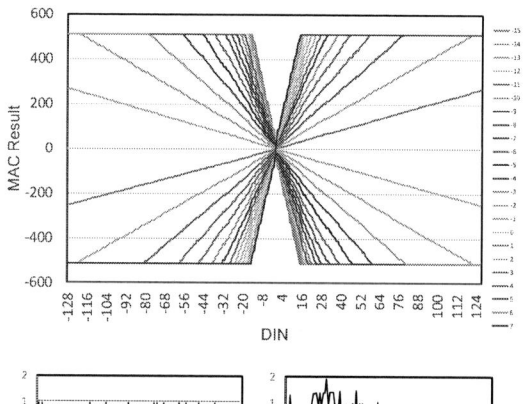

Fig.11 Basic input-output characteristics of a single bidirectional GRO MAC.

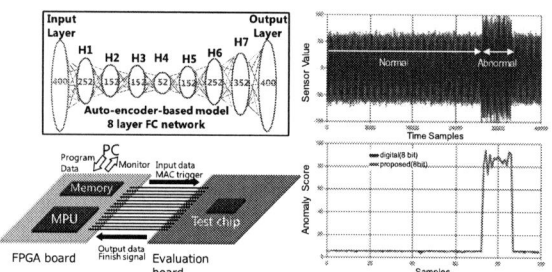

Fig.12 Evaluation setup and anomaly detection demonstration.

Fig.11 plots the measured MAC results at each input data Din and weight W. Good linearity is confirmed as the DNL and INL were no more than 1 LSB and 2 LSB, respectively.

MNIST and anomaly detection task sets were performed in the test setup shown in Fig.12. The host MPU implemented on the FPGA loads MAC tasks in a maximum of 256 groups to the test chip, and outputs the recognition and/or detection results. Fig.12 summarizes the auto-encoder-based anomaly detection network and the execution results assuming IoT edge. The network trained in a normal condition has 400 inputs as series sampling data of the target waveform, and predicts the next 400 samples[6]. The anomaly score is extracted based on the difference between the predicted waveform and the actual

Table I. Performance comparison with previous analog MACs.

	This Work		[1]		[2]	[3]
Domain	Phase		Phase		Charge	Time
Process	28 nm		28 nm		40 nm	65 nm
Resolution	8 bit		8 bit		3 bit	1 bit
MAC Area [μm²]	960(*)		1200		12000	13000
Application	MNIST	Anomaly Detection	MNIST	Anomaly Detection	CIFAR 10	MNIST
MAC rate [MHz]	753	675	780	700	1000	N.A.
Efficiency [TOPS/W]	12.4	10.3	14	11.6	8.77	77
Efficiency [TOPS/W*Bit]	99.2	82.4	112	92.8	26.3	77
Area Efficiency [TOPS/μm²]	1563	1406	1300	1167	167	N.A.

*Calculated area from gate count needed to implement

one. As shown in the waveform, the bidirectional 8bit PMAC outputs closely resemble the anomaly score with the 8 bit fully-digital execution results as a reference.

Table I compares recently reported analog MACs. Compared with the unidirectional PMAC, the bidirectional architecture saves 20% gate counts in exchange for 3.5% throughput degradation due to the larger parasitic load capacitances. The sharing topology of the delay nodes shown in Fig.3(b) increments both gate and drain originated capacitances and significantly degrades the oscillation frequency, however, its impact on the overall throughput is small as described above because there exist other common delay overheads related to read and write operations. In addition, the silicon area efficiency of proposed PMAC is higher than conventional arts.

V. CONCLUSIONS

Power efficient, small gate count, precise phase domain MAC circuit was demonstrated in 28 nm CMOS for DNN inference applications. The proposed bidirectional GRO-based architecture reduces gate count for the whole architecture by 20%. Moreover, the potential phase error due to leakage current is fundamentally eliminated, since bidirectional GRO continues operation for both positive and negative inputs. These advantages of the proposed scheme are retained in further-deep-scaled technologies whereas they are unavailable in the conventional analog MACs. Two other key features of the proposed scheme, namely, asynchronous readout and 2-step DTC techniques are presented.

REFERENCES

[1] K. Yoshioka, et al, "PhaseMAC: A 14 TOPS/W 8bit GRO based Phase Domain MAC Circuit for In-Sensor-Computed Deep Learning Accelerators," Symposium on VLSI Circuits, pp.263-264, 2018.

[2] E. Lee et al, "A 2.5GHz 7.7TOPS/W Switched-Capacitor Matrix Multiplier with Co-designed Local Memory in 40nm," IEEE ISSCC, pp.418-419, 2016.

[3] D. Miyashita, et al, "A Neuromorphic Chip Optimized for Deep Learning and CMOS Technology With Time-Domain Analog and Digital Mixed-Signal Processing," IEEE JSSC, Vol. 52, No.10, pp.2679-2689, 2017.

[4] N. Jouppi, et al, "In-Datacenter Performance Analysis of a Tensor Processing Unit," IEEE ISCA, pp.1-12, 2017.

[5] M. Straayer et al, "A multi-path gated ring oscillator TDC with first-order noise shaping," IEEE JSSC, Vol. 44, No.4, pp.1089-1098, 2009.

[6] S. Chauhan et al, "Anomaly detection in ECG time signals via deep long short-term memory networks," IEEE DSAA, pp.1-7, 2015.

2-2 (8095)

IEEE Asian Solid-State Circuits Conference
November 5 - 7, 2018/Tainan, Taiwan

A Power and Area Efficient 2.5-16 Gbps Gen4 PCIe PHY in 10nm FinFET CMOS

Shenggao Li, Fulvio Spagna, Ji Chen, Xiaoqing Wang, Luke Tong, Sujatha Gowder, Wenyan Jia, Roan Nicholson,
Sita Iyer, Rui Song, Lily Li, Meng-hung Chen, Amanda Tran, Michael De Vita, Deepar Govindrajan, Marcus
Pasquarella, Dave Bradley*, Frank Verdico*, Matt Duwe*, Eric Lee*, Michelle Wigton*
Intel Corporation, Santa Clara, CA, United States, *Fort Collins, CO, United States
Shenggao.Li@Intel.com

Abstract— **This paper presents a 2.5-16 Gbps Gen4 PCIe transceiver with 3-tap Tx EQ, and 8-tap Rx DFE in a 10nm FinFET CMOS technology. A low latency digital CDR is designed supporting a flexible timing recovery scheme. The CDR uses a 3-stage ring DCO, with a low-noise cascaded NMOS voltage regulator to provide 40dB PSRR with an operating temperature range from -40C to 125C. A senary-weighted hybrid C-DAC in the DCO achieves a monotonic fine tuning characteristic with a measured DNL of 0.3 LSB. The power efficiency is 7.5pJ/b, with the smallest reported footprint of 129.6x945.0 um² per link.**

Keywords—Gen4 PCIe, DCO, DFE, CTLE, CDR

I. INTRODUCTION

The world is embracing a data centric era. The growth in computing power, thanks to several decades of semiconductor advancement following Moore's Law, and the exponential growth of data, attributed to the fast proliferation of Internet of Things, Social Media, Mobile Internet, Machine Learning and Artificial Intelligence, form a virtuous cycle of economic growth. To this end, there is a growing need for more IO bandwidth between ICs, from die to die, socket to socket, machine to machine, and building to building. As of this writing, plenty of work has been done to tackle the design challenges around speed, power, and performance at per pin data rate as high as 100Gbps in the Ethernet families. In the micro-processor space, the pace of per pin data rate is trending at 16Gbps (Gen4 PCIe), with 32Gbps (Gen5 PCIe) on the horizon. The per pin data rate is practically constrained by the platform form factor and cost structure of the eco-system.

This paper focuses on a Gen4 PCIe design at 16Gbps link rate [1-3]. The PCIe family is back compatible, which means a Gen4 device must support Gen1/2/3 rates of 2.5, 5.0, and 8.0Gbps. The targeted insertion loss of the channel (including package loss at each end of the channel) is 5-30dB at 8GHz.

Section II below provides a top-level view of the Rx/Tx architecture. Section III focuses on individual design challenges and circuit techniques. Section IV presents silicon results with performance comparison to similar designs.

II. RX/TX ARCHITECTURE

Fig. 1 shows the Rx architecture. It consists of a RC based all-pass tunable structure to provide on-die AC coupling capability while creating a low frequency zero-pole pair for tail

ISI cancellation, followed by a 4-stage CTLE/VGA block for gain control and peaking control with a high-frequency zero-pole pair. Non-integrating summers are used in the odd/even interleaves to implement an 8-tap Decision Feedback Equalizer (DFE). Each interleave includes 3 samplers which can be configured as data/error samplers for baud-rate timing recovery, or data/edge samplers for 2x-sampling timing recovery. The timing recovery logic controls a DCO (digital CMOS ring oscillator) followed by a duty cycle correction circuit to generate a differential sampling clock. Complex functionalities, including offset calibration, channel adaptation, timing recovery, dfx, and power management, are implemented with standard digital logic to ensure design robustness.

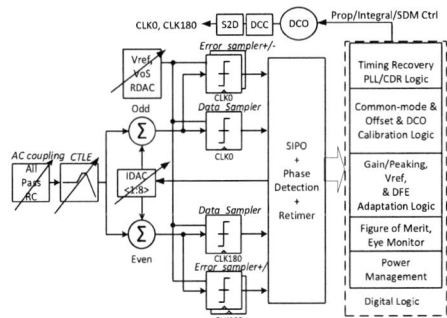

Fig. 1. Rx Architecture

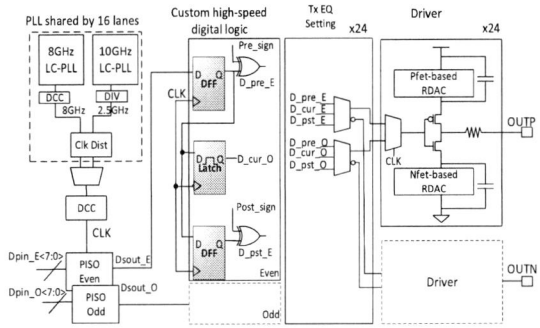

Fig. 2. Voltage Mode Tx with 3 Tap EQ

Fig. 2 shows the Tx PLL, PISO and driver. Two LC VCO based PLLs operating at 8GHz and 10GHz are used to support

978-1-5386-6414-8/18 $31.00 © 2018 IEEE

the 4 PCIe data rates simultaneously. CMOS clock distribution tree is adopted for low power operation. The Tx is a voltage mode driver with 24 identical slices configurable with up to - 9.5dB de-emphasis and 6dB pre-shoot. The Tx output impedance is calibrated symmetrically in each slice through the use of Pfet and Nfet DACs, together with a linear metal resistor. The high frequency output of the Tx is further boosted by adding capacitors in parallel with fixed resistance elements within the driver stack.

III. CIRCUIT SOLUTIONS

A. Cpad reduction at Rx/Tx front-end

Parasitic capacitance, resulting from ESD devices, programmable termination resistors, metal routes, and intrinsic device caps seen at the Rx/Tx front-end pads, affects the insertion loss, and return loss performance. 16Gbps is an inflection point where Tcoils or Pi-coils are deemed necessary to mitigate the Cpad impact. In this design, we chose not to use coils to avoid mutual coupling and route congestions. Design/layout optimization is the key in reducing the pad capacitance to meet 250V CDM ESD spec while satisfying the -6dB return loss specifications without Tcoils.

B. Co-designed CTLE & VGA

An all pass RC circuits preceding a continuous time linear equalizer is shown in Fig. 3 to provide on-die AC coupling and low frequency Rx equalization. It uses a simple feed forward C_1 with a series R_2, C_2 load to implement a low frequency zero-pole pair for cancelling the long tail ISI not cancelled by the 8-tap DFE. R0, R1 provides a DC path to mitigate the baseline wander. They are sized to provide a flat DC-AC transition. For

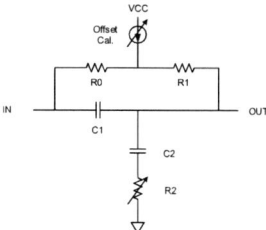

Fig. 3 All-pass RC filter (half-circuit) at Rx front-end

simplicity of deriving a first order transfer function, we can ignore the effect of the relatively large R_0 and R_1. The above structure provides a zero at $1/(2\pi R_2 C_2)$ and a pole/zero ratio of $1+C_2/C_1$. A current DAC between R_0 and R_1 is used for offset correction.

Fig. 4 shows the 4-stage CTLE that provides a high frequency zero-pole pair to compliment the TXEQ and DFE capabilities. It achieves a nominal peaking of 8dB relative to DC at Nyquist frequency. It also provides a nominal DC gain range of [-18, 7] dB to cover a wide variety of OEM channels. The high frequency zero-pole pair is implemented in stage 3 with a

parallel RC as source degeneration. The attenuation is implemented in multiple stages in the forms of source degeneration (stage1 and 4) and load tuning (stage2). Offset cancellation is done at the output of stages 1 and 3 to keep residual offset of each stage to a small range. Continuous time offset cancellation (CTOC) is also implemented to handle the offset change due to voltage, temperature drift and gain setting change during amplitude adaptation.

The upper half of Fig. 4 shows the generation of the biasing voltage, Vgate_hv for the NMOS loads in each stage. A replica circuit mirroring the diff-pairs and tail current sources in each stage is included to track the temperature and voltage drift in functional mode.

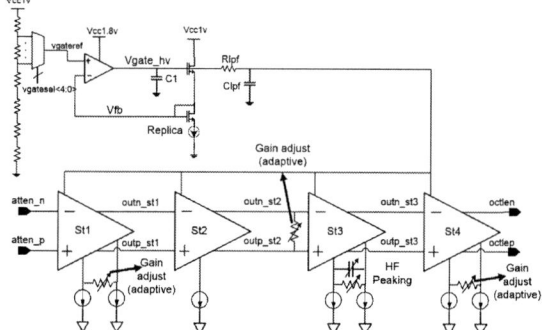

Fig. 4 CTLE topology

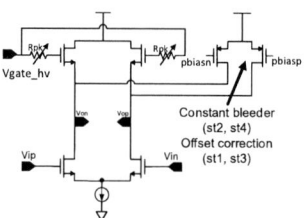

Fig. 5 CTLE single stage schematic

Each stage is a classic NMOS differential pair (Fig. 5) with source degeneration, except that stage 2 has no source degen to achieve extra gain. In order to avoid using spiral inductors to save area, active inductive peaking load is implemented in each stage for extra peaking and bandwidth extension. The load device is an NMOS source follower with its gates biased through a peaking resistor, $R_{pk,}$ implemented with PMOS devices operating in linear region. A bleeder circuitry is used for gain boosting and offset cancellation.

C. DFE summer nonlinearity and sampler kickback noise

A non-integrating summer with 8-tap DFE is used in each interleave to drive 3 samplers (Fig. 6). Similar to the CTLE, active inductor loads are used in the non-integrating summer for bandwidth extension. Unlike a linear resistor load, the output impedance of the active inductor load is a function of the summer output signal. Normally, this nonlinearity adds no significant burden to the system during the data sampling instant when the summer outputs are equalized by the DFE feedback at a positive clock transition. However, at the negative clock transition, when the samplers entering its reset phase, DFE feedback is not at all optimal at this instant. The common-mode kickback noise will be converted by the non-symmetric output impedance to diff-mode noise, which represents as unsettled data-dependent ISI. To mitigate this residue ISI, diode-connected NFETs has been used in the summer loads to counter the nonlinearity of the inductive load. Further, a current injection circuit is added to the sampler to shift the operating voltage of the sampler internal nodes to minimize charge-sharing which is the source of kickback noise.

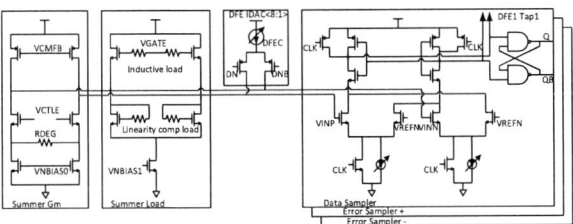

Fig. 6 Linearized DFE summer with 8 Tap DFE

D. All Digital DCO with Low Noise Regulator

The clock and data recovery (Fig. 7) is fully digital with proportional, integral, and sigma-delta paths that control the cap elements in a 3-stage inverter based CMOS ring-oscillator operating between 2.5 and 8 GHz. A cascaded 2-stage NMOS source follower is implemented to provide supply noise rejection with up to 40dB PSRR at 10MHz. Low thermal noise and high PSRR are the primary consideration for choosing this specific regulator topology. The RC filters (BW=100kHz) on the NMOS gate provide the needed PSRR. Given the open-loop nature of this NMOS regulator, multi-rounds of regulator calibration are necessary to tune the regulator to its optimal operating condition. The diode connected NMOS, along with digital temperature tracking allows the DCO to adapt to temperature changes during normal operation or aggressive CPU power management events.

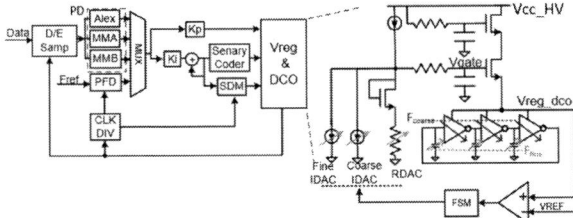

Fig. 7 Conceptual all digital CDR with CMOS 3-stage ring oscillator

The voltage sensing loop, consisting a comparator, a FSM, a coarse/fine IDAC, a RDAC, and a selectable VREF reference, maintains Vreg_dco within a preset range (not too low to cause jitter degradation, and not too high to cause reliability risk). The RDAC and coarse IDAC work together to compensate for process and voltage variations at power on. The fine IDAC (8-bit thermometer DAC) provides fine granularity voltage control on the fly. The coarse frequency tuning is done by varying gm of each stage, whereas the fine frequency tuning (including integral, proportional, fractional) is achieved through a hybrid C-DAC. Table *1* illustrates the coding scheme of the C-DAC. It consists of 42 bit of thermometer coded cap arrays (of 6 unit caps each, evenly distributed on 3 inverter stages), and 3 bit of binary weighted cap arrays (of 4, 2, 1 unit caps). Each "C" stands for one unit cap. This senary (base-6, only 0-5 values are valid) arrangement of the unit cap cells simplifies the physical implementation of the C-DAC.

Table 1. DCO C-DAC array with unit cap C for fine frequency tuning

	Thermo-code			Binary-code		
	t41	...	t0	b2	b1	b0
stg1	CC	...	CC	C		C
stg2	CC	...	CC	CC	C	
stg3	CC	...	CC	C	C	

E. Programmable CDR with Low Loop Latency

The CDR has 3 Phase-Detection (PD) functions (Alexander, Mueller-Muller type-A and type-B [4]) to handle different use scenarios. When operating in Alexander mode, the PD basic timing function can, at higher data rate, be modified to compensate for pre-cursor ISI by shifting the sampling point so as to maximize the difference between the cursor and the pre-cursor (therefore minimizing the effect of the uncorrected pre-cursor ISI). When operating in baud-rate mode the two available PD timing functions are Mueller-Muller type-A or type-B; the type-A algorithm which, because of the DFE, needs to satisfy a zero first pre-cursor requirement, is better suited for lower IL channels. At high IL where satisfying a zero first pre-cursor would require large amount of TX Equalization pre-emphasis (and would lead to a significant second pre-cursor), the type-B is better suited and has been found to yield very symmetrical eye. The ability to choose between different timing functions allows for an additional degree of freedom to optimize the receiver vertical and horizontal margins when operating at different data rates and over a large variety of interconnects and Tx equalization capabilities. Particular care has been paid to the CDR loop latency which, left unchecked, could result in limit cycle or limit the CDR loop bandwidth. A fast phase error path (excluding integral & SDM elements) is designed in custom logic to minimize the clock cycles along Prop. Gain path to DCO tuning cap arrays. The total latency (from sampling to DCO cap change) is 12 UI at Alexander Phase detector mode and 24UI at Mueller-Muller Phase

Detector modes. (see [3] regarding latency on the P/I adder path which is a limitation of phase interpolator based CDR).

IV. MEASUREMENT RESULTS & CONCLUSIONS

Fig. 8 shows the die photo of a Gen4 PCIe x16 port. Fig. 9 is the Tx eye with no EQ applied. The measured Rj is 0.4 ps. One measured fine frequency tuning curve and the DNL of the Rx DCO is shown in Fig. 10. The DNL is within +/-0.3 LSB, indicating a monotonic tuning curve. Fig. 11 shows the DCO Pnoise at 8GHz in open-loop setup. The measured Pnoise is -72dBc/Hz at 1MHz offset and -109dBc/Hz at 10MHz offset. The jitter tolerance result [9] is shown in Fig. 12, a healthy jitter margin is achieved at BER of 1e-12 with 30dB insertion loss.

This work is compared to prior state of the art NRZ transceivers at similar speeds (Table 2). This work achieved robust performance with smallest area due to process scaling and circuit choices (i.e., minimal use of passive inductors). Channel characteristics and targeted product yield may affect power efficiency of a design. The power of this work is 7.5pJ/bit (Rx =78mW, Tx = 36mW, amortized PLLs/Clock=6.6mW). With more aggressive power delivery and clock distribution schemes, as demonstrated by [6] (4.3pJ/bit), and [7-8] (7.3-7.5pJ/bit), power efficiency improvement is possible.

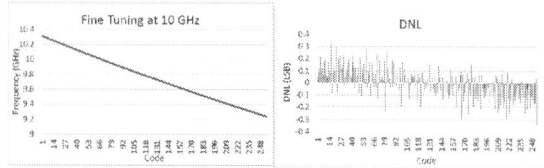

Fig. 8 PCIe Gen4 x16 port die photo

Fig. 9 Measured Tx eye (Rj = 0.4ps)

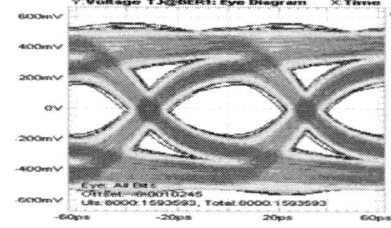

Fig. 10 Measured DCO fine frequency tuning curve and DNL

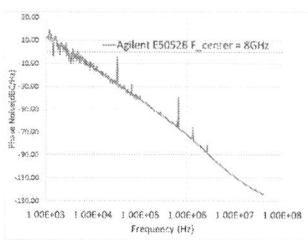

Fig. 11 Measured RX DCO Phase Noise

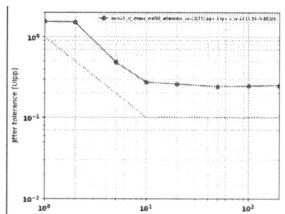

Fig. 12 Gen4 Rx JTOL test

Table 2 Performance comparison

Reference	This work	[1]	[2]	[5]	[6]
Data Rate	2.5-16 Gbps	0.5-16.3 Gbps	16 Gbps	28 Gbps	8.5-13 Gbps
Process	10nm FinFET CMOS	16nm FinFET CMOS	45nm SOI	28nm CMOS	28nm CMOS
Modulation	NRZ	NRZ	NRZ	NRZ	NRZ
Power	7.5 pJ/bit	13.4 pJ/bit	24.1 pJ/bit	10.5 pJ/bit	4.3 pJ/bit
Area	0.122 mm²/link	4.34 mm²/Quad	0.272 mm²/link	N/A	0.22 mm²/link
Insertion Loss	30 dB/Nyquist	28 dB/Nyquist	32 dB/Nyquist	40 dB/Nyquist	35 dB/Nyquist
Rx EQ	8 tap DFE	11 tap DFE w loop unroll	12 tap DFE w loop unroll	14 tap DFE w loop unroll	5 tap DFE
Tx EQ	3	3	3	5	4
Supply	0.8/1.0/1.8	0.9/0.9/1.2	N/A	1/1.25	1

REFERENCES

[1] M. Erett et al., "A 0.5–16.3 Gbps Multi-Standard Serial Transceiver With 219 mW/Channel in 16-nm FinFET," in JSSC, pp. 1783-1797, July 2017

[2] G. R. Gangasani et al., "A 16-Gb/s Backplane Transceiver With 12-Tap Current Integrating DFE and Dynamic Adaptation of Voltage Offset and Timing Drifts in 45-nm SOI CMOS Technology," in JSSC, pp. 1828-1841, Aug. 2012

[3] P Francese, et al., "A 16 Gb/s 3.7 mW/Gb/s 8-Tap DFE Receiver and Baud-Rate CDR With 31 kppm Tracking Bandwidth", in JSSC, pp2490-2502, Nov. 2014

[4] K. H. Mueller and M. Muller, "Timing recovery in digital synchronous data receivers," IEEE Trans. Commun, vol. COM-14, pp. 516-530, May 1976

[5] B Zhang, "A 28Gb/s Multistandar Serial Link Transceiver for Backplane Applications in 28nm CMOS", JSSC 2015

[6] N Kocaman, "A 3.8mW/Gbps Quad-Channel 8.5-13Gbps Serial Link With a 5 Tap DFE and 4 tap Transmit FFE in 28nm CMOS", JSSC 2016

[7] H Miyaoka, "A 28.3 Gb/s 7.3 pJ/bit 35 dB backplane transceiver with eye sampling phase adaptation in 28 nm CMOS", VLSI 2016

[8] S Parikh, "A 32Gb/s Wireline Receiver with a Low-frequency Equalizer, CTLE and 2-Tap DFE in 28nm CMOS", ISSCC 2013

[9] PCIe 4.0, http://www.pcisig.com

40-nm 64-kbit Buffer/Backup SRAM with 330 nW Standby Power at 65°C Using 3.3 V IO MOSs for PMIC less MCU in IoT Applications

Yoshisato Yokoyama, Tomohiro Miura, Yukari Ouchi, Daisuke Nakamura,
Jiro Ishikawa, Shunya Nagata, Makoto Yabuuchi, Yuichiro Ishii and Koji Nii

Renesas Electronics Corporation, Tokyo, Japan
{yoshisato.yokoyama.jx, koji.nii.uj}@renesas.com

Abstract— **An effective standby power reduction of buffer/backup SRAM in MCU is proposed for power module IC (PMIC) less edge system in IoT applications. The proposed SRAM macro is implemented using 3.3 V thick-gate-oxide IO MOSs for effectively reducing the leakage power with source bias control techniques. Four multiples interleave wordline circuitry is also introduced to reduce read and write operating power. A test chip with 64-kbit SRAM macro is designed and fabricated using 40-nm technology. The measured data show that the leakage power is 330 nW at 65°C (47 nW at 25°C), which is 1/140 of other works. The read/write power is reduced by 60% by interleave wordline circuitry.**

Keywords—Thick-gate transistor, SRAM, low leakage, standby power, dynamic power, retention

I. INTRODUCTION

Recently, in the rapidly growing IoT market, it is expected that all applications will be connected wirelessly. Thus, there are strong demands for extremely low-power operation for longer battery life. An edge equipment consists of some sensors and low-power micro-controller-unit (MCU) with embedded non-volatile-memory (NVM) and RF circuit. To reduce the total BOM cost, the edge system is required to remove the power module IC (PMIC), which is often used in the mobile system for suppling multi-voltages to SoCs/MCUs. Fig. 1 shows required single power supply system without PMIC in the IoT applications, whereas high performance mobile systems require the power management system with PMIC. To implement the single power supply system with a buttery, MCUs typically have voltage down converters (VDC) internally. On the intermitted operation in the edge equipment, the MCU is computing along with the obtained data from sensors in the active mode. Meanwhile almost systems are shut down in the standby/sleep mode. However, the latest data from sensors buffered in the embedded memory of MCU are needed to be retained before sleeping. The embedded NVM has a good data retention characteristic with less leakage power, however the endurance has limitations and writing/programing power and speed are not met the target of such embedded buffer memories.

In the work, we propose the buffer/backup (BB) SRAM using 3.3 V IO MOSs for PMIC less low-power MCU in IoT applications. Using volatile memories with low leakage current

have been reported [1,2,5,7,11,12], however, those are required either an additional process steps or custom process to fabrication. increasing the chip cost. proposed BB SRAM macro can be implemented without any process customizations.

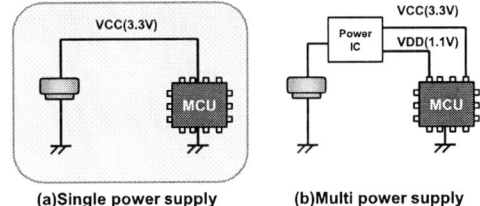

(a)Single power supply **(b)Multi power supply**

Fig. 1 Power-supply for MCU.

II. BUFFER/BUCKUP SRAM WITH 3.3 V THICK-GATE-OXIDE IO MOSS

Fig. 2 (a) and 2(b) show the conventional power management system and proposed one using on-die VDCs. In the conventional system, main blocks are shut-down by power gating and no standby power in the corresponding VDC, becoming almost zero standby power in these blocks. However, in the other always power-on block, the VDC typically has several μA standby current [3] to generate the core voltage of 1.1 V in 40-nm technology to retain the SRAM data and wakeup/sleep system control logic. On the other hand, in the proposed power management system, always power-on blocks including BB SRAM and logics are implemented in the 3.3 V IO MOS regions without any VDCs as shown in Fig. 2 (b). In this case, there is no leakage power of VDC, only consuming the leakage powers in BB SRAM and system logics.

To reduce the total standby power of BB SRAM in MCU, we propose a 6T single-port SRAM bitcell using 3.3 V thick-gate-oxide IO MOSs. Fig. 3(a) illustrates the circuit of 6T SRAM bitcell. It has separated sources which are shown in Fig. 3(a) as VDD/ARVDD and VSS/ARVSS, respectively. The 6T SRAM bitcell is optimized by changing both gate length (L) and gate width (W) for each pull-up (PU) PMOS, pull-down (PD) NMOS, and pass-gate (PG) NMOS to reduce the leakage current keeping with acceptable area and access speed. Optimized L/W ratio of each PU, PD and PG is shown in Fig. 3(b). Fig. 3(c) illustrates the layout of proposed 6T SRAM bitcell, which area is 2.888

978-1-5386-6414-8/18 $31.00 © 2018 IEEE

μm^2. Source bias control techniques [1, 4-6] are effective for reducing standby leakage power in SRAM bitcells. In this work, we apply 3.3 V supply voltage for proposed SRAM macro including bitcell arrays, it has more room for source bias control in the operating margin point of view. Then, both VDD source bias and VSS source bias techniques are introduced to effectively reduce the leakage power in SRAM bitcell array. Fig. 4 shows the circuit diagram of the bias controller in the proposed BB SRAM macro. If input signal "RS" is set to "H", the SRAM macro transits to resume standby mode that can hold the stored data with less leakage current than normal standby mode. The signal "NMA[2:0]" are pins to adjust the bias levels of ARVDD and ARVSS in resume standby mode.

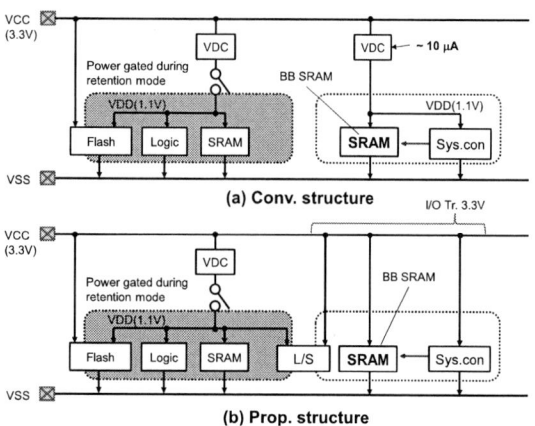

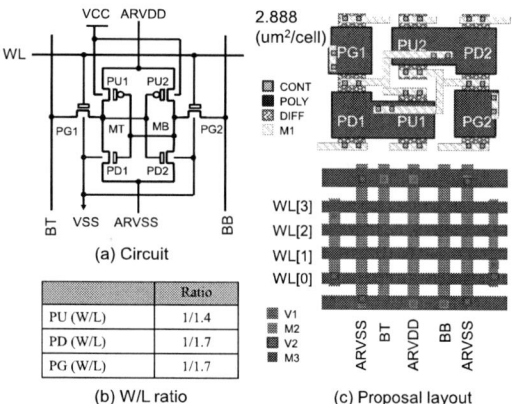

Fig. 2 Power management system using on-die voltage down converter (VDC).

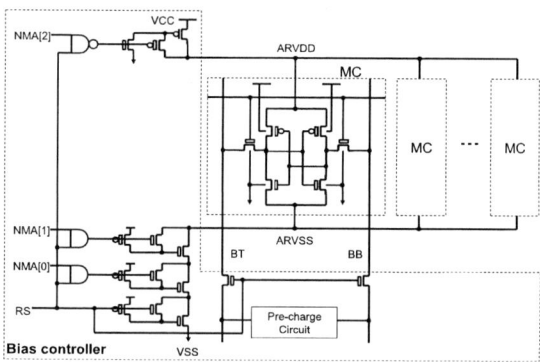

Fig. 3 6T SRAM bit cell using 3.3 V thick- gate-oxide IO MOSs

Fig. 5 shows the simulation result of the leakage current reduction. The bias controller lowers the voltages of ARVDD and bitline (BL), and raises the voltage of ARVSS, and the leakage current is reduced by 98%. All simulations are executed at the worst condition of FF (NMOS:Fast and PMOS:Fast) process corner, 3.6 V supply voltage (+10%), and 65°C. Fig. 6(a) shows the temperature dependencies of the read static noise margin (SNM) [8], write margin (WM) defined by write-trip-

point [8] of proposed SRAM bitcell at -6.0 σ, 2.7 V and each worst condition. Fig. 6(b) shows simulation results of the retention margin obtained by [8] in the both case of normal standby (w/o source bias) and resume standby (w/ source bias) conditions. The biased voltages generated by Fig. 4 are reproduced and reflected in the result of Fig. 6(b). The simulation results show that the read SNM and WM have enough margin, and the retention margin becomes worse at the -40°C, but still has margin.

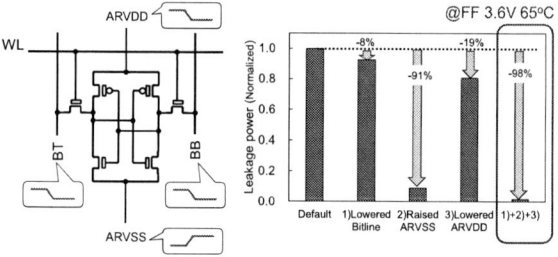

Fig. 4 Circuit diagram of bias controller in the proposed BB SRAM. macro.

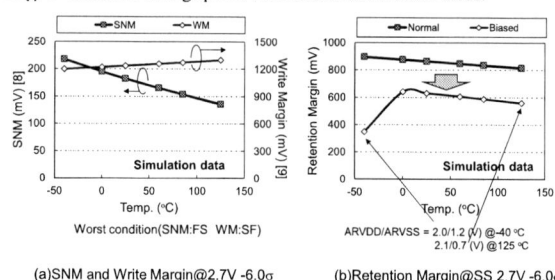

Fig. 5 Simulated leakage power reduction in the retention mode.

Fig. 6 Simulated temperature dependences of the static-noise margin (SNM), write margin (WM) and retention margin.

III. DYNAMIC POWER REDUCTION

The proposed buffer/backup (BB) SRAM with 3.3 V IO MOSs can reduce the standby leakage as described in section II, however the dynamic power increases compared to typical SRAMs with 1.1 V core MOS. The techniques to reduce dynamic power are needed. The reported technique to select column address with wordline (WL) using the advantage of

978-1-5386-6414-8/18 $31.00 © 2018 IEEE

wiring area is effective[1]. As shown in Fig. 3 (a), proposed cell has four WLs in a cell, then not only row address but also column address can be selected by WLs. Fig. 7 indicates a simple block diagram of designed SRAM macro. The signal of RA and CA in Fig. 7 are row addresses and column addresses respectively. WL decoder can select a unique address specified by RA and CA. Furthermore, column MUXs in I/O block can select column address, as a result, it is possible to reduce the power consumption for unselected BLs. Fig. 8(a) and Fig. 8(b) show SPICE simulation waveforms at the worst process-voltage-temperature (PVT) conditions in the write operation and read operation, respectively. The signals of MT/MB in Fig. 8 (a) are internal complemental nodes in memory cell, and CTR/CBR in Fig. 8(b) are differential inputs for a sense amplifier shown in Fig. 7. The implemented 64-kbit BB SRAM macro has enough write/read operation margin considering with 6-sigma local variations. Fig. 8(c) shows the SPICE simulation waveforms which transmit from normal standby to resume standby mode and from resume standby mode to normal standby mode. Waveforms with different colors indicate variations of NMA[2:0]=000~111. The voltages of ARVDD and ARVSS are adjusted using NMA[2:0]. Then, power consumption to transit from resume standby mode to normal mode is reduced by adjusting NMA[2:0]. That power consumption is 1083 μW/MHz if NMA[2:0]=111, and it becomes to 521 μW/MHz if NMA[2:0]=000. The setting of NMA[2:0]=000 is effective when switching normal standby/ resume standby mode frequently.

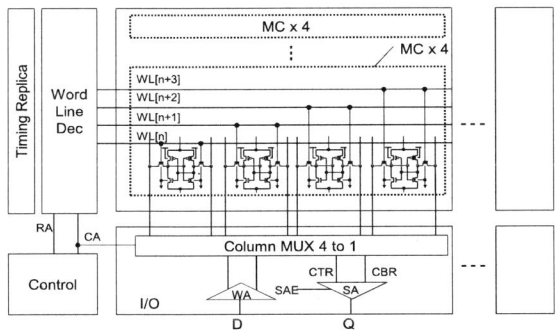

Fig. 7 Block diagram of proposed Retention SRAM.

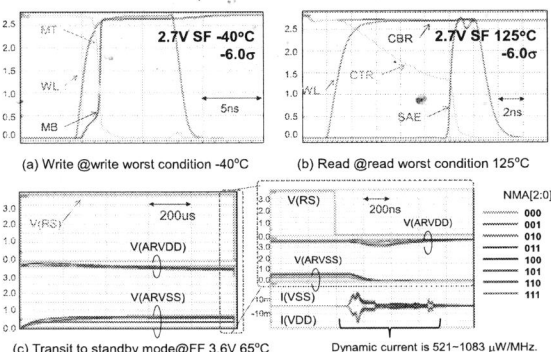

Fig. 8 Simulation waveforms of 64-kbit BB SRAM (2048-word x 32-bit).

IV. DESIGN AND EVALUATION OF TEST CHIP

Fig. 9 portrays a die photograph of a test chip using 40 nm CMOS technology [10]. 64-kbit BB SRAM macro which has supposed circuit is implemented in the test chip. Table 1 presents a summary of the test chip features. We observed full read/write functions at temperatures of -40°C to 125°C. Fig. 10 presents a typical SHMOO plot at 125°C, showing the minimum operating voltage (V_{min}) vs. the access time. The access time of the measured SRAM macro is 16.7 ns at the typical supply voltage of 3.3 V. It includes the delay of level shifter. Fig. 11(a) and Fig. 11(b) show a cumulative distribution functions (CDFs) of write/read V_{min} and retention V_{min} at RS mode, respectively at temperatures of -40°C to 125°C. The total number of measured dies is 25. The median of write/read V_{min} for typical process is 1.28 V and the median of retention V_{min} is 1.19 V, confirmed enough operation margin for typical supply voltage of 3.3 V. Fig. 12(a) shows CDF of standby leakage powers in RS modes at 65°C and 125°C. The median of leakage power of 64-kbit SRAM is 6.0 μW at 125°C and 0.33 μW at 65°C, respectively. The leakage power at 25°C is lower than measurement accuracy of tester. Therefore, it is extrapolated as 47 nW from that at 65°C and simulated dependency of leakage current on temperature by SPICE simulation.

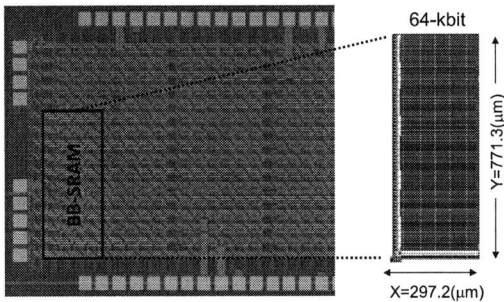

Fig. 9 Microphotograph of the Test chip and layout plots of proposed retention SRAM macro using 40-nm technology.

Table 1 Features of the test chip.

	Features
Technology	40-nm Embedded Flash process
Macro configuration	64-kbit (2048 word x 32 bit)
Macro size (included level shifter)	297.2 μm x 771.3 μm 225655 μm²@64-kbit (Non-rectangular)
Bit density	0.277 Mbit/mm²
Access time @typ	16.7 ns (included level shifter)
Cycle Time @typ	23.8 ns(42 MHz)
Dynamic power @typ	Read:174 μW/MHz Write:180 μW/MHz
Standby power @64-kbit	6.0 μW @typ 125°C 0.33 μW @typ 65°C

Fig. 12(b) shows the measured dynamic power consumptions of write and read operations *vs* supply voltage. The measured dynamic power consumptions with proposed circuitry are 174 μW/MHz for read operation and 180 μW/MHz for write operation at the typical process, 3.3 V and 125°C, respectively, obtained good correlations with simulation results. The dynamic power with multi-interleave WL scheme can be reduced to 40% compared to without scheme, that read power is

426 µW/MHz. Table 2 shows a comparison with previous reports. Here, we assume the VDC has 2 µA of standby current at least, which is expected 1/3.5 of latest report by [3]. Proposed BB SRAM using 3.3 V IO MOS does not needed such VDC leakage power, having advantage of total standby leakage power reduction to 1/140, as shown in Table 2.

Table 2 Comparison with previous reports.

		ISSCC'14 [1]	VLC'17 [2]	This work
Standby power	SRAM(*1)	2.6 nW @25°C	0.86 nW @25°C	330 nW@65°C
	SRAM+VDC(*2)	6.60 µW @25°C	6.60 µW @25°C	47 nW@25°C
Techology node		65nm	65nm	40nm
Voltage		1.2V	0.75V	3.3V
Advanced or additional Process		Yes	Yes	No
Transistor		thick-gate	thin-gate	thick-gate
Cell Area		2.159 µm2	0.5408 µm2	2.888 µm2
Capacity		128 kbit	128 kbit	64 kbit
Dynamic power		25 µW/MHz	6.3 µW/MHz	174 µW/MHz

(*1):Normalized to 64-kbits.
(*2):Assumed VDC has 2 µA standby current at 3.3 V.

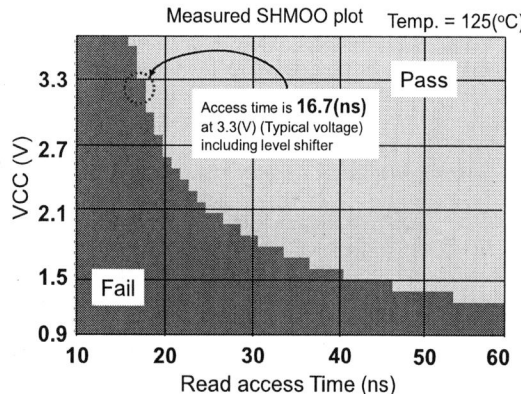

Fig. 10 SHMOO plot V_{min} vs. the access time at 125°C.

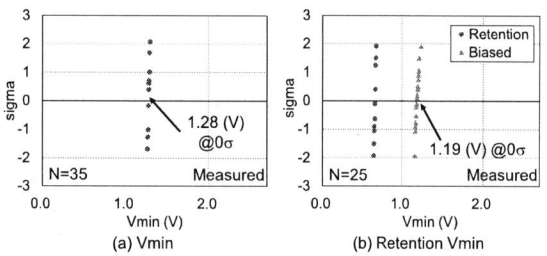

Fig. 11 Distribution of measured V_{min} at temperatures of -40°C to 125°C.

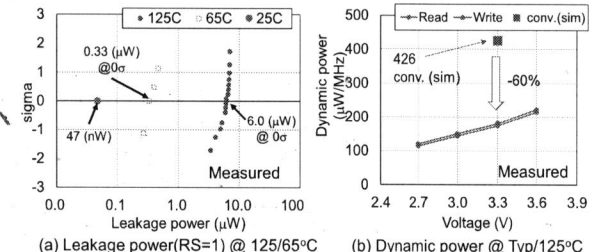

Fig. 12 Distribution of measured leakage power and voltage dependency of dynamic power.

V. CONCLUSION

We proposed an effective standby power reduction of buffer/backup SRAM in MCU for PMIC less edge system in IoT applications. It is implemented using 3.3 V thick-gate-oxide IO MOSs without on-die VDC for effectively reducing the leakage power. Four multiples interleave WL circuitry was also introduced to reduce the dynamic power. A test chip with 64-kbit SRAM macro was designed and fabricated using 40-nm technology. From the measured data, we obtained that the leakage power was 330 nW at 65°C (47 nW at 25°C), which is less 1/140 than other works. The read/write dynamic power is reduced by 60% by interleave WL circuitry.

REFERENCES

[1] Toshikazu Fukuda et. al, "13.4 A 7ns-access-time 25µW/MHz 128kb SRAM for low-power fast wake-up MCU in 65nm CMOS with 27fA/b retention current", ISSCC, pp.236 - 237, 2014.

[2] Makoto Yabuuchi et. al, "A 65 nm 1.0 V 1.84 ns Silicon-on-Thin-Box (SOTB) embedded SRAM with 13.72 nW/Mbit standby power for smart IoT", 2017 Symposium on VLSI Circuits, pp.220 - 221.

[3] Qianneng Zhou et. al, "Embedded DC-DC Voltage Down Converter for Low-Power VLSI Chip", APCCAS, pp.670 – 673, 2006.

[4] T. Kamei et. al, "A resume-standby application processor for 3G cellular phones", ISSCC, pp.336-531, Vol.1, 2004

[5] N. Maeda et. al "A 0.41 µA Standby Leakage 32 kb Embedded SRAM with Low-Voltage Resume-Standby Utilizing All Digital Current Comparator in 28 nm HKMG CMOS", JSSC, pp.917-923, Vol.4, 2013

[6] Patent US7087942, 2006

[7] Yoshisato Yokoyama et. al, "40nm Ultra-low leakage SRAM at 170 deg.C operation for embedded flash MCU", ISQED, pp.24-31, 2014.

[8] E. Seevinck, F. J. List, and J. Lohstroh, "Static-Noise Margin analysis of MOS SRAM cells," IEEE J. Solid-State Circuits, vol. SC-22, no. 5, pp. 748-754, Oct. 1987.

[9] R. Heald and P. Wang, "Variability in sub-100 nm SRAM designs," ICCAD Digest, pp. 347-352, 2004.

[10] Yoshisato Yokoyama, et. al, "A cost effective test screening method on 40-nm 4-Mb embedded SRAM for low-power MCU", A-SSCC, Proc. Industry2-3, Nov. 2015

[11] Koji Nii, et. al, "A dynamic/static SRAM power management scheme for DVFS and AVS in advanced automotive infotainment SoCs", 2016 Symposium on VLSI Circuits, pp.220 – 221

[12] Takahiko Ishizu, et. al, "A 140 MHz 1 Mbit 2T1C gain-cell memory with 60-nm indium-gallium-zinc oxide transistor embedded into 65-nm CMOS logic process technology", 2017 Symposium on VLSI Circuits, pp.C162 – C163

Logic Process Compatible 40nm 256Kx144 Embedded RRAM with Low Voltage Current Limiter and Ambient Compensation Scheme to Improve the Read Window

Chien-An Lai, Chung-Cheng Chou, Chi-Hsiang Weng, Zheng-Jun Lin, Pei-Ling Tseng, Chien-Fan Wang, Chih-Chen Wang, Chin-I Su, Wei-Chi Chen, Yu-Cheng Lin, Tong-Chern Ong, Chi Chang, Yu-Der Chih, Tsung-Yung Chang

Taiwan Semiconductor Manufacturing Company, Hsin-Chu, Taiwan

Abstract—In this paper, we present a low voltage current limiter that can effectively confine the filament size by limiting the write current to a preset compliance level after forming or SET operations. In addition, a word-line (WL) location-aware and a temperature compensation schemes are also proposed to deal with the ambient variations and tighten cell current distribution. As a result, the silicon data measured from a 36Mb of RRAM test chip demonstrates a 9.5uA of read window after 10K cycles and 85C 10 years of retention test in 40nm logic process.

Keywords—RRAM, forming, compensation, read window

I. INTRODUCTION

RRAM technology has made a significant progress in the past decay as a competitive candidate for the next generation non-volatile memory. RRAM is suitable for the embedded applications due to its full compatibility with logic process. A fresh RRAM element (RE) requires an one-time process, called forming to become filamentary which is stochastic and critical to write performance. Several techniques like write termination and current limiting were proposed to tighten cell current distribution [1][2]. Write termination circuit detects the write current flowing through the cell and generates a control signal to turn off write bias when forming or SET success. Consequently, the SET stimulus pulse is cut from the written cell to avoid over-set and save write power for fast cells. However, the circuit consumes a substantial area and voltage overhead to implement the detection circuit. Besides, the sudden termination of write operation may not grow a dense filament, even for fast cells. Another method to prevent over-SET is current limiting. A current limiter not only can prevent over-stress from SET or forming but also can grow a denser filament if write stimulus pulse width is long enough. However, the voltage overhead from the conventional current limiter is significant due to its operation at MOS's saturation region. In addition, the cell current variation resulted from the location dependency and temperature variation was not dealt with before. As a result, a wider cell current distribution after forming and SET is observed. In this paper, we target to develop a current limiter with a low voltage overhead and present several techniques to compensate the ambient variations and tighten cell current distribution after forming and SET operations. As a result, a bigger tail-to-tail cell current separation - valid read window - is measured from silicon.

II. LOW VOLTAGE CURRENT LIMITER (LVCL)

The proposed current limiter is a stack of two NMOS, M2 and M4 depicted in Fig.1. M2 is the footer device shared with the multiple columns and M4 is the selected column mux. A bias generator comprised of M3 and M1 to replicate the stack and generate the gate biases VY and VG. Through the closed loop of the OPamp, VY and VG will bias the drain of M1 and M2 at a preset voltage, 0.2V for the example, as the write current reaches the compliance current level Iref. So the voltage overhead of the current limiter is the drain-source voltage of M3 + 0.2V.

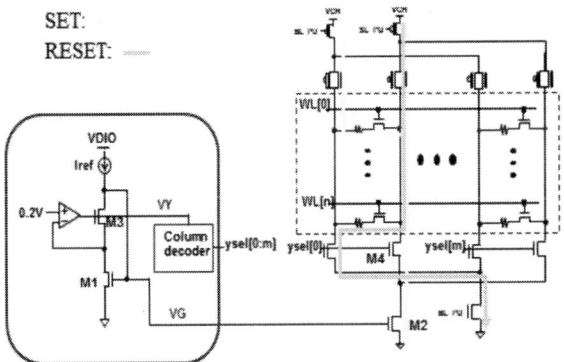

Fig.1 The schematic of the proposed LVCL

If we well design the size of M3 and make M3 operate in linear region at the compliance current level, the voltage overhead of the current limiter can be less than 300mV. To realize the OPamp to accommodate a low input voltage like 0.2V, a two stage of OAamp is adopted [3]. The PMOS type of differential pair is used as the input stage to accommodate low input voltage and the second stage provides the gain to minimize the offset. The output of the OPamp generates the voltage source VY for the column decoder. Since VG and VY fix the gate and drain bias of M2, the write current can be confined to a constant value and a fixed filament size can be achieved. The drain voltage of M4 will increase and prevent RRAM element from over stress.

978-1-5386-6414-8/18 $31.00 © 2018 IEEE

III. WL LOCATION-AWARE COMPENSATION (WLAC)

In write operation, RRAM element (RE) will flow a substantial current level at the end of SET or the onset of RESET. The high cell current induces a significant IR-drop from the parasitic resistance. Instead of the dedicate source-line (DSL) architecture, we adopt the common source-line (CSL) architecture which allows two columns share one source line, so that SL can be implemented with a wider metal width comparing to the DSL. Therefore SL resistance can be reduced accordingly. Besides, the metal rule can be relaxed due to the less metal track and make a smaller bit-cell more feasible than the DSL architecture. CSL and DSL architectures are depicted as Fig 2. For CSL, equalizers are required to prevent half-selected cells from read or write disturb.

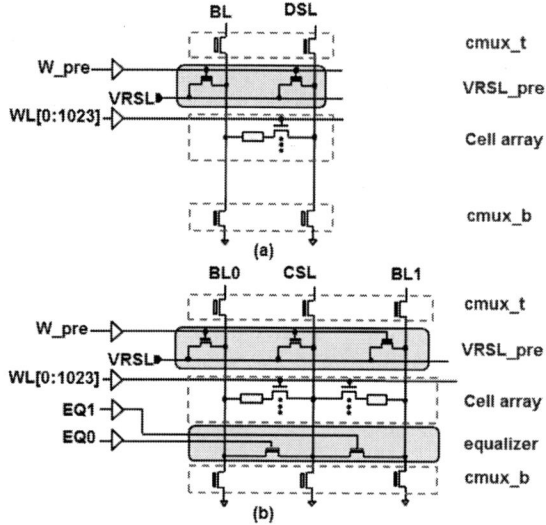

Fig.2 Array architecture: (a) DSL; (b) CSL.

Although IR drop can be reduced through CSL architecture, RE still experiences different bias at different WL location, WL<0> ~ WL<1023>, due to BL/SL resistances (RBL/RSL). In addition, the on-state resistance of select transistor would be modulated by the body effect resulted from the RBL/RSL [4].

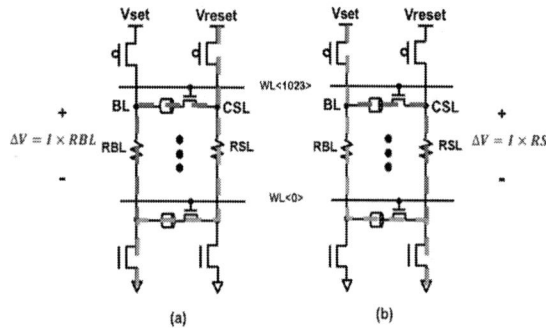

Fig.3 One column of write path for (a) RESET (b) SET. (Only half of the CSL structure is shown for simplicity)

That would cause the variation of resistance state after write operations. For example, cells located at WL<1023> encounter a higher IR drop than cells located at WL<0> by Icell x RSL for SET or Icell x RBL for RESET, shown as Fig. 3. Since we would like to keep the RRAM cells with similar write conditions to improve write cycles, we don't alter the limiting current to compensate the location variation [5]. To compensate this effect, a self-adjusting word-line activation voltage (VWL) generator is presented as Fig. 4.

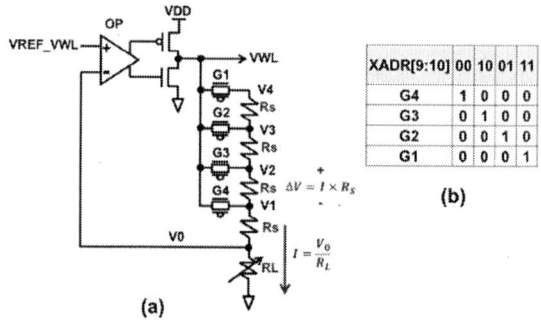

Fig.4 WLAC scheme: (a) VWL generator; (b) Truth table to steer G1~G4 with corresponding WL location.

A column is divided into four segments according to WL location, 256 WLs are grouped in a same segment for this case. The four segments can be decoded by the two MSB of the addresses XADR[9:10]. Rs is the resistance to generate the compensating voltage $\Delta V = I x Rs$. For a balanced RBL and RSL, the compensating voltage ΔV is the same for SET and RESET. That can simplify the VWL generator design. RL is a knob to fine tune the bias current I and the compensating voltage.

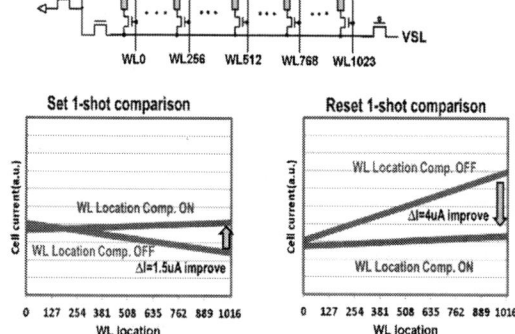

Fig.5 The measured cell current from 1 shot of SET(left) and 1 shot of RESET(right) versus WL locations.

V1~V4 are the word-line activation voltage corresponding to the four segments to compensate the body effect, so that the effective IR-drop along the write path can be remained the same to apply effective write bias on each bit-cell along BL/SL. Silicon measurement demonstrates the cell current variation along a column is reduced by 1.5u/4uA for SET/RESET operation, shown as Fig. 5. The measurement was taken from one shot of SET or RESET, so that the cell current levels would not be influenced by the verification levels of retry. Each cell number corresponds to different WL location is the average of 2304 bit-cells.

IV. TEMPERATURE COMPENSATION FOR READ CURRENT

The parasitic resistance effect not only influences write operation but also read operation. To read out the data stored in RRAM element (RE), a read voltage Vread will be applied on RRAM cell to generate the cell current. The gap between the high bound of Ir0 distribution and the low bound of Ir1 distribution decides the read window. The read current of a bit-cell can be approximated as $\frac{Vread}{R_{on}(1T)+RRE+Rpar}$. Where Ron(1T) is the on-state resistance of select transistor, Rpar is the lumped parasitic resistance and RRE is the state resistance of RE. At high temperature, both Ron(1T) and Rpar become higher because of the mobility degradation and a smaller cell current will be observed. The temperature effect is more sensitive to the low resistance state of RE and causing an uneven cell current shift to high and low resistance states. That impacts the tail-to-tail cell current separation. To maintain the read margin over temperature, an elevated VWL is required at high temperature region. We propose a piece-wise temperature compensation scheme for the purpose which is shown as Fig. 6. Vmax is the high bound of VWL to prevent reliability concern; Vmin is the minimum to ensure correct read operation.

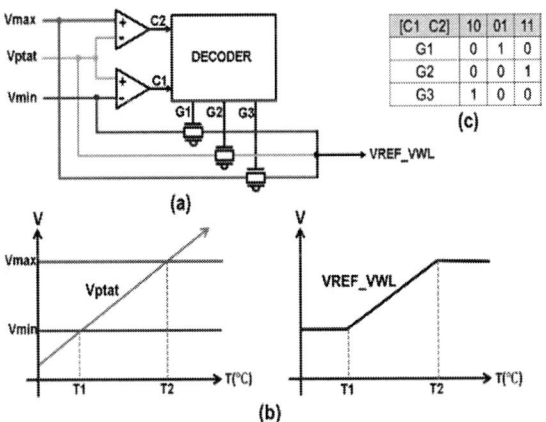

Fig.6 Temperature-adjusting VWL generation: (a). function diagram. (b). Superposition of Vptat with Vmax and Vmin. (c). The truth table of the steering logic.

A V_{PTAT} (Proportional To Absolute Temperature) is generated from the band gap reference acting as a temperature detection [6]. V_{PTAT} intersects with Vmin and Vmax at T1 and T2. Two voltage comparators are used to compare V_{PTAT} with Vmax and Vmin and define three operating temperature regions. VWL increases from T1 and T2 to reduce Ron(1T)

and compensates mobility degradation effect. Silicon data shown as Fig. 7, about 10% improvement of read window at 85C and 125C is achieved.

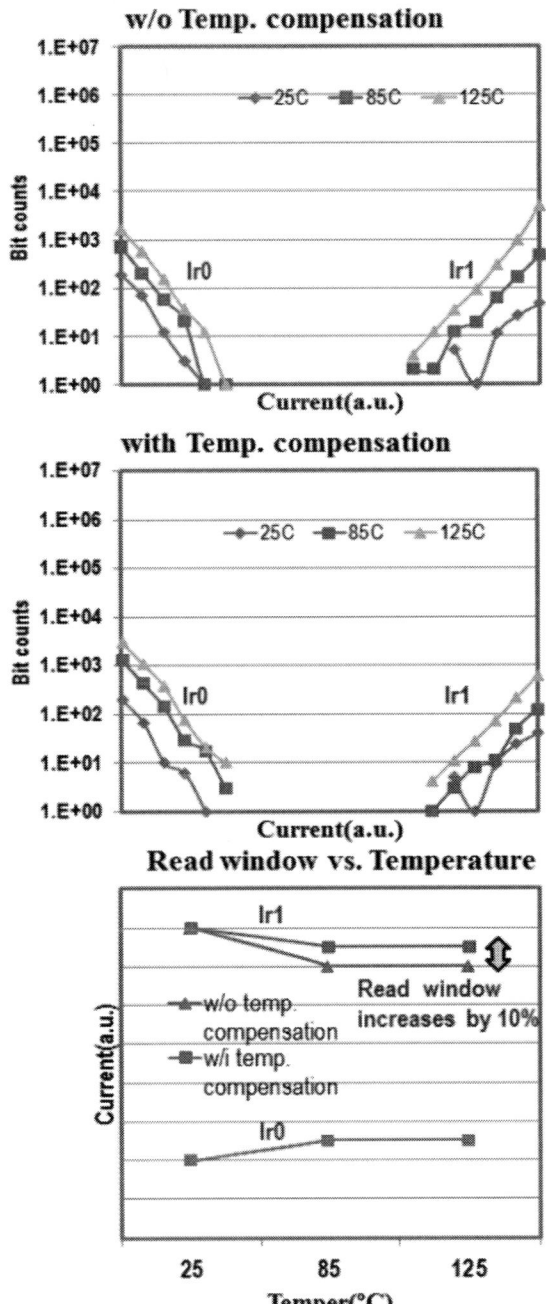

Fig.7 Read window is improved by the temperature compensation. Ir0 would be increased at high temperature because of the increased sub-threshold leakage.

V. MEASUREMENT RESULT

In our previous work [7], retention test after 1K cycles (1KC RAC) was taken with an 11Mb of memory density. We further move on to the reliability test for 10K cycles of endurance check at -40C, 125C and 25C, then bake those samples at 175C for 11hrs, termed as 10KC RAC (retention after cycles). The read window is measured and compared with our previous test chip which without those proposed features, shown in Fig. 8.

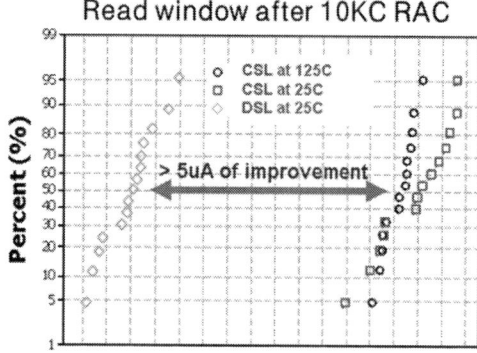

Fig. 8 Read window comparison: the work vs. prior work.

VI. CONCLUSION

A 36Mb of RRAM test chip is designed and manufactured in 40nm technology. Fig. 9 shows the die photo of the test chip. The significant IR-drop reduction is acquired by the presented CSL architecture. The presented WL location-aware compensation scheme reduces the variation resulted from the location dependency and LVCL effectively improves the forming uniformity, shown as Fig. 10. As a result, a more tightened cell current distribution after forming can be achieved. The improved forming quality can benefit SET and RESET performance, achieve a faster write time and more concentrated Ir0 and Ir1 distributions. The temperature-adjusting VWL further reduces the variation resulted from temperature effect. As a result, a robust 9.5uA [Fig.8] of read window after 10KC RAC is achieved.

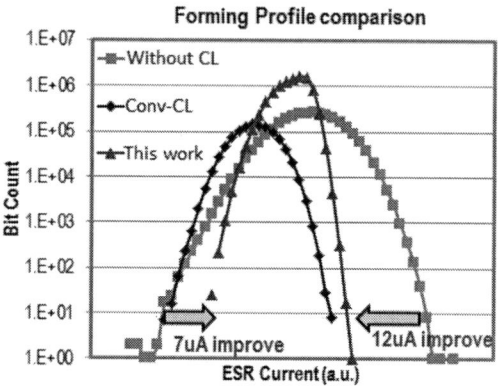

Fig.10 Cell current distributions are measured after forming. Comparison between without current limiter (red dots), the conventional current limiter (black dots) and the work which includes LVCL and WLAC (blue dots)

REFERENCES

[1] Xiaoyong Xue,et al., " A 0.13um 8Mb Logic-Based Cu Si O ReRAM with Self-Adaptive Operation for Yield Enhancement and Power Reduction", IEEE JSSCC, Vol.48, No. 5, May 2013.

[2] Y.L. Song,et al., "Reliability Significant Improvement of Resistive Switching Memory by Dynamic Self-adaptive Write Method", VLSI Technology Symposium. pp. T102-T103, 2013.

[3] Behzad Razavi. "Design of Analog CMOS Integrated Circuits." McGraw-Hill Higher Education. p.309, 2001

[4] Sedra, A.S. & Smith, K.C. "Microelectronic circuit." New York: oxford. p.552, 2004

[5] Hyung-rok Oh,et al.,"Enhanced write performance of a 64-mb phase change random access memory", ISSCC, pp.48-49, 2005

[6] Wenguan Li,et al., "A Low Power CMOS Bandgap Voltage Reference with Enhanced Power Supply Rejection", IEEE ASICON, 2009

[7] Chung-Cheng Chou,et al., "An N40 256K×44 embedded RRAM macro with SL-precharge SA and low-voltage current limiter to improve read and write performance", ISSCC, pp.478-480, 2018

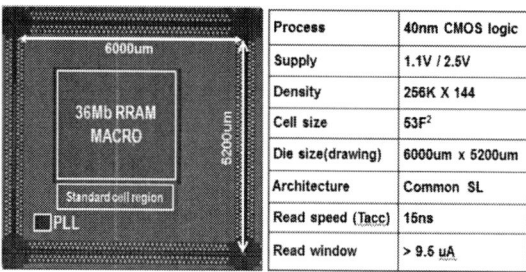

Process	40nm CMOS logic
Supply	1.1V / 2.5V
Density	256K X 144
Cell size	53F^2
Die size(drawing)	6000um x 5200um
Architecture	Common SL
Read speed (Tacc)	15ns
Read window	> 9.5 uA

Fig. 9 Die photo and key parameters

A 10-Bit 1026-Channel Column Driver IC with Partially Segmented Piecewise Linear Digital-to-Analog Converters for Ultra-High-Definition TFT-LCDs with One Billion Color Display

Chih-Wen Lu*, You-Gang Chang*, Xing-Wei Huang*, Jhih-Siou Cheng†, Po-Yu Tseng†, and Chih-Hsien Chou†

*Department of Engineering and System Science, National Tsing Hua University, Taiwan, R.O.C.
Email: cwlu@mx.nthu.edu.tw; franknb4088@hotmail.com; fredwilly11@gmail.com
†Novatek Microelectronics Corp., Taiwan, R.O.C.

Abstract—Herein, a 10-bit 1026-channel column driver IC with partially segmented piecewise linear digital-to-analog converters (DACs) is designed, simulated, and prototyped. This device is intended to improve the color depth and gamut of ultra-high-definition thin-film transistor liquid-crystal displays. The proposed column driver can output a precise gamma-corrected transfer curve without loss of effective bit resolution. An area-efficient two-voltage selector, a level shifter with a two-to-four decoder function, and a linearity-enhancing DAC-embedded operational amplifier are included in the design to minimize the overall chip area. The die dimensions of the 1026-channel column driver IC are only 18.14 mm × 1.2 mm. The prototype achieves a maximum settling time of 5.6 μs for driving a load with 5-KΩ resistance and 300-pF capacitance within 10-mV tolerance of the final voltage.

Keywords—*LCD column driver, DAC, ultra-HD, TFT-LCD, and DAC-embedded op-amp.*

I. INTRODUCTION

Rapid progress in display technology has involves an increase in resolution, improved brightness, and enhanced color range [1]. The global market for ultra-high-definition (UHD)/4K panel displays is expected to grow in the near future. Further development of display driver technology can improve the color depth and gamut of such displays. Fig. 1 shows the typical driver architecture for driving a large liquid-crystal display (LCD) panel. The timing controller receives and processes incoming image data and sends these data to column driver ICs. The column driver ICs drive rows of subpixels as they are selected by the row drivers. The resolution of these column driver ICs is critical for achieving wide color gamut and precise color depth. A large panel needs several column driver ICs. For example, 12 1026-channel column driver ICs are needed to drive a UHD (3,840 × 2,160) or 4K (4,096 × 2,160) panel. Column driver ICs process digital image codes, convert these digital codes to analog levels, and drive the LCD panel accordingly with high output voltages. Fig. 2 shows a typical LCD column driver with four signal-processing blocks: an image data-reading circuit, level shifters, digital-to-analog converters (DACs), and output buffers. To reduce the power consumption and silicon die area of the digital circuit, the data-reading circuit is supplied with low voltage; the level shifters, DACs, and output buffers are all high-voltage circuits. Therefore, level shifters are placed between the digital reading circuit and DACs to translate each logical "1" signal from low voltage to high voltage. The reading circuit receives the image

codes serially from the timing controller and sends one-row digital codes to the DACs through the level shifters. The DACs then convert the digital codes to analog levels, and the corresponding output buffers drive the display panel. The quality of the DACs determines the gray level of the display.

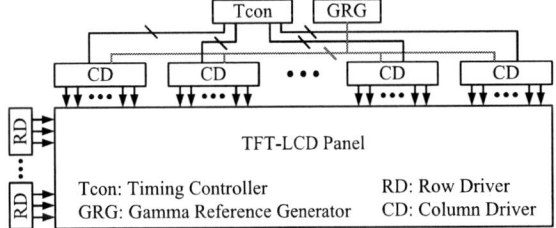

Fig. 1 Typical driver architecture for driving a large LCD panel.

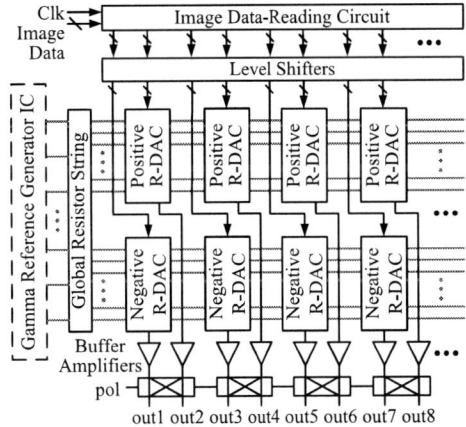

Fig. 2 Typical LCD column driver with four signal-processing blocks: an image data-reading circuit, level shifters, DACs, and output buffers.

To obtain channel-to-channel uniformity across the display, resistor-string DACs (R-DACs) are usually used in column drivers. To perform one/two-dot inversion, two adjacent channels take turns to drive a pair of the LCD panel's data lines. One channel drives the negative gamma voltage, and the other drives the positive gamma voltage. Each pair of DACs consists of a negative DAC and a positive DAC, which perform D/A conversions for the negative and positive gamma voltages, respectively. The global resistor string provides gamma voltages for all the channel DACs. These gamma voltages are designed to compensate for the nonlinearity of the LC response.

978-1-5386-6414-8/18 $31.00 © 2018 IEEE

Because several column driver ICs are required to drive a high-resolution display panel, the gamma reference generator IC provides several gamma voltages for each global resistor string to achieve chip-to-chip uniformity in the display (Figs. 1 and 2). To achieve a deep-color display with 10 bits each for red, green, and blue, one 10-bit R-DAC is required for each output channel. Consequently, DACs use the largest proportion of the die area in column driver ICs for deep-color displays.

Alternatively, linear DACs can be used in column driver ICs [2-5]. In these DACs, the inverse of the LC nonlinear response is fitted digitally, which requires a lookup table in the timing controller. This means that two extra bits are needed for LC compensation [4]. For example, a 10-bit color depth image would need column drivers with 12-bit linear DACs. Further, 12-bit DACs will occupy a large area on column driver ICs. To implement a compact column driver for UHD/4K thin-film transistor (TFT) LCDs with 10-bit color depth, we designed, fabricated, and tested a column driver architecture built from partially segmented piecewise linear DACs. The column driver can output a precise gamma-corrected transfer curve without loss of the effective bit resolution. An area-efficient two-voltage selector, a level shifter with a logic function, and a DAC-embedded operational amplifier (op-amp) also reduce the overall chip area in the proposed design.

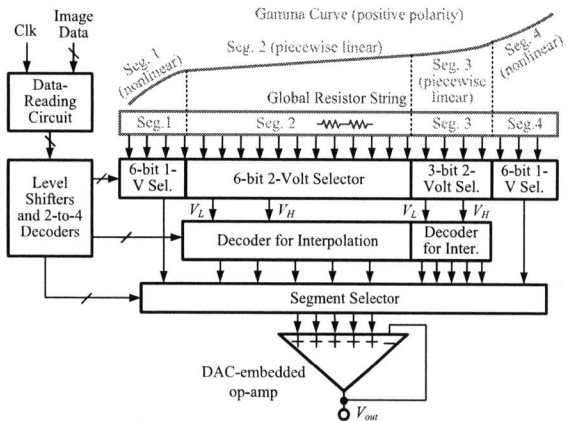

Fig. 3 Architecture of the proposed 10-bit column driver IC, with one positive-polarity channel illustrated in detail.

II. PROPOSED COLUMN DRIVER IC

This study proposes a 10-bit 1026-channel column driver IC with partially segmented piecewise linear DACs. Fig. 3 shows the proposed design, with one positive-polarity channel illustrated in detail. In each output channel, a data-reading circuit samples the image data and relays one subpixel code to each input of the level shifters. The level shifters then boost the voltage of the logic signals. The DAC generates a voltage corresponding to the subpixel code, and the output buffer drives the data line of the LCD panel. To generate a 10-bit color depth nonlinear gamma voltage, the resistor values of the global resistor string are varied and the resistor string is divided into four segments following the curvature of the gamma curve. Because Segments 1 and 4 have larger curvatures, no voltage interpolation is used in them to maintain precise gamma correction. Two 6-bit single-voltage selectors choose one voltage each from the gamma voltages of Segments 1 and 4

according to the six LSBs ($b_5b_4b_3b_2b_1b_0$). Since the middle gamma curve is smoother, the interpolation technique is used in Segments 2 and 3 to generate additional gamma voltages, which reduces the die area of the DAC. A 6-bit two-voltage selector and a 3-bit two-voltage selector choose two adjacent voltages (V_H and V_L) from Segments 2 and 3 according to the image data $b_9b_8b_7b_6b_5b_4$ and $b_5b_4b_3$, respectively, and then pass these voltages to the following decoder for voltage interpolation using a segment selector and a DAC-embedded op-amp. The DAC-embedded op-amp generates 832 ($52 \times 2^4 = 832$) voltage levels from Segment 2 and 64 ($8 \times 2^3 = 64$) voltage levels from Segment 3 of the global resistor string using 4-bit and 3-bit interpolations, respectively. Therefore, the 10-bit DAC can generate 1024 voltage levels ($2^6 + 52 \times 2^4 + 8 \times 2^3 + 2^6 = 1024$).

To reduce the die area of the DAC, we designed a compact two-voltage selector. Fig. 4 shows a schematic of the proposed 3-bit tree-type two-voltage selector that chooses two adjacent voltages from the voltages V_0, V_1, ..., V_8 of Segment 3 of the global resistor string according to image data $b_5b_4b_3$. The switching matrix of this selector consists of three stages connected in a cascade. The first stage is controlled by b_5. According to the truth table, a group of voltages, (V_0, V_1, ..., V_4) or (V_4, V_5, ..., V_8), is chosen by the first-stage switches when b_5 is 0 or 1, respectively. The first-stage switch network consists of five pairs of complementary switches so that it can choose five voltages from the resistor string. Successive stages can be arranged using the same rule. The second stage involves choosing three voltages from the outputs of the first stage. The final stage then outputs two adjacent voltages from these three voltages. Higher resolution two-voltage selectors can be extended accordingly. Compared with the state-of-the-art two-voltage selector [5], the required number of switches for a 6-bit two-voltage selector is reduced from 190 to 138.

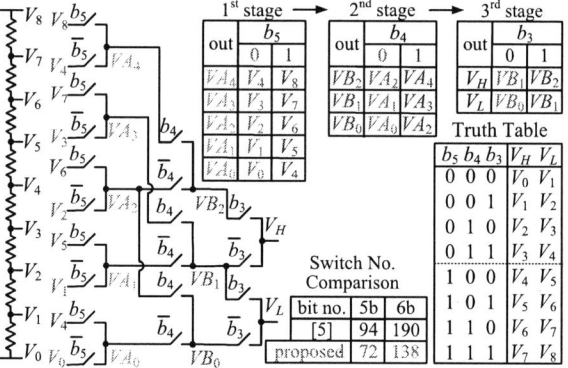

Fig. 4 Schematic of the proposed 3-bit tree-type two-voltage selector.

As an alternative, hybrid-type DACs can be used in the column driver to reduce propagation delay. The 3-bit tree-type two-voltage selector can be extended to a 6-bit hybrid-type three-stage two-voltage selector for choosing two adjacent voltages from Segment 2 of the global resistor string. Fig. 5 shows a schematic of the proposed 6-bit hybrid-type two-voltage selector that we designed for choosing two adjacent voltages from the voltages V_0, V_1, ..., V_{52} of the global resistor string according to the image data $b_9b_8b_7b_6b_5b_4$. The first stage is controlled by two MSBs (b_9b_8). Four groups of voltages, (V_0,

$V_1, \ldots, V_{16}$), ($V_{16}, V_{17}, \ldots, V_{32}$), ($V_{32}, V_{33}, \ldots, V_{48}$), and ($V_{48}, V_{49}, \ldots, V_{64}$), are chosen by the first-stage switches if $b_9 b_8$ are 00, 01, 10, or 11, respectively. The second and third stages use the same voltage selection scheme. Because 52 voltages ($V_0, V_1, \ldots, V_{52}$) must be interpolated to generate 832 voltage levels in Segment 2, the voltages $V_{53}, V_{54}, \ldots, V_{64}$ are not connected to the 6-bit two-voltage selector. Three 2-to-4 decoders are required to provide control signals to the switch network. These decoders can be constructed from four high-voltage two-input NOR gates. In the present work, these decoders are combined with level shifters to minimize the die area. Fig. 6 shows a schematic of the proposed four-output level shifter with a 2-to-4 decoder. The pull-down network comprises two inverters and four NAND gates, with transistors M13–M16 forming a NOR function for each output. The two inverters, four NAND gates, and transistors M13–M16 are all low-voltage devices. The power supply to the inverters and NAND gates is 1.8 V. The pull-up network (M1–M12) is connected in a quadruple cross-couple pair to raise one of the four outputs to V_{DDH} (18 V). The bias devices Mbp1–Mbp4 and Mbn1–Mbn4 limit the short-circuit current. The transistors M1–M12, Mbn1–Mbn4, and Mbp1–Mbp4 are high-voltage devices. The truth table for the level shifter with 2-to-4 decoder function is also shown in Fig. 6.

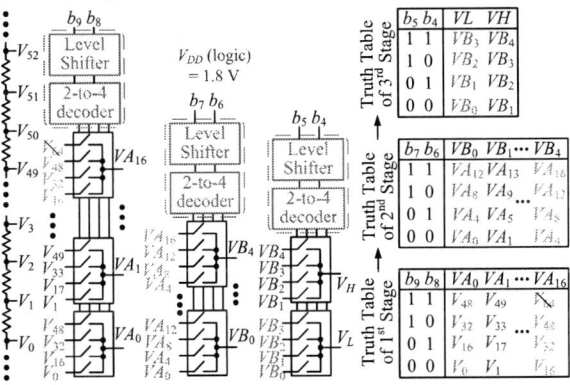

Fig. 5 Schematic of the proposed 6-bit hybrid-type two-voltage selector.

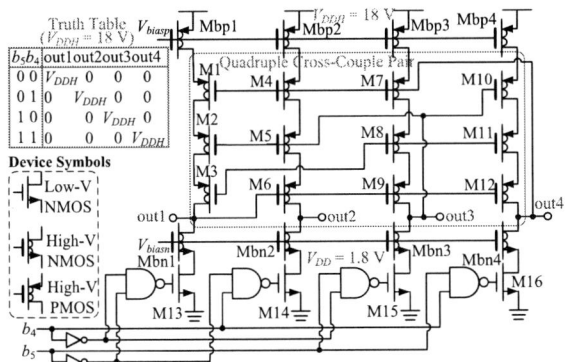

Fig. 6 Schematic of the proposed four-output level shifter with a 2-to-4 decoder.

Recently, DAC-embedded op-amps have become popular for use in the column drivers of small LCD panels. These devices handle voltage interpolation and drive the display panels. The small linear range of the I–V characteristics of the differential pair limits the resolution of voltage interpolation, making DAC-embedded op-amps unsuitable for driving high-voltage displays. To make the DAC-embedded op-amp appropriate for use in high-voltage display driver ICs, we designed a differential op-amp with enhanced linearity (Fig. 7). To achieve 4-bit voltage interpolation, five differential pairs are required in the DAC-embedded op-amp. The ratios of the transistor sizes of the five differential pairs and their tail currents are 1:1:2:4:8. To extend the linear region of the I–V curve of the differential pairs, two adaptive source degeneration devices (M3 and M4 as well as M7 and M8) are inserted between the sources of the differential pairs and their tail currents. Because two different voltage levels are applied to the inputs of each differential pair, the resistances of these two degeneration devices are unequal. The larger the differential input voltage, the greater the difference between the source degeneration resistances. This increases the amount of negative feedback to the differential pair with source degeneration for a larger input differential voltage, thereby extending the linear range of the I–V curve. Simulated transfer characteristics and transconductances of the proposed and conventional differential pairs are shown in Fig. 8 to demonstrate how the proposed design accomplishes linear enhancement. The variations in the transconductance of the proposed differential pair are much smaller than those of the conventional circuit in the input voltage range of ± 0.3 V, which indicates that linear enhancement was successful.

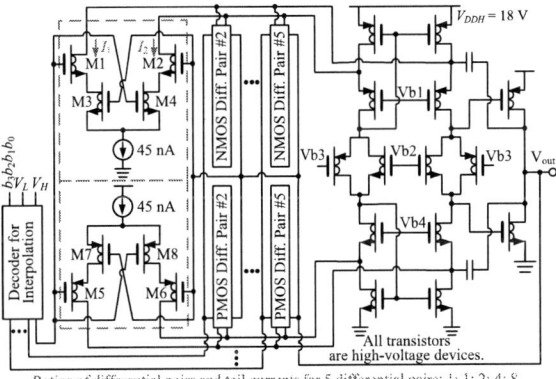

Fig. 7 A 4-bit DAC-embedded op-amp with the proposed linear-enhanced differential pairs.

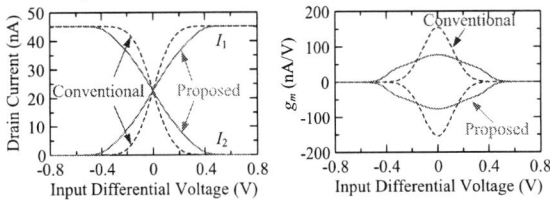

Fig. 8 Simulated transfer characteristics and transconductances of the proposed and conventional differential pairs.

III. EXPERIMENTAL RESULTS

Using the 0.18-μm 1P4M 1.8V/9V/18V CMOS technology, a 1026-channel column driver was fabricated. Fig. 9 shows the gamma-corrected transfer curves and histograms of the output voltage deviations measured from five column driver ICs through all channels. The histogram of output voltage deviations was calculated from 10,506,240 measurements. Without offset cancellation, the maximum inter-chip DVO is 16 mV. Fig. 10 shows the output waveforms measured from two adjacent output channels with a 5-KΩ resistance and 300-pF capacitance load sweeping across the digital input data. The maximum settling time to reach the final voltage with a tolerance of 10 mV was 5.6 μs. Fig. 11 shows the die micrograph and a comparison of the chip areas. The proposed 1026-channel column driver IC covers only 18.14 × 1.2 mm² of the die area, taking up 64% of the area of the 10-bit 642-channel column driver IC fabricated using the 0.1-μm 1P5M 1.5V/5V CMOS technology [4]. Table I presents the performance summary. Fig. 12 is a picture displayed on the TFT-LCD panel. The performance of the proposed column driver IC was successfully verified for use in commercial products.

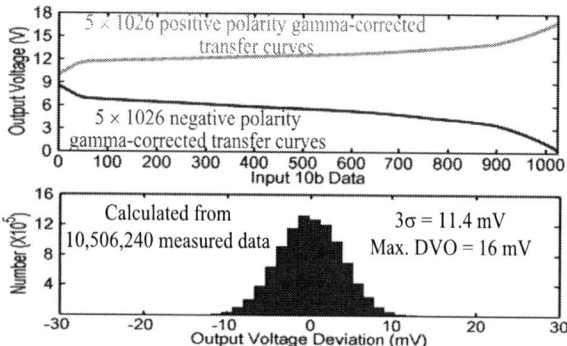

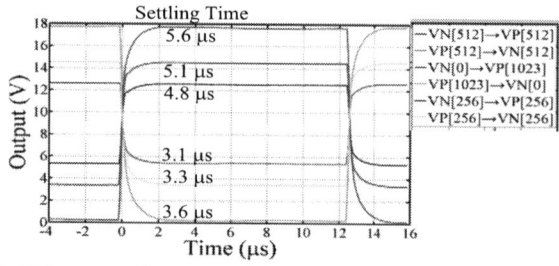

Fig. 9 Gamma-corrected transfer curves and histograms of the deviations in output voltage measured from five column driver ICs through all channels.

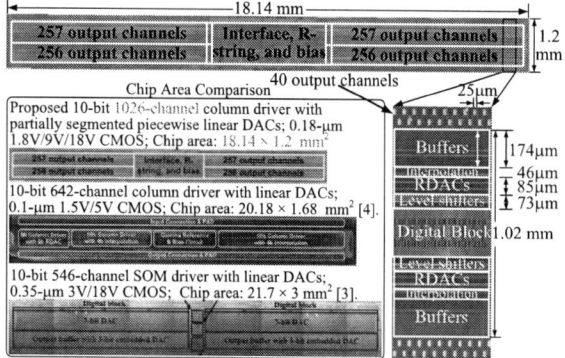

Fig. 10 Output waveforms measured from two adjacent output channels with a 5-KΩ resistance and 300-pF capacitance load sweeping across the input data.

Fig. 11 Die micrograph and a comparison of the chip areas.

IV. CONCLUSION

This paper discussed the design of a 10-bit 1026-channel column driver IC with partially segmented piecewise linear DACs. This DAC design can be used in a UHD TFT-LCD with 10-bit color depth. The proposed column driver outputs a precise 1024-level gamma-corrected transfer curve. The 1026-channel column-driver IC covers only 18.14 mm × 1.2 mm of a silicon die. The maximum settling time for driving a 5-KΩ-resistance and 300-pF-capacitance load within 10-mV tolerance is 5.6 μs. With no offset cancellation, the maximum inter-chip DVO is 16 mV.

Table I Performance summary and comparison.

	[3]	[4]	This work
Tech.	0.35-μm 2P4M	0.1-μm 1P5M	0.18-μm 1P4M
Color depth	< 10 bit	< 10 bits	10 bits
Output Range	6V (10 bits; linear)	0.25V to 4.75V (10 bits; linear)	0.2V to 8.8V (10 bits) 9.2V to 17.8V(10 bits)
Max. DVO	2.7 mV	20 mV (measured from 16 channels (4 chips))	16 mV (measured from 1026 x 5 channels; 10,506,240 measurements)
Static Current	N.A.	1 μA/buffer	7 μA/buffer
Settling Time	1.93 μs (C_L = 40pF, 4.5V ~ 10.5V)	N.A.	5.6 μs (R_L = 5 K, C_L = 300pF, 0.2V ~ 17.8V)
Die area	21.7×3 mm² (546 channels)	20.18×1.68mm² (642channels)	18.14×1.2 mm² (1026 channels)

Fig. 12 A picture displayed on the TFT-LCD panel.

REFERENCES

[1] D. McCartney, "Tuning LCDs to Tune TV," Information Display, Vol. 20, No. 10, Oct. 2004, pp. 14-17.

[2] H.-S. Kim, et al, "A 5.6mV Inter-Channel DVO 10b Column-Driver IC with Mismatch-Free Switched-Capacitor Interpolation for Mobile Active-Matrix LCDs," ISSCC Dig. Tech. Papers, pp. 392-393, Feb. 2013.

[3] J.-S. Kang, et al, "A 10b Driver IC for a Spatial Optical Modulator for Full HDTV Applications," ISSCC Dig. Tech. Papers, pp. 138-139, Feb. 2007.

[4] H.-M. Lee, et al, "A 10b Column Driver with Variable-Current-Control Interpolation for Mobile Active-Matrix LCDs," ISSCC Dig. Tech. Papers, pp. 266-267, Feb. 2009.

[5] C.-W. Lu, et al, "A 10-b Two-Stage DAC with an Area-Efficient Multiple-Output Voltage Selector and a Linearity-Enhanced DAC-Embedded Op-Amp for LCD Column Driver ICs," IEEE JSSC, vol. 48, no. 6, pp. 1475–1486, Jun. 2013.

A WDR CMOS Image Sensor Employing In-pixel Capacitive Variation using a Re-configurable Source Follower for Low Light Applications

Neha Priyadarshini, Chandani Anand, and Mukul Sarkar
Department of Electrical Engineering, Indian Institute of Technology, Delhi, India-110016
Email: eez168078@iitd.ac.in, msarkar@ee.iitd.ac.in

Abstract—A wide dynamic range image sensor that uses a 5-transistor pixel architecture to provide user-programmable conversion gain enhancement for detection of low illumination signals is presented. The dynamic range enhancement is a result of periodic variation of the gate-bulk capacitance of the in-pixel source follower by re-configuring it as a MOS capacitor. The pixel is capable of providing standard integrated photo-signal in high illumination and amplified signal in low illumination. A prototype sensor containing a 64×64 array of the proposed pixels has been fabricated in AMS 0.35 μm, 3.3 V CMOS technology to verify the operation. The $10 \times 10 \ \mu m^2$ pixel has a 20.25% fill factor. The in-pixel signal amplification results in a dynamic range enhancement of 18 dB under low illumination condition and a conversion gain enhancement of 160 $\mu V/e^-$.

I. INTRODUCTION

Over the past decade, CMOS image sensors for vision systems are required to display an excellent contrast between high exposure and low exposure regions of a scene. The ability to image this contrast defines the dynamic range (DR) of the image sensor. There is an increasing demand for enhancing the DR for low illumination levels since it is required in medical imaging, satellites and star tracking, security and surveillance systems and other applications. It is advantageous if the DR enhancement is achieved by changing the standard pixel architecture as minimally as possible. Typically, CMOS image sensors have a DR of 50 - 60 dB which is less for certain applications.

Some important DR enhancement architectures reported in literature include logarithmic pixels [1], in-pixel lateral overflow capacitors [2], multiple exposure techniques [3] and signal charge multiplication. Logarithmic pixels are capable of detecting several orders of photocurrent by non-linearly compressing the signal. But due to subthreshold conduction, they suffer from image lag and poor SNR under low illumination. Their low illumination response is improved using lin-log pixels [4]. But they require additional transistors to behave in a dual mode making their pixel structure complex. In-pixel lateral overflow capacitors can achieve as high as 185 dB DR [5] but they severely reduce the fill factor. Dual and multiple exposure techniques do not change the standard pixel architecture. However, they often require memory elements in the column, consume high power and pose a stringent requirement on the bandwidth of the column circuitry [6]. Charge multiplication is performed for low illumination ap-

plications using impact ionization [7]. This method requires high electric fields obtained from high supply voltages (10 - 15 V). High voltage operation has drawbacks in terms of power consumption making this method unfeasible for various applications.

This paper presents a wide DR pixel architecture for detection of low illumination signals in which the in-pixel source follower is periodically transformed as a MOS capacitor. A MOS capacitor's capacitance can be parametrically varied [8] and the variation is user-programmable. According to the change in capacitance, the pixel conversion gain (CG) is increased resulting in amplification of the voltage signal. Due to the reconfiguration, no extra capacitor is required to achieve the amplification and the fill factor is not compromised. The amplification is performed in the charge domain and does not involve any active components or resistors. Thus, the amplification itself is noiseless [8]. Moreover, the proposed technique amplifies the signal in the early stage of the image sensor signal chain making the noise sources appearing afterwards less noticeable at the sensor's output. In [9], the authors have presented the device physics and the TCAD simulation results of this technique. This work concentrates on the hardware implementation of the technique and the experimental results. The amplification of low illumination signals yields an 18 dB increase in the sensor DR and 160 $\mu V/e^-$ enhancement in the pixel CG using the proposed technique.

II. PIXEL ARCHITECTURE AND OPERATION

The architecture of the proposed pixel with its column circuitry is shown in Fig. 1(a). The pixel consists of a p-sub/n-well photodiode, a sample and hold switch (SH), a reset switch (RST), transistor MS, a row select switch (RSEL) and an additional switch (S_{PA}). The column consists of switches S_{P1}, S_{P2} and the CDS circuit. The timing diagram is shown in Fig. 1(b). In the proposed pixel, transistor MS serves a dual purpose. It works as a source follower when S_{PA} is off and as a three terminal MOS capacitor when S_{PA} is on. The shorted source-drain terminal (control terminal) of MS can be switched to V_{DD} or V_{PA} by controlling the switches S_{P1} and S_{P2}. Initially, PD and SD nodes are reset and the reset signal V_{RST} is readout and stored on the capacitor C_{RST}. Then SH is opened and signal charges are integrated on node PD. Under low illumination condition, switches S_{PA} and S_{P1} are

978-1-5386-6414-8/18 $31.00 © 2018 IEEE

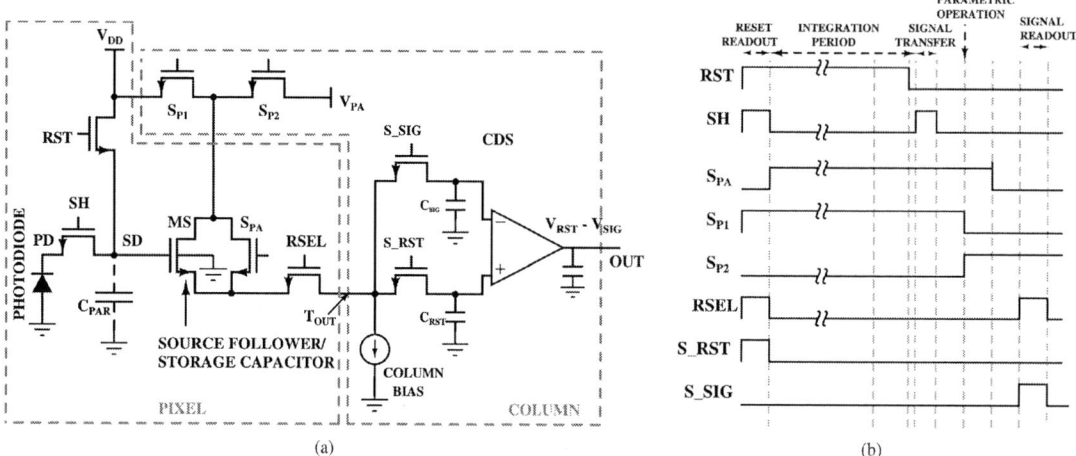

Fig. 1: (a) Proposed 5-transistor pixel and column circuit (b) Timing Diagram

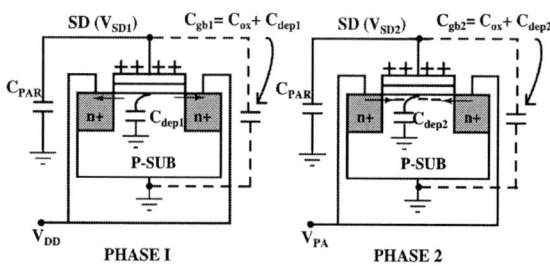

Fig. 2: A pictorial representation showing variation in the depletion capacitance of transistor MS when the control terminal is switched from V_{DD} (Phase I) to V_{PA} (Phase II)

closed and S_{P2} is open so that MS becomes a MOS capacitor and holds the charges at its gate. After exposure, integrated electrons from the node PD are sampled onto the parasitic capacitance C_{PAR} at node SD. So, the signal charge is held at the floating node SD and the control terminal potential is held at V_{DD}.

A positive gate potential due to the signal charge is unable to create inversion in MS since the free electrons in the channel of MS sink into the control terminal due to its higher potential. Thus, a depletion region with its equivalent capacitance, C_{dep1} is created in MS as shown in phase I of Fig. 2. The gate-bulk capacitance of MS, C_{gb} is a series combination of the oxide capacitance C_{ox} and depletion capacitance, C_{dep}. Thus, the sampled charge in phase I can be written as

$$Q_{G1} = V_{SD1}C_{gb1} + (V_{SD1} - V_{DD})(C_{gs1} + C_{gd1}) + V_{DD}(C_{sb1} + C_{db1}) \quad (1)$$

where V_{SD1} is the sampled voltage on node SD. C_{gs1}, C_{gd1}, C_{sb1} and C_{db1} are the gate-source, gate-drain and substrate $p-n$ junction capacitances respectively.

Once the charge is held, lowering the control terminal potential from V_{DD} to V_{PA} makes some electrons re-enter the channel. This situation is shown as parametric operation

in Fig. 1(b) and phase II in Fig. 2. The new charges on node SD are

$$Q_{G2} = V_{SD2}C_{gb2} + (V_{SD2} - V_{PA})(C_{gs2} + C_{gd2}) + V_{PA}(C_{sb2} + C_{db2}) \quad (2)$$

where V_{SD2} is the new gate potential balancing the substrate charges. C_{gs2}, C_{gd2}, C_{sb2} and C_{db2} are the gate-source, gate-drain, and substrate $p-n$ junction capacitances respectively.

The depletion capacitance is changed from C_{dep1} to C_{dep2} due to the potential change at the control terminal. Thus, only the potential of SD should change in order to balance the change in capacitance since charge remain the same. In Fig. 2, the potentials of SD before and after parametric operation are V_{SD1} and V_{SD2} respectively. Equating $Q_{G1} = Q_{G2}$ due to charge conservation, V_{SD2} can be written as

$$V_{SD2} = \frac{V_{SD1}(C_{gb1}+C_{gs1}+C_{gd1})}{C_{gb2}+C_{gs2}+C_{gd2}}$$
$$- \frac{V_{DD}(C_{gs1}+C_{gd1}+C_{sb1}+C_{db1}) - V_{PA}(C_{gs2}+C_{gd2}+C_{sb2}+C_{db2})}{C_{gb2}+C_{gs2}+C_{gd2}} \quad (3)$$

The first term in (3) is a function of C_{gb1}/C_{gb2} whereas the second is an offset that remains constant for a fixed V_{PA}

$$V_{SD2} = \frac{V_{SD1}(C_{gb1})}{C_{gb2}} - V_{offset} \quad (4)$$

According to (4), V_{SD2} is lower than V_{SD1} since C_{gb2} is greater than C_{gb1}. For the same light input, reset signal V_{RST} remains unchanged whereas the light dependent signal, V_{SIG} is lowered. After amplification, MS is converted back into a source follower by switching S_{PA} off and V_{SIG} is readout and stored on capacitor C_{SIG}. Thus, V_{RST}-V_{SIG} is amplified without amplifying the noise of the readout electronics making low illumination signals detectable by amplifying in the charge domain at the SD node itself.

Under high illumination, switches S_{PA} and S_{P2} remain off whereas S_{P1} is closed through out the operation as there is

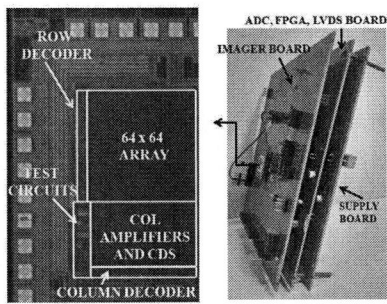

Fig. 3: Chip Micrograph and Customized test boards

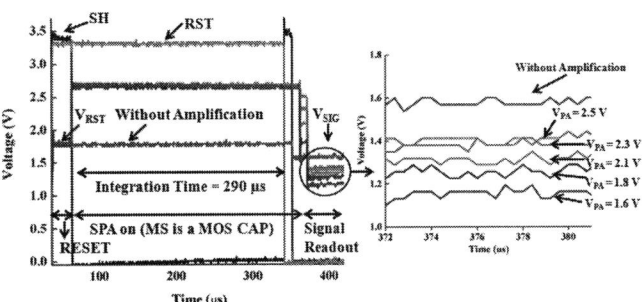

Fig. 4: CRO results of a test pixel. The sampled output is shown for V_{PA} = 2.5 V, 2.3 V, 2.1 V, 1.8 V and 1.6 V at a fixed illumination level of 3 lux

no need for amplification. The signal is read via the source follower buffer similar to a standard pixel

III. EXPERIMENTAL RESULTS

A CMOS image sensor has been realized in 0.35 μm AMS Opto process. The chip micrograph and test boards are shown in Fig. 3. The resolution is 64 × 64. The output analog signal from the chip is processed by an off chip ADC (AD9822) and transferred via a frame grabber (NI PCI-PXI-1424) to the PC for image capture. An FPGA (EP1C3T144C8) provides the external clocks.

The CRO results in Fig. 4 display the pixel output for six operating modes at a fixed illumination of 3 lux. The mode labeled 'Without Amplification' has switches S_{PA} and S_{P2} open and S_{P1} closed so that V_{PA} is disconnected from the pixel. The amplification modes follow the timing in Fig 1(b). Varying V_{PA} amplifies the voltage change due to the signal electrons at the SD node by different gains according to the change in capacitance as explained in (4). The output voltages according to change in V_{PA} are displayed in the magnified version of Fig. 4.

Fig. 5 shows the CG measured from the photon transfer curve of a single pixel. Without amplification, the CG is measured as 72 $\mu V/e^-$. For a V_{PA} of 2.4 V, 2.0 V and 1.6 V, the CG is 147.2 $\mu V/e^-$, 188.7 $\mu V/e^-$ and 232 $\mu V/e^-$. In [10], pump-gate jot devices (jots) capable of low light imaging report a CG as high as 432 $\mu V/e^-$. Jots modify the potential profile in the charge transfer path by creating a storage well underneath the TG gate to eliminate the overlap capacitance to FD node in 4T pixel. However, the method reduces the full well capacity (FWC). Thus, in comparison to the proposed method in which both, CG and DR increase, jots enhance CG but DR remains low.

The photo-electric conversion characteristics without amplification and for three values of V_{PA}: 2.4 V, 2.0 V and 1.6 V are shown in Fig. 6. The sensor is exposed to a uniform fixed illumination of 13 lux from a 550 nm DC light source. The 12 bits digital output is shown with respect to varying integration time. Each data point is an average of 50 frames. The offset in (4) is compensated by dark frame subtraction. The sensitivity of the imager without amplification is 3771 LSB/lux.s. Increasing $V_{DD} - V_{PA}$ increases the slope. The sensitivities of the modes for V_{PA} = 2.4 V, 2.0 V and 1.6

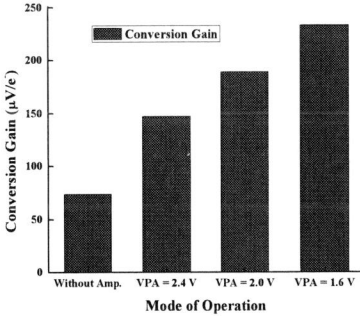

Fig. 5: Measured pixel conversion gain for different modes of operation

V are 4698 LSB/lux.s, 5193 LSB/lux.s and 5873 LSB/lux.s, respectively. Thus, the proposed method provides the capability of tailoring the conversion characteristics according to the ambient illumination by controlling V_{PA}. V_{PA} is not reduced further than 1.6 V, since it becomes the drain voltage of transistor MS in the subsequent readout phase. During readout, MS is a source follower (S_{PA} is off) and has to be maintained in saturation.

The modes displayed in Fig. 6 extend the DR upto 18 dB. The inherent 59 dB DR obtained without amplification is extended to 77 dB for low illumination. The measured SNR is shown in Fig. 7. The SNR at the switching point from the amplification mode with V_{PA} = 1.6 V to V_{PA} = 2.0 V is decided by the read noise noise floor. The measured results show an SNR dip of 1.6 dB. The SNR at the switching point from V_{PA} = 2.0 V to V_{PA} = 2.4 V is decided by the photon shot noise of the V_{PA} = 2.4 V amplification mode and shows a dip of 2.8 dB. The third switching point between V_{PA} = 2.4 V to the mode without amplification is dominated by the photon shot noise of the mode without amplification and shows the largest SNR dip of 13 dB.

Sample images for the modes plotted in Fig. 6 are shown in Fig. 8(a-d). Without amplification, the details of the image where there is lack of sufficient illumination are not displayed. As V_{PA} decreases or V_{DD}-V_{PA} increases, keeping an illumination of 3 lux and exposure time of 300 μs fixed, the sensitivity increases. However, for V_{PA} = 1.6 V, parts of the

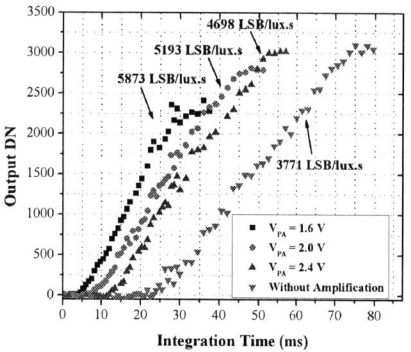

Fig. 6: Photo-electric conversion characteristics

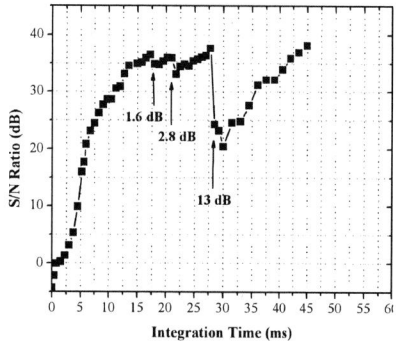

Fig. 7: Measured Signal-to-Noise Ratio at switching points

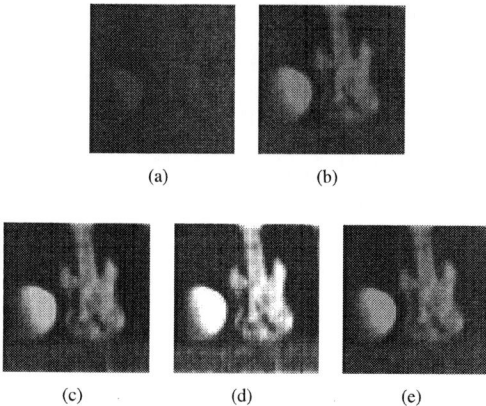

Fig. 8: Sample Images at a fixed illumination of 3 lux and integration time of 300 μs (a) Without amplification (b) V_{PA} = 2.4 V (c) V_{PA} = 2.0 V (d) V_{PA} = 1.6 V (e) Reconstructed Image

in a user-controlled way. Based on the proposed sensor, a self adapting low light WDR camera can be developed and SNR can be further optimized. Threshold variation of the source follower increases the FPN in the pixel which can be removed by using a dark row for calibration. The technique is compatible with 4T pixels having a pinned photodiode and an inherent DR of 60 - 70 dB, making the attainable DR, after amplification, 80 - 90 dB.

REFERENCES

[1] M. Loose, K. Meier, J. Schemmel, "A self-calibrating single-chip CMOS camera with logarithmic response," IEEE J. of Solid-state circuits, vol. 36, no. 4, pp.586-596, 2001.

[2] N. Akahane, S. Sugawa, S. Adachi, K. Mori, T. Ishiuchi, and K. Mizobuchi, "A sensitivity and linearity improvement of a 100 dB dynamic range CMOS image sensor using a lateral overflow integration capacitor," IEEE J. of Solid State Circuits, vol. 41, no. 4, pp. 851-858, 2006.

[3] J. H. Park, M. Mase, S. Kawahito, M. Sasaki, Y. Wakamori and Y. Ohta, "A 142dB dynamic range CMOS image sensor with multiple exposure time signals," IEEE Conf. on Asian Solid-State Circuits, pp. 85-88, 2005.

[4] M. Vatteroni, P. Valdastri, A. Sartori, A. Menciassi, and P. Dario, "Linear logarithmic CMOS pixel with tunable dynamic range," IEEE Trans. on Electron Devices, vol. 58, no. 4, pp. 1108-1115, 2011.

[5] N. Ide, W. Lee, N. Akahane, and S. Sugawa, "A wide DR and linear response CMOS image sensor with three photocurrent integrations in photodiodes, lateral overflow capacitors, and column capacitors," IEEE J. of Solid-State Circuits, vol. 43, no. 7, pp. 1577-1587, 2008.

[6] M. Mase, S. Kawahito, M. Sasaki, Y. Wakamori, and M. Furuta, "A wide dynamic range CMOS image sensor with multiple exposure-time signal outputs and 12-bit column-parallel cyclic A/D converters," IEEE J. of Solid-State Circuits, vol. 40, no. 12, pp. 2787-2795, 2005.

[7] R. Shimizu, M. Arimoto, H. Nakashima, K. Misawa, K. Suzuki, T. Ohno, Y. Nose, K. Watanabe, T. Ohyama, K. Tanis, "A Charge-Multiplication CMOS Image Sensor Suitable for Low-Light-Level Imaging," IEEE Int. Solid-State Circuits Conf., pp. 50-51, 2009.

[8] S. Ranganathan and Y. Tsividis. "A MOS capacitor-based discrete-time parametric amplifier with 1.2 V output swing and 3/spl mu/W power dissipation," IEEE Int. Solid-State Circuits Conf., pp. 406-502, 2003.

[9] G. Musalgaonkar, N. Priyadarshini and M. Sarkar, "An HDR Pixel With Over 60-dB Dynamic Range Enhancement Using In-Pixel Parametric Amplification," IEEE Trans. on Electron Devices, vol. 65, no. 2, pp. 555-563, 2018.

[10] J. Ma, D. Starkey, A. Rao, K. Odame, and E. R. Fossum, "Characterization of quanta image sensor pump-gate jots with deep sub-electron read noise," IEEE J. of the Electron Devices Society, vol. 3, no. 6, pp. 472-480, 2015.

image start getting saturated. Thus, a reconstructed image to display the total DR obtained from all the modes is shown in Fig. 8(e). The sensor characterization results are summarized in table I.

TABLE I: Performance Summary

Technology	0.35 μm
Pixel Size	10 μm x 10 μm with 20.25% Fill Factor
Pixel Array	64 x 64
Sensor Output	12 bit digital output
Full Well Capacity	24000 (Without amplification)
Power Consumption	3.55mW with 3.3 V Power Supply
DR Extension	18 dB (In low illumination)
Temporal Noise	$25e^-$ (Without amplification) $3.3e^-$ (Amplification mode with V_{PA} = 1.6 V)
FPN	2.9% (Without amplification) 5% (Amplification mode with V_{PA} = 1.6 V)

IV. CONCLUSION AND FUTURE PERSPECTIVES

A novel technique for obtaining wide DR in CMOS image sensors for low light applications has been implemented. It attains 18 dB DR extension featuring a good SNR for low illumination levels. Measurement results display the ability of the technique to amplify the low illumination signal in the charge domain at the pixel level. Thus, the DR can be enhanced using only one additional transistor switch in the pixel compared to standard architectures. There is no requirement of changing the material or using high supply voltages. Various sensitivity characteristics can be obtained

3-3 (8147)

A CMOS Imager for Reflective Pulse Oximeter with Motion Artifact and Ambient Interference Rejections

Hsiang-Lin Chen, Sung-En Hsieh, Tzu-Hsiang Hsu, Chih-Cheng Hsieh

Department of Electrical Engineering, National Tsing Hua University, Hsinchu, Taiwan

cchsieh@ee.nthu.edu.tw

Abstract—This paper presents a dual-mode CMOS imager for reflective pulse oximeter and fingerprint capturing capability. The same sensing array with 3-transistor active pixel sensor (3T-APS) is used in both modes for area and cost efficiency. In the oximeter mode, a coarse-fine ADC is applied to extend the system resolution. A periodical tracking and subtracting (PTS) technique is proposed to prevent the physiological signal (photoplethysmogram, PPG) from saturation caused by the motion artifact. The 60Hz time-varying ambient light from power line noise is suppressed by the post processing of the moving averaging filter with a sampling frequency (f_S) of 120Hz. A prototype with 64×64-pixel array and 20×20um^2 pixel pitch was fabricated in TSMC 0.18um CMOS technology. The measured result shows a motion rejection capability of ±199× AC swing, a motion tracking speed of 61.4kHz, and an ambient interference rejection capability of -36dB. The achieved SpO$_2$ accuracy is ±0.51% (@SpO$_2$ = 97%) at a power consumption of 380uW (without LED driver).

Keywords—ambient light rejection, bio-signal sensors, fingerprint sensor, motion artifact rejection, reflective pulse oximeter

I. INTRODUCTION

Health-managing monitors based on photoplethysmogram (PPG) are highly demanded including oxygen saturation (SpO$_2$), heart rate, blood pressure, and respiratory rate applications. The SpO$_2$ is the amount of oxygen dissolved in blood. According to Beer's law, SpO$_2$ is calculated by the concentration of oxygenated hemoglobin (HbO$_2$) and deoxygenated hemoglobin (Hb) that responses to the red (R) and infrared (IR) light respectively. It is defined as (1), where R$_{PPG}$ is the signal ratio of AC/DC in red and infrared illuminations as shown in (2). Normally, the ratio between AC (pulsatile) and DC (static) is around 0.5% to 4%. The AC is defined as the trough-to-peak value of the pulsatile part of PPG at a frequency between 0.1Hz to 5Hz.

$$S_pO_2 = \frac{0.81 - 0.18R}{0.73 + 0.11R} \quad (1) \qquad R_{PPG} = \frac{AC_{RED}/DC_{RED}}{AC_{IR}/DC_{IR}} \quad (2)$$

For the commercial pulse oximeters, the error tolerance of SpO$_2$ is specified to be less than ±2% within the detectable range of 70%~100%. To achieve an improved SpO$_2$ error of ±1%, the required relative error of R$_{PPG}$ in (2) needs be less than ±3%. To achieve the target accuracy with a AC/DC signal ratio of 0.5%, the total required resolution of PPG signal acquisition is around 15 bits. Therefore, a high-resolution analog-to-digital converter (ADC) is usually required for directly readout of PPG [1] with penalties of area and power. To reduce the required dynamic range and resolution of quantizer, several works with static DC-component cancellation and dual feedback loops of LED current control have been reported [2-3].

However, the error of residue signal (after DC cancellation) occurs when the intensity of the LED changes within a PPG cycle. The static DC cancellation can be implemented using least-mean-square adaptive filter [4-5], but the achievable

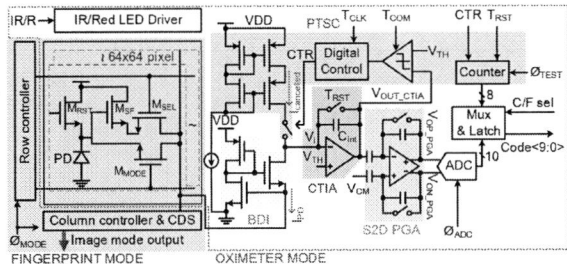

Fig. 1 The proposed architecture.

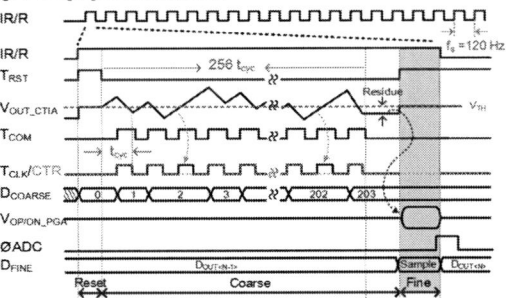

Fig. 2 Timing diagram of the proposed pulse oximeter.

motion rejection bandwidth is limited by the filter convergence speed and feedback loop latency. A DC level detection and calibration scheme was reported [6] using a current DAC without low-frequency filter. Nonetheless, the low DC updating rate results in a signal clip error from motion artifact and ambient light interference.

To solve the mentioned issues, this paper presents a CMOS imager for reflective pulse oximeter with motion artifact and ambient interference rejection. A coarse-fine ADC with periodical tracking and subtraction circuit (PTSC) is proposed to achieve DC cancellation and motion artifact rejection. The moving averaging (MA) filter with a sampling frequency proportional to utility frequency is used to remove the ambient interference from power line. Moreover, the proposed CMOS image sensor (CIS) also supports a dual-mode operation [6] for anti-counterfeit optical fingerprint application using the same sensing array for area and cost efficiency. The rest of this paper is organized as follows. Section II describes the system architecture and the operation. Section III presents the measurement results. Section IV gives the conclusion.

II. PROPOSED OXIMETER ARCHITECTURE

Fig. 1 shows the system architecture of the proposed dual-mode CIS. In oximeter mode, M$_{MODE}$ is turned on in each pixel to connect all the diodes together. The SpO$_2$ signal sensing path consists of sensing array, 8-bit (256-time) periodical tracking and subtracting circuit (PTSC), buffer direct injection (BDI) circuit, capacitive trans-impedance amplifier (CTIA), single-ended to differential programmable gain amplifier (S2D PGA), 10-bit SAR ADC, and IR/RED LED driver. With the proposed

978-1-5386-6414-8/18 $31.00 © 2018 IEEE   25

PTSC, the signal drift induced by motion artifact can be compensated in every sample using a maximum 256-time charge subtraction to avoid signal saturation. The adopted BDI frontend [7] provides the photodiode a stable reverse bias and releases the gain requirement of CTIA about 40dB by reducing the equivalent impedance of the pixel array. In fingerprint mode, all the pixel acts like a conventional 3T-APS array with enabling of the corresponding peripheral readout circuit including the array controller and column-shared correlated double sampling (CDS) circuit.

As mentioned, to achieve a ±1% accuracy of SpO2, the required dynamic range of PPG signal acquisition is at least 15 bits for a AC/DC signal ratio of 0.5% and 7-bit resolution for AC signal (AC swing = 128LSB). In this work, with one additional bit for motion artifact tolerance, the targeted system resolution is 16 bits. An 8-bit PTSC and a 10-bit SAR ADC are implemented as the coarse and fine ADCs respectively with 2-more bits for motion artifact rejection and coarse-fine error redundancy. Fig. 2 shows the timing diagram of pulse oximeter mode (M_{MODE} = 1). The AC/DC signals of IR and R are readout interleaving with corresponding LED enabling controlled by IR/R. When IR/R = 1, the infrared and red LEDs turn on and off respectively and vice versa. Each readout operation (IR/R=1) can be divided into Reset, Coarse, and Fine phases. After resetting CTIA by T_{RST} ("Reset" phase), 256 cycles (t_{cyc}) of periodical tracking and subtracting (PTS) operation is enabled by T_{CLK} ("Coarse" phase). When the integrated output of CTIA (V_{OUT_CTIA}) is larger than a specific threshold V_{TH}, CTR is high to enable a fixed-amount charge ($Q_{LSB/Coarse} = I_{cancelled} \times 0.5 \times t_{cyc}$) subtraction. By counting the "1" number of comparator's output CTR (enabled by T_{COM}), the conversion of 8-bit coarse code (D_{COARSE}) is accomplished. After 256-t_{cyc} conversions, the residue is then amplified by PGA and digitized (D_{FINE}) by the following 10-bit SAR ADC in "Fine" phase to achieve an effective conversion resolution of 17 bits. With the proposed PTS technique, a real-time tracking and subtracting operation is implemented to avoid signal saturation caused by motion artifact. The achieved motion tracking speed is 61.4kHz (2×120Hz×256) and the tolerance range of motion artifact is ±199× of AC swing (@DC level = 0.5× full swing, AC/DC = 0.5%) at the sampling rate of 120Hz. For rejection of the ambient light interference from utility power (60Hz in Taiwan), the moving averaging (MA) filtering function with a window (W) of 2 is implemented by selecting a double sampling frequency (120Hz) to realize a sinc function with a frequency response notch at 60Hz.

III. MEASUREMENT RESULTS

A prototype is fabricated in TSMC 0.18um CMOS technology and measured. Fig. 3 shows the measured PPG signal w/i (in blue) and w/o (in red) MA filtering in time domain and frequency domain under indoor lighting. It shows the inference tone at 60Hz has been rejected by 36dB (from -65dB to -101dB). Fig. 4 shows the comparison of reconstructed PPG waveforms (in blue) w/i and w/o motion artifact rejection and reported coarse code (in red). Under the test with motion artifact, compared to the distorted signal (saturation by motion) in prior art [6] due to the slow coarse-code update speed (1Hz), it shows there is no signal clipping using the proposed oximeter with a real-time (61.4kHz) motion artifact tracking capability. Table I summarizes the performance comparison of the state-of-the-art works. The measured SNR_{AC} is 33.21dB to 51.27dB with a AC/DC signal ratio range of 0.5% to 4%. Since the SNR_{AC} improves with a larger photocurrent I_{PD} and AC/DC at a penalty of increased LED power consumption, a FoM defined as $I_{PD} \times (AC/DC)/10^{SNR/20}$ is used to provide a fair benchmarking. It shows this work achieves the best FoM performance (>3× better) and motion artifact tracking speed

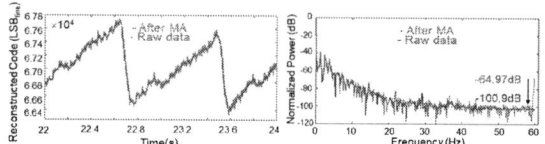

Fig. 3 Ambient interference rejection in time domain and frequency domain.

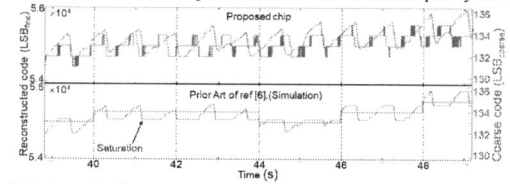

Fig. 4 Motion rejection of proposed oximeter and simulated prior art.

Table. I. Comparison table

	TBioCAS 2013 [2]	TBioCAS 2015 [3]	JSSC 2015 [4]	JETCAS 2017 [5]	ASSCC 2016 [6]	This work
Technology	0.35um	0.18um	0.18um	0.18um	0.18um	0.18um
Supply Voltage(V)	3.3	1.8	1.2	5/1.5	3.3/1.8	3.3/1.8
Array Size	Off-chip component	Off-chip component	ECG SoC	Organic sensor	64×128	64×64
Pixel Pitch (um)			sensor		20	20
Fingerprint Captured	NO	NO	NO	NO	YES	YES
Oximeter Type	Trans-missive	Reflective		Reflective	Reflective	Reflective
Sampling Frequency (Hz)	100	185	32k	50~5k	0.5(DC)/500(AC)	120
Equivalent Resolution (bit)	17	17	13.5	4(DC)/10(AC)	12.3~18.3	17
SNR_{AC} (dB) w/ & w/o Filter	37 @fc=10Hz (AC/DC=2.8%)	48* @fc=10Hz (AC/DC=0.5%)	N/A	30	35.9 @fc=10Hz (AC/DC=1.3%)	33.21~51.27 */** (w/o filter) 36.76~54.83 */** @fc=10Hz
I_{PD} (uA)	0.044	30		0.3	0.328	0.056
FoM of Oximeter (pA) $I_{PD} \times (AC/DC) / 10^{SNR/20}$	174	597.16		379.5	68.36	4.67
Motion-rejection/AC (@DC level=0.5x full swing, AC/DC = 0.5%)	N/A	±10×	±13.3×	N/A	±1×	±199×
Motion-tracking Speed (Hz)	N/A	185	10	N/A	1	61.4k
Power (mW) (w/o LED)	0.528	0.216	0.075	0.233	0.0998	0.38

*when DC level is 0.5x full swing of the system **AC/DC=0.5%~4%

(>5× better) compared to the state-of-the-art works. According to the experimental R-to-SpO2 characteristic transfer function, the prototype sensor achieves a SpO2 accuracy of ±0.51% without any post signal processing (low-pass filter) under a normal condition of 97% SpO2 concentration and the total power is 380uW excluding the LED driver. In optical fingerprint mode, the fingertip image is successfully captured with recognizable detail (sweat pole) for biometric ID application.

IV. CONCLUSION

This paper presents a dual-mode CMOS imager for reflective pulse oximeter and fingerprint capturing capability. With the proposed PTS and MA schemes, the prototype demonstrates a low-power oximeter solution with motion artifact rejection capability of ±199 times of AC and ambient interference rejection capability of -36dB. The achieved accuracy of SpO2 detection is ±0.51% (@SpO2=97%) without any post signal processing at a power consumption of 380uW (w/o LED).

ACKNOWLEDGMENT

This work is supported by National Chip Implementation Center and Ministry of Science and Technology, Taiwan under the contract 104-2221-E-007-103-MY3.

REFERENCES

[1] Texas Instruments, AFE4403 datasheet, SBAS650B, May 2014.

[2] K. N. Glaros and E. M. Drakakis, "A sub-mW fully-integrated pulse oximeter front-end," in *TBioCAS*, vol. 7, no. 3, pp. 363-75, Jun. 2013.

[3] E. S. Winokur *et al.*, "A Low-Power, Dual-Wavelength Photoplethysmogram (PPG) SoC With Static and Time-Varying Interferer Removal," in *TBioCAS*, vol. 9, no. 4, pp. 581-589, Aug. 2015.

[4] N. Van Helleputte *et al.*, "A 345 µW Multi-Sensor Biomedical SoC With Bio-Impedance, 3-Channel ECG, Motion Artifact Reduction, and Integrated DSP," in *JSSCC*, vol. 50, no. 1, pp. 230-244, Jan. 2015.

[5] Y. Lee *et al.*, "Sticker-Type Hybrid Photoplethysmogram Monitoring System Integrating CMOS IC With Organic Optical Sensors," in *JETCAS*, vol. 7, no. 1, pp. 50-59, March 2017.

[6] A. Y.-C. Chiou *et al.* "An integrated CMOS optical sensing chip for multiple bio-signal detections," in *A-SSCC*, pp. 197-200, Toyama, 2016.

[7] N. Bluzer, R. Stehlik, "Buffered direct injection of photocurrents into charge-coupled devices", in *JSSC*, vol. SC-13, pp. 86-92, 1978.

3-4 (8205)

A 4-channel 5.04 μW 0.325 mm^2 Orthogonal Sampling-Based Parallel Neural Recording System

Reza Ranjandish[1], and Alexandre Schmid[2]

[1,2]Microelectronic Systems Laboratory, Swiss Federal Institute of Technology (EPFL), Switzerland

reza.ranjandish@epfl.ch

Abstract—**The application of orthogonal sampling for parallel neural recording is presented in this paper. Orthogonal sampling enables reducing the number of the ADCs in conventional recording systems into one single unit. Consequently, the ADC bandwidth and dynamic range is effectively employed and shared between all the channels without any loss in the temporal information of the channels during sampling, which is not the case of time-multiplexed ADCs. A 4-channel neural recording system based on orthogonal sampling is implemented in a 0.18 μm technology with a power consumption of 1.26 μW per channel supplied at 0.8 V to validate the proposed methodology.**

Index Terms—**Neural recording, orthogonal sampling, capacitively-coupled instrumentation amplifier, neural activity.**

I. INTRODUCTION

In order to analyze data that is transferred between populations of neurons, a recording system is needed to sense and amplify the delicate neural signals. Better understanding of the brain network topology and functionality requires a recording system with a large number of channels. Over the past years, the number of simultaneous recording channels in neural interface has doubled every seven years [1]. As the number of recording channels is growing, significant improvements are required at system level to support the stringent constraints. One major challenge in multichannel recording consists of reducing temporal information loss during the recording [2]. Hence, either a dedicated ADC or a system-level technique should be used to avoid temporal information loss that takes place in time-multiplexed ADCs such as in [3]. In addition, the bandwidth of the ADC of time-multiplexed recording systems is not effectively used. Considering a recording system with N channels, assuming that each channel is sampled at the rate of SR, then the ADC should perform at a sampling rate of $N \times SR$. The architectures proposed in [4], [5] use a dedicated ADC in each recording channel. The recording channels proposed in [6] are implemented using rail-to-rail differential $\Delta^2\Sigma$ neural recording channels. Using this technique a low-power and area implementation of the neural recording channels with dedicated ADC per channel is possible for low-bandwidth (*i.e.* 500 Hz) recording using oversampling technique. However, by increasing the recording bandwidth, the power consumption per channel increases linearly and makes the recording system power hungry. Compressing data using multi-channel compressive sensing (MCS) is one of the methods used to share the ADC among multiple channels [7], [8]. However, the reconstruction of the compressed signal

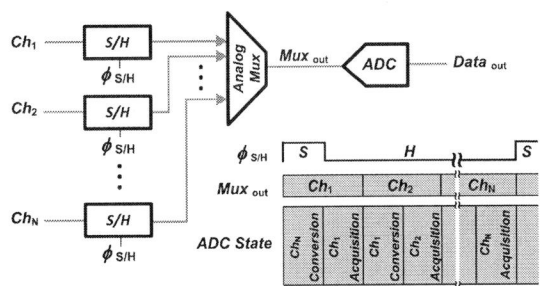

Fig. 1. parallel-sampling time-multiplexing method for parallel neural recording.

introduces latency, complexity and loss in the SNR of the reconstructed signal [9]. A straightforward method to perform parallel neural recording using a single ADC consists of simultaneously sampling all the channels, hold and digitize them respectively in time as shown in Fig. 1. This architecture needs one sample-and-hold circuit per channel. Furthermore, due to the droop rate of the sample-and-hold unit, information loss in Channel N is significantly larger than it is in Channel 1, during digitization. In [1], parallel recording is performed by sharing a single ADC. Using a Walsh-Hadamard coding technique, samples are encoded and digitized using a single ADC, as summrized in Fig. 2. Employing this technique in an N-channel parallel recording, each sample of the channels are held using a track and hold (T/H) for a period of N successful modulation and sub-conversion cycles of the ADC. Moreover, the required ADC resolution (B_{ADC}) prescribes ($log_2 N$) bits in addition to the resolution of the individual channel (B_{Ch}) [1]:

$$B_{ADC} = B_{Ch} + log_2 N \qquad (1)$$

Hence, although this method eliminates the droop rate issue of the parallel-sampling time-multiplexing method, one track-and-hold unit is still required per channel. Furthermore, Walsh-Hadamard multiplication performs employing one G_m cell per channel, hence increasing the required power and area with respect to the parallel-sampling time-multiplexing method.

In this paper a new parallel neural recording system is presented in which the signals are modulated using orthogonal frequencies at the sample-and-hold stage. Therefore, the

978-1-5386-6414-8/18 $31.00 © 2018 IEEE

27

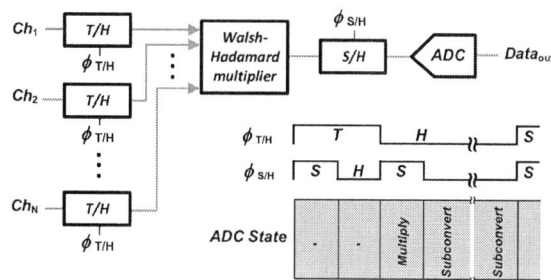

Fig. 2. Overview of the method proposed in [2] for parallel neural recording.

sample-and-hold unit is shared among all the channels. In addition, the proposed method does not require additional resolution for the ADC in contrast to the method presented in [2].

II. System Architecture

The proposed system architecture is shown in Fig. 3. Each analog front-end (AFE) consists of a dual-stage capacitive-coupled (CC) fully differential low-noise amplifier and a low-pass filter. In order to limit the flicker noise at G_1, the PMOS input pair is designed large while smaller transistors can be employed in the input pair of G_2. The analog output of each channel is modulated using orthogonal frequencies, summed up and simultaneously sampled. In addition, the gain at the summing stage is programmable to provide sufficient amplification. The output of the orthogonal sampler is then digitized using a 10-bit successive approximation register ADC (SAR ADC).

To ensure the orthogonality between each of the modulation frequencies, the first four Walsh functions are used to generate modulation frequencies. Wal_0, Wal_1, Wal_2 and Wal_3 are the first four Walsh functions, which are generated by the Walsh

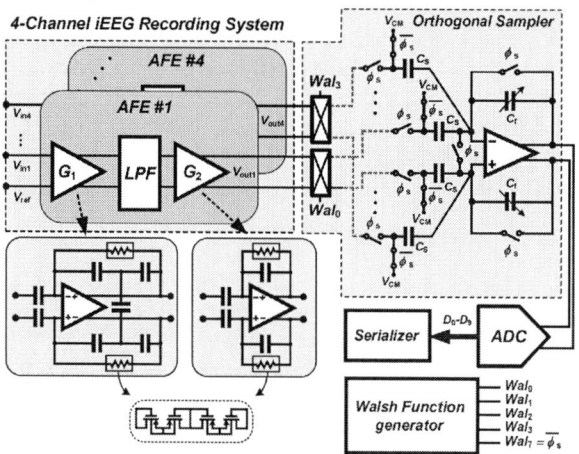

Fig. 3. Architecture of the 4-channel parallel recording system implementing the orthogonal sampling technique.

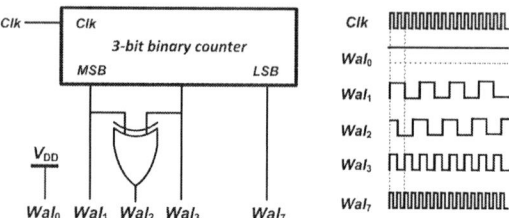

Fig. 4. Circuit-level implementation of the Walsh function generators.

function generator. Walsh functions are orthogonal functions that only take two values of ±1. Definition of the Walsh functions is based on the Rademacher functions which are defined as follows:

$$R_j : [0, T] \rightarrow \{-1, 1\}$$
$$R_0 = 1 \tag{2}$$
$$R_j(x) = \text{sgn}(\frac{\sin(2^j \pi x)}{T}), \ j \geq 1$$

$$R_j(x + T) = R_j(x) \tag{3}$$

where

$$\text{sgn}(x) = \begin{cases} -1 & if \ x \leq 0 \\ +1 & otherwise \end{cases} \tag{4}$$

Walsh functions can be generated using the following iterative process:

$$Wal_n(x) = \prod_{j=0}^{m} R_{j+1}(x)^{g_j} \tag{5}$$

where $(b_{m-1}b_{m-2}...b_0)_2$ is the binary representation of n and $g_i = b_i \oplus b_{i+1} (b_m = 0)$ is its corresponding Gray code, and $\oplus$ is the exclusive-or (XOR) operator. The circuit-level implementation of Walsh function generators is shown in Fig. 4 which consists of a 3-bit binary counter and an exclusive OR logic gate.

The simplified block diagram of the 4-channel recording system is shown in Fig. 5 including the timing diagram of the first four Walsh functions as well as Wal_7, which is used for the sampling. As a benefit of this technique, for an N-channel recording system, each channel is sampled with an oversampling factor of N thanks to the orthogonal sampling. In order to clarify this method operation, we assume a 2-channel recording system as shown in Fig. 6. The first channel is not modulated while the second channel is modulated at a frequency of F_M, which is equivalent to the modulation with Walsh functions Wal_0 and Wal_1. Wal_1 has a period of $1/F_M$. Subsequently, the outputs are summed up. Due to the modulation, the spectrum of channel two spreads in the frequency domain as shown in Fig. 6, while the spectrum of channel one does not change. If the output of the summer is sampled at a frequency of $2F_M$ (Wal_3), then all the harmonics higher than F_M fold back into a frequency range smaller than F_M. As a result, each channel is sampled and digitized at $2F_M$ that has an oversampling ratio of 2.

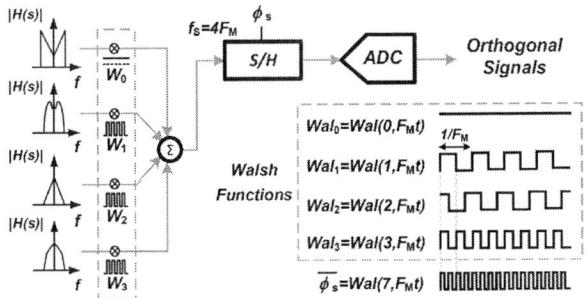

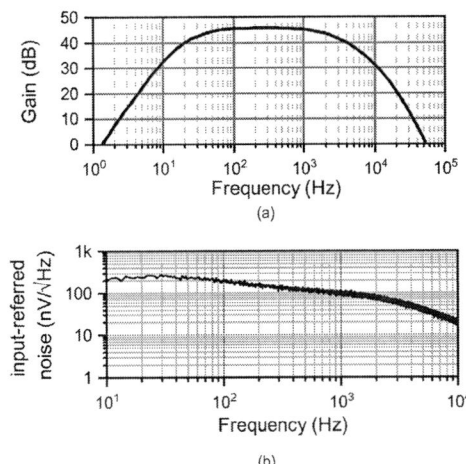

Fig. 5. simplified block-level illustration and timing operation of the 4-channel parallel neural recording system.

At the receiver side, the data stream is demodulated by the same Walsh functions in the digital domain (Walsh-Hadamard matrix), and is consequently filtered and decimated to extract the corresponding channel. As a benefit of this method, all the channels are oversampled by a factor of N in an N-channel parallel recording system. In order to avoid crosstalk among the recording channels, high order of filtering is required at the receiver side. Comparing with the time-multiplexed recording systems, the orthogonal sampling efficiently use the bandwidth of the ADC to perform parallel recording using a single ADC and it requires one sample and hold block shared among the all recording channels. In addition, oversampling with a factor of N allows extracting additional $log_2\sqrt{N}$ bits using decimation filtering. Hence, in contrast to [1], the ADC used in orthogonal sampling does not need additional number of bits and the ADC resolution (B_{ADC}) is equal to the resolution of the individual channel (B_{Ch}):

Fig. 7. Measured (a) frequency response and (b) input-referred-noise of AFE.

$$B_{ADC} = B_{Ch} \qquad (6)$$

The presented method is not limited to 4-channel and can be developed to an N-channel recording system ($N = 4n$, $n > 1$), in which Wal_0-Wal_{N-1} should be used for modulation and Wal_{2N-1} should be used to perform sampling.

III. EXPERIMENTAL RESULTS

The individual front-end channel draws 1 μA from a 0.8 V supply. The front-end shows a mid-band gain of 45.3 dB, a high-pass corner of 30 Hz and a low-pass corner of 2.35 kHz (Fig. 7a). The input-referred noise of the individual front-end is depicted in Fig. 7b, which exhibits an integrated input-referred noise of 4.6 μVrms (30 Hz-2.35 kHz). The total power consumption of the system including the ADC, summer, Walsh and bias current generators is equal to 5 μW which yields a noise efficiency factor (NEF) of 4.6 per channel. To validate the performance of the system, each channel is stimulated with a single-tone sinusoid using attenuators. The input sinusoid

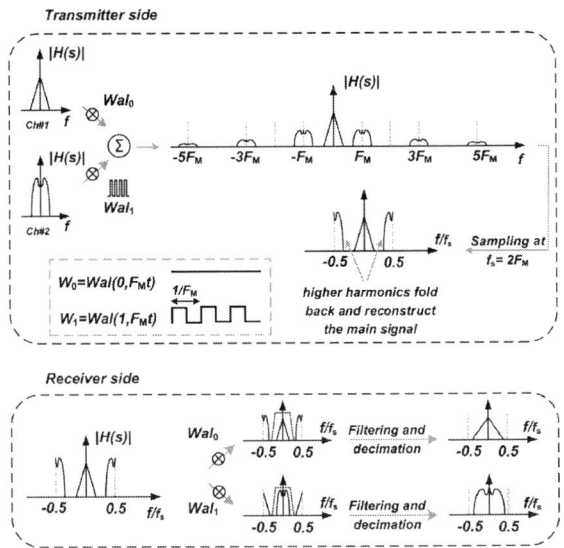

Fig. 6. Illustration of the orthogonal sampling performance in frequency domain using two orthogonal channels.

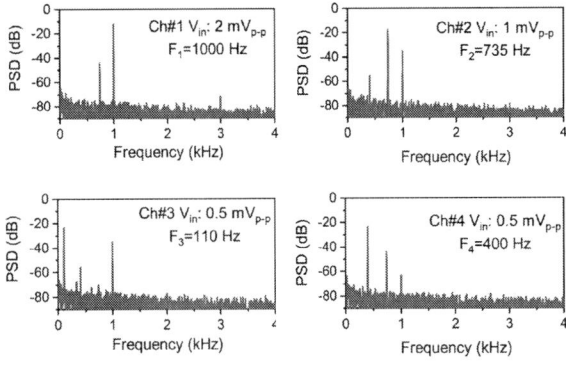

Fig. 8. Spectrum of the demodulated, filtered and decimated signals for each channel at the receiver using 3rd order FIR filter.

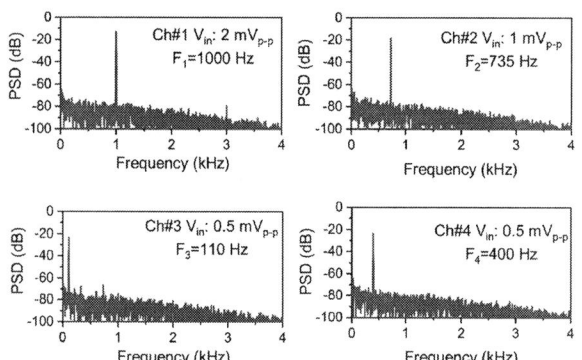

Fig. 9. Spectrum of the demodulated, filtered and decimated signals for each channel at the receiver using 10th order FIR filter.

signals are a 2 mV$_{pp}$ sinusoid at 1 kHz for channel number one, a 1 mV$_{pp}$ sinusoid at 735 Hz for channel number two, a 0.5 mV$_{pp}$ sinusoid at 110 Hz for channel number three and 0.5 mV$_{pp}$ sinusoid at 400 Hz for channel number four.

The modulated samples with sampling rate of 32 kS/s are serially captured and demodulated, filtered and decimated in Matlab which are depicted in Fig. 8 and Fig. 9. Fig. 8 shows the results filtered by a 3rd order FIR low-pass filter and Fig. 9 shows the results filtered by a 10th order filter. As discussed earlier, a low-pass filter of higher-order favors lower crosstalk. By comparing the results shown in Fig. 8 and Fig. 9, the importance of using higher-order filter is recognizable. According to Fig. 9, the worst crosstalk is measured at -51 dB using a 10th order FIR low-pass filter. Fig. 10 shows the die photo of the 4-channel parallel neural recording system including the size of the individual front-end. An individual AFE occupies 0.04 mm^2 and the porposed 4-channel parallel recording system occupies 0.325 mm^2 resulting in 0.08 mm^2 per channel. Table 1 summarizes the performance of the proposed system and compares it with the prior arts, using uncompressed parallel recording systems which employ conventional neural amplifiers.

IV. CONCLUSION

A 4-channel parallel neural recording system based on orthogonal sampling is presented. A parallel recording technique based on frequency modulation is proposed to prevent temporal data loss in the information acquired from multiple channels. Neural signals are modulated using Walsh functions

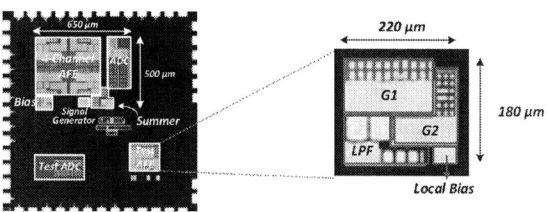

Fig. 10. Die micrograph (left) and one recording channel (right).

and summed up at the sample-and-hold stage. This technique provides an orthogonality between the information of different channels, while enabling sharing one sample-and-hold among all channels. This technique provides parallel neural recording although sharing a single ADC among all recording channels. The proposed method is implemented as a 4-channel parallel neural recording system in 0.18 μm technology.

ACKNOWLEDGMENT

This work has been supported by the Swiss NSF under grant number 200020-175790.

Table 1. Performance Summary and Comparison

Parameter	[2]	[4]	[5]	This Work
Technology	0.18 μm	0.13 μm	0.18 μm	0.18 μm
Supply Voltage	1.2 V	1.2 V	0.5 V	0.8 V
Power/Channel [1]	22.4 μW	68 μW	3.05 μW [2]	1.26 μW
Architecture	Single ADC	ADC/Ch	ADC/Ch	Single ADC
Feature	Coding	Windowed integrator	Δ-ΔΣ	Orthogonal sampling
ADC resolution	$B_{ch}+log_2\sqrt{N}$	B_{ch}	B_{ch}	B_{ch}
Bandwidth (Hz)	11.5-56 k	280-10 k	0.4-10.9 k	30-2.35 k
LNA Gain	41-60 dB	56 dB	38.5 dB	45.6 dB
Input ref. noise	4.1 μV$_{rms}$ [3]	2.2 μV$_{rms}$	3.32 μV$_{rms}$	4.6 μV$_{rms}$
NEF	5.6	6.4	3	4.6
Area/Channel [4]	0.12 mm^2	0.26 mm^2	0.07 mm^2 [5]	0.08 mm^2

[1] System total power consumption/No. of channels.
[2] AFE+ADC.
[3] from 10 Hz to 100 kHz
[4] System total area/No. of channels.
[5] Including digital interface. In this system, capacitors are placed on top of the active circuitry.

REFERENCES

[1] I. H. Stevenson and K. P. Kording, "How advances in neural recording affect data analysis," *Nature neuroscience*, vol. 14, no. 2, p. 139, 2011.

[2] V. Majidzadeh, A. Schmid, and Y. Leblebici, "A 16-channel, 359 μw, parallel neural recording system using walsh-hadamard coding," in *Custom Integrated Circuits Conference (CICC), 2013 IEEE*. IEEE, 2013, pp. 1–4.

[3] J. Dragas, V. Viswam, A. Shadmani, Y. Chen, R. Bounik, A. Stettler, M. Radivojevic, S. Geissler, M. E. J. Obien, J. Müller *et al.*, "In vitro multi-functional microelectrode array featuring 59 760 electrodes, 2048 electrophysiology channels, stimulation, impedance measurement, and neurotransmitter detection channels," *IEEE journal of solid-state circuits*, vol. 52, no. 6, pp. 1576–1590, 2017.

[4] H. Gao, R. M. Walker, P. Nuyujukian, K. A. Makinwa, K. V. Shenoy, B. Murmann, and T. H. Meng, "Hermese: A 96-channel full data rate direct neural interface in 0.13μ m cmos," *IEEE Journal of Solid-State Circuits*, vol. 47, no. 4, pp. 1043–1055, 2012.

[5] S.-Y. Park, J. Cho, K. Na, and E. Yoon, "Toward 1024-channel parallel neural recording: Modular δ-$\delta\sigma$ analog front-end architecture with 4.84 fj/cs· mm 2 energy-area product," in *VLSI Circuits (VLSI Circuits), 2015 Symposium on*. IEEE, 2015, pp. C112–C113.

[6] H. Kassiri, M. T. Salam, M. R. Pazhouhandeh, N. Soltani, J. L. P. Velazquez, P. Carlen, and R. Genov, "Rail-to-rail-input dual-radio 64-channel closed-loop neurostimulator," *IEEE Journal of Solid-State Circuits*, vol. 52, no. 11, pp. 2793–2810, 2017.

[7] M. Shoaran, M. Shahshahani, M. Farivar, J. Almajano, A. Shahshahani, A. Schmid, A. Bragin, Y. Leblebici, and A. Emami, "A 16-channel 1.1 mm 2 implantable seizure control soc with sub-μw/channel consumption and closed-loop stimulation in 0.18 μm cmos," in *VLSI Circuits (VLSI-Circuits), 2016 IEEE Symposium on*. Ieee, 2016, pp. 1–2.

[8] M. Shoaran, C. Pollo, K. Schindler, and A. Schmid, "A fully integrated ic with 0.85-μw/channel consumption for epileptic ieeg detection," *IEEE Transactions on Circuits and Systems II: Express Briefs*, vol. 62, no. 2, pp. 114–118, 2015.

[9] H. Hosseini-Nejad, A. Jannesari, and A. M. Sodagar, "Data compression in brain-machine/computer interfaces based on the walsh–hadamard transform," *IEEE transactions on biomedical circuits and systems*, vol. 8, no. 1, pp. 129–137, 2014.

4-1 (8144)

An Integrated DC-DC Converter with Segmented Frequency Modulation and Multiphase Co-Work Control for Fast Transient Recovery

U-Fat Chio, Kuo-Chih Wen, Sai-Weng Sin[1], Chi-Seng Lam, Yan Lu, Franco Maloberti[2], R. P. Martins[1,3]

State Key Laboratory of Analog and Mixed-Signal VLSI (http://www.amsv.umac.mo)
1 - Dept. of ECE, Faculty of Science and Technology, University of Macau, Macao, China
2 – Department of Electrical, Computer and Biomedical Engineering, University of Pavia, Pavia, Italy
3 – On leave from Instituto Superior Técnico/Universidade de Lisboa, Portugal
E-mail: terryssw@umac.mo

Abstract— **This paper presents a fully integrated VCO-based switched-capacitor (SC) DC-DC converter in 65 nm CMOS. We propose two transient-enhancement techniques: segmented frequency modulation (SFM) and multiphase co-work control (MCW) to reduce the latency of the VCO-based control loop and shorten the SC DC-DC converter response time. We design a 15-phase interleaved converter to support an output voltage of 1 V from a 2.4 V input supply, delivering up to 138 mA of load current, which takes only 25/29 ns for output voltage recovering to the steady state from heavy-to-light/light-to-heavy load transients, respectively. It obtains a peak efficiency of 82.8 % and keeps the efficiency above 80 % from 31 mA to the maximum load current. The SC DC-DC converter chip occupies 0.61 mm², and its output power density is 240 mW/mm².**

Keywords— *SC DC-DC converter, fully integrated, switched-capacitor, voltage-controlled oscillator (VCO).*

I. INTRODUCTION

Fully integrated switched-capacitor (SC) DC-DC converters have increasingly proven their advantages while achieving high-conversion efficiency and generating different supply voltage domains on-chip with few shared voltage supply pins [1]-[5]. Multiphase interleaving of SC DC-DC converters can adjust the number of switching phases along with a wide range of load currents. Besides, multiple phases allow the control-loop to respond at every phase that is a fraction of the switching period [3], thus having a strong ability to achieve a fast load transient response. However, the response of the control loop is not able to provide the fast recovery requested to the converter output during load transition. As shown in [2], the VCO, whose frequency is proportional to the charge pump integrator voltage, determines the clock frequency of the multiphase generator. To accomplish both the rapid detection and fast recovery, [2] uses an additional bypassing control path to handle the load transition specifically. The frequency of the VCO-based clock generator was set to the maximum immediately after detecting an undershoot. However, for limited light-to-heavy load transition, such brute-force action caused the excessive switching of the SC array that exceeds the frequency required by the corresponding steady-state load current. Consequently, after the undershoot is recovered, a new overshoot is generated due to the overcharged voltage of the charge pump integrator which prolongs the total load transient recovery time [2].

The circuit described in this paper uses a segmented frequency modulation technique (SFM) and a multiphase co-

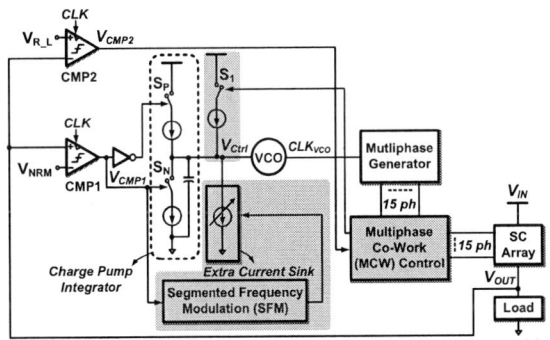

Fig. 1. Simplified circuitry of the VCO-based SC DC-DC converter with proposed load transient enhancement control techniques.

work scheme (MCW) for the multi-phase SC DC-DC converter to improve the transient recovery speed during light-to-heavy and heavy-to-light load transients, thus achieving smooth and fast control at the same time. The details will be described next.

II. PROPOSED LOAD TRANSIENT RECOVERY SCHEMES

Fig. 1 depicts the simplified block diagram of the proposed VCO-based SC DC-DC converter. It is based on the implementation in [4], with the CMP1, the charge-pump integrator and the VCO forming the primary control path to drive the multi-phase clock generator, and the CMP2 providing the brute-force path for undershoot detection. This paper introduced two extra blocks for the load transient enhancement control, i.e., a segmented frequency modulation (SFM) control with extra current sink control and a multiphase co-work (MCW) control circuit. These two additional techniques relaxe the limitation outlined in [2].

A. Segmented Frequency Modulation (SFM)

Fig. 2 shows the circuitry of the proposed SFM modulator. It consists of a pulse counter, a 2-bit decoder, three switches S_{X0}, S_{X1}, S_{OV} and three binary weighted extra sink current branches. The pulse counter accumulates the output states of the CMP1 in each comparison cycle. If the load current changes from heavy-to-light, it generates an overshoot at the output that switches the CMP1 to 1 (V_{NRM} is the detection threshold). When the sum of pulse counts is larger than a given number, an overshoot detection signal V_{SIG} changes from low to high and turns on the switch S_{OV} which increases the sink current of the charge pump integrator. Depending on the degree

---

This research work was financially Supported by Research Committee of University of Macau and Macao Science and Technology Development Fund SKL/AMS-VLSI/SSW/FST, SKL/AMS-VLSI/WMC/FST & 120/2016/A3.

978-1-5386-6414-8/18 $31.00 © 2018 IEEE   31

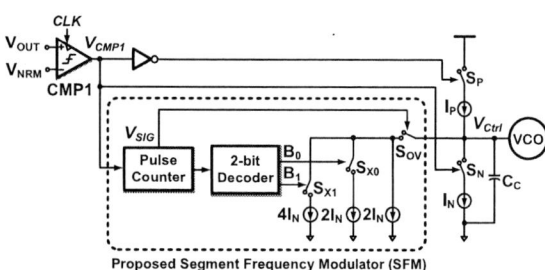

Fig. 2. Circuitry of the proposed segment frequency modulator (SFM).

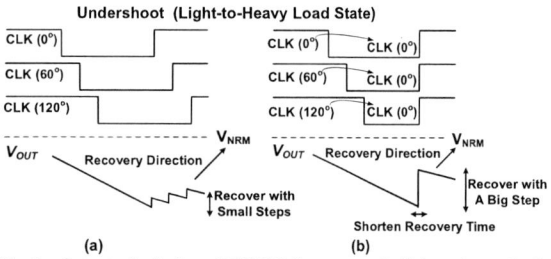

Fig. 3. Output of a 3-phase SC DC-DC converter in light-to-heavy load transient with (a) normal phase interleaving, and (b) MCW schemes.

of overshooting, up to an 8x additional sink current can be supported. This additional sink current can discharge the VCO control voltage V$_{CTRL}$ quickly. Actually, the switching frequency of a VCO-based SC DC-DC converter scales with the load current. Therefore, a fast recovery time denotes speed in turning down the switching frequency for load step-down.

B. Multi-phase Co-Work (MCW) Technique

The multi-phase interleaving technique can reduce the output ripples of the DC-DC converter while avoiding a dedicated output decoupling capacitor [3]. Fig. 3(a) gives an example of the output voltage of a 3-phase interleaved SC DC-DC converter. The ripples of V_{OUT} decrease by a factor of 3 because of 3 channels interleaving, which also imposes a slow transient recover due to the reduced voltage steps. This work uses the MCW technique depicted in Fig. 3(b), which allows simultaneously ripple reduction and fast transient recovery in light-to-heavy load transition. The solution aims to synchronize the selected interleaving multiple-phase SC channels to provide an immediate larger energy step to the load. Indeed, by increasing the energy supply of the multi-phase SC network over time, the load transient recovery speed can be enhanced within a short period.

III. MEASUREMENT RESULTS

The proposed DC-DC converter is implemented in a 65 nm CMOS process, with an active area of 0.61 mm^2 (Fig. 4(a)). An on-chip load current generator implemented with a programmable PMOS array can adjust the equivalent width of the PMOS devices to obtain correct I-V measurements during the load transient.

In the measurement, we set the nominal voltage V$_{NRM}$ =1 V, and the undershoot-detection threshold V$_{R\_L}$ =0.9 V. The loop response time is within 5 ns, with the VCO switching

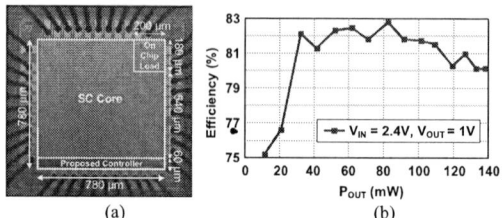

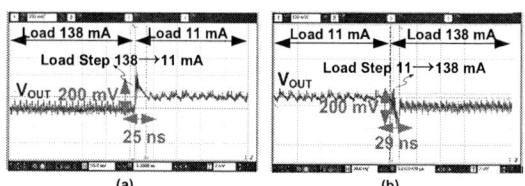

Fig. 4. (a) Chip photo, (b) Measured efficiency vs. output power.

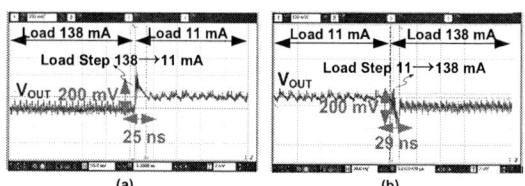

Fig. 5. Measured output voltage waveforms during (a) heavy-to-light load transition and (b) light-to-heavy load transition.

TABLE I: PERFORMANCE SUMMARY

Technology	65 nm	Max. out Power	138 mW
Topology	1/2 SC	Active Area	0.61 mm^2
Interleave Phase	15	Power Density	240 mW/mm^2
Capacitor Type	MOS	Peak Efficiency	82.8%
C$_{fly}$ / C$_{out}$	2.56 nF / 0	Load Transient Step	11<->138 mA
SC Frequency	1 - 250 MHz	Light-to-Heavy Recovery	25ns, 5.1mA/ns
V$_{IN}$ / V$_{OUT}$	2.4 V / 1 V	Heavy-to-Light Recovery	29ns, 4.38mA/ns
Ripple	55 mV	Controller Area Overhead	7.7%

frequency ranging from 1 MHz to 250 MHz for the entire load range, while the clock frequency of the comparators is 1.5 GHz. Fig. 4(b) shows the measured efficiency of the converter when V_{IN} = 2.4 V and V_{OUT} = 1 V. The load current sweeps from 11 mA to 138 mA. The power efficiency is higher than 75 % at light load. From 31 mA to 138 mA the efficiency exceeds 80 % with a peak of 82.8 % at 80 mA.

Fig. 5(a) shows the output voltage waveforms during heavy-to-light load (138 to 11 mA) transient. The SFM technique helps to improve the recovery speed as low as to 25 ns. Fig. 5(b) shows the waveforms during light-to-heavy load with a transient from 11 mA to 138 mA. The MCW helps to improve the recovery speed to 29 ns. These results confirm the effectiveness of the proposed SFM and MCW control techniques. Table I shows the performance summary. Without the use of deep-trench capacitors, the circuit attains a fast recovery and the output power density of 240mW/mm^2.

REFERENCES

[1] T. Souvignet et al., "A Fully Integrated Switched-Capacitor Regulator With Frequency Modulation Control in 28-nm FDSOI," *IEEE Trans. Power Electron.*, vol. 31, No. 7, pp. 4984-4994, Jul. 2016.

[2] H. P. Le et al., "A Sub-ns Response Fully Integrated Battery-Connected Switched-Capacitor Voltage Regulator Delivering 0.19W/mm^2 at 73% Efficiency," *ISSCC Dig. Tech. Papers*, pp. 372-373, Feb. 2013.

[3] Y. Lu, J. Jiang et al., "123-phase DC-DC converter-ring with fast-DVS for microprocessors," *ISSCC Dig. Tech. Papers*, pp. 364-365, Feb. 2015.

[4] S. Bang et al., "A Low Ripple Switched-Capacitor Voltage Regulator Using Flying Capacitance Dithering," *IEEE J. Solid-State Circuits*, vol. 51, No.4, pp. 919-929, Apr. 2016.

[5] B. Zimmer et al., "A RISC-V Vector Processor With Simultaneous-Switching Switched-Capacitor DC–DC Converters in 28 nm FDSOI," *IEEE J. Solid-State Circuits*, vol. 51, No.4, pp. 930-942, Apr. 2016.

978-1-5386-6414-8/18 $31.00 © 2018 IEEE

4-2 (8096)

An 88% Efficiency 2.4µW to 15.6µW Triboelectric Nanogenerator Energy Harvesting System Based on a Single-Comparator Control Algorithm

Karim Rawy[1], Ruchi Sharma[1], Hong-Joon Yoon[2], Usman Khan[2], Sang-Woo Kim[2], Tony T. Kim[1]

[1]School of EEE, Nanyang Technological University, Singapore, [2]Sungkyunkwan University, Suwon, Korea

Karim002@e.ntu.edu.sg

Abstract— **This paper presents an energy harvesting system (EHS) based on a triboelectric nanogenerator (TENG). A novel TENG HDL spice model was developed to optimize the proposed ultra-low power (ULP) EHS. The proposed TENG-EHS utilizes a novel single-comparator-control (SCC) algorithm for improving the power conversion efficiency (PCE). It modulates the switching frequency of the implemented switched capacitor charge pump (SCCP) in proportion to the load condition at a given applied vibration frequency (i.e. excitation frequency). Moreover, a novel hysteresis control technique was introduced. It regulates the input voltage at the maximum possible power point without IC breakdown, and adopts a dropout excess charge storage technique. The fabricated test chip in 65-nm CMOS technology achieves a peak PCE of 88% with 2.4 µW to 15.6 µW input power and power density of 39.59 µW/mm².**

I. Introduction

The rapid growth of ultra-low power applications including IoT devices, implantable sensors, and wireless electronics raises the call for fully-autonomous power management circuit (PMC) design. In these applications, energy harvesters aim to assist or remove batteries as primary energy sources for PMC [1, 2]. Various energy harvesting devices such as photovoltaic (PV) cells, thermoelectric generators (TEG), and piezoelectric devices have been explored. Piezoelectric devices have been widely utilized for harvesting mechanical energy.

Recently, as a newly introduced mechanical energy scavenger, TENG has been excessively studied from the material engineering point of view. It has been demonstrated to harvest energy from horizontal, vertical, and rotational vibrations. Compared with piezoelectric transducers, which suffer from complex fabrication and high-cost [3], TENG has the features of simple fabrication, low cost, high flexibility, low weight, and small size [3], which makes it a promising power source. Moreover, it supports several harvesting modes [3, 4], allowing it to be employed as an adequate candidate for EHS for IoT applications. Fig. 1 illustrates how a TENG device operates. It generates electric power based on the coupling of triboelectrification and electrostatic induction using two materials with relatively opposite triboelectric (TE) properties, as explained in [3, 4]. The first TENG-EHS in an integrated circuit form was presented in [5]. It proposes a dual-input rectifier with a maximum power point tracking (MPPT) circuit. Although it achieves a 97% tracking efficiency, the control circuit suffers from modest power conversion efficiency (PCE)

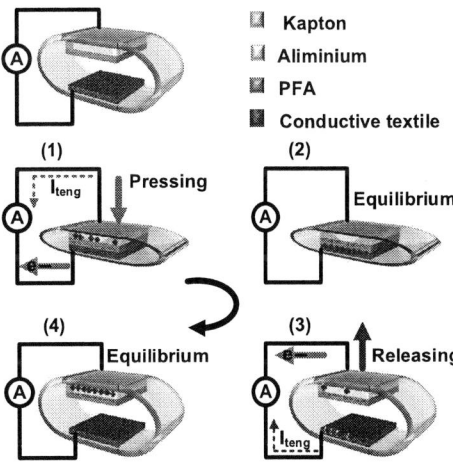

Fig. 1. The fabricated vertical contact TENG harvesting mechanism.

of 51.1%. This limited PCE stems from the high internal impedance (hundreds of MΩ) and the limited output current (<5 µA) of the TENG device. These undesirable characteristics entails the design of a smart control circuit that minimizes the intolerable conduction and switching losses within the implemented switched capacitor charge pump (SCCP) with minimum power consumption.

This work presents a novel TENG-EHS for harvesting energy from a fabricated vertical contact mode TENG device (Fig. 1). A conduction loss reduction technique is developed using an ULP single-comparator-control (SCC) algorithm. It follows a voltage conversion ratio (VCR) self-correction algorithm by modulating the SCCP switching frequency (f_{sw}) with respect to the load condition. The control circuit shows a 36% PCE improvement compared with [5]. Furthermore, a hysteresis-input-regulation-control (HIRC) technique is proposed for excess charge storage and input voltage regulation around the maximum possible power point. This minimizes tracking efficiency degradation, and prevents the EHS from breakdown due to high voltage. In addition, a HDL-based spice model of the employed TENG device has been developed. The developed spice model facilitates the optimization of the proposed TENG-EHS at various operating points. Experimental results demonstrate that the developed spice model shows high accuracy.

978-1-5386-6414-8/18 $31.00 © 2018 IEEE

II. TENG CHARACTERIZATION AND SPICE MODEL

Fig. 2 shows the TENG output AC waveforms (V_{teng}) and their uneven peak values (i.e. V_{pos} and V_{neg}) which depend on the pressure and the TE materials, while V_{teng} frequency depends on f_{ex}. A characterization circuit was developed, as depicted in Fig. 2, to study the output characteristics of the TENG with the variation of f_{ex} under different load values (i.e. Z_{load}). In this work, a conventional full-wave rectifier (FWR) was employed to convert V_{teng} to a regulated DC-value (V_{rect}). The P-V characteristic curves, shown in Fig. 3, can be generated by varying Z_{load} from zero to ∞. It can be shown that the TENG acts as a current limited voltage source. In addition, at a certain f_{ex}, the maximum power point voltage (V_{mpp}), at which the tracking efficiency is maximized, exceeds 30 V, which mandates high voltage (HV) technology process to regulate V_{rect} near V_{mpp}.

By investigating the theoretical model (Fig. 4 (a)) presented in [6], V_{teng} can be obtained as follows.

$$V_{teng}(t) = -\frac{Q}{S\varepsilon_o}\left[\frac{d_1}{\varepsilon_{r1}} + x(t)\right] + \frac{\sigma}{\varepsilon_o}x(t) \qquad (1)$$

Here, d_1 and S are the dielectric thickness and area respectively. While 'ε_{r1}' and 'ε_o' are the dielectric constant and the air permittivity respectively. Q and $x(t)$ are the charges transferred and the time-variant distance between the two electrodes, respectively. V_{teng} can be further simplified based on the conventional model in [7] (Fig. 4 (b)) as follows

$$V_{teng}(t) = -\frac{Q}{C_{TENG}\big(x(t)\big)} + V_{oc}\big(x(t)\big) \qquad (2)$$

where C_{TENG} and V_{oc} are the time-variant capacitor and the open-circuit voltage within the TENG, respectively. Based on (2), V_{oc} and C_{TENG} are time-variant components and function of $x(t)$, defining V_{teng} value at a given time. Utilizing this information for TENG spice modelling and following (1) and (2), Fig. 4 (c) demonstrates the proposed TENG HDL-based spice model. Firstly, $x(t)$ is generated and varies with f_{ex}. Secondly, C_{TENG} and V_{oc} are calculated simultaneously based on (1) and using $x(t)$, as shown in Fig. 4 (c), to generate proper values for V_{teng} and TENG output current (I_{cap}). The experimental and the simulated results of the fabricated TENG and its proposed spice model, respectively, shown in Fig. 5, verify the accuracy of the deduced HDL spice model.

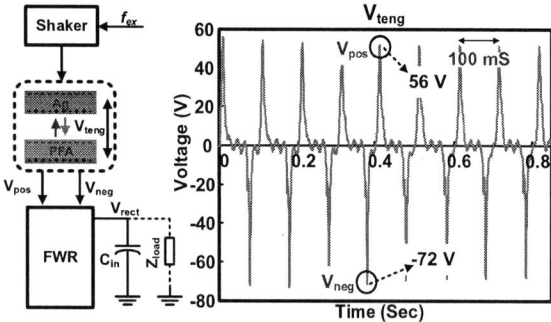

Fig. 2. TENG characterization circuit and AC output waveforms.

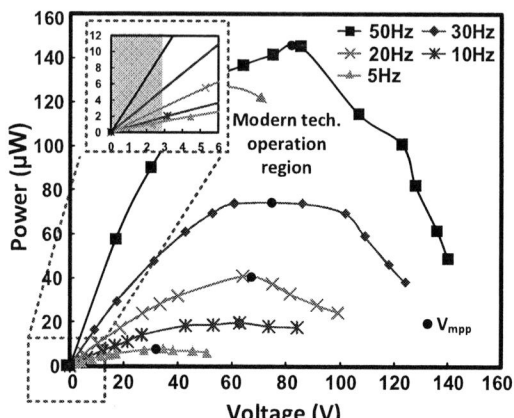

Fig. 3. Experimental P-V characteristic curves at different f_{ex}.

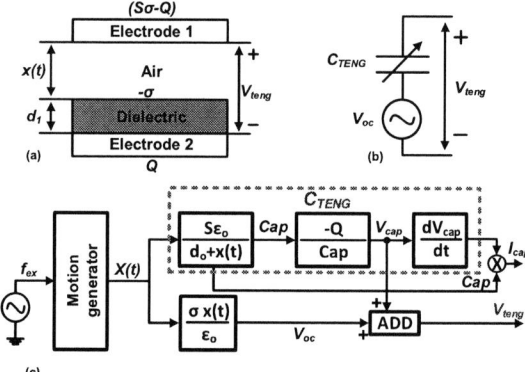

Fig. 4. TENG block diagram of (a) theoretical model, (b) conventional electrical model, and (c) proposed HDL spice model.

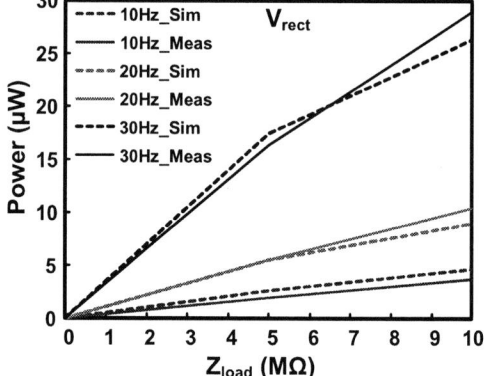

Fig. 5. The measured and simulated V_{rect} across different Z_{load}.

III. TENG-EHS ARCHITECTURE AND CIRCUIT IMPLEMENTATION

The proposed TENG-EHS architecture, shown in Fig. 6 comprises four main blocks, namely a doubler SCCP, dynamic latched comparators for the SCC algorithm and HIRC implementation, an excess energy storage circuit, and a digital core for executing the proposed SCC and HIRC algorithms. The

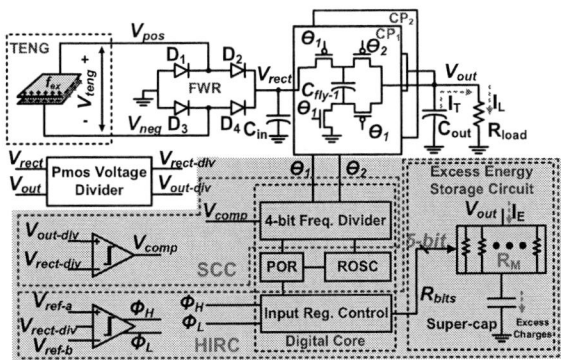

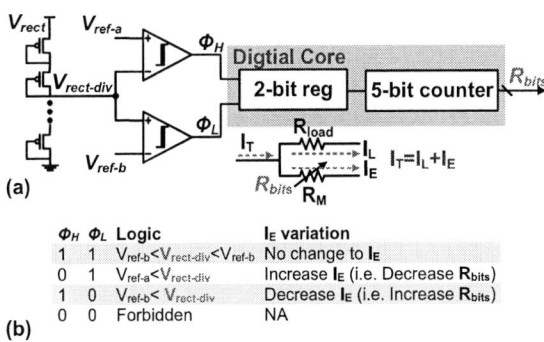

Fig. 6. Proposed TENG-EHS tope architecture.

HV limitation of the employed 65-nm CMOS technology prevents V_{rect} from operating near V_{mpp} described in Fig. 3. Hence, the proposed TENG-EHS adopts a HIRC technique that maintains V_{rect} at the maximum power point within the technology operation region ($V_{mpp-max}$) depicted in Fig. 3. Moreover, HIRC reduces the harvested energy loss by developing an excess energy storage circuit. For a non-isolating TENG-EHS, the V_{rect} value is defined by the total output current (I_T) and f_{ex}. I_T is composed of load current (I_L) and excess charges storage current (I_E) (Fig. 6). The I_E flow rate is controlled by altering the total resistance R_M, which consists of a binary-weighted resistor matrix set by 5-bit digital signal (R_{bits}). Hence, the proposed HIRC regulates V_{rect} around $V_{mpp-max}$ by indirectly modulating I_E through R_{bits} with respect to I_L and f_{ex} variations. Fig. 7 (a) describes the proposed HIRC block diagram. The outputs of the two latched comparators (Φ_H and Φ_L) are stored in a 2-bit register and fed to a 5-bit counter. The later controls the R_M value through a 5-bit digital signal (R_{bits}). Following the logic table in Fig. 7 (b), at light I_L and/or high f_{ex}, $V_{rect-div}$ goes above V_{ref-a} (i.e. $\Phi_H=0$, and $\Phi_L=1$), hence the 5-bit counter decrements R_{bits} so that R_M decreases (i.e. I_E increases) to compensate for the decrease in I_L or the increase in f_{ex}. Likewise, at high I_L and/or low f_{ex}, $V_{rect-div}$ drops below V_{ref-b}, then R_{bits} increments to decrease I_E. Finally, at both scenarios, I_E is used to store the excess charges in a supper-capacitor. A part of the stored energy is lost in R_M due to the HV limitation of the process technology while maintaining V_{rect} near $V_{mpp-max}$.

A doubler SCCP with a gain of 0.5 is implemented to provide a regulated output voltage (V_{out}) at ~1.2 V. Due to the limited power around $V_{mpp-max}$ (Fig. 3), an SCC algorithm was proposed to minimize the conduction and switching losses within the SCCP, and improve the PCE. Basically, for efficient harvested power conversion, the VCR (i.e. 0.5) must be kept constant across various load conditions. When the VCR deviates from 0.5 due to a variation in I_L, the SCCP f_{sw} must be adjusted to avoid unnecessary high f_{sw} at light loads (i.e. switching loss) and insufficient f_{sw} (i.e. conduction loss) at heavy loads. Thus, the SCC algorithm adopts a proposed event/time driven technique. It consists of a single-comparator that compares V_{out} with V_{rect}

Fig. 7. HIRC (a) block diagram, and (b) truth and logic table.

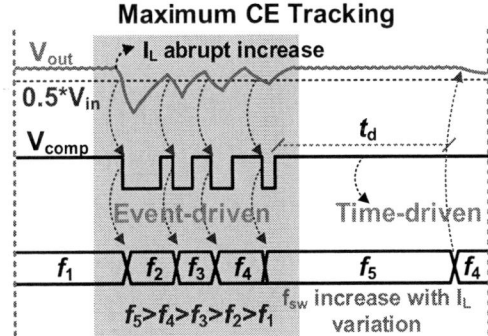

Fig. 8. Proposed single-comparator control timing diagram.

after an appropriate voltage division (Fig. 6), followed by a 4-bit frequency divider. The 4-bit frequency divider divides the ring oscillator (ROSC) output frequency by a 4-bit digital value 'N'. Finally, 'N' is set, according to 'V_{comp}' (Fig. 6) to modulate f_{sw} in proportion to I_L variation, adjusting the VCR to 0.5. Following the timing diagram depicted in Fig. 8, if $V_{out} < 0.5V_{rect}$, 'N' is decremented one bit per V_{comp} negative edge (event-driven), increasing f_{sw}. Likewise, If $V_{out} > 0.5V_{rect}$, f_{sw} is decreased by incrementing 'N' one bit after a defined time 't_d' (time-driven). At steady state, V_{out} oscillates around $0.5V_{rect}$. Thus, PCE along with the voltage conversion efficiency (VCE) can be increasingly improved.

IV. MEASUREMENTS RESULTS

The proposed TENG-EHS was designed and fabricated in 65-nm CMOS process. Fig. 9 shows the performance of the proposed SCC algorithm. When I_L increases, the comparator transmits its pulses to the 4-bit frequency divider (Fig. 6), to modulate f_{sw} accordingly, maintaining the VCR around 0.5. f_{sw} increases one step per V_{comp} negative edge to restore V_{out} to its regulated value (Fig. 8), improving the PCE as well as the VCE.

The PCE values at different f_{ex} across the output power range of the proposed TENG-EHS are recorded in Fig. 10. The tested TENG-EHS chip achieves a peak PCE of 88% at f_{ex} 30 Hz. The PCE results validates the implemented SCC algorithm for minimizing the conduction and the switching losses within the SCCP. Moreover, the ultra-low power consumption of the

978-1-5386-6414-8/18 $31.00 © 2018 IEEE

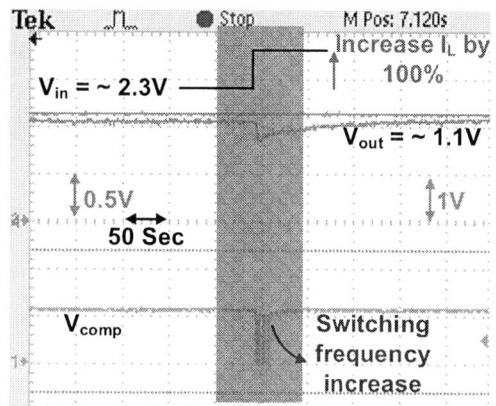

Fig. 9. The proposed SCC algorithm response at I_L variation.

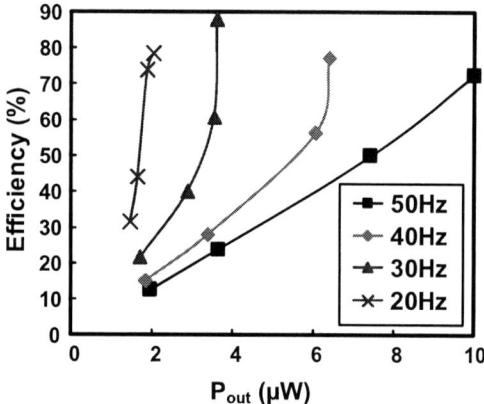

Fig. 10. Power conversion efficiency at different f_{ex} across the load range

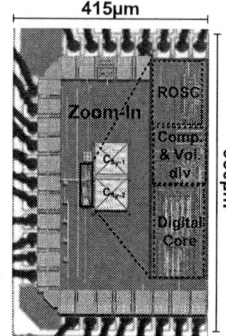

Parameters	[5] ISSCC '18	This work
Technology	0.18µm BCD	65nm CMOS
Type of converter	Buck	Switched capacitor
Harvester type	Triboelectric	Triboelectric
Output voltage	1V-5V	1.2V
Input voltage	<70V	2.5V**
Input power	4.5µW-16µW	2.5µW-15.6µW
Peak conversion efficiency	51.1%	88%
Monolithic area	2.2 mm²	0.394 mm²

**Maximum input voltage due to the technology limitation

Fig. 11. TENG-EHS die photo and comparison table with Prior Arts.

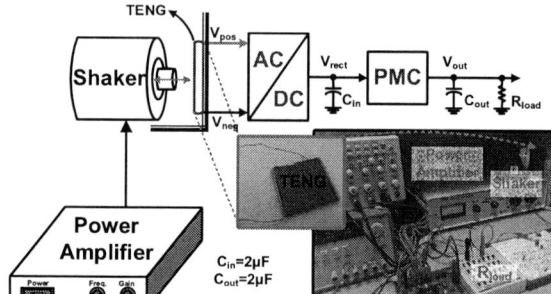

Fig. 12. Measurement setup of the tested TENG-EHS.

dynamic comparators, the voltage dividers, and the digital core boosts the PCE across the output power range.

Fig. 11 shows the test chip microphotography along with a comparison table with the TENG-EHS prior art in literature [5]. The proposed TENG-EHS occupies a silicon area of 0.394 mm² and exhibits a peak PCE of 88%, showing an almost a 36% improvement compared with [5]. In addition, it shows a 39.59 µW/mm² power density while the prior work in [5] achieves 7.3 µW/mm². Fig. 12 illustrates the measurement setup of the proposed TENG-EHS. A power amplifier is used to accurately control the f_{ex} and the pressure of the vibration generated by a commercial mini-shaker. It agitates one plate of the TENG while the other plate is fixed on a horizontal surface.

V. CONCLUSION

This work proposes a TENG-EHS with a ULP control circuit. Firstly, a TENG spice model is developed for optimizing the proposed EHS. In addition, a novel SCC algorithm is proposed to reduce the losses within the SCCP, and improve the PCE and VCE. The proposed TENG-EHS achieves a peak PCE of 88% with 36% improvement over prior art. Finally, a HIRC algorithm was developed to prevent the fabricated chip from breakdown and store the excess TENG harvested energy. The implemented digital core is synthesized and applicable to any CMOS technology.

REFERENCES

[1] K. Rawy, T. Yoo, and T. T. H. Kim, "An 88% Efficiency 0.1-300-µW Energy Harvesting System With 3-D MPPT Using Switch Width Modulation for IoT Smart Nodes," *IEEE Journal of Solid-State Circuits*, pp. 1-12, 2018.

[2] X. Liu and E. Sánchez-Sinencio, "An 86% Efficiency 12 µW Self-Sustaining PV Energy Harvesting System With Hysteresis Regulation and Time-Domain MPPT for IOT Smart Nodes," *IEEE Journal of Solid-State Circuits*, vol. 50, pp. 1424-1437, 2015.

[3] R. Hinchet, W. Seung, and S.-W. Kim, "Recent Progress on Flexible Triboelectric Nanogenerators for SelfPowered Electronics," *ChemSusChem*, vol. 8, pp. 2327-2344, 2015.

[4] Z. L. Wang, "Triboelectric Nanogenerators as New Energy Technology for Self-Powered Systems and as Active Mechanical and Chemical Sensors," *ACS Nano*, vol. 7, pp. 9533-9557, 2013/11/26 2013.

[5] I. Park, J. Maeng, D. Lim, M. Shim, J. Jeong, and C. Kim, "A 4.5-to-16µW Integrated Triboelectric EnergyHarvesting System Based on High-Voltage Dual-Input Buck Converter with MPPT and 70V Maximum Input Voltage," in *2018 IEEE International Solid - State Circuits Conference - (ISSCC)*, 2018, pp. 146-148.

[6] S. Niu, S. Wang, L. Lin, Y. Liu, Y. Sheng Zhou, Y. Hu, *et al.*, *Theoretical study of contact-mode triboelectric nanogenerators as an effective power source* vol. 6, 2013.

[7] S. Niu, Y. Liu, Y. S. Zhou, S. Wang, L. Lin, and Z. L. Wang, "Optimization of Triboelectric Nanogenerator Charging Systems for Efficient Energy Harvesting and Storage," *IEEE Transactions on Electron Devices*, vol. 62, pp. 641-647, 2015.

A Wide-Range Capacitive DC-DC Converter with 2D-MPPT for Soil/Solar Energy Extraction

I-Che Ou[1], Jia-Ping Yang[1], Chia-Hung Liu[1], Kai-Jie Huang[1], Kun-Ju Tsai[2], Yu Lee[2], Yuan-Hua Chu[2], and Yu-Te Liao[1]

[1] Department of Electrical and Computer Engineering, National Chiao Tung University, Hsinchu City, Taiwan
[2] Industrial Research and Technology Institute, Hsinchu, Taiwan

Abstract—This paper presents a capacitive DC-DC converter with adaptive DC-DC conversion ratios and maximum power point tracking (MPPT) for soil and solar energy extraction. To overcome the varying input power ranges of the soil/solar energy sources, a two-dimension power tracking loop with time-based current slope detection was employed. The design was fabricated in a 0.18-μm CMOS process, achieving >80% efficiency in a throughput power range of 360μW to 25mW in the soil mode and from 400μW to 10mW in the solar mode while the peak system efficiency is 89.5%.

I. INTRODUCTION

Ambient energy harvesting from the environment for powering small, low-cost devices that perform extraordinary tasks is key to providing broad-scale practicality for wireless environment sensing. However, ambient energy harvesting technology has still far been limited by the instability of energy production in varying environments. To overcome the varying energy supply for remote sensing applications, a complementary solar and soil energy harvesting system is proposed in this paper. Solar cells provide energy during sunny days, while soil energy [1] can be used in rainy or cloudy conditions. Harvesting energy from multiple sources can provide a more constant energy supply to wireless sensor nodes, but the wide input power range must be accommodated in the power management circuits. Recently, designs combined multiple-dimension power extraction methods can extend the high-efficiency range over a wide input range [2] [3]. Some of these works only focus on the optimal efficiency of the DC–DC converter while neglecting the power loss due to voltage conversion, which results in a limited input range at a regulated output voltage. This paper presents a wide-range, high-efficiency converter that uses reconfigurable conversion ratios for the different input voltages of soil and solar energy harvesters to meet the desired regulated output voltage and a time-domain sensing mechanism that eliminates the demand for a current sensor or other analog circuits and effectively reduces power consumption.

II. DESIGN OF WIDE-RANGE DC-DC CONVERTER

Fig. 1 shows the schematic of the proposed capacitive DC–DC converter for the soil/solar energy extraction. The design consists of one forward path for energy delivery, one feedforward control path for voltage regulation, and two control paths for MPPT: one for conversion ratio (CR) selection (feedforward) and the other for switching frequency tuning (feedback). A start-up circuit, including a five-stage charge pump and ring oscillator, initiates the operation of the power management unit while connecting to the energy transducer. Once the output voltage reaches 1.6V, the main high-efficiency charge pump takes over the operations, boosts

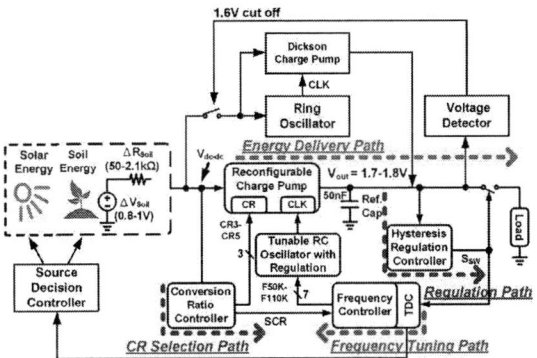

Fig. 1. The proposed soil/solar energy harvesting system architecture

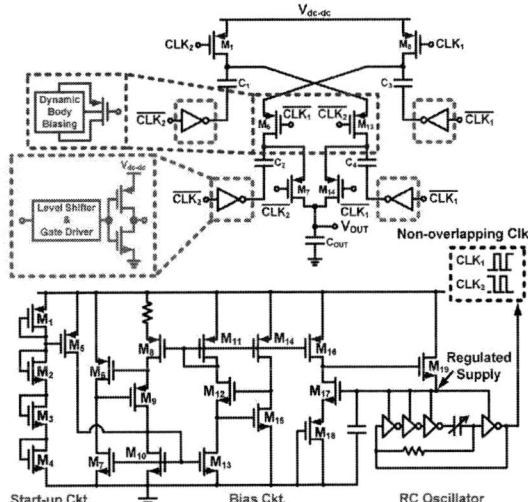

Fig. 2. Schematic of the main-path charge pump

the input voltage, and delivers the extracted power to the load under the control of a hysteresis regulation.

Fig. 2 shows the schematic of the main-path differential voltage tripler [4]. An adaptive body-biasing technique is used in the switching transistors to reduce the power loss in conduction and isolation. Furthermore, the level shifter is adopted in the switch driver to enhance conductance at a low input voltage. To reduce the effects of supply fluctuations, the clock generation is implemented by a regulated RC oscillator with a supply-insensitive voltage reference, making the oscillation frequency only related to the value of the resistor and capacitor.

978-1-5386-6414-8/18 $31.00 © 2018 IEEE

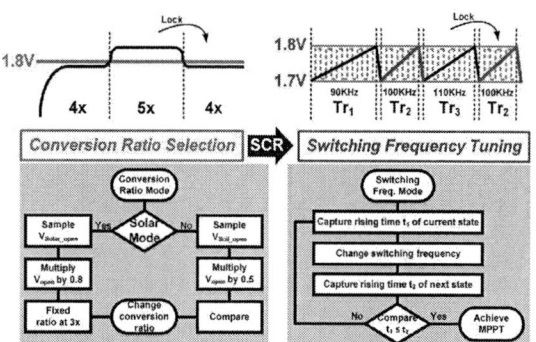

Fig. 3. The proposed 2-D maximal power tracking flow diagram for soil/solar power extraction

To cover the wide input range over environment variations, a two-dimension MPPT architecture, including conversion-ratio and frequency optimization, is proposed. Fig. 3 shows the proposed 2D maximal power tracking flow for soil and solar power extraction. First, the conversion ratio is determined according to the input voltage to maintain the output voltage in the desired range. The input voltage is sampled by the capacitive divider, which sets the voltage ratio of 0.8 and 0.5 in the solar mode and the soil mode, respectively. The sampled voltage is compared to reference voltages to decide the conversion ratio (3x, 4x, 5x). By controlling the conversion ratio, the output voltage approaches the regulated voltage (1.8V) after conversion, reducing the charge redistribution losses [5] and making the converter operate efficiently over a wide input range in both solar and soil modes. Once the conversion ratio mode is complete, the maximum power tracking controller starts frequency tuning to find the optimal frequency to compromise the switching loss and conduction loss in the charge pump. By tuning the capacitor values, the RC-oscillator frequency can be programmed from 50kHz to 110kHz. The feedback-control frequency tuning architecture with current and time-domain sensing mechanisms is applied by sampling the output voltage of the DC–DC converter and controlling the input impedance at the interface. The frequency optimization uses the comparisons of charging time of the capacitor at the output. Instead of pulse-width modulation, constant-on-time power delivery is used to reduce design complexity, the need for a high-speed comparator, and the output ripples within a required range. Since the frequency of pulse depends on the input power or charging time of a fixed capacitor, a time-based counter can be used to measure the output current slope by counting the number of pulses in a set voltage range (1.7-1.8V). The output digital codes from the counter adjust the frequency controller until the optimal power delivery is found. In addition, the mode selection between the soil mode and the solar mode is decided by comparing the counter's final digital codes after running the MPPT algorithm for each mode. The combination of voltage regulation and two-dimension maximal power tracking circuits reduces the power consumption and keeps high efficiency in a wide input range.

III. MEASUREMENT RESULTS

The proposed adaptive conversion-ratio charge pump with an MPPT controller for dual-mode power extraction was fabricated in a 0.18-μm CMOS process and occupies an area of 0.75mm². The measured output voltage of the charge pump is boosted to a range between 1.7V and 1.8V, as set by the

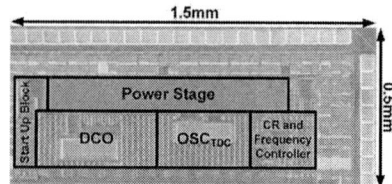

Fig. 4. Chip micrograph

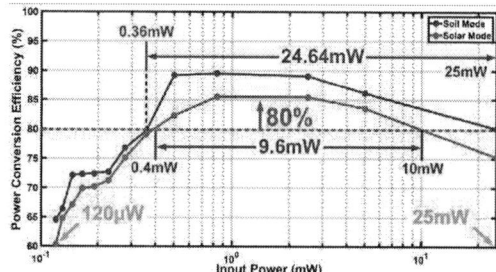

Fig. 5. Measured efficiency over a wide range input power

Table 1 Performance Summary and Comparisons

	[2]	[3]	[4]	This Work
Tech. (nm)	65	180	180	180
Architecture	Capacitive	Inductive	Capacitive	Capacitive
V_{IN} (V)	0.35-1	2.6 and 3	1.1–1.5	0.4–0.5
V_{OUT} (V)	1	1, 1.8 and 3	3.3	1.8
Peak Eff. (%)	88	83	86.4	89.53
Throughput Power (μW)	0.1–300 (>60%)	1–10000 (>60%)	<21 (>30%)	120–25000 (>60%)
Regulation	Yes	Yes	Yes	Yes
MPPT	Yes	Yes	Yes	Yes
Area (mm²)	0.54	4.6225	2.25	0.75

hysteresis controller. Fig. 4 shows the chip micrograph. Fig. 5 shows the measured conversion efficiency of the proposed dual mode DC–DC converter. The measured input range of conversion efficiency above 80% is from 400μW to 10mW in the solar mode and 360μW to 25mW in the soil mode. The peak efficiency of this design is 89.5%. Table 1 shows the performance comparisons to other works. By using the proposed 2D MPPT tracking architecture, the design can track wide-range input variations and achieve high power conversion efficiency.

REFERENCES

[1] F. T. Lin, et. al. "A Self-Powering Wireless Environment Monitoring System Using Soil Energy," in *IEEE Sensors Journal*, vol. 15, no. 7, pp. 3751-3758, July 2015.

[2] K. Rawy, et al. "An 88% Efficiency MPPT for PV Energy Harvesting System with Novel Switch Width Modulation for Output Power 100nW to 0.3mW," *IEEE ASSCC*, Seoul, 2017, pp. 117-120.

[3] G. Yu, et al. "A 400 nW Single-Inductor Dual-Input–Tri-Output DC–DC Buck–Boost Converter With Maximum Power Point Tracking for Indoor Photovoltaic Energy Harvesting," in *IEEE Journal of Solid-State Circuits*, vol. 50, no. 11, pp. 2758-2772, Nov. 2015.

[4] X. Liu and E. Sánchez-Sinencio, "An 86% Efficiency 12μW Self-Sustaining PV Energy Harvesting System With Hysteresis Regulation and Time-Domain MPPT for IOT Smart Nodes," in *IEEE Journal of Solid-State Circuits*, vol. 50, no. 6, pp. 1424-1437, June 2015.

[5] X. Liu and E. Sanchez-Sinencio, "A 0.45-to-3V reconfigurable charge-pump energy harvester with two-dimensional MPPT for Internet of Things," *IEEE International Solid-State Circuits Conference (ISSCC) Digest of Technical Papers*, San Francisco, CA, 2015, pp. 1-3.

4-4 (8036)

A 13.56 MHz 88.7%-PCE Voltage Doubling Rectifier Using Adaptive Delay Time and Pulse-Width Control

Ye-Sing Luo, Hsing-Hung Lin, and Shen-Iuan Liu

Graduate Institute of Electronics Engineering & Department of Electrical Engineering
National Taiwan University, Taipei, Taiwan 10617, R.O.C.
E-mail: lsi@ntu.edu.tw

Abstract—**A voltage doubling rectifier using an adaptive pulse controller is presented to receive the low input amplitude. The adaptive pulse controller adjusts the delay time and pulse-width of the pulses in the background to control the power switches. By keeping the output voltage as high as possible, the power conversion efficiency (PCE) of this voltage doubling rectifier is enhanced. This work is fabricated in a 0.18μm CMOS process. For the input amplitude of 0.8V, an input frequency of 13.56MHz, and a load resistor of 140Ω, this rectifier achieves a PCE of 88.7%. For the input amplitude of 1.3V, the rectifier has a peak PCE of 89.3%.**

I. INTRODUCTION

For the implantable medical devices, wireless sensors, and energy harvesting circuits, the wireless power transmission is receiving significant attention. For these applications, an AC-to-DC rectifier [1-4] is often expected to work under a low received AC voltage and a high power conversion efficiency (PCE) over a wide input range. A comparator-based rectifier is widely adopted [1-4]. An active diode is realized by a comparator and a power transistor. When the turn-on delay time $t_{D\text{-}ON}$ of the comparator is increased and the conduction time of the power transistor is reduced. Similarly, when the turn-off delay time $t_{D\text{-}OFF}$ of the comparator is increased. It induces a reverse leakage current [1-4]. The turn-on and turn-off delay times dramatically degrade the PCE of a rectifier.

In [3], both turn-on and turn-off delay times are compensated by adding an equivalent input offset voltage in the comparators. However, the off-chip control signals may be required to program these offset voltages. In [4], an on-chip calibration is used to compensate the delay times of the comparator. However, while the input amplitude of a rectifier is lower than 1.3V [4], the rectified supply voltage is low. The analog comparators and the error amplifiers may not work in such a low supply voltage.

In this work, the power switches of a rectifier are controlled by using pulses, which delay time and pulse-width are adjusted in the background to enhance the output voltage and the PCE. Since the pulses and its control circuits are mostly realized by the digital circuits, this rectifier works even the input amplitude voltage of a rectifier is as low as 0.8V.

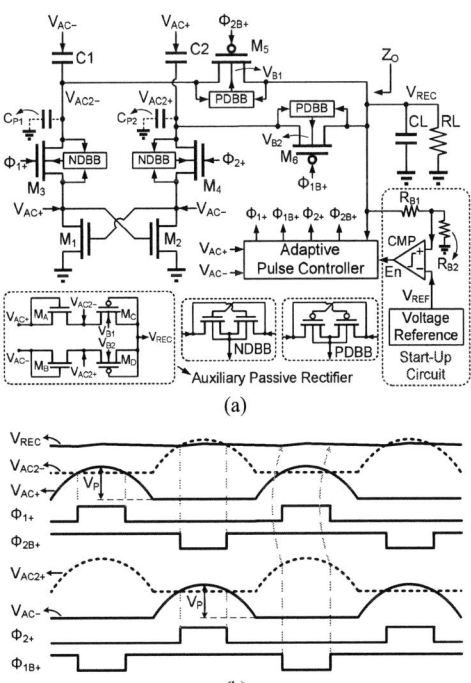

(a)

(b)

Fig. 1. (a) A rectifier with voltage doubling and its timing diagram while the adaptive pulse controller is (b) enabled in the steady state.

II. CIRCUIT DESCRIPTION

Fig. 1(a) shows the proposed rectifier with voltage doubling. It consists of six power switches $M_1 \sim M_6$, an auxiliary passive rectifier, two n-type dynamic body bias (NDBB) circuits [3], two p-type dynamic body bias (PDBB) circuits [3], a start-up circuit, an adaptive pulse controller, a resistor RL, and three capacitors C1, C2, and CL. To rectify the low-swing AC voltage, the nMOS cross-coupled pair is used because its positive feedback [2] helps to turn on M_1 and M_2, respectively. To double the output voltage, a voltage doubler [5] is adopted by using the pumping capacitors,

978-1-5386-6414-8/18 $31.00 © 2018 IEEE   39

$C1\sim C2$, and the pMOS load switches, $M_5\sim M_6$, which avoid a large V_{TH} drop for low voltage applications [6]. To eliminate the body effect of the power switches, the PDBB (NDBB) circuit connects the body of the PMOS (NMOS) to the high (low) voltage. An auxiliary passive rectifier is composed of four diode-connected transistors, $M_A\sim M_D$. The substrates of M_C and M_D are connected to V_{B1} and V_{B2}, respectively, which are provided by two PDBB circuits. The start-up circuit is supplied by V_{REC}. It is composed of a voltage reference [7], a comparator (CMP), and two resistors R_{B1} and R_{B2}. The timing diagram is shown in Fig. 1(b). Assume that the on-resistances of M_k (k=3~6) are neglected and the peak amplitude of V_{AC+} and V_{AC-} are equal to V_P. When $V_{AC+}>V_{AC2-}$, a short pulse Φ_{1+} turns on M_3. Since $V_{AC-}\approx 0$ and $V_{AC2}\approx V_{AC+}$, the voltage across $C1$ is approximated to V_P. As the rectified voltage $V_{REC}<V_{AC2-}$, a short pulse Φ_{2B+} goes low to turn on M_5. Due to a parasitic capacitor C_{P1}, the voltage V_{AC2-} is expressed as

$$V_{AC2-} \approx V_{AC-} \cdot \frac{C1}{C1+C_{P1}} + V_P \qquad (1)$$

Assuming $C1>>C_{P1}$ and M_5 is turned on for a short time, the peak amplitude of V_{AC2-} is approximated to $2V_P$. It leads to $V_{REC}\approx V_{AC2-}\approx 2V_P$. Similarly, as $V_{AC2+}>V_{REC}$, Φ_{1B+} goes low to turn on M_6 and it leads to $V_{REC}\approx V_{AC2+}\approx 2V_P$. Thus, this rectifier with voltage doubling is achieved.

To properly turn on/off $M_3\sim M_6$, the delay times and pulse-widths of four pulses are controlled by an adaptive pulse controller as shown in Fig. 2(a). V_{REC} serves as a power supply for two sample-and-hold circuits, SAH1 and SAH2, two latch comparators, CMP1 and CMP2, a timing generator, the level shifters [9] & buffers, and a regulator [9]. V_{SUP} is generated by a regulator, and it is a supply voltage of two D-flip-flops, DFF1 and DFF2, a controller, a comparator CMP3, a 7-bits digitally-controlled delay line (DCDL) [10], and a 6-bits pulse width modulator (PWM). V_{AC+} and V_{AC-} are converted into a clock CK_{IN} by using a comparator CMP3. While En is high, the timing generator, DFF1, DFF2, controller, DCDL, and PWM will be enabled. The timing diagram is shown in Fig. 2(b). The code $D[6;0]$ controls the 7-bits DCDL to alter the delay time t_D. The code $PW[5;0]$ controls the 6-bits PWM to adjust the pulse-width t_{PW}. To deglitch CK_D and $\Phi_1\sim\Phi_{2B}$, $D[6;0]$ and $PW[5;0]$ are synchronized to CK_{IN} and CK_D by a controller, respectively. To lower the power of the controller, a timing generator uses a relaxation oscillator and digital circuits to generate four non-overlapping clocks, $P_1\sim P_4$, for SAH1, SAH2, CMP1, CMP2, and a controller. The SAH1 and SAH2 sample V_{REC} to be V_{S1} and V_{S2}, respectively. If V_{REC} is rising, $V_{S1}<V_{S2}$. When P4 goes high to latch CMP1, V_{O1} becomes low. To deglitch V_{O1}, DFF1 is used to re-sample V_{O1} by P4, where the output Q_1 is sent to the controller. If V_{REC} is rising, $V_{S2}<V_{S1}$. When P2 goes high to latch CMP2, V_{O2} goes low. To deglitch V_{O2}, DFF2 is used to re-sample V_{O2} by P2, and the output Q_2 is sent to the controller. The state diagrams of a controller are shown in Fig. 2(c). Initially, $D[6;0]=[00...0]$ and $PW[5;0]=[00...0]$. When P_2 goes low, this controller starts from the state S_{D1} and increases $D[6;0]$ by one LSB to adjust the delay time first. When $Q_1=0$, the controller stays in S_{D1} and increases $D[6;0]$ by one LSB again until $V_{S1}>V_{S2}$. The controller will switch from the state S_{D1} to S_{D2}. When the delay time is short, V_{REC}

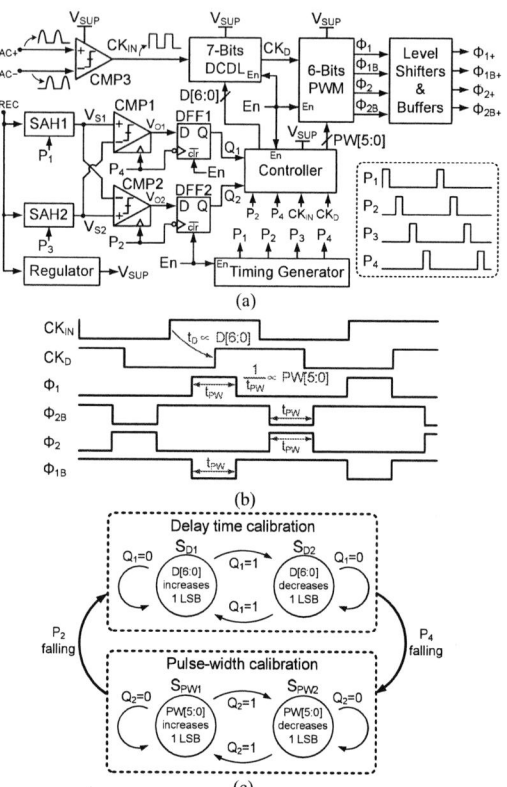

Fig. 2. (a) An adaptive pulse controller, (b) its timing diagram, and (c) the state diagram.

becomes low due to the leakage current. Then, the controller increases the delay time, and V_{REC} will be increased. When the delay time is increased too much, V_{REC} will be decreased. In the steady state, the controller switches between the states S_{D1} and S_{D2}, and the delay time is adjusted in background. Similarly, when P_4 goes low, the pulse-width is calibrated by $PW[5;0]$ with the same algorithm as shown in Fig. 2(c).

A. Timing Generator

Let Z_O be the total output impedance of the proposed rectifier which is approximated as

$$Z_O \approx r_{on,M5} + \frac{1}{j\omega \cdot C1} \qquad (2)$$

where $r_{on,M5}$ is the on-resistor of the transistor M_5 in Fig. 1(a). Since the capacitor $C1$ is 100nF with an angular frequency ω of $2\cdot\pi\cdot 13.56$ Mrad/s, the impedance of $C1$ is pretty small and Z_O is dominant by $r_{on,M5}$. The transient response of the output voltage V_{REC} is modeled as a first-order RC circuit. Assuming the voltage error of the step response is less than 1%, the settling time of V_{REC} is given as

$$t_S \geq 4.7 \cdot \left(RL // r_{on,M5}\right) \cdot CL. \qquad (3)$$

While RL=140Ω, CL=100nF, and $r_{on,M5}$=0.87Ω, the required settling time t_S should be longer than 0.4μs by (2). To correctly sample V_{REC}, the settling time is chosen as 3.2μs. So, the required period of $P_1\sim P_4$ should be 12.8μs and their corresponding frequency is chosen as 78.125kHz. The timing

978-1-5386-6414-8/18 $31.00 © 2018 IEEE

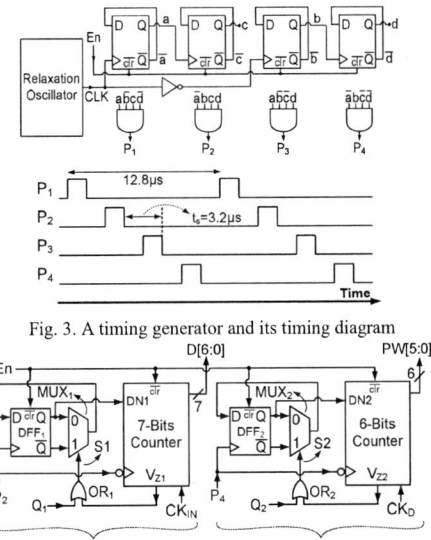

Fig. 3. A timing generator and its timing diagram

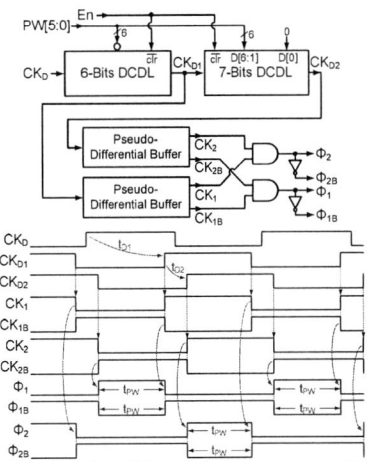

Fig. 4. A controller for the delay time and the pulse-width calibrations.

generator is shown in Fig. 3. A relaxation oscillator generates a clock CLK of 312.5kHz. While En is high, CLK is divided by 4 to realize four non-overlapping clocks, $P_1 \sim P_4$.

B. Controller

To control the 7-bits DCDL [10] and the 6-bits PWM, a controller is shown in Fig. 4. It is composed of two D-flip-flops, DFF_1 and DFF_2, two multiplexers, MUX_1 and MUX_2, two OR gates OR_1 and OR_2, a 7-bits counter, and a 6-bits counter. To deglitch the digital codes D[6:0] and PW[5:0], CK_{IN} and CK_D are used to re-sample P_2 and P_4 in the 7-bits and a 6-bits counters, respectively. While the signal DN1 (DN2) is low, the output code D[6:0] (PW[5:0]) of the counter is increased. When DN1 (DN2) is high, the code D[6:0] (PW[5:0]) is decreased. For the delay time calibration, Initially, En, V_{Z1}, S1, and DN1 are high, Q_1 is low, and D[6:0]=0. As P_2 goes high, DN1 becomes low. While P_2 goes low, D[6:0] is increased by one LSB, and both V_{Z1} and S1 become low. Since DN1 is low, D[6:0] will be increased by one LSB as P_2 goes low. When Q1 becomes high, S1 will go high. While P_2 goes high, DN1 becomes high. When P_2 goes low, D[5:0] will be decreased by one LSB. For the pulse-width calibration, in the beginning, En, V_{Z2}, S2, and DN2 are high, Q_2 is low, and PW[5:0]=0. When P_4 becomes high, DN2 goes low. As P_4 goes low, PW[5:0] is increased by one LSB. Then, V_{Z2} and S2 become low. Since DN2 is low, PW[5:0] is increased by one LSB while P_4 becomes low. As Q2 goes high, S2 will become high. When P_4 goes high, DN2 becomes high. As P_4 goes low, PW[5:0] is decreased by one LSB.

C. Pulse Width Modulator

To control the pulse-width t_{PW} of Fig. 2(b), a 6-bits PWM in Fig. 5 is used. It is realized by a 6-bits DCDL[10], a 7-bit DCDL, two pseudo-differential buffers, two AND gate and two inverters. While En is high, the 6-bits and 7-bit DCDLs are enabled. A clock CK_{D1} is generated by CK_D and the 6-bits DCDL. The delay time t_{D1} is inversely proportional to PW[5:0]. Then, CK_{D2} is generated by CK_{D1} and the 7-bits DCDL with a delay time of t_{D2}, which is proportional to

Fig. 5. A 6-bits pulse width modulator and its timing diagram.

CK_{D2} into the differential ones CK_1, CK_{1B}, CK_2, and CK_{2B}, respectively. By using an AND gate, CK_{2B} and CK_{1B} are used to generates Φ_1. The time interval between the rising edge of CK_{2B} and the falling one of CK_{1B} determines the pulse-width of Φ_1, which is t_{PW}. Similarly, CK_2 and CK_1 are used to generates Φ_2 by using an AND gate. The pulse-width of Φ_2, t_{PW}, is determined by the time interval between the rising edge of CK_2 and the falling edge of CK_1.

III. EXPERIMENTAL RESULTS

The proposed voltage doubling rectifier is fabricated in a 0.18μm CMOS process. While an input amplitude is 0.8V and RL is 100Ω, the measured transient responses of V_{REC}, V_{SUP}, PW[5:0], D[6:0], and En are shown in Fig. 6. Initially, V_{REC} starts from 0V. While V_{REC}>0.46V, En goes high. Then, the adaptive pulse controller will be enabled to alter the digital codes PW[5:0] and D[6:0], alternatively. The measured settling time is 1.28ms. There is a trade-off between the controller's power and the settling time. The final V_{REC} and V_{SUP} are 0.9V and 0.67V, respectively. The measured V_{REC} is lower than the calculated V_{REC} in Section II. It is because the measured V_{REC} has to provide the additional power for the adaptive pulse controller and start-up circuit. The measured input amplitude versus V_{REC} with the different loading resistors is shown in Fig. 7(a). While a load resister RL is 100Ω and the input amplitude is 0.8V, the output voltage V_{REC} is 0.98V. While the input amplitude is 1.3V and RL=140Ω, a peak V_{REC} is 1.85V and its output power is 24.4mW. Fig. 7(b) shows the measured input amplitude versus PCE with the different loading resistors. Fig. 8(a) shows the measured output voltage V_{REC} and PCE versus the different loading resistors RL while the input amplitude is 1V. When RL=1000Ω, the peak V_{REC} is 1.72V. As RL=300Ω, the peak PCE is 89.3%. Since the power of the adaptive pulse controller is constant, the PCE is reduced in the light load. The auxiliary passive rectifier, the comparator CMP3, the 7-bits DCDL, the 6-bits PWM, level shifters, and buffers use the low-V_{TH} nMOS and pMOS transistors. The controller uses both low-V_{TH} and normal transistors. Other circuits use the normal transistors. A die photo is shown in Fig. 8(b) and its area is 1.44mm².

TABLE I. PERFORMANCE SUMMARY AND COMPARISON

	[1]	[2]	[3]	[4]	[11]	This work
Technology (nm)	0.35	0.5	0.18	0.065	0.35	**0.18**
Rating Voltage (V)	3.3	5	1.8	2.5	3.3	**1.8**
Frequency (MHz)	1.5	13.56	13.56	13.56	13.56	**13.56**
Voltage Doubling	1X	1X	2X	1X	1X	**2X**
Loading	100Ω	500Ω	100Ω	100Ω	500Ω	**140Ω**
Peak PCE @ Input Amplitude	87% @2.4V	80.2% @3.8V	85% @1.19V	94.6% @2.5V	91.4%	**89.3% @1.3V**
VCR* @ Input Amplitude	0.87 @2.4V	0.82 @3.8V	1.67 @1.19V	0.952 @2.5V	0.924 @3.6V	**1.42 @1.3V**
Peak Output Power @ Input Amplitude	43.6mW @2.4V	30.4mW	44.8mW @1.35V	57.1mW @2.5V	22.1mW @3.6V	**24.4mW @1.3V**
Lowest Input Amp.	1.2V	NA	1.15V	1.3	1.8V	**0.8V**
Input Amplitude Range (PCE>80%)	1.2V	NA	0.2V	1.2V	1.8V	**0.5V**

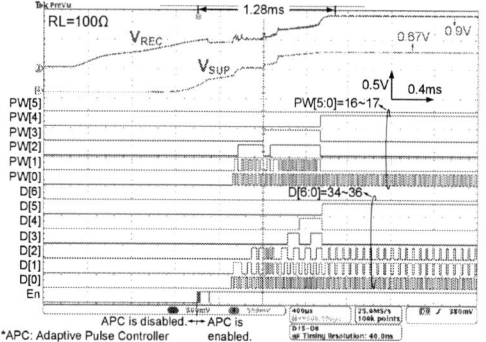

Fig. 6. Measured output voltage V_{REC}, V_{SUP}, PW[5:0], D[6:0], and En.

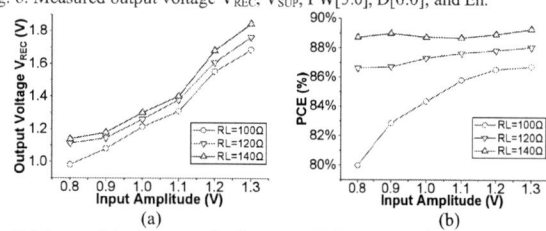

Fig. 7. Measured the input amplitude versus (a) the output voltage V_{REC} and (b) PCE with the different loading resistors.

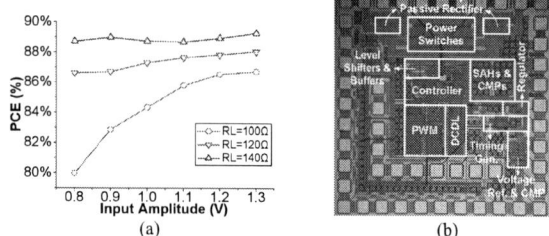

Fig. 8. (a) The measured PCE with different loading resistors and (b) Die photo of this voltage doubling rectifier.

IV. CONCLUSIONS

A low-input-swing voltage doubling rectifier using adaptive delay time and pulse-width control is presented. The performance summary and comparison are shown in Table I. This work achieves the lowest input amplitude of 0.8V with a PCE of 88.7%. Comparing with 2X voltage doubling structures [3], this work achieves the lowest input amplitude

of 0.8V, and a widest input amplitude range of 0.5V for PCE>80%.

ACKNOWLEDGEMENT

The authors would like to thank National Chip Implementation Center, Ministry of Science and Technology, Taiwan, and Donation Grant FD105012.

REFERENCES

[1] S. Guo, and H. Lee, "An efficiency-enhanced CMOS rectifier with unbalanced-biased comparators for transcutaneous-powered high-current implants," *IEEE J. Solid-State Circuits*, vol. 44, no. 6, pp. 1796–1804, June 2009.

[2] H. M. Lee and M. Ghovanloo, "An integrated power-efficient active rectifier with offset-controlled high speed comparators for inductively powered applications," *IEEE Trans. Circuits Syst. I, Reg. Papers*, vol. 58, no. 8, pp. 1749–1760, Aug. 2011.

[3] C. Y. Wu, et. al., "A 13.56 MHz 40 mW CMOS high-efficiency inductive link power supply utilizing on-chip delay-compensated voltage doubler rectifier and multiple LDOs for implantable medical devices," *IEEE J. Solid-State Circuits*, vol. 49, no. 11, pp. 2397–2407, Nov. 2014.

[4] C. Huang, T. Kawajiri, and H. Ishikuro, "A near-optimum 13.56 MHz CMOS active rectifier with circuit-delay real-time calibrations for high-current biomedical implants," *IEEE J. Solid-State Circuits*, vol. 51, no. 8, pp. 1797–1809, Aug. 2016.

[5] P. Favrat, et. al, "A high-efficiency CMOS voltage doubler," *IEEE J. Solid-State Circuits*, vol. 33, no. 3, pp. 410–416, Mar. 1998.

[6] J. Kim, P. K. T. Mok, and C. Kim, "A 0.15 V input energy harvesting charge pump with dynamic body biasing and adaptive dead-time for efficiency improvement," *IEEE J. Solid-State Circuits*, vol. 50, no. 2, pp. 414–425, Feb. 2015.

[7] B. Razavi, *Design of Analog CMOS Integrated Circuits*. Boston, MA: McGraw-Hill, 2001.

[8] B. Zhai, et. al., "Energy-efficient subthreshold processor design," *IEEE Trans. Very Large Scale Integr. (VLSI) Syst.*, vol. 17, no. 8, pp. 1127–1137, Aug. 2009.

[9] P. Li and R. Bashirullah, "A wireless power interface for rechargeable battery operated medical implants," *IEEE Trans. Circuits Syst. II, Exp. Briefs*, vol. 54, no. 10, pp. 912–916, Oct. 2007.

[10] R. J. Yang and S. I. Liu, "A 40–550 MHz harmonic-free all-digital delay-locked loop using a variable SAR algorithm", *IEEE Journal of Solid-State Circuits*, vol. 42, no. 2, pp. 361-373, Feb. 2007.

[11] L. Cheng, et. al., "Adaptive on/off delay-compensated active rectifiers for wireless power transfer systems", *IEEE Journal of Solid-State Circuits*, vol. 51, no. 3, pp. 712-723, Mar. 2016.

4-5 (8091)

A 6800-μm² Resistor-Based Temperature Sensor in 180-nm CMOS

Jan Angevare, Kofi A.A. Makinwa
Electronic Instrumentation Laboratory
Delft University of Technology
Delft, the Netherlands
j.a.angevare@tudelft.nl

Abstract—A resistor-based temperature sensor has been realized in 180 nm CMOS for SoC thermal management applications. Occupying only 6800 μm², it is the smallest resistor-based temperature sensor ever reported. This is achieved by employing a compact highly-digital VCO-based ADC. After a 2-point trim, the sensor achieves an inaccuracy of ±0.35 °C (3σ) in a temperature range from -35 °C to 125 °C. By achieving a resolution of 0.12 °C (rms) at 2.8 kSa/s, it can track the fast thermal-transients in SoCs.

Keywords—Temperature sensor, Thermal Sensing, Wien-Bridge

I. INTRODUCTION

As the dimensions of CMOS devices continue to shrink, their energy density goes up, leading to severe self-heating issues in large Systems-on-Chips (SoCs). This can degrade performance, reduce life-time and even cause permanent damage. To prevent this, SoCs typically employ dynamic thermal management (DTM) [1, 2]. This involves using on-chip temperature sensors to monitor die temperature and then taking action to ensure that it does not rise above pre-determined limits. Possible actions range from reducing the clock frequency and/or the supply voltage to complete shutdown. To account for temperature sensor errors, however, DTM limits must include safety margins. Since such margins directly translate into unspent power, and therefore to reduced performance, the accuracy of temperature sensors for thermal management applications is paramount and should be less than ±1 °C at 70 °C [3].

In modern SoCs, local die temperatures can rise significantly in only a few milliseconds [3]. As such, local hotspots may be created, which must be monitored to avoid reliability risks. Because such hotspots are typically caused by the activity of large digital blocks, their precise location is difficult to predict at design time [3]. As a result, on-chip temperature sensors will often be located some distance away from the actual hotspot, leading to significant errors in its estimated temperature, and thus requiring even larger safety margins. To minimize such errors, large SoCs will typically contain multiple temperature sensors. These should be as small as possible to maximize area efficiency and facilitate flexible placement in digital blocks [3]. Furthermore, they should sample fast enough (> 1 kSa/s) to detect thermal transients.

Most temperature sensors proposed for thermal management applications belong to one of three categories. The most common are bandgap-based temperature sensors, which exploit the temperature dependence of parasitic bipolar transistors (BJTs) [3]. Although very accurate, their voltage headroom requirements do not scale with technology. As a scaling-friendly alternative, sensors based on gate-delay have been proposed. Although they can be quite small, they are inherently sensitive to power supply variations [5, 6] and are prone to MOSFET aging. Recently, thermal-diffusivity (TD) sensors have been proposed. These exploit the well-defined temperature-dependent propagation of heat through Silicon. Although they are naturally small, accurate and scale well with technology, they are comparatively power hungry [7-9].

As an alternative, the temperature dependence of on-chip resistors can be exploited to realize resistor-based temperature sensors [10]. Even though they are not as area efficient as transistors, the moderate accuracy and resolution requirements of DTM means that they can be quite compact. In [11], however, the sensor outputs a temperature dependent frequency, which must still be digitized by extra circuity. In this paper, a proof-of-concept resistor-based temperature sensor is proposed with a VCO-based read-out circuit that provides a direct digital output and results in an even smaller sensor.

II. ARCHITECTURE

A. Wien-Bridge

In general, temperature-to-digital converters (TDCs) can be realized with either time-domain or amplitude domain techniques. Since advanced processes typically support only low supply voltages but offer very fast circuitry, TDCs in such processes can best be realized with time-domain techniques. This requires a temperature-dependent signal in the time-domain. An on-chip RC filter can generate a temperature-dependent time delay (or phase shift). This delay is mainly due to the resistors, since on-chip (MIM) capacitors are relatively stable. Such RC filters should minimize area while maximizing temperature sensitivity. 1st order low-pass (LP), Wien-Bridge (WB) [10] or Poly-Phase (PP) filters [11] can be employed. The LPF has the smallest area, but its phase shift also exhibits the lowest sensitivity to changes in resistance, and hence in temperature. This sensitivity can be doubled by using a WB or

978-1-5386-6414-8/18 $31.00 © 2018 IEEE

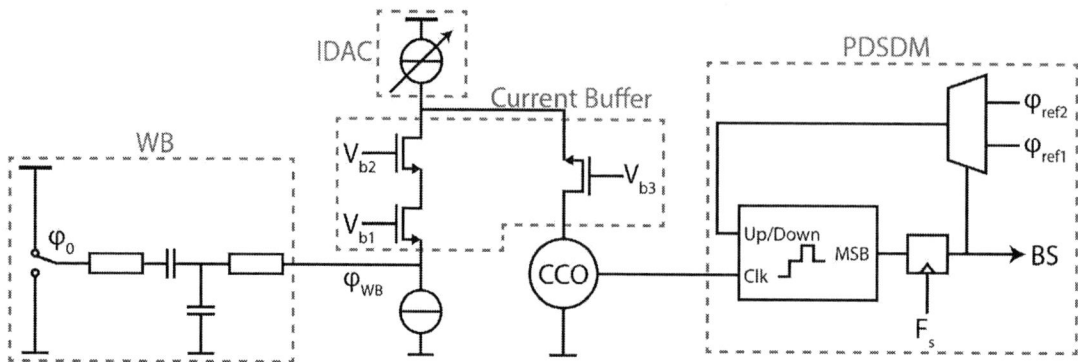

Fig. 1. Complete system schematic.

a PPF, but at the expense of more resistors and capacitors [10, 11]. However, when driven by rail-to-rail square-waves, e.g. generated by standard logic, the output voltage of a PPF exceeds these rails and thus requires the use of thick-oxide devices. The WB does not suffer from this problem, and so was chosen for use in this work. To save area, a single-ended configuration was used (Fig. 1).

The design of the WB involves a number of trade-offs. A lower center frequency (f_{WB}) results in larger phase delay, which in turn minimizes the errors caused by other sources of delay, and thus increases accuracy. However, it requires larger components. In this design, R = 28 kΩ (silicided poly) and C = 1.8 pF (MIM) were chosen, resulting in f_{WB} = 3 MHz. The resulting WB then occupies about 60% of the sensor's total area. This should scale well, since process scaling is usually accompanied by an increase in capacitance density.

B. Phase-Domain Sigma-Delta Modulator

A phase-domain sigma-delta Modulator (PDSDM) is used to readout the phase shift of the WB [8]. It uses a chopper as a phase detector, and the feedback loop of the PDSDM forces the integrator input to be 0 on average (Fig. 1). This occurs when the WB output phase (φ_{in} in Fig. 8) is 90° phase shifted with respect to the phase of the chopping waveform ($\varphi_{in}+90°$).

In [10] and [11], the analog readout circuitry was realized with the help of large integration capacitors. In this work, a highly digital VCO-based PDSDM is used [7]. By converting the signal of interest into the frequency-domain (by means of a VCO), a digital up/down counter can be used as a time-domain

integrator, thus obviating the need for integration capacitors. This has three advantages: it is more area efficient, it scales well with technology, and, once designed, it can be readily synthesized.

Conceptually, a counter can be seen as an integrator of digital pulses. So instead of integrating current, a counter integrates frequency. The output current of the WB is processed by a current buffer that feeds a current-controlled oscillator (CCO) used to transform the WB signal into the frequency domain (Fig. 1). The CCO output is fed into the up/down counter. Toggling the counter's up/down signal inverts its integration polarity, effectively chopping the input signal and realizing the phase detector function. The comparator can also be efficiently implemented by sampling the counter's most-significant bit (MSB).

C. Counter

In this design the quantization noise associated with the use of a counter as a discrete-time integrator [12] is the dominant noise source. It has a white spectrum and can be minimized by maximizing the ratio between the CCO's oscillation frequency and the chopping frequency (up/down frequency), and by maximizing the CCO's output swing so as to make optimal use of the counter's dynamic range. However, a higher CCO output frequency requires more counter bits to prevent overflow. As a compromise, a 7-bit counter with a maximum clock frequency of 800 MHz was designed.

To reduce the counter's area and power consumption, it is split into a 2-bit fine and 5-bit coarse counter (Fig. 3). The fine counter uses gray-coding to reduce circuit complexity and employs clock-gating to reduce the frequency in the coarse counter by a factor 4, and reduce its power consumption. To

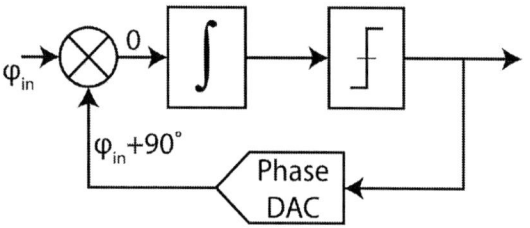

Fig. 2. Block diagram of the Phase-Domain Sigma-Delta Modulator.

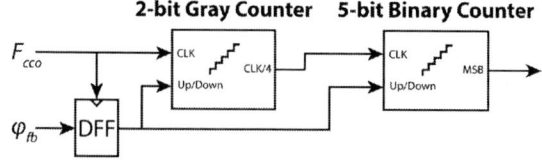

Fig. 3. Block diagram of the counter.

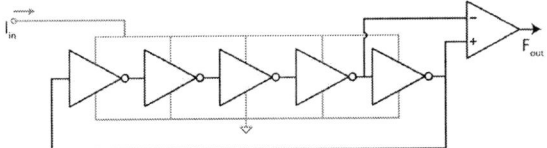

Fig. 4. Schematic of the current-controlled oscillator and levelshifter.

prevent meta-stability issues, the up/down signal is re-clocked by the counter clocking signal. The complete digital circuit, including the 2-bit gray and 5-bit binary counter, was designed to operate from -55 to 125 °C.

D. Current Controlled Oscillator

The combination of the WB and the CCO were designed to generate a frequency swing of about 600 MHz peak-to-peak, leaving some margin for process spread. An additional (6-bit) current DAC allows for a one-time trim of the mean CCO frequency, ensuring that it is close to the middle of the desired frequency range, i.e., 400 MHz.

The CCO consists of a 5-stage ring-oscillator (Fig. 4). Since its output amplitude varies significantly over PVT, a level-shifter is used to generate a rail-to-rail output signal. However, this must be quite fast to avoid introducing extra delay. In [9], the level shifter consisted of a CM sensing circuit and a differential pair. In this design, both inputs are simply connected to consecutive stages of the ring-oscillator.

For accuracy, the output of the WB should be connected to a low impedance node to avoid altering f_{WB} and causing extra phase spread. Because the CCO has a significant input impedance (~20 kΩ), a folded-cascode current buffer (Fig. 3) is used to create a better virtual ground (~600 Ω). The folded cascode requires a bias current of ~100 µA in order to reduce the phase-shift contribution. Due to the finite output impedance of the various current sources, the gain of the current buffer is less than unity. However, this is not a problem, since the PDSDM measures phase information, and is insensitive to amplitude variation.

III. MEASUREMENT RESULTS

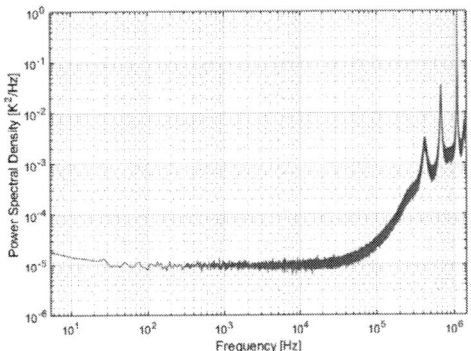

Fig. 5. PSD of 2^{19} bitstream samples, averaged over 100 cycles.

The sensor has been realized in a 180-nm CMOS process and occupies 6800 µm². Each chip consists of 20 temperature sensors and shared bias and reference phase generators. The latter supplies a selected sensor with a zero-phase WB drive signal and two reference phases for the PDSDM feedback. Each sensor dissipates 1.6 mW, 60% of which is consumed by the counter.

The PDSDM is operated at 2.9 MHz and each conversion consist of 1024 clock cycles resulting in a conversion rate of 2.8 kSa/s. After decimation by a sinc¹ filter, the sensor's output is processed by a 1st order polynomial whose coefficients are derived by a two-point calibration (at 5 °C and 125 °C). The resulting systematic non-linearity (Fig. 6) is then removed by a fixed 3rd order polynomial. For flexibility, both the decimation filter and the polynomial correction were implemented off-chip. As shown in Fig. 7, the sensor achieves an inaccuracy of ±0.35 °C in a temperature range from -35 °C to 125 °C. The resulting relative accuracy is the best in class and is only surpassed by [10], which requires a much larger. Fig. 5 shows the PSD of the bitstream, which indicates that the sensor is white noise limited up to 10 kHz. Fig. 8 shows a chip photograph of the sensor. In Table I the sensor's performance is compared to other state-of-the-art temperature sensors intended for DTM applications.

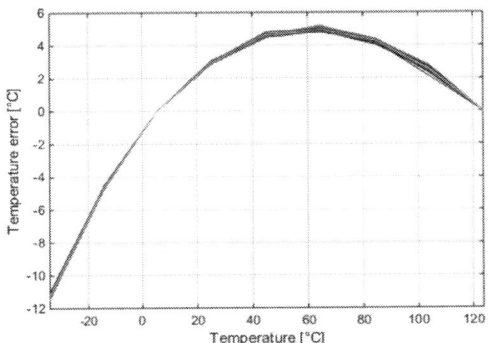

Fig. 6. Measured temperature error, before systematic non-linearity removal.

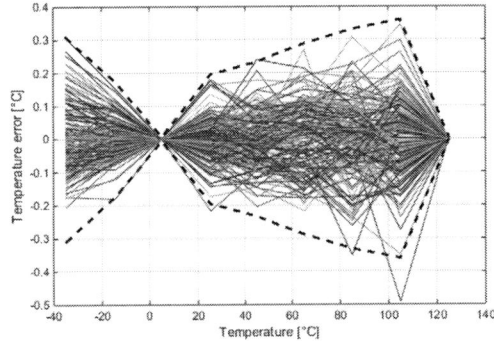

Fig. 7. Measured temperature error for 8 measured chips (160 sensors) after a two-point trim. Black dashed lines indicate the 3σ error.

TABLE I. PERFORMANCE SUMMARY AND COMPARISON WITH STATE-OF-THE-ART.

Publication	This Work	U. Sonmez [9]	S. Pan [10]	W. Choi [11]	Y-C Hsu [13]	M. Eberlein [14]
Year	2018	2016	2017	2018	2017	2017
Type	Resistor	TD	Resistor	Resistor	PNP	NPN
Technology	180nm	40nm	180nm	65nm	28nm	28nm
Area (μm^2)	6800	1650	720000	7000	9460	3800
Inaccuracy (3σ, °C)	-	1.4	-	-	0.87	3.6
Inaccuracy, 1-pt trim (3σ, °C)	-	0.75	0.3	-	-	-
Inaccuracy, 2-pt trim (3σ, °C)	0.35	-	0.075	0.35	-	-
Relative inaccuracy (3σ, m°C/°C)	2.19	8.48 / 4.55	2.4 / 0.6	2.8	6.96	24
Temperature Range (°C)	-35 to 125	-40 to 125	-40 to 85	-40 to 85	0 to 125	-20 to 130
Resolution (°C)	0.12	0.36	0.00017	0.0028	0.15	0.5
Speed (kSa/s)	3	1	0.1	1	0.15	0.5
Supply Voltage (V)	1.8	1	1.8	0.85-1.05	1.8	1.1
Power (mW)	1.6	2.5	0.18	0.068	0.0188	0.0176

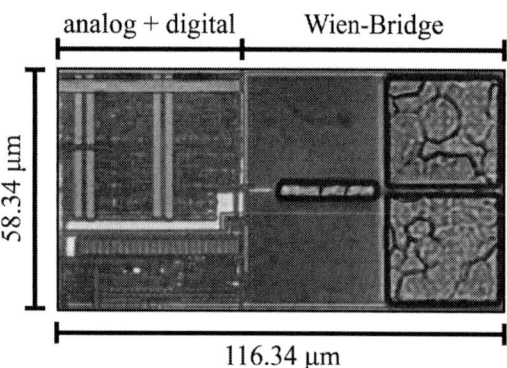

Fig. 8. Chip photograph.

IV. CONCLUSION

A compact resistor-based temperature sensor has been proposed. Compared to other resistor-based temperature sensors aimed at thermal management, this proof of concept design achieves the lowest area and best relative accuracy, despite being realized in a mature 180nm process. The reason for its low area is the use of a highly digital VCO-based ADC. Since this design scales well with technology, its performance is expected to improve when ported to more advanced processes.

REFERENCES

[1] D. Brooks and M. Martonosi, "Dynamic thermal management for high-performance microprocessors," *International Symposium on High-Performance Computer Architecture (HPCA)*, January 2001.

[2] E. Rotem, et al., "Temperature measurement in the Intel Core Duo processor," *Proc. THERMINIC*, pp. 23-27, Sep. 2006.

[3] J. Shor and K. Luria, "Miniaturized BJT-based thermal sensor for microprocessors in 32- and 22-nm technologies," *IEEE J. Solid-State Circuits*, Vol. 48, no. 11, pp. 2860-2867, Nov. 2013.

[4] C. Lu, et al, "An 8b subthreshold hybrid thermal sensor with ±1.07°C inaccuracy and single-element remote-sensing technique in 22nm FinFET," *Dig. ISSCC*, pp. 318-320, Feb. 2018.

[5] M. Cochet, et al., "A 225 μm^2 probe single-point calibration digital temperature sensor using body-bias adjustment in 28 nm FD-SOI CMOS," *IEEE SSC-L*, pp. 14-17, Vol. 1, Issue 1, Jan. 2018.

[6] K. Yang, et al., "A 0.6nJ -0.22/+0.19°C inaccuracy temperature sensor using exponential subthreshold oscillation dependence," *Dig. ISSCC*, pp. 160-161, Feb. 2017.

[7] R. Quan, et al., "A 4600μm^2 1.5°C (3σ) 0.9kS/s thermal-diffusivity temperature sensor with VCO-based readout," *Dig. ISSCC*, pp. 488-489, Feb. 2015.

[8] C. Vroonhoven and K. Makinwa, "A CMOS temperature-to-digital converter with an inaccuracy of 0.5° C (3σ) from -55 to 125° C," *Dig. ISSCC*, pp. 576-577, Feb. 2008.

[9] U. Sonmez, F. Sebastiano and K. Makinwa, "1650μm^2 thermal-diffusivity sensors with inaccuracies down to ±0.75°C in 40nm CMOS," *Dig. ISSCC*, pp. 206-207, Feb. 2016.

[10] S. pan, et al., "A resistor-based temperature sensor with a 0.13pJ·K^2 resolution FOM," *Dig. ISSCC*, pp. 158-159, Feb. 2017.

[11] W. Choi, et al., "A 0.53pJK^2 7000μm^2 resistor-based temperature sensor with an inaccuracy of ±0.35°C (3σ) in 65nm CMOS," *Dig. ISSCC*, pp. 322-323, Feb. 2018.

[12] U. Sonmez, F. Sebastiano and K. Makinwa, "Analysis and design of VCO-based phase-domain $\Sigma\Delta$ modulators," *Trans. Circuits and Systems I*, pp. 1075-1084, Sep. 2017.

[13] Y. Hsu, et al., "An 18.75μW dynamic-distributing-bias temperature sensor with 0.87°C(3σ) untrimmed inaccuracy and 0.00946mm^2 area," *Dig. ISSCC*, pp. 102-103, Feb. 2017.

[14] M. Eberlein and I. Yahav, "A 28nm CMOS ultra-compact thermal sensor in current-mode technique," *Proc. VLSI Circuits*, pp. 1-2, Jun. 2016.

FPGA-based CNN Processor with Filter-Wise-Optimized Bit Precision

Asuka Maki, Daisuke Miyashita, Kengo Nakata, Fumihiko Tachibana, Tomoya Suzuki, and Jun Deguchi

Toshiba Memory Corporation

Kawasaki, Japan

Abstract—Many efforts have been made to improve the efficiency for inference of deep convolutional neural network. To achieve further improvement of the efficiency without penalty of accuracy, we propose filter-wise optimized quantization with variable precision and the hardware architecture that fully supports it; as the bit precision for operations is reduced by granularly optimizing weight bit precision filter-by-filter, the execution time is reduced proportionally to the total number of computations multiplied with the number of weight bit. We implement the proposed architecture on FPGA and demonstrate that ResNet-50 run with 5.3× less execution cycles without penalty of accuracy.

Keywords—deep learning; convolutional neural network; quantization; variable bit width; FPGA

I. INTRODUCTION

The effectiveness of deep convolutional neural network (CNN) has been demonstrated for wide variety of applications such as image recognition [2], [5], [7], object detection [12], and semantic segmentation [4]. In line with the rapid increase of algorithmic researches, the importance of hardware-related researches is growing since the aid of energy efficient hardware is essential to execute those algorithms that consist of huge amount of computations with practical power consumption and execution time.

In this paper, in order to improve the energy efficiency for CNN inference, we employ the quantization of weights and activations, which is the best suited to massively parallel computing since it reduces the number of computations without deteriorating computational intensity unlike other approaches such as pruning and separable convolution [6]. Although extremely low bit quantization approaches such as 1 bit [11] and 2 bit [3] are attractive in terms of efficiency, significant recognition accuracy penalty is inevitable, they report 13.8% and 9.9% drops, respectively, compared to the full precision AlexNet [7] top-5 accuracy of 83% on ImageNet [2]. On the other hand, in order to maintain the accuracy with *uniform* bit width, more than 8 bit precision is required in general.

Recent works have revealed that layer-wise optimized quantization is quite effective for reducing the total bit number [9], [15]. Actually, for CNN, the required bit width for computation to maintain the accuracy varies not only between layers but also between pixels (i.e. spatial positions), channels, and filters. Then we have opportunities to reduce the number of computations by granularly controlling bit width. However, in order to exploit this feature, i.e. granularly optimized bit

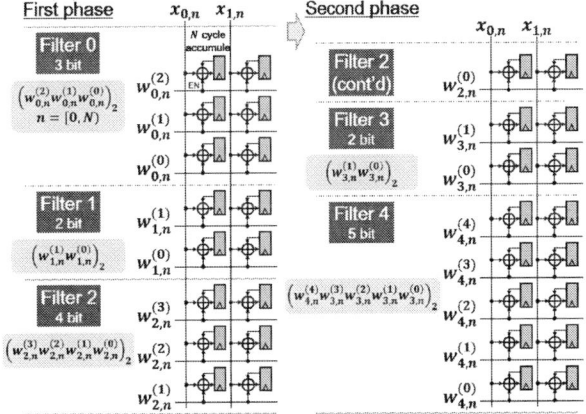

Fig. 1. The concept of the proposed parallel multiply-accumulate (MAC) processor with variable bit precision.

precision, a specific hardware is required. Ideally, the efficiency is proportionally improved, or the number of execution cycle is proportionally reduced, with the number of bit × the number of operation. In this paper, we propose to adopt filter-wise optimized bit precision for the weights, and propose the hardware architecture that fulfils the above-mentioned requirements. Furthermore, we demonstrate the validity of the proposed architecture by implementing it on FPGA and executing image classification task.

II. PROPOSED ARCHITECTURE

A. MAC processor with filter-wise variable bit precision

Fig. 1 shows the concept of the proposed MAC processor with filter-wise variable bit precision. This architecture is designed to execute following algorithm. A signed B-bit weight is represented as

$$w_n = -2^{B-1}w_n^{(B-1)} + 2^{B-2}w_n^{(B-2)} + \cdots + 2^1 w_n^{(1)} + 2^0 w_n^{(0)},$$
(1)

where $w_n^{(b)}$ is a binary value (0 or 1). Using this representation, a MAC computation is described as

$$\sum_{n=0}^{N-1} w_n x_n = \sum_{n=0}^{N-1} \left(-2^{B-1}w_n^{(B-1)}x_n + \cdots + 2^0 w_n^{(0)}x_n \right)$$
(2)

978-1-5386-6414-8/18 $31.00 © 2018 IEEE

$$= -2^{B-1} \sum_{n=0}^{N-1} w_n^{(B-1)} x_n + \cdots + 2^0 \sum_{n=0}^{N-1} w_n^{(0)} x_n, \quad (3)$$

where N is the number of accumulation *length*, which is the product of the number of channels, kernel height and kernel width for normal convolution computation or the number of channels for Winograd [8] convolution computation. Therefore a MAC computation with multiple-bit weights is able to be computed by summing properly bit shifted results of MAC computations using 1-bit weights. We refer to this process as *bit-synthesis*.

As shown in Fig. 1, we unroll the MAC computation units for 1-bit weight in parallel. Since the weight is 1 bit, an accumulator with enable (EN) is employed as MAC unit, i.e. only when the value of weight is 1, the accumulator is enabled and it accumulates the input activation.

Input activations are broadcast vertically and reused in 8 MAC units for multiplied with 8 different 1-bit weights. On the other hand, weights are broadcast horizontally and reused in 2 MAC units for multiplied with 2 different input activations. This 2-D reuse strategy minimizes the amount of loading of weights and activations from buffer for executing a certain number of MAC computations [10].

In the example shown in Fig. 1, we have 8 units in a column and we would like to compute the MAC computations for 5 filters. The numbers of bit width for each filter are $\{3, 2, 4, 2, 5\}$. At the first phase, MAC computations for filter 0 and filter 1, and a part of filter 2 (upper 3 bits out of 4 bits) are executed.

In every phase, N cycle accumulation is executed in each unit without storing and/or loading intermediate results (output stationary in [1]), which reduces the output bandwidth. After completing N cycle accumulation, *bit-synthesis* is executed. Note that this change of the order of the computation from Equation (2) to Equation (3), i.e. *bit-synthesis after accumulation* instead of *accumulation after bit-synthesis* which are employed in [13], enables to reduce the required dynamic range of the each accumulator as well as reduce the number of addition from $B \times N$ to $B + N$.

At the second phase, MAC computations for the remaining 1 bit of filter 2, filter 3 and filter 4 are executed. After accumulation, the *bit-synthesis* for filter 2 is executed with the registered intermediate result computed in the previous phase.

Consequently, the proposed architecture enables not only to maximize the number of filters executed simultaneously but also to keep the resource utilization close to 100% even if the numbers of bit width for filters are not aligned with the number of implemented MAC units. Due to this feature, the proposed *bit-parallel* architecture is better suited to parallel computing with the filter-wise optimized bit precision than *bit-serial* architecture [9]. In *bit-serial* architecture, when computations for multiple filters are executed in parallel, the execution time will vary depending on the bit precision of filters. This deteriorates resource utilization.

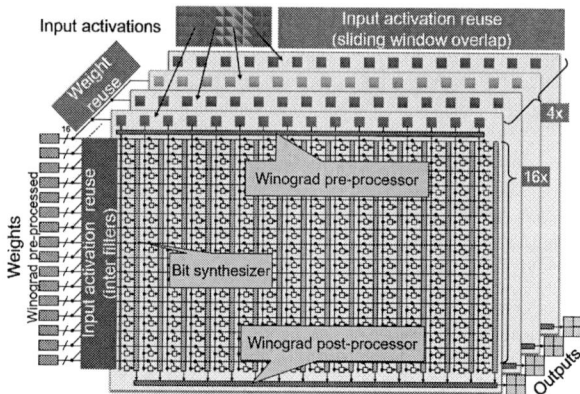

Fig. 2. The detailed architecture of the implemented MAC processor with filter-wise bit precision.

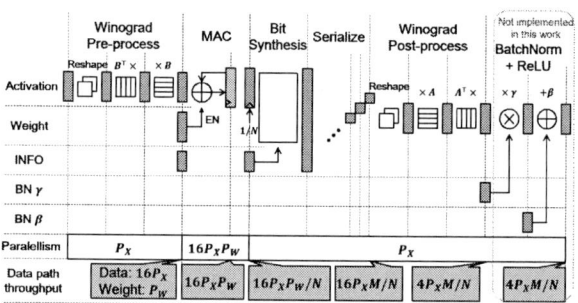

Fig. 3. Pipeline strategy of the proposed CONV engine. $P_X = 4$ and $P_W = 16$ in this work. N and M are the number of input channels and the number of filters, respectively.

B. Detail of the implemented architecture

Fig. 2 shows the detail of the MAC processor implemented in this work. Our implementation supports Winograd convolution. Weights are pre-processed for Winograd convolution in advance and stored in memory. Winograd pre-processor converts 4×4 input activations to 4×4 pre-processed input activations. These 16 pre-processed input activations are multiplied with 16 different 1-bit weights and accumulated in parallel (represented with 16 columns in the figure). As explained in the previous section, we employ further 2-D parallelism, i.e. $16 \times$ weights for input activation reuse (represented with 16 rows in the figure) and $4 \times$ input activations for weight reuse (represented with 4 planes in the figure). Besides, by exploiting the fact that 4×4 input activations before Winograd pre-processing are overlapped as shown in the top of Fig. 2, we reduce the bandwidth for loading activations. A *bit synthesizer* is placed in each column, a Winograd pre- and post-processor are placed in each plane.

Fig. 3 shows the pipeline strategy in the proposed CNN processor. In order to increase the throughput, all the operations are pipelined.

At first, activations are pre-processed for Winograd [8] and

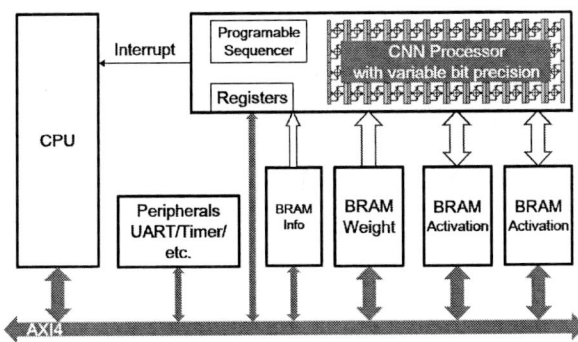

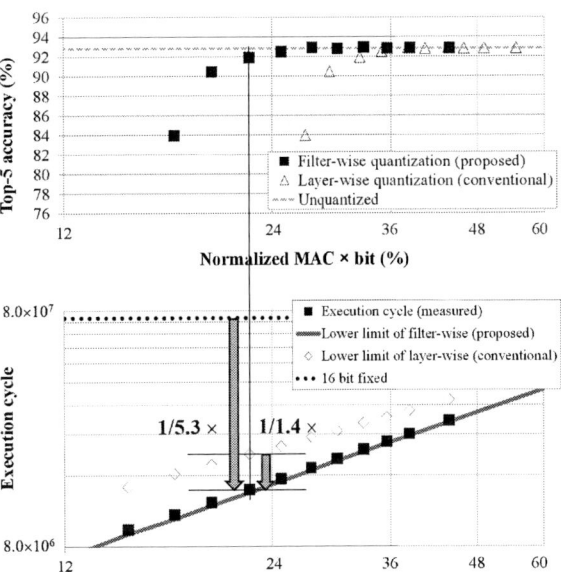

Fig. 4. The system architecture of the proposed convolution processor implemented on FPGA.

fed to the proposed MAC array. Weights are pre-processed for Winograd in advance and stored in the block RAM (BRAM) buffer, then directly fed to the MAC array. After completing MAC computations, the outputs are synthesized to obtain the result for the multiple-bit weight using additional information of bit widths which are also loaded from on-chip buffer. Next, the post-processing for Winograd is applied. Although BatchNorm and ReLU are not implemented in this work in order to focus on the convolution function, they can also be embedded in this pipeline as shown in the figure.

After the MAC computations, since results are reduced by summation and the throughput after that stage are decreased by the number of input channels N, the MAC outputs computed in parallel are serialized for the following processes such as Winograd post-processing, BatchNorm, and ReLU to exploit the temporal capacity with small hardware resource utilization. The parallelism and data throughput for each stage are shown in the bottom of Fig. 3. For normal convolution, Winograd pre- and post-processes are skipped.

III. IMPLEMENTATION

We implement the proposed architecture on Xilinx ZCU102 Zync UltraScale+ MPSoC board.

Fig. 4 shows the system architecture implemented in this work. Inputs (weights and activations) to the proposed CNN processor are loaded from external DRAM to BRAM buffers in advance. Then, controlled by the programmable sequencer, the weights and activations are fed to the CNN processor in proper order.

Despite that our design includes 1024 parallel 1-bit MAC processors, the FPGA gate utilization is only around 20 %. Note that DSP48 instances are not used at all.

IV. EXPERIMENTS

In order to validate the effectiveness of the proposed architecture, we evaluate the execution time of the proposed implementation with that of conventional fixed 16-bit weights implementation. First, we quantize the pre-trained weights of ResNet-50 [5] with filter-wise optimized bit precision. As shown in the upper figure of Fig. 5, compared to layer-wise

Fig. 5. Upper: top-5 accuracy vs. normalized MAC × bit. The horizontal axis shows the number of computations × the number of weight bits normalized to the number before applying our quantization method (16 bit for weight). Top-5 accuracy is measured on GPU with 10000 images in the validation dataset. The image size is 256 × 256. Lower: the number of execution cycles vs. the normalized MAC × bit. The number of cycles is counted by the counter implemented on FPGA and does not include cycles for inter-layer data movement between BRAM and DRAM.

optimization, the number of MAC operations × the number of bit of weights (referred to as MAC×bit hereinafter) is significantly reduced under the same penalty of top-5 accuracy. In the lower figure of Fig. 5, the gray solid line shows the theoretical lower limit of the execution cycles when we assume the hardware with 64 MAC units for $16b$ weight and the execution cycles are ideally proportional to MAC×bit. As shown in the figure, the measured execution cycles of our implementation are very close to the theoretical limits. The average utilization of the MAC units reaches > 97%. Compared to the case of $16b$ fixed bit precision and the case of layer-wise optimized one, the number of execution cycles are reduced by 5.3× and 1.4×, respectively.

Fig. 6 shows the comparison of the number of execution cycles of each layer of the proposed architecture and that of a conventional architecture. The numbers of execution cycles of the proposed architecture are significantly reduced compared to that of conventional one.

Table I shows the summery of the implementation and performance. Our design fully supports the convolution computation with filter-wise optimized weight bit precision, which is useful to reduce the computations without accuracy penalty.

Fig. 7 shows the experimental setup. We run ResNet-50 for the image classification task on the ZCU102 board and all 52 convolutional layers except for the first 7 × 7 one are executed

978-1-5386-6414-8/18 $31.00 © 2018 IEEE

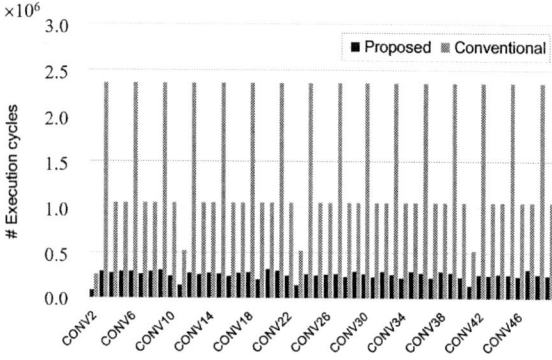

Fig. 6. Comparison of the number of execution cycles of the proposed architecture vs. conventional one. As a conventional architecture, we assume 64 parallel architecture for 16b-fixed MAC, which consumes similar hardware resources to the proposed one.

TABLE I
SUMMARY.

	[14]	[3]	Our design
weight precision	32b float	2b	**1-16b**
act. precision	32b float	2b	8b
Frequency	100 MHz	200 MHz	100 MHz
FPGA chip	Vertex7 VX485T	Zynq XC7Z020	Zynq XCZU9EG
kLUTs used	186	44	57
DSP used	2240	89	0
BRAM used	1024	105.5	900
Peak performance (GMACS×bit)	-	-	102.4
Dataset	ImageNet	ImageNet	ImageNet
CNN type	AlexNet	AlexNet	ResNet-50
Top-5 Accuracy	-	73.1 %	**91.9** %
Performance (GMACS)	30.81	205.11	47.7

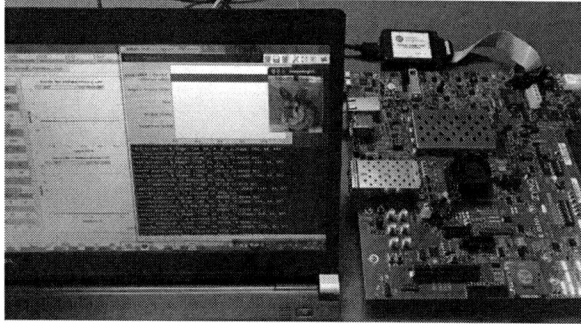

Fig. 7. Experimental setup.

on FPGA where the proposed architecture is implemented. The other operations such as element-wise add in ResNet-50 and data movement between external DRAM and FPGA are executed by CPU on Zynq. The PC screen shows an image of rabbit to be classified and the bar plot indicating that it is correctly inferred to be "wood rabbit, cottontail, cottontail rabbit."

V. CONCLUSION

In this paper, we proposed filter-wise quantization of weights to reduce the number of computations for the inference of convolutional neural networks without penalty of the accuracy. Furthermore, we proposed a hardware architecture that fully exploits the reduced precision, i.e., as the number of bit of operation is reduced, the number of execution time is reduced proportionally. We implemented the proposed architecture on FPGA and demonstrated that when ResNet-50 on ImageNet run on the FPGA, the number of execution cycles is reduced by $5.3\times$ without penalty of top-5 accuracy compared to 16b fixed case. Implementation of some functions which are not implemented in this work such as DMAs is left as future work.

ACKNOWLEDGMENT

The authors would like to thank Shinichi Sasaki, Takahisa Kaihotsu, Takeshi Kumagaya, Ryuichi Fujimoto, and Yoshio Masubuchi for their supports.

REFERENCES

[1] Y.-H. Chen, J. Emer, and V. Sze, "Eyeriss: A Spatial Dataflow for Energy-Efficient Dataflow for Convolutional Neural Networks," In *43rd Annual International Symposium on Computer Architecture (ISCA)*, pages 367-379, 2016.

[2] J. Deng, W. Dong, R. Socher, L.-J. Li, K. Li, L. Fei-Fei, "ImageNet: A Large-Scale Hierarchical Image Database," *CVPR09*, 2009.

[3] L. Jiao, C. Luo, W. Cao, X. Zhou, L. Wang, "Accelerating low bit-width convolutional neural networks with embedded FPGA," *2017 27th International Conference on Field Programmable Logic and Applications (FPL)*, 2017.

[4] K. He, G. Gkioxari, P. Dollár, and R. Girshick. "Mask R-CNN," *ArXiv:1703.06870*, Mar. 2017.

[5] K. He, X. Zhang, S. Ren, and J. Sun, "Deep Residual Learning for Image Recognition," *ArXiv:1512.03385*, Dec. 2015.

[6] A. G. Howard, M. Zhu, B. Chen, D. Kalenichenko, W. Wang, T. Weyand, M. Andreetto, and H. Adam, "MobileNets: Efficient Convolutional Neural Networks for Mobile Vision Applications," *ArXiv:1704.04861*, Apr. 2017.

[7] A. Krizhevsky, I. Sutskever, H.E. Hinton, "Imagenet classification with deep convolutional neural networks," In *Advances in neural information processing systems*. pages 1097-1105, 2012.

[8] A. Lavin and S. Gray, "Fast Algorithms for Convolutional Neural Networks," *ArXiv:1509.09308*, Sept. 2015.

[9] J. Lee, C. Kim, S. Kang, D. Shin, S. Kim, H. J. Yoo, "UNPU: A 50.6TOPS/W unified deep neural network accelerator with 1b-to-16b fully-variable weight bit-precision," *2018 IEEE International Solid - State Circuits Conference - (ISSCC)*, Feb. 2018.

[10] B. Murmann, D. Bankman, E. Chai, D. Miyashita, and L. Yang, "Mixed-Signal Circuits for Embedded Machine-Learning Applications," *Asilomar Conference on Signals, Systems and Computers*, Nov. 2015.

[11] M. Rastegari, V. Ordonez, J. Redmon, A. Farhadi, "XNOR-Net: ImageNet classification using binary convolutional neural networks," in *Computer Vision ECCV*, pages 525-542, 2016.

[12] S. Ren, K. He, R. Girshick, and J. Sun, "Faster R-CNN: Towards Real-Time Object Detection with Region Proposal Networks," *ArXiv:1506.01497*, June 2015.

[13] H. Sharma, J. Park, N. Suda, L. Lai, B. Chau, J. K. Kim, V. Chandra, H. Esmaeilzadeh, "Bit Fusion: Bit-Level Dynamically Composable Architecture for Accelerating Deep Neural Networks," In *45th Annual International Symposium on Computer Architecture (ISCA)*, June 2016.

[14] C. Zhang, P. Li, G. Sun, Y. Guan, B. Xiao, J. Cong, "Optimizing FPGA-based accelerator design for deep convolutional neural networks," *Proceedings of the 2015 ACM/SIGDA International Symposium on Field-Programmable Gate Arrays*, pages 161-170, 2015.

[15] Y. Zhou, S.-M. Moosavi-Dezfooli, N.-M. Cheung, and P. Frossard, "Adaptive Quantization for Deep Neural Network," *ArXiv:1712.01048*, Dec. 2017.

An Asynchronous Energy-Efficient CNN Accelerator with Reconfigurable Architecture

Weijia Chen*, Hui Wu*, Shaojun Wei*[†], Anping He[‡], Hong Chen*[†]
*Institute of Microelectronics, Tsinghua University,
[†]Beijing National Research Center for Information Science and Technology, Beijing, China
[‡]School of Information and Technology, Lanzhou University, Lanzhou, China
Email: hongchen@tsinghua.edu.cn

Abstract—In this paper, we introduce an asynchronous energy-efficient convolutional neural network (CNN) accelerator with reconfigurable architecture including six computing cores, each of which contains 5x5 processing elements. With the dynamically reconfigurable architecture, the data path, the calculation method, the activation function, and the pooling way and size can be modified according to the configurable information for different CNN models. In the computing cores, the global clock is replaced by the local pulse signals from Click elements. An asynchronous pipeline formed by Click elements enables the circuits to work in pipeline mode without any sacrifice of speed because of the self-timed characteristic of asynchronous circuits. Each computing core has a 5x5 registers array that is fully connected by an asynchronous Mesh network, by which the input data can be fully reused. A novel computing pattern called convolution-and-pooling-integrated computing, which combines convolution and pooling computing together, is proposed to reduce the access to the intermediate data. These yield an 88% decrease of the access to off-chip memory, which significantly reduces energy consumption. A CNN model, LeNet-5, is implemented in our accelerator with the FPGA of Xilinx VC707. The asynchronous computing core has 84% less dynamic power than that of the synchronous core. The efficiency achieves 30.03 GOPS/W, which is 2.1 times better than that of previous works.

Keywords—CNN; Energy-Efficient; accelerator; Asynchronous

I. INTRODUCTION

Convolutional Neural Networks (CNN) have been widely used in the field of computer vision and show its great advantages in image classification, object detection and video surveillance [1]. The inference of CNNs is usually realized by CPU and GPU. However, the CPU has limited computing resources and parallelism. Although GPU outperforms CPU in the inference of CNNs because it is designed for parallel computing of large-scale data, but GPU consumes too much power (for example: 33W for NVIDIA GTX840M, and 235W for NVIDIA Tesla K40 [2,3]). Hence, CNN accelerators require a trade-off between flexibility and energy efficiency. As we know, ASIC design can obtain the best power efficiency but only a certain CNN model can be implemented in ASIC circuit because of its worst flexibility. FPGA shows acceptable performance but its fine-grained computing and routing resources limit the power efficiency and runtime reconfiguration for different CNNs. To obtain a better flexibility and energy efficiency, some CNN accelerators adopt a coarse-grained dynamic reconfigurable architecture (CGRA) such as the Eyeriss from MIT [4] and the Thinker [1] with high performance

and flexibility. On the other hand, the asynchronous circuits are characterized by their local data- or control-driven flow of operations, which differs from the global clock-driven flow of synchronous designs. This character enables the different portions of the asynchronous circuits to operate at their individual ideal "frequencies"—or rather to operate and idle as needed, consuming energy only when and where needed. Clock gating has a similar goal—enabling registers only when needed—but does not address the power drawn by the centralized control and clock-tree buffers [5]. As a result, asynchronous logic has been advocated as a means of reducing power consumption in a number of applications [6,7]. IBM's TrueNorth which successfully implements Spiking Neuron Networks (SNN) has only 65mW power [8], and the dataflow processing unit (DPU) from Wave Computing company with asynchronous processing element achieves 181 TOPS. In this paper, we propose an asynchronous accelerator with dynamic reconfigurable architecture to achieve great flexibility, low power and high energy efficiency.

II. DESIGN OF THE CNN ACCELERATOR

A. Architecture of the accelerator

The top-level architecture is shown in Fig. 1, in which the input data is stored in the off-chip DRAM. The configuration information from the controller will be input into the computation array including six cores together with the registers array. According to the configuration information, the activation function of processing element (PE) in each core, pooling way and size, and the direction of data flow in the registers array for input data reuse will be determined. The computation array is responsible for convolution and pooling computing layer by layer. The computation results will be stored into DRAM through the output buffer.

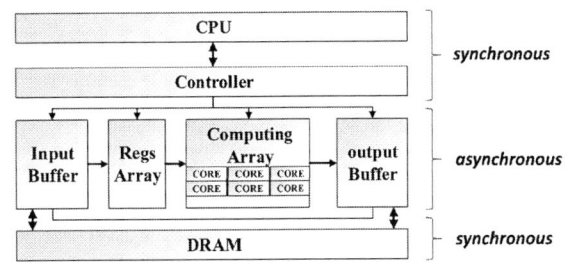

Fig. 1. The top-level architecture of the accelerator.

978-1-5386-6414-8/18 $31.00 © 2018 IEEE

B. Design of the computing core

A computing core that contains twenty-five PEs together with the registers array is shown in Fig. 2. Each PE includes a register and a multiplier. As a result, each PE has the function of both multiplication and data storage. All the PEs are fully connected through an asynchronous Mesh network, by which each PE can receive data from its neighbor PE from any direction (such as top, bottom, left and right). Besides, this fully connected network make it possible to reuse the input data. The size and data path in the core can be configured according to the requirement of different layers or CNN models. That is, each PE can work independently or all PEs can work together. Instead of global clock, handshake protocol is used to make sure the data flow properly between PEs. Because of the "event-driven" feature of asynchronous handshake protocol, a PE will be completely turned off when it does not receive a request signal, therefore its power consumption is reduced. All the results from each multiplier are added up, and the request signal will be generated and sent to the reconfigurable pooling unit for next computation.

C. Design of the PE.

We adopt bundled data and two-phase asynchronous handshake protocol with Click element in the PE design. As shown in Fig. 4(a), with the control of request and acknowledgment signals, C_n and C_{n+1} use the bundled data (single-rail) datapath, which is similar to those used in synchronous design [8]. The four-phase and two-phase protocols are shown in Fig. 4(b) and Fig. 4(c) respectively. In four-phase protocol, the request and acknowledgment signals need to return to zero while in two-phase protocol each transition of request and acknowledgment represent an event. As a result, the two-phase outperform four-phase in speed and power

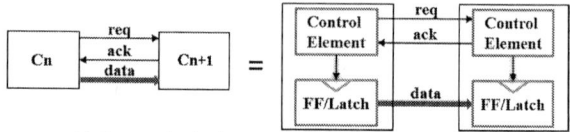

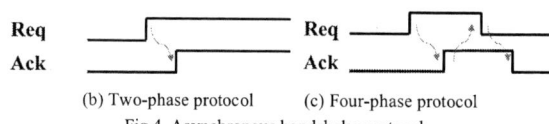

(a) Communication between asynchronous circuit modules

(b) Two-phase protocol (c) Four-phase protocol

Fig.4. Asynchronous handshake protocol.

because of the lower signal transition frequency. We choose Click element shown in Fig.3 for the control circuit. Once a request signal arrives, Click generates a local pulse signal 'fire', which enables the FF or Latch.

The designed PE circuit is shown in Fig. 5. Three Clicks connected in series form a three-stage asynchronous pipeline. The delay match circuits (not shown in Fig. 5) between Clicks make each stage self-timed regardless of the critical path of the whole circuits by matching the delay of the combination logic between DFFs. Besides, if there is no request signal, the whole asynchronous circuits are completely turned off and there will be no dynamic power. When a request signal arrives, the configuration data will be written into 'DFF1' when the signal 'fire1' goes high. Then, the Click in the middle generates 'fire2', by which 'DFF2' read the input data from one of the directions and send it to the multiplier as an operand at the same time. Finally, the request signal comes to the Click on the right and 'fire3' goes to high which enables 'DFF3' to store the input data so that the neighbor PEs can reuse it.

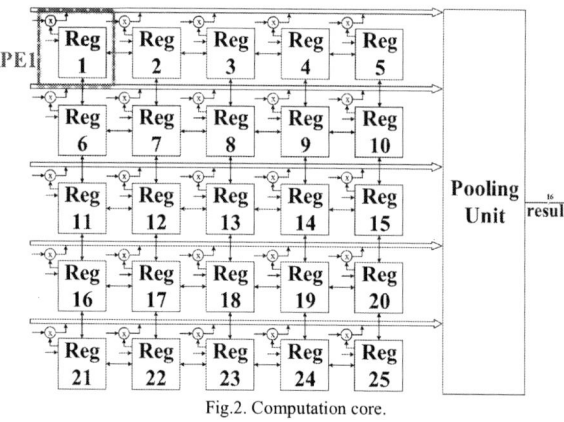

Fig.2. Computation core.

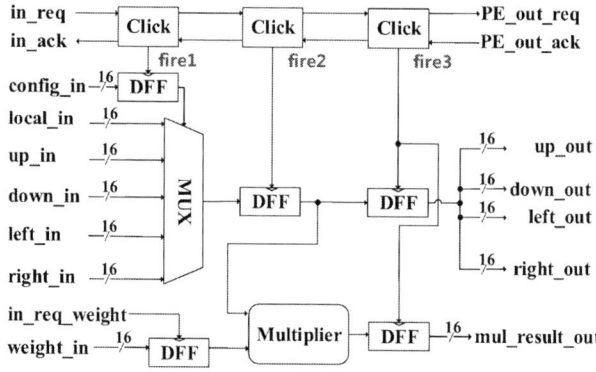

Fig.5. Processing element circuit.

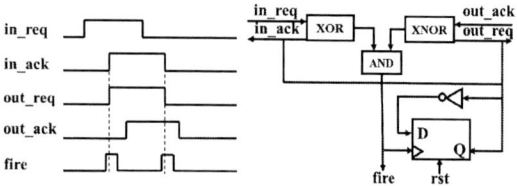

(a) Waveforms of Click element (b) Structure of Click element

Fig.3. Click element.

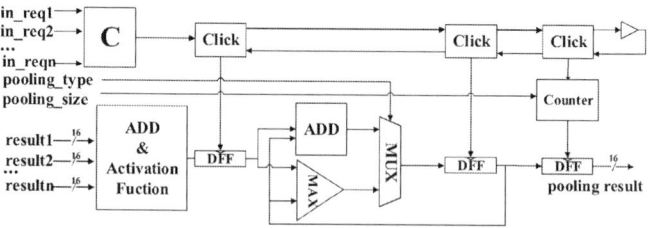

Fig.6. Pooling unit circuit.

D. Design of the pooling unit (PU)

The designed PU circuit is illustrated in Fig. 6. All the out request signals of PEs from one core are connected to the Muller C element for completion detection. In other words, only when all the request signals arrive will the Muller C generate a next request signal, which ensures all the multiplication results of the PEs are ready and then the PU starts to work. The activation function, pooling way (maximum or average) and pooling size can be dynamically modified according to the configuration information. Some typical activation functions, such as Sigmoid, tanh and ReLU are widely used. It is easy to realize ReLU, but difficult for Sigmoid and tanh, which cost too many hardware resources. Therefore, we adopt a 256KB ROM to store all the activation results for Sigmoid and tanh. The input data with a fixed-point number can be considered as an address to find the output value after activation. This method also makes computing faster because the complex calculations are avoided.

III. THE CONVOLUTION-AND-POOLING-INTERGRATED COMPUTING AND DATA REUSE METHODS

For CNN accelerators, the frequent access to off-chip memory is always the bottleneck of energy optimization. To reduce the off-chip memory access, we put forward novel methods of convolution-and-pooing-integrated computing and input data reuse, which yield an 88% decrease of the access to off-chip memory, and as a result the power consumption decrease significantly.

A. The convolution-and-pooling-intergrated computing

To verify our accelerator, we use an improved LeNet-5 model (shown in Fig. 7), which has five layers: two convolution layers, two pooling layers, and one full connection layer. Generally, the output data of the previous layer is the input data of the next layer, and the computing resources can be reused. However, the access to the intermediate data wastes too much power, especially when necessary access to off-chip memory happens.

The proposed convolution-and-pooing-integrated computing avoids the access to the intermediate data. As shown in Fig. 8, we give an example (2x2 convolution with 2x2 pooling) to show the different convolution kernel moving method. In the traditional accelerator, the convolution kernel needs to read the input data from left to right and top to bottom in sequence. The convolution results have to be stored temporarily and wait for pooling. In contrast, with the "event-driven" feature of asynchronous circuit, in the convolution-and-pooling-integrated computing, the direction of reading input data matches the generation of each pooling result. For example, after four convolutions, we obtain the four results A, B, C, and D in Fig.8(a). Because the first pooling computing needs A, B, H, G as the operands, A and B have to be stored until H and G are ready. While in the convolution-and-pooing-integrated computing shown in Fig. 8(b), four operands for pooling computing are ready at the same time because of different moving way of the convolution kernel. In other words, there is no intermediate data to be stored, every pooling result will be calculated out just after the four convolutions When performing the maximum pooling, the lager convolution result will replace the smaller one, and when the average pooling, we only need to

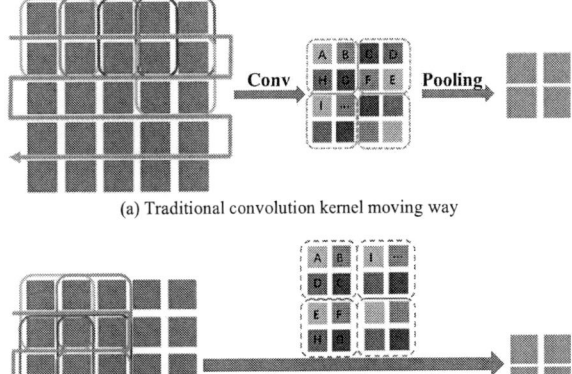

(a) Traditional convolution kernel moving way

(b) The moving way of the convolution kernel
in convolution-and-pooing-integrated computing

Fig. 8. Comparison of the moving way of the convolution kernel.

accumulate values one by one with the weight value, which has been divided by the size before accumulation. Thus, the final result of accumulator is the average pooling result. In general, the access to intermediate data in the pooling layer can be reduced to zero by this way.

B. The input data reuse

Take the 2x2 convolution (shown in Fig. 9) for example, we find the same input data in two adjacent convolution operations can be reused. For once convolution, the weight values (W0-W4) are fixed and only need to be read once, so we choose weight-stationary way. Four results will be obtained after four convolutions (i.e., O1-O4). Since the four PEs are connected by the Mesh Network, so once one convolution of data in the red frame (i.e., I0, I1, I3, and I4) in Fig. 9 is completed, the next convolution needs the data in the yellow frame (i.e., I1, I2, I4, and I5). We find that the new input data of this convolution are I2 and I5, and I1 and I4 can be read from the near PE. That is, I1 and I4 can be reused instead of being read again. Table I shows the amount of the access data in each layer of the improved LeNet-5. Without the reuse and convolution-and-pooing-integrated computing, the total data access amount s 66.4KB, However, with our methods of integration computing and input data reuse, it becomes 7.8 KB, which has an 88% decrease. For more complicated CNNs, more off-chip memory access will be saved.

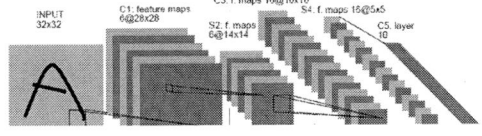

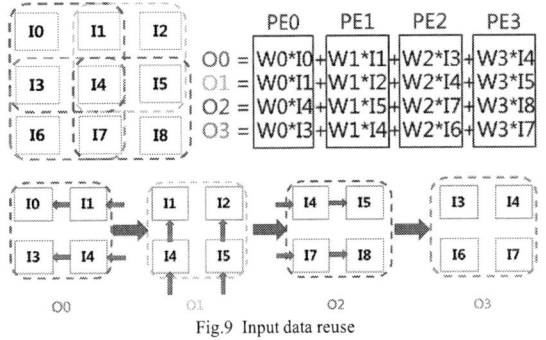

Fig.9 Input data reuse

TABLE I
The amount of data access of each layer of LeNet-5

LeNet-5 Layer	1	2	3	4	5	total
Function	conv	pooling	conv	pooling	full connection	
Input data (KB)	1.568	6.912	1.728	2.048	0.512	12.8
Access amount without our methods (KB)	28.8	13.824	19.2	4.096	0.512	66.4
Access amount with our methods (KB)	5.8	0	1.48	0	0.512	7.8

IV. EXPERIMENT RESULTS

We implement the improved LeNet-5 with Xilinx FPGA VC707. To compare the performance of power, we design a synchronous accelerator with the same function. From the Xilinx Xpower Analyzer tool, the dynamic power of the synchronous core is 45.96mW, but only 7.25mW for the asynchronous core. The experiment platform is shown in Fig. 10. The comparison results between our accelerator and previous works (synchronous CNN accelerators for LeNet-5) are presented in Table II. Our accelerator's efficiency achieves 30.03 GOPS/W, more than 2 times better than that in [10] and [11], and 9 times better than that in [9]. Besides, ten thousand images in MINST data have been tested by our circuits and the error rate is 2%.

The Demo link is https://vimeo.com/273064365 (password: **asscc2018demowjc**).

V. CONCLUSION

In this paper, we first apply the asynchronous circuits to a reconfigurable CNN accelerator and propose novel methods of the input data reuse and the convolution-and-pooling-intergraded computing to decrease the access to off-chip memory. Experimental results show that our accelerator achieves flexibility, 88% decrease of data access, and doubled energy efficiency. More CNN models will be tested on our platform, and the computing pattern of asynchronous PEs and delay- match circuits will be optimized in the future.

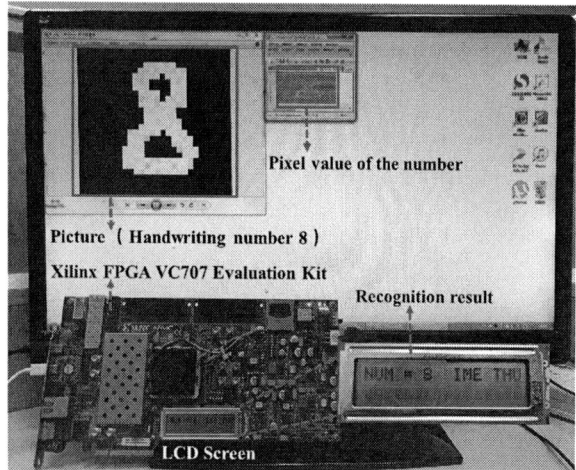

Fig.10. The experiment platform.

TABLE II
Experiment Results with comparison of previous works

Benchmark LeNet-5	FPGA 2015[10]	ICCAD 2016[11]	FPGA 2016[12]	This work
Platform	VC7VX 485T	VC5VX 690T	ZYNQ Z-7045	VC7VX 485T
Precision	16bit-fixed	16bit-fixed	16bit-fixed	16bit-fixed
Clock (MHz)	100	150	150	
DSP Used	2240	2833	780	406
BRAM Used	1024	1248	486	101
LUT Used	186251	350892	182616	75221
FF Used	205704	311904	127653	38577
Performance (GOPS)	61.62	354	136.97	20.3
Power(W)	18.61	26	9.63	0.676
Efficiency (GOPS/W)	3.31	13.62	14.22	30.03

ACKNOWLEDGMENT

This work is supported by National Natural Science Foundation of China (No. 61674090), xpartly supported by Beijing National Research Center for Information Science and Technology (No. 042003266) and Beijing Engineering Research Center (No. BG0149).

REFERENCES

[1] F. Tu, S. Yin, P. Ouyang, S. Tang, L. Liu and S. Wei, "Deep Convolutional Neural Network Architecture With Reconfigurable Computation Patterns," in IEEE Transactions on Very Large Scale Integration (VLSI) Systems, vol. 25, no. 8, pp. 2220-2233, Aug. 2017.

[2] Q. V. Le, "Building high-level features using large scale unsupervised learning," in IEEE International Conference on Acoustics, Speech and Signal Processing (ICASSP). IEEE, 2013, pp. 8595–8598.

[3] NVIDIA, "Tesla k40 gpu active accelerator," Report, 2013.

[4] Y. H. Chen, T. Krishna, J. S. Emer and V. Sze, "Eyeriss: An Energy-Efficient Reconfigurable Accelerator for Deep Convolutional Neural Networks," in IEEE Journal of Solid-State Circuits, vol. 52, no. 1, pp. 127-138, Jan. 2017.

[5] ABeerel, Peter A, and M. E. Roncken. "Low Power and Energy Efficient Asynchronous Design." Journal of Low Power Electronics 3.3(2007):234-253.

[6] I. E. Sutherland, "Micro pipelines." Communications of the ACM, vol. 32, no. 6, pp. 720-738, 1989.

[7] Andreas Steininger et al. "Exploring the state dependent SET sensitivity of asynchronous logic - The muller-pipeline example." Computer Design (ICCD), pp. 61-67, 2014.

[8] F. Akopyan et al., "TrueNorth: Design and Tool Flow of a 65 mW 1 Million Neuron Programmable Neurosynaptic Chip," in IEEE Transactions on Computer-Aided Design of Integrated Circuits and Systems, vol. 34, no. 10, pp. 1537-1557, Oct. 2015.

[9] I. E. Sutherland, Micropipelines. Communications of the ACM 32,720 (1989)

[10] Chen Zhang, Peng Li, Guangyu Sun, Yijin Guan, Bingjun Xiao, and Jason Cong.Optimizing fpga-based accelerator design for deep convolutional neural networks.In FPGA, pages 161–170, 2015.

[11] Chen Zhang, Zhenman Fang, Peipei Zhou, Peichen Pan, and Jason Cong. Caffeine:towards uniformed representation and acceleration for deep convolutional networks. In ICCAD, page 12, 2016.

[12] Jiantao Qiu, Jie Wang, Song Yao, Kaiyuan Guo, Boxun Li, Erjin Zhou, Jincheng Yu, Tianqi Tang, Ningyi Xu, and Huazhong Yang. Going deeper with embedded fpga platform for convolutional neural network. In FPGA, pages 26–35, 2016

Hardware Architecture for Fast General Object Detection using Aggregated Channel Features

Koichi Mitsunari, Jaehoon Yu, and Masanori Hashimoto
Graduate School of Information Science and Technology, Osaka University
Email: {k-mitunr,yu.jaehoon,hasimoto}@ist.osaka-u.ac.jp

Abstract—For embedded system applications, high detection accuracy and fast detection must be achieved within a limited power budget. This paper proposes an embedded system-oriented hardware accelerator for object detection with aggregated channel features (ACF). The proposed accelerator consists of hardware architectures dedicated for HOG features, quantization, and boosted decision trees, and they contribute to 2006X speed-up and 601X memory reduction. Our FPGA implementation result shows that the proposed accelerator can detect pedestrians at 170 fps for Full HD images, and 6-class traffic objects at 78 fps for Full HD images.

I. INTRODUCTION

The primary goal of embedded object detection systems aiming at driver assistance and autonomous robots is to achieve fast and accurate detection with low power consumption. These applications are time-critical, and missed detection can be a threat to human life. Therefore, fast and accurate detection is indispensable and a social requirement. However, accurate detection with novel algorithms, for example, rich object representations, generally demands massive computations, which prevents fast detection and low power implementation. On the other hand, early works on object detection (e.g. [1]) use simple object representation, and the accuracy required for critical applications is not satisfied. We need to construct an accurate detection system with sophisticated algorithms and develop their efficient hardware implementations for attaining low latency and power consumption.

The automatic braking system in driver assistance needs to detect objects 33 meters ahead when driving speed is 30 km/h. In this case, Full HD 60 fps image processing is essential. Here, let us compare three well-known algorithms adopted for hardware implementation; support vector machine (SVM), convolutional neural network (CNN), and aggregated channel features (ACF) [2]. SVM is a linear classifier, and it is suitable for parallel implementation due to its simple structure [1]. However, it suffers from low detection accuracy, and it is not suitable for critical applications. CNN is drawing a lot of attention since it achieves high detection accuracy. However, due to its inherent tremendous amount of computation, low power consumption and fast detection are hardly achievable. Recently, CNN hardware is widely studied (e.g. [3]), but an implementation that satisfies high accuracy, low latency, and low power consumption is not presented yet. ACF achieves good detection accuracy. ACF uses a boosted decision tree (BDT) classifier, which requires a small amount of computation, and achieves fast detection in software implementation.

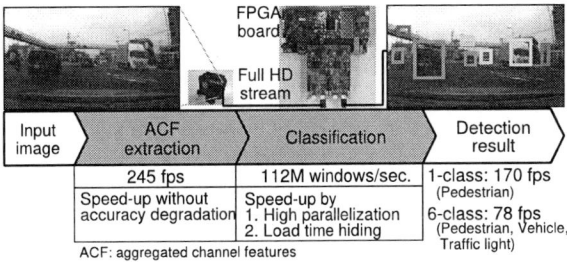

Fig. 1. Proposed Object Detection System Overview.

TABLE I
PROBLEMS IN ACF HARDWARE IMPLEMENTATION.

	Memory	Area	Speed
HOG feature extraction		✗	
Feature representation	✗	✗	
Parallel classification using BDT			✗

Thanks to this, [4] achieves the fast classification of 480p30 even with a serial hardware implementation. However, its parallel hardware implementation pursuing higher throughput is not straightforward since the memory access depends on input data, and it prevents parallel implementation.

This work uses ACF as a baseline to exploit its reasonably high accuracy and low computational cost. For enhancing throughput and minimizing hardware cost, we have performed algorithm-hardware co-optimization and improved the compatibility of ACF with the hardware implementation [5]–[7]. This paper proposes a general object detection system shown in Fig. 1 and presents its FPGA implementation. ACF extraction is speeded up by adopting a hardware-oriented feature descriptor which extracts equivalent information in a small amount of computation [5], and 245 fps is achievable. BDT classification is speeded by parallel implementation and hiding load time of coefficients [7], and 112M windows/sec is attained. In addition, a quantization method which is robust to accuracy degradation [6] is adopted for memory saving and power reduction. Consequently, the proposed system can detect multi-objects of pedestrians, vehicles, and traffic signals in 1080p60, which satisfies the above-mentioned requirement for the automatic braking system.

978-1-5386-6414-8/18 $31.00 © 2018 IEEE

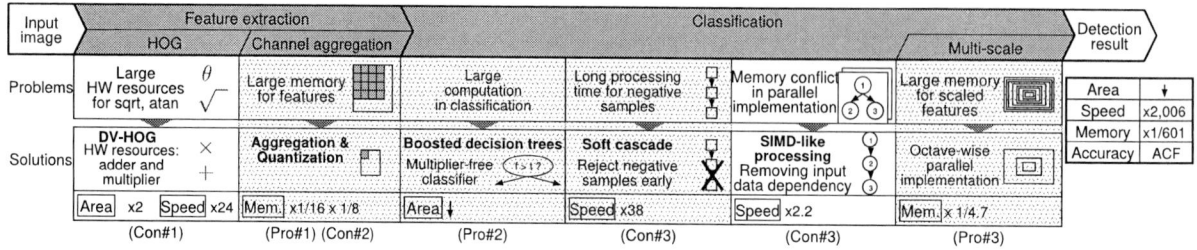

Fig. 2. Problems and Solutions for ACF Hardware Implementation.

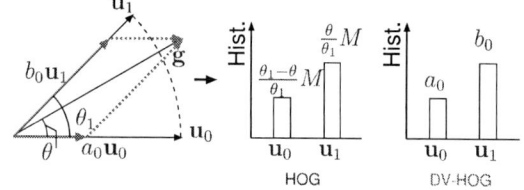

Fig. 3. HOG Feature Extraction.

II. PROBLEMS OF ACF HARDWARE IMPLEMENTATION

ACF is originally developed for software implementation, and the good accuracy with small computational cost is demonstrated in [2]. The following discusses the compatibility of ACF with hardware implementation and points out problems to be addressed for high-throughput ACF implementation.

ACF has high affinity with the hardware implementation regarding the following three points. In feature extraction, ACF aggregates feature maps by 4x4, and it reduces the required memory capacity to one-sixteenth (Pro#1). In the classification step, the BDT classifier does not require multipliers, and then the necessary hardware resource is small (Pro#2), where BDT classifier consists of multiple decision trees, and at each tree, the algorithm selects a leaf node by recursive comparison between a threshold and a feature from the root node. Also, BDT can use soft cascade, which rejects negative samples early, for fast detection (Pro#3) since a cascade structure represents BDT.

However, ACF has incompatibilities with hardware implementation, which are listed in TABLE I. First, we need to extract HOG features from an input image, but this feature extraction process includes expensive computations such as trigonometric function and square root, which demands large hardware resource (Con#1). Second, a large memory is necessary to store features during feature extraction (Con#2) and classification. Finally, as explained, the parallel hardware implementation of BDT is difficult (Con#3) since the memory access patterns depend on input data, and straightforward memory segmentation cannot avoid memory access collision. We need to resolve these incompatibilities.

III. PROPOSED HARDWARE ACCELERATOR ARCHITECTURE

This section presents the proposed architecture for ACF-based object detection. Fig. 2 summarizes the problems of

object detection accelerator and their solutions. Each solution contributes to at least either of area reduction, speed-up, or memory reduction. The advantages of ACF (Pro#1) to (Pro#3) provide memory reduction in channel aggregation, area reduction and fast classification in classification, respectively. (Con#1) to (Con#3), on the other hand, are resolved by the proposed architecture, and the following subsections explain the proposed hardware solutions one by one. As a result, the proposed accelerator achieves area reduction, 2,006-times speed-up, and memory reduction to 1/601 while keeping the detection accuracy almost identical to the original software ACF implementation.

A. Decomposed Vector Histograms of Oriented Gradients (DV-HOG)

HOG feature extraction calculates the magnitude and orientation of edges in an input image and generates their histogram. Gradient vector $\mathbf{g} \in \mathbb{R}^2$ is calculated by differentiating the input image along x and y-axes, and its magnitude and angle are defined as

$$M = \sqrt{g_x^2 + g_y^2}, \quad \theta = \tan^{-1}\left(\frac{g_y}{g_x}\right). \quad (1)$$

Histogram bins are linearly spread from 0 to π radian. Given $\mathbf{g}$, it votes M to adjacent bins using interpolation in terms of angle as shown in Fig. 3. HOG extraction includes the trigonometric function and square root calculation, and it requires a large amount of hardware resource.

To solve the problem, [5] proposes a complex computation-free interpolation based on vector decomposition as shown in Fig. 3. This DV-HOG regards $\mathbf{g}$ as the weighted sum of adjacent unit vectors and adopts L1 norm for magnitude calculation. DV-HOG is calculated only with multiplication with constant and addition, and it reduces the circuit area to 1/12 without accuracy degradation compared with interpolation in terms of angle.

B. Quantization for Decision Tree Classifiers

Classification using decision trees compares a feature and a threshold at each decision node. To reduce the memory requirement for decision tree classifiers, [6] proposed a quantization method focusing on the classifier's threshold range. Namely, child node selection in BDT is based on the comparison result between feature and threshold values, and the difference between them is not used for classification.

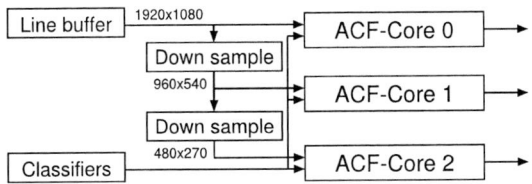

Fig. 4. Multi-scale ACF Hardware Architecture.

TABLE II
PARAMETERS USED IN THE EVALUATION.

Module		Parameter	Value
Feature extraction		Parallelism	24
		Bit-width	4bit
Classification	ACF-Core0	Parallelism	8x16x8
	ACF-Core1		8x4x8
	ACF-Core2		8x2x8

Therefore, fine quantization is applied only to the range of threshold values for memory saving. [6] reports that if 2% accuracy degradation on INRIA Person Dataset is allowed, the numerical precision is reduced from 32bit to 2bit, which indicates that memory requirement is reduced to one-sixteenth.

C. Hardware Architecture for BDT Classifier

Classification using a BDT classifier includes complicated memory access because selected decision nodes depend on input data. Thus, parallel processing is difficult due to memory access conflict. To solve the problem, [7] visits all the decision nodes and enables SIMD-like processing. When each channel has a memory bank, channel-wise parallel implementation is feasible. With the processing order scheduling and 1,024-parallel classification, the hardware can classify 350 fps multi-scale Full HD stream. Also, this method is compatible with soft cascade, and [7] reports that 33-times speed-up is obtained by soft cascade for pedestrian detection.

D. Hardware Architecture for ACF

Thanks to the above hardware solutions, the feature extraction and classification are implemented in parallel for fast detection. For detecting multiple sizes of objects, on the other hand, multi-scale detection, which scales the input image or applies classifiers of different window sizes, must be implemented. In this case, the processing time for the feature extraction is the bottleneck compared with that of classification. To reduce the feature extraction time, this architecture adopts an approach proposed by Benenson et al. [8]. Fig. 4 shows the overview, where the scale octave is 1/2 and classifiers dedicated for each scaled image are applied to ACF-Core 0 to 2 in parallel. In the case of Full HD image, there exist three octaves.

IV. EVALUATION ON FPGA BOARD

A. Implementation

The proposed hardware architecture is implemented on FPGA. The evaluation uses Xilinx ZC706 evaluation board

(a) Pedestrian detection   (b) Traffic object detection

Fig. 5. Detection Results.

and two FMC cards for HDMI input and output. The FPGA board has programmable logic (PL) and ARM core, and the ACF-HW module is implemented at register transfer level using Verilog HDL in PL. To handle the 1080p60 input stream and overlay the bounding boxes on the output stream, we used the Vivado IPs of video input, video output, and video timing controller. The trained classifier is stored in the SD card, and loaded from software running on the ARM core.

TABLE II shows the parameters used in the implementation, and the target frequency for the ACF-HW module is 100MHz. TABLE III shows the resource utilization, which shows the balanced usage of resources. We can see that DV-HOG modules occupy 8% of LUTs in total. It should be noted, on the other hand, that the original computation of Eq. (1) requires 12x resources, which makes a single FPGA implementation infeasible.

B. Object Detection Performance Evaluation

The processing performance is evaluated on pedestrian detection and traffic object detection. The evaluation using pedestrian detection aims to compare the performance with the related approaches. In traffic object detection, multi-class classification is evaluated considering practical applications.

1) Pedestrian Detection: The evaluation uses twelve classifiers whose window size ranges from 48x96 to 92x184, and each classifier consists of 2,048 depth-two decision trees. Training process uses INRIA Person Dataset [10]. Fig. 5(a) shows the detection result. For quantitative analysis, software simulation is used to count clock cycles for each step. The result shows that feature extraction and classification consume 4.08 and 1.80 (= 0.15 × 12(scales)) milliseconds, respectively. Consequently, the proposed accelerator can process 170 fps of Full HD. TABLE IV shows the processing performance comparison. [1] and [9] achieve Full HD 60 fps processing. However, they suffer from the higher log-average miss rate (MR) of 46% and 20% on INRIA Person Dataset, respectively, whereas the log-average MR of the proposed architecture is 17%. For a fair comparison between ACF-based architectures, we use the processing number of windows in a second as an evaluation metric. The table indicates that the proposed accelerator achieves 57 times speed-up compared with ACF hardware [4].

2) Traffic Object Detection: As a multi-class evaluation, pedestrian, vehicle, and traffic light detection are performed on FPGA. TABLE V summarizes the classifiers. The detection candidate area is limited as shown in TABLE V to reduce the number of false positives and speed-up. Fig. 5(b) shows the

TABLE III
FPGA RESOURCE UTILIZATION.

Module	Slice		LUT (Logic)		LUT (Memory)		LUT (FF)		32Kb BRAM		DSP	
ACF-HW	124,770	(57%)	124,476	(57%)	294	(0%)	128,626	(59%)	356.5	(65%)	122	(14%)
→ ACF-Core0	70,755	(32%)	70,657	(32%)	98	(0%)	72,719	(33%)	204.5	(38%)	41	(5%)
→ ImageBuf	195	(0%)	195	(0%)	0	(0%)	601	(0%)	9	(2%)	0	(0%)
→ FeatGen	6,928	(3%)	6,831	(3%)	97	(0%)	8,215	(4%)	0	(0%)	0	(0%)
→ DV-HOG	6,168	(3%)	6,168	(3%)	0	(0%)	7,457	(3%)	0	(0%)	0	(0%)
→ AggCube	2,021	(1%)	2,021	(1%)	0	(0%)	2,878	(1%)	8.5	(2%)	40	(4%)
→ ACFCube	23,891	(11%)	23,891	(11%)	0	(0%)	24,420	(11%)	170	(31%)	0	(0%)
→ LeafCube	35,144	(16%)	35,143	(16%)	1	(0%)	36,839	(17%)	16	(3%)	0	(0%)
→ OutputBuf	39	(0%)	39	(0%)	0	(0%)	1,086	(0%)	1	(0%)	0	(0%)
→ ACF-Core1	26,127	(12%)	26,119	(12%)	98	(0%)	27,158	(12%)	84.5	(16%)	41	(5%)
→ ACF-Core2	18,859	(9%)	18,761	(9%)	98	(0%)	19,686	(9%)	64.5	(12%)	40	(4%)
Total	139,215	(64%)	137,939	(63%)	1,276	(2%)	149,129	(68%)	389	(71%)	128	(14%)

TABLE IV
DETECTION PERFORMANCE COMPARISON.

Method	Speed	Method	log-avg. MR	#win. / sec.
[1]	Full HD 60 fps	SVM	46%	6284k
[9]	Full HD 60 fps	DPM	20%	3,975k
[4]	VGA 30 fps	ACF	17%	1,972k
Ours	Full HD 170 fps	ACF	17%	112,501k

TABLE V
CLASSIFIERS FOR TRAFFIC OBJECTS.

Target	Pedestrian	Vehicle (Front, Rear)	Traffic light (Green, Yellow, Red)
Depth	2	2	2
#weak classifier	2,048	512	512
Window size	[48, 96], ..., [92, 184]	[48, 48], ..., [92, 92]	[48, 16], ..., [84, 28]
#classifier	12	12	4
Total #classifier	12	24	12
Area	Lower 2/3	Lower 2/3	Upper half

detection result, and TABLE VI summarizes the processing time. We can see soft cascade contributes to 17.6 times speed-up on average. The system can process 78 fps of full HD frames, which is enough for the driving assistance system at 30 km/h. It should be noted that the proposed accelerator does not use any domain-specific knowledge and hence it is applicable to any object detection applications, although the required fps for the driving assistance system is exemplified above.

V. CONCLUSION

Embedded object detection systems need to simultaneously achieve high detection accuracy, fast detection, and low power

TABLE VI
CLASSIFICATION SPEED EVALUATION.

		# cycle		Speed-up
		Soft cascade		
		Off	On	
Vehicle	Front	148,677	13,626	10.9x
	Rear	153,143	15,193	10.1x
Pedestrian		575,037	14,996	38.3x
Traffic Light	Green	156,133	9,481	16.5x
	Yellow	116,308	10,716	10.9x
	Red	140,734	9,403	15.0x
Total		1,290,032	73,415	17.6x

consumption, and its design is highly challenging. To solve the issue, this paper proposed a hardware architecture for general multi-class object detection using ACF. The proposed hardware architecture makes use of the advantages of the ACF algorithm itself and incorporates multiplier-free DV-HOG, aggressive quantization and BDT parallel computation architecture with the overall accelerator architecture. In total, the system is speed-up by 2,000 times and reduced memory to 1/600. FPGA implementation result showed that the proposed system could detect pedestrians at 170 fps for a Full HD image, and 6-class traffic objects at 78 fps for Full HD, which satisfied the requirement for the automatic braking system.

ACKNOWLEDGMENT

This work was supported by JSPS KAKENHI Grant Number JP16K16085.

REFERENCES

[1] A. Suleiman and V. Sze, "An energy-efficient hardware implementation of HOG-based object detection at 1080HD 60 fps with multi-scale support," J. Signal Process. Syst., vol. 84, no. 3, pp. 325–337, Sep. 2016.

[2] P. Dollár et al., "Fast feature pyramids for object detection," IEEE Trans. Pattern Anal. Mach. Intell., vol. 36, no. 8, pp. 1532–1545, Aug. 2014.

[3] R. Zhao et al., "Optimizing cnn-based object detection algorithms on embedded FPGA platforms," in Applied Reconfigurable Comput., 2017, pp. 255–267.

[4] H. Song et al., "Hardware implementation of aggregated channel features for ADAS," in Proc. Int. SoC Des. Conf., Oct. 2016, pp. 167–168.

[5] K. Mitsunari et al., "Decomposed vector histograms of oriented gradients for efficient hardware implementation," IEICE Trans. Fundam. Electron., Commun. and Comput. Sci. (conditional acceptance).

[6] K. Mitsunari and J. Yu, "Influence of numerical precision on machine learning and embedded systems," in Proc. Int. Workshop Smart Info-Media Syst. Asia, Sep. 2016, pp. 164–169.

[7] K. Mitsunari et al., "Hardware architecture for high-speed object detection using decision tree ensemble," IEICE Trans. Fundam. Electron., Commun. and Comput. Sci. (to appear).

[8] R. Benenson et al., "Pedestrian detection at 100 frames per second," in Proc. IEEE Comput. Soc. Conf. Comput. Vis. and Pattern Recognit., Jun. 2012, pp. 2903–2910.

[9] A. Suleiman et al., "A 58.6 mW 30 frames/s real-time programmable multiobject detection accelerator with deformable parts models on full HD 1920 × 1080 videos," IEEE J. Solid-State Circuits, vol. 52, no. 3, pp. 844–855, Mar. 2017.

[10] N. Dalal and B. Triggs, "Histograms of oriented gradients for human detection," in Proc. IEEE Comput. Soc. Conf. Comput. Vis. and Pattern Recognit., Jun. 2005, pp. 886–893.

A Neural Network Accelerator With Integrated Feature Extraction Processor for a Freezing of Gait Detection System

Val Mikos*, Chun-Huat Heng*, Arthur Tay*, Shih-Cheng Yen*, Nicole Shuang Yu Chia[†], Karen Mui Ling Koh[†],
Dawn May Leng Tan[‡] and Wing Lok Au[†]

*Department of Electrical and Computer Engineering, National University of Singapore
[†]Department of Neurology, National Neuroscience Institute, Singapore
[‡]Department of Physiotherapy, Singapore General Hospital, Singapore

Abstract—Parkinson's disease patients are at risk of falls due to freezing of gait (FoG). Wearable detection systems providing biofeedback for aid rely on accurate FoG classifiers, but such algorithms have yet to propose a dedicated hardware implementation. This paper is a first proposal of a dedicated hardware for a real-time FoG feature extractor and classifier on a single chip. The neural network classifier and FoG feature extractor exploit an FoG system's inherent time-sharing affinity to reduce area consumption which is crucial in wearable applications. The design has been fitted onto a small form-factor 10x10 mm² Artix-7 FPGA. It consumes 6.716 μJ per classification which surpasses non-dedicated FoG system implementations in terms of efficiency, while remaining competitive with 91.32% and 93.64% in sensitivity and specificity, respectively.

I. Introduction

Parkinson's disease patients experience sudden, episodic inabilities to maintain regular gait known as freezing of gait (FoG). Such an abrupt disruption in gait can result in falls and has hence incited research to develop wearable FoG detection systems that provide biofeedback cues for aid in overcoming FoG. The design of accurate FoG classifiers consequently became a crucial aspect of a detection system's viability. Unfortunately, a substantial part of today's potent FoG classifiers do not provide a hardware implementation [1]–[4]. The wearable hardware implementations that are capable of real-time detection either lack the needed processing power to run sophisticated FoG classification algorithms [5], or are bulky and power hungry due to wireless communication between a computational processor and a decentralized inertial measurement unit (IMU) [6], [7]. As noted in [6], a miniaturization of the bulky FoG classifier's hardware in form of dedicated hardware could significantly reduce power consumption and augment wearability. This paper is a first attempt at providing such a dedicated hardware. It proposes a system-on-chip (SoC) solution that accommodates all central aspects of an FoG detection system: acquisition of sensor data, feature extraction, and a sophisticated classification algorithm for biofeedback activation. The proposed design has been evaluated on an Artix-7 FPGA for its power consumption and hence feasibility for battery-powered, wearable, small form-factor PCBs.

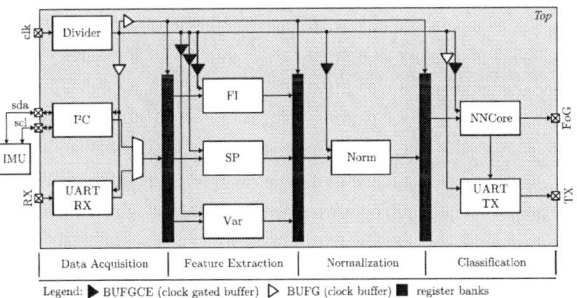

Fig. 1. The proposed SoC containing a feature extraction processor and neural network accelerator that exploit an FoG system's inherent time-sharing affinity for low area consumption.

The paper first provides an overview of the overall FoG detection system in Section II. The proposed feature extraction processor and neural network classifier are then illustrated in Sections III and IV. Finally, Section V evaluates the FPGA implementation before the paper concludes in Section VI.

II. Freezing of Gait Detection System

FoG detection systems predominantly rely on classification of data from an IMU worn on a subject's lower limbs. A miniaturized system would hence integrate an IMU, a feature extractor and a classification algorithm on the same PCB. Our proposed system combines the latter two into a single SoC as illustrated in the overall architecture in Fig. 1. Data from the IMU is first acquired through I²C communication. The feature extraction processor then extracts the three features freeze index (FI), stride peak (SP) and variance (Var) from the acquired data. After normalizing the feature vectors, a neural network core classifies the current sample and outputs the classification result for biofeedback activation, using UART if necessary. The transition between two successive computational sections is enabled through register banks acting as serial to parallel interfaces. Since battery-powered, wearable systems are subject to strict power constraints, the processing blocks are clock gated to reduce the dynamic power consumption during idling periods. Regular clock buffers are then used to balance the

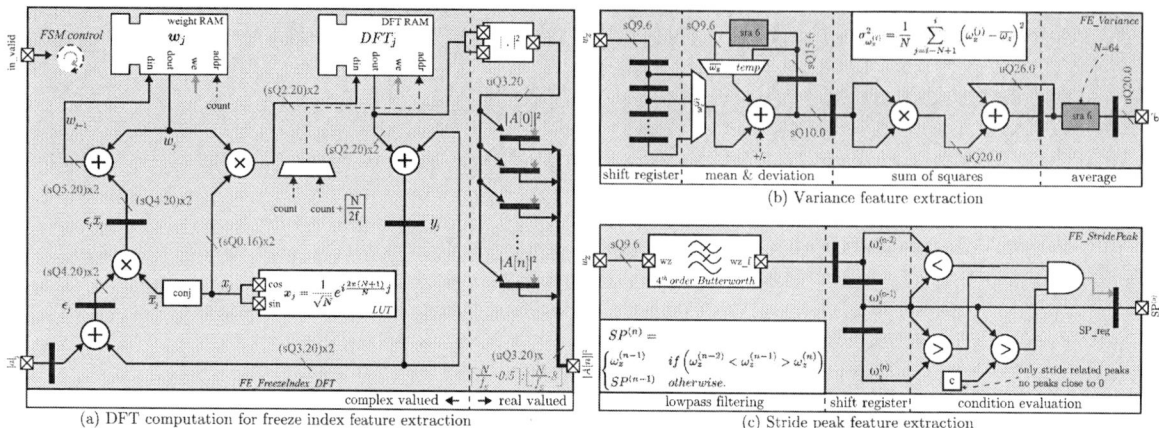

(a) DFT computation for freeze index feature extraction

(b) Variance feature extraction

(c) Stride peak feature extraction

Fig. 2. Feature extraction processor showing (a) DFT computation for extracting the FI, (b) variance computation and (c) SP computation.

clock tree for the circuitry that is always clocked, ensuring a minimal clock skew between any two flipflops.

III. FEATURE EXTRACTION PROCESSOR

After data acquisition, the processor is tasked with the extraction of the three aforementioned features as follows.

A. Freeze Index (FI) via Discrete Fourier Transform (DFT)

The FI is a highly established feature for detecting FoG and is defined as

$$FI = \frac{\int_3^8 |A(f)|^2 \, df}{\int_{0.5}^3 |A(f)|^2 \, df} = \frac{\sum_{n=\lceil \frac{N}{f_s} \cdot 3 \rceil}^{\lfloor \frac{N}{f_s} \cdot 8 \rfloor} |A[n]|^2}{\sum_{n=\lceil \frac{N}{f_s} \cdot 0.5 \rceil}^{\lfloor \frac{N}{f_s} \cdot 3 \rfloor} |A[n]|^2}. \quad (1)$$

Here, $A(f)$ is the Fourier transform of the acceleration signal $|a|$, $A[n]$ the corresponding N-point DFT and f_s the sampling frequency. The DFT is obtained through a least mean squares (LMS) spectrum analyzer [8] with the proposed implementation shown in Fig. 2(a). For a new IMU sample $|a|_j$ with time index j, the LMS spectrum analyzer attempts to replicate that value through a weighted sum y_j of the harmonic phasor vector $\boldsymbol{x}_j$ with weights $\boldsymbol{w}_j$, as in

$$y_j = \boldsymbol{x}_j^T \boldsymbol{w}_j, \quad \boldsymbol{x}_j = \frac{1}{\sqrt{N}} \begin{bmatrix} 1 & e^{i\frac{2\pi}{N}j} & \dots & e^{i\frac{2\pi(N+1)}{N}j} \end{bmatrix}. \quad (2)$$

In other words, it will adapt the weight vector $\boldsymbol{w}_j$ to take on values that minimize the complex error ϵ_j

$$\epsilon_j = |a|_j - y_j = |a|_j - \boldsymbol{x}_j^T \boldsymbol{w}_j. \quad (3)$$

The weights are then adapted by

$$\boldsymbol{w}_j \leftarrow \boldsymbol{w}_j + \epsilon_j \overline{\boldsymbol{x}}_j. \quad (4)$$

Per time index j, two LMS iterations are conducted. The weighted sum of the harmonic phasors $\boldsymbol{x}_j$ finally resolves the input into its Fourier components, as in

$$DFT_j = \boldsymbol{x}_j \boldsymbol{w}_j. \quad (5)$$

The computation of the N-point DFT via (2)-(5) requires $3N$ complex adders and $2N$ complex multipliers. This can quickly

overwhelm small form-factor FPGAs with limited digital signal processing (DSP) blocks at their disposal. Fortunately, the sampling frequency f_s of the IMU can be chosen relatively low in FoG systems, resulting in ample time for the DFT computation to finish before a new sample is supplied ($t = 1/f_s$). This is exploited in the proposed VLSI implementation by time-sharing 3 complex adders and 2 complex multipliers, making the number of incurred DSPs independent of N as long as the FPGA's clock is fast enough to complete the computation within $1/f_s$. Time-sharing requires the weight vector $\boldsymbol{w}$ and DFT bins to be stored in block RAM, while the phasor vector $\boldsymbol{x}$ can be implemented using a Look-up Table (LUT). The absolute value of the DFT bins is only computed for those that lie within $[0.5, 8]$ Hz as the rest are not required. The summation of the absolute values and ratio computation is subsequently computed based on (1).

B. Variance and Stride Peak (SP)

The variance and SP feature extractors are shown in Fig. 2(b) and (c). Both of their computations are based on the IMU's angular velocity signal in the subject's frontal plane ω_z. For SP extraction, ω_z is first low-pass filtered before the conditions stated in Fig. 2(c) are evaluated. For variance extraction, a window length of 64 samples is chosen such that the averaging reduces to a simple bitshift operation, highlighted as "sra 6".

IV. NEURAL NETWORK ACCELERATOR

Once the features are extracted, they are normalized and routed to the neural network core for classification. Fig. 4 illustrates a generic, fully connected neural network which takes the three extracted features as input and classifies them by means of forward propagation. This requires the induced field $z_j^{(l)}$ and the corresponding activation $a_j^{(l)}$ to be computed for every j^{th} unit of every l^{th} layer. The fields can be computed by the expressions

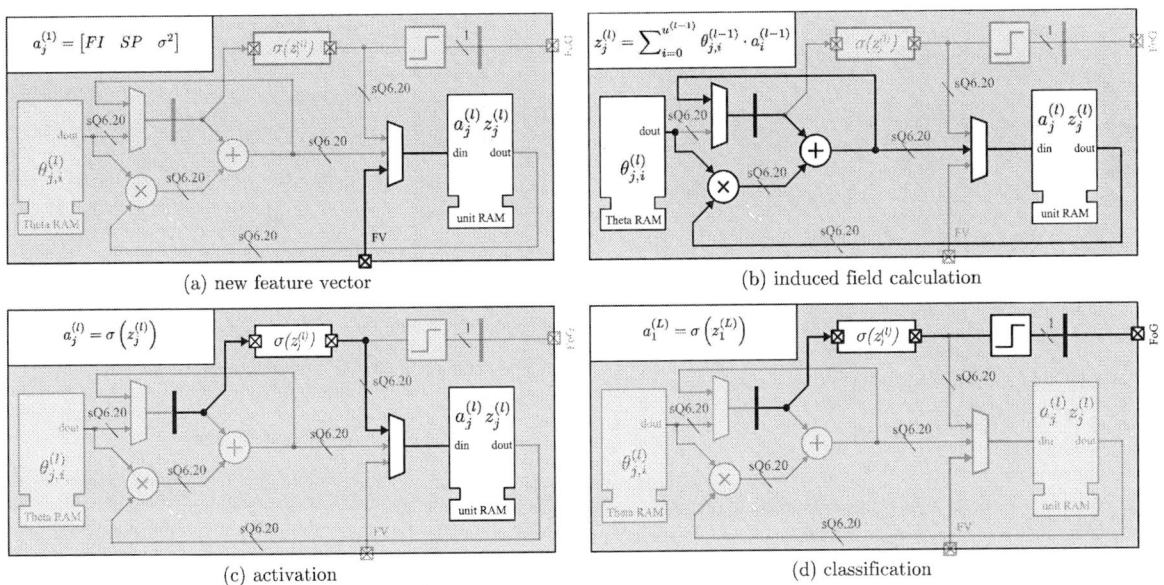

(a) new feature vector

(b) induced field calculation

(c) activation

(d) classification

Fig. 3. Forward propagation of a feature vector by (a) writing the new feature vector into RAM, (b) computing the induced fields, (c) activating the computed induced field using a LUT and finally (d) outputting the classification result.

$$z_j^{(l)} = \sum_{i=0}^{u^{(l-1)}} \theta_{j,i}^{(l-1)} \cdot a_i^{(l-1)}, \quad \forall j \in [1, u^{(l)}] \tag{6}$$

$$a_j^{(l)} = \sigma\left(z_j^{(l)}\right), \qquad \forall j \in [1, u^{(l)}] \tag{7}$$

where $u^{(l)}$ denotes the number of units in layer l, $\sigma(.)$ the sigmoid function and $\theta_{j,i}^{(l-1)}$ the neural network parameter from unit i in layer $l-1$ to unit j in layer l. Note that every zeroth unit is a bias unit, as in $a_0^{(l)} = 1$ for all layers l. The proposed VLSI implementation computing (6) and (7) for classification is shown in Fig. 3. The new feature vector is first stored within a unit RAM as $a_j^{(1)}$ (Fig. 3 (a)), writing one feature per clock cycle. With the first layer hence stored in unit RAM, the second layer's induced fields $z_j^{(2)}$ are computed

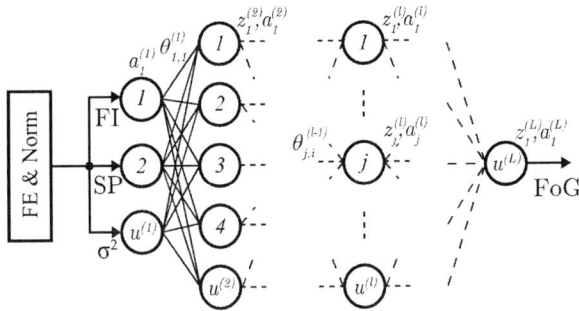

Fig. 4. A generic, fully connected neural network. Three input units accommodate the three extracted features, while only one output unit is required in binary classification problems.

next (Fig. 3 (b)). The required neural network parameter $\theta_{j,i}^{(1)}$ are obtained from Theta RAM, which is initialized with the appropriate values. The resulting induced field is written into unit RAM and retained within the accumulation register. A LUT implementing the sigmoid function is then enabled to compute the activation of the induced field using the accumulation register as input (Fig. 3 (c)). Steps (b) and (c) are repeated until all units of all layers have been computed. The activated function of the final layer L is outputted as a classification result for biofeedback activation after rounding the LUT result to 1 bit (Fig. 3 (d)).

As for the feature extraction, the proposed neural network exploits an FoG system's time-sharing affinity by employing only one real adder and one real multiplier, making the number of incurred DSPs independent of the neural network's architecture ($u^{(l)}$ and depth L), as long as the FPGA's clock is fast enough to compute the classification within $1/f_s$. That is why the sigmoid activation function is stored in a LUT, as it allows for a single-cycle computation of (7). For the proposed implementation to be viable, the neural network architecture hence needs to be constrained in size to not exceed the available computation time. This can be expressed by

$$t_{NNCore} = t_{clk}\left(3 + \sum_{l=2}^{L} u^{(l)} \cdot \left(u^{(l-1)} + 1\right)\right) \leq 1/f_s, \tag{8}$$

where the factor '3' represents the three cycles required to initially store the feature vector, and the factor '1' represents the cycle required for the LUT to activate a computed induced field. Evaluation of (8) indicates that even at a 1 MHz clock, the neural network could still be $L = 46$ layers deep with $u^{(l)} = 20$ for all layers, since f_s can easily be set to 50

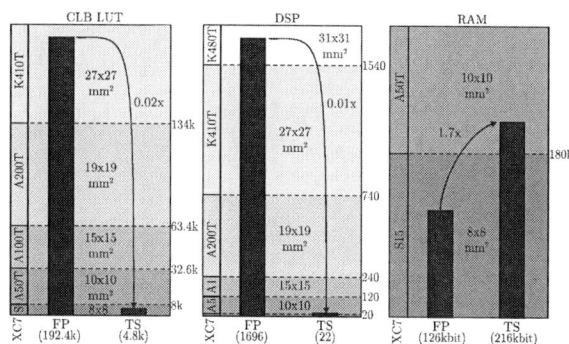

Fig. 5. Area savings for our time-shared (TS) implementation as compared to a fully-parallel (FP) one. The most resourceful 7-Series (XC7-) FPGAs for a given package form factor are displayed.

Hz to conform with Nyquist's theorem when computing the DFT for (1). In fact, the proposed implementation could even compute deep neural network algorithms proposed for FoG classification [1], while keeping area consumption at a minimum.

V. IMPLEMENTATION AND RESULTS

The neural network architecture implemented had 5 units per layer ($u^{(l)}$) and a depth of $L = 3$. Deeper networks did not indicate any improvement in classification accuracy, hence a narrow neural network was chosen for superior power efficiency. For the feature extractor, a 128-point DFT was computed and the IMU was sampled with $f_s = 50$ Hz. The design was mapped onto Xilinx's 7-Series FPGAs (XC7-). Fig. 5 illustrates how the exploitation of an FoG system's low sampling frequency through time-sharing significantly reduces the LUT and DSP blocks required to fit the design, while keeping the RAM required to store the intermediate results within bounds. A fully-parallel implementation would fit only on a Kintex-7 (XC7K480T, 31x31mm²), while the time-shared design allows for an Artix-7 (XC7A35T, 10x10mm²). The Spartan-7 (XC7S15, 8x8mm²), however, is not powerful enough to run our algorithm. Ultimately, we mapped our design onto the Artix-7 (XC7A35T, 10x10mm²) running on a 12 MHz clock. It consumes 6 mW dynamic power, 36 mW static power and achieves an overall classification efficiency of 6.716 μJ per classification. Despite the FPGA's routing overhead, this efficiency nearly equals those of ASIC based classifiers for temporal signals [9]. The lack of any feature extractor on that particular ASIC design makes the achieved classification efficiency especially noteworthy. With respect to FoG detection systems, our design significantly outperforms non-dedicated hardware implementations that consume 23.4-31.2 mJ/c [6]. The comparison with these designs is summarized in Table I. As indicated there, the accelerator achieved 91.32% sensitivity and 93.64% specificity, which is comparable with the performance of recent FoG detection algorithms that lack a hardware implementation [1]–[4].

	This work	Meth. Inf Med [6]	ESSCIRC 2017 [9]
Application	FoG	FoG	Temporal Signals
Classifier	NN	non ML	RNN
Feature Extractor	✓	✓	✗
IC	XC7A35T	XScale μC	ASIC
Technology Node	28 nm	180 nm	65 nm
Voltage	1.0 V	✗	1.2 V
Frequency	12 MHz	✗	400 MHz
Efficiency	6.716 μJ/c	23.4-31.2 mJ/c	4.9 μJ/c
N-point DFT	128-point	256-point	✗
Sensitivity	91.32%	73.10%	✗
Specificity	93.64%	81.60%	✗

TABLE I
SUMMARY OF THE PROPOSED DESIGN AND COMPARISON TO EXISTING
FoG AND ASIC SYSTEMS.

VI. CONCLUSION

A first dedicated hardware accelerator for wearable FoG detection systems has been proposed. The architecture time-shares most circuitry for low area consumption which allows for the employment of small form-factor FPGAs. Coupled with a classification efficiency of 6.716 $\mu J/c$, the proposed design becomes highly suitable for battery-powered, wearable, small form-factor PCBs. Compared to current FoG detection systems, the proposed hardware significantly improves classification efficiency and augments wearability through miniaturization.

ACKNOWLEDGMENT

This research is supported in part by NUS-NNI 2016 Grant R263000C36133 and in part by NMRC/CISSP/2014/2015.

REFERENCES

[1] J. Camps *et al.*, "Deep learning for freezing of gait detection in parkinson's disease patients in their homes using a waist-worn inertial measurement unit," *Knowledge-Based Systems*, vol. 139, pp. 119–131, 2018.

[2] P. Tahafchi *et al.*, "Freezing-of-gait detection using temporal, spatial, and physiological features with a support-vector-machine classifier," in *2017 39th Annual International Conference of the IEEE Engineering in Medicine and Biology Society (EMBC)*, July 2017, pp. 2867–2870.

[3] C. Ahlrichs *et al.*, "Detecting freezing of gait with a tri-axial accelerometer in parkinson's disease patients," *Medical & biological engineering & computing*, vol. 54, no. 1, pp. 223–233, 2016.

[4] K. Nazarzadeh, S. P. Arjunan, D. K. Kumar, and D. P. Das, "Non-invasive detection of the freezing of gait in parkinson's disease using spectral and wavelet features," in *2016 38th Annual International Conference of the IEEE Engineering in Medicine and Biology Society (EMBC)*, Aug 2016, pp. 876–879.

[5] D. Ahn *et al.*, "Smart gait-aid glasses for parkinson's disease patients," *IEEE Transactions on Biomedical Engineering*, vol. 64, no. 10, pp. 2394–2402, Oct 2017.

[6] M. Bächlin, M. Plotnik, D. Roggen, N. Giladi, J. M. Hausdorff, and G. Tröster, "A wearable system to assist walking of parkinsons disease patients," *Methods Inf Med*, vol. 49, pp. 88–95, 2010.

[7] S. Mazilu *et al.*, "Online detection of freezing of gait with smartphones and machine learning techniques," in *Pervasive Computing Technologies for Healthcare (PervasiveHealth), 2012 6th International Conference on*, May 2012, pp. 123–130.

[8] B. Widrow, P. Baudrenghien, M. Vetterli, and P. Titchener, "Fundamental relations between the lms algorithm and the dft," *IEEE Transactions on Circuits and Systems*, vol. 34, no. 7, pp. 814–820, July 1987.

[9] C. Chen *et al.*, "Ocean: An on-chip incremental-learning enhanced processor with gated recurrent neural network accelerators," in *ESSCIRC 2017 - 43rd IEEE European Solid State Circuits Conference*, Sept 2017, pp. 259–262.

978-1-5386-6414-8/18 $31.00 © 2018 IEEE

A 0.25-27Gb/s Wideband PAM4/NRZ Transceiver with Adaptive Power CDR for 8K System

Yoshihide Komatsu, Akinori Shinmyo, Masami Funabashi, Shuji Kato

Kazuya Hatooka, Kenji Tanaka, Mayuko Fujita, and Kouichi Fukuda

Panasonic Corporation

Kyoto, Japan

Abstract - **A multimodal PAM4/NRZ transceiver, including adaptive ultra-wide range receiver and power noise stabilized transmitter, is proposed. The multimodal receiver features wide-range phase detector to optimize jitter tolerance for each speed while supporting both clock-forward and clock-embedded mode. An adaptive CDR controller selects the optimum CDR mode according to transmitter jitter performance to reduce power consumption while ensuring interoperability. The multimodal transmitter features PAM4 emphasis driver and stabilizer, which are applied for reducing distortion impact and power induced jitter. A test chip was fabricated in 28nm CMOS process and achieved 0.25-27Gb/s wide-range operation, and it achieves 33% power reduction compared to conventional architecture without the adaptive controller, resulting in 3.8pJ/bit energy efficiency for TX and 6.2pJ/bit for RX.**

I. INTRODUCTION

Ultra-high-resolution 8K display solutions revolutionize our life in the field of not only consumer electronics, but also industrial businesses such as broadcasting, medical, digital signage, and so on. In the industrial area, optical fiber cable is suitable for video data transportation due to the necessity for long-distance connection across large workspace. PAM4 has been widely introduced for efficient use of link bandwidth [1-3]. A PAM4 transmitter with standalone PLL has been developed [1] and it achieved a low jitter, because by assembling in one component, jitter may be increased by power noise which is induced by the pre-driver. ADC based multimodal receiver has been also developed [2], however heavy input capacitance is always required which degrades both return loss and insertion loss, and then it requires increasing input amplitude and increasing power consumption. On the other hand, in the consumer electronics, metal cable is mandated in order to satisfy backward compatibility which enables us to connect legacy source devices with very low operating speed.

In accordance with these market requirements, this paper presents a novel multimodal transceivers capable of operating from 0.25G to 27Gb/s while supporting both forwarded-clock and embedded-clock mode for optical/metal hybrid transmission system. As shown in Fig.1, a metal/optical hybrid connector is applied for this system, which enables us to connect both legacy metal cable and optical cable. A new stabilizing circuit allows the transmitter to reduce the power induced jitter and output clean eye pattern for both optical and metal system. In addition, the CTLE based receiver features a reconfigurable phase detector and data picker in order to cover wide range operation. This RX also has automatically offset-calibrated AFE and samplers to achieve high jitter tolerance for PAM4. The transceiver is fabricated in 28nm process, and consumes 102mW (3.8pJ/bit) for TX, and 167mW (6.2pJ/bit) for RX at 27Gb/s.

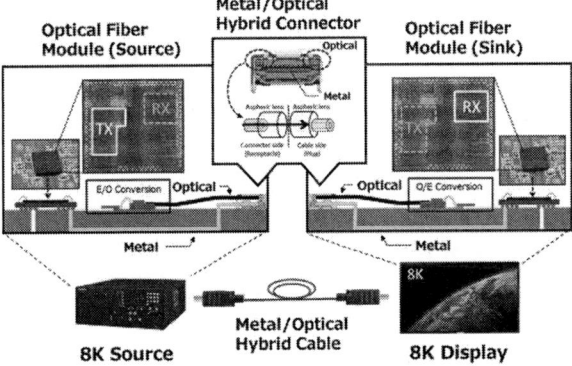

Fig.1: Optical/Metal hybrid transmission system

II. WIDE-RANGE TRANSMITTER WITH STABILIZER

Fig.2 shows the configuration diagram of PAM4/NRZ multi-mode transmitter. This contains three features such as ultra-high speed by PAM4, compatibility with multi-mode, and correction for waveform distortion of the optical signal. For optical transmission, it is assumed to use in PAM4 mode, and activates all serial converters and drivers to output PAM4 signal with the data rate of 18G to 27Gbps. For metal transmission, it is assumed to be used in NRZ mode and only one serial converter and one PAM4 driver is activated to output the NRZ signal. It covers the range of 0.25G to 13.5Gbps and realizes backward compatibility.

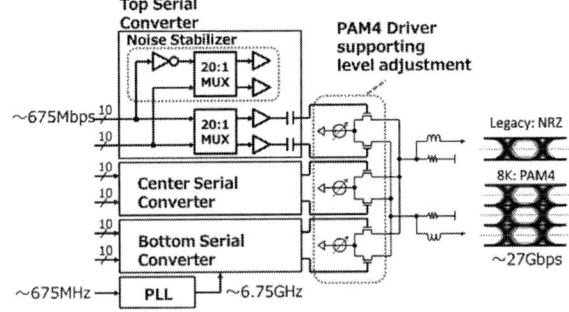

Fig.2: PAM4/NRZ multi-mode Transmitter

To realize the faster transmitter rate, it is necessary to reduce the jitter of serial differential output. The random jitter is caused by a thermal noise inside the PLL and by a flicker noise. On the other hand, the period jitter is almost caused by buffer delay fluctuation induced by a power supply ripple noise.

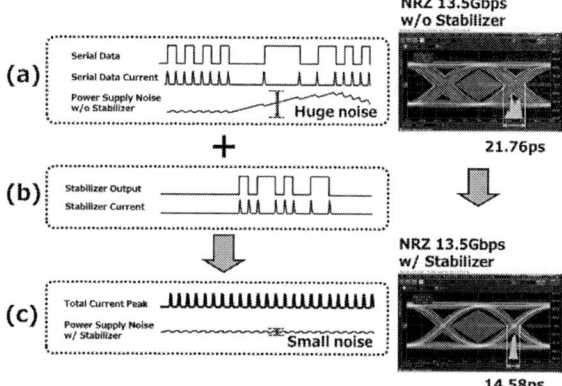

Fig.3: Jitter improvement with noise stabilizer

We propose a noise stabilizer circuit as shown in Fig.2. The jitter caused by power supply noise is successfully reduced with this stabilizer. Fig.3(a) shows the output waveform when the stabilizer is disabled. Since the pattern-dependent spike current causes power supply noise, this noise shifts the buffer latency. And then, the output jitter is increased. The noise stabilizer reduces a period jitter by stabilizing power supply noise. As shown in Fig.3(b), this stabilizer generates a dense pattern at the point where a sparse data pattern exists, and generates a sparse pattern at the point where a dense pattern exists. After adding the spike current of the noise stabilizer to the original current, as shown in Fig.3(c), the spike current density becomes constant regardless of the data pattern. The amount of deterministic period jitter is calculated by

$$DJ[ps] = Z(f) \cdot I(f) \cdot S\left(\frac{t}{v}\right) \quad (1)$$

where "Z" is the VDD impedance, and "I" is instantaneous current as shown in Fig.4, and "S" is the delay sensitivity. According to (1), the jitter is roughly calculated by the difference between 2 components of integrated power profile; one is a dense pattern (the data run length is $1T$) and the other is a sparse pattern (the data run length is nT). This equation is expressed as below.

$$DJ[ps] = \left(\int_{BW}^{\infty} Z(f) \cdot I_{1T}(f)\, df - \int_{BW}^{\infty} Z(f) \cdot I_{nT}(f)\, df \right) \times \left(\frac{\Delta Delay}{\Delta V} \right) \quad (2)$$

Applying stabilizer, the jitter is reduced because it cancels the pattern dependency between $1T$ and nT, then the pattern dependent components are almost canceled. Remained jitter components are only random jitter with window function (3) and duty-cycle distortion (4) as shown in following equations, where "G" is phase noise spectrum.

$$RJ^2{}_{RMS}[ps] = 4 \cdot \int_{BW}^{\infty} G_{\Delta\phi}(f)\, \sin^2\left(\frac{2\pi \cdot f}{2f_o}\right) df \quad (3)$$

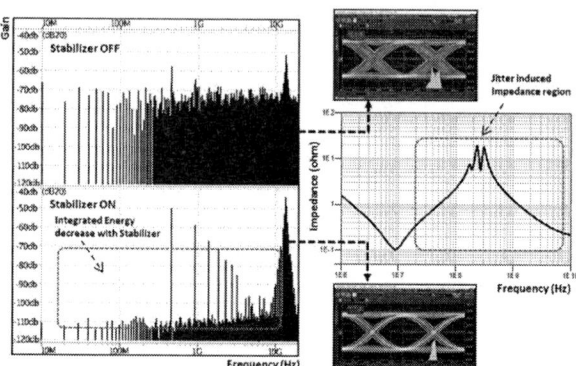

Fig.4: Instantaneous Current FFT with Stabilizer and VDD impedance

$$DCD[ps] = \frac{\delta(t - \Delta t_{rise})}{2} + \frac{\delta(t - \Delta t_{fall})}{2} \quad (4)$$

Therefore, the jitter without noise stabilizer was 21.76ps, while the jitter with noise stabilizer could be reduced to 14.58ps.

III. CURRENT OPTIMIZATION FOR DISTORTION

In the case of transmitting and receiving optical signals, mutual conversion between electric signals and optical signals is indispensable. As shown in Fig.5(a), the mutual conversion is performed with VCSEL (Vertical Cavity Surface Emitting Laser) Driver, VCSEL, PD (Photo Diode), TIA (Trans Impedance Amplifier). In this photoelectric conversion process, although linear transformation is desired, as shown in Fig.5(b), the nonlinearity of each element is a factor and distortion occurs in the waveform. In order to solve this problem, we devised a driver that can adjust the current for each data of Top/Center/Bottom. Fig.5(c) shows the eye evaluation result after optimally adjusting the current value of each driver. Before the current adjustment, Top eye was closing, whereas if the current was adjusted optimally, 78.1mV/37ps eye opening could be achieved.

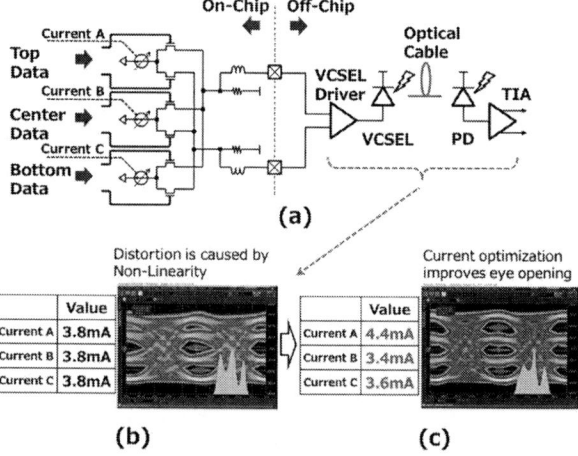

Fig.5: PAM4 eye improvement with current optimization

978-1-5386-6414-8/18 $31.00 © 2018 IEEE

Additionally, Fig.6 illustrates PRBS7 and the proposed 8B5Q encoding for PAM4. Transmission efficiency is not degraded with 8B5Q compare to 8B10B. With this 8B5Q data encoding, the PAM4 data full-swing transition is restricted, then particularly the TOP and BOTTOM eye openings increase more for optional PAM4 waveform control. Eye opening is improved about 33% to 40% for both "time" and "amplitude".

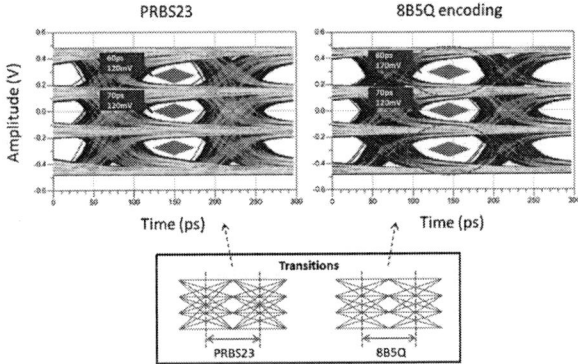

Fig.6: 8B5Q encoding for PAM4

IV. PAM4/NRZ MULTI-MODE RECEIVER

Fig.7 shows the proposed RX with the adaptive CDR. The received signal through an AFE is sampled by samplers, and then input to a DeMUX. Adjusted offset voltage is applied to both the AFE and samplers to distinguish Top/Center/Bottom data. The DeMUX parallelizes the sampled data and outputs signals to subsequent control logic and phase detector (PD). The PD compares the data edge with sampling clock edge, then it judges whether the phase of the PLL clock should be UP or DOWN to control the PLL. The adaptive CDR controller receives parallelized data and controls PLL, PD settings and the number of activated samplers.

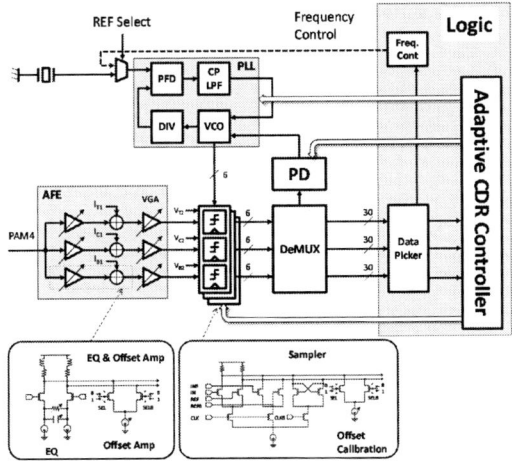

Fig.7: Proposed PAM4-RX Architecture

Our proposed RX supports three selectable CDR modes as shown in Fig.8. In Mode(1), the number of parallelism for data processing is two. This means that the clock frequency is 6.75GHz. UP signal is created by executing XOR of P1 & P2, P4 & P5. DN is created by P0 & P1, P3 & P4. The data sampled by P1, P4 are picked up as main data. In this case, jitter tolerance is maximized since the delay time from the detection of phase difference to proper phase adjustment, 2.3UI, is minimized. However, the power consumption is higher than other modes since the clock frequency is high. On the other hand, in Mode(2), the number of parallelism is three. UP signal is created by executing XOR of P0 & P1, P4 & P5, and DN are created by P1 & P2, P3 & P4. The data sampled by P0, P2, P4 are picked up as main data. In this case, the reduced clock frequency (4.5GHz) reduces power consumption but degrades jitter tolerance. For further improvement of power efficiency, Mode(3) activates the limited number of samplers. In this mode, the PD outputs UP/DN only once every 3bit.

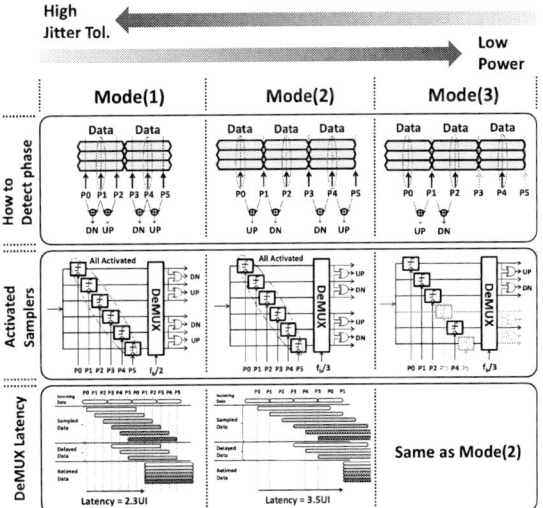

Fig.8: PAM4 Selectable CDR mode

This proposed CDR concept is to reduce power consumption as much as possible under the condition that the jitter tolerance exceeds the amount of TX jitter. When TX jitter amount is huge, RX is required to operate as power-consuming mode since DeMUX feedback latency need to be decreased to enhance jitter tolerance. On the other hand, clean TX jitter case enables RX to operate at low power mode since required jitter tolerance can be relaxed. Optimizing jitter tolerance therefore allows RX to reduce power consumption while ensuring interoperability.

Fig.9(a) shows the relationship between simulated jitter tolerance of each mode and various TX jitters. The adaptive CDR controller switches the mode if TX jitter exceeds the RX jitter tolerance. A sequential flow for mode change is shown in Fig.9(b). The CDR initially starts from Mode(3). Adaptive CDR controller checks whether incoming data is error-free or not. If error detected, the controller changes CDR mode until error-free.

978-1-5386-6414-8/18 $31.00 © 2018 IEEE

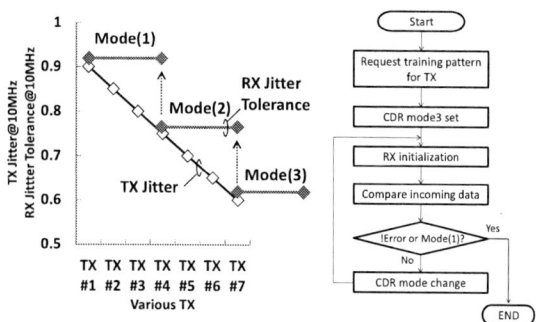

Fig.9 (a): Simulated jitter tolerance@10MHz
(b): Sequential flow for mode change

Fig.10 shows the evaluation results of the multi-mode RX. Jitter tolerance was measured for both NRZ (13.5Gbps) and PAM4 (27Gbps). Jitter tolerance target is 0.2UI for both NRZ and PAM4. As shown in Fig.10, it was confirmed that the target tolerance was satisfied with proposed phase detector in MODE(1) through MODE(3).

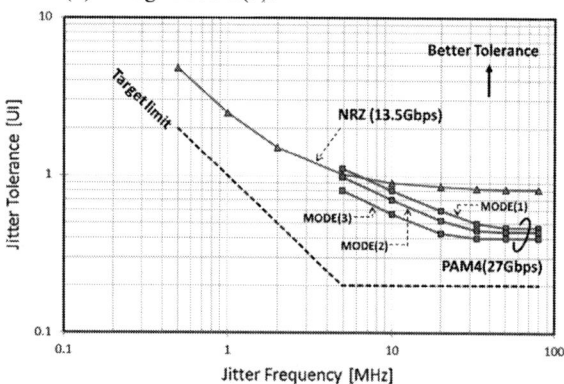

Fig. 10: Jitter Tolerance Measurement Result

Adaptive Power consumption is shown in Fig.11, Mode(1) shows the highest jitter tolerance as 0.91UI@10MHz, and consumes 250mW. On the other hand, Mode(2) and Mode(3) with adaptive CDR controller reduces power by 19%, and 33% compared to the conventional architecture, respectively.

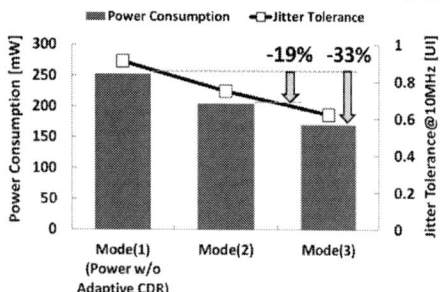

Fig.11: Adaptive Power Measurement results

To evaluate this hybrid transceiver, we produced a test chip containing PAM4/NRZ TX in the 0.66mm^2 area and RX in the 0.52mm^2 area as shown in Fig.12, and we measured the characteristics of each PAM4/NRZ mode.

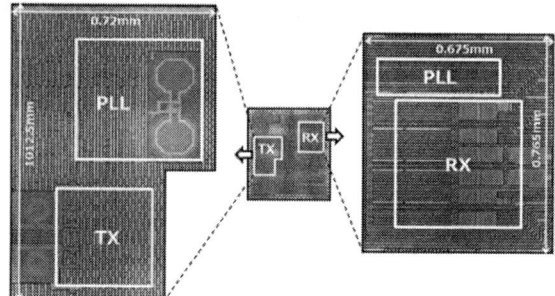

Fig. 12: CHIP micrograph

Table.1 shows the comparison result with conventional results. This proposed RX is suitable for production due to its wide operating range and small power and area.

TABLE I
Performance summary and Comparison

Design	TX [1] ISSCC2017	TX This work	RX [2] ISSCC2016	RX This work
Technology	40nm	28nm	28nm	28nm
Data Format	NRZ/PAM4	NRZ/PAM4	NRZ/PAM4	NRZ/PAM4
Data rate [Gb/s]	56	0.25-27	32	0.25-27
Power [mW]	220	102	320	167
Power Efficiency [pJ/bit]	3.93	3.78	10.00	6.19
Area [mm²]	1.39	0.66	0.89	0.52

V. CONCLUSION

This paper described a wideband PAM4/NRZ Transceiver based on the stabilizer transmitter with distortion adjustments and wide-range phase detector adjustments according to operating data rate with adaptive CDR control. In addition, it reached an industrial top class speed improvement using PAM4/NRZ transmission; it could reduce a distortion impact of optical transmission and ensure backward compatibility by wide range operation. We investigated the power induced jitter theory and we proved to reduce the jitter amount with the stabilizer function. This proposed transceiver consumes average energy efficiency about 3.8pJ/bit for TX and 6.2pJ/bit for RX at PAM4 mode.

REFERENCES

[1] P.Peng, J.Li, L.Chen, J.Lee "A 56Gb/s PAM-4/NRZ Transceiver in 40nm CMOS" in ISSCC, pp.110-112, 2017
[2] D.Cui, H.Zhang, N.Huang, A.Nazemi, B.Catli, H.G.Rhew, B.Zhang, A.Momtaz, J.Cao "A 320mW 32Gb/s 8b ADC-Based PAM-4 Analog Front-End with Programmable Gain Control and Analog Peaking in 28nm CMOS" in ISSCC, pp.58-59, 2016
[3] Aurangozeb, A.D.Hossain, M.Hossain "Channel Adaptive ADC and TDC for 28 Gb/s PAM-4 Digital receiver" in CICC, 2017
[4] T.Shibasaki, T.Danjo, Y.Ogata, Y.Sakai, H.Miyaoka, F.Terasawa, M.Kudo, H.Kano, A.Matsuda, S.Kawai, T.Arai, H.Higashi, N.Naka, H. Yamaguchi, T.Mori, Y.Koyanagi, H.Tamura "A 56Gb/s NRZ-Electrical 247mW/lane Serial-Link Transceiver in 28nm CMOS" in ISSCC, pp.64-66, 2016
[5] T.J.Yamaguchi, M.Soma, D.Halter, R.Raina, J.Nissen, M.Ishida "A method for measuring the cycle-to-cycle period jitter of high-frequency clock signals" in VLSI Test Symposium, pp.102-110, 2001

6-2 (8221)

A Fully-Integrated 25Gb/s Low-Noise TIA+CDR Optical Receiver designed in 40nm-CMOS

Juncheng Wang[1], Xuefeng Chen[2], Shang Hu[1], Yaxin Cai[2], Rui Bai[2], Xin Wang[2], Yuanxi Zhang[2], Shenglong Zhuo[1], Liu Chang[1], Bozhi[1] Yin, Jianxu Ma[2], Hao Yan[2], Jiangao Xuan[2], Milton Lu[2], Tao Xia[2], Qi Nan[3], and Patrick Yin Chiang[1,2]

Email: pchiang@fudan.edu.cn
[1]Fudan University, Shanghai, China
[2]PhotonIC Technology, Shanghai, China
[3]Chinese Academy of Sciences, Beijing, China

Abstract—A fully-integrated 25Gbps low-noise optical receiver is presented that integrates a Transimpedance Amplifier (TIA), Continuous-Time Linear Equalizer (CTLE), high gain and high bandwidth Limiting Amplifier (LA), and Clock and Data-Recovery (CDR) into a single die. The TIA employs an inverter-based pseudo differential TIA with Cross-Coupled Negative Gm pair and a Negative capacitor to increase the signal bandwidth, while a MOSFET Corner Compensation (MCC) circuit compensates for the CMOS corner variation. An Automatic Gain Control (AGC) scheme is proposed that solves the group delay issue caused by TIA input impedance variation from small input to overload current. Finally, a 2x-oversampling CDR using a bang-bang phase detector is included. The 850nm VCSEL-based full-link measurement results show that the optical receiver achieves a sensitivity (BER<1e-12) of 52uApp (RSSI Current=51uA, ER=4.89dB) with a 150fF Photodiode from the 3.3V and 1.3V supplies, respectively.

Keywords—low noise, optical receiver, corner compensation, auto gain control, CDR

I. INTRODUCTION

As electrical I/Os approach the bandwidth limit due to severe loss at high-frequency, optical fiber communication becomes a promising technology for energy-efficient, high-bandwidth and low-cost interconnects. The possibility of an all-in-CMOS link is especially attractive compared with a SiGe-based optical transceiver as it is compatible with the low-cost standard CMOS technology and enables integration with large-scale digital signal-processing. For the optical-electrical (O/E) conversion, a low-noise CMOS-based optical receiver front-end is limited by both the bandwidth and intrinsic noise. Recent CMOS-based optical receivers have been proposed [1-3], but have not integrated the CDR with the TIA. This work presents an all-CMOS optical receiver that integrates both the TIA+CDR into a single CMOS RX.

II. CIRCUIT DESCRIPTION

The low noise optical receiver is shown in Fig. 1. The receiver mainly has two parts: analog front-end and clock and data recovery. For the normal 25Gb/s operating mode, the front end includes a low-bandwidth TIA with a CTLE to compensate

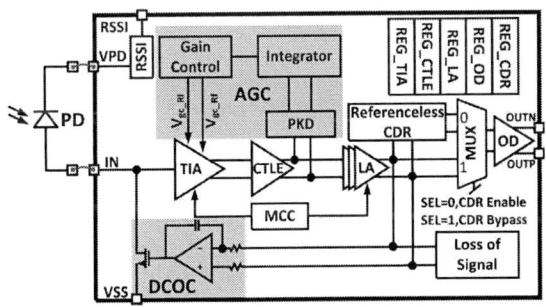

Fig. 1. Optical Receiver system block diagram.

for the limited bandwidth, a high gain (>40dB) and high bandwidth (>30GHz) LA that amplifies the current to meet the CDR sensitivity requirement (BER<1e-12). An adaptive loop such as a DCOC is needed to insure the signal path circuits have suitable DC operating point. The Low Dropouts (LDOs) provide a stable power supply and filter out noise of the signal path's supply.

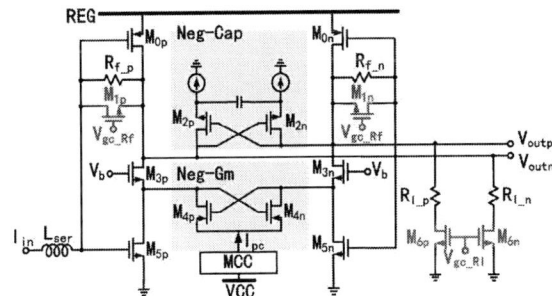

Fig. 2. Proposed inverter-based TIA schematic

A. Signal Path

A low-bandwidth inverter-based TIA (Fig. 2), is used to convert the optical current into voltage and a CTLE [1] to increase the TIA bandwidth from 6.3GHz to 17GHz. The TIA

978-1-5386-6414-8/18 $31.00 © 2018 IEEE

67

employs a negative capacitor and negative-gm pair to boost the bandwidth, such that TIA output bandwidth is improved efficiently across different negative-gm setting. The negative-gm compensation current comes from the proposed MOSFET Corner Compensation (MCC), compensating for the corner variation. The TIA+CTLE group delay and front-end gain variation is markedly reduced after corner compensation is enabled. The LA, which has 41dB gain and 30GHz bandwidth, amplifies the CTLE output to saturation for a total front-end gain of 64K-ohms.

A DC Offset Cancellation (DCOC) loop is also integrated that can sink the common-mode current of the TIA input, insuring that the signal path has a suitable DC operation point across a large dynamic range of optical input power.

B. Automatic Gain Control (AGC)

An auto-gain control loop, which includes a peak-detector (PKD), an integrator and a gain control, is used to adaptively control the TIA gain across different input currents, such that the reliability and overload issue can be avoided.

Another issue for the AGC is the TIA gain control scheme. The TIA input LRC network is sensitive to TIA input impedance which may cause ISI and bit errors when the TIA operates at large input current. In order to keep the TIA input impedance stable, M_{1p}, M_{1n} and M_{6p}, M_{6n} are adopted to adjust the feedback resistor R_{f_p} and inverter gain simultaneously.

C. Clock and Data Recovery Unit

A 25Gb/s clock & data-recovery (CDR) block is also integrated, which retimes the data and recovers the clock-phase. A digital referenceless-CDR algorithm extracts the frequency from the input signal without requiring a separate frequency detector, and is continuously robust from 24.3Gb/s to 28.1Gb/s with no deadbands or false-locking. After data-recovery, a high-speed 2:1 serializer sends the 25Gb/s data to a two-tap pre-emphasis output driver and equalizer.

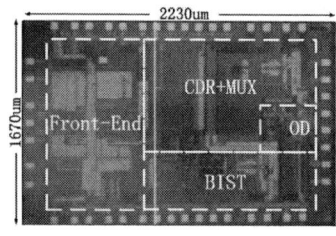
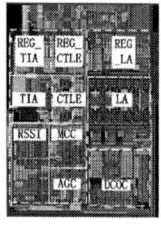

Fig. 3. (a) RX Chip photo (b) Front-End layout.

III. MEASUREMENT RESUTLS

The proposed optical receiver was fabricated in a 40nm CMOS process (Fig. 3a). The fabricated RX front-end (Fig. 3b) and integrated CDR was connected to a variable optical source which is consists of a Variable Optical Attenuator (VOA) and a CMOS-based 25Gbps VCSEL Transmitter. The PRBS31 data is generated from a ML-4009 BERT. The sensitivity is measured across different negative-gm setting as seen in Fig. 4a. The

25Gb/s full link measurement results show that the optical receiver achieves a sensitivity (BER<1e-12) of 52uA$_{pp}$ (RSSI Current=51uA, ER=4.89dB) using a 150fF Photodiode,

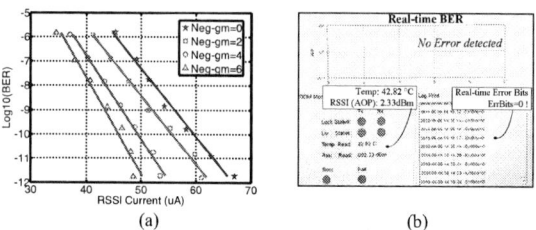

Fig. 4. Measurement results: (a) BER at different negative-gm setting, (b) BER test at I_{OMA}=0.85mA$_{pp}$

dissipating 0.62mA and 194mA (TIA+LA: 70mA, CDR: 100mA, MUX+OD: 24mA) current from 3.3V and 1.3V supplies, respectively. The RSSI is unity current gain from 5uA to 2.4mA. Fig. 4b show the AGC measurements results. The TIA+CDR RX can operate without bit error at nearly 0.9mA$_{pp}$ overload current.

TABLE I. PERFORMANCE SUMMARY

	[1]	[2]	[4]	This work
Technology	65nm CMOS	28nm CMOS	130nm BiCMOS	40nm CMOS
Architecture	TIA+LA	TIA+LA	TIA/LA+CDR	TIA/LA+CDR
Data Rate(Gb/s)	25	25	25.78	25.78
PRBS	2^{31}-1	2^{15}-1	2^{31}-1	2^{31}-1
AGC	N/A	N/A	N/A	Yes
Input referred noise current (uA$_{rms}$)	1.8	N/A	N/A	2.41
Sensitivity(uA$_{pp}$)	59	25*	63**	52
Photo diode capacitance (fF)	80	<25	N/A	150
Supply(V)	1/1.8	N/A	2.5/3.3	1.3/3.3
Power consumption (mW)	93	4.3	290	254

* Emulated photodiode test
** Calculated based on photodiode responsibility = 0.5A/W.

REFERENCES

[1] D. Li, G. Minoia, M. Repossi, D. Baldi, E. Temporiti, A. Mazzanti, and F. Svelto, "A Low-Noise Design Technique for High-Speed CMOS Optical Receivers," *IEEE J. Solid-State Circuits*, vol. 49, no. 6, pp 1437-1447, June 2014.

[2] S. Saeedi, S. Menezo, G. Pares, and A. Emami, "A 25 Gb/s 3D-integrated CMOS silicon-photonic receiver for low-power highsensitivity optical communication," *J. Lightw. Technol.*, vol. 34, no. 12,pp. 2924–2933, Jun. 15, 2016.

[3] A.Cevrero, et. al., "A 64 Gb/s 1.4 pJ/b NRZ Optical-Receiver Data-Path in 14 nm CMOS FinFET", *IEEE ISSCC Dig. Tech. Papers*, pp. 482-483,Feb. 2017.

[4] Takayuki Shibasaki,et.al."4×25.78Gb/s Retimer ICs for Optical Links in 0.13μm SiGe BiCMOS" *IEEE ISSCC Dig. Tech. Papers*, pp. 412-413, Feb. 2015.

A Low Input Referred Noise and Low Crosstalk Noise 25 Gb/s Transimpedance Amplifier with Inductor-Less Bandwidth Compensation

Akitaka Hiratsuka[1], Akira Tsuchiya[2], Kenji Tanaka[3], Hiroyuki Fukuyama[3],
Naoki Miura[3], Hideyuki Nosaka[3], Hidetoshi Onodera[1]

[1]Dept. Communications and Computer Engineering, Kyoto University, Kyoto, Japan
[2]Dept. Electronic Systems Engineering, The University of Shiga Prefecture, Hikone, Shiga, Japan
[3]NTT Device Technology Laboratories, NTT, Atsugi, Kanagawa, Japan
tsuchiya.a@e.usp.ac.jp

Abstract—This paper presents a low-noise and high-speed transimpedance amplifier (TIA) for optical interconnection. For high density parallel integration of optical receiver, small area and crosstalk mitigation are important as well as high speed, low power and so on. We propose an inverter TIA (INV-TIA) with inductor-less bandwidth compensation circuits and a passive crosstalk filter. Since these additional circuits are composed with resistance and capacitance, the area overhead and the power overhead are much smaller than existing techniques. We fabricated a 25 Gb/s TIA in a 65-nm CMOS. Compared to the reference design, the proposed circuit reduces the input referred noise by 60% and the crosstalk noise by 25%.

Index Terms—optical communication, transimpedance amplifier, input referred noise, bandwidth compensation, crosstalk

I. INTRODUCTION

The optical communication is a key technology for high-capacity data transfer. The application is not only long-distance communication, but also short-distance called optical interconnection. Transimpedance amplifier (TIA) needs both high-speed operation and low noise because it is the first block of the optical receiver. For higher data capacity, high density parallel integration of optical receivers is emerging. Thus, small area and crosstalk mitigation are becoming important as well as high speed, low power and low noise.

In TIA design, there is a trade-off between the bandwidth and the input referred noise current. To achieve both high-speed operation and low input referred noise, bandwidth enhancement techniques are necessary. Inductive peaking is a widely used bandwidth enhancement. However the area of on-chip inductor is much larger compared to other components, MOS FET, resistor and capacitor. Especially in multi-stage TIA, inserting peaking inductor to each stage consumes huge area and the area overhead might not be acceptable.

Considering interference among neighboring optical receivers, crosstalk noise via the power/ground line is a problem. When multiple receivers share the power/ground line, di/dt

This work was supported by JSPS KAKENHI Grant Number 16K18092, and VLSI Design and Education Center (VDEC), The University of Tokyo with the collaboration with Synopsys, Cadence, Mentor Graphics and Keysight Technologies.

noise degrades signal integrity. Ref. [1] proposed a techniques to cancel the power/ground noise. However adding active noise canceler consumes extra power, then the power efficiency becomes worse.

This paper proposes a low input referred noise and low crosstalk-noise sensitivity 25 Gb/s TIA. The basic topology is 3-stage inverter-TIA (INV-TIA) and the target transimpedance gain is 50 dBΩ. For low input referred noise current, bandwidth compensation circuits are employed. One is a capacitive degeneration to compensate the frequency characteristic, another is a negative capacitance to enhance the bandwidth of the first stage. For crosstalk mitigation, an RC low-pass filter in inserted to the power line of the first stage. These three additional circuits are constructed by only resistance and capacitance, thus the power overhead is almost zero and the area overhead is much smaller than on-chip inductor. The proposed TIA was fabricated in a 65-nm CMOS. The measurement results verify 50 dBΩ and 25 Gb/s operation and show that the input referred noise current is 18.1 pA/$\sqrt{\text{Hz}}$ in 12.6 GHz −3-dB bandwidth. Compared to the reference design, the input referred noise current is 60% smaller. The crosstalk noise from a neighboring TIA is also reduced by 25%. The area is 68 μm × 94 μm (0.0064 mm^2) and the power consumption is 3.96 mW.

II. TARGET SYSTEM

Fig. 1 shows the target system. A number of TIAs are integrated into one chip and they share the power/ground line. The di/dt noise due to the inductance of the bonding wires causes interference among TIAs. In this paper, we write this interference via the power/ground line as "crosstalk noise". To improve the power integrity, decoupling capacitor is a common practice. However for high density implementation, the area for each channel should be small. The challenge of this work is to develop a small-area and low-power noise reduction without sacrificing the gain and the bandwidth. The target specification of TIA is 50 dBΩ transimpedance gain and 25 Gb/s/ch operation speed. The TIA topology is 3-stage INV-TIA.

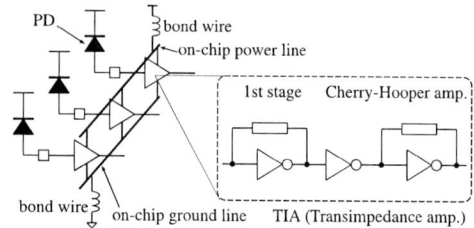

Fig. 1. Target system and the 3-stage INV-TIA under discussion.

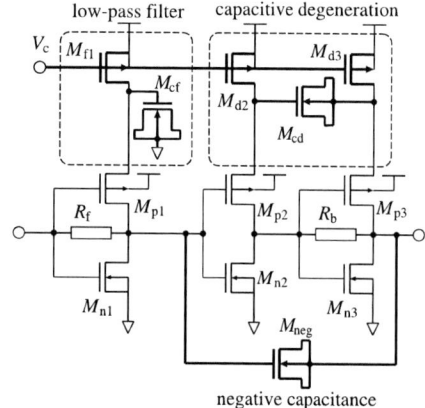

Fig. 2. Schematic of the proposed circuit.

III. LOW-NOISE DESIGN AND BANDWIDTH COMPENSATION

Here we explain the key ideas of the proposed circuit. Fig. 2 shows the circuit schematic of the proposed TIA. The added components are shown in thick line.

A. Input Referred Noise Reduction and Bandwidth Compensation

In TIA design, the input referred noise current is approximately in inverse proportional to the feedback resistor R_f. On the other hand, higher feedback resistance degrades the bandwidth. We employ two bandwidth compensation techniques to achieve both low input referred noise and high bandwidth. Fig. 3 shows the design concept. Larger feedback resistor in the first stage makes the gain higher, the bandwidth lower, and the input referred noise smaller. Then the 2nd and 3rd stages (Cherry-Hooper amplifier) are designed to have a peak in the gain curve to compensate the bandwidth degradation in the first stage.

Capacitive degeneration is applied to the second and the third stage of the TIA. The MOS FETs M_{d2} and M_{d3} are resistors and they compose source degeneration circuit. The MOS FET M_{cd} is a capacitor to bypass the resistors M_{d2} and M_{d3} in high frequency. This configuration suppresses the gain in lower frequency, and produces a peak in the gain curve. As shown in Fig. 3, this circuit compensates the characteristics of the first stage.

Negative capacitance is used to enhance the bandwidth of the first stage. As shown in Fig. 2, a MOS capacitor M_{neg} is

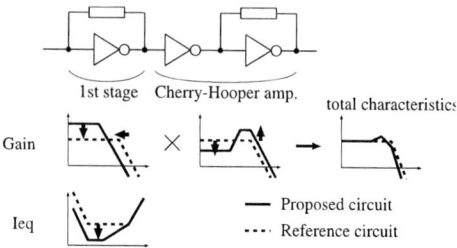

Fig. 3. Design concept for low input referred noise.

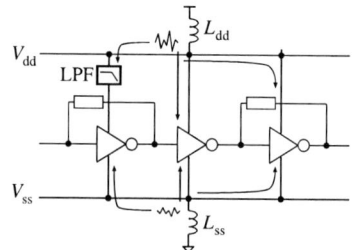

Fig. 4. Design concept for low crosstalk noise.

connected between the output of the first stage and that of the third stage. Since the set of the second and the third stages (Cherry-Hooper amplifier) is a non-inverting amplifier, the effective capacitance at the output of the first stage becomes negative by Miller effect. The negative capacitance reduces the load of the first stage and it results bandwidth enhancement of the first stage. On the other hand, for the third stage, the negative capacitance is an additional load. Thus, the capacitance value should be chosen carefully not to degrade the bandwidth of the whole circuit.

B. Low Crosstalk Noise Design

To mitigate the crosstalk noise from the power/ground line, adding a replica circuit to cancel di/dt on the power/ground line [1] has been proposed. However active noise canceler needs extra power. For area efficiency and power efficiency, we use only one low-pass filter as shown in Fig. 2. Fig. 4 shows the design concept. We assume the number of bonding wires for power supply is less than that for ground. Then the inductance L_{dd} is larger than L_{ss}, and di/dt noise in the power line is larger. Obviously, the most noise-sensitive part in TIA is the first stage, thus the low-pass filter is inserted to the power supply of the first stage of the TIA.

IV. MEASUREMENT RESULTS

This section describes the design details and the measurement results.

A. Design Detail

To reveal the effect of the proposed design, we fabricated a reference TIA and the proposed TIA. The reference TIA is a basic 3-stage INV-TIA to achieve 50 dBΩ and 25 Gb/s. The proposed TIA is the circuit shown in Fig. 2. The design parameters of each circuits are shown in Table. I. The reference

TABLE I
Design parameters.

Channel width	M_{p1}	M_{n1}	M_{p2}	M_{n2}	M_{p3}	M_{n3}
Reference	$10\mu m$	$4\mu m$	$16\mu m$	$8\mu m$	$16\mu m$	$12\mu m$
Proposed	$40\mu m$	$6\mu m$	$28\mu m$	$20\mu m$	$32\mu m$	$14\mu m$

Channel width	M_{f1}	M_{d2}	M_{d3}	M_{cf}	M_{cd}	M_{neg}
Reference	—	—	—	—	—	—
Proposed	$40\mu m$	$80\mu m$	$80\mu m$	$960\mu m$	$576\mu m$	$20\mu m$

Resistance	R_f	R_b
Reference	278 Ω	1112 Ω
Proposed	968 Ω	242 Ω

(Channel length of all MOS FETs: 60 nm)

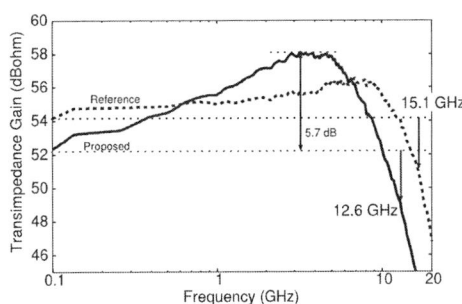

Fig. 6. Measured transimpedance gains.

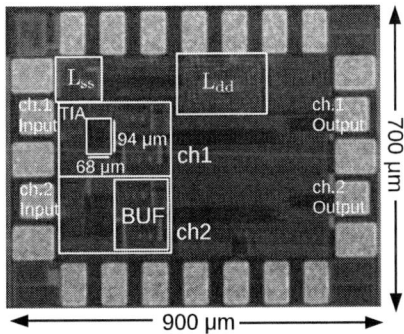

Fig. 5. Chip microphoto of two channels of TIA and an output buffer.

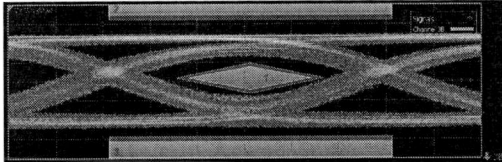

Fig. 7. Eye-diagram of the reference TIA. (49.2 mV/div, 6.47 ps/div)

circuit uses relatively low feedback resistance R_f to obtain higher bandwidth. On the other hand, the R_f of the proposed circuit is 968 Ω to reduce the input referred noise. The resistors R_f and R_b are realized by poly resistors. Fig. 5 shows the chip-microphoto fabricated in a 65-nm CMOS with 1.2 V supply voltage. The proposed TIA occupies 68 μm × 94 μm area and followed by a 50-Ω output buffer. To evaluate the crosstalk noise, two channels are integrated and share the power/ground line. Since the measurement is done by on-wafer measurement, we add inductors L_{dd} and L_{ss} to mimic the power/ground noise due to bonding wires. We assume that the number of bond wire for ground is more than that for V_{dd}, then we set the inductance L_{dd} to 0.68 nH, and L_{ss} to 0.14 nH.

B. Transimpedance Gain

Fig. 6 shows the measured transimpedance gain of the reference TIA and the proposed TIA. The transimpedance gain is calculated from s-parameter measurement. The gain of the reference TIA is 55.0 dBΩ at low frequency, and its −3-dB bandwidth is 15.1 GHz. The low frequency gain of the proposed TIA is 52.3 dBΩ, and the −3-dB bandwidth is 12.6 GHz.

C. Eye-diagram

Figs. 7 and 8 are the measured eye-diagram of the reference TIA and the proposed TIA, respectively. The evaluation is done by electrical measurement with $2^{31} - 1$ PRBS input signal. The overlaid eye-mask is for 100GBASE-SR4 Rx SEC. Table II describes the performance of the eye-diagrams. These

results verify that the proposed TIA can achieves 25 Gb/s operation as well as the reference TIA with bandwidth-aware configuration.

D. Input Referred Noise

Fig. 9 shows the input referred noise. The noise is calculated from the noise spectrum measured by a spectrum analyzer and the measured transimpedance gain in Fig. 6. As shown in Fig. 9, the input referred noise current is reduced because the larger feedback resistance of the first stage. The input referred noise current of the proposed TIA is 18.1 pA/$\sqrt{Hz}$ and it is 60% smaller than that of the reference TIA, 44.3 pA/$\sqrt{Hz}$.

E. Crosstalk Noise

To evaluate the crosstalk noise, we input 25 Gb/s, 200 μA peak-to-peak PRBS to the input of channel 1 and observe the output of channel 2. The input of channel 2 is quiet (continuous 0). The peak-to-peak voltage of channel 2 output corresponds the noise voltage induced by the channel 1. The

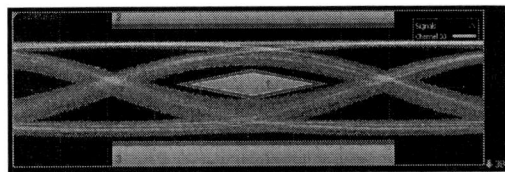

Fig. 8. Eye-diagram of the proposed TIA. (43.2 mV/div, 6.46 ps/div)

TABLE II
Summary of eye-diagram measurement.

	V_{swing}	eye height	eye width	Mask margin
Conv.	180 mV	87.8 mV	26.8 ps	17.0%
Proposed	153 mV	62.1 mV	25.2 ps	18.0%

V_{swing} : voltage swing between zero level and one level.

TABLE III
PERFORMANCE SUMMARY AND COMPARISON.

	Proposed	Reference	[2]	[3]	[4]	[5]	[1]
Technology	65 nm	65 nm	65 nm	65 nm	80 nm	65 nm	65 nm
Z_T [dBΩ]	52.3	55.0	42.0	41.0	74.0	83.0	76.8
Bit-rate [Gb/s]	25	25	30	25	20	25	28
−3dB BW [GHz]	12.6	15.1	21.0	21.0	9.3	16.1	21.0
Power [mW]	3.96	3.95	5.0	4.8	13.8	41	43.5
EPB [pJ/bit]	0.158	0.158	0.167	0.192	0.69	1.61	1.56
I_{eq} [pA/√Hz]	18.1	44.3	15.8	—	—	7.2	17.8
Area [mm²]	0.0064	0.0020	—	—	0.0080	0.35	—

Z_T : transimpedance gain, EPB : Energy per Bit, I_{eq} : input referred noise current spectral density

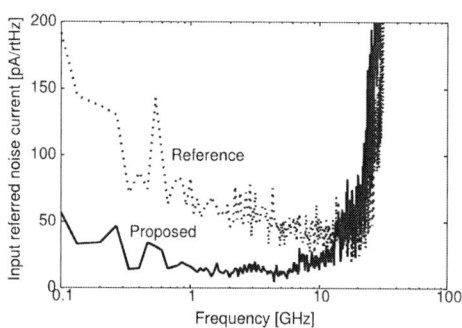

Fig. 9. Measured input referred noise current spectral density.

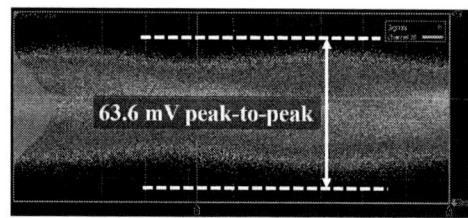

Fig. 10. Measured crosstalk noise of the reference TIA. (10.4 mV/div, 6.67 ps/div)

measured waveforms are shown in Figs. 10 and 11. The peak-to-peak voltage of the reference TIA is 63.6 mV and that of the proposed circuit is 48.0 mV. The proposed circuit reduces the crosstalk noise by 25%.

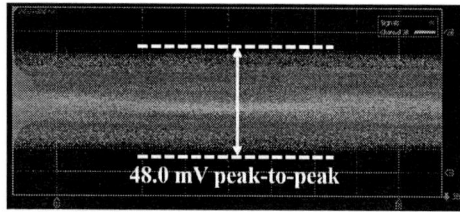

Fig. 11. Measured crosstalk noise of the proposed TIA. (10.4 mV/div, 4.76 ps/div)

F. Performance Summary

Table III shows the performance summary and comparison with published works. Compared to the reference design, the proposed TIA reduces the input referred noise current by 60%. Since the proposed circuit does not use active circuit, the power overhead is zero and the power consumption is 3.96 mW. The proposed TIA realizes the highest energy efficiency (the lowest energy per bit) though it has a crosstalk mitigation circuit. The additional circuit needs some area overhead and the proposed TIA is larger than the reference TIA, however the proposed TIA is the smallest compared to other works. From discussion above, the proposed design realizes area-efficient and energy efficient noise reduction for both the input referred noise and the crosstalk noise.

V. CONCLUSION

This paper proposes a low input referred noise and low crosstalk noise design of multi-stage INV-TIA. We employ a capacitive degeneration and a negative capacitance for bandwidth compensation, and a low-pass filter for crosstalk noise mitigation. Realizing all additional circuits by resistors and capacitors, the proposed technique suppresses the area-overhead and the power-overhead. The proposed circuit reduce the input referred noise by 60% and the crosstalk noise by 25% without power overhead. The inductor-less bandwidth enhancement realizes 25 Gb/s operation and 0.0064 mm² area. This technique contributes to realize high density integration of optical receiver analog front-end.

REFERENCES

[1] T. Takemoto, H. Yamashita, T. Yazaki, N. Chujo, Y. Lee, and Y. Matsuoka, "A 25-to-28 Gb/s high-sensitivity (−9.7 dBm) 65nm CMOS optical receiver for board-to-board interconnects," *IEEE Journal of Solid-State Circuits*, vol. 49, no. 10, pp. 2259–2276, 2014.

[2] Q. Pan, Y. Wang, Z. Hou, L. Sun, Y. Lu, W.-H. Ki, P. Chiang, and C. P. Yue, "A 30-Gb/s 1.37-pJ/b CMOS receiver for optical interconnects," *IEEE Journal of Solid-State Circuits*, vol. 33, no. 4, pp. 778–786, Feb. 2015.

[3] J.-Y. Jiang, P.-C. Chiang, H.-W. Hung. C.-L. Lin, T. Yoon, and J. Lee, "100Gb/s ethernet chipsets in 65nm CMOS technology," in *International Solid-State Circuits Conference*, 2013, pp. 120–122.

[4] L. Szilagyi, R. Henker, and F. Ellinger, "20 Gbit/s ultra-compact optical receiver front-end with variable gain transimpedance amplifier in 80 nm CMOS," in *IEEE MTT-S Latin Ameria Microwave Conference*, 2016.

[5] D. Li, G. Minoia, M. Repossi, D. Baldi, A. Ghilioni, E. Temporiti, and F. Svelto, "A 25Gb/s 3D-integrated silicon photonics receiver in 65nm CMOS and PIC25G for 100GbE optical links," in *International Symposium on Circuits and Systems*, 2016, pp. 2334–2337.

6-4 (8012)

A 10-Gb/s, 0.03-mm², 1.28-pJ/bit Half-Rate All-Digital Injection-Locked Clock and Data Recovery with Maximum Timing-Margin Tracking Loop

Min-Seong Choo, Han-Gon Ko, Sung-Yong Cho, Kwangho Lee, and Deog-Kyoon Jeong

Department of Electrical and Computer Engineering and Inter-University Semiconductor Research Center
Seoul National University, Seoul, Korea
Email: mschoo@isdl.snu.ac.kr, dkjeong@snu.ac.kr

Abstract— A 10-Gb/s, 0.03-mm², 1.28-pJ/bit half-rate all-digital injection-locked clock and data recovery (ILCDR) with a path mismatch tracking (PMT) loop is presented. When injection timing is not perfectly matched with the local oscillator, the timing margin of the data sampler is reduced, resulting in the degradation of jitter tolerance (JTOL) performance. By simply de-serializing the error information from the phase detector in the conventional phase-locked loop (PLL) based CDR with respect to the polarity of the data transition, the proposed ILCDR achieves robust injection behavior over path mismatch variations. Fabricated in 28-nm CMOS technology, the proposed ILCDR occupies 0.03 mm² and consumes 12.8 mW at 10 Gb/s with a 0.9-V supply voltage. The measured JTOL is 1 UIpp at 31 MHz with the target bit error rate of 10^{-12} in the presence of the initial path delay mismatch.

Keywords—Clock and data recovery (CDR); injection-locked oscillator (ILO); injection-locked CDR (ILCDR); path mismatch tracking (PMT); jitter tolerance (JTOL); half rate

I. INTRODUCTION

Recently, an injection-locked clock and data recovery circuit (ILCDR) has widely been adopted in wireline receivers since it shows excellent jitter tolerance (JTOL) performance and low power consumption with a minimal hardware [1]–[8]. Typically, to achieve a superior JTOL performance in the conventional phase-locked loop (PLL) or phase interpolator (PI) based CDRs, large power consumption should be required for a better phase noise of the oscillator and the higher sensitivity of the sampler. However, an injection-locked oscillator (ILO) have a large bandwidth because it directly forwards the transition of data to the local oscillator. In other words, an ILO tracks the phase of the input data stream rapidly, which results in higher JTOL performance. In spite of its superior JTOL performance, the design of ILCDR has a critical issue to be resolved, which is a frequency offset between the input data stream and the local oscillator. When the offset is not completely eliminated, the timing margin of the receiver is reduced significantly. Thus, the reduced timing margin causes a bit error when consecutive identical digits show up. Besides, if the offset exceeds the locking range of an ILO, it could fail to lock. For these reasons, to employ an ILO in CDR applications, a timing calibrator that cancels out the offset is necessary for robust operation over the process, supply voltage, temperature (PVT) variations.

In recently published papers, some efforts to calibrate the frequency offset have been made [2]–[8]. In [2]–[5], a replica oscillator embedded in a PLL forwards the control voltage to an ILO. Thus, the free-running frequency of an ILO, f_{osc}, is set within the locking range of an ILO and the phase is aligned instantly on the input data stream. However, this architecture is vulnerable to the PVT variations due to their structural limitation of replica scheme. In addition, the frequency difference between the input data stream and the external reference clock also causes the frequency mismatches between the two oscillators. Even if these mismatches are small, their impact on performance is critical, resulting in the degradation of the timing margin. To mitigate such performance degradation, an ILCDR using a single oscillator is presented [6]. However, the mismatches are still present as f_{osc} is set from the reference clock, not from the input data stream. The phase locking in [7] and [8] is performed without a reference clock and adopts the conventional PLL-based architecture, which has two-phase tracking paths. One is the direct injection to the oscillator, and the other is through the phase detector. However, when the two path delays are not properly calibrated, the frequency offset is inevitably occurred, degrading JTOL performance.

This work presents new tracking loop and phase detector as the solution of the inherent two-phase modulation in PLL-based ILCDR. Through the proposed tracking loop that continuously tracks the input data stream, great JTOL performance is achieved. Furthermore, by embedding a delay element which is controlled by the proposed phase detector before injecting the input data stream, it is possible to track and compensate the delay mismatch between the two paths, guaranteeing maximum timing margin of the sampler over PVT variations.

II. PROPOSED ARCHITECTURE AND IMPLEMENTATION

A. Proposed ILCDR and ILDCO architecture

Fig. 1 shows the proposed ILCDR architecture. The output clock is modulated by the two paths as mentioned above. When the sum of the two path delays, $\tau_{inj} + \tau_{clk}$, equals to the integer multiples of a bit time, the free-running frequency of an ILO (f_{osc}) is equal to the target frequency, f_{target}. In this work, f_{target} is 5 GHz since the input data rate is 10 Gb/s, and half-rate clock recovery scheme is adopted. The injection-path delay, τ_{inj}, is

978-1-5386-6414-8/18 $31.00 © 2018 IEEE

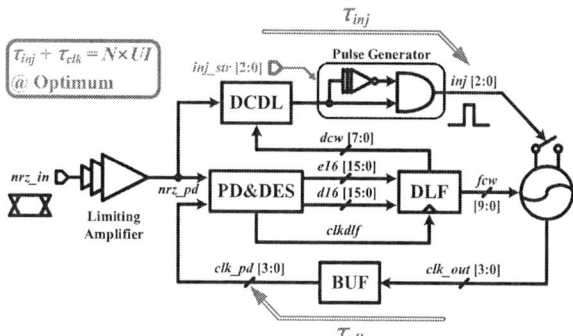

Fig. 1. Block diagram of the proposed ILCDR.

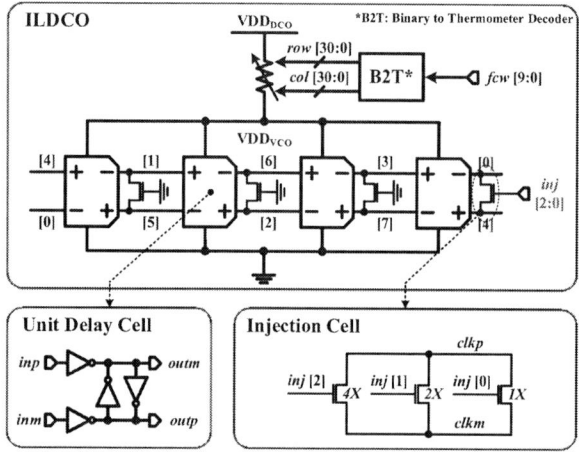

Fig. 2. Block diagram of the ILDCO.

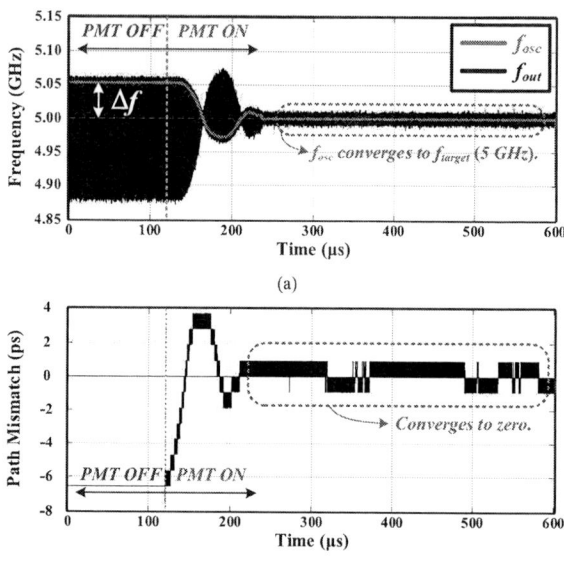

(a)

(b)

Fig. 3. Behavioral transient simulations of the proposed ILCDR. (a) Frequency versus time. (b) Path mismatch versus time. In this example, injection strength (β) is 0.8, and initial path mismatch is set to –6.56 ps with 10-Gb/s, 2^7–1 PRBS input data pattern.

continuously adjusted by the 8-bit digitally controlled delay line (DCDL), satisfying the condition using the proposed tracking loop. Since this work utilizes only the information of the traditional 2X-oversampling phase detection in CDR, it is implemented with a minimal hardware overhead. The detailed operation of the proposed phase detector for extracting the path mismatch would be covered in following Section II.B.

Fig. 2 shows the block diagram of the injection-locked digitally controlled oscillator (ILDCO). It consists of four-stage pseudo-differential ring oscillator for quadrature-phase clocks to operate in half-rate clock recovery. Injection cells are designed to short differential nodes in a binary-weighted array for various injection strengths. When an injection cell is disabled, inj is tied to low from the pulse generator. The dummy injection switches are equally implemented and tied to low for the rest of the differential nodes to generate accurate quadrature-phase clocks. Frequency tuning is performed using 10-bit digitally controlled resistor array as described in [11].

Fig. 3 shows behavioral transient simulations demonstrating the operation of the proposed path mismatch tracking (PMT) loop. In this simulation, initial path delay mismatch is set to –6.56 ps, and injection strength, β, defined in [9] is 0.8. The input data rate is 10 Gb/s with $2^7 - 1$ PRBS pattern, which is identical

to the test condition in Section III. In Fig. 3(a), when the tracking loop is off (<100 µs), the timing mismatch forces the free-running frequency of the oscillator (f_{osc}) to deviate from the target frequency (f_{target}), showing the two-point modulation problem in a conventional structure. The amount of deviation in frequency is a function of the timing mismatch and the injection strength. In other words, the output frequency (f_{out}) is determined by not only the free-running oscillator but also the injection effect. After the PMT loop is enabled (>100 µs), the initial path mismatch is tracked, and finally converges to zero in average as shown in Fig. 3(b). Since the path mismatch is diminished owing to the loop, the corresponding free-running frequency of the oscillator (f_{osc}) also goes to the target value (f_{target}), and the amount of variation in the output frequency (f_{out}) is minimized. The timing margin of the sampler is maximized to the half of the unit data interval (0.5 UI) which is the optimal point for error-free operation. In addition, since the free-running frequency of the oscillator converges to the target frequency, it is much more tolerant to the consecutive identical digits than the typical PLL-based ILCDR with path mismatch.

B. Path Mismatch Tracking (PMT) for optimum behavior

Fig. 4 depicts the conceptual timing diagram under path delay mismatch when the proposed tracking loop is disabled. In this example, the path delay deviation causes the frequency error ($\Delta f = f_{osc} - f_{target} < 0$), and the inj pulse forces the oscillator to move forward in phase. For simplicity, it is assumed that the injection strength is strong enough to replace the original edges, and the CDR loop is also strong enough to correct the edges in every cycle. At the time t_0, the first edge of clk_pd [0] is aligned to the rising edge of nrz_pd by the CDR loop, and $\Delta\phi$ is zero. However, the next edge is replaced by the pulse injection, and therefore it is not aligned to the falling edge of nrz_pd due to the

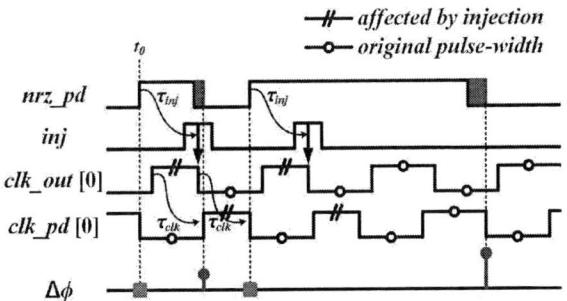

Fig. 4. Conceptual timing diagram under path delay mismatch when the PMT loop is disabled.

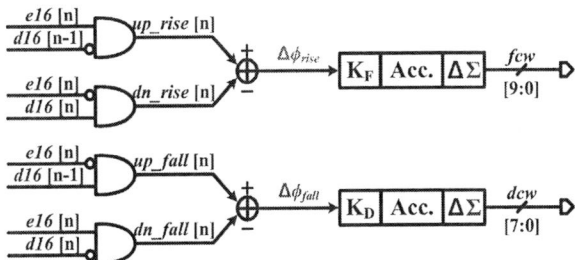

Fig. 5. Block diagram of the proposed phase detector in digital loop filter that detects the path mismatch.

delay deviation. With the two assumptions and the fact that the injection is applied only at the rising edges of the input data stream, the proposed phase detector embedded in the digital loop filter extracts the delay deviation on the two paths as $\Delta\phi_{fall}$. Fig. 5 shows the block diagram of the proposed phase detector in the digital loop filter. To separate the error information, the XOR gates alone in the conventional phase detector are replaced by the AND gates with minimal hardware additives. The errors are divided into $\Delta\phi_{rise}$ and $\Delta\phi_{fall}$ according to the direction of the input data stream, in other words, whether the injection is applied or not. These separated errors are integrated and applied to the ILDCO and DCDL, respectively.

III. MEASUREMENT RESULTS

The proposed ILCDR fabricated in 28-nm CMOS technology occupies an active area of 0.03 mm^2, and consumes 12.8 mW at 10 Gb/s with a 0.9-V supply voltage. Fig. 6 shows a chip photomicrograph and description of the building blocks with power breakdown. Each power consumption is measured by the digital power source, Agilent B2926A. Fig. 7 shows the measured jitter histograms of the 2.5-GHz recovered clock (divided by two) when the PMT loop is on and off. When the PMT loop is off, the delay control word is fixed to the edge of the locking range, satisfying the bit error rate less than 10^{-12}. As shown in Fig. 7(a), the histogram shows the irregular distribution with two peak tones. One is from the free-running frequency of the oscillator (f_{osc}), and the other is from the injection operation. The frequency offset caused by the path mismatch makes deterministic noise, and root-mean-square (rms) and peak-to-peak jitter are 10.3 ps and 54.8 ps, respectively. On the other hand, with the PMT loop on, regular Gaussian distribution is obtained as shown in Fig. 7(b), and rms and peak-to-peak jitter are 3.59 ps and 26.8 ps, respectively. From this result, it is found that the frequency offset made by the path mismatch is successfully attenuated using the proposed tracking loop.

To investigate the effect of the path mismatch and verify the PMT loop, an extra test is performed as shown in Fig. 8. In this test, a significant amount of the random jitter (about 1 UIpp) equipped in JBERT, Agilent N4903A, is intentionally added to the input data stream to figure out the operation of the PMT loop, not to estimate the absolute tolerance of the proposed ILCDR.

In Fig. 8(a), the separated power domains of ILDCO and all other blocks are tied together, and it is swept using the digital power source. Since the supply voltage determines the amount of path mismatch, the typical PLL-based ILCDR without the PMT loop (blue line with triangular markers) shows much narrower locking range than the proposed one (red line with circular markers). In addition, to find out the locking range of the ILDCO itself, an injection is applied to the oscillator with all other feedback loops disabled (black line with diamond markers). Since the frequency of the oscillator is controlled solely by the supply voltage in this test, there is a single point that shows the minimum bit error rate.

In order to further validate the operation of the PMT loop, the delay control words are varied in the same test condition, and the corresponding bit error rates are measured. With the PMT loop on, the initial delay control word is gradually adapted to the optimum value with the flat bit error rate, which is almost equal to the lowest bit error rate of the disabled PMT. In addition, the locking range with respect to the delay control words is much wider than the disabled one about ten times.

Fig. 9 shows the measured JTOL for $2^7 - 1$ PRBS pattern at 10 Gb/s with several different test conditions when the target bit error rate is 10^{-12}. The injection strengths and channels of the input data are varied in this test. When the size of the injection

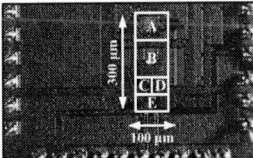

	Block Description	Power (mW)
A	Digital Loop Filter	2.6
B	ILDCO	3.38
C	Pulse Gen. & DCDL	6.83
D	PD & Deserializer	
E	Limiting Amplifier	

Fig. 6. Chip photomicrograph, block description, and separated power consumption at 10 Gb/s with a 0.9-V supply voltage.

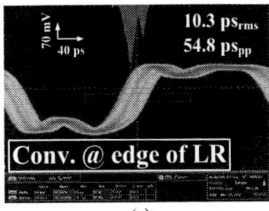
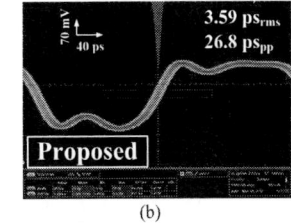

Fig. 7. Measured jitter histograms of the 2.5-GHz recovered clock for (a) the conventional PLL-based ILCDR at the edge of the locking range and (b) the proposed ILCDR, satisfying the bit error rate less than 10^{-12}.

978-1-5386-6414-8/18 $31.00 © 2018 IEEE

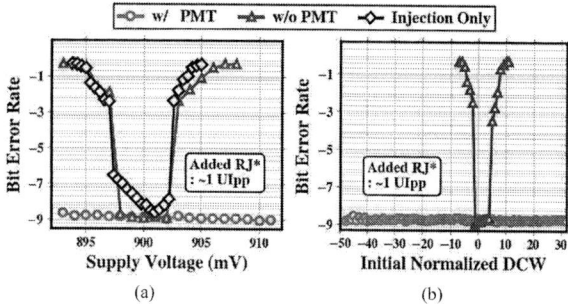

*Note that a significant amount of *Random Jitter* (~ 1 UIpp) is intentionally added.

Fig. 8. Measured bit error rate at 10 Gb/s with several different ILCDR with various (a) supply voltage and (b) delay control words. In this test, the random jitter of about 1 UIpp is intentionally added for degradation of bit error rate.

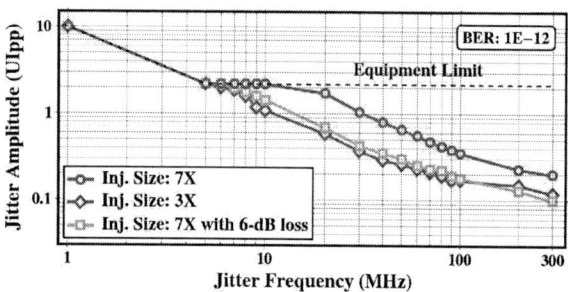

Fig. 9. Measured jitter tolerance of the proposed ILCDR for $2^7 - 1$ PRBS pattern at 10 Gb/s with the bit error rate of 10^{-12} in several different test conditions.

cells is 7X as explained in Fig. 2, the proposed ILCDR tolerates 1-UIpp sinusoidal jitter amplitude at the frequency of 31 MHz (red line with circular markers). In the other condition, the injection size is changed to 3X, and it shows the degraded JTOL performance compared with the strong injection since the injection strength determines the bandwidth of the overall architecture. In addition, to investigate the effect of inter-symbol interference, lossy channel (6-dB loss at the Nyquist frequency) is used with 7X injection size (yellow line with square markers). With 6-dB lossy channel, it achieves over 10-MHz jitter frequency when 1-UIpp sinusoidal jitter amplitude is applied.

Table I summarizes the performance and compares with other recently published ILCDRs. This work shows the best JTOL performance between the works [1]–[8] satisfying the bit error rate less than 10^{-12} as 1-UIpp amplitude at the sinusoidal jitter frequency of 31 MHz. In addition, the proposed ILCDR achieves the highest energy efficiency of 1.28 pJ/bit among the fully functional ILCDR chips published in the literature.

IV. CONCLUSION

This paper describes a new injection-locked clock and data recovery circuit (ILCDR) that continuously tracks the frequency offset between the input data stream and the local oscillator. The proposed ILCDR adjusts the conventional phase detector with respect to the polarity of the data transition. By employing the strong injection at the rising edges of the input data stream and the path-tracking loop, the proposed ILCDR achieves the exceptional jitter tolerance performance and remarkable energy efficiency compared with the existing ILCDR architectures.

TABLE I. PERFORMANCE SUMMARY AND COMPARISON.

	ISSCC'08 [6]	JSSC'15 [7]	JSSC'16 [8]	This Work
Technology (nm)	250	90	28	28
Architecture	Full rate	Full rate	Half rate	Half rate
Oscillator Type	GVCO (RING)	ILO (RING)	ILO (RING)	ILO (RING)
Supply Voltage (V)	3.3 / 1.8	1.2	0.9	0.9
Data Rate (Gb/s)	10.3125	2.2	1 – 12	10
Power (mW)	856	6.1	$22.9^{3)} / 11.0^{4)}$	12.8
Bit Error Rate	$< 10^{-12}$	$< 10^{-12}$	$< 10^{-9}$	$< 10^{-12}$
Jitter Tolerance	1.0 UIpp @ 20 MHz$^{1)}$ 0.27 UIpp @ 80 MHz	1.0 UIpp @ 4 MHz$^{1)}$ 0.3 – 0.6 UIpp @ 10 MHz	1.09 UIpp @ 100 MHz 0.56UIpp @ 300 MHz$^{5)}$	1.0 UIpp @ 31 MHz 0.2 UIpp @ 300 MHz
Active Area (mm²)	9.0	0.44	-	0.03
Energy Efficiency (pJ/bit)	83.01	$2.77^{2)}$	$1.9^{3)} / 0.9^{4)}$	1.28
Path Mismatch Tracking	No	No	Yes	Yes

$^{1)}$ Estimated from JTOL results.

$^{2)}$ Excluding limiting amplifier.

$^{3)}$ Including analog front end, de-multiplexer, and excluding digital logics.

$^{4)}$ Excluding analog front end, de-multiplexer, and digital logics.

$^{5)}$ Measured by the internal calculation circuits, not external equipment.

REFERENCES

[1] J.-H. C. Zhan, J. S. Duster, and K. T. Kornegay, "Full-rate injection-locked 10.3Gb/s clock and data recovery circuit in a 45GHz-fT SiGe process," in *Proc. IEEE Custom Integrated Circuits Conference (CICC)*, Sep. 2005, pp. 552–555.

[2] M. Nogawa *et al.*, "A 10 Gb/s burst-mode CDR IC in 0.13 μm CMOS," in *IEEE Integrated Solid-State Circuits Conference (ISSCC) Dig. Tech. Papers*, Feb. 2005, pp. 228-595.

[3] J. Lee and M. Liu, "A 20-Gb/s burst-mode clock and data recovery circuit using injection-locking technique," *IEEE J. Solid-State Circuits*, vol. 43, no. 3, pp. 619–630, Mar. 2008.

[4] K. Maruko *et al.*, "A 1.296-to-5.184Gb/s transceiver with 2.4mW/(Gb/s) burst-mode CDR using dual-edge injection-locked oscillator," in *IEEE Integrated Solid-State Circuits Conference (ISSCC) Dig. Tech. Papers*, Feb. 2010, pp. 364–365.

[5] Y. Take, N. Miura, and T. Kuroda, "A 30 Gb/s/Link 2.2 Tb/s/mm² inductively-coupled injection-locking CDR for high-speed DRAM interface," *IEEE J. Solid-State Circuits*, vol. 46, no. 11, pp. 2552–2559, Nov. 2011.

[6] J. Terada, K. Nishimura, S. Kimura, H. Katsurai, N. Yoshimoto, and Y. Ohtomo, "A 10.3125Gb/s burst-mode CDR circuit using a ΔΣ DAC," in *IEEE Integrated Solid-State Circuits Conference (ISSCC) Dig. Tech. Papers*, Feb. 2008, pp. 226–609.

[7] W.-S. Choi, T. Anand, G. Shu, A. Elshazly, and P. K. Hanumolu, "A burst-mode digital receiver with programmable input jitter filtering for energy proportional links," *IEEE J. Solid-State Circuits*, vol. 50, no. 3, pp. 737–748, Mar. 2015.

[8] T. Masuda *et al.*, "A 12 Gb/s 0.9 mW/Gb/s wide-bandwidth injection-type CDR in 28 nm CMOS with reference-free frequency capture," *IEEE J. Solid-State Circuits*, vol. 51, no. 12, pp. 3204–3215, Dec. 2016.

[9] S. Ye, L. Jansson, and I. Galton, "A multiple-crystal interface PLL with VCO realignment to reduce phase noise," *IEEE J. Solid-State Circuits*, vol. 37, no. 12, pp. 1795–1803, Dec. 2002.

[10] A. Elkholy, M. Talegaonkar, T. Anand, and P. K. Hanumolu, "Design and analysis of low-power high-frequency robust sub-harmonic injection-locked clock multipliers," *IEEE J. Solid-State Circuits*, vol. 50, no. 12, pp. 3160–3174, Dec. 2015.

[11] H. Song, D.-S. Kim, D.-H. Oh, S. Kim, and D.-K. Jeong, "A 1.0–4.0-Gb/s all-digital CDR with 1.0-ps period resolution DCO and adaptive proportional gain control," *IEEE J. Solid-State Circuits*, vol. 46, no. 2, pp. 424–434, Feb. 2011.

A 28.16-Gb/s Area-Efficient 60GHz CMOS Bi-Directional Transceiver for IEEE 802.11ay

Jian Pang, Korkut Kaan Tokgoz, Shotaro Maki, Zheng Li, Xueting Luo, Ibrahim Abdo,
Seitarou Kawai, Hanli Liu, Bangan Liu, Makihiko Katsuragi, Kento Kimura, Atsushi Shirane, Kenichi Okada
Department of Electrical and Electronic Engineering, Tokyo Institute of Technology
2-12-1-S3-28, Ookayama, Meguro-ku, Tokyo 152-8552, Japan
Email: pangjian@ssc.pe.titech.ac.jp

Abstract—This paper introduces a 60-GHz CMOS transceiver designed for IEEE 802.11ad/ay featuring the area-efficient bi-directional operation. The proposed bi-directional PA-LNA occupies for less than half on-chip area while staying a similar performance with the conventional standalone PA and LNA. The measured noise figure in RX mode is 4.8dB at 62.56GHz, and the measured EVM is -26dB in TX mode with an output power of -4.2dBm. Thanks to the compact PA-LNA, this work realizes a maximum data-rate of 28.16Gb/s in 16QAM with only 3mm². The power consumption is 105mW in TX mode and 128mW in RX mode.

Index Terms—60 GHz, CMOS, transceiver, bi-directional,

I. INTRODUCTION

The exponential growth in data traffic demands a much higher wireless communication speed. To further boost the data-rate, MIMO is supposed to be supported in IEEE 802.11ay. The latest research on 60-GHz transceivers [1–4] realized a data-rate up to 42.24Gb/s. However, all of the works mentioned above implement TX and RX separately, which consume sizable on-chip area. Considering the MIMO configuration, the manufacturing cost of the communication system will be increased along with the physical size. To realize a low-cost, compact system, the bi-directional transceiver offers an area-efficient solution by replacing the standalone PA and LNA with a bi-directional amplifier. This work presents a bi-directional transceiver chip designed for IEEE 802.11ay MIMO application. Compared with the conventional 60-GHz transceivers, the proposed transceiver reduce the required on-chip area by more than half while staying a similar performance. A measured maximum data-rate of 28.16Gb/s in 16QAM is realized with a 4-channel bonding bandwidth.

II. IMPLEMENTATION OF THE BI-DIRECTIONAL TRANSCEIVER

Fig. 1(a) shows the block diagram of the proposed 60-GHz bi-directional transceiver. Direct-conversion topology is adopted in this work due to the reduced system complexity. Entirely two transceiver cells are integrated into the same chip, which is capable of supporting a MIMO or beam-forming configuration. Each of the transceiver cells consists of a 5-stage bi-directional PA-LNA, an I/Q bi-directional mixer, an I/Q 2-stage baseband amplifier, and bypass switches for TX mode. The LO in this work is realized with a quadrature injection-locked oscillator with an off-chip injection. It covers

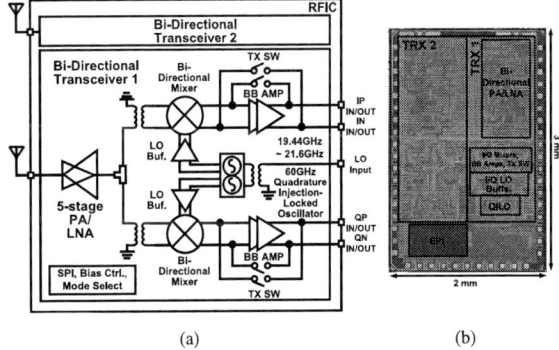

(a)   (b)

Fig. 1.   (a) Block diagram of the proposed bi-directional transceiver. (b) Die micrograph.

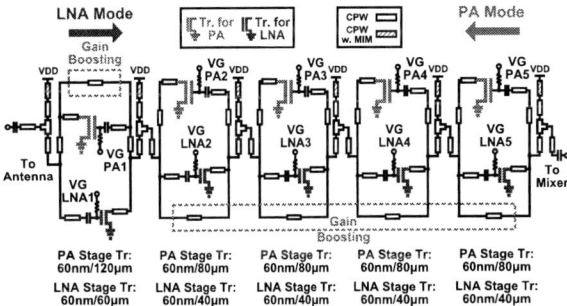

Fig. 2.   Circuit Schematic of the proposed bi-directional amplifier.

all of the required LO frequency for channels defined in IEEE 802.11ad/ay. The I/Q phase mismatch calibration is realized by tuning the free-run frequency of the I oscillator.

Fig. 2 shows the circuit schematic of the proposed bi-directional PA-LNA. Conventional bi-directional amplifiers utilize switches to change the operation mode between PA and LNA [5]. However, the required area is still large due to the separate inter-stage matching. The proposed bi-directional PA-LNA utilizes five bi-directional units, each of which consists of a PA stage transistor (Tr.) and an LNA stage Tr.. Compared with conventional bi-directional amplifier, the required area of this work is significantly reduced by sharing the inter-stage matching networks. A shunt gain-boosting transmission-line (TL.) is utilized in each stage for improved RF gain performance. In most circumstances, the system has different

978-1-5386-6414-8/18 $31.00 © 2018 IEEE   77

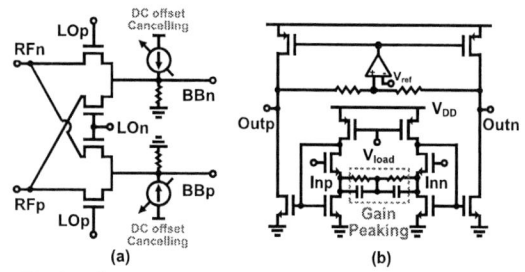

Fig. 3. Circuit Schematic of the mixer and the baseband amp.

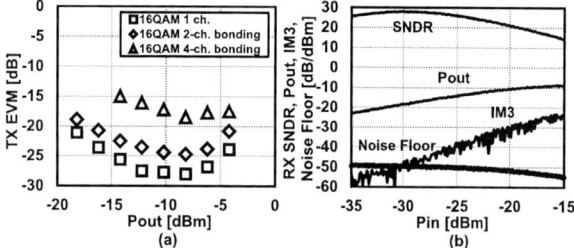

Fig. 4. Measured (a) EVM of the TX and (b) SNDR of the RX.

Carrier freq.	61.56GHz	61.56GHz	61.56GHz
BW.	(1-ch.)	(4-ch. bonding)	(4-ch. bonding)
Modulation	64QAM	QPSK	16QAM
Data rate*	10.56Gb/s	14.08Gb/s	28.16Gb/s
Constellation**			
Spectrum**			
TX EVM**	-27.3dB	-19.3dB	-18.5dB
TX-to-RX EVM***	-24.0dB	-17.7dB	-17.6dB

*The roll-off factor is 0.25. The bandwidth is 2.16GHz except for the channel bonding.
**Constellation, Spectrum, and Tx EVM are measured with an external downconverter.
***Tx-to-Rx EVM is measured through Tx and Rx, which is equal to −SNR(MER).

Fig. 5. Measured constellation and performance summary.

	This Work	Intel [1]	Tokyo Tech [4]	Tokyo Tech [3]	Broadcom [2]
Data rate/ Modulation	28.16Gb/s +28.16Gb/s 16QAM	27.8Gb/s (H TRX+V TRX) 16QAM	21.12Gb/s +21.12Gb/s 64QAM	42.24Gb/s 64QAM	4.6Gb/s 16QAM
Pout /ant. path	-4.2dBm TX EVM=-26dB	N.A.	7.0dBm TX EVM=-22dB	7.3dBm TX EVM=-22dB	-2.0dBm TX-to-RX EVM=-22dB
NF	4.8dB	8.0dB	N.A.	N.A.	<7.0dB
Integration	65nm, direct conversion, 2TX, 2RX, 2LO, w/o PLL.	28nm, digital-pol. TX, direct-con. RX, 2TX, 2RX, 2LO, w/o PLL.	65nm, direct conversion, 2TX, 2RX, 2LO.	65nm, direct conversion, TX, RX, LO.	40nm, 144-array, heterodyne, TX, RX, LO, analog&dig. BB.
Area	6.0mm² (2TRX)	4.3mm² (2TRX)	17.6mm² (2TRX)	6.0mm²	21.9mm² (RFIC only) (12TRX)
Data rate /Element Area	9.4 Gb/s/mm²	6.4 Gb/s/mm²	2.4 Gb/s/mm²	7.0 Gb/s/mm²	2.5 Gb/s/mm²
Power Consumption	TX:210mW RX:256mW 2TRX	TX: 210mW RX: 110mW 2TRX	TX: 544mW RX: 432mW 2TRX	TX: 169mW RX: 139mW	TX: 8400mW RX: 6600mW 144-element

Fig. 6. Performance comparison of 60-GHz CMOS transceivers.

requirements on PA and LNA. As a result, larger Tr. size is chosen for PA than for LNA. In this work, the PA-LNA shares the antenna port in an area-efficient switchless manner. The antenna sharing network is designed with small insertion loss together with the PA last stage Tr. and LNA first stage Tr.. The measured NF in LNA mode is 4.2dB at 61.56GHz, while the measured output P_{1dB} is 0dBm in PA mode.

Fig. 3 shows the circuit schematic of the bi-directional mixer and the baseband amplifier. In addition to the PA-LNA, the area is further saved with an I/Q bi-directional mixer. This work utilizes a passive double-balanced mixer for both up and down conversion. Not only the mixer but also the entire I/Q LO chain could be completely shared between the TX and RX. In this work, the carrier leakage is canceled by two current sources at the baseband side. Considering the noisy received signal, a 2-stage flipped-voltage-follower (FVF) based baseband amplifier is applied in RX. The baseband amplifier is optimized for flat gain response together with high linearity.

III. MEASUREMENT RESULTS AND CONCLUSION

The proposed bi-directional transceiver is fabricated in a 65-nm CMOS process. Fig. 1(b) shows the die photo of this work. The chip size is 3mm x 2mm, which includes two transceiver elements. The area required for a single-element transceiver is less than 3mm². Fig. 4(a) shows the measured TX mode EVM in 16QAM at ch. 2.5 (center frequency: 61.56GHz). A maximum EVM of -28dB is achieved at -8.2dBm output power with 1-ch. bandwidth. Fig. 4(b) shows the measured P_{out}, IM_3, and noise floor in RX mode at 62.56GHz. The measured NF is 4.8dB. The calculated SNDR in RX mode with a 4-ch. bonding bandwidth is also shown in Fig. 4(b). A maximum SNDR of 28.0dB is achieved with an input power of -29.3dBm. Fig. 5 summarize the measured constellation and EVM performance. All of the measured spectrums satisfy the spectrum mask from IEEE 802.11ay. 64QAM could be supported with 1-ch. bandwidth by this work with a TX-to-RX EVM of -24dB. The measured maximum data-rate is 28.16Gb/s in 16QAM with a 4-bonded channel. The communication distance in this condition is 1cm with two 14-dBi horn antennas. The maximum communication distance is 0.54m in QPSK. Fig. 6 compares this work with some state-of-the-art 60-GHz transceivers. Thanks to the bi-directional operation, this work achieves a data-rate of 28.16Gb/s per element TRX with an area of only 3mm². The two-element bi-directional transceiver chip is capable of supporting a high-performance, low-manufacturing-cost transceiver array system.

ACKNOWLEDGMENT

This work is partially supported by the MIC/SCOPE #175003017, STAR and VDEC in collaboration with Cadence Design Systems, Inc., Synopsys, Inc., and Mentor Graphics, Inc.

REFERENCES

[1] S. Kang, et al., "A 40Gb/s 6pJ/b RX baseband in 28nm CMOS for 60GHz polarization MIMO," ISSCC, pp. 164–166, Feb 2018.

[2] T. Sowlati, et al., "A 60GHz 144-element phased-array transceiver with 51dBm maximum EIRP and ±60° beam steering for backhaul application," ISSCC, pp. 66–68, Feb 2018.

[3] J. Pang, et al., "A 128-QAM 60GHz CMOS transceiver for IEEE802.11ay with calibration of LO feedthrough and I/Q imbalance," ISSCC, pp. 424–425, Feb. 2017.

[4] R. Wu, et al., "64-QAM 60-GHz CMOS Transceivers for IEEE 802.11ad/ay," IEEE Journal of Solid-State Circuits, vol. 52, no. 11, pp. 2871–2891, Nov 2017.

[5] E. Cohen, et al., "A bidirectional TX/RX four element phased-array at 60GHz with RF-IF conversion block in 90nm CMOS process," RFIC, pp. 207–210, June 2009.

978-1-5386-6414-8/18 $31.00 © 2018 IEEE

A 77-GHz Mixed-Mode FMCW Generator Based on a Vernier TDC with Dual Rising-Edge Fractional-Phase Detector

Jianxi Wu[1], Zipeng Chen[1], Wei Zheng[2], Yibo Liu[1], Shufu Wang[2], Nan Qi[3] and Baoyong Chi[1]

1. Institude of Microelectronics, Tsinghua University, Beijing, 100084, China

2. Radarchip Technology Co., Ltd., Beijing, 100084, China

3. Institute of Semiconductors, Chinese Academy of Sciences, Beijing, 100084, China

Email: chibylxc@tsinghua.edu.cn

Abstract—A 77-GHz mixed-mode frequency-modulated continuous-wave (FMCW) generator based on a Vernier time-to-digital converter (TDC) is presented. By utilizing a coarse-fine segmented digital-to-analog converter (DAC), the generator supports variable slopes and sawtooth-type FMCW chirp modulation. A dual rising-edge fractional-phase detector is proposed to overcome the issue caused by the rising and falling time mismatch of the delay chain in TDC. A self-calibrated delay chain is employed in the TDC to overcome process, voltage and temperature (PVT) variations. A divider-less frequency error estimator is proposed to save area and power. Sampling-edge-selection is utilized to avoid frequency glitches during the retiming process. An LC voltage-controlled oscillator with split varactors is proposed to improve the linearity tuning range. Implemented in 65nm CMOS, the generator consumes 99 mW power and 0.56 mm² chip area. The measurement results show that the generator achieves 0.6 to 10 ms FMCW chirp period (triangle-type), 250 MHz to 4 GHz bandwidth, and less than 251 kHz root-mean-square (RMS) frequency error. And the phase noise from 78.7 GHz carrier is -87.4 dBc/Hz at 1 MHz offset.

Keywords—Frequency-modulated continuous-wave (FMCW); radar; mm-wave; PLL; Vernier TDC; Glitch; Segmented DAC

I. INTRODUCTION

The automotive millimeter-wave (mm-wave) radars have drawn a lot of attention from the vehicle transportation industry [1]. The frequency-modulated continuous-wave (FMCW) chirp generator is a key component in the automotive radar transceiver. Fractional-N analog phase-locked loop (PLL) and all-digital PLL (ADPLL) are typically employed to implement FMCW generator and present competitive performances on linearity and power consumption. Compared to the fractional-N analog PLL, ADPLL has more flexible loop dynamics to generate the FMCW chirp signal with reconfigurable slopes. However, the wide-tuning-range digitally-controlled oscillator (DCO) in the ADPLL needs complicated SRAM-based calibrations [2]. A mixed-mode PLL compromises between the analog and digital PLLs to generate reconfigurable chirps without complicated calibration [3]. Without utilizing a programmable frequency divider, the counter-assisted mixed-mode PLL guarantees a better phase-noise performance, but it suffers from the glitch and the rising and falling time mismatch of the delay chain in time-to-digital convertor (TDC) [4]. Furthermore, the digital-to-analog converter (DAC) of the mixed-mode PLL needs wide dynamic range to meet the goal of variable slopes and sawtooth-

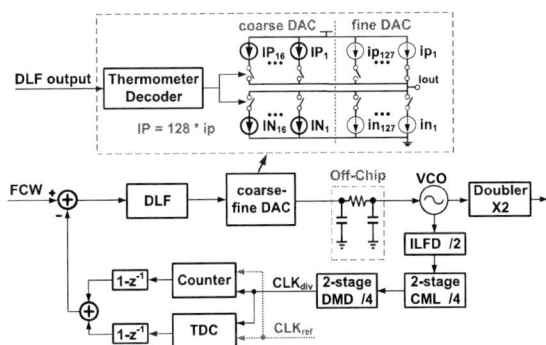

Fig. 1. Block diagram of the proposed 77-GHz mixed-mode FMCW generator.

type FMCW chirp modulation, which increases the design complexity. A divide-operation during the frequency error estimation also consumes large area in both the all-digital and counter-assisted mixed-mode PLLs.

This paper presents a reconfigurable 77-GHz counter-assisted mixed-mode FMCW generator with 40 MHz reference clock. To overcome the issue caused by the rising and falling time mismatch of the delay chain in TDC, a dual rising-edge fractional-phase detector is proposed, and a wide detection range Vernier TDC with automatically calibrated delay chain is employed. A coarse-fine segmented DAC and a divider-less frequency error estimator are proposed to reduce the area and power consumption. To remove glitches in the retiming process, a sampling-edge-selection technique is utilized. An LC voltage-controlled oscillator (VCO) with split varactors is proposed to improve the linearity tuning range. The FMCW generator prototype is implemented in 65nm CMOS.

II. CIRCUIT DESCRIPTIONS

A. System Architecture

The system architecture of the proposed mixed-mode FMCW generator is shown in Fig. 1. A frequency doubler is employed to simplify the local-oscillation (LO) generation and enhance the modulation bandwidth. The output of the VCO is divided by an injection-locked frequency divider (ILFD), a 2-stage current-mode logic (CML) frequency divider and a 2-stage D flip-flop-based (DFF-based) dual-modulus divider (DMD).

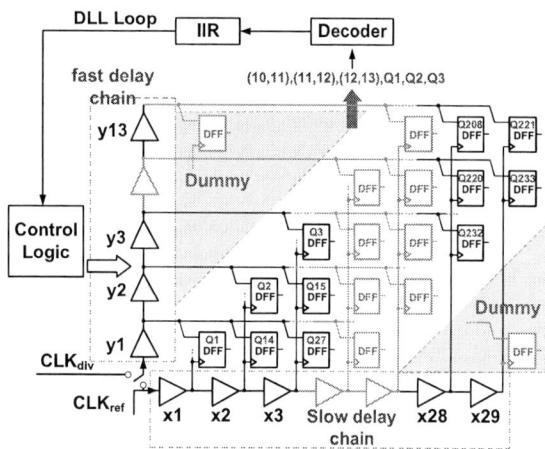

Fig. 2. Proposed 2-dimensional Vernier TDC with an automatic calibration DLL.

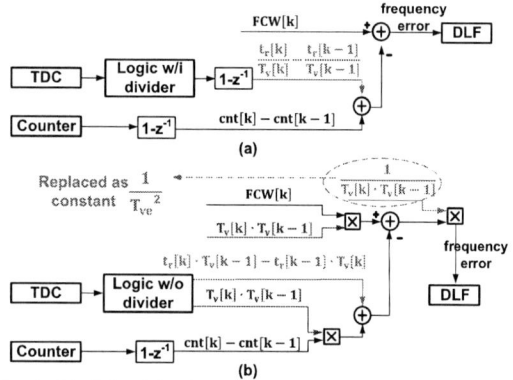

Fig. 3. Principle of the frequency error estimator in (a) conventional topology with divider and (b) proposed divider-less topology.

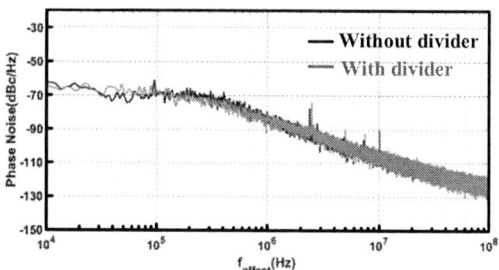

Fig. 4. Simulated phase noise of the proposed mixed-mode FMCW generator at 76.18 GHz, when the TDC resolution is set to 10 ps.

rising edges are detected in the proposed fractional-phase detector for higher TDC linearity.

B. A Wide Range Vernier TDC With Automatic Calibration

The detection range of the TDC should be doubled to ensure two consecutive CLK_{div}'s rising edges be observed in one sampling process. As shown in Fig. 2, a 2-dimensional Vernier TDC [5] is utilized to achieve wide detection range and high resolution with low power and area consumption. In the conventional Vernier TDC, an arbiter array and a register bank controlled by digital logic are employed to obtain and hold the edge detection information. The arbiter-register block is replaced by a DFF array from the standard cell library in the proposed architecture with less area and power consumption. The setup time of the DFF introduces a constant offset during the phase estimation and is removed by the differentiator. Dummy DFFs are inserted in the array to provide the same capacitive load to all the stages in the delay lines. The delay unit is realized as two cascade CMOS inverters. Ideally, the unit delay in the fast delay chain must be 12/13 of that in the slow delay chain, and they are designed as 120 and 130 ps. The total TDC range can achieve 2.3 ns with 13 stages in the fast delay chain and 29 stages in the slow delay chain, which makes it possible to cover two CLK_{div} cycles in one sampling process.

As shown in Fig. 2, a delay locked loop (DLL) in the Vernier plane is employed to ensure the 12/13 ratio of the unit delays and enhance the linearity of the TDC. During the calibration, the CLK_{ref} is fed into both chains, and some dedicated outputs of the DFF array around the position (column-12, row-13) are decoded, accumulated in the IIR filter, and fed back to correct the ratio. The rising edges of two reference clocks meet at the position (column-12, row-13) while the calibration is done. A 7-bit switched-capacitor array with 0.5 ps step size is inserted into each delay unit to cover a range of 60 ps. The measurement result shows that the resolution can be adjusted from 8.8 to 12.6 ps by using the DLL-based automatic calibration.

C. Divider-Less Frequency Error Estimator

The normalized time difference between the rising edges of CLK_{div} and CLK_{ref} ($t_r[k]$), as well as the normalized CLK_{div} period ($T_v[k]$) can be obtained from a decoder following the TDC. As shown in Fig. 3, a divider-less frequency error estimator is proposed to reduce the power and area consumption. Instead of differentiating the fractional phase, the denominator $T_v[k] * T_v[k-1]$ is moved to the forward path while the FCW and the counter output ($cnt[k]$) are multiplied by the same factor.

The divider output clock (CLK_{div}) and the reference clock (CLK_{ref}) are sent into a counter and a TDC to obtain the integer and fractional phase information normalized by the CLK_{div} period. The frequency converted by the differentiator is compared with the chirp frequency command word (FCW), and the difference between the two values is filtered by a digital loop filter (DLF). The DLF consists of 2 IIR filter stages and a FIR filter (including a proportional path and an integral path). The IIR filters are utilized to reduce the DAC input range and suppress the out-of-band noise. A thermometer 7-bit DAC is employed to provide the reconfigurable FMCW slopes from 250 MHz/5ms to 4 GHz/0.3ms, while a coarse 4-bit DAC establishes fast frequency-ramping-down for the sawtooth-type FMCW chirp modulation without consuming much area and power. The output current of the DACs is integrated by an off-chip capacitor to provide the input tuning voltage for the VCO. An RC filter is inserted to provide the anti-aliasing filtering.

In the conventional all-digital or counter-assisted mixed-mode PLL, the time difference between the CLK_{div}'s rising and falling edges is detected by the TDC to obtain the CLK_{div} period information. But the mismatch between the rising and falling time in the delay chain significantly deteriorates the TDC performance. To overcome this issue, two consecutive CLK_{div}'s

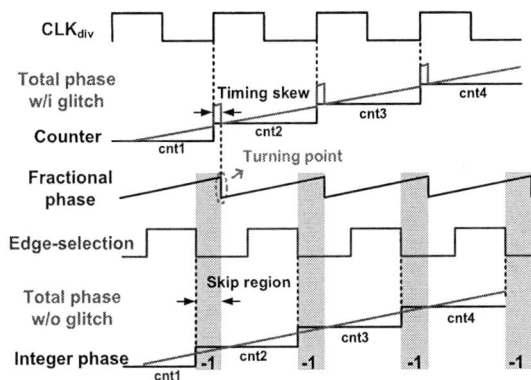

Fig. 5. Principle of the proposed sampling-edge selection to remove glitches in the frequency error estimator.

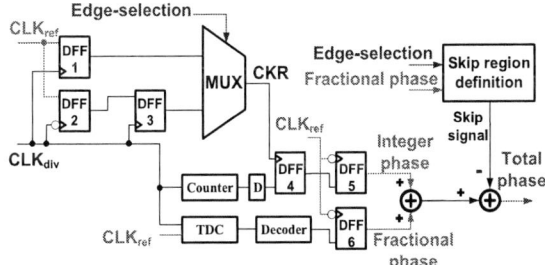

Fig. 6. The implementation of proposed sampling-edge selection.

Then, the denominator in the forward path is replaced by T_{ve}^2. Since T_{ve} is the constant predefined CLK_{div} period, the divider can be avoided. The transmission function G(s) in the proposed model is multiplied by $T_v[k] * T_v[k-1]/T_{ve}^2$ whose value only influences the bandwidth and is thus negligible. Behavioral simulation in Fig. 4 shows that the simplified estimator achieves the same phase noise as that with the divider, while the area consumption of the digital logic could be decreased by 1/3.

D. Glitch Removing With Sampling-Edge Selection Technique

Depicted in Fig. 5, due to the loading mismatch of the counter and TDC, a timing skew is introduced between their outputs. As a result, the estimated frequency would have glitches, which significantly deteriorate the TDC performance. To remove glitches, an additional correction logic is proposed in [4], with the cost of ignoring any frequency steps above the predefined threshold. In this work, a sampling-edge selection technique is proposed and implemented (Fig. 6) to remove the glitch in the frequency error estimator without ignoring any frequency steps.

An edge-selection signal is required to define a time window centered at the fractional phase turning point, which lasts for half of the CLK_{div} period. Some bit of the TDC outputs which corresponds to the quantized 1/4 CLK_{div} cycle is utilized as the edge-selection signal. This signal is then used to control the MUX to select the CKR, from either the CLK_{ref} sampled by the CLK_{div} rising edge (edge-selection=1), or that of the falling edge (edge-selection=0). During each edge-selection cycle when it is '0', the fractional phase will always be added by the counter result after the current CLK_{div} rising edge. Therefore, the integer phase keeps constant around the point where the fractional phase

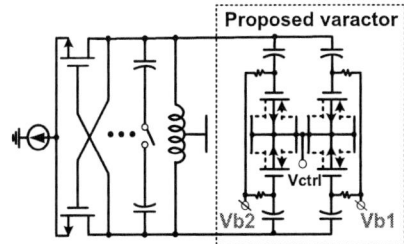

Fig. 7. The wide tuning range VCO.

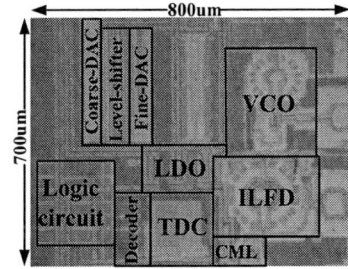

Fig. 8. Chip microphotograph of the presented FMCW generator.

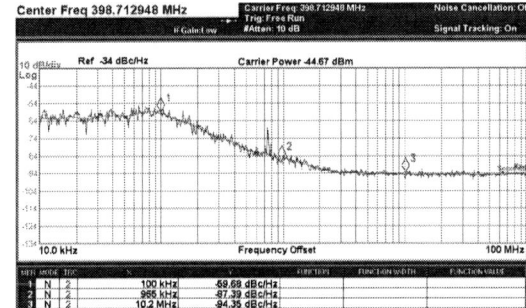

Fig. 9. Measured frequency doubler output phase noise at 78.7 GHz.

switches from 1 to 0, and thus the time skew can be neglected. There is a skip region when the CLK_{ref}'s rising edge leads the fractional phase's turning point and the selection signal is 0. The fractional phase distributes within [0.75, 1] or [0, 0.25] when selection signal is 0, and in [0.25, 0.75] when selection signal is 1. So, the skip region is defined as the overlap when edge-selection signal is 0 and the fractional phase is above 0.5. If the sampling happens within the skip region, the phase would be subtracted by 1, which removes the glitches completely. There is no timing skew between the integer phase and skip region, because the integer phase is controlled by the edge-selection signal through edge-selection retiming. When CLK_{ref}'s rising edge locates in the region where the selection signal is 1, the CLK_{ref} will be resampled by the rising edge of CLK_{div} normally.

E. Wide Tuning Range VCO

A wide tuning range LC VCO is utilized to generate the oscillation signal around 38.5 GHz. The linear tuning range of VCO is limited due to the nonlinear C-V characteristics of the accumulation-mode MOS varactor. Fig. 7 shows the VCO schematic integrated with the proposed varactor. The varactor is split into two branches in parallel with different bias voltage applied for an improved linearity performance. The optimized

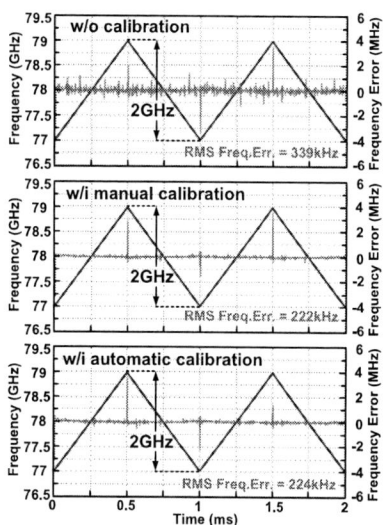

Fig. 10. Measured FMCW chirp waveforms and the frequency error including TAPs with 1ms period and 2 GHz bandwidth.

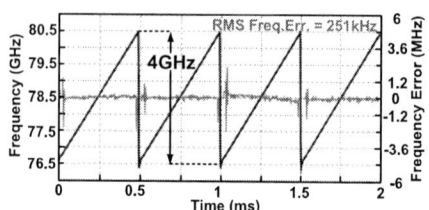

Fig. 11. Measured FMCW sawtooth chirp and frequency error with 0.5ms period and 4 GHz bandwidth.

TABLE I
FMCW GENERATOR PERFORMANCE COMPARISON

	This work				[1]	[2]
Architecture	Mixed-mode PLL				Fractional-N PLL	ADPLL
Frequency (GHz)	76.5-80.5				76.92-78.85	61.6-62.6
Reference (MHz)	40				200	40
BW/Period(GHz/ms)	0.25/10	2/1	4/0.6	4/0.5**	1.93/2	1/0.42
RMS Freq. Error (kHz)	37	224	181*	251	674*	384*
PN @ 1MHz(dBc/Hz)	-87.4				-81	-90
Technology(nm)	65				65	65

* Without the turn-around points
** Sawtooth-type FMCW chirp modulation

C-V curve features higher linearity, since the flat part in the C-V curve of one branch neutralizes the steep part in that of the other branch. The proposed 2.5 V varactor enables 2 GHz tuning range within one single tuning curve. By employing a 2-bit switched-capacitor array, the total tuning range covers from 37.2 to 42.0 GHz with enough overlap between adjacent tuning curves to ensure the continuous frequency modulation.

III. MEASUREMENT RESULTS

The presented FMCW generator is implemented in 65nm CMOS. The chip microphotograph is shown in Fig. 8 occupying a core die area of 0.56 mm^2. The total power consumption is 99 mW, in which the VCO, divider and doubler dissipate 77.5 mW, and the TDC, DAC and logic circuits dissipate 21.5 mW.

Fig. 9 shows the phase noise measured at the frequency doubler output with Keysight N9010A signal analyzer and harmonic mixer. The phase noise from 78.7 GHz carrier is -87.4 dBc/Hz at 1 MHz offset, even though the phase noise of the harmonic mixing LO from the instrument is -94.35 dBc/Hz. The CLK$_{div}$ is sampled by Keysight MSO9254A and demodulated with Keysight 89601B, which is multiplied by 64 to get the FMCW chirp waveform and frequency error. Fig. 10 shows the measurement results in different calibration modes. The RMS frequency error including the turn-around points (TAPs) is 339 kHz without calibration and is improved to 222 kHz and 224 kHz with the optimized manual calibration and automatic calibration, respectively, which shows that the automatic calibration can work as good as the optimized manual calibration. As shown in Fig. 11, a sawtooth-type chirp with 4 GHz bandwidth and 0.5 ms period is also supported by the proposed architecture. Compared to the results in [3], a much faster modulation slope is obtained with small RMS frequency error. The fast frequency-ramping-down process takes 5 us and an extra 15 us is needed before the loop stabilizes again. Table I summarizes the performance of the proposed FMCW generator

and makes a comparison with the state-of-the-art. It shows the presented mixed-mode generator achieves reconfigurable modulation period from 0.6 to 10 ms (triangle-type) and bandwidth from 250 MHz to 4 GHz with smaller RMS frequency error.

IV. CONCLUSION

A 77-GHz mixed-mode FMCW generator is presented in this paper. A 2.3 ns detection range, 10 ps resolution Vernier TDC with automatic calibration, a coarse-fine segmented DAC and a divider-less frequency error estimator are employed in the generator. A sampling-edge selection technique is utilized to remove the glitch. An LC VCO with split varactors is proposed. The presented FMCW generator is implemented in 65nm CMOS and less than 251 kHz RMS frequency error is achieved.

ACKNOWLEDGMENT

This work is supported in part by the National Natural Science Foundation of China (No. 61331003 and No. 61774093).

REFERENCES

[1] Haikun Jia, et al. "A 77 GHz Frequency Doubling Two-Path Phased-Array FMCW Transceiver for Automotive Radar," IEEE J. Solid-State Circuits, vol. 51, no. 10, pp. 2299-2311, Oct. 2016.

[2] Wanghua Wu, et al. "A 56.4-to-63.4 GHz Multi-Rate All-Digital Fractional-N PLL for FMCW Radar Applications in 65-nm CMOS," IEEE J. Solid-State Circuits, vol. 49, no. 5, pp. 1081-1096, May 2014.

[3] Y. Wang, et al. "A Ku-Band 260mW FMCW Synthetic Aperture Radar TRX with 1.48GHz BW in 65nm CMOS for Micro-UAVs," ISSCC, pp. 240-241, Feb. 2016.

[4] M. Lee, et al. "A Low-Noise Wideband Digital Phase-Locked Loop Based on a Coarse-Fine Time-to-Digital Converter With Subpicosecond Resolution," IEEE J. Solid-State Circuits, vol. 44, pp. 2808-2816, Oct. 2009.

[5] A. Liscidini, L. Vercesi, R. Castello, "Time to digital converter based on a 2-dimensions Vernier architecture," CICC Sept. 2009.

7-3 (8184)

IEEE Asian Solid-State Circuits Conference
November 5 - 7, 2018/Tainan, Taiwan

A CMOS 76-81 GHz 2TX 3RX FMCW Radar Transceiver Based on Mixed-Mode PLL Chirp Generator

Taikun Ma[1], Zipeng Chen[1], Jianxi Wu[1], Wei Zheng[2], Shufu Wang[2], Nan Qi[3], Baoyong Chi[1]

1. Institute of Microelectronics, Tsinghua University, 100084, Beijing, China
2. Radarchip Technology Co., Ltd., 100084, Beijing, China
3. Institute of Semiconductors, Chinese Academy of Sciences, 100084, Beijing, China
Email: chibylxc@tsinghua.edu.cn

Abstract—**A fully integrated 76-81 GHz frequency-modulated continuous-wave (FMCW) radar transceiver (TRX) in 65nm CMOS is presented. Two transmitters (TXs) and three receivers (RXs) are integrated for MIMO processing. A 38.5 GHz mixed-mode PLL with reconfigurable loop bandwidth and frequency doubling scheme are employed to generate the reconfigurable FMCW chirp waveforms. Passive voltage-mode down-conversion is utilized to improve the RX linearity against TX leakage. A bottom-switching PA is proposed to realize the Bi-Phase modulation, and the magnetically-coupled resonator technique is widely used to effectively expand the link bandwidth. Measurement results show that the FMCW TRX could generate reconfigurable chirps with the bandwidth from 250 MHz to 4 GHz and the period from 600 us to 10 ms. The root-mean-square (RMS) frequency error is less than 251 kHz. The TX maximum output power is 13.4 dBm and is adjustable within 3 dB by reconfiguring its LDO output voltage. The RX achieves 15.3 dB noise figure and -8.5 dBm RF input-referred P1dB. Real-time experiments are carried out using the proposed TRX chip, in which the measured average distance error is 10 cm. The overall power consumption is 921mW with 2-TXs and 3-RXs powered on.**

Keywords— *Radar; frequency-modulated continuous-wave (FMCW); mm-wave; transceiver; PLL; CMOS;*

I. INTRODUCTION

In recent years, high-performance millimeter-wave (mm-wave) automotive radars have been promising research topics. Among them, frequency-modulated continuous-wave (FMCW) radars have drawn many attentions because of its simplicity, powerful performances and functionalities [1-3]. In practical applications, the TX leakage may saturate the RF front-end and deteriorate the RX sensitivity. However, few works can be found focusing on it. In addition, some of the works only have one TX or RX channel. Compared to those works, MIMO radar can achieve higher resolution and sensitivity. Moreover, most of the works employ fractional-N analog PLLs, whose loop bandwidth cannot be flexibly reconfigured to support various FMCW chirps with different slope.

This paper presents a 76-81 GHz 2TX 3RX FMCW radar

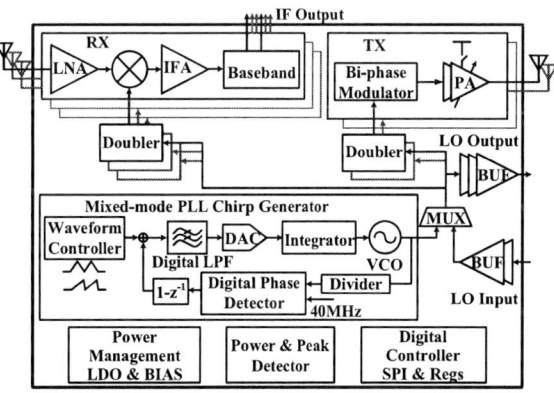

Fig. 1. System block diagram of the presented FMCW transceiver.

transceiver. To improve the TX leakage resilience, a high-linearity receiver is employed. The PA that support Bi-Phase modulation and on/off keying is proposed for MIMO application. A mixed-mode PLL with flexible loop bandwidth configuration is utilized to generate the FMCW chirp.

II. SYSTEM DESCRIPTIONS

Fig. 1 illustrates the architecture of the proposed 76-81GHz FMCW radar transceiver. The system consists of two TXs and three RXs for MIMO operation. A mixed-mode PLL chirp generator around 38.5 GHz is integrated, which is distributed in the LO network to multiple branches. The LO network includes 5 frequency doublers dedicated for each TX/RX channel, as well as a pair of input/output LO buffers for multiple chip cascading. In the chirp generator, a programmable waveform controller is utilized to generate triangular or sawtooth waveform control code. Each RX consists of a low noise amplifier (LNA), a mixer, an IF baseband amplifier (IFA) and a programmable analog baseband while each TX contains a PA with a Bi-Phase modulation stage.

To realize a high linearity receiver front-end against TX leakage, a single-stage low-gain LNA and a voltage-mode passive mixer are employed. An IF baseband amplifier is utilized to eliminate the leakage-induced dc-offsets. Due to the

978-1-5386-6414-8/18 $31.00 © 2018 IEEE

83

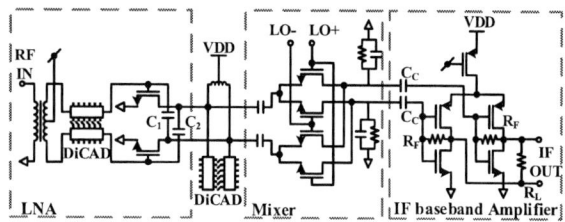

Fig. 2.   Schematic of the receiver front-end.

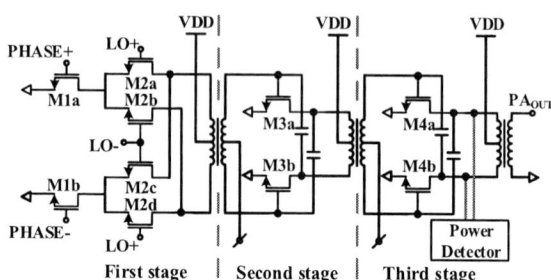

Fig. 3.   Schematic of the power amplifier

degraded noise-suppression capability of the low-gain LNA, low-noise design is involved in the mixer and IF baseband amplifier. For the MIMO application, 2 TXs and 3 RXs are integrated in this work. In each TX, a bottom-switching PA that can be turned on and off quickly is proposed to realize Bi-Phase modulation and on/off keying. To generate the FMCW chirp, a mixed-mode PLL with flexible loop bandwidth configuration is utilized in this work. Compared to all-digital PLLs, it avoids the complicated digitally-controlled oscillator (DCO) design by replacing the DCO with a voltage-controlled oscillator (VCO) and a current DAC followed by an integrator.

III.   CIRCUIT IMPLEMENTATIONS

A.   High Linearity Receiver

Fig. 2 shows the schematic of the receiver front-end, including a LNA, a mixer and an IF baseband amplifier. The LNA adopts the single-stage common-source topology, which provides large output swing at low supply voltage. A pair of cross-coupled capacitors (C_1, C_2) are used to neutralize the transistor gate-drain parasitic capacitance [4] to improve the reverse isolation. In the LNA input matching network, a transformer and a digitally controlled artificial dielectric (DiCAD) T-line [5] are utilized to realize single-to-differential conversion, as well as the matching frequency tuning. Compared to the MIM switched-capacitor array, the DiCAD introduces less loss to the network.

A double-balanced passive voltage-mode mixer is adopted in the front-end with several advantages. Firstly, the passive mixer has high linearity, which is important to resist the TX leakage. Secondly, no static current flows through the passive mixer, which contributes less flicker noise and saves power.

In the IF baseband amplifier, the resistor (R_F) and the coupling capacitor (C_C) form a high-pass filter (HPF), which eliminates the dc-offsets introduced by TX leakage. To address the poor noise-suppression of the LNA, low-noise design must be involved in the IF baseband amplifier. By employing the current reuse technique and inserting R_F into the feedback, this amplifier is self-biased, and the noise contribution from R_F as well as the transistors are minimized.

An analog baseband is integrated after the IF baseband amplifier, composing of a three-stage PGA with 0~52.6 dB gain, two low-pass filters with 2MHz bandwidth, and a HPF whose bandwidth can be reconfigured from 200 to 500 kHz.

B.   Transmmiter with Bi-phase Modulation

The 3-stage PA is illustrated in Fig. 3, in which the first stage is used as a Bi-Phase modulator, and the second and third stages are common-source amplifiers with neutralization capacitors.

There are two methods to implement Bi-Phase modulation, bottom-switching and top-switching (PHASE+ and PHASE- are differential control signals). The top-switching uses M_{2a-d} for the Bi-Phase modulation and $M_{1a,b}$ for LO input. The parasitic capacitance of M_{2a-d} makes it difficult to realize wideband impedance matching, while the turn-on resistance increases the loss of the matching network. In the proposed bottom-switching structure, the transistors $M_{1a,b}$ are used as the tail switches that switch in the modulation. Larger size tail transistors can be used to obtain higher conductance without loading the output node, thereby reducing the loss without degrading the matching bandwidth.

To achieve wideband performance, magnetically-coupled resonators (MCR) [6] are used to realize the inter-stage and output stage impedance matching. Magnetically-coupled resonator form a fourth-order network, and by controlling the resonant frequencies at both primary and secondary coils as well as their amplitudes, wideband response with acceptable ripple could be achieved.

The PA output power can be adjusted from 10.3 to 13.4 dBm by reconfiguring its LDO output voltage. Two TXs can be turned on and off rapidly to support the dual-channel time-division operation. According to the simulation results, the switching time is less than 30ns. A power detector is used to monitor the output power of the PA [7].

C.   LO generation and mixed-mode PLL

Fig. 4 depicts the building blocks in the LO generation and distribution, including a waveform controller, a PLL and a LO network. The mixed-mode PLL is adopted to generate LO because its loop bandwidth can be flexibly programmed on-chip, which is useful to configure the FMCW chirp slope. Compared to all-digital PLLs, it replaces the complicated DCO-based solution with a simple VCO and a DAC.

In the chirp generator, the waveform controller generates triangular or sawtooth waveform control code with reconfigurable bandwidth and slope. In the mixed-mode PLL, a LC VCO works at around 38.5 GHz, whose output is divided by 32 and sent to a digital phase detector (DPD). The DPD

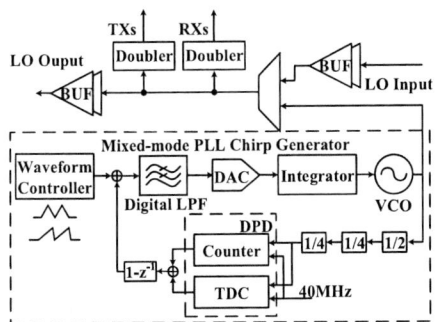

Fig. 4.   Scheme of mixed-mode PLL chirp generator and LO distributions.

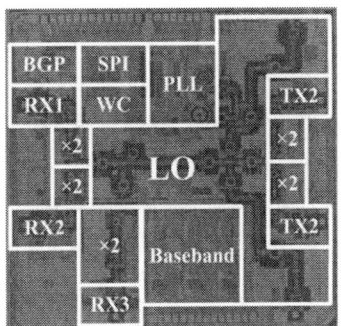

Fig. 5.   Chip photograph (×2 = doubler, WC = Waveform Controller).

consists of a fine TDC and a coarse counter, detecting phase error between the divided clock and the external reference clock. A digital differentiator ($1-z^{-1}$) converts the phase error into frequency and adds to the waveform control code. It travels through the digital loop filter (DLF) and current DAC before being integrated to a voltage, which then controls the VCO frequency.

In the DPD, a TDC is used to detect the period of the divided clock. However, the mismatch between the rising and falling times in the TDC degrades its performance. To address this issue, a dual-rising edge DPD including a wide range 2-dimension Vernier TDC is employed. A delay locked loop (DLL) is proposed in the TDC for self-calibration. Thus, the TDC achieves 2.3 ns detection range and 10 ps resolution.

The PLL is designed to work at around 38.5 GHz, whose output then gets frequency-doubled to 77 GHz, acting as the LO signal in TX transmission and RX down-conversion. This frequency doubling scheme simplifies the VCO and LO buffer design, reduces the transition loss and extends the LO network bandwidth. At maximum, the LO frequency slope and bandwidth can achieve 13.3 MHz/us and 4 GHz respectively. The falling time of the sawtooth wave reaches as short as 5 us.

IV.   EXPERIMENTAL RESULTS

The proposed 76~81 GHz 2TX 3RX FMCW radar TRX has been implemented in 65nm CMOS. As depicted in Fig. 5, the entire chip area is 7.29 mm^2 including all bonding pads.

The VCO measurement shows that it achieves 4.8 GHz frequency tuning range, from 37.2 GHz to 42.0 GHz. The measured PLL phase noise is -87.4 dBc/Hz at 1MHz offset from 78.7 GHz carrier.

The integrated chirp generator could generate reconfigurable chirps with the bandwidth from 250 MHz to 4 GHz and the period from 600 us to 10 ms. The root-mean-square (RMS) frequency error is less than 251 kHz. Fig. 6 shows the measured demodulated chirp waveform and frequency error, where the chirp bandwidth is set to 4 GHz and the period is set to 600 us and 1 ms, respectively. The RMS frequency error is 181 kHz without considering turning points.

The conversion gain of the receiver path (RF+IF) is programmable from 26.2 to 78.8 dB. The minimum receiver gain and corresponding noise figure are measured by fixing a

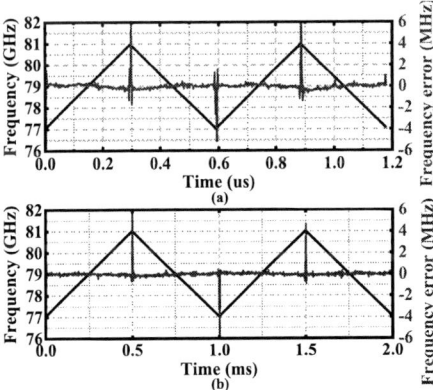

Fig. 6.   Measured demodulated FMCW waveform and frequency error with (a) 600us period. (b) 1ms period (Bandwidth is set to 4 GHz).

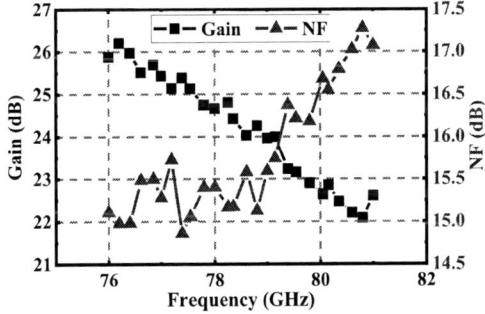

Fig. 7.   Measured receiver minimum conversion gain and noise figure.

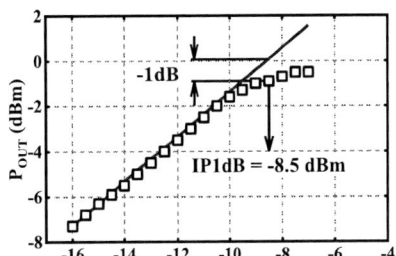

Fig. 8.   Measured IP1dB of the receiver front-end.

600kHz IF and sweeping the RF frequency from 76 to 81 GHz. As shown in Fig. 7, the noise figure is 15.3dB at 78 GHz. The

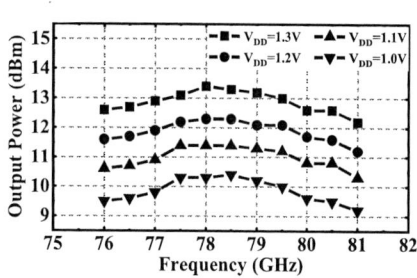

Fig. 9. Measured PA output power versus the operation frequency

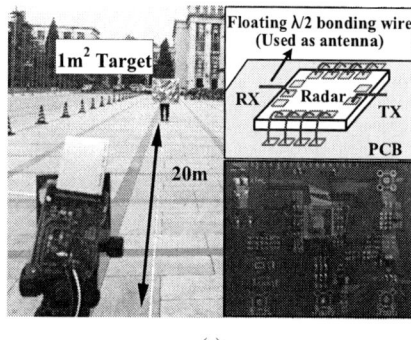

(a)

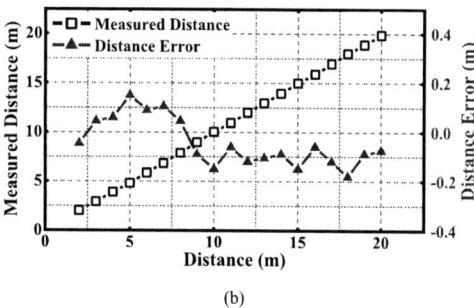

(b)

Fig. 10. Real-time experiment. (a) Setup. (b) Result.

center frequency of the receiver is somewhat lower than the desired one, due to inaccurate modelling.

Fig. 8 shows that the input 1dB compression point of the receiver front-end is -8.5 dBm. Since the output power of the PA is 13.4 dBm, the receiver can resist the TX leakage once the TX-RX isolation is better than 21.9 dB.

Fig. 9 shows the measured PA output power with different power supplies by configuring the LDO. The maximum value is 13.4 dBm at 79 GHz with the PAE of 13.5%. The 1dB BW is greater than 5 GHz, covering the entire operating band.

Fig. 10a shows the real-time experimental setup, in which simple half-wavelength bond-wires are used as antennas. A 1 m² aluminum foil is used as the target, and the output IF signal of the receiver is sent to the spectrum analyzer to calculate the distance. The maximum detection range is measured as 20 m, which is limited due to the poor antenna gain. The average detection error is 10 cm, and can be further improved through digital signal processing.

The performances are summarized and compared to similar works in Table I. This works achieves better linearity in the case of the same chirp bandwidth and slope.

Table. 1. COMPARISON

	[1]	[2]	[3]	This Work
Process	65nm CMOS	65nm CMOS	45nm CMOS	65nm CMOS
Channel	RX+TX	2RX+TX	4RX+3TX	3RX+2TX
Chirp BW (GHz)	0.7	1.93	4	4
Chirp slope (MHz/us)	1.4	1.93	100	13.3
RMS error (kHz)	65	674	-	181
Phase noise (dBc/Hz)	-85.33	-81	-94	-87.4
P_{OUT}(dBm)	5.1	12.9~13.2	10.8	13.4
RX gain(dB)	38.7	47.8~100.7	-	26.2~78.8
NF(dB)	7.4@1M	5.04@3.3M	18@1M	15.3@1M
Area(mm²)	1.04	4.64	22	7.29
Power (W)	0.243	0.343	3.4	0.921

V. CONCLUSION

A 76-81GHz 2TX 3RX FMCW radar transceiver based on a mixed-mode PLL chirp generator in 65nm CMOS is presented, enabling multi-chip cascading operating for MIMO processing. A mixed-mode PLL is utilized to generate the FMCW chirp signal. High linearity LNA and passive mixer are employed to resist the TX leakage. A bottom-switching PA is proposed for Bi-Phase modulation. Measurement results show 4 GHz FMCW bandwidth, 13.3 MHz/us chirp slope, 13.4 dBm PA output power and -8.5dBm RX input P1dB, at the cost of 921 mW power consumption in total.

REFERENCES

[1] J. Lee, Y. A. Li, M. H. Hung, and S. J. Huang, "A Fully-Integrated 77-GHz FMCW Radar Transceiver in 65-nm CMOS Technology," IEEE Journal of Solid-State Circuits, vol. 45, pp. 2746-2756, Dec 2010.

[2] H. K. Jia, L. X. Kuang, W. Zhu, Z. P. Wang, F. Ma, Z. H. Wang, et al., "A 77 GHz Frequency Doubling Two-Path Phased-Array FMCW Transceiver for Automotive Radar," IEEE Journal of Solid-State Circuits, vol. 51, pp. 2299-2311, Oct 2016.

[3] K. S. Brian P. Ginsburg1, Sreekiran Samala1, "A Multimode 76-to-81GHz Automotive Radar Transceiver with Autonomous Monitoring," ISSCC, 2018.

[4] W. L. Chan and J. R. Long, "A 58-65 GHz Neutralized CMOS Power Amplifier With PAE Above 10% at 1-V Supply," IEEE Journal of Solid-State Circuits, vol. 45, pp. 554-564, Mar 2010.

[5] B. C. Haikun Jia, Lixue Kuang, Zhihua Wang, "Simple and robust self-healing technique for millimetre-wave amplifiers," IET Circuits, Devices & Systems, 2015.

[6] H. K. Jia, C. C. Prawoto, B. Y. Chi, Z. H. Wang, and C. P. Yue, "A 32.9% PAE, 15.3 dBm, 21.6-41.6 GHz Power Amplifier in 65nm CMOS Using Coupled Resonators," 2016 IEEE Asian Solid-State Circuits Conference (A-SSCC), pp. 345-348, 2016.

[7] S. Kawai, T. Mitomo, and S. Saigusa, "A 60GHz CMOS Rectifier with-27.5dBm Sensitivity for mm-Wave Power Detection," 2012 IEEE Asian Solid State Circuits Conference (A-SSCC), pp. 281-284, 2012.

A 25 fps 32 × 24 Digital CMOS Terahertz Image Sensor

Tong Fang[1,2], Run-jiang Dou[1,2], Li-yuan Liu[1,2], Jian Liu[1,2] and Nan-jian Wu[1,2]

[1]State Key Laboratory of Superlattices and Microstructures, Institute of Semiconductors,
Chinese Academy of Sciences, Beijing 100083, China
[2]University of Chinese Academy of Sciences, Beijing 100049, China
Email: liujian@semi.ac.cn and liuly@semi.ac.cn

Abstract—A 32×24 digital CMOS terahertz image sensor for 25 fps imaging at 860 GHz is presented. Pixel with DRC free structure is proposed and column parallel readout architecture is employed. The sensor array and the readout circuits are fully integrated into one single silicon chip in 180 nm CMOS process. Measured averaged room-temperature responsivity of single pixel is 1.28 kV/W at 860 GHz. The THz image sensor can operate up to 25 fps under continuous-wave illumination, and we acquired a video of a rotating blade chopping the backward wave oscillator (BWO) source beam, and the rotating blade can be clearly distinguished in the video.

Keywords—Terahertz imaging; CMOS Terahertz image sensor; DRC free patch antenna; Column parallel readout

I. INTRODUCTION

Terahertz (THz) radiation with frequency range from 0.3 to 3 THz (1 THz=10^{12} Hz) has many properties: non-ionizing penetration through many dry, non-metallic and non-polar materials such as plastics, paper sheets or common clothing, higher spatial resolution in comparison to millimeter waves, and specific spectral fingerprints for many substances. Those properties are suitable for some applications such as security screening [1], bio-sensing [2], and drug and explosives analysis [3]. Among these, security screening and bio-sensing have been drawn more attentions. Thus, cost effective, real-time, high sensitive, low-noise and room temperature THz imaging systems are strongly demanded. Single THz detector has now progressed towards detector array for reduced imaging time. Dues to the CMOS advantage of high yield and low-cost, particular attention has been devoted to the CMOS integrated THz detector array [4]-[8]. Unfortunately, none of them has integrated on-chip readouts circuits, external commercial amplifiers and ADCs could be used to process the analog outputs from the chip, but the system is still complicated and bulky. Although flip-chip technique could be used to hybridize the detector array and the CMOS readout integrated circuit (ROIC) [9], it dramatically increases the cost.

In this paper, we report a 32×24 digital CMOS image sensor for 25 fps THz imaging at 860 GHz. To reduce system-level complexity and save costs, the sensor array and the readout circuits are fully integrated into a single silicon chip in 180 nm CMOS process. It can operate up to 25 fps under continuous-wave illumination, which results in a digital CMOS THz imaging chip for many promising applications.

This work was supported by National Key R&D Program of China (Grant No. 2016YFA0202202) and the National Science and Technology Major Projects (Grant No. 2016ZX03001002-002) of the Ministry of Science and Technology of China, National Natural Science Foundation of China (Grant Nos. 61474108, 61331003) and Youth Innovation Promotion Association Program, Chinese Academy of Sciences (No.2016107).

II. CHIP ARCHITECTURE

Fig. 1 shows the chip architecture diagram. A 32×24 pixel array is located in the middle of the chip. Column parallel readout circuits array for odd- and even-numbered channels are put on the top and bottom respectively. Each readout channel consists of a chopping stabilization amplifier and an incremental A/D converter (I-ADC). Digital control logic on the left can generate all the control signals including row selection time and the parameter configuration for on-chip amplifiers and ADCs. There are two modes for chip operation. In normal mode, the pixel array together with readout circuits work in rolling-shutter manner and output imaging data at a frame rates up to 25 fps. In test mode, every pixel can be addressed and measured individually. In addition, pixel out is also able to be accessed directly by bypassing the readout channel. Bias circuit is on the right. It distributed all the bias current to readout channels.

A. Pixel with DRC Free Patched Antenna

Fig.2 shows the diagram of the THz pixel similar to our previous work [10]. THz radiation is received by on-chip patch antenna. A matching network is designed to couple THz

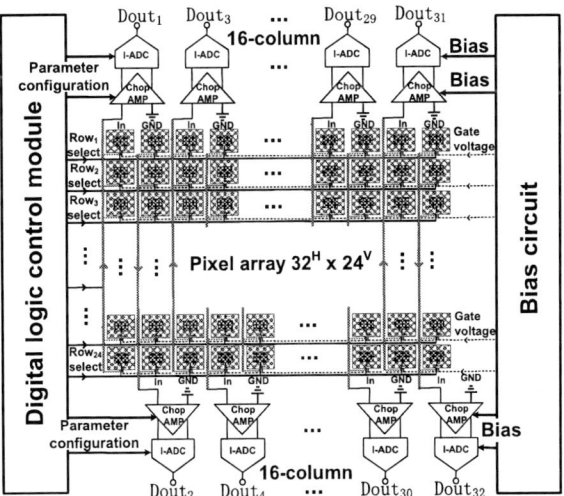

Fig. 1. Chip architecture diagram.

signal from the antenna to the source terminal of an NMOS transistor (M1). To obtain optimum responsivity (R_v), M1 is sized to be 240 nm/180 nm, the minimum allowable dimension in the technology [6]. A notch filter by metal transmission line is also designed and connected to the gate of M1. The notch filter can provides ac ground at the corresponding THz frequency for the gate, thus to eliminate the influence of the gate bias supply line [10].

For 860 GHz radiation, our patch antenna is composed of a top metal plane (metal 6) with 100 μm × 67 μm and a bottom metal plane (metal 1) with 220 μm × 200 μm as shown in the Fig. 2. Such large metal planes are usually not allowed because of reliability issue due to unreleased metal stress. To deal with it, we deliberately make some periodically placed square holes on both top and bottom plane. The size of hole at the patch plane is 1.5 μm × 1.5 μm and at the ground plane is 1.0 μm × 0.4 μm. By doing this, large plane can be segmented into small ones and metal stress can be released. Due to small size of holes, antenna performance is still guaranteed according to our electromagnetic (EM) simulation.

Another design issue with patch antenna design is that the vertical distance between the top and the bottom plane should be maximized to obtain high radiation efficiency. In this work, metal 1 and metal 6 are used as bottom and top plane. Other metal layers from metal 2 to 5 should not exist in the space right between metal 1 and metal 6. As a result, metal densities of these layers become extremely low in this area. This will

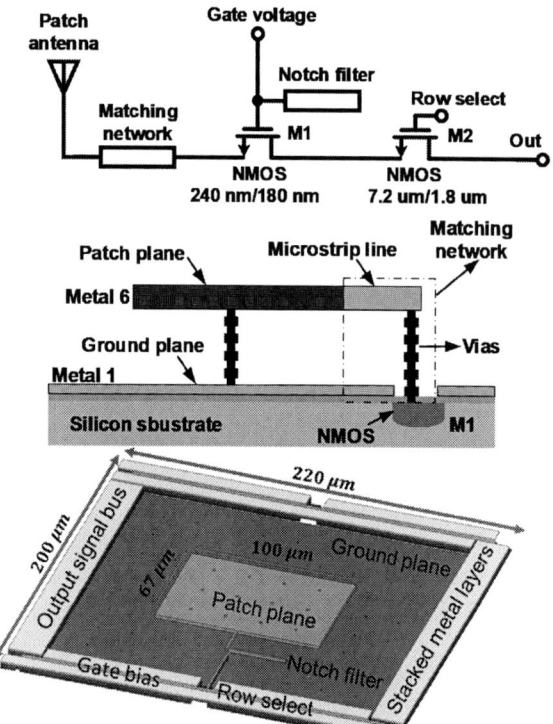

Fig. 2. Diagram of the THz pixel and the simulation model of the single pixel.

also result in manufacture failure related to process steps such as metal etching. In this design, stacked dummy metal layers are adopted and put around the edge of our patch antenna to ensure sufficient metal density. The dummy shape and distance to the top patch are carefully chose and simulated in order to guarantee the antenna's performance.

B. Column Parallel Readout Scheme

The THz pixel adopts a passive detection method, the output magnitude of which is determined by the radiation power level. Since Backward Wave Oscillation (BWO) source can only emit 860 GHz radiation with power of less than 150 μW, the pixel output is roughly on the order of several millivolts even if the pixel has moved to the center of the focused source beam; for pixels at the edge of the source beam, outputs are much weaker. To achieve enough dynamic range of imaging, the readout circuits must be capable to readout very weak signal and to ensure sufficiently low input-refer noise.

Readout noise includes thermal noise and flicker noise. Flicker noise can easily be removed by chopping technique, so that we concern more about the thermal noise. Thermal noise is signal band related. For a sampling system, suppose a total thermal noise power of N_t, sampling rate of f_s and signal update rate of f_u, the input refer noise of a readout channel can be expressed as $N_{input}=N_t \times f_u/f_s$. Compared to reducing N_t by increasing the capacitor size with extended power consumption, a more suitable way is keeping the ratio of f_u/f_s as small as possible. This implies that employing oversampling system with high oversampling ratio. Since increasing sampling frequency f_s leads to difficulties such as generation and distribution of high frequency clock, higher power dissipation, more signal coupling and disturbance, we turn to lowering down f_u. Since a target frame rate has specified, we finally reach the column parallel readout scheme as shown in Fig.1. By choosing a column parallel architecture, each channel can operate under a very slow update rate and low thermal noise is achieved. Meanwhile frame rate is still guaranteed thanks to the acceleration by parallel fashion.

III. CHIP DESIGN

An array of 32×24 pixels was designed. Each pixel occupied area of 220 μm × 200 μm. Selection transistor was sized to be 7.2 μm/1.8 μm for small on resistance. The output of each pixel is connected to column readout bus through the selection transistor. This bus is then connected to the positive input terminal of the readout channel whose negative terminal is grounded. The readout channel is composed of a preamplifier followed by a 3-rd order feed forward cascade integrator feedback (CIFB) I-ADC [10]. The systematic clock of our sensor is 1MHz. To achieve 25fps imaging rate, each row was exposed for 1664 clock period. Simultaneously the ADC will gather 1664 samples and output a stream of 1s and 0s with the same length. The stream is directly sent to off-chip FPGA for post processing.

The preamplifier adopts a two-stage Miller compensated instrumentation topology [10]. Default gain is set to 16. Miller capacitor is 2 pF here. Sampling capacitor of the I-ADC is

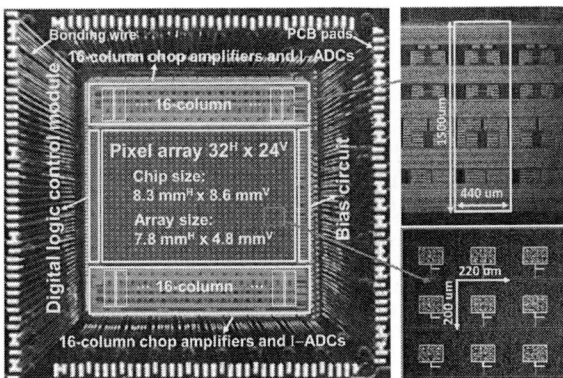

Fig. 3. Die photos of the CMOS THz image sensor and a close-up photograph of the readout circuit and the single pixel

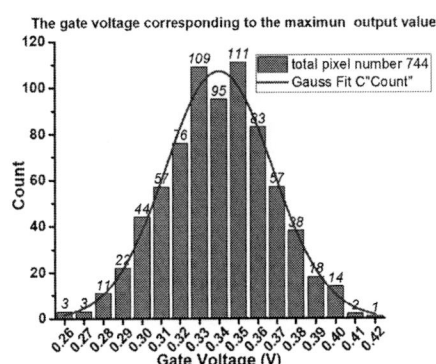

Fig. 4. Histogram of gate voltage corresponding to the maximum output value in an array (744 pixels).

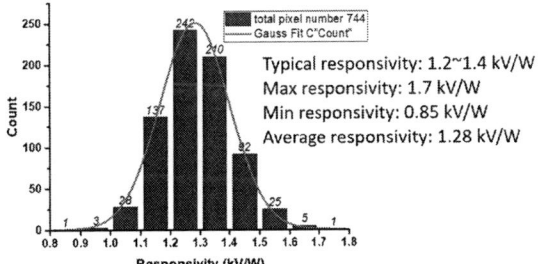

Fig. 5. Histogram of the responsivity in an array (744 pixels).

designed to be 4 pF. These capacitor sizes are large enough for thermal noise below 10 μV_{RMS} with above decimation ratio. To eliminate flicker noise, both the preamplifier and amplifier inside I-ADC are chopped. For preamplifier, its chopping frequency is set to 125 kHz which is only 1/8 of f_s and hence the chopping component lies far away from the cutting edge of the SINC filter and can be removed completely. Amplifiers inside I-ADC are also chopped, but the chopping frequency is as high as 1/2 f_s.

For test purpose, each read channel can be powered down and the output of pixel is directly connected to bonding pads through switch.

Fig. 3 shows the chip die photograph. It is fabricated in 1P6M 180 nm standard CMOS technology. The array size is $7.8 \times 4.8\ mm^2$ with pixel pitch of 220 μm × 200 μm, the chip size is $8.3 \times 8.6\ mm^2$, the column readout circuits size is $1500 \times 440\ \mu m^2$, all circuits operate under 1.8 V.

IV. EXPREIMENTAL RESULTS

We let the sensor chip work at test mode to measure the performance of each pixel individually, pixel-out can be connected to the lock-in amplifier directly by bypassing the readout channel. The scattered output THz radiation from the BWO was focused on the sensor by two off-axis parabolic mirrors, the pixel analog output was connected to the lock-in amplifier, and the BWO was chopped at a frequency of 177 Hz which also provides a reference signal for the lock-in amplifier, the data from the lock-in amplifier was collected by the personal computer through the GPIB card. We measured 744 pixels' output voltage versus the gate voltage (the last column was designed for other measurements). Fig. 4 shows the histogram of the maximum output value corresponding to the gate voltage, which closed to a Gaussian distribution.

Fig. 5 shows the histogram of measured responsivity of 744 pixels. It is close to Gaussian distribution with an average responsivity of 1.28 kV/W.

Fig. 6 shows the measurement set-up for real-time imaging of the source beam which is chopped by a motor controlled blade. The THz non-penetrated blade was put near the THz wave outlet of the BWO and was driven by a motor rotating at low speed. The chopped THz wave was focused on the THz

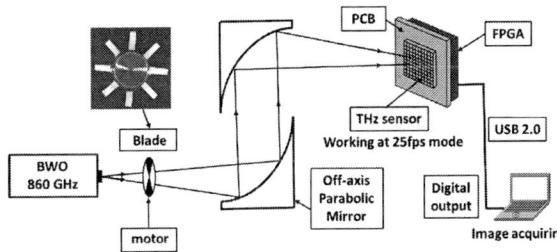

Fig. 6. Measurement set-up of real-time imaging of the source beam chopped by a rotating blade.

sensor chip by two off-axis parabolic mirrors. The sensor chip received the THz wave and immediately processed by the ROICs which amplified the weak analog signal and converted into digital signal. Then the FPGA captures the digital outputs from the sensor chip and transfers them to a computer. Fig. 7(a) shows actual measurement set-up of Fig. 6. In this experiment the sensor chip worked at 25 fps mode. Fig. 8 shows the video sequence about a rotating blade (anticlockwise) chopping the source beam. It can be seen that the source beam had been chopped roughly three times in 1 second. Table I is the performance comparison between this work and other THz focal plane array technology. Our sensor achieves a relative high responsivity excluding the gain of the integrated amplifier. Compared to these who uses external commercial amplifiers and ADCs to readout or flip-chip technology to hybridize the detector array and the CMOS ROIC, our chip applied on-chip readout circuits, thus reduces system-level complexity and saves costs. Our results demonstrate a promising application in low-cost digital CMOS THz camera.

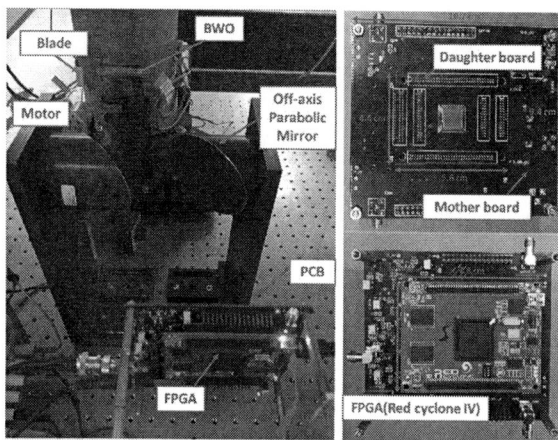

Fig. 7. Left: Actual measurement set-up of Fig. 6, right: the front side and back side of the measurement board.

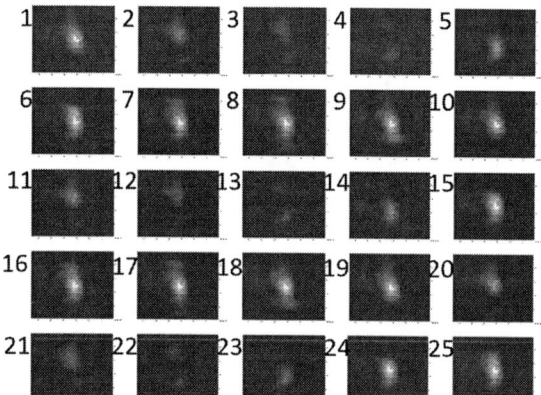

Fig. 8. Video sequence of the source beam chopped by a rotating blade (anticlockwise).

V. CONCLUSIONS

We report a 32×24 digital CMOS THz image sensor for 25 fps imaging at 860GHz. Pixel with DRC free structure was proposed and consideration of choosing column parallel

readout scheme was completed. The sensor array and the readout circuits are integrated into a single silicon chip in 180 nm CMOS process, the chip size is $8.3 \times 8.6 \ mm^2$ including bonding pads. The pixels were achieved an average responsivity of 1.28 kV/W. The chip can operate up to 25 fps. And a video of chopped BWO source beam is captured and the rotation of the chopper blade can be clearly distinguished

REFERENCES

[1] D. L. Woolard, J. O. Jensen, R. J. Hwu, and M. S. Shur, Terahertz Science and Technology for Military and Security Applications. Singapore: World Scientific, 2007.

[2] P. de Maagt, P. H. Bolivar, and C. Mann, , K. Chang, Ed., "Terahertz science, engineering and systems—From space to earth applications," in Encyclopedia of RF and Microwave Engineering. NewYork: Wiley, 2005, pp. 5175–5194.

[3] Hoshina, H., Sasaki, Y., Hayashi, A., Otani, C., and Kawase, K., "Noninvasive mail inspection system with terahertz radiation," Applied spectroscopy 63, 81-86 (2009).

[4] R. Al Hadi, et al., "A 1 k-Detector Video Camera for 0.7–1.1 Terahertz Imaging Applications in 65-nm CMOS," IEEE J. Solid-State Circuits, Vol. 47, no. 12, Dec. 2012.

[5] D. Y. Kim, S. Park, R. Han, Kenneth K. O. "820-GHz imaging array using diode-connected NMOS transistors in 130-nm CMOS." Vlsi Circuits IEEE, 2013:C12-C13.

[6] E. Öjefors, U. Pfeiffer, A. Lisauskas, and H. Roskos, "A 0.65 THz focal-plane array in a quarter-micron CMOS process technology," IEEE J. Solid-State Circuits, vol. 44, no. 7, Jul. 2009.

[7] J. Zdanevičius, M. Bauer, S. Boppel, V. Palenskis, A. Lisauskas, V. Krozer, H. G. Roskos. "Camera for High-Speed THz Imaging." Journal of Infrared Millimeter & Terahertz Waves, 2015, 36(10):986-997.

[8] A. Boukhayma, A. Dupret, J. P. Rostaing and C. Enz. "A Low-Noise CMOS THz Imager Based on Source Modulation and an In-Pixel High-Q Passive Switched-Capacitor N-Path Filter"[J]. Sensors, 2016, 16(3):325

[9] G. C. Trichopoulos, H. L. Mosbacker, D. Burdette, and K. Sertel. "A Broadband Focal Plane Array Camera for Real-time THz Imaging Applications." IEEE Transactions on Antennas & Propagation, 2013, 61(4):1733-1740.

[10] Z. Y. Liu, L. Y. Liu, J. Yang, and N. J. Wu, "A CMOS Fully Integrated 860-GHz Terahertz Sensor." IEEE Transactions on Terahertz Science and Technology, vol. 7, no. 4, Jul. 2017.

TABLE I.   COMPARISON BETWEEN THE TERAHERTZ FPA TECHNOLOGY

Technology	BW/Freq. [THz]	Pixel [x × y]	Max R_v [V/W]	Min NEP [pW/Hz$^{1/2}$]	Frame Rate[fps]	Optics	On-chip readout circuits	Ref.
180 nm CMOS	0.86	32×24	[1]1.28k	-	25	-	Yes	This work
65nm bulk CMOS	0.79-0.96	32×32	[2]140k, 0.86 THz	100, 0.86THz	25	Si lens	No	[4]
150 nm CMOS	0.6	24×24	300	43	450	-	No	[7]
130 nm CMOS	0.2, 0.27, 0.6	31×31	[3]300k, 0.27THz	18.7	100	-	No	[8]
Sb-HBD	0.6-1.0	80×32	600(optical), 0.7THz	850	100	Si lens	No	[9]

[1]Average responsivity of 744 pixels.
[2]Including an on-chip amplifier with a 50-dB open-loop gain and a 5-dB VGA gain.
[3]Include a 58-dB on-chip closed-loop gain.

10-1 (8105)

IEEE Asian Solid-State Circuits Conference
November 5 - 7, 2018/Tainan, Taiwan

A 3.9μW, 81.3dB SNDR, DC-coupled, Time-based Neural Recording IC with Degeneration R-DAC for Bidirectional Neural Interface in 180nm CMOS

Hyuntak Jeon[1], Jun-Suk Bang[2], Yoontae Jung[1], Taeju Lee[1], Yeseul Jeon[1],
Seok-Tae Koh[1], Jaesuk Choi[1], Doojin Jang[1], Soonyoung Hong[3], Minkyu Je[1]

[1]School of Electrical Engineering, KAIST, Daejeon, Korea, [2]Samsung Electronics, Hwaseong, Korea, [3]DGIST, Daegu, Korea

Abstract—**This paper presents a 5-bit VCO-based neural recording IC, which directly quantizes the input signal and achieves a large dynamic range (DR) to process the small-amplitude neural signal in the presence of the large-amplitude stimulation artifact (SA). A feedback-controlled source degeneration is applied to the input transconductor circuit ($G_{m,in}$) by using a resistor DAC (R-DAC). It mitigates the circuit nonlinearity, resulting in a large signal-to-noise-and-distortion ratio (SNDR) and a high input impedance (Z_{in}). The implemented neural recording IC achieves 81.3dB SNDR over 200Hz signal bandwidth and 200mV$_{PP}$ maximum allowable input range while consuming 3.9μW per channel.**

I. INTRODUCTION

The bidirectional neural interface system for the closed-loop interaction between the human nervous system and the human-made machine should provide a large DR performance for neural recording function. There should be no loss or distortion of the recorded neural signal even when the strong SA and the significant DC electrode offset (DEO) are present. If we target to record the local field potential (LFP) signal with direct DC coupling, both the DEO and SA can be as large as several tens of mV in the signal band, as shown in Fig. 1. More than 80dB of DR is therefore required to record the LFP signal with a good SNR while accommodating the DEO and SA.

The conventional neural recording circuit employing an instrumentation amplifier (IA) followed by an ADC as in [1] has difficulty in achieving a high DR. To overcome this limit, a recording structure consisting of a $G_{m,in}$ and a quantizer based on a current-controlled oscillator (CCO) is proposed in [2]. However, due to its inherently nonlinear characteristics, a digital calibration circuit needs to be added, resulting in a significant power and area overhead even when implemented in 40nm technology, in addition to the inevitable calibration cost. In [3], a VCO-based DSM is introduced with employing a capacitor DAC (C-DAC) to feedback the output to the AC-coupled input, achieving a large DR without suffering from the nonlinearity issue. However, its low Z_{in} renders it suboptimal for the neural recording application.

In this work, by applying a feedback-controlled degeneration R-DAC to the VCO-based DSM, a large DR, as well as a high Z_{in}, are obtained, so that the neural signal can be recorded concurrently with large interferences such as SA and DEO.

II. PROPOSED CONCEPT AND CIRCUIT IMPLEMENTATION

As shown in Fig. 2, the $G_{m,in}$ implemented by a pair of PMOS transistors converts the differential input voltage signal, $V_{in}^+ - V_{in}^-$, into the imbalanced branch currents, I_{in}^+ and I_{in}^-, which are used to control the frequency of the ring oscillators, CCO$_P$ and CCO$_N$, respectively. The output phase difference of these two ring oscillators, $\phi_P - \phi_N$, is converted to 5-bit digital output by the phase quantizer. The digital output is fed back to control the degeneration R-DAC so that the branch currents become balanced and the frequency difference of the two ring oscillators converges to zero in the steady state. By doing so,

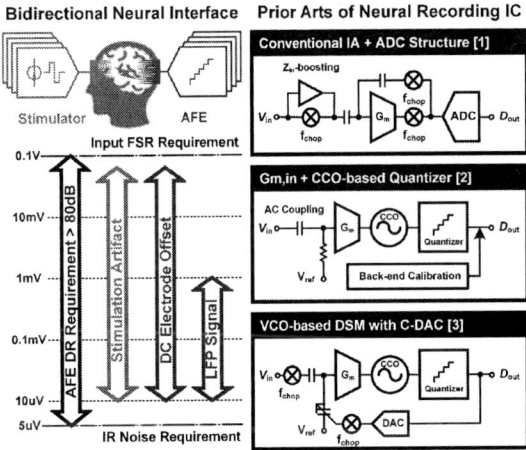

Fig. 1. DR requirement of bidirectional neural interface system and previously reported neural recording circuit structures

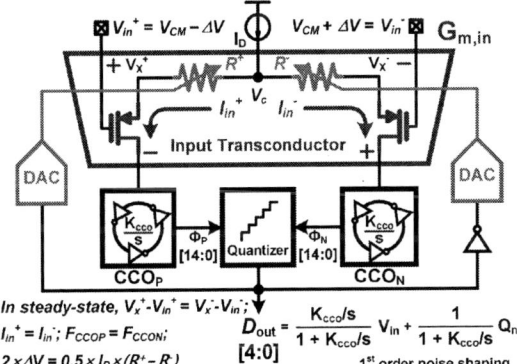

Fig. 2. Key concept of proposed structure

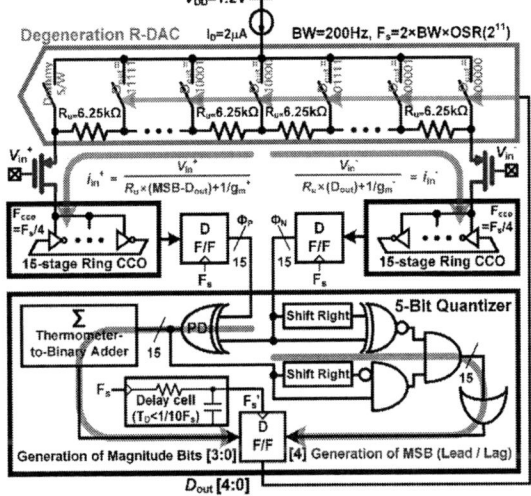

Fig. 3. Detailed circuit implementation

978-1-5386-6414-8/18 $31.00 © 2018 IEEE

91

the characteristics of the $G_{m,in}$ and CCOs are linearized, and the targeted large DR is obtained. Since the input is directly DC-coupled and the large DR absorbs the DEO, the circuit can also achieve a very high Z_{in} in the LFP band. Note that the CCO can be modeled as an integrator in the negative feedback loop, introducing the benefit of the 1st-order noise-shaping effect.

Fig. 3 shows the detailed circuit implementation. The differential input voltage is converted to the differential branch currents by the $G_{m,in}$, and each of them is fed to the current input port of a 15-stage inverter-based ring oscillator. From the magnitude of the difference between the output phases of the two CCOs, the phase detector (PD) implemented by an array of XOR gates generates 15 thermometer bits, which in turn are converted to 4 binary bits ($D_{out}[3:0]$) by a thermometer-to-binary adder. In addition, to generate the MSB ($D_{out}[4]$) representing the relative lead/lag information between the two CCO outputs, the structure used for the phase-extended quanizer in [4] is employed to minimize the implementation complexity and power consumption of the quantizer.

III. Measurement Results and Performance Comparison

The proposed neural recording IC is fabricated in 180nm CMOS process. The prototype IC shown in Fig. 6 has 4 channels and each channel occupies an area of 0.45×0.5mm^2, implying that a 100-channel neural recording system can be realized within an area of 5×5mm^2 using 180nm process. Fig. 4 shows the output spectrum measured with applying 200mV$_{pp}$ 59.375Hz sinusoidal voltage input, and the SNDR measured with a BW of 200Hz and a sampling frequency of 819.2kHz. The measurement results exhibit the performance of 81.3dB SNDR and 91.2dB SFDR.

Fig. 5 shows the LFP waveform measured through the phosphate-buffered saline (PBS) solution under the setup depicted in Fig. 6. The measurement was carried out using the prerecorded LFP waveform and the instrument-generated SA waveform as signal sources. The middle and bottom plots of Fig. 5 present the LFP waveform contaminated by the modeled SA signal. Thanks to the large DR of the proposed recording circuit, the original LFP waveform is well preserved even in the presence of large artifacts without any saturation or significant distortion. Table I summarizes the performance and compares it with that of other state-of-the-art designs. The proposed design achieves very large SNDR and allowable input range, as well as nearly infinite Z_{in} in the LFP band, with low power consumption.

Acknowledgement

This research was supported by the Convergence Technology Development Program for Bionic Arm (2017M3C1B2085296) and the Brain Research Program (2017M3C7A1028859) through the National Research Foundation of Korea (NRF) funded by the Ministry of Science & ICT.

References

[1] H. Chandrakumar, et al., "A 2.8uW 80mV$_{pp}$-Linear-Input-Range 1.6GΩ-Input Impedance Bio-Signal Chopper Amplifier Tolerant to Common-Mode Interference up to 650mV$_{pp}$," ISSCC, pp. 448–449, 2017.

[2] W. Jiang, et al., "A ±50mV Linear-Input-Range VCO-Based Neural-Recording Front-End with Digital Nonlinearity Correction," ISSCC, pp. 484–485, 2016.

[3] C. –C. Tu et al., "A 0.06mm2 ± 50mV Range -82dB THD Chopper VCO-based Sensor Readout Circuit in 40nm CMOS," SOVC, pp. C84–C85, 2017.

[4] S. Li, et al., "A 174.3dB FoM VCO-Based CT ΔΣ Modulator with a Fully Digital Phase Extended Quantizer and Tri-Level Resistor DAC in 130nm CMOS," ESSCIRC, pp. 241–244, 2016.

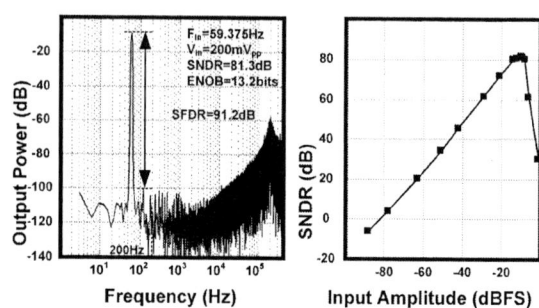

Fig. 4. Measured output spectrum and SNDR performance

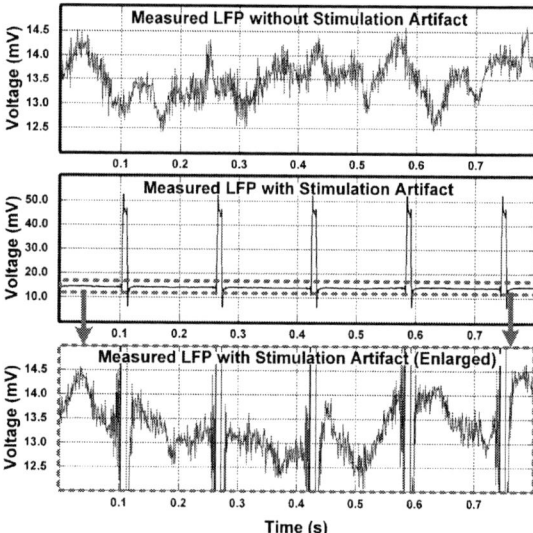

Fig. 5. LFP waveform measured *in vitro*: with and without contamination by large stimulation artifacts

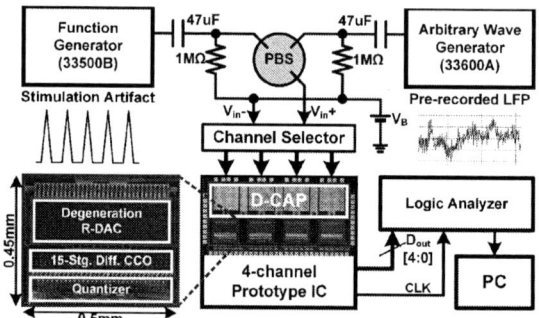

Fig. 6. Die micrograph and *in-vitro* measurement setup

Table I. Performance summary and Comparison

	ISSCC 2017 [1]	ISSCC 2016 [2]	VLSI 2017 [3]	This Work
Process	40nm	40nm	40nm	180nm
Topology	IA+ADC	VCO-based Quantizer	VCO-based DSM + C-DAC	VCO-based Quantizer + R-DAC
Target	LFP+AP	LFP	Hall Sensor	LFP
Power/Channel	2.8μW	7μW	21μW	3.9μW
Area/Channel	0.069mm²	0.135mm²	0.06mm²	0.225mm²
Peak Input	40mV$_{PP}$	100mV$_{PP}$	100mV$_{pp}$	200mV$_{PP}$
Bandwidth	10kHz	200Hz	2kHz	200Hz
F$_{sample}$	N/A	1kHz	1MHz	819.2kHz
Dynamic Range	SNR: 74dB (LFP) SNR: 81dB (AP)	SFDR: 79dB	SNDR: 74.9dB	SNDR: 81.3dB SFDR: 91.2dB
Z$_{in}$	1.6GΩ	∞	N/A	∞
Input Referred Noise	1.8μV$_{rms}$ (LFP) 5.3μV$_{rms}$ (AP)	6μV$_{rms}$	N/A	6.08μV$_{rms}$
Noise Efficiency Factor	*7.4 (LFP) *4.4 (AP)	N/A	N/A	**23

*Calculated only in IA, **Calculated for total recording interface.

A Second-Order Purely VCO-Based CT ΔΣ ADC Using a Modified DPLL in 40-nm CMOS

Yi Zhong[1,2], Shaolan Li[1], Arindam Sanyal[3], Xiyuan Tang[1], Linxiao Shen[1], Siliang Wu[2] and Nan Sun[1]

[1]The University of Texas at Austin, Austin, TX, USA, [2]Beijing Institute of Tech., Beijing, China, [3]The State University of New York at Buffalo, Buffalo, NY, USA.

Email: zhongyi@utexas.edu, slliandy@utexas.edu, nansun@mail.utexas.edu

Abstract—**This paper presents a power-efficient purely VCO-based 2nd-order CT ΔΣ ADC featuring a modified DPLL structure. It combines a VCO with an SRO-based TDC, which enables 2nd-order noise shaping without any OTA. The nonlinearity of the front-end VCO is mitigated by putting it inside a closed loop. A multi-PFD scheme reduces the VCO center frequency and power. The proposed architecture also realizes an intrinsic tri-level DWA. A prototype ADC in 40-nm CMOS process achieves a Schreier FoM of 170.3 dB with a DR of 72.7 dB over 5.2-MHz BW, while consuming 0.91 mW under 1.1-V supply.**

I. INTRODUCTION

OTAs are widely used to build integrators in classic ΔΣ ADCs. However, technology scaling makes them difficult to design under low power supply and small transistor intrinsic gain. VCO-based integrators, which are simple, power-efficient, and highly digital, are promising candidates to replace power-hungry active RC integrators in advanced processes. Most existing purely VCO-based ΔΣ ADCs are limited to only 1st-order noise shaping [1]. Therefore, it presents a need to realize high-order VCO-based ΔΣ ADC. Ref. [2], [3] combines the VCO with classic OTA-based integrators to achieve high-order shaping, but they sacrifice the scaling friendliness and the power efficiency due to the use of OTA. The work in [4] demonstrated a low-power passive RC and VCO hybrid. It however renders only 1st-order in-band noise suppression, and thus, has a higher in-band noise floor compared to classic 2nd-order ΔΣ ADCs. A 3rd-order purely VCO-based ADC reported in [5] leverages up-down counters to cascade multiple VCOs. Nevertheless, it suffers from VCO nonlinearity as the first VCO integrator operates in open loop and experiences the full signal swing, thereby requiring nonlinearity calibration.

This paper presents an OTA-free 2nd-order purely VCO-based CT ΔΣ ADC featuring a modified digital phase-locked loop (DPLL) structure. It is inspired by the fact that a DPLL intrinsically exhibits not only the mechanism of ΔΣ modulation, but also a purely time-domain oriented structure. The proposed ADC employs two VCOs, one serving as a phase-domain translator and the other as a switched ring oscillator (SRO)-based time-digital converter (TDC), *to realize true 2nd-order noise shaping by using logic gates and simple analog components*. The outputs of the TDC readily facilitate a tri-level DAC control pattern and maintain an *intrinsic* data-weighted averaging (DWA) capability, removing the need for an explicit DEM block. Compared to existing OTA-free VCO-based ADC, this work is able to not only extend the scaling-friendly merits to high-order, but also maintains both high linearity and 2nd-order in-band noise shaping performance simultaneously without needing calibration.

II. PROPOSED 2ND-ORDER PURELY VCO-BASED ΔΣ ADC

The key concept of the proposed ADC is to leverage the loop mechanism of a DPLL [6], [7], that the control voltage will be enforced to cancel any phase perturbation in the system. As graphically represented in Fig. 1, under this idea, the input (V_{in}) is inserted as an intentional perturbation to the VCO, which triggers the phase variation. The phase difference is detected by the phase/frequency detector (PFD), which is assumed to have a reference phase of 0, and subsequently digitized by the TDC. Since the control voltage tracks the perturbation and is mapped directly to the TDC code, the TDC output (D_{out}) therefore can readily serve as a digital representation of the input signal. In this work, the TDC is designed to provide one extra order of noise shaping through using a reset-less SRO followed by a digital differentiator. Based on impulse-invariance transform [8], the noise transfer function (NTF) of the system is expressed as:

$$NTF = \frac{2 \cdot (1 - z^{-1})^2}{2 + (k_{v1} k_{v2} \alpha G T_s^2 - 2) \cdot z^{-1} + k_{v1} k_{v2} \alpha G T_s^2 \cdot z^{-2}} \quad (1)$$

where k_{v1} and k_{v2} are the gain of VCO and SRO respectively, α is the gain of PFD, G is the DAC gain, and T_s is the sampling period. Based on (1), as G increases, the NTF poles move out of the unit circle. For a fixed G, increasing k_{v1} and k_{v2} also leads to instability of the ADC. Therefore, k_{v1}, k_{v2} and G should be carefully chosen to ensure stable operation. Equation (1) also indicates the in-band quantization error is 2nd-order shaped. The proposed ADC has lower in-band quantization error than the technique of [4] which only has 1st-order noise shaping at low frequency. Compared to [5], the proposed ADC has feedback at the back-end stage which makes the signal swing at the first VCO integrator input very small, and thus, it is immune to VCO nonlinearity.

The schematic of the proposed 2nd-order purely VCO-based CT ΔΣ ADC is presented in Fig. 2. The input currents are injected into two differential current controlled oscillators (CCOs) whose phase difference is quantized by PFD. The two CCOs serve as reference to each other, which removes the need for an externally derived reference in the PFD. Note that if a single PFD is used as in a classic PLL, the CCO's free running frequency would have to be higher than the sampling frequency as the phase information needs at least one rising edge within each sampling period to update, leading to high power penalty. By contrast, the proposed work utilizes 6 PFDs to speed up the phase information update, and thus, relaxes the CCO free-running frequency requirement. Despite using multi-PFD introduces both V/I mismatch error and SRO non-linearity in the back-end stage, these non-idealities are strongly attenuated by the front-end CCO-based integrator when input referred. Based on the post-layout simulation result, compared with the single

PFD scheme, the power consumption of the first stage CCOs drops from 616 μW to 210 μW with a power penalty of only 61uW for 5 extra PFDs. Although the discrete nature of the CCO and PFD incurs a quantization behavior at the first stage output, the absence of sampling allows the artifact energy to concentrate at high frequencies. It brings negligible impact on the in-band noise as the back-end TDC filters it before sampling [9].

The back-end stage consists of a pair of pseudo-differential SRO-based TDC with 15 quantization levels. This structure provides immunity to even-order non-linearity. However, the asymmetry between PFDs' "UP" and "DN" pulses would introduce common mode modulation if applied to the SROs separately, causing harmonic leakage. This work addresses this concern by cross feeding "UP" and "DN" to ensure the TDC inputs are fully differential. By using the XOR-based differentiator, the TDC provides an intrinsic data-weighted averaging (DWA) behavior [2]. A tri-level DAC control can be derived easily through the tri-level generator in Fig 2. Fig. 3(a) intuitively illustrates an example that tri-level DAC is split into 2 DWA pattern outputs. The behavioral simulation spectrum in Fig. 3(b) proves that this tri-level DAC retains the intrinsic DWA mechanism as 1st-order element mismatch shaping.

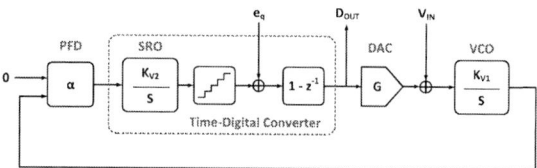

Fig. 1. Architecture of proposed purely VCO-based 2nd-order ΔΣ ADC.

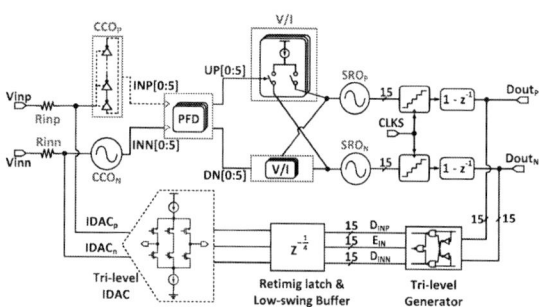

Fig. 2. Schematic of proposed purely VCO-based 2nd-order ΔΣ ADC.

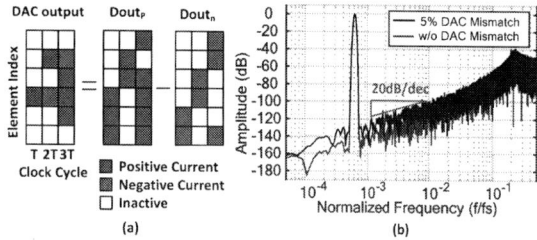

Fig. 3. (a) Tri-level generator element selection pattern example and (b) behavioral simulation results showing 1st-order mismatch error shaping.

III. MEASUREMENT RESULTS

The proposed 2nd-order VCO-based CT ΔΣ ADC occupies an area of 0.086 mm^2 in 40-nm CMOS process. Die photo is shown in Fig. 4(a). It operates at 260MS/s under 1.1-V supply and consumes 906 μW in total, where 376 μW for digital logics, 368 μW for two VCO-based integrators and 162 μW for DAC. The measured SNDR and SNR across input amplitude is shown in Fig. 4(b). This work achieves a DR of 72.7 dB and Schreier FoM of 170.3 dB with the full scale of 1.0 V$_{p-p}$ (equivalently 500

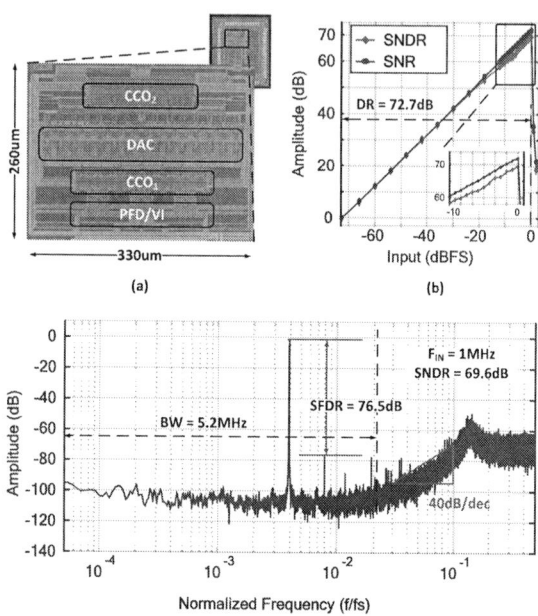

Fig. 4. (a) Die photo. (b) SNDR/SNR vs input amplitude. (c) Measured ADC output spectrum.

TABLE I. PERFORMANCE COMPARISON

	This work	[3]	[4]	[5]	[10]
Loop order	**2nd**	4th	2nd	3rd	3rd
Purely VCO	√	×	×	√	×
OTA-free	√	×	√	√	×
Calibration-free	√	√	√	×	√
Process(nm)	**40**	65	40	65	65
Fs(MS/s)	**260**	1200	330	1600	1280
Power (mW)	**0.91**	54	0.524	3.7	38
BW (MHz)	**5.2**	50	6	10	50
SNDR (dB)	**69.6**	71.5	68.6	65.7	64
DR (dB)	**72.7**	72	70.8	71	75
FoM_s* (dB)	**170.3**	161.7	171.4	165.3	166.2

*$FoM\_s = DR + 10*log10\ (BW/Power)$

μA$_{p-p}$) in 5.2-MHz BW. The measured output PSD shown in Fig. 4(c) proves 2nd-order noise shaping. At 5.2-MHz BW, the ADC achieves SNDR and SNR of 69.6 dB, 72.1 dB, respectively. As shown in Table I, the proposed ADC demonstrates in-line performance to other state-of-the-art high-order VCO-based ΔΣ ADCs. To the best knowledge of the authors, this work is the first to demonstrate a calibration-free high-order purely VCO-based CT ΔΣ ADC with chip measurement results.

REFERENCES

[1] S. Li, *et.al.*, *JSSC*, pp. 1940-1952, July 2017.
[2] M. Z. Straayer and M. H. Perrott, *JSSC*, pp. 805-814, April 2008.
[3] K. Reddy, *et.al.*, *VLSI Symp.*, 2015, pp. C256-C257.
[4] S. Li and N. Sun, *VLSI Symp.*, 2017, pp. C36-C37.
[5] A. Babaie-Fishani and P. Rombouts, *JSSC*, pp. 2141-2153, Aug. 2017.
[6] A. Sanyal and N. Sun, *Electronics Letters*, pp. 1204-1205, 2016.
[7] V. Prathap, *et.al.*, *MWSCAS*, 2017, pp. 687-690.
[8] J. A. Cherry and W. M. Snelgrove, *TCAS-I*, pp. 376-389, Apr 1999.
[9] Y. Yoon, *et.al.*, *ISCAS*, 2014, pp. 926-929.
[10] B. Young, *et.al.*, *VLSI Symp.*, 2014, pp. 1-2.

An 11b 1GS/s Time-Interleaved ADC with Linearity Enhanced T/H

Yan Zhu[1], Chi-Hang Chan[1], R.P. Martins[1,2]

[1]State-Key Laboratory of Analog and Mixed Signal VLSI, University of Macau, Macau, China, ivorchan@umac.mo
[2]On leave from Instituto Superior Técnico, Universidade de Lisboa, Portugal

Abstract - **This paper presents a 1GS/s 11 bit 4 × Time-Interleaved (TI) ADC employing the proposed Track-and-Hold (T/H) to enhance the sampling linearity and avoid timing skews. A dual auxiliary Source-Follower (SF) T/H provides signal double boosting to suppress the sampling distortion while maintaining good power efficiency. We present a dynamic SF-based switch boosting technique, providing a fast signal boost up and less signal attenuation to maximize the tracking duration and V_{gs} of the sampling switch. A prototype in 65nm CMOS achieves SNDR of 55.8dB @Nyquist input and ERBW up to 2×Nyquist input with total 22mW power.**

I. INTRODUCTION

High broadband applications dictate the need for low-power ADCs with high speed and high resolution. To implement GS/s sampling rates TI multiple power efficient ADCs [1][2] are the common scheme to obtain good FoM. However, the interleaving spurs degrade the performance. The offset and gain mismatches can be effectively corrected, while timing and bandwidth errors demand heavy calibration efforts limiting the performance of those converters [1][2]. The TI ADCs with active T/H [3][4] avoid the timing errors and isolate the input from varying impedances and kick-backs from back-end interleaving channels, thus improving the performance robustness and reliability. Although those TI ADCs [3][4] achieve either better sampling linearity or higher bandwidth, the use of active T/Hs limits the FoM to several hundred fJ/conv.step level. To obviate this the paper presents a *dual auxiliary source follower* (DASF) assisted T/H that improves the sampling linearity for a wideband. A 1GS/s 4-way TI 11-bit ADC implements with DASF in a 65nm CMOS prototype achieving a SNDR of 55.8dB @Nyquist input and ERBW up to the 2nd Nyquist zone resulting a FoM of 43.7 fJ/conv.step. The ADC performs only the offset calibrations for the sub-ADCs that are all implemented on-chip.

II. ADC ARCHITECTURE

Fig.1 depicts the overall ADC architecture, incorporating a single DASF T/H followed by 4 TI sub-ADCs. The DASF T/H samples the input signal and buffers it to the sub-ADCs for an 11 bit conversion. The DASF T/H output drives the sub-ADCs through a 1 to 4 demux. The T/H and the sub-ADCs sample at 1 GS/s and 250MS/s, respectively. The single T/H eliminates the time spurs that would arise from the interleaved sampling scheme. The sub-ADC uses a two-stage pipelined-SAR architecture that quantizes 5 and 7 bits in the 1st and 2nd stages, respectively, with an amplification gain of 4. We use one bit redundancy to relax the

This work was financially supported by NSFC with funding code: 61604180, Macao Science and Technology Development Fund under Grant ((077/2017/A2)) and Research Grants of University of Macau with funding code number MYRG2018-00104-AMSV.

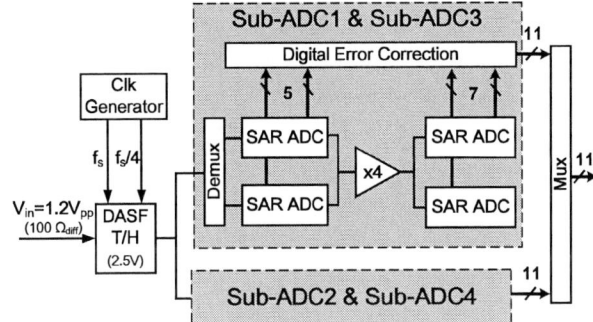

Fig.1: Overall ADC architecture.

the 1st stage quantizer accuracy and apply an opamp-sharing technique to 2 sub-ADC pairs for power and area efficiency. The gain mismatches are tolerated by intrinsic capacitor matching, while the offset mismatches are corrected on-chip. We reuse both coarse and fine SAR ADCs in two stages to measure and compensate the offsets, which saves significantly the digital circuit overhead.

III. PROPOSED DUAL AUXILIARY SOURCE FOLLOWER

Fig. 2 shows the simplified schematic of the DASF T/H. The sampling network employs a bootstrapped switched-capacitor topology and a simple cascade source follower with DASFs to improve the sub-ADCs sampling linearity. In the track mode, the sampling capacitor C_S follows the input signal, while powering down the Auxiliary SF (ASF) isolates the buffer output from its loading capacitance CDAC. In hold mode, both ASFs are activated to buffer the input signal to the CDAC for back-end quantization.

To enhance the linearity of the SF the channel length modulation of M1 must be suppressed. An auxiliary p-type source-follower provides a level-shifted input signal to M2 who duplicates the input signal at its source terminal producing much smaller V_{ds} variation in M1. This suppresses the device M1 non-linearity due to low and non-linear output resistance (r_o) and provides more V_{ds} headroom for M1 to further enlarge r_o. A larger r_o is highly desired, as it improves the DC gain and consequently a better SNR. The achieved DC gain is -0.24dB. When compared with the switched-capacitor boosting technique [5] the proposed solution avoids extra clocking which is more efficient in a high speed implementation. As the auxiliary SF drives only the small gate capacitance of M2, it has negligible overhead in power and area. The current variation due to the channel length modulation in the current source device causes non-linearity, which can be tolerated and reduced through design optimization.

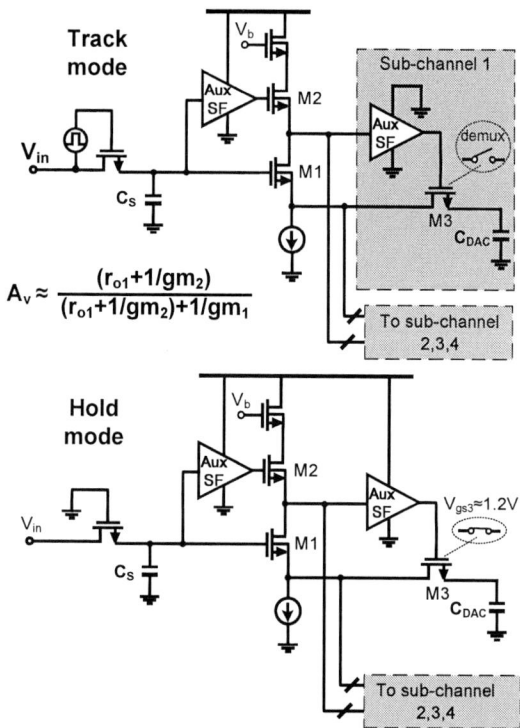

Track mode

$$A_v \approx \frac{(r_{o1}+1/gm_2)}{(r_{o1}+1/gm_2)+1/gm_1}$$

Sub-channel 1

demux

To sub-channel 2,3,4

Hold mode

$V_{gs3} \approx 1.2V$

To sub-channel 2,3,4

Fig. 2 Simplified schematic of the DASF Track-and-hold circuit.

The hold mode allows stringent T/H settling time (~400ps) to drive a large load capacitance (~1pF) designed according to the overall noise requirement for 11 bit accuracy. The demux before the sub-ADC has a variable V_{gs} that causes the distortion due to limited tracking duration. The implementation of a conventional switched-capacitor bootstrap appears as the extra loading capacitor at the main T/H's output, which would be quite large due to 4× interleaving. Its startup time is related with the T/H's driving capability and routing parasitics in the signal path. To improve both linearity and bandwidth we use a dynamic ASF, which duplicates and shifts up the signal at the M2 source terminal producing another boosting level of $V_{in}+2\times V_{gs}$ to the gate of M3. The input parasitcs of the ASF are negligible, minimizing the burden to the main T/H. The conventional bootstrap has boosting signal attenuation due to the parasitics of the sampling switch M3 that is sized large for better BW. Using the proposed active boosting can avoid this problem. The DASF T/H including the biasing circuit consumes only 4.3mA current that is less than half of the overall ADC power consumption.

IV. MEASUREMENT RESULTS

The ADC implemented in a 65nm CMOS exhibits an active area of 0.128mm² (Fig.3). Fig. 4 shows the spectral plots at Nyquist and 2×Nyquist inputs, where the ADC achieves SNDRs of 55.8dB and 54.3dB, respectively. The offset mismatches are calibrated off-chip and their spurs are all below –70dB. Fig. 5 depicts the ADC performance at 1GS/s sampling rate versus the input frequencies where the SNDRs are 57.1dB @19.4MHz input and staying above 50dB up to near the 4th Nyquist zone. The ADC consumes 22mW from 1.2/2.5V supplies at 1GS/s, including

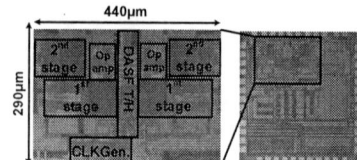

Fig. 3: Die microphotograph of the ADC.

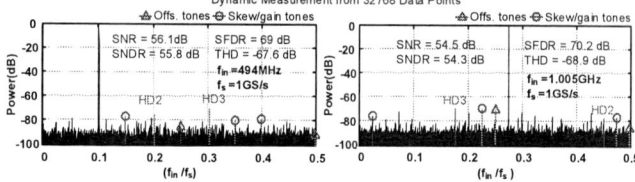

Fig. 4: Measured FFT @Nyquist and 2×Nyquist inputs (decimated by 125)

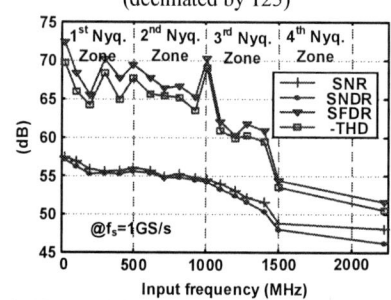

Fig.5: Measured dynamic performance v.s. input frequency

Table I Performance summary & comparisons with state-of-the-art designs

	This Work	[2] ISSCC' 14	VLSI' 16 L. Y. Zhu	ISSCC' 15 H.K. Hong	ISSCC' 15 N. L. Dortz
Architecture	TI Pipelined-SAR	TI-SAR	TI-SAR	TI-SAR	TI-SAR
Technology (nm)	65	65	16	45	65
Resolution (bit)	11	10	10	10	10
Sampling Rate (GS/s)	1	1	1.6	1.7	1.62
Supply Voltage (V)	1.2/2.5	1	0.95	1	1.1
Input Swing (V$_{pp}$)	1.2	N/A	1	N/A	1
SFDR @ 1st Nyq. (dB)	69	60	61	62	61
SNDR @1st Nyq. (dB)	55.8	51.4	50.3	51.2	48
SNDR @2nd Nyq. (dB)	54.3	N/A	N/A	46	N/A
ERBW (GHz)	>1	0.5	<0.7	0.6	0.78
Area (mm²)	0.128	0.78	0.23	0.057	0.83
Power (mW)	22	18.9	8.2	15.4	93
FoMW@Nyq. (fJ/conv.step)	43.7	62.3	19.2	30.5	283
FoMS@Nyq. (dB)	159.4	155.6	160.2	158.6	147.4
Timing Cal.	No	Yes	Yes	No	Yes

17mW analog power and 5mW digital power. The DNL/INL is 0.45/0.79LSBs. The ADC attains Schreier and Walden FoM$_{,hf}$ of 159.4dB and 43.7fJ/conv.step, respectively. Table I shows the performance summary and a comparison with state-of-the-art TI SAR ADCs with timing-calibration. This work demonstrates better performance for higher input bandwidth, while the achieved FoM is still competitive to those fully dynamic ADCs.

REFERENCES

[1] C. Y. Lin, et al., *ISSCC Dig. Tech. Papers*, pp. 468-469, Feb. 2016.
[2] S. Lee, et al., *ISSCC Dig. Tech. Papers*, pp. 384-385, Feb. 2014.
[3] C. C. Hsu, et al., *ISSCC Dig. Tech. Papers*, pp. 464-465, Feb. 2007.
[4] K. Doris, et al., *ISSCC Dig. Tech. Papers*, pp. 180-181, Feb. 2011.
[5] A. M. A. Ali, et al., *ISSCC Dig. Tech. Papers*, pp. 482-483, Feb. 2014.

978-1-5386-6414-8/18 $31.00 © 2018 IEEE

10-4 (8162)

A 1-V 3.1-ppm/°C 0.8-μW Bandgap Reference with Piecewise Exponential Curvature Compensation

Hongrui Luo, Quan Sun, Ruizhi Zhang, Hong Zhang

School of Electronic and Information Engineering, Xi'an Jiaotong University, Xi'an, 710049, China
Email: hongzhang@xjtu.edu.cn

Abstract—This paper presents a low-power precision BGR, which is realized with a current-mode BGR compensated by a piecewise exponential current generated by three simple differential MOS pairs. Fabricated in a 0.18-μm CMOS process, measured results show that the BGR achieves average and best TCs of 7.08 and 3.1 ppm/ °C after trimming, respectively, in the range of -40 to 125°C, and a line sensitivity of 0.03%/V in a supply voltage ranges from 1 to 3V. The current consumption is 0.8 μA under a 1-V supply.

Keywords—bandgap reference; low voltage; high-precision; piecewise; temperature curvature compensation

I. INTRODUCTION

Conventional first-order compensated BGRs show limited temperature stability due to the temperature curvature in the V_{BE} of a BJT. Many high-order curvature compensation techniques have been employed to achieve a BGR with very low temperature coefficient (TC) [1]-[3]. Inherent curvature compensation is used in [1], achieving a lowest TC of 3.9 ppm/°C, at the cost of high power consumption. In [2], the velocity saturation index depending on the operation region of the NMOS is applied to implement high-order compensated reference with best TC of 5.6ppm/°C, but, the ZTC-mode structure shows relatively high power consumption and reference offset. In [3], better TC is achieved with piecewise linear curvature compensation, however, its compensation circuit is complex and the BGR shows relatively high power consumption.

This work presents a precision BGR with piecewise exponential curvature compensation. Three differential pairs are combined to generate a piecewise exponential temperature compensation current for a current-mode BGR core, resulting in a very low TC with sub-1μW power consumption.

II. OPERATION PRINCIPLE AND CIRCUIT IMPLEMENTATION

The conventional first-order compensated BGR exhibits curvature in the output because the V_{BE} of a BJT is a complex function of T containing many high-order terms. Therefore, an exponential function could be more efficient than a linear function for curvature compensation. However, it is not feasible to realize an exponential voltage using simple circuit structure with acceptable power consumption. In this paper, a simple differential pair operated in sub-threshold region is used to generate a current exponentially varying with T. Then, three differential pairs are combined to generate a piecewise exponential current to cancel the curvature of a current-mode BGR.

A. Proposed Exponential Compensation Current Generator

The proposed exponential current generator is realized by

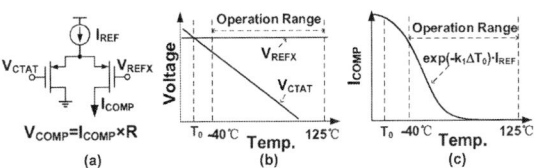

Fig. 1. Proposed exponential current generator: (a) circuit structure; (b) ideal temperature dependence of V_{REFX} and V_{CTAT}; (c) I_{COMP} curve.

a simple differential pair, as shown in Fig. 1(a). It is assumed that the temperature-independent current I_{REF} is so low that the two PMOS always operate in sub-threshold region. Then, if V_{CTAT} is lower enough than V_{REFX}, I_{COMP} can be obtained approximately as an exponential function of $(V_{CTAT}-V_{REFX})$:

$$I_{COMP} \approx \exp(\frac{V_{CTAT} - V_{REFX}}{mV_T}) \cdot I_{REF} \quad (\text{for } V_{CTAT} \ll V_{REFX}) \quad (1)$$

where m is the sub-threshold slope factor, V_T is the thermal voltage. In order to translate (1) into a function of temperature difference, V_{CTAT} and V_{REFX} are selected as complementary-to-absolute-temperature (CTAT) voltage and temperature-independent voltage, respectively, as shown in Fig. 1(b). The relationship between V_{REFX} and V_{CTAT} can be expressed as $V_{CTAT} = -\alpha_{CTAT}(T-T_0) + V_{REFX}$, where α_{CTAT} is the TC of V_{CTAT} and T_0 is the temperature at the extrapolated crossover point of V_{CTAT} and V_{REFX}.

Therefore, the condition for (1) can be fulfilled if T_0 is much lower than the operational range (–40~125 °C for this work). Then, I_{COMP} can be obtained as

$$I_{COMP} \approx \exp(-k_1\Delta T) \cdot I_{REF}, \quad k_1 = \frac{\alpha_{CTAT}}{mV_T} \bigg\} \text{ for } T \gg T_0 \quad (2)$$

where $\Delta T = T - T_0$. Therefore, neglecting the slight temperature dependence of V_T, I_{COMP} indeed decreases exponentially as ΔT increases in the operation range, as shown in Fig. 1(c). As implied by Fig. 1(a), I_{COMP} can be converted into V_{COMP} with a resistor in a practical BGR.

B. BGR Architecture

The overall BGR circuit is shown in Fig. 2. The low-voltage current-mode BGR in [4] is used as the core BGR for the proposed work. Two temperature-independent voltages, V_{REFX1} and V_{REFX2}, and two CTAT voltages, V_{CTAT1} and V_{CTAT2} are picked up from the core circuit for the 3 differential pairs in the compensation circuit.

With a proper design of V_{REFX1}, V_{REFX2}, V_{CTAT1} and V_{CTAT2} as that given in Fig. 3(a), each differential pair generates an exponential current in a given piece of the operational temper-

978-1-5386-6414-8/18 $31.00 © 2018 IEEE

97

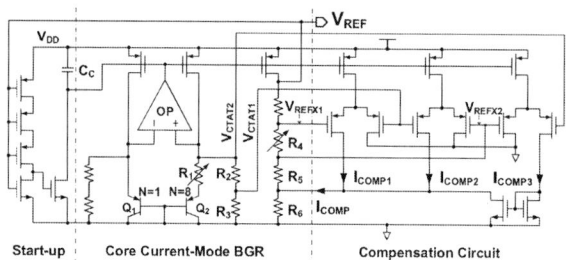

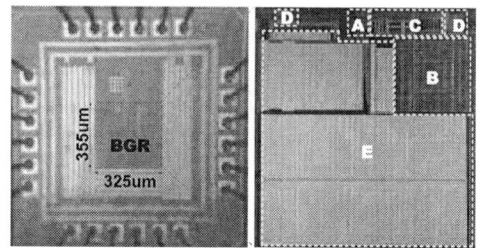

Fig. Chip microphoto and layout (A: start-up circuit; B: BJTs; C: MOS transistors; D: trimming circuits; E: resistors).

Fig. 2. Schematic of the proposed BGR.

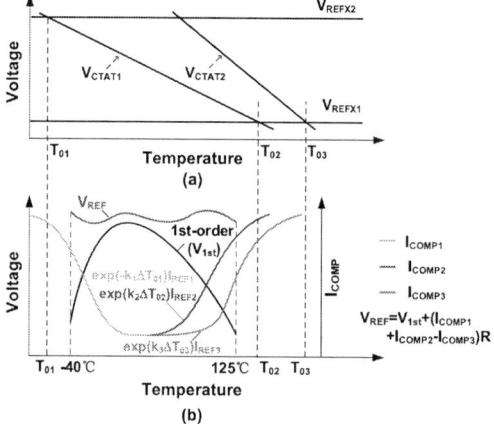

Fig. 3. Operation principle: (a) voltages selected from the core BGR for the three differential pairs; (b) conceptual compensation result.

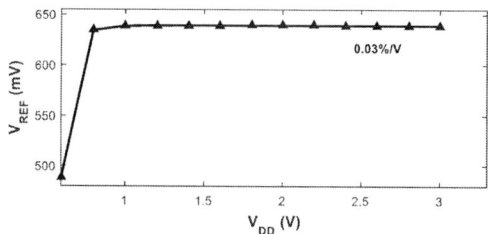

Fig. 5. Measured reference voltage versus supply voltage.

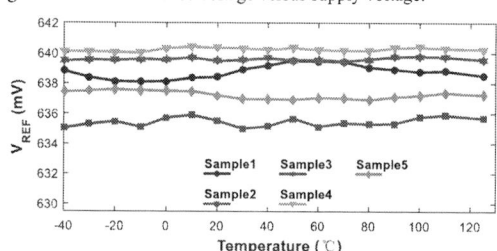

Fig. 6. Measured reference voltage for 5 samples after trimming (average and best TCs are 7.08 and 3.1 ppm/ °C, respectively).

ature range, as shown in Fig. 3(b). The three currents are then combined as a total current (I_{COMP}), which flows into R_6 of the core BGR to compensate the curvature of V_{REF}, resulting in a compensated output $V_{REF} = V_{1st} + (I_{COMP1} + I_{COMP2} - I_{COMP3})R$.

The compensation circuit consumes very low current because the differential pairs are required to be in sub-threshold region. Simulation shows that a very low TC of 0.45ppm/°C is realized with 0.8μA current for the total BGR.

III. MEASUREMENT RESULTS

The proposed BGR is fabricated in a 0.18μm CMOS process, with a core area of 0.115 mm², as shown in Fig. 4. Fig. 5 shows the measured untrimmed V_{REF} varying with supply voltage at 25 °C, showing a very low line sensitivity of 0.03%/V in a supply voltage range of 1 ~ 3V.

The TCs of five samples after trimming in a temperature range from -40 °C to 125 °C under 1.8-V V_{DD} are given in Fig. 6, with calculated TCs of 8.3, 6.4, 13.6, 3.1, 4ppm/ °C, respectively, resulting in an average TC of 7.08ppm/ °C. The trimming is implemented by adjusting the resisters R_1 and R_4 in Fig. 2. The BGR consumes 0.8μA when the V_{DD} is 1V.

The performance is summarized in Table I with comparison to other precision BGRs. The FoM in [5] is selected for a general comparison (FoM=$(T_{max}-T_{min})^2/(TC×Power×Area)$), showing that the proposed BGR achieves the highest FoM in Table I.

Table I
PERFORMANCE SUMMARY AND COMPARISON

Parameter	This work	[1]	[2]	[3]
Technology (nm)	180	350	65	180
Supply Voltage (V)	1-3	2.5	0.8	1.3-1.8
V_{REF} (V)	634.7m	617.7m	428m	547m
Temp. Range (°C)	-40-125	-15-150	-40-125	-40-140
best TC (ppm/ °C)	3.1	3.9	5.6	1.67
Line sensitivity	0.03%/V	0.04%/V	0.39%/V	0.08%/V
Current Diss. (μA)	0.8 @1V	38	16.25	28
FoM (°C³/(W×mm²))	9.546 ×10¹⁶	0.072 ×10¹⁶	3.6 ×10¹⁶	5.67 ×10¹⁶
Active area (mm²)	0.115	0.1019	0.0104	0.0094

REFERENCES

[1] C. M. Andreou, et al. "A Novel Wide-Temperature-Range, 3.9 ppm/°C CMOS Bandgap Reference Circuit," in *IEEE Journal of Solid-State Circuits*, vol. 47, no. 2, pp. 574-581, Feb. 2012.

[2] J. Jiang, et al. "A 5.6 ppm/°C Temperature Coefficient, 87-dB PSRR, Sub-1-V Voltage Reference in 65-nm CMOS Exploiting the Zero-Temperature-Coefficient Point," in *IEEE Journal of Solid-State Circuits*, vol. 52, no. 3, pp. 623-633, March 2017.

[3] H. M. Chen, et al. "A Sub-1 ppm/°C Precision Bandgap Reference With Adjusted-Temperature-Curvature Compensation," in *IEEE Transactions on Circuits and Systems I: Regular Papers*, pp. 1308-1317, June 2017.

[4] H. Banba *et al.* "A CMOS bandgap reference circuit with sub-1-V operation," in *IEEE J. of Solid-State Circuits*, pp. 670-674, May 1999.

[5] A.C. de Oliveira, et al, "Picowatt, 0.45-0.6 V Self-Biased Subthreshold CMOS Voltage Reference," in *IEEE Transactions on Circuits and Systems I: Regular Papers*, pp. 3036-3046, Dec. 2017.

A 40nW, Sub-1V Truly 'Digital' Reverse Bandgap Reference Using Bulk-Diodes in 16nm FinFET

Matthias Eberlein *, Georgios Panagopoulos *, and Harald Pretl [†]

Communication and Device Group, * Intel Germany, Neubiberg, Germany; [†] Intel Austria, Linz, Austria

Email: matthias.eberlein@intel.com

Abstract—We present a novel, simple concept to generate a robust voltage reference, which is based on capacitive bias of pn-junctions. The respective PTAT and CTAT signals are sampled from the voltage-decay by means of different timings, and combined through charge sharing. This provides for precise current ratios of N >10000, resulting in exceptionally large PTAT and reverse-bandgap levels. Here, for the first time, the Nwell/Psub diode of a standard CMOS process is utilized in replacement of parasitic BJTs. The measured samples, in 16nm FinFet on 2200µm² active area, achieve an untrimmed accuracy of ±0.82% (3σ) at 235mV output. Line sensitivity is 0.7mV/100mV, operating at a minimum supply of 0.85V with 47nA power drain. The compact Bandgap circuit is digital to that effect that no amplifiers, resistors, biasing or matching currents are required, neither is it impacted by any analog transistor performance.

Keywords— Bandgap voltage reference; FinFet process; bulk-diode; switched-capacitor

I. Introduction

Classic bandgap references require analog components like amplifiers, MOS current mirrors, resistors etc., which need to meet certain performance for accuracy, e.g. matching. They are difficult to implement in a digital process and do not scale well with future nodes. Specifically the ratio of BJTs or currents is limited to values of typically N=8...20 (Fig. 1), which results in a small PTAT-voltage, sensitive to offset and mismatch errors.

In FinFet process there is an additional challenge with the availability of bipolar transistors: Traditionally the parasitic PNP is utilized, built from the drain/source, bulk and substrate junctions of a standard PMOS device. However, the critical base-emitter junction is squeezed inside the fin (Fig. 1), with various non-uniformity issues, while the base-width Wb is large and fluctuating. This results in very low beta (<<1), large spread and overall poor linearity.

Most prior art solutions do not perform at very low biasing conditions, like Vdd < 1.0V, or currents << 1uA, which would be required for wearables and IOT applications. Recent publications in that area have achieved good accuracy/power ratios through switch-cap techniques [4, 5], and also in FinFet [3]. But they still depend on critical matching of analog components, voltage doublers, or use extra process options [2].

In this paper an innovative concept of 'capacitive bias' is applied to a single pn-junction, which allows to determine the current density precisely by pure timing control. The circuit contains only capacitors and switches, driven by a pulse generator, and is therefore compatible to digital design and FinFet. We also introduce the profitable usage of bulk-diode in forward-bias, which further improves robustness towards future scaling.

II. Capacitive Bias of PN-Junctions

Essential part of the idea is the utilization of this specific characteristic in cap-voltage decay: When a capacitor – pre-charged to e.g. Vdd – discharges through a diode (Fig. 2), the resulting voltage over time can be calculated [1]:

$$V_D(t) = -mV_T \cdot ln\{1 - [1 - exp\left(\frac{-Vdd}{mV_T}\right)] \cdot exp\left(\frac{-Is}{C \cdot mV_T} \cdot t\right)\}$$

where m is diode ideality factor, V_T is the thermal voltage and Is = diode saturation current (series-R is neglected). For times after an initial short period (~30ns), which is characterized by series resistance and initial conditions, this formula can be simplified to:

$$V_D(t) = -mV_T \cdot ln\left(\frac{Is}{C \cdot mV_T} \cdot t\right) \qquad (1)$$

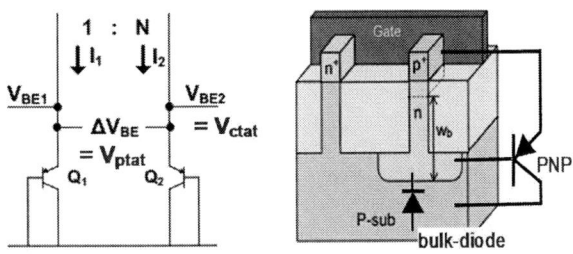

Fig. 1. Classic bandgap core with PNPs (left) and parasitic bipolar junctions of FinFet section (right).

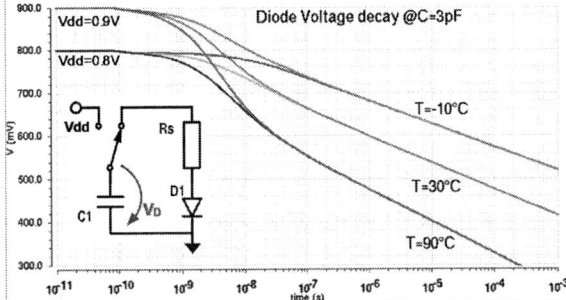

Fig. 2. Simulated diode voltage for different Vdd during capacitor discharge versus time and temperature

978-1-5386-6414-8/18 $31.00 © 2018 IEEE

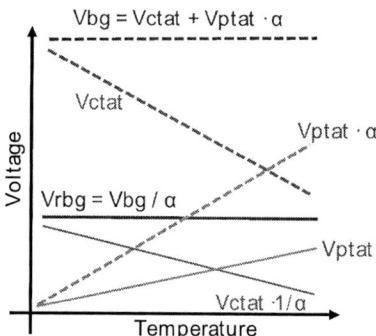

Fig. 3. Composition of conventional bandgap (Vbg, dashed) and reverse bandgap (Vrbg, solid) reference from scaled PTAT and CTAT voltages.

Most important and confirmed by simulation (Fig. 2): For several decades of time (t) the diode voltage follows a strict logarithmic function, with no relevant dependence from initial bias. Consequently the diode current is also well defined and proportional to 1/t. Here we utilize this relation to generate precisely current ratios of N ≫1000, down to very low levels (<10nA). Hence, any parasitic resistors have negligible effect.

A. Application to reverse bandgap principle

A bandgap reference is basically generated by addition of two scaled signals with positive (PTAT) and negative (CTAT) temperature coefficient (Fig. 3). While Vctat is simply a diode voltage (Vd2), the PTAT part is $\Delta Vd = Vd1 - Vd2 = mV_T \cdot \ln(N)$, where N is the ratio of current densities. For the classic bandgap, only the PTAT voltage is multiplied by a factor k and therefore most sensitive to errors. Formulated with diode voltages:

$$Vbg = Vd2 + k \cdot \Delta Vd = (k+1) \cdot Vd2 - k \cdot Vd1$$

The 'reverse bandgap' creates a fraction of Vbg and effectively applies scaling to CTAT only:

$$Vrbg = \frac{Vbg}{k+1} = Vd2 - \left(\frac{k}{k+1}\right) \cdot Vd1 \qquad (2)$$

While ln(N) is ~2.1, the factor k required to achieve temperature compensation is typically ~9, yielding Vrbg ~ 120mV. However, with above technique of capacitive bias, much larger values of N can be achieved, resulting in smaller k and a more robust reference level (≫200mV).

The presented invention [6] uses capacitor switching and charge sharing, realized in 4 phases with pulses tc / ts1 / ts2 / ta as explained in Fig. 4: Initially (during tc) C1 and C2 are

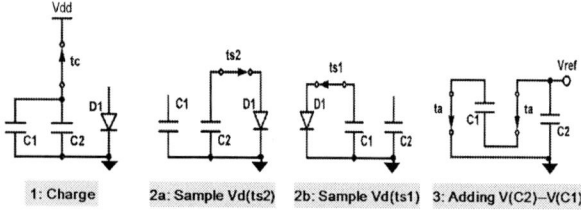

Fig. 4. Capacitor switching phases to generate a bandgap voltage Vref with sampling timings tc, ts1, ts2 and ta.

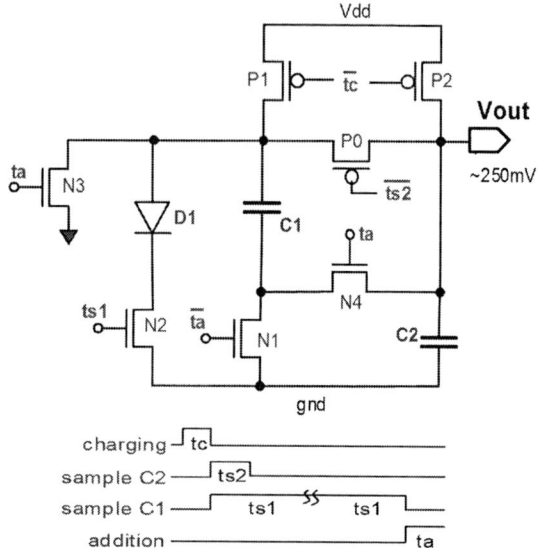

Fig. 5. Basic time-controlled bandgap example with a single floating diode, and corresponding control pulses.

connected to Vdd, which is any voltage > ~0.8V. In phase 2a/b, each capacitor is discharged through a diode for a time ts1 and ts2, respectively. During phase 3 (ta) the sampled voltages Vd1=Vd(ts1) and Vd2=Vd(ts2) are subtracted, with a weighting according to respective capacitors.

The resulting reference output implements a reverse bandgap as per formula (2). However, the voltage level is reduced by a damping factor as a result of charge sharing in phase 3:

$$Vref = \frac{C2}{C1+C2} \cdot \left[Vd2 - \frac{C1}{C2} \cdot Vd1 \right] \qquad (3)$$

III. CIRCUIT REALIZATION

Fig. 5 shows a straight forward implementation of above principle, together with certain sequence of control signals. Switch devices are arranged for optimum on/off performance and simplicity. For good results C1, C2 and D1 require accurate models. Phase 2a/2b are merged, therefore overlapping the pulses ts1 and ts2. In Fig. 6 typical waveforms of capacitor voltage are explained, here with an extreme pulse-ratio of N=2·10⁴. Notably, also trimming may be accomplished simply through adaption of control timings.

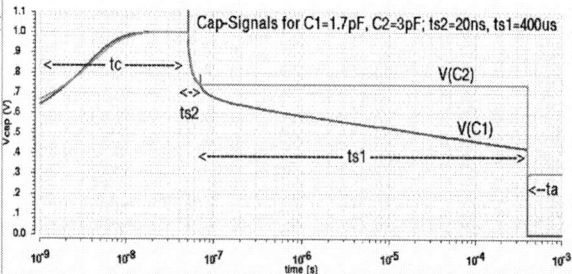

Fig. 6. Example circuit: Voltage across C1 and C2 during switching phases.

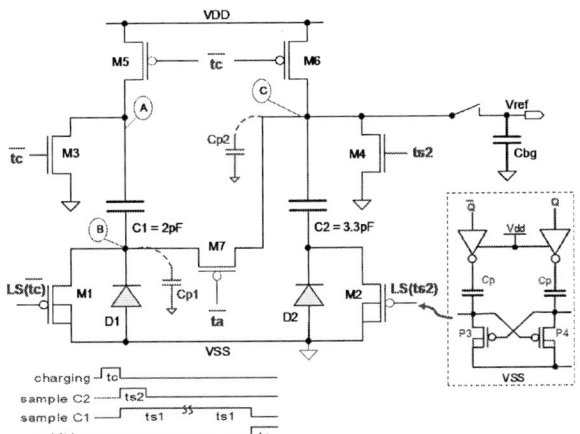

Fig. 7. Final bandgap circuit (simplified), with negative level-shifter (right) to drive the LS-gates. Control pulses are generated on-die by simple counter.

This simple realization, however, requires a floating (true) diode, which is not available in digital FinFet process. Instead, we introduce here the use of Nwell/Psub diode (Fig. 1), biased through a negative charge-pump. This junction is expected to have superior characteristic over conventional drain/source regions, since buried in the clean substrate and not affected from scaling.

Fig. 7 shows the final implementation of the bandgap circuit, together with a simple switch driver for negative level-shifting. This is required to control PMOS M1, M2 by signals LS($\overline{tc}$) and LS(ts2), during the charge- and add-phase. We chose a configuration with two bulk diodes D1 and D2, to simplify the switch design. Initially at pulse tc, both capacitors charge to Vdd. During sampling times ts1 and ts2, NMOS devices M3, M4 are closed, respectively, while M1 and M2 are open. This shifts the bottom plate of C1 and C2 to -Vdd, applying forward-bias to the diodes. In the last phase (addition) M2 and M7 are closed, and the charge Q(C2) is merged with –Q(C1). A continuous output signal can be achieved through repeated operation of phase 1-3, and sampling of Vout to another capacitor (Cbg) or buffer. Only standard elements from digital process flow are used, like metal-finger capacitors.

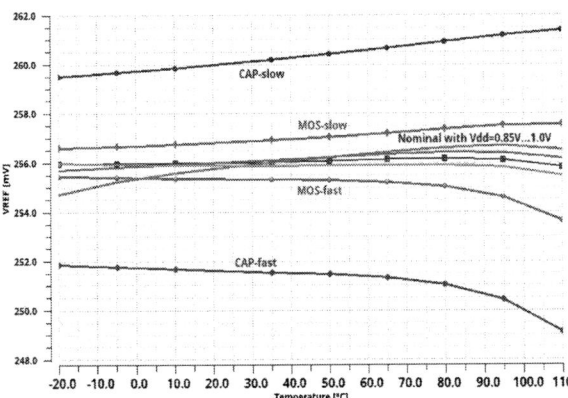

Fig. 8. Simulated Vref output vs. Vdd, MOS corners and capacitor spread

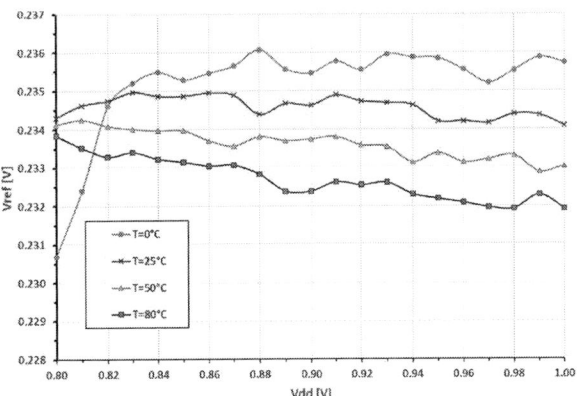

Fig. 9. Measured Vref vs. Vdd and across temperatures for a typical sample.

For accurate calculations we must consider the effective sampling cap C1+C2 during ts2, due to time overlapping. Combining with formula (1) into (3) therefore yields the reference output:

$$Vout = \frac{C2 \cdot V_T}{C1+C2} \cdot \left[ln \frac{(C1+C2) \cdot V_T}{Is2 \cdot ts2} - \frac{C1}{C2} \cdot ln \frac{C1 \cdot V_T}{Is1 \cdot ts1} \right] \quad (4)$$

With selection of D1/D2 = 8, and ts1/ts2 = 100µs/50ns, the circuit realizes a total current ratio of N = 16000. Such parameters induce extremely low current levels in the diodes, increasing sensitivity to leakage. For that reason some switches were implemented with IO-devices. Another challenge are parasitic capacitances, with dominant impact from components at node B and C (see Fig. 7): These add to the effective value of C1, C2, and cause additional scaling of the sampled voltage V(C2) during the last phase. Also the junction capacitance of D1 must be considered. The impact of parasitics can be handled by careful extraction of layout parameters.

IV. SIMULATION AND MEASUREMENT RESULTS

The structure was analysed in pre-silicon simulations across full temperature range and process corners (Fig. 8). The reference shows low sensitivity towards pre-charge conditions, which is Vdd, as predicted by the capacitive bias concept. Also the impact of switch performance, reflected by MOS process corners, is small. Assuming a low variability of the Nwell-diodes, the major error contribution results from capacitors, predicting ~ 3mV spread for Vref. However, this is still 50% lower than traditional bias (of BJTs) based on poly resistors.

The circuit was implemented in 16nm FinFet, and 24 samples from a single wafer are evaluated. Fig. 9 shows measurements versus supply for various temperatures. This confirms the excellent PSRR of +/-0.19%/200mV (at 25°C), and accurate operation down to Vdd=0.85V. The variability of the output voltage, including 3σ-values, is displayed in Fig. 10. This results reveal a residual temperature coefficient and offset from simulated values. We found through device extraction that the bulk diode, specifically junction capacitance, was not modelled correctly, which accounts for this deviation. Fig. 11 illustrates the statistical behavior of all 24 samples at room temperature.

978-1-5386-6414-8/18 $31.00 © 2018 IEEE

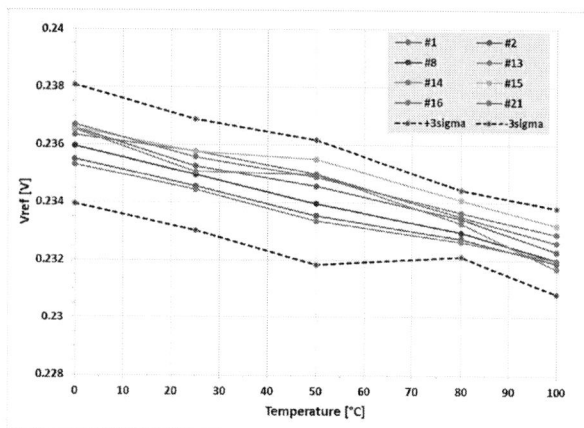

Fig. 10. Output voltage vs. temperature for 8 random samples, at Vdd = 0.90V

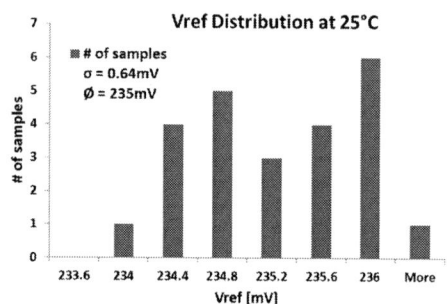

Fig. 11. Distribution of reference output at ambient and Vdd = 0.85V

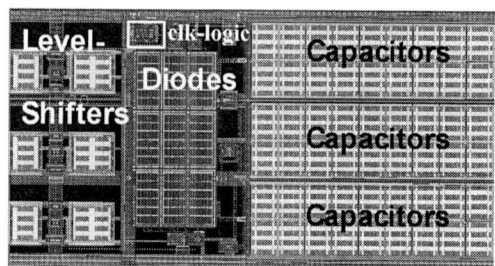

Fig. 12. Layout of the 'digital' bandgap (incl. logic) in 16nm FinFet

V. CONCLUSION

We propose a compact nano-power reverse bandgap, which controls PTAT and CTAT signals precisely by simple pulse timings. The switch-cap structure includes several innovations: For the first time the Nwell-diode is used in forward bias with superior stability over BJT. The sampling concept enables extreme current ratios of N >>1000, supporting large and precise PTAT voltages. Further, this circuit requires no analog components, except capacitors and switches with optimized on/off performance. Therefore it is especially suitable for FinFet technology, digital systems and future scaling. Convenient calibration is possible by timing adaption.

On 2200µm² silicon area (Fig. 12), the untrimmed bandgap achieves a 3σ-accuracy of 0.82%, with good supply rejection down to 0.85V. The bias current is ~ 47nA, including a major contribution from the switch drivers. Table 1 summarizes the performance and compares with previous related work.

TABLE I.   PERFORMANCE COMPARISON WITH RELATED WORK

	This work	**[2] (Kumar W.)**	**[3] (Chang)**	**[4] (Ivanov)**	**[5] (Shrivastava)**
Process	16nm FinFet	16nm FinFet	16nm FinFet	130nm	130nm
VDD	0.85-1.0V	1.8V	1.8V	0.75V	0.5V
Power	40nW	340µW	-	170nW	32nW
Vref	235mV	1.22V	615mV	185mV	423mV
Spread (3σ) w/o trim	+/- 0.82%	+/-1.3%	+/-1.67%	+/-0.9%	+/-2.0%
PSRR	~ - 44 dB	-	-	~ - 53dB	- 40dB
Area	0.0022mm²	0.011mm²	0.0023mm²	0.07mm²	0.0264mm²
# Samples	24	-	63	~ 300	6
Type	Reverse BG	NPN BG	Current-mode BG	Reverse BG	Fractional SC BG

REFERENCES

[1] E. Hellen, "Verifying the diode-capacitor circuit voltage decay", American Journal of Physics, vol. 71, August 2003.

[2] S. K. Wadhwa and N. Chaudhry, "High accuracy, multi-output bandgap reference circuit in 16nm FinFet", VLSID 2017, pp. 259-262.

[3] C. H. Chang, J. J. Horng, A. Kundu, C. C. Chang and Y. C. Peng, "An ultra-compact, untrimmed CMOS bandgap reference with 3σ inaccuracy of +0.64% in 16nm FinFET," 2014 IEEE Asian Solid-State Circuits Conference (A-SSCC), KaoHsiung, 2014, pp. 165-168.

[4] V. Ivanov, R. Brederlow and J. Gerber, "An Ultra Low Power Bandgap Operational at Supply From 0.75 V," in IEEE Journal of Solid-State Circuits, vol. 47, no. 7, pp. 1515-1523, July 2012.

[5] A. Shrivastava, K. Craig, N. E. Roberts, D. D. Wentzloff and B. H. Calhoun, "5.4 A 32nW bandgap reference voltage operational from 0.5V supply for ultra-low power systems," 2015 IEEE International Solid-State Circuits Conference - (ISSCC) Digest of Technical Papers, San Francisco, CA, 2015, pp. 1-3.

[6] M. Eberlein, "Bandgap reference circuit with capacitive bias", US20170285680A1, 2017.

10-6 (8031)

A 0.6V 1.63fJ/c.-s. Detective Open-Loop Dynamic System Buffer for SAR ADC in Zero-Capacitor TDDI System

Yao-Sheng Hu, Li-Yu Huang and Hsin-Shu Chen

Department of Electrical Engineering and Graduate Institute of Electronics Engineering, National Taiwan University

Taipei, Taiwan 10617, R. O. C.

hschen@ntu.edu.tw

Abstract - A low-power detective open-loop dynamic (DeOLD) system buffer co-designed with a SAR ADC without using any off-chip decoupling capacitor is proposed. It is pertinent for the zero-capacitor touch with display driver integration (TDDI) system. The system buffer detects the ADC input signal actively to compensate its loss charge back quickly. The dynamic charge sharing technique instead of the well-known pump squeezing technique is adopted to reduce 42.4% power of charger compensators in the buffer. Without the off-chip decoupling capacitor, the DeOLD system buffer co-designed with the 400kS/s 10b SAR ADC consumes 334nW under a 0.6V supply. Its active area is only 0.0098mm². It achieves an SNDR of 55.96dB, which results in an FoM$_W$ of 1.63fJ/c.-s.

Index Terms - SAR ADC, open-loop, dynamic operation, power system, zero-capacitor TDDI technology.

I. INTRODUCTION

Touch is the most intuitive operation in human machine interfaces. While the touch panel gains its popularity, the usage of a low-cost touch with display driver integration (TDDI) system using the zero-capacitor technology is an inevitable trend for the mobile devices. The zero-capacitor technology allows the TDDI system to get rid of the off-chip decoupling capacitor, C$_{OFDC}$, on the flexible printed circuit (FPC) board. It not only reduces the cost of system, but also enables the borderless display module.

SAR ADCs are known for their compact-area and low-power characteristics, which is advantageous to the zero-capacitor TDDI system. However, the huge dynamic current from the SAR ADC, I$_{L\_AD}$, offers its power system a real challenge, especially without any C$_{OFDC}$. An unclean ADC power supply voltage, V$_{DD\_AD}$, deteriorates the overall SNR performance of the TDDI system. It is a direct method by using the conventional LDO [1] to regulate V$_{DD\_AD}$. However, its static close loop passively sensing ΔV$_{DD\_AD}$ is slow and power-hungry as shown in Fig. 1. A large off-chip C$_{OFDC}$ is also required to reach enough phase margin. There are many reference buffer designs to replace the conventional LDO. The capacitor reservoirs method [2] can isolate disturbance from V$_{DD\_AD}$. The signal-dependent I$_{L\_AD}$ still decreases its accuracy. The reference neutralizing method [3] is fast, but its static reference consumption is not energy-efficient.

This paper proposes an energy-efficient detective open-loop dynamic (DeOLD) system buffer co-designed with a SAR ADC in the zero-capacitor TDDI system as illustrated in Fig. 2. The system buffer supplies clean voltages for the

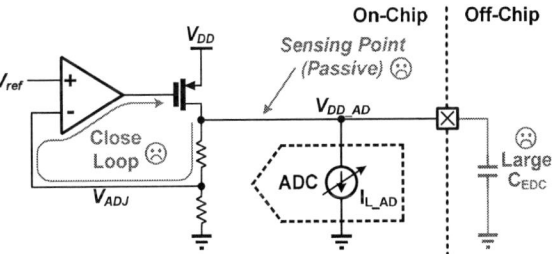

Fig. 1. The simplified block diagram of the conventional LDO.

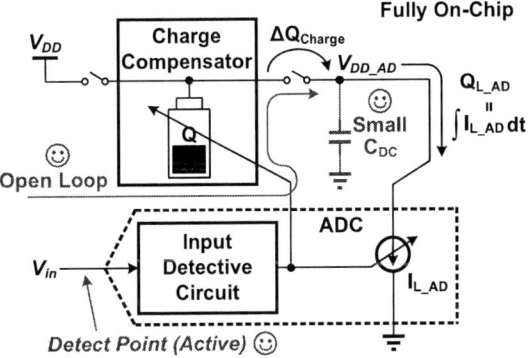

Fig. 2. The concept of the proposed DeOLD system buffer.

reference, the analog and the digital supply voltages within the whole system. The input signal, V$_{in}$, determines the ADC charge loss, Q$_{L\_AD}$. Instead of sensing the disturbance of ΔV$_{DD\_AD}$ passively, the proposed DeOLD system buffer actively detects V$_{in}$ to derive the signal-dependent Q$_{L\_AD}$, which is consumed by the dynamic current I$_{L\_AD}$ of the ADC. Then, a tunable charge compensator is utilized to compensate the dissipated charge, ΔQ$_{Charge}$, to V$_{DD\_AD}$. The asynchronous dynamic charge sharing technique is adopted to save 42.4% power consumption in comparison with the usage of the well-known pump squeezing method [3]. The open-loop approach without opamps can be faster and more energy-efficient than the close-loop feedback with the consideration of phase margin. The opamp-free design is also suitable for the low-voltage environment. Moreover, it only needs a small on-chip decoupling capacitor, C$_{DC}$, rather than a large C$_{OFDC}$. As a result, the FoM$_W$ performance of the DeOLD system buffer with a 400kS/s SAR ADC achieves as low as 1.63fJ/c.-s. under a 0.6V supply.

978-1-5386-6414-8/18 $31.00 © 2018 IEEE

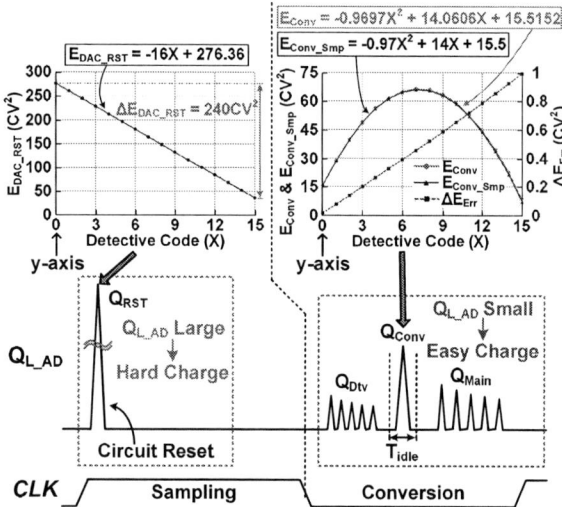

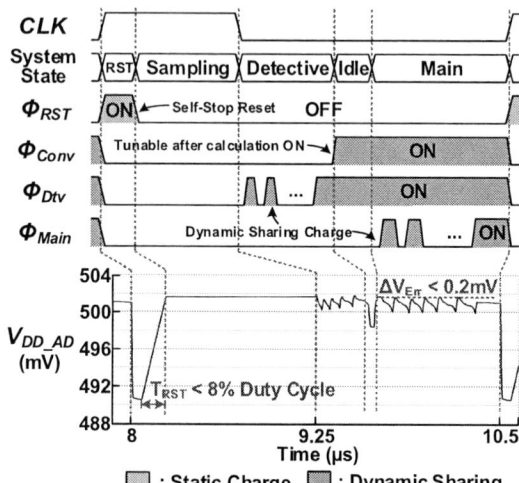

Fig. 5. The timing diagram of this work.

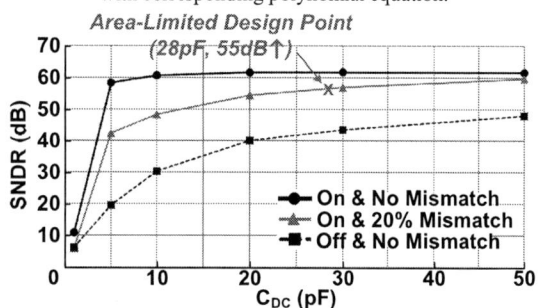

Fig. 4. The SNDR simulation versus the value of C_{DC}, charge compensator on/off and process variation.

II. ALGORITHM OF THE PROPOSED SYSTEM BUFFER

The detective circuit, which predicts the ADC switching energy, is the key block of the proposed DeOLD system buffer. Our system buffer takes advantage of the coarse sub-ADC, which is already existed in the subranging SAR ADC [4], as the detective circuit. In addition, the skipping switching method of the subranging SAR ADC is also chosen for this work. Not only its relationship between Q_{L_AD} and V_{in} can be estimated precisely with a polynomial equation, but also its skipping state reduces the DAC switching energy. Furthermore, its timing arrangement merges the switching phase of the MSB main capacitors to the same interval, which simplifies the charge compensator circuits of V_{DD_AD} and decreases working times of the DeOLD system buffer.

Fig. 3 shows the Q_{L_AD} dissipation versus different phases of the subranging SAR ADC. Q_{Conv} is caused by the ideal DAC switching energy during the conversion phase. Its energy dissipation, E_{Conv}, is symmetrical to the middle of the detective code, which is depicted as the gray solid-line curve at the right-up side of Fig. 3. The detective code, X, from the detective circuit represents the digital format of V_{in}, which is related to E_{Conv}. The left-hand side of X is not shown here for

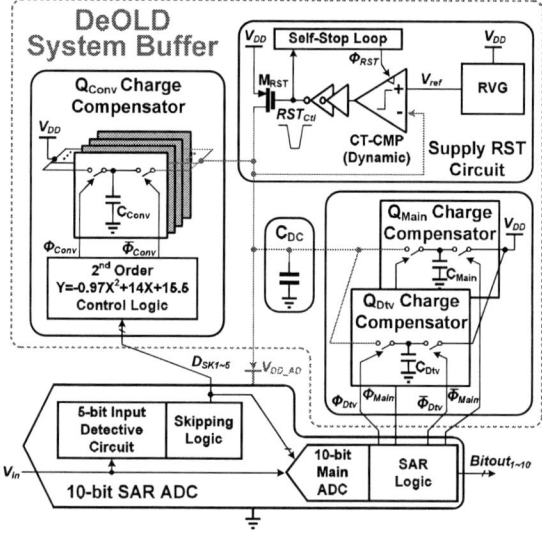

Fig. 6. The block diagram of the DeOLD buffer and the ADC.

simplicity. The right-hand side of E_{Conv} can be modeled as a 2nd order polynomial equation by a leastz-square error approximation method, but it is hard to be implemented. Thus, in order to be more feasible, the coefficients of E_{Conv} are simplified as

$$E_{Conv\_Smp} = -0.97X^2 + 14X + 15.5 , \qquad (1)$$

The energy of detective error, ΔE_{Err}, is the difference between E_{Conv} and E_{Conv_Smp}. It is small enough to be neglected. Other small spikes of Q_{L_AD} caused by the energy loss from the detective circuit, Q_{Dtv}, and the main ADC, Q_{Main}, can also be supplemented easily.

At the beginning of the sampling phase, the large Q_{RST} is resulted from the DAC reset energy, E_{DAC_RST}. It can be modeled as

$$E_{DAC\_RST} = -16X + 276.36 . \qquad (2)$$

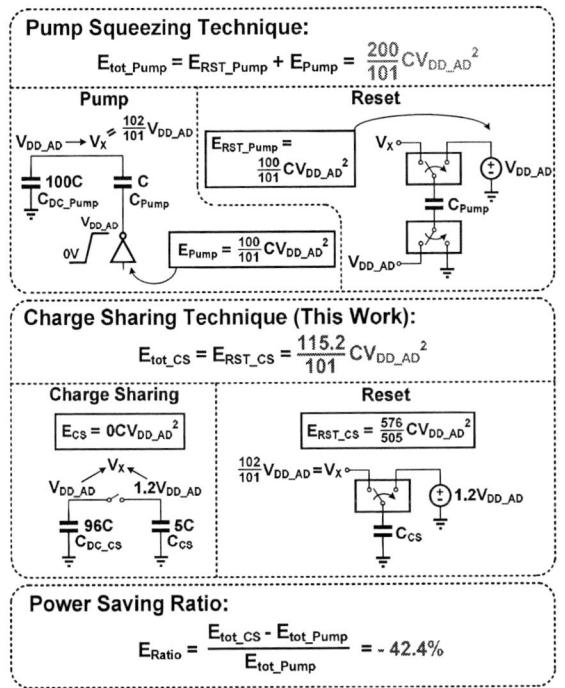

Pump Squeezing Technique:

$$E_{tot\_Pump} = E_{RST\_Pump} + E_{Pump} = \frac{200}{101}CV_{DD\_AD}^2$$

Pump

$$V_{DD\_AD} \rightarrow V_X \approx \frac{102}{101}V_{DD\_AD}$$

$100C$ $\quad$ C
C_{DC_Pump} $\quad$ C_{Pump}

$$E_{Pump} = \frac{100}{101}CV_{DD\_AD}^2$$

Reset

$$E_{RST\_Pump} = \frac{100}{101}CV_{DD\_AD}^2$$

C_{Pump}

Charge Sharing Technique (This Work):

$$E_{tot\_CS} = E_{RST\_CS} = \frac{115.2}{101}CV_{DD\_AD}^2$$

Charge Sharing

$$E_{CS} = 0CV_{DD\_AD}^2$$

V_X

V_{DD_AD} $\quad$ $1.2V_{DD\_AD}$

$96C$ $\quad$ $5C$
C_{DC_CS} $\quad$ C_{CS}

Reset

$$E_{RST\_CS} = \frac{576}{505}CV_{DD\_AD}^2$$

$$\frac{102}{101}V_{DD\_AD} = V_X$$

$1.2V_{DD\_AD}$

C_{CS}

Power Saving Ratio:

$$E_{Ratio} = \frac{E_{tot\_CS} - E_{tot\_Pump}}{E_{tot\_Pump}} = -42.4\%$$

Fig. 7. The energy dissipation of the pump squeezing technique and the charge sharing technique in the charge compensator.

Although E_{DAC_RST} can be modeled as a simple 1st order polynomial equation, ΔE_{DAC_RST} is so large that it is difficult to design an effective charge compensator. Instead, this work overlaps the reset phase of the DeOLD system buffer to that of the ADC. It only needs to charge the slight energy loss from Q_{Dtv}, Q_{Main} and the signal-dependent Q_{Conv}. This way both reduces the number of charge compensators and increases the compensation accuracy as well.

Fig. 4 illustrates the SNDR simulation versus Q_{Conv}, Q_{Dtv} and Q_{Main} charge compensators turned on or off with the value of C_{DC}. The larger C_{DC} is, the more system PVT tolerance range is. The accuracy dramatically decreases when the charge compensators are turned off. By considering 20% of the process mismatch and the limitation of area, 28pF of C_{DC} is chosen to get 55dB or more of SNDR in this work.

III. ARCHITECTURE AND CIRCUIT IMPLEMENTATION

The timing diagram and block diagram of the proposed DeOLD system buffer with the ADC are shown in Fig. 5 and Fig. 6, respectively. It consists of the ADC, which includes a detective circuit, skipping logic, and a main ADC, and the DeOLD buffer, which includes a C_{DC}, [Q_{Conv} Q_{Dtv} Q_{Main}] charge compensators and a supply reset, RST, circuit. The DeOLD system buffer provides a clean V_{DD_AD} of 0.5V for the ADC.

At first, the 5-bit detective circuit roughly detects V_{in} and transfers the results to the skipping logic to generate $D_{SK1\sim5}$. The C_{Dtv} keeps sharing charges to supplement the successive

energy loss from the detective circuit, Q_{Dtv}, at this time. The charge compensators use dynamic charge sharing technique to compensate charge from V_{DD} to C_{DC}. $D_{SK1\sim5}$ goes to the 2nd order control logic to generate Φ_{Conv}, which charges back the signal dependent Q_{Conv} loss during the idle mode. It also makes the main ADC skip the unnecessary switching for power-saving. The beginning of the main ADC LSB conversions is sensitive to the disturbance of V_{DD_AD}. Hence, the SAR logic sends Φ_{Main} to keeps C_{Main} sharing its charge to compensate the energy loss of the main ADC conversion, Q_{Main}, which suppresses ΔV_{DD_AD} under 0.2mV at the starting point of every comparison phase. After sharing, C_{Dtv}, C_{Main} and C_{Conv} in these charge compensators become as a part of C_{DC} to reduce the voltage ripple of V_{DD_AD}.

The charge loss of E_{DAC_RST}, Q_{RST}, is too large to be accurately compensated. Therefore, V_{DD_AD} should be reset by the supply RST circuit during the short RST phase. It also removes the accumulation error caused by the mismatch of the charge compensators. The comparator of the supply RST circuit controls M_{RST} to charge the V_{DD_AD} back to about 0.5V, according to the value of V_{ref}. When the reset signal, Φ_{RST}, of the DeOLD system buffer comes, the continuous comparator (CT-CMP) is turned on. RST_{Ctl} starts from low-level, so M_{RST} is turned on to pull V_{DD_AD} up. Until V_{DD_AD} and V_{ref} cross, RST_{Ctl} goes back to high-level and turns off M_{RST}. Then, RST_{Ctl} goes through the self-stop loop to gate the large static current wasted from the CT-CMP. The overshoot and offset of the CT-CMP does not affect the accuracy of this work. It only causes the slight offset of V_{DD_AD}. By doing this, it supplements the large energy loss from Q_{RST}. The active time of the CT-CMP only occupies 8% of the duty cycle, which makes the DeOLD system buffer more energy-efficient.

This work adopts the dynamic charge sharing technique to save more energy than that of the well-known pump squeezing technique. Fig. 7 illustrates the energy consumption of the pump squeezing technique and the dynamic charge sharing technique. Assume these two methods use the same total capacitance, C_{tot}, and get the same final voltage value, V_X, which is equal to $101C$ and $\frac{102}{101}V_{DD\_AD}$, respectively. C is assumed to be a unit capacitor for simplicity. The pump squeezing technique requires $\frac{100}{101}CV_{DD\_AD}^2$ of energy, E_{Pump}, to pump C_{Pump} from 0V to V_{DD_AD}. It also needs another $\frac{100}{101}CV_{DD\_AD}^2$ of energy, E_{RST_Pump}, to reset C_{Pump}. Its total energy dissipation, E_{tot_Pump}, is equal to $\frac{200}{101}CV_{DD\_AD}^2$. On the other hand, the charge sharing technique shares the charge from C_{CS} to C_{DC_CS}. During the charge sharing phase, both of C_{CS} and C_{DC_CS} are not connected to the power supply, so its energy consumption, E_{CS}, is zero. It only takes $\frac{115.2}{101}CV_{DD\_AD}^2$ of energy, E_{CS_RST}, to charge C_{CS} back to $1.2V_{DD\_AD}$. Without energy dissipation during the charge sharing phase, the total energy of the dynamic charge sharing technique, E_{tot_CS}, saves 42.4% of energy than that of the pump squeezing technique. Hence, it is utilized in this work.

IV. MEASUREMENT RESULTS

Fig. 8 shows the chip micrograph of the ADC fabricated in 40nm CMOS. The core circuit occupies an area of 0.0098mm² (94μm×104μm). The system works at 0.6V with 0.5V of V_{ref}. Thanks for the dynamic and open-loop features of the DeOLD system buffer, the total power of this system is only 334nW at 400kS/s. It can be broken down as follows: The Q_{Conv} charge compensator uses only 16.1%; The Q_{Dtv} and Q_{Main} charge compensators use 46.3%; The Supply RST circuit uses 37.6%, including 17% of CT-CMP with the self-stop loop. The power efficiency of this work is 71.3%, which is the ratio between the simulated power of ADC and the measured total power dissipation of the whole system.

Fig. 9 illustrates the measured DNL and INL plots at the conversion rate of 400kS/s. The unit capacitor of ADC capacitor array is 1fF. The DNL and INL are +0.27/-0.37 and +0.39/-0.39LSB, respectively. Fig. 10 shows the dynamic performance versus input frequency. It gets better SFDR with a band pass filter in front of the input signal, such as 100kHz and 200kHz. The Nyquist rate FFT plot is shown in Fig. 11. The measured SNDR, SFDR, SNR, and THD at Nyquist rate achieve 55.96dB, 65.47dB, 57.23dB, and 60.94dB, respectively. The ADC is isolated from the power supply except the RST phase, so the PSRR performance is limited by the supply RST circuit. Without any large C_{OFDC}, the FoMw achieves 1.63fJ/c.-s., including the power consumption of both the DeOLD system buffer and the ADC. Table I shows the performance summary and comparison table of the energy-efficient ADCs in recent years.

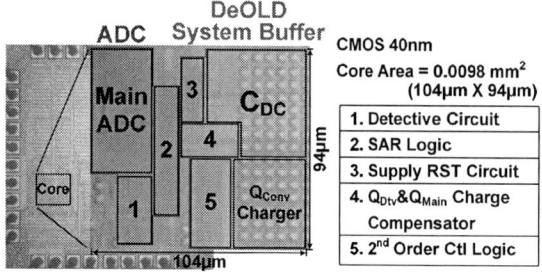

Fig. 8. Chip micrograph.

Table I. Performance summary and comparison.

	VLSI'14 Y.-J Chen	ISSCC'14 H.-Y Tai	VLSI'16 S.-E Hsieh	ISSCC'15 M. Liu	This Work
Technology (nm)	90	40	90	65	40
Resolution (bits)	10	10	11	10	10
Active Area (mm²)	0.04	0.0065	0.035	0.26	0.0098
Supply (V)	0.4	0.45	0.3	0.8	0.6
Fs (kS/s)	250	200	600	80	400
DNLmax (LSB)	0.43	0.44	0.63	0.60	0.37
INLmax (LSB)	0.67	0.45	0.72	0.94	0.39
Power (nW)	200	84	0.187	106	334
Including Power System	No	No	No	Yes	Yes
CEDC	-	-	-	10μF	No
ENOBNyquist (bit)	8.6	8.95	9.46	9.1	9
FoMW (fJ/c.-s.)	2.02	0.85	0.44	2.4	1.63

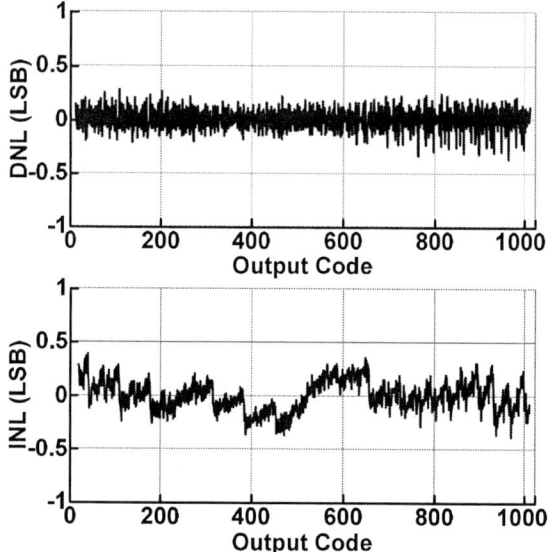

Fig. 9. Measured static performance.

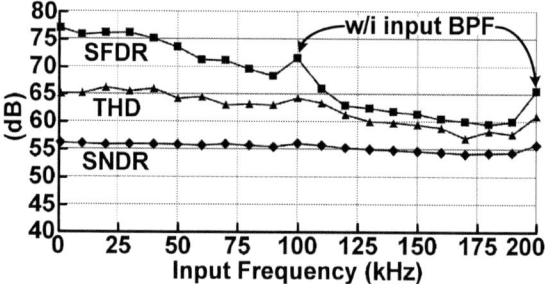

Fig. 10. Measured dynamic performance versus input frequency.

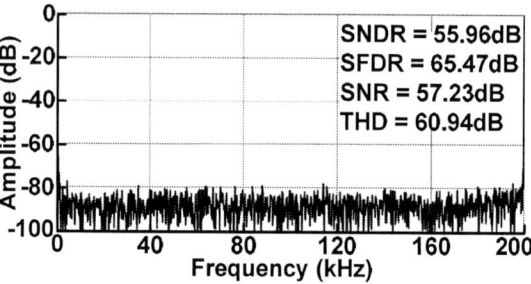

Fig. 11. Measured FFT plot at the Nyquist rate.

REFERENCES

[1] M. Liu, et al., "A 0.8V 10b 80kS/s SAR ADC with duty-cycled reference generation," *IEEE ISSCC Dig. Tech. Papers*, pp. 278-279, Feb. 2015.

[2] J. Shen, et al., "A 16-bit 16MS/s SAR ADC with On-Chip Calibration in 55nm CMOS," *IEEE Symp. VLSI Circuits*, pp. C282-C283, June 2017.

[3] Y.-Z Lin, et al., "A 8.2-mW 10-b 1.6-GS/s 4× TI SAR ADC with Fast Reference Charge Neutralization and Background Timing-Skew Calibration in 16-nm CMOS," *IEEE Symp. VLSI Circuits*, pp. C204-C205, June 2016.

[4] H.-Y. Tai, et al., "A 0.85fJ/conversion-step 10b 200kS/s Subranging SAR ADC in 40nm CMOS," *IEEE ISSCC Dig. Tech. Papers*, pp. 196-197, Feb. 2014.

Design of a 2.45-GHz RF Energy Harvester for SWIPT IoT smart sensors

Pengcheng Xu, Denis Flandre and David Bol

ECS group, ICTEAM Institute
Université catholique de Louvain, Belgium
{pengcheng.xu, denis.flandre, david.bol}@uclouvain.be

Abstract—Simultaneous wireless power and information transfer (SWIPT) is a flexible and cheap way to supply IoT smart sensors avoiding battery replacement. In this paper, we study the design of a 2.45-GHz RF energy harvester (RFEH) system based on a discrete-component matching network, a custom 65nm CMOS cross-coupled rectifier coupled with an off-the-shelf storage-charging power management unit (PMU) performing rectifier output voltage regulation with maximum power point tracking (MPPT). For 2.45-GHz operation, we propose a parasitic-aware sizing of π matching network. This introduces a maximum bound on the real impedance at the rectifier input to ensure good impedance matching in practice. MPPT is used to help reaching this target real impedance at given input RF power, as the rectifier input impedance is a function of both its input and output voltages. Measurement results show a sensitivity as low as -17.1 dBm with a peak power harvesting efficiency (PHE) of 48.3% at -3 dBm 2.45-GHz input RF power. Comparison between simulation and measurement results demonstrate that these results are limited by the parasitic capacitance and that PHE around 45% can be obtained down to -10 dBm with PCB/package improvement for lower parasitic capacitance.

Keywords—RF energy harvester, 2.45-GHz, maximum power point tracking, power harvesting efficiency, impedance matching

I. INTRODUCTION

Wireless power transfer (WPT) is a flexible way to supply IoT smart sensors without battery replacement. This is usually done with sub-GHz WPT using a specific power transmitter added to the building infrastructure. As mainstream short-range low-power communications like WiFi or BLE uses 2.45-GHz band, sub-GHz WPT also requires a second specific antenna on the smart sensor for receiving the RF power. Therefore, WPT at 2.45-GHz is an interesting alternative for more compact and cheap smart sensors, also re-using BLE/WiFi infrastructure for simultaneous wireless power and information transfer (SWIPT). As an example, such a 2.45-GHz SWIPT occupancy-detection smart system is demonstrated in [1].

Fig. 1 shows a typical RF-supplied IoT smart sensor for SWIPT. The single antenna is used for BLE communications and RF energy harvesting (RFEH). The RFEH system collects the AC electromagnetic power of the RF wireless link and converts it into a DC electrical power to supply the ultra-low-power (ULP) sensor platform. A typical RFEH system is composed of an impedance matching network (MN), an AC/DC rectifier, a power management unit (PMU) and a rechargeable energy storage (battery or supercapacitor). The matching network matches the rectifier input impedance to

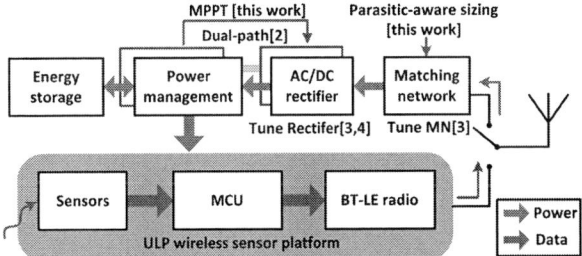

Fig. 1. SWIPT IoT smart sensor and RFEH state-of-art techniques.

the antenna impedance, typically 50/75Ω, to avoid reflecting incident RF power. After AC/DC conversion by the rectifier, the PMU generates a stable voltage to supply the ULP platform regardless of the variable incident RF power, while managing the storage element.

Because of strong path loss, the incident RF power is limited in a range from -25 to 0 dBm [1]. Therefore, it is a key challenge for SWIPT system to achieve good harvesting efficiency over this incident power range. Solutions in the literature include dual-path rectifiers, which consists of a low-power path using low-V_{th} transistors for high efficiency at low RF power and a high-power path using high-V_{th} transistors for high efficiency at high input power [2]. Another solution is reconfigurable rectifiers based on multiple identical primary cells connected in series/parallel depending on the input RF power level [3][4]. In order to compensate rectifier impedance variation with input power, a tunable MN based on a capacitor bank was also implemented in [3]. However, in all these solutions, the configuration switches introduce power losses due to their series resistance. Large switches can be selected for low series resistance. However, as we will study in this paper, any parasitic capacitance (e.g. by these switches) added at the rectifier input becomes critical for the MN design at 2.45-GHz. In this paper, we study the design of an RFEH system at 2.45-GHz without reconfiguration switches. We demonstrate the impact of the parasitic capacitance at the rectifier input and validate the interest of rectifier output voltage regulation with maximum power point tracking (MPPT). The proposed RFEH system is presented in Section II. The parasitic-aware sizing of the MN at 2.45-GHz is discussed in Section III. The characteristics of the rectifier operated at its MPPT are studied in Section IV with measurement results in Section V.

978-1-5386-6414-8/18 $31.00 © 2018 IEEE

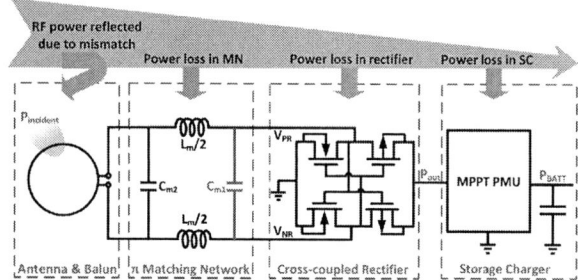

Fig. 2. Proposed RFEH system and its power losses classification.

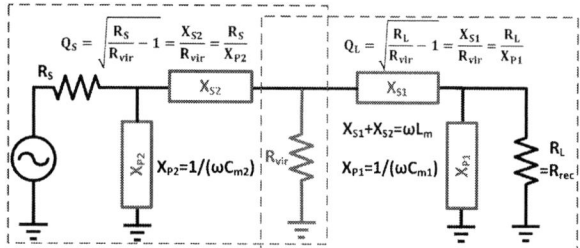

Fig. 3. Π matching network design (the virtual resistor R_{vir} is for sizing but not real component and $R_{vir} \leq \min(R_S, R_L)$).

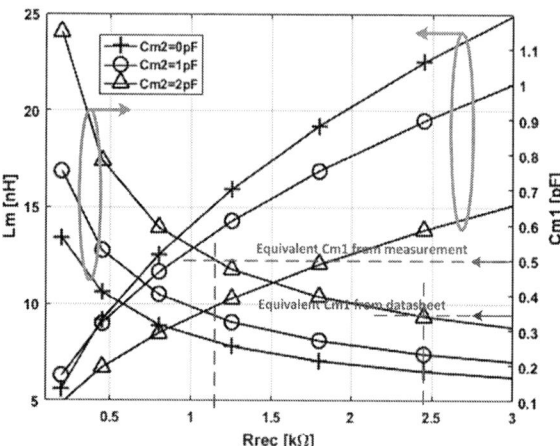

Fig. 4. Sizing of the π-MN elements vs. R_{rec} at 2.45-GHz with 50-Ω antenna (blue curve with $C_{m1}=0.35$pF and red curve with $C_{m1}=0.50$pF).

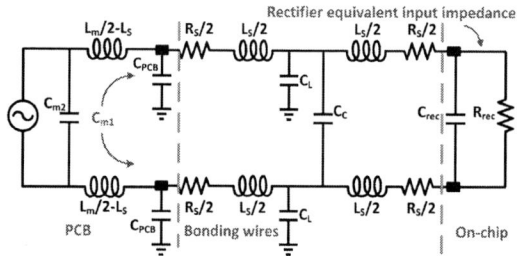

Fig. 5. Model of PCB/package parasitics (5×5mm QFN32L with $R_S=46$mΩ, $L_S=1.11$nH, $C_L=202$fF, $C_C=39$fF).

II. PROPOSED RFEH SYSTEM

The proposed 2.45-GHz RFEH system is depicted in Fig. 2. It is built around a custom cross-coupled rectifier designed in 65nm CMOS. This rectifier architecture allows dynamic self-V_{th} cancellation but as its input is differential, it requires the addition of a Balun to the single-ended antenna. The MN is based on the π topology to help cancel the rectifier parasitic input capacitance coming from PCB and chip package, as will be explained in Section III. The rectifier output voltage is regulated by the storage-charging path inside the PMU, which performs MPPT to preserve good power conversion efficiency (PCE) in the rectifier and limit its input impedance variation over the RF incident power range. The PMU also boosts the voltage to charge the storage element.

There are four main sources of RF power losses in RFEH system: RF power reflected back over the air due to impedance mismatch, power loss in the MN, power loss in the rectifier and power loss in the PMU. The power harvesting efficiency (PHE) of RFEH system is the ratio between the extracted DC output power P_{out} delivered to the load (here in Fig. 2 we consider the PMU as the load) and the RF incident electromagnetic power on the receiver antenna, which would be collected with a perfect impedance match, $P_{incident}$ in short. PHE can be expressed as:

$$PHE = P_{BATT}(\text{or } P_{out} \text{ without considering PMU})/P_{incident}.$$

Because of the wide RF incident power range, preserving good PHE over this range is a challenge [1]. Let us mention that the PCE metric defined as $PCE = P_{out}/P_{absorbed} = P_{out}/(P_{incident} - P_{reflected})$ in [5][6] is a limited metric for RFEH systems as it neglects the reflected power due to impedance mismatch. A high PCE is thus a required condition but not a sufficient condition for high performance

of RFEH systems. This is why we use the PHE metric as a performance metric in this paper.

III. PARASITIC-AWARE SIZING OF THE MATCHING NETWORK

To extract maximum power from an RF source, the load impedance must be equal to the complex conjugate (i.e. identical real impedance with opposite reactance) of the source impedance. As shown in Fig. 5, the rectifier input impedance Z_{rec} can be modeled as a resistor R_{rec} and capacitor C_{rec} in parallel for representing its real and imaginary parts. In two-element L-MN, the quality factor (Q) is fixed when the source and load impedances are determined. In contrast, π-MN can be used for narrow-band high-Q application [7]. As described in Fig. 3, it can be separated into two L networks with a virtual resistor R_{vir} whose resistance must be smaller than either antenna impedance R_S or load resistor R_L (here R_L is the input resistance R_{rec} of the rectifier operated at its MPPT) for sizing π-MN parameters. Fig. 4 shows the required L_m and C_{m1} values to achieve a perfect match as a function of the rectifier input resistance R_{rec} and the C_{m2} value. If $C_{m2}=0$, the π-MN network is actually an L-MN and the value of its elements L_m and C_{m1} follows the L-MN sizes. Let us mention already that R_{rec} varies with the RF incident power. This comes from the varying voltage amplitude at the rectifier input combined with the highly non-linear characteristics of the transistors at low gate voltage.

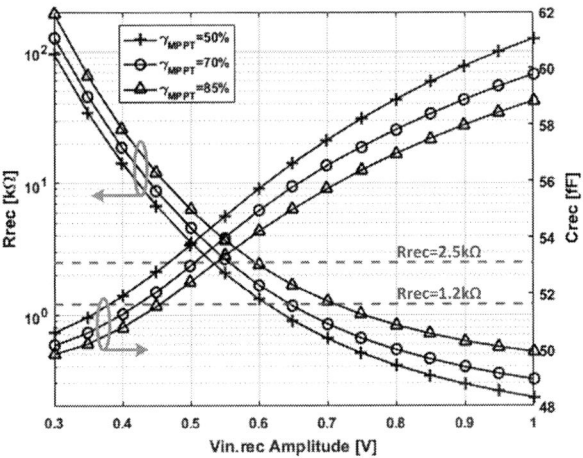

Fig. 6. Rectifier input impedance with four rectifiers in parallel (NMOS 5um/60nm, PMOS 10um/60nm) at 2.45-GHz, γ_{MPPT}=V$_{rec.out}$/V$_{rec.out.OC}$ (V$_{rec.out.OC}$ is V$_{rec.out}$@rectifier output open circuit).

At 2.45-GHz, PCB and package parasitics have a magnified impact compared to sub-GHz. The value of the C_{m1} capacitance in the π-MN has a minimum value coming from the combination of the PCB (C_{PCB}) and package (C_L and C_C) parasitics as well as the rectifier equivalent input capacitance (C_{rec}). From the datasheet of QFN package model in Fig. 5, the total parasitic from two input RF bonding wires and PCB is around 0.30pF (=C_L/2+C_C+C_{PCB}/2) if C_{PCB}=0.30pF (with a trace length of 5.5mm and width of 0.7mm in FR4 PCB). Adding C_{rec}, C_{m1} minimum value is estimated at 0.35pF. As depicted in Fig. 4, the 0.35pF C_{m1} value requires an equivalent input resistance of the rectifier below 0.5kΩ when C_{m2} is 0pF (corresponding to an L-MN). This is a difficult target for sizing the rectifier at low RF incident power. Therefore, using C_{m2}=2pF, the R_{rec} requirement is relaxed to 2.50kΩ. However, in real measurement, due to extra parasitic capacitor from PCB arbitrary soldering and extra bonding wire length for die position, which depends on fabrication technique and is variable case by case, the equivalent C_{m1} is estimated at 0.50pF. The corresponding matched R_{rec} is close to 1.20kΩ with C_{m2}=2pF and L_m=10nH, which results in an explicit L_m of 7nH considering the parasitic inductance L_S of the bonding wires. This means the best matching occurs when the RF incident power results in an R_{rec} of 1.20kΩ. Let us mention here that assuming sub-GHz WPT at 900MHz, the same 0.50pF C_{m1} equivalent parasitic capacitance would require a target R_{rec} of 4kΩ, which is much easier to obtain at low input RF power.

IV. RECTIFIER CHARACTERISTIC UNDER MPPT OUTPUT VOLTAGE REGULATION

The proposed RFEH uses a custom rectifier design in 65nm LP CMOS with LVT transistors. It is based on the dynamic self-V_{th}-cancellation cross-coupled topology [8], which achieves a higher rectifier PCE than the conventional rectifier, at low input power conditions. In this differential scheme, as shown in Fig. 2, the gate of transistors is actively biased by

Fig. 7. Proposed RFEH system implementation.

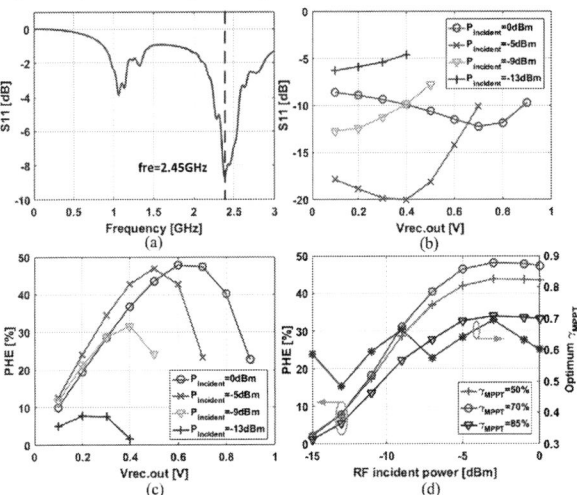

Fig. 8. RFEH measurement results. (a) RFEH operation frequency band. (b) S11 vs. Vout at different input RF power. (c) PHE vs. Vout at different RF input power. (d). Impact of γ_{MPPT} ratio on PHE.

a dynamic differential-mode signal. When V_{PR} is negative, which corresponds to the forward bias condition for the MN1 diode, the gate voltage of MN1, which is V_{NR}, is positively biased and effectively decreases the turn-on voltage of MN1, resulting in a small ON-resistance. On the other hand, when V_{PR} becomes positive, which corresponds to the reverse bias condition, the gate voltage rapidly decreases, which effectively reduces the reverse leakage current.

For a given sizing of the rectifier, Z_{rec} is a function of the rectifier input $V_{rec.in}$ and output $V_{rec.out}$ voltages as well as the RF frequency:

$$Z_{rec}=R_{rec} \parallel C_{rec} =f(V_{rec.in}, V_{rec.out}, \text{frequency})$$

Post-layout simulations show in Fig. 6 that, the rectifier input resistance R_{rec} increases greatly at low $V_{rec.in}$ as transistors enter the near-/sub-threshold regime. However, its capacitance C_{rec} is less dependent on rectifier input and output terminal voltage, as the transistor gate capacitance (C_{gs} and C_{gd}) variations are cancelled out due to the two differential input voltages V_{PR} and V_{NR}.

The selected battery-charging PMU operates with pseudo MPPT. Indeed, it periodically disconnects the rectifier output and senses its open circuit voltage (in this design the V$_{rec.out.OC}$ is close to 0.82 V$_{rec.in}$ amplitude) and then regulates the

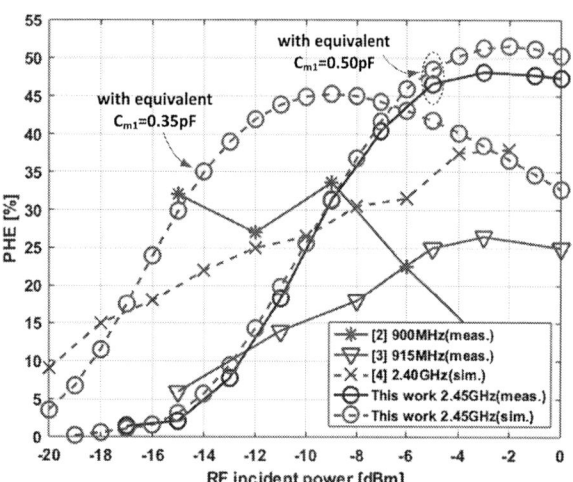

Fig. 9. PHE (=$P_{out}/P_{incident}$ with γ_{MPPT}=70%) compared to the state-of-art.

rectifier output voltage by forcing $V_{rec.out}$ to a configurable ratio γ_{MPPT} of $V_{rec.out.OC}$. Fig. 6 shows that γ_{MPPT} affects R_{rec} and the lowest R_{rec} is reached with γ_{MPPT} of 50%.

V. MEASUREMENT AND ANALYSIS

The cross-coupled RF rectifier was designed and fabricated in 65nm LP CMOS process with 5μm×25μm area. As depicted in Fig. 7, the full RFEH system also includes a Balun, a discrete-component MN and the AEM30940 PMU from e-peas semiconductors for storage charging with pseudo-MPPT. It allows configuring the γ_{MPPT} ratio at 50, 70 or 85%.

The measured S11 parameter at the MN input for -5 dBm incident power with open load shows excellent matching at 2.45 GHz as depicted in Fig. 8(a). As this matching depends on R_{rec}, it is sensitive to $V_{rec.in}$ and $V_{rec.out}$. Fig. 8(b) shows that optimum matching is reached for different $V_{rec.out}$ values. MPPT thus helps preserving a good matching from 0 dBm to -10 dBm. However, below -10 dBm, the mismatch becomes serious as R_{rec} becomes significantly higher than the target 1.2kΩ for which the MN was sized. Fig. 8(c) clearly shows that the PHE is maximized for a $V_{rec.out}$ voltage that varies from 0.25V to 0.65V as a function of the RF input power. This really validates the need for $V_{rec.out}$ MPPT regulation. The γ_{MPPT} ratio to select is analyzed in Fig. 8(d) with the PHE obtained for 50, 70 and 85% ratios. Best PHE is reached at γ_{MPPT} ratio of 70% as a trade-off between an optimum PCE at higher γ_{MPPT} ratio and a better matching (thanks to lower R_{rec}) at lower γ_{MPPT}.

Fig. 9 show that PHE measurement results are in excellent agreement to the simulation results with all parasitic elements taken into account. The proposed RFEH has excellent performance from 0 dBm to -10 dBm thanks to the parasitic-aware MN sizing methodology and the absence of reconfiguration switches. Below -10 dBm, impedance mismatch shows up. The PHE suffers from both the reflected power loss and the associated $V_{rec.in}$ drop, which results in a low PCE for the rectifier. One 900-MHz RFEH [2] offers a better PHE at

TABLE I. TECHNIQUE AND PERFORMANCE COMPARISON.

	This work	[2] 2017 TCAS-II	[3] 2017 JSSC	[4] 2016 ISCAS(sim.)
Process	65nm	65nm	0.18μm	0.18μm
Frequency	2.45GHz	900MHz	915MHz	2.40GHz
Topology	Cross coupled	Cross coupled	Greinacher doubler	Greinacher doubler
Configuration Switch	No	Yes	Yes	Yes
Rectifier Load	MPPT PMU	147kΩ Resistor	DC/DC Boost	Charged 1V Capacitor
Sensitivity	-17.1dBm	-17.7dBm	-14.8dBm	-22dBm
Peak PHE	48.3% @-3dBm	34.5% @-9dBm	26% @-3dBm	38.4% @-2dBm

low RF incident power, which can be explained by the lower $R_{rec.in}$ requirement at sub-GHz frequency given the parasitic capacitance value (as discussed in Section III). One 2.4-GHz RFEH [4] also obtain slightly better PHE at low $P_{incident}$ but it is based only on simulations without mentioning PCB/package parasitics.

To demonstrate that the PHE of the proposed RFEH system is limited at low $P_{incident}$ by the PCB/chip parasitic capacitance, we simulated the RFEH assuming that this equivalent capacitance could be brought down to 0.35pF. Results in Fig. 9 show that in this case, the PHE can be kept higher than 25% down to -16 dBm with sensitivity below -20 dBm.

VI. CONCLUSION

A fully 2.45-GHz RFEH system with MPPT for SWIPT IoT smart sensors is presented in this paper. The proposed π-MN sizing methodology fully considers parasitic components, which are more critical at 2.45-GHz compared to sub-GHz. MPPT helps to improve the PHE by tuning the rectifier input impedance. Measurement results in Table I show a sensitivity of -17.1 dBm and peak PHE of 48.3% at -3 dBm. Simulation shows that reducing the parasitic PCB/package capacitance would result in 45% PHE down to -10 dBm with a sensitivity as low as -22.7 dBm.

ACKNOWLEDGMENT

This work was supported by the Brussels Institute for Research and Innovation (INNOVIRIS) under the COPINE-IoT project.

REFERENCES

[1] R. Dekimpe and et al., "A Battery-less BLE IoT Motion Detector Supplied by 2.45-GHz RF Energy Harvesting," *PATMOS*, 2018.
[2] Y. Lu and et al., "A wide input range dual-path CMOS rectifier for RF energy harvesting," *TCAS-II*, 2017.
[3] M. A. Abouzied and et al., "A fully integrated reconfigurable self-startup RF energy-harvesting system with storage capability," *JSSC*, 2017.
[4] Z. Zeng and et al., "A WLAN 2.4-GHz RF energy harvesting system with reconfigurable rectifier for wireless sensor network," *ISCAS*, 2016.
[5] T. Soyata and et al., "RF energy harvesting for embedded systems: A survey of tradeoffs and methodology," *IEEE Circuits and Systems Magazine*, 2016.
[6] C. R. Valenta and et al., "Harvesting wireless power: Survey of energy-harvester conversion efficiency in far-field, wireless power transfer systems," *IEEE Microwave Magazine*, 2014.
[7] C. Bowick and et al., *RF Circuit Design*, 2nd ed. Newnes, 2008.
[8] K. Kotani and et al., "High-efficiency differential-drive CMOS rectifier for UHF RFIDs," *JSSC*, 2009.

A 6.78–200 MHz Offset-Compensated Active Rectifier with Dynamic Logic Comparator for mm-size Wirelessly Powered Implants

Jianming Zhao and Yuan Gao

Institute of Microelectronics, A*STAR (Agency for Science, Technology and Research), Singapore
zhaojm@ime.a-star.edu.sg; gaoy@ime.a-star.edu.sg

Abstract— **This paper presents an active rectifier with wideband operation frequency range from 6.78 MHz to 200 MHz for mm-size wireless powered implantable medical device. An energy efficient high speed dynamic logic comparator is proposed to operate up to 200 MHz. In addition, a sample and hold (S/H) offset compensation loop is designed to automatically compensate the intrinsic delay of the dynamic logic comparator. Implemented in a 65nm CMOS process, the proposed active rectifier achieved measured peak voltage conversion ratio of 95% and the measured peak power conversion efficiency (PCE) is 94% at 6.78 MHz and 84% at 200 MHz, respectively.**

Keywords— *active rectifier; dynamic logic comparator; high frequency rectifier*

I. INTRODUCTION

Wireless power transfer (WPT) is a key enabling technology for the miniaturized medical implant [1, 2]. WPT allows the recharging of battery without physical access to the implant device. Hence, the battery size can be reduced to miniaturize the device dimensions without compromising the device life time. More importantly, the miniaturized device can be implanted in the area closer to the target organ or nerve with minimally invasive surgery, which will lead to better therapeutic outcomes. WPT usually works at the license-free ISM frequency bands such as 6.78 MHz and 13.56 MHz. Recently, to further miniaturize the implant device to millimeter (mm)-size for the emerging electroceutical applications, there are rising interests to explore WPT at RF frequencies [3, 4].

The rectifier is a critical building block in the WPT system. Its main function is to convert the AC wireless power input signal to DC output. The simplest rectifier is a full-wave diode bridge rectifier (DBR). However, DBR circuit has low power conversion efficiency (PCE) due to the V_{th} drop across the diode. Various types of active rectifier have been proposed to improve the PCE [5-10]. With proper control of the switches on/off timing, rectification function can be achieved. The schematic of a conventional active rectifier is shown in Fig. 1 to illustrate the operation mechanism. PM_{1-2} are the cross-coupled PMOS transistors and NM_{1-2} are the NMOS transistors driven by the comparators (CMP_{1-2}) as active diodes. The comparator is utilized to compare the input AC signal with reference ground (GND) to generate the gate drive signals DR+ and DR- to turn on/off NM_{1-2}. Ideally, when AC signal is

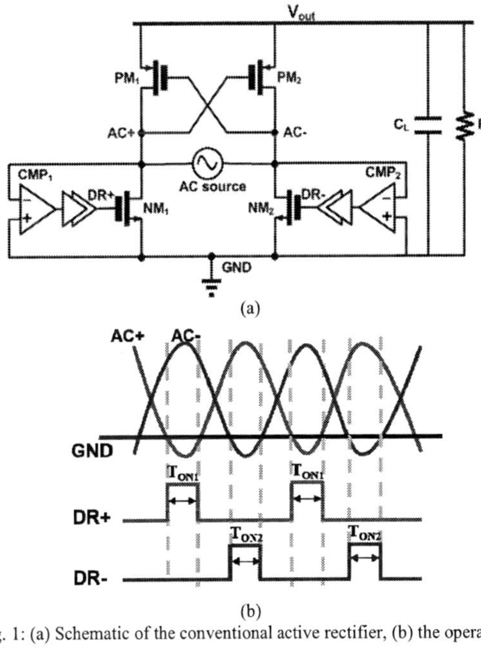

Fig. 1: (a) Schematic of the conventional active rectifier, (b) the operation timing diagram.

crossing GND, the comparator is expected to switch instantaneously. But in reality, it takes time for comparator to respond. In addition, the intrinsic comparator offset will shift the switching point, leading to degraded PCE performance. It consumes high power for the conventional comparator to achieve high speed switching with low offset. On the other hand, when AC signal is far from GND, the comparator maintains the output and no high speed operation is required. However, the conventional comparator usually requires constant bias current, therefore significant power is wasted. Energy efficient high speed comparator is a major bottleneck for high frequency active rectifier design.

Different techniques have been proposed to improve the active rectifier PCE. There are mainly two groups of solutions to compensate the comparator delay, including delay lock loop [3, 8] and sample loop [6]. However, it remains a great challenge to design a robust high efficiency active rectifier that works up to RF frequencies [4].

This work is funded by A*STAR (Agency for Science, Technology and Research) BMRC (Biomedical Research Council), Singapore under the grant no. IAF311022.

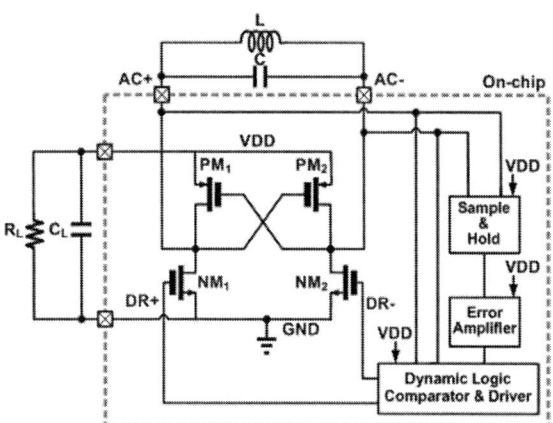

Fig. 2: Block diagram of the proposed active rectifier

In this paper, a wideband active rectifier with high speed dynamic logic comparator and offset-compensation loop is proposed for mm-size wirelessly powered implants. A novel high speed energy efficient dynamic logic comparator is proposed. The comparator offset is compensated by the low speed background offset compensation loop. With these two novel design techniques, the circuit is able to operate in wide frequency range from 6.78 MHz to 200 MHz with PCE in the range of 94% − 84%. Reliable self-startup is also achieved with an additional switch control circuit in the feedback loop error amplifier.

II. SYSTEM ARCHITECTURE

Fig. 2 shows the block diagram of the proposed active rectifier. An off-chip LC tank is formed by a coil and a capacitor to resonate at the operation frequency to receive the wireless power. The active rectifier circuit includes a cross-coupled power transistor core, a delay compensation loop, high speed comparators and the power transistor drivers.

The rectifier core includes cross-coupled PMOS transistors PM_{1-2} and NMOS transistors NM_{1-2} driven by the corresponding control signal DR+ and DR-. At the switching point of NM_{1-2}, the input AC voltage is sampled and compared with GND by an error amplifier. The amplifier output is feedback to control the NM_{1-2} switching on/off time by adjusting the dynamic logic comparator's bias current.

It should be noted that the feedback loop works at a much lower speed compared to the RF carrier frequency through the sample and hold operation. The background feedback loop will converge to a steady state to compensate the comparator offset and in turn enhance the rectifier conversion efficiency.

III. KEY CIRCUIT BLOCKS

A. Dynamic Logic Comparator and Switch Driver

The schematic of the proposed dynamic logic comparator and switch driver is shown in Fig. 3 (a). The comparator has two branches to generate the switch turn on signal V_{ON} and

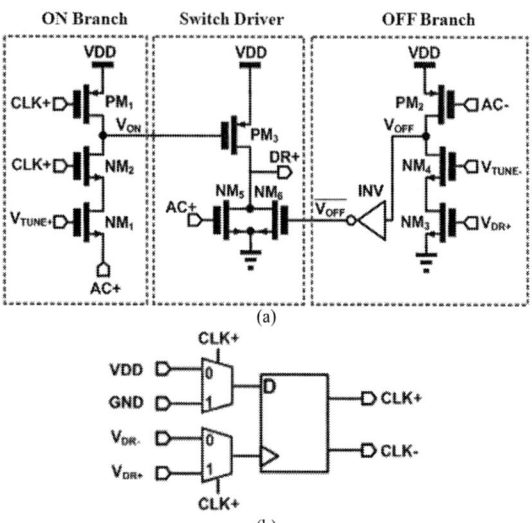

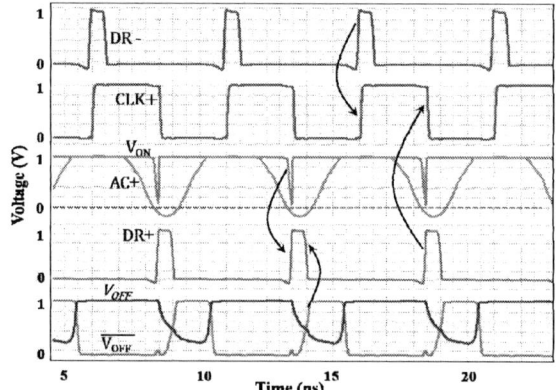

Fig. 3: The schematic of (a) the proposed dynamic logic comparator with switch driver (b) clock generator.

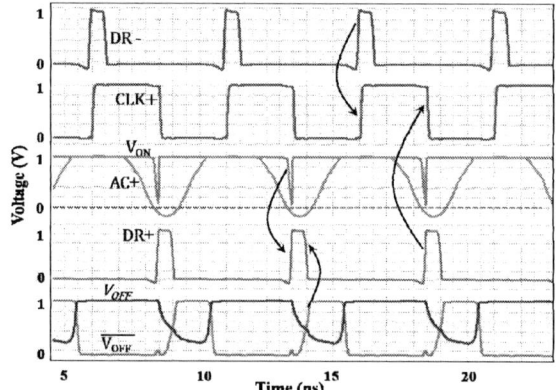

Fig. 4: Timing diagram of DR+ branch nodes.

turn off signal V_{OFF} separately and then combined at the switch driver. Fig.3 (b) shows the associated clock generator.

The operation principle of the dynamic logic comparator is illustrated with an example of DR+ timing diagram shown in Fig. 4. When AC+ decreases towards GND, the ON branch NM_1 starts to turn on as its V_{GS} increases. Because CLK+ is HIGH, therefore NM_2 is on and PM_1 is off, V_{ON} is hence pull down to LOW and DR+ starts to rise. Meanwhile, due to the symmetrical nature of the differential signals, AC- rises towards the peak. In the OFF branch, PM_2 V_{GS} decreases to turn it off. DR+ turns on NM_4, hence V_{OFF} is pulled down slowly with the discharging slop controlled by V_{TUNE-}. As shown in the clock generator in Fig. 3 (b), the rising edge of DR+ flips the DFF output CLK+ to LOW, V_{ON} is HIGH and PM_3 is turned off.

After AC+ passes the bottom peak, when it rises close to GND, the decreased V_{OFF} finally flips the inverter and NM_6 is turned on to pull down DR+. The turn on and off delay is

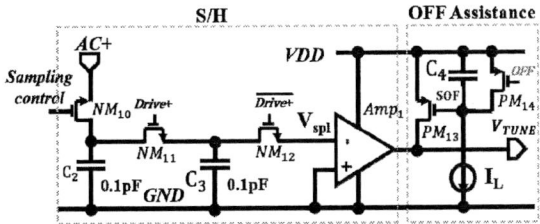

Fig. 5. Schematic of sample and hold circuit and error amplifier.

controlled by the loop feedback voltage V_{TUNE+} and V_{TUNE-}, which control the bias current of ON and OFF branches separately. DR- works in the same manner for the NM_2 control.

Compared to the state-of-art push-pull comparator [6] and current starved delay cell [3] designs, the proposed dynamic logic comparator depends on the parasitic leakage time, which eliminates the feed-through current in the previous two types of comparators and allows the rectifier to work in higher frequency.

B. S/H Circuit and Error Amplifier

Fig. 5 shows the schematic of the sample and hold (S/H) circuit and the error amplifier in the feedback control loop. In order to reduce the power consumption, the sampling block is enabled once for every 64 RF carrier cycles and lasts for 4 continuous carrier cycles. Then the sampling block will be disabled and wait for the next sampling enable signal. For sampling at the turning on edge, NM_{10}'s control signal is set to HIGH by DR- and set to LOW by DR+. In this manner, AC+ voltage is sampled just at the DR+ rising edge. On the other hand, for sampling at the turn off edge, DR+ directly controls the gate of NM_{10}. Because the voltage on C_2 is not stable during sampling when NM_{10} is on, therefore C_3 is added to smooth the sampled voltage after MN_{10} is off. The error amplifier AMP_1 compares the sampled voltage with GND to generate V_{TUNE} to control the delay by tuning dynamic logic comparator bias current.

In addition, self-startup is an important feature for the active rectifier so that the system can restore normal operation even when the system power supply is already drained. Before the active diode restore normal operation, the rectification is through the body diode of NMOS transistor. To avoid energy leakage, the NMOS switches should be fully turned off. To prevent this turn-off failure happens, OFF assistance circuit in Fig. 5 is added. When the voltage feedback to pull down the gate voltage of NM_1 in Fig. 2 is too low to generate an effective off signal, PM_{14} is always off. Hence, the feedback voltage for off is pushed to VDD by PM_{13}. An effective OFF signal is guaranteed to turn off power switch during the period of start-up. After the startup period, when VDD is charged to sufficient high, PM_{14} charges the node SOF to VDD, so that PM_{13} is turned off and leaves the control of error amplifier AMP_1 back to the feedback loop.

IV. MEASUREMENT RESULTS

The proposed active rectifier has been implemented in a standard 65nm CMOS process. Fig. 6 shows the chip

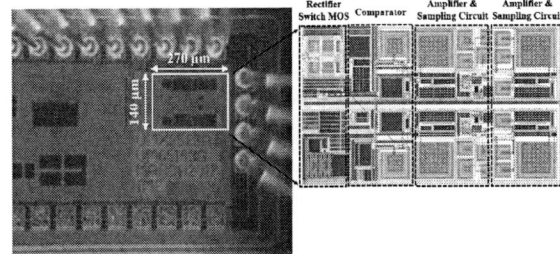

Fig. 6. Chip micrograph.

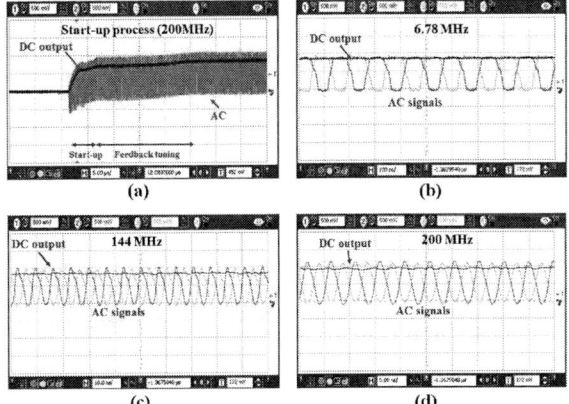

Fig. 7. Measured system waveforms (a) at 200 MHz for start-up, (b) at 6.78MHz, (c) 144 MHz, (d) 200 MHz.

micrograph of the active rectifier and the key function blocks layout. The core active area occupies 140 µm × 270 µm. Simulation results show that the circuit consumes 10µA and 80µA at 6.78 MHz and 200 MHz, respectively, when the rectifier DC output is 1V. For chip measurement, a commercial wideband transformer (Model: Mini circuits TT1-6) is used to convert the single-ended output from signal generator into fully differential signals. Fig. 7 shows the measured rectifier waveforms under different conditions. Fig. 7 (a) shows the system cold startup waveforms at 200MHz. It can be observed that the system can self-start and reach steady operation state after the feedback tuning process. Fig. 7 (b)-(d) show the measured input AC and DC output at 6.78 MHz, 144 MHz and 200 MHz, respectively. The same load, a 1 kΩ resistor in parallel with a 1 nF capacitor, is used in all the measurements. These results prove that, with the dynamic logic comparator technique, the proposed active rectifier is able to operate in a wide range of frequency from 6.78 MHz up to 200 MHz.

Fig. 8 shows the rectifier's voltage conversion ratio and power conversion efficiency vs. RF carrier frequency and rectifier DC output. The peak voltage conversion ratio is 95% at 6.78 MHz and 84% at 200 MHz. The rectifier achieves 94% peak power efficiency at 6.78 MHz and 84% efficiency at 200 MHz. It can be observed that the measured power conversion efficiency fluctuates between 6.78 MHz and 200 MHz. This could be due to the parasitic resonance caused by the PCB/packaging parasitics. This matter will be further investigated in the follow-up study.

978-1-5386-6414-8/18 $31.00 © 2018 IEEE

TABLE I. PERFORMANCE SUMMARY

Year	[3]	[4]	[6]	[8]	[9]	[10]	This Work
Description	Passive/active Rectifier	Passive Rectifier	Active Rectifier	Active Rectifier	Active Diode	Passive Rectifier	Active Rectifier
Process Technology	65nm CMOS	0.18μm SOI	0.35μm CMOS	0.18μm CMOS	0.18μm CMOS	0.35μm BCD	65nm CMOS
Frequency (MHz)	300	144	13.56	13.56	13.56	6.78	6.78-200
Input range (V)	~0.5	1-1.2	1.8-3.6	4V$_{max}$	~1.5-3.1	5 (6W)	0.7-1.1
Load cap (nF)	4	1	2	4	12.2	N.A.	1
Load resistor (Ω)	~8k	~10k	500	50-5k	~500	N.A.	1k
Voltage conversion ratio	N.A.	80% - 100%	90% - 92%	71% - 94%	61% - 79%	N.A.	95%@6.78MHz 95%@13.56MHz 85%@144MHz 83%@200MHz
Power conversion efficiency (%)	70	65*	89-91	96	78	86	94@6.78MHz 94@13.56MHz 85@144 MHz 84@200MHz

*Simulation result.

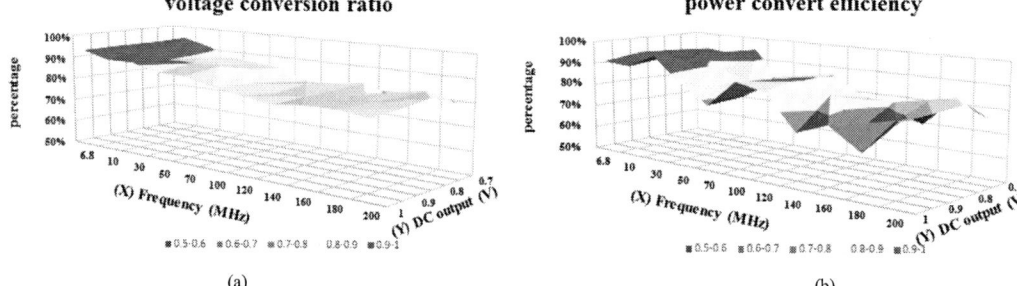

(a)   (b)

Fig. 8. Measured results of the rectifier (a) voltage conversion ratio and (b) power convert efficiency vs. RF frequency (X axis) and the rectifier DC output (Y axis).

The active rectifier measurement results are summarized in Table I. Compared with the other state-of-art designs, the proposed rectifier demonstrated the widest operation frequency range from 6.78 MHz to 200 MHz without any external tuning. In the low frequencies range at 6.78 MHz and 13.56 MHz, this design achieved one of the best PCE performance. It also outperform the other reported designs at 144 MHz range and achieves 84% PCE at 200MHz.

V. CONCLUSIONS

In this paper, a wideband high speed active rectifier with dynamic logic comparator and offset-compensation loop is proposed. With these two design techniques, the proposed active rectifier is able to work from 6.78 MHz to 200 MHz with PCE between 95% and 84%. The next step is to incorporate this active rectifier design into an mm-size wirelessly powered implantable medical device system.

REFERENCES

[1] K.A. Ng, et al., "A low power, high CMRR neural amplifier system employing CMOS inverter-based OTAs with CMFB through supply rails," IEEE J. of Solid-State Circuits, vol. 51, no. 3, pp. 724-737, Mar. 2016.

[2] L. Yao, et al., "A 20V-compliance implantable neural stimulator IC with closed-loop power control, active charge balancing, and electrode

impedance check," in IEEE Asian Solid-State Circuits Conference, Nov. 2014, pp. 201-204.

[3] R. Muller, et al., "A minimally invasive 64-channel wireless μEOG implant," IEEE J. of Solid-State Circuits, vol. 50, no. 1, pp. 344-359, Jan. 2015.

[4] C. Kim, et al., "A 144 MHz fully integrated resonant regulating rectifier with hybrid pulse modulation for mm-sized implants," IEEE J. of Solid-State Circuits, vol. 52, no. 11, pp.3043-3055, Nov. 2017.

[5] J. Zhao, et al., "An integrated wireless power management and data telemetry IC for high-compliance-voltage electrical stimulation applications," IEEE Trans. on Biomedical Circuits and Systems, vol. 10, no. 1, pp. 113-124, Feb. 2016.

[6] C. Huang, et al., "A near-optimum 13.56 MHz CMOS active rectifier with circuit-delay real-time calibrations for high-current biomedical implants," IEEE J. of Solid-State Circuits, vol. 51, no. 8, pp. 1797-1809, Aug. 2016.

[7] S. Guo, et al., "An efficiency-enhanced CMOS rectifier with unbalanced-biased comparators fro transcutaneous-powered high-current implants," IEEE J. of Solid-State Circuits, vol. 44, no. 6, pp. 1796-1804, Jun. 2009.

[8] H. Cruz, et al., "A 13.56 MHz, 162 mW magnetically coupled digital rectifier with 94% VCR, 96% PCE over 50-to-5kΩ load range, and embedded 80 kbos DBPSK demodulator for biomedical applications," in IEEE Asian Solid-State Circuits Conference, Nov. 2016, pp. 209-212.

[9] H. Shinohara, et al., "A ZVS CMOS active diode rectifier with voltage-time-conversion delay-locked loop for wireless power transmission," in IEEE Asian Solid-State Circuits Conference, Nov. 2015, pp. 1-4.

[10] J.-H. Choi, et al., "Resonant regulating rectifiers (3R) operating for 6.78 MHz resonant wireless power transfer (RWPT)," IEEE J. Solid-State Circuits, vol. 48, no. 12, pp. 2989–3001, Dec. 2013.

Photovoltaic-Assisted Self-Vth-Cancellation CMOS RF Rectifier for Wide Power Range Operation

Ren Usami[1], Takao Komiyama[2], Yasunori Chonan[2], Hiroyuki Yamaguchi[2], and Koji Kotani[2]

[1]Graduate School of Systems Science and Technology, Akita Prefectural University
[2]Faculty of Systems Science and Technology, Akita Prefectural University
84-4, Aza Ebinokuchi, Tsuchiya, Yurihonjo City 015-0055, Japan

Abstract—In this study, a photovoltaic (PV)-assisted self-Vth-cancellation CMOS RF rectifier for energy harvesting from ambient radio waves was developed. Because the threshold voltage (Vth) of MOSFETs in the rectifier was effectively compensated for by a DC bias voltage generated from not only the on-chip PV cells but also the output voltage of the rectifier itself, the power conversion efficiency (PCE) under low input power conditions was improved. In addition, a bias voltage limiting mechanism was newly adopted to appropriately suppress the excess bias voltage for Vth compensation and to improve the PCE under high input power conditions. As a result, the rectifier operates with a higher efficiency than previously reported rectifiers over a wide input power range, from as low as −25 dBm to higher input power without degradation of the PCE.

Keywords—energy harvesting, photovoltaic, radio waves, rectifier

I. INTRODUCTION

Energy harvesting from ambient energy sources is attracting much attention as a power supply technology for IoT devices such as remote sensors. Although there are various ambient energy sources including light, heat, electromagnetic waves, and vibrations, this study focused on electromagnetic waves. Energy harvesting from ambient radio waves is carried out by converting a received radio frequency (RF) signal to DC power using a rectifier. It has been demonstrated that TV broadcasting radio waves were received and converted into DC power, and a temperature sensor with a liquid crystal display was driven [1]. Assuming an IoT device application, a high RF to DC power conversion efficiency (PCE) at the input power level, ranging from −30 dBm to 0 dBm, is required.

The PCE of the rectifier is defined and roughly approximated by the following equation [2].

$$\mathrm{PCE} \equiv \frac{P_{\mathrm{OUT}}}{P_{\mathrm{IN}}} = \frac{P_{\mathrm{OUT}}}{P_{\mathrm{OUT}} + P_{\mathrm{LOSS}}}$$
$$\cong \frac{V_{\mathrm{DC}}}{V_{\mathrm{DC}} + V_{\mathrm{TO}}} \cong \frac{V_{\mathrm{RF}} - V_{\mathrm{TO}}}{V_{\mathrm{RF}}}, \quad (1)$$

where P_{OUT}, P_{IN}, P_{LOSS}, V_{DC}, V_{TO}, and V_{RF} are the DC output power, the RF input power, the power loss of the rectifier, the output DC voltage, the turn-on voltage of the diode, and the peak voltage amplitude of the applied RF signal, respectively. In general, in CMOS integrated circuit technology, a diode-connected MOSFET is used as a rectifying diode, for which the turn-on voltage is equal to the threshold voltage (Vth) of the MOSFET. Therefore, lowering the effective Vth of the MOSFET is essential for improving the PCE. Based on this understanding, a number of high efficiency rectifiers, in which Vth is effectively compensated for by various means, have been proposed [2-8].

One such high performance rectifier is the self-Vth-cancellation (SVC) CMOS rectifier [2]. In this rectifier, a high PCE can be obtained by compensating for the Vth of the MOSFETs by the DC output voltage generated by the rectifier itself. However, under low input power conditions, where the output voltage is not yet enough, the Vth is not sufficiently compensated for and a high PCE is not obtained. In addition, the Vth is excessively compensated for with a high output voltage under high input power conditions. In this case, the PCE deteriorates due to the increase in the reverse leakage current. Therefore, the input power range in which a high PCE is obtained is limited to the medium input power region for this rectifier.

The photovoltaic (PV)-assisted CMOS rectifier [3-5], in which the Vth is compensated for by the DC bias voltage of PV cells (solar cells), has been proposed as an example of the "synergistic ambient energy harvesting" concept. Although it is based on a conventional voltage doubler rectifier, on-chip PV cells are inserted in the gate bias paths of diode-connected nMOS and pMOS transistors. Compared with a simple voltage doubler, the PCE is improved in the entire input power range. However, because the output voltage of a single on-chip PV cell is small, the Vth is not sufficiently compensated for. Therefore, the PCE is still not enough under low input power conditions. In addition, it is difficult to increase the bias voltage by using a series-connection of PV cells. Although it is possible to realize a single unit of series-connected PV cells with CMOS technology [9], two electrically isolated units, which are required in this application, cannot be integrated in the same substrate.

In this paper, we propose a photovoltaic-assisted self-Vth-cancellation CMOS rectifier by combining both the SVC technique and the PV-assisted Vth-compensation technique. The Vth of the MOSFETs was compensated for by the DC bias voltage jointly generated by the PV cell and the rectifier itself. In addition, a voltage limiter mechanism was newly adopted to appropriately suppress the excess bias voltage for Vth compensation and to prevent degradation of the PCE under high input power conditions.

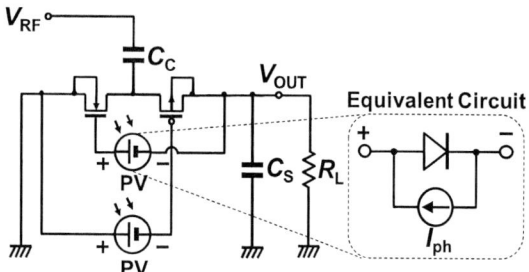

Fig. 1 PV-assisted SVC CMOS rectifier.

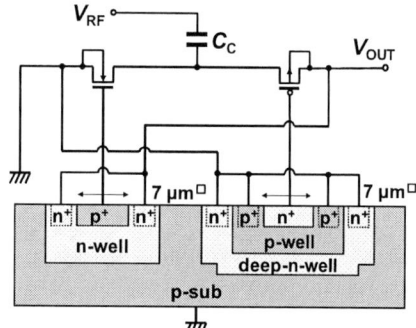

Fig. 2 PV cell structure.

II. PV-ASSISTED SVC CMOS RECTIFIER

The proposed PV-assisted SVC CMOS rectifier is shown in Fig. 1. Based on the SVC CMOS rectifier [2], on-chip PV cells consisting of on-chip pn junctions were inserted in the gate bias paths of nMOS and pMOS. The Vth of the MOSFETs was compensated for by the DC bias voltage generated not only from the PV cell, but also from the rectifier itself. Therefore, the PCE under low input power conditions was greatly improved. As shown in the inset in Fig. 1, the PV cell under light irradiation is modeled by the parallel connection of a diode and a current source. Unlike conventional power generation by PV cells which requires a large cell area, a high level of light irradiation, and adaptive load impedance control to maintain the PV cells at the maximum power point condition, a small cell area and a low level of light irradiation were sufficient for this application because the purpose of the PV cells in this case was not power generation but voltage generation. Because the output of the PV cell was connected to the high resistance MOS gate, the PV cell operated under almost open-circuit conditions. The photocurrent generated in the pn junction in the PV cell flowed into the pn junction itself, resulting in the bias voltage generation. The PV cells in the SVC mechanism could provide appropriate voltage and effectively bias the MOSFETs even with a small PV cell area and under low irradiance conditions.

A test circuit for the PV-assisted SVC CMOS rectifier was designed and fabricated with a 0.18-μm, 5-ML CMOS process. The channel width/length of nMOS and pMOS transistors used

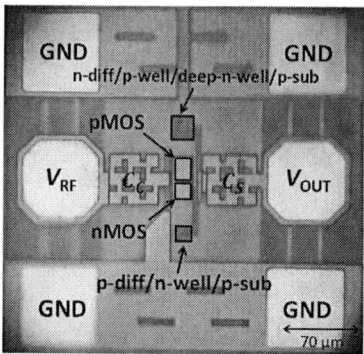

Fig. 3 Photomicrograph of the PV-Assisted SVC CMOS rectifier test circuit.

in the rectifier was 3.6 μm/0.18 μm and 10.8 μm/0.18 μm, respectively. The structure of PV cells integrated on a substrate is shown in Fig. 2. For nMOS biasing, the PV cell is composed of the p^+-diffusion/n-well/p-substrate structure. For pMOS biasing, the PV cell is composed of the n^+-diffusion/p-well/deep-n-well/p-substrate structure. PV cells acting for the gate bias were, however, only a junction between the surface diffusion layer and the underlying well (p^+-diffusion/n-well for nMOS and n^+-diffusion/p-well for pMOS). Because the terminals of the PV cells were connected to nodes having different potentials, these structures were needed to electrically isolate them. Furthermore, by this configuration, the gate bias voltage of the PV cells became symmetrical [4]. The n-well/p-substrate junction of the PV cell for nMOS biasing was reverse biased and irradiated by light. Photo-generated current flowed from the DC output node of the rectifier to the ground. This current (in nA) was, however, much smaller than the output load current of the rectifier (in μA), and it could therefore be ignored from the viewpoint of rectifier efficiency. The PV cell area (surface diffusion layer) was 7 × 7 μm. The transistor area was covered by the top two metal layers (ML4 and ML5) for shading. Both coupling capacitor C_C and output smoothing capacitor C_S were designed to be 1.13 pF. Fig. 3 shows a photomicrograph of the fabricated test circuit.

The PCE was evaluated by on-wafer measurements. Measurements were performed at an RF input frequency of 1 GHz and an output load resistance of 10 kΩ. A vector network analyzer (VNA, Agilent N5242A) was used to apply RF power to the rectifier and measure the reflection coefficient. The PCE was calculated by

$$\text{PCE} \equiv \frac{P_{\text{OUT}}}{P_{\text{IN}}} = \frac{\dfrac{V_{\text{OUT}}^2}{R_{\text{L}}}}{P_{\text{SRC}}(1 - |S_{11}|^2)}, \qquad (2)$$

where V_{OUT}, R_{L}, P_{SRC}, and S_{11} are the output voltage of the rectifier, the output load resistance, the source power supplied from the VNA, and the measured reflection coefficient, respectively. An LED lamp was used as the light source to emulate a normal indoor environment.

978-1-5386-6414-8/18 $31.00 © 2018 IEEE

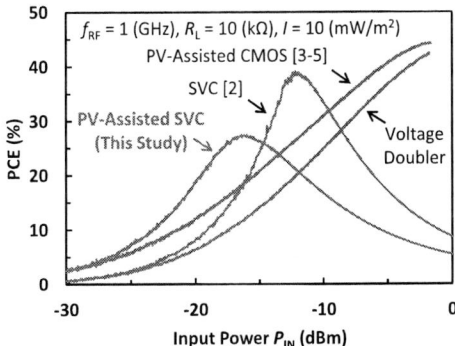

Fig. 4 Measured PCE as a function of P_{IN}.

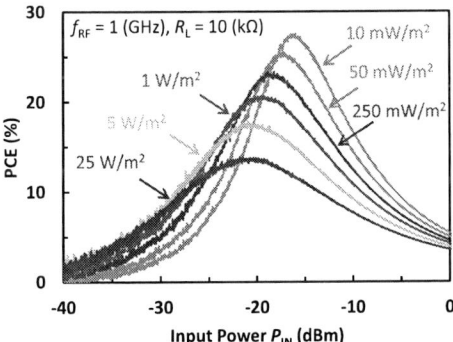

Fig. 5 Dependence of the irradiance.

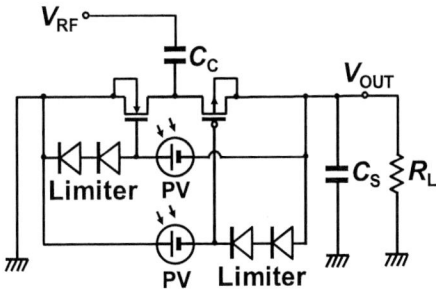

Fig. 6 PV-assisted SVC CMOS rectifier with limiter.

conditions are shown in Fig. 5. As the irradiance increased, the PCE for low input power conditions improved. This is because the bias voltage increased as the irradiance increased, and the Vth of the MOSFETs was more compensated for. On the other hand, as the irradiance increased, the peak value of the PCE decreased. This is because the Vth compensation was increased and the influence of the reverse leakage current by the SVC became large, even under low input power conditions.

III. PV-ASSISTED SVC CMOS RECTIFIER WITH BIAS VOLTAGE LIMITER

As described in the previous section, the reason for the PCE degradation under high input power conditions was the excessive Vth compensation. Therefore, if the gate bias voltage was effectively limited to an appropriate value, a higher PCE could be expected even under high input power conditions. For this purpose, a voltage limiter was newly adopted to appropriately suppress the gate bias voltage and to improve the PCE under high input power conditions.

The newly proposed PV-assisted SVC CMOS rectifier equipped with a bias voltage limiter is shown in Fig. 6. Based on the PV-assisted SVC CMOS rectifier, bias voltage limiters consisting of diodes (pn junction) were inserted between the nMOS gate and ground and between the pMOS gate and V_{OUT}. Note that by connecting two diodes in series, the current-voltage (I-V) characteristic of the limiter was adjusted for realizing an appropriate limiting voltage. For fine adjustments, the area of the diode could be further tuned.

The structures of the PV cells and voltage limiter integrated on the same substrate are shown in Fig. 7. For limiting the nMOS bias, the limiter was composed of a series connection of

The PCE measurement results are shown in Fig. 4. When the light irradiance was 10 mW/m², a high PCE that has not been realized by previously reported rectifiers [2-5] was obtained for the input power range from −25 dBm to −15 dBm. Specifically, at an input power of −16 dBm, the PCE was 27.5%. Note that although the light irradiance of typical indoor light conditions is about 3 W/m², a high PCE could be obtained even with a very small light irradiation of 10 mW/m². However, the problem associated with high input power still existed. The PCE deteriorated under high input power conditions due to the excessive threshold compensation, as was the case with the SVC CMOS rectifier [2].

PCE measurement results under various light irradiation

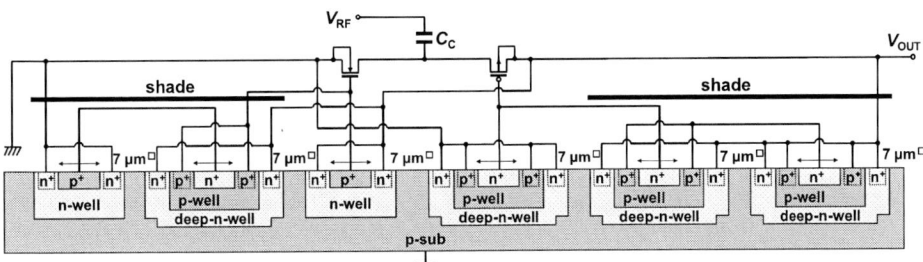

Fig. 7 PV cell and limiter structures.

978-1-5386-6414-8/18 $31.00 © 2018 IEEE

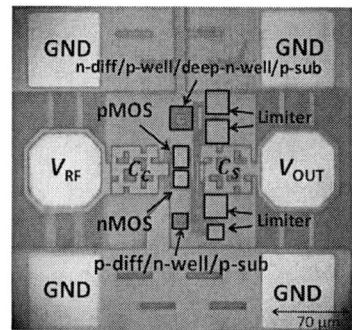

Fig. 8 Photomicrograph of the PV-Assisted SVC CMOS rectifier test circuit with limiter.

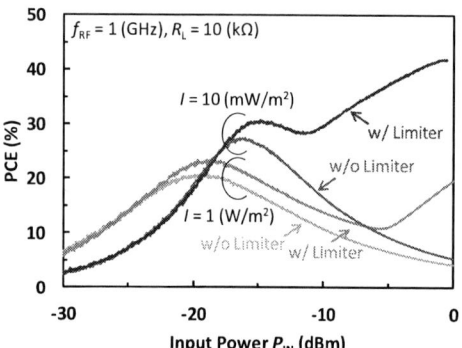

Fig. 9 Measured PCE as a function of P_{IN}.

the n$^+$-diffusion/p-well/deep-n-well/p-substrate structure and the p$^+$-diffusion/n-well/p-substrate structure. This irregular structure and connection were used for the upper-side diode structure. This configuration was utilized to prevent photocarriers generated in the non-shading region from flowing into the limiter. Although photocarriers generated in a non-shaded area flowed into the deep-n-well of the upper-side diode, it did not affect the operation of the limiter because the photocurrent was supplied from the output terminal of the rectifier. For limiting the pMOS bias, the limiter was composed of a series connection of two identical diode structures composed of the n$^+$-diffusion/p-well/deep-n-well/p-substrate structure. The limiter areas (surface diffusion layer) were 7×7 µm. The transistor area and limiter area were covered by the top two metal layers (ML4 and ML5) for shading. Fig. 8 shows a photomicrograph of the fabricated test circuit.

Fig. 9 shows the PCE measurement results for light irradiances of 10 mW/m^2 and 1 W/m^2. Compared with the PV-assisted SVC CMOS rectifier without the limiter, a high PCE was realized at a high input power level. Excessive Vth compensation was prevented by the effect of limiters. In more detail, when V_{OUT} increased due to the increase in P_{IN} and the gate bias voltage became higher than the turn-on voltage of the limiter, the photocurrent flowed into the limiters, and the appropriate bias voltage was maintained. However, compared

with that of 10 mW/m^2, the photocurrent in the case of 1 W/m^2 was large, so that the limiter operation threshold was shifted to too high in the P_{IN} region and the effect of the limiter became weak. This was due to the insufficient optimization of the limiter. Incidentally, because the current flowing through the limiters was supplied from the output of the rectifier, some amount of power may have been lost. However, because the limiter and the PV cell were connected in series and the current was regulated by the photocurrent of the PV cell, which was in the "pA" range, it could be neglected with respect to the "µA" range of the output load current of the rectifier. Therefore, the influence on the PCE could be ignored.

IV. CONCLUSION

A high efficiency photovoltaic-assisted self-Vth-cancellation CMOS rectifier for energy harvesting from ambient radio waves was developed. The bias voltage jointly generated by the PV cells and the rectifier itself compensated for the Vth of the MOSFETs. Moreover, by adding a voltage limiter mechanism, the rectifier could operate efficiently in a wide input power range.

ACKNOWLEDGMENT

The VLSI chip in this study was fabricated as part of the chip fabrication program of the VLSI Design and Education Center (VDEC), University of Tokyo, in collaboration with Rohm Corporation and Toppan Printing Corporation.

REFERENCES

[1] A. Sample, and J. R. Smith, "Experimental Results with two Wireless Power Transfer Systems," in IEEE Radio Wireless Symp., pp. 16-18, Jan. 2009.

[2] K. Kotani and T. Ito, "Self-Vth-Cancellation High-Efficiency CMOS Rectifier Circuit for UHF RFIDs," IEICE Transactions on Electronics., Vol. E92-C, No. 1, pp. 153-160, Jan. 2009.

[3] K. Kotani, T. Bando, and Y. Sasaki, "A Photovoltaic-Assisted CMOS Rectifier for Synergistic Energy Harvesting from Ambient Radio Waves," IEICE Trans. Electron., Vol. E97-C, No.4, pp. 245-252, Apr. 2014.

[4] K. Kotani, "Efficiency Improvement in Photovoltaic-Assisted CMOS Rectifier with Symmetric and Voltage-Boost PV-Cells," IEICE Transactions on Fundamentals, Vol. E98-A, No.2, pp. 500-507, Feb. 2015.

[5] K. Kotani, "Improvement of power conversion efficiency in photovoltaic-assisted UHF rectifiers by non-silicide technique applied to photovoltaic cells," Japanese Journal of Applied Physics, Vol. 54, 04DE02, 2015.

[6] G. Papotto, F. Carrara, and G. Palmisano, "A 90-nm CMOS Threshold-Compensated RF Energy Harverster," IEEE Journal of Solid-State Circuits, Vol. 46, No. 9, pp. 1985-1997, Sep. 2011.

[7] M. Stoopman, S. Keyrouz, H. J. Visser, K. Philips, and W. A. Serdijn, "Co-Design of a CMOS Rectifier and Small Loop Antenna for Highly Sensitive RF Energy Harvesters," IEEE Journal of Solid-State Circuits, Vol. 49, No. 3, pp. 622-634, Mar. 2014.

[8] Yan Lu, Haojuan Dai, Mo Huang, Man-Key Law, Sai-Weng Sin, Seng-Pan U, and Rui P. Martins, "A Wide Input Range Dual-Path CMOS Rectifier for RF Energy Harvesting," IEEE Transactions on Circuits and Systems II: Express Briefs, Vol. 64, No. 2, pp. 166-170, Feb. 2017.

[9] F. Horiguchi, "Integration of Series-Connected On-Chip Solar Battery in a Triple-Well CMOS LSI," IEEE Trans. Electron Devices, Vol. 59, No. 6, pp. 1580-1584, Jun. 2012.

Stable, Self-Biased and High-Gain Organic Amplifiers with Reduced Parameter Variation Effect

Masoud Seifaei[1], Daniel De Dorigo[1], David Ingvar Fleig[1], Matthias Kuhl[1]
Ute Zschieschang[3], Hagen Klauk[3], Yiannos Manoli[1,2]

[1]Fritz Huettinger Chair of Microelectronics, IMTEK, University of Freiburg, Germany
[2]Hahn-Schickard, Villingen-Schwenningen, Germany
[3]Max Plank Institute for Solid State Research, Stuttgart, Germany
Email: masoud.seifaei@imtek.uni-freiburg.de

Abstract—This paper presents the design and implementation of differential and self-biased single-ended high-gain amplifiers based on p-channel organic thin-film transistors (TFTs) and thin-film carbon resistors (TFCRs). Using resistive bias circuitry, the effect of mobility degradation due to aging is significantly reduced from 43% to 2.7% and 7.9% during 2 and 7 months, respectively. In addition, the bias-point instability due to bias-stress-induced threshold-voltage shifts, which can have large time constants, is limited to 9.6%. The circuits become also nearly insensitive to light. Differential and self-biased single-ended amplifiers with resistive load and bias circuitry are implemented having a gain of 29.4 dB and 23.1 dB and a gain-bandwidths (GBW) of 6.9 kHz and 2.4 kHz, respectively. An input-referred offset (V_{OS}) of 2 mV is achieved. Due to the small footprint of the TFCRs, the circuit area is significantly reduced.

I. INTRODUCTION

Organic electronic devices can be fabricated at relatively low temperatures on large-area flexible substrates, which makes them attractive for many applications, such as bio-signal acquisition, smart sensors and smart textiles. Implementation of signal processing units with sensing elements on the same substrate can lead to low-cost, flexible, and integrated systems. The signal chain from the source to the digital signal processing unit contains analog signal conditioning and converter circuits (namely amplifiers, filters and analog-to-digital-converters). Up to date these blocks are not reliably implemented on flexible substrates due to three main issues:

- The technology is still in the development stage with various not fully understood phenomena.
- Different physics and limitations of organic TFTs, such as low carrier mobility ($\sim 2 \frac{cm^2}{Vs}$), high contact resistance ($\sim 300\,\Omega.cm$), low operating frequency ($\sim 1\,MHz$) and large device mismatch ($\sim 10\%$).
- Time-varying transistor parameters, namely, degradation of the carrier mobility over time (aging), and bias-stress-induced threshold-voltage shifts [1], [2] which cause unstable circuit operating point.

Thus, circuit design techniques should be adapted in order to overcome these limitations.

One of the most important circuit building blocks in the signal chain that plays a crucial role in the overall system performance is the amplifier. The few organic amplifier implementations already reported [3]–[7] do not adequately address

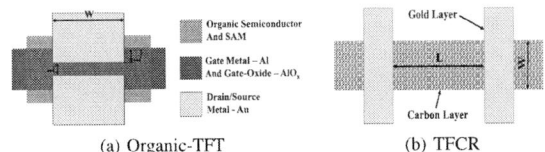

(a) Organic-TFT   (b) TFCR

Fig. 1. Physical layout of the organic TFTs (a) and TFCRs (b).

aging and threshold-voltage shift and a stable integrated bias circuit is not reported. On the other hand, due to the absence of high-performance complementary organic transistors, low carrier mobility, and poor matching, amplifiers are still the bottleneck of the system performance.

In this paper, TFCRs with high unit resistance ($1\,M\Omega/\square$) are used in the bias circuitry, as amplifier load and feedback elements. The layout of a thin-film carbon resistor is shown in Fig. 1(b), which consists of a 30 nm-thick carbon layer deposited between two 30 nm-thick gold contacts [8]. Considering insulating substrates, TFCRs have negligible parasitic capacitance and their frequency response is flat over the active operating frequency range of the TFTs.

One of the most important achievements of this work is a stable integrated bias circuit which reduces the effect of the time varying parameters and sensitivity to light. In addition, the highest gain-per-stage, bandwidth-over-power performance and smallest area among organic technologies published to date is achieved.

In section II, the amplifiers and bias circuits are explained in detail. Implementations and measurement results are given in section III and the conclusion is presented in section IV.

II. ORGANIC AMPLIFIERS AND BIAS SCHEME ARCHITECTURES

The circuits were fabricated using p-channel organic TFTs based on the vacuum-deposited small-molecule semiconductor dinaphtho[2,3-b:2',3'-f]thieno[3,2-b]thiophene (DNTT) [2] and thin-film resistors based on vacuum-deposited carbon [8], fabricated on flexible polyethylene naphthalate (PEN) substrates using polyimide shadow masks [2].

A major objective in amplifier design is to achieve high gain and GBW with low power consumption. The limiting factor

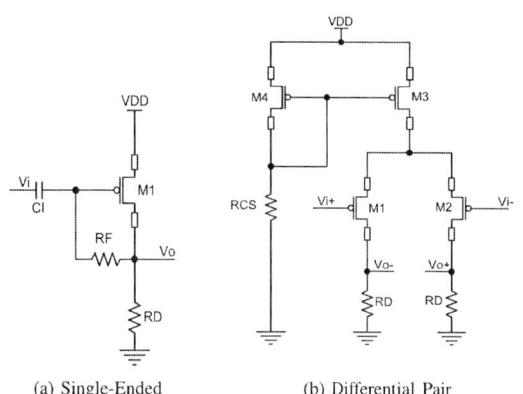

(a) Single-Ended   (b) Differential Pair

Fig. 2. The schematic of TFCR-based differential and self-biased single-ended organic amplifiers. The resistor R_F provides feedback to the input gate and reduces the influence of transistor parameter variations.

of the organic-TFT technology for achieving high gains is the absence of high-performance n-channel organic TFTs, which makes current source load elements unfeasible. The diode-connected-load amplifiers not only exhibit low gains but also suffer from a high parasitic capacitance on the output node as the employed organic TFTs have a large gate capacitance per unit area ($C_I = 600 \frac{\text{nF}}{\text{cm}^2}$) [2]. The TFCRs have negligible parasitic capacitance and high resistance which make them an attractive candidate for load elements. The schematics of resistive organic amplifiers are shown in Fig. 2. Another drawback of using diode-connected loads is the variation of the output DC voltage level due to changes in threshold voltage [1] and carrier mobility [2] even in the presence of an ideal external current source. Therefore, the TFCRs are used not only to increase the gain and GBW of the amplifiers but also to reduce the detrimental effect of the transistor parameter variations as they do not age and are time invariant.

A. Self-Biased Single-Ended Resistive Amplifier

The schematic of a self-biased single-ended amplifier is shown in Fig. 2(a). As the effect of the contact resistance can be translated to lower effective carrier mobility (μ_{eff}) [9] the current equation of the circuit can be written as follows:

$$V_{DD} = R_D I_D + \sqrt{\frac{I_D}{\kappa}} - V_{th} \qquad (1)$$

where $\kappa = \frac{1}{2}\mu_{\text{eff}} C_I \frac{W}{L}$ and the drain current can be derived as follows:

$$I_D = \frac{V_{DD} + V_{th}}{R_D} + \frac{1}{2\kappa R_D^2} \\ - \frac{1}{R_D\sqrt{\kappa}}\sqrt{\frac{1}{4\kappa R_D^2} + \frac{V_{DD} + V_{th}}{R_D}} \qquad (2)$$

where $\mu_{\text{eff}} C_I = 1 \frac{\mu A}{V^2}$, $V_{th} = -1\,V$ and R_D in MΩ range. It can be seen from equation (2) that the first term is not dependent

on μ_{eff} but on V_{th} where due to bias stress, V_{th} may change by up to 40% [1]. By choosing V_{DD} five times larger than V_{th}, the influence of threshold shift will be limited to around 10%. The second and third terms are heavily dependent on μ_{eff}. By designing R_D and $\frac{W}{L}$ sufficiently large, the effect of μ_{eff} on the operating point can be decreased significantly.

The operation principle of the bias circuit is based on the high-impedance feedback R_F, that is designed to affect the DC operating point but not the AC performance. As the organic TFT ages, μ_{eff} decreases and causes a decrease in the drain current. Any decrease in drain current causes an increase in the source-gate voltage (V_{SG}) of the transistor resulting in an increase in the drain current. This same argument applies to changes of V_{th}. This mechanism compensates changes in μ_{eff} or V_{th} by adjusting V_{SG}.

One of the characteristics of organic TFTs subjected to aging which affects the circuit performance is the current-gain cutoff frequency f_T [9]:

$$f_T = \frac{\mu_{\text{eff}}(V_{SG} + V_{th})}{2\pi L(2L_{ov} + L)} \qquad (3)$$

As μ_{eff} decreases the feedback path causes an increase in V_{SG}, therefore the change in f_T and the frequency response of the amplifier can also be compensated. L_{ov} and L are the gate-source/drain overlap and channel lengths, respectively, as shown in Fig. 1(a).

By performing an AC analysis on the circuit shown in Fig. 2(a), it can be seen that R_F is a Miller resistance, which loads the input of the amplifier by $R_F \frac{1}{1+A}$ and the output by $R_F \frac{A}{1+A}$. Assuming a gain higher than 20 dB and $R_F \gg R_D$, the effect of R_F on the gain of the amplifier is not noticeable. The input has to be AC-coupled because the bias point of the amplifier is determined by R_F and should not be changed by the input. In order to have a low cutoff frequency, a large input capacitor is required. The measured integrated MIM capacitors have a unit-area capacitance of $5.3\,\text{nF}/\text{mm}^2$. Together with an R_F in the MΩ range, it results in a cutoff frequency of less than 1 Hz.

B. Differential Resistive Amplifier with Self-Biased Current-Source

By extending the principle of section II-A to differential amplifiers and assuming that all transistors have similar aging rate and bias-stress-induced V_{th} shifts, we can achieve the same results for the operation point stability in differential amplifiers because equation (2) is also valid for the current source of Fig. 2(b). The only design consideration is to keep V_{DS} of M_4 and M_3 as close as possible in order to have the same bias-stress effect. This can be achieved by designing the input common-mode voltage (V_{iCM}) around that of the output and choosing the aspect ratio of M_3 equal to M_4 and double those of M_1 and M_2.

As the TFCRs have smaller footprints than the TFTs, they can be placed very close to each other in order to match well. Measurements have shown that closely placed TFCRs have smaller mismatch (less than 5%) than organic TFTs, leading to small offset voltages in differential amplifiers.

978-1-5386-6414-8/18 $31.00 © 2018 IEEE

III. IMPLEMENTATION AND MEASUREMENT RESULTS

Based on the fact that the output resistance of the organic TFTs is much larger than R_D, the gain of the amplifier can be calculated as follows:

$$|A_{DC}| = g_m . R_D$$
$$g_m = 2\sqrt{\kappa I_D}$$
$$R_D = \frac{V_{DCout}}{I_D} = \frac{V_{DD}}{2I_D} \qquad (4)$$
$$|A_{DC}| = \frac{V_{DD}}{\sqrt{\frac{I_D}{\kappa}}} = \frac{V_{DD}}{V_{SG} + V_{th}}$$

According to the discussions in section II, the following design strategies can be deduced:

- $V_{DD} = 5\,V$
- I_D is determined by the output capacitive load and the operating frequency ($I_D \propto f.C_L$)
- $\frac{W}{L}$ of the organic TFTs is calculated based on the desired gain of the amplifier from equation (4).
- Calculation of $R_D = \frac{V_{DD} - (V_{SG} + V_{th})}{I_D}$ and $R_F > 10 R_D$

The design parameters of two differential amplifiers with $A_V > 20$dB are given in Table I. The simulated and measured frequency responses are depicted in Fig. 3, showing a good agreement between design, simulation and measurements, except for the operating current which is caused by absolute variation of resistors, μ_{eff} and V_{th}. Gains of 26.1 dB and 23.1 dB along with GBWs of 0.92 kHz and 2.4 kHz are achieved for single-stage differential pairs. The differential amplifiers have $V_{OS} \leq 2$ mV due to small relative error of resistors. The values of R_D are given as aspect ratio in Fig. 3 and can be calculated with the unit resistance of ~ 1 M$\Omega/\square$ but as seen in the results this value has an absolute variation of $\sim 30\%$ which explains for the difference between design and measurement.

TABLE I
DESIGN PARAMETERS OF TWO DIFFERENTIAL AMPLIFIERS.

	Diff. Amp.1	Diff. Amp.2
I_{D1}	90 nA	475 nA
W_4	2000 μm	1000 μm
L_4	20 μm	20 μm
R_D	22 MΩ	4 MΩ
R_{CS}	22 MΩ	4 MΩ
A_{DC}	36 dB	26 dB

In order to see the effect of aging , Diff. Amp.2 (Table I) and a single-ended amplifier with $R_D = 4$ MΩ, $R_F = 60$ MΩ and $\frac{W}{L} = \frac{500\,\mu m}{20\,\mu m}$ have been measured seven and two months after fabrication respectively. The simulation of a single-ended amplifier with expected reduction of 40% in μ_{eff} [2] results in a bias-current reduction of 43% without high-impedance feedback, but only in 2.2% with it. The measurement results are shown in Fig. 4.

The effect of bias stress and light have also been investigated using a single-ended amplifier having $R_D = 2$ MΩ, $\frac{W}{L} = \frac{500\,\mu m}{20\,\mu m}$ and $R_F = 60$ MΩ and without the feedback resistance but constant V_{iCM}. The experiment has been performed in the following stages:

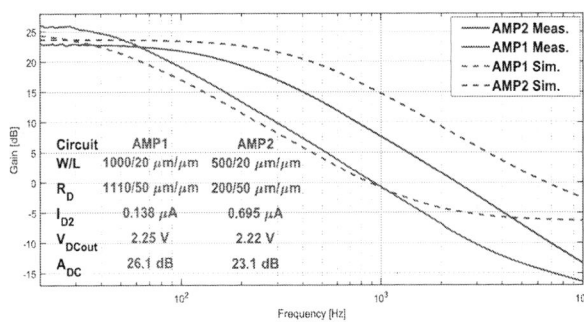

Fig. 3. Simulated and measured frequency response of the resistive differential amplifiers with an output load of 20 pF. The measurement results show high gain-per-stage, high GBW and low power consumption, which are in good agreement with the design goals set in Table I.

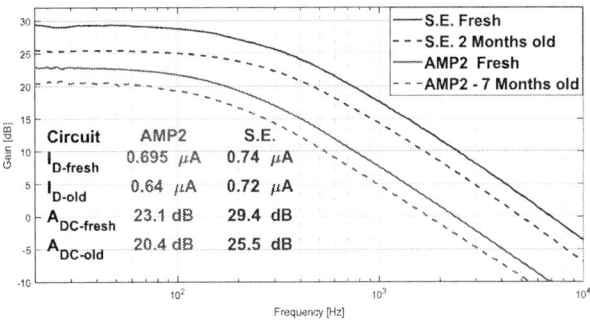

Fig. 4. The frequency response comparison of 7-months old differential and 2-months old single-ended amplifiers with fresh samples loaded by 20 pF and 10 pF, respectively. The bias currents have changed by 2.7% and 7.9% after 2 and 7 months, where more than 43% was expected.

- Circuit start-up in darkness and measurement of I_D
- Measurement of I_D after biasing for 60 minutes in darkness
- The light exposure with intensity of $0.167\,\frac{mW}{mm^2}$
- Measurement of I_D after one minute and two minutes of biasing in light

The experiment timing was chosen by considering the large time constant of the bias-stress effect [1] and the fast response of organic TFTs to light [10]. The results of this experiment are reported in Table II, showing that the bias point is very stable against changes in V_{th} due to bias stress and illumination.

TABLE II
THE EFFECT OF BIAS-INDUCED THRESHOLD SHIFT AND LIGHT.

Drain Current	Amp. with R_F	Amp. without R_F
$I_{D,Start-up}$	1.20 μA	0.02 μA
$I_{D,60\,mins-Dark}$	1.09 μA (-9.2%)	0.75 μA (+3650%)
$I_{D,1\,min-Light}$	1.09 μA (<1%)	0.79 μA (+5.3%)
$I_{D,2\,mins-Light}$	1.09 μA (<1%)	0.81 μA (+2.5%)

With the feedback resistor the rate of change in the current becomes negative, because at start-up $V_{SG} = V_{DD}$ and V_{th} is at their maximum value. As I_D increases, V_{SG} decreases and leads to a decrease in I_D. Meanwhile, V_{th} decreases to its

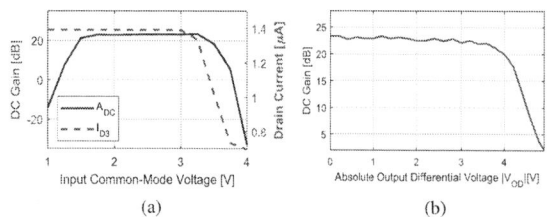

(a)   (b)

Fig. 5. Output swing and V_{iCM} range of the diff. amp.2 of Table I. (a) V_{iCM} range is 1.5 V to 3.5 V. The supply current is constant for $V_{iCM} < 3$ V, which is due to low channel-length modulation of organic TFTs. (b) The amplifier has an output swing ($|V_{OD}| = |V_{O+} - V_{O-}|$) of ± 4 V with $V_{DD} = 5$ V.

final value and, as the effect of this change is in the opposite direction to the change in I_D, leads to a more stable bias point.

The measurement results of the output swing and V_{iCM} range of Diff. Amp.2 are depicted in Fig. 5. An output swing of ± 4 V has been achieved with $V_{DD} = 5$ V. Since the organic TFTs have a relatively large channel length ($20\,\mu$m) and a relatively large contact resistance ($\sim 300\,\Omega$.cm) which acts as source degradation, the supply current is nearly insensitive to V_{iCM}, Fig. 5(a). The amplifier does not experience performance degradation with V_{iCM} in the range of 1.5 V to 3.5 V, Fig. 5(b).

The single-ended and Diff. Amp.2 have an input-referred noise voltage of $8.4\,\mu V_{rms}$ and $28.3\,\mu V_{rms}$, respectively. The 1/f noise is the dominant noise source, because the noise-corner of the amplifiers are around 5 kHz which is much higher than their noise-bandwidth.

The performance comparison of the amplifiers to state-of-the-art organic-TFT-based amplifiers is presented in Table III. The presented amplifiers not only have high gains but also have the smallest area and the best GBW-over-power performance which are achieved by using TFCRs. The area of the single-ended self-biased amplifier is just 1 mm² using the similar technology as reported in [5], where the area is approximately 75 mm². Photographs of the single-ended and differential amplifiers (Diff. Amp.2) are shown in Fig. 6.

IV. CONCLUSION

Resistive differential/single-ended amplifiers with 2.4/6.9 kHz GBW, 23.1/29.4 dB gain and 7/3.6 μW power consumption occupying 3.85/1 mm² area are presented, which to the best knowledge of the authors have the best GBW-over-power performance reported to date. The input-referred offset voltage of differential pairs is reduced to less than 2 mV. The resistive bias circuit successfully reduced the effect of transistor aging on the circuit bias point within 2 and 7 months to 2.7% and 7.9%, where more than 43% was expected. The effect of bias-stress induced threshold voltage shift is limited to 9.6% and the circuits are nearly insensitive to light illumination.

ACKNOWLEDGMENT

This work is supported by the German Research Foundation (DFG) under Grant Number MA 2193/17-1 and KL 2223/6-1.

TABLE III
COMPARISON WITH STATE-OF-THE-ART ORGANIC AMPLIFIERS.

Reference	[6]	[5]	[4]	[7]	This Work		
Topology	Positive Feedback	S.E. 1-stage	Diff. 1-stage	Diff. 3-stage	S.E. 1-stage	Diff. 1-stage	
A_{DC}	27 dB	28 dB	10 dB	24.6 dB	29.4 dB	26.1dB	23.1dB
GBW	70 Hz	2 kHz	3 kHz	2 kHz	6.9 kHz	0.9 kHz	2.4 kHz
Power	-	30 μW	49 μW	-	3.6 μW	1.4 μW	7 μW
V_{DD}	60 V	2 V	10 V	$\pm$15 V	5 V	5 V	5 V
Area	962 mm²	75* mm²	-	10.4 mm²	1 mm²	5 mm²	3.85 mm²
CMRR	-	-	-	-	-	-	68.4 dB
NEF	-	-	-	-	19.5	-	62.7
V_{OS}	-	-	-	-	-	< 2 mV	< 2 mV
FOM $= \dfrac{GBW}{Power}$	-	66.7 $\frac{Hz}{\mu W}$	61.2 $\frac{Hz}{\mu W}$	-	1924 $\frac{Hz}{\mu W}$	667 $\frac{Hz}{\mu W}$	345 $\frac{Hz}{\mu W}$

* Estimated according to the substrate photo.

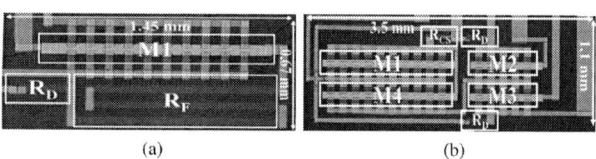

(a)   (b)

Fig. 6. Photographs of the single-ended self-biased amplifier (a) and differential amplifier with resistive current source (b).

REFERENCES

[1] U. Zschieschang, R. T. Weitz, K. Kern, and H. Klauk, "Bias stress effect in low-voltage organic thin-film transistors," *Applied Physics A*, vol. 95, no. 1, pp. 139–145, Apr 2009.

[2] U. Zschieschang and H. Klauk, "Low-voltage organic transistors with steep subthreshold slope fabricated on commercially available paper," *Organic Electronics*, vol. 25, pp. 340 – 344, 2015.

[3] M. Raja, D. Donaghy, L. Gonzalez-Macia, and A. J. Killard. "Design and simulation of a high-gain organic operational amplifier for use in quantification of cholesterol in low-cost point-of-care devices," *IET Circuits, Devices Systems*, vol. 11, no. 5, pp. 504–511, 2017.

[4] M. Torres-Miranda, A. Petritz, A. Fian, C. Prietl, H. Gold, H. Aboushady, Y. Bonnassieux, and B. Stadlober, "High-speed plastic integrated circuits: Process integration, design, and test," *IEEE Journal on Emerging and Selected Topics in Circuits and Systems*, vol. 7, no. 1, pp. 133–146, March 2017.

[5] H. Fuketa, K. Yoshioka, Y. Shinozuka, K. Ishida, T. Yokota, N. Matsuhisa, Y. Inoue, M. Sekino, T. Sekitani, M. Takamiya, T. Someya, and T. Sakurai, "1 μm-thickness ultra-flexible and high electrode-density surface electromyogram measurement sheet with 2 V organic transistors for prosthetic hand control," *IEEE Transactions on Biomedical Circuits and Systems*, vol. 8, no. 6, pp. 824–833, Dec 2014.

[6] J. Chang, X. Zhang, T. Ge, and J. Zhou, "Fully printed electronics on flexible substrates: High gain amplifiers and dac," *Organic Electronics*, vol. 15, no. 3, pp. 701 – 710, 2014.

[7] D. Kim, Y. Kim, K. Y. Choi, D. Lee, and H. Lee, "A solution-processed operational amplifier using direct light-patterned a-ingazno tfts," *IEEE Transactions on Electron Devices*, vol. 65, no. 5, pp. 1796–1802, May 2018.

[8] H. Ryu, D. Kaelblein, O. G. Schmidt, and H. Klauk, "Unipolar sequential circuits based on individual-carbon-nanotube transistors and thin-film carbon resistors," *ACS Nano*, vol. 5, no. 9, pp. 7525–7531, 2011, pMID: 21870841.

[9] T. Zaki, *Short-channel organic thin-film transistors: fabrication, characterization, modeling and circuit demonstration*. Springer, 2015.

[10] J. Milvich, T. Zaki, M. Aghamohammadi, R. Rdel, U. Kraft, H. Klauk, and J. Burghartz, "Flexible low-voltage organic phototransistors based on air-stable dinaphtho[2,3-b:2',3'-f]thieno[3,2-b]thiophene (DNTT)," *Organic Electronics*, vol. 20, pp. 63 – 68, 2015.

978-1-5386-6414-8/18 $31.00 © 2018 IEEE

11-5 (8043)

An Encryption-Authentication Unified A/D Conversion Scheme for IoT Sensor Nodes

Vinod. V. Gadde,　Hiromitsu Awano,　Makoto Ikeda,

Dept. of Electrical Engineering and Information Systems, The University of Tokyo, Tokyo, Japan
E-Mail: vinod@silicon-utokyo.ac.jp,　awano@vdec.u-tokyo.ac.jp,　ikeda@silicon.u-tokyo.ac.jp

Abstract — Widely distributed sensors, in the IoT era, are vulnerable to attacks, including data sniffing and spoofing. Tamper resistance, data encryption and authentication, which are essential aspects of security of all IoT devices, are becoming ever more critical. We have proposed an Analog-to-Digital Conversion scheme, based on slope A/D conversion, involving two randomized slopes, to realize resistance to side channel attacks and perform data encryption-authentication during the A/D conversion process. We have designed and fabricated the proposed encryption-authentication unified ADC in 0.18μm CMOS process and demonstrated an ENOB of 7.64 bits, DNL = +/- 0.6 LSBs and INL = +0.5/-0.4 LSBs at 24KS/s.

Keywords — *ADCs, side channel attacks, information security encryption, authentication.*

I.　INTRODUCTION

With the advent of IoT, the volume, in units, of sensor nodes and the complexity of wireless sensor networks created by these nodes are growing exponentially. Assuring the security of these sensor nodes, and hence the end-to-end security of the networks connecting them, is also growing into a very complex problem. A typical case of a generic sensing scheme and its associated security measures are illustrated in Fig.1. Tamper proofing/resistance is built into the element for physical security and encryption-authentication of the digitized data is facilitated for digital information security.

A. Hole in the End-to-End Security

While tamper proofing/resistance of sensing element and encryption and authentication of the digitized data seem sufficient security countermeasures for an isolated sensor node with physical access restrictions, ubiquitous sensor nodes, in the open, may pose an entirely different security challenge, as they are vulnerable to side channel attacks, like estimation of data from electromagnetic leakage and tampering of A/D conversion process by physical abuse of electronic circuitry. This security hole is depicted in Fig.1.

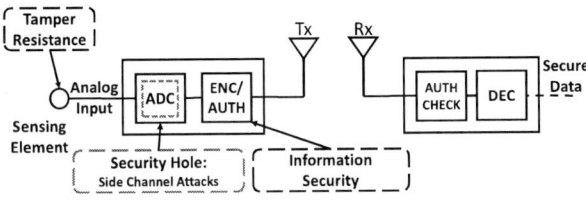

Fig.1, A typical sensing scheme and hole in the end-to-end security due to sniffing/spoofing attack vulnerability of ADC.

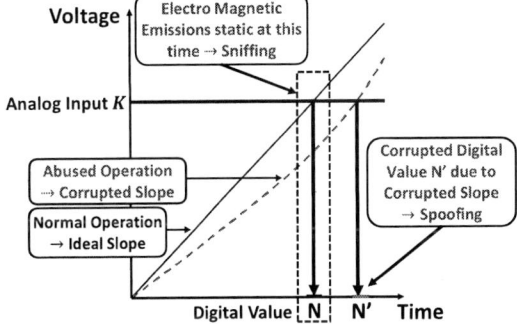

Fig.2. Single Slope ADC: Sniffing attacks during ideal operation and spoofing attacks by abusing operation.

B. Attack vulnerability – Case of Single Slope ADC

To understand the implication of the mentioned security hole, let us consider the case of a single slope ADC (Fig.2). It operates by generating a ramp, comparing it with the analog input, and measuring the time from the start of ramp to the crossing of comparator threshold using a digital counter. The digital count (N) is proportional to the analog input (K), and hence is the digital equivalent of the analog value. Considering an adversary with electromagnetic estimation capabilities, as the time of the comparator trigger is static w.r.t counter start time, the digital value can be estimated. Considering an adversary with circuit abuse capabilities, like noise injection or jamming, corruption of digital value can be effected, and cannot be detected, as every digital output from ADC is a valid value. These scenarios are illustrated in Fig.2. This vulnerability is generic to any ADC architecture and the experimental support for this suspicion is based on [1-4].

C. Problem Statement

(a) The electromagnetic emission sequence for a given input is static in time and makes the ADC vulnerable to estimation, causing confidentiality compromise of the post processed encryption.

(b) The circuit abuse causing corrupt digitization, if detection is unfacilitated, causes integrity compromise of the post-processed authentication.

Emerging trends in ADCs, like Compressive Sensing (CS) based secure ADCs[5-7] with random sampling, may addresses the issue (a), but lacks generality, as these operate only on specific sets of signals that meet some sparsity criteria. Besides, (b) may still be a concern for such CS based ADCs.

978-1-5386-6414-8/18 $31.00 © 2018 IEEE　　　　123

II. Proposed Encryption-Authentication unified A/D Conversion Scheme

We propose an A/D conversion scheme involving multiple randomized non-linear mapping of input voltage, followed by digitization and inverse mapping, as depicted in Fig.3. The data acquisition flow is detailed in the Fig.4. The mapping functions are chosen at random from a set of function pairs using a randomization parameter r. If x represents the normalized analog input, f_r and g_r represent normalized non-linearly mapped output, then, for successful A/D conversion, the mapping function pair $(f_r(x), g_r(x))$ must meet a certain mathematical criterion as defined by equation (1) below:

$$max\left(|\dot{f}_r|, |\dot{g}_r| \right) \geq 1 \quad \forall x \in (0,1) \ ;$$

$$\text{where} \quad \dot{f}_r = \frac{\partial f_r(x)}{\partial x} \ ; \ \dot{g}_r = \frac{\partial g_r(x)}{\partial x}$$

$$(1)$$

A. Solution Set { (f_r, g_r) } and the working procedure

A set of solutions for the equation (1), that can be used to realize the A/D conversion scheme as described in Fig.3, is as detailed in equation (2). A function pair for two different values of r, $(r = r1)$ and $(r = r2)$, has profiles as illustrated in Fig.5(a).

$$f_r = \frac{x}{(1-r)-x} \quad \forall x \in \left(0, \frac{1-r}{2}\right) ; \ \frac{1-x}{r+x} \quad \forall x \in \left(\frac{1-r}{2}, 1\right);$$

$$g_r = \frac{x}{(2-r)-x} \quad \forall x \in \left(0, \frac{2-r}{2}\right) ; \ \frac{1-x}{x-(1-r)} \quad \forall x \in \left(\frac{2-r}{2}, 1\right);$$

$$(2)$$

The concept of randomization of the A/D conversion operation (randomization steps) and the attack detection check function (verification step) of Fig.4 are explained in Fig.5 and 6 respectively. Choosing a random mapping function pair, on a per sample basis, as illustrated in Fig.5(a), results in randomization of the digital pair (N_1, N_2), hence the randomization of the A/D conversion process itself, and the electromagnetic emissions there-of, and addresses the issue (a) of the problem statement. Under proper circuit operation condition, a given analog input K must produce a mapped pair (N_1, N_2) which lies on a parametrized curve as depicted in the fig.5(b), failing which is an indication of altered circuit behavior. By some means, if the circuit operation is altered, causing the mapping pair functions to deviate from ideal, as illustrated in Fig.6(a), the generated digital pair also deviate from the ideal expected point on the verification curve, as depicted in Fig.6(b). A verification of the presence of the generated point, on the verification curve, can thus be used as a check against circuit operation abuse. In practical application however, a decision margin need to be determined depending on the ideal state noise performance of the system implementing this scheme, and the attack detection sensitivity of the scheme and ideal state noise performance of the system form a tradeoff relationship. This addresses point (b) of the problem statement. These features are enabled at a cost, of realizing non-linear mapping functions (f_r, g_r) in analog domain, inverse mapping and verification functions in digital domain, generation of r synchronized at both Tx and Rx ends (for example by using a synchronized pseudo-random digital bit stream and a DAC at Tx) and one extra ADC.

B. Circuit Implementation

The solution set described above can be implemented by physically realizing four variable slope straight line equations as depicted in the Fig.7(a). The solutions of the intersection of the straight lines (y_1, y_2), with y_3 and y_4, yield the mappings f_r and g_r, respectively. Fig.7(b) is the circuit implementation of this using integrators and level shifters and the intersection points are obtained using comparators C1 through C4 and digital counters.

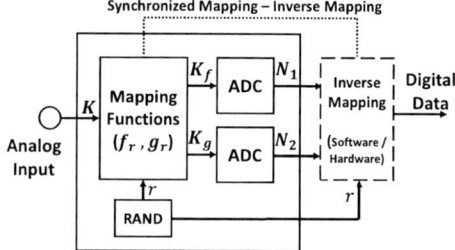

Fig.3. Proposed Encryption-Authentication unified A/D conversion scheme: randomized non-linear mapping of analog input followed by digitization and inverse mapping.

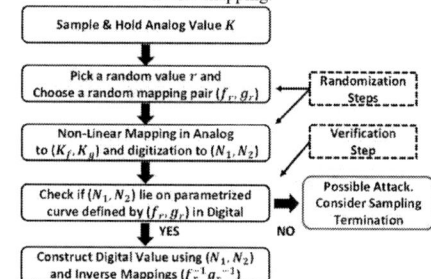

Fig.4. Data acquisition flow of the proposed A/D conversion scheme.

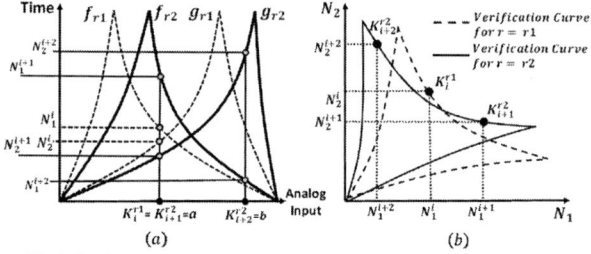

Fig.5. Fundamental operation: (a) Randomization: Random (N_1, N_2) are generated for input sequence K=(a,a,b,…) by changing r=(r$_1$, r$_2$, r$_2$, …) on per sample basis. (b) Verification: Verification of point (N_1, N_2) on respective curves defined by r.

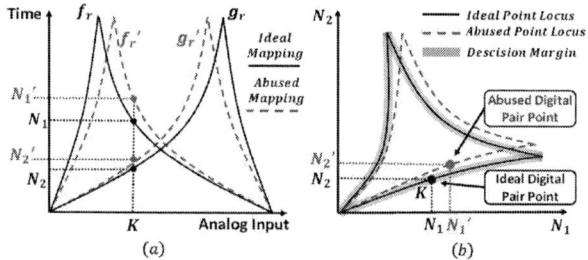

Fig.6. Attack Detection: (a) During attack on circuit, abused mapping function pair leads to corrupted digital pair (N_1', N_2'). (b) Corresponding point fall-out of the verification curve and the locus of points formed by such abused digital pairs, along with a pass/fail decision boundary.

978-1-5386-6414-8/18 $31.00 © 2018 IEEE

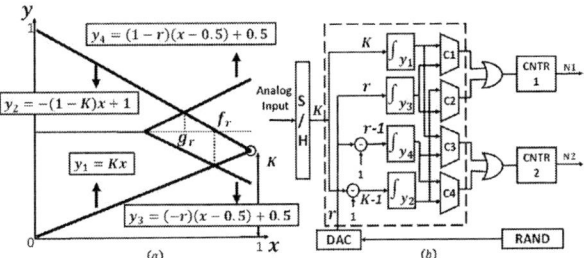

Fig.7. (a) Formulating non-linear mapping (f_r, g_r) as solutions of intersection of straight line pairs. (b) Architecture diagram of a circuit implementing the process defined in (a).

C. A 0.18um CMOS Realization of the Analog Domain

The analog section of the proposed solution architecture, depicted by the dashed rectangle of Fig.7(b), was realized in 0.18um process CMOS. The chip microphotograph is shown in Fig.8. The integrators, implemented as switched capacitor circuits, realize the straight lines y_1 through y_4, as a 256 step ramps, to facilitate an 8-bit A/D conversion. The analog input values of K and r are fed into the chip and the four comparator outputs are post-processed in software (including counters). The inverse equations used for the reconstruction of the digital data using the values N_1 and N_2 and the randomization parameter r, are detailed in (3). The parameters SC_x and OC_x are the Slope Correction and Offset Correction factors for each pairs of straight lines, of which f_r and g_r are the solutions.

$$f_r^{-1} = \frac{[1-(r+SC_1)]N_1}{1+N_1} - OC_1 \qquad \forall (N_1 \le N_2)$$

$$\frac{[2-(r+SC_2)]N_2}{1+N_2} - OC_2 \qquad \forall (N_2 > N_1)$$

$$g_r^{-1} = \frac{1-(r+SC_3)N_1}{1+N_1} - OC_3 \qquad \forall (N_1 \le N_2)$$

$$\frac{[1-(r+SC_4)]N_2+1}{1+N_2} - OC_4 \quad \forall (N_2 > N_1)$$

$$Digital\ Value = f_r^{-1} \ \forall \ (\dot{f}_r > \dot{g}_r) \ ; \ g_r^{-1} \ \forall (\dot{g}_r > \dot{f}_r) \tag{3}$$

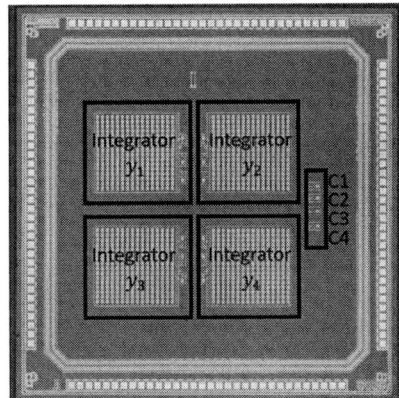

Fig.8. Microphotograph of the 0.18um CMOS chip realizing proposed Encryption-Authentication unified ADC.

III. MEASUREMENT RESULTS

A. Basic ADC Operation

The chip measurement summary and a comparison with two different slope based ADCs are consolidated in Table I, and the supporting graphs are consolidated in Fig.9. The usable digital range, post calibration, was found to be from $(20-240)$, and the results are for corresponding sub-full-scale analog input.

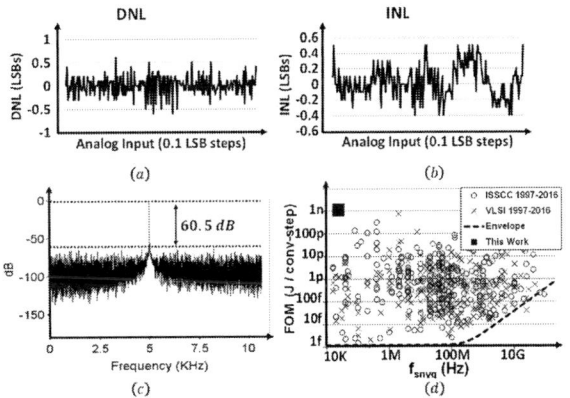

Fig.9. (a) Measured DNL plot; (b)Measured INL plot.
(c) FFT spectrum for a 5kHz input sinusoid.
(d) Comprehensive FoM comparison plot [8].

B. Side Channel Attack Resistance and Attack Detection

The proposed ADC works by selecting random mapping function pairs on a per sample basis. Fig.10 demonstrates a set of 32 mapping function pairs generated on chip by sweeping the randomization parameter r over normalized range 0 to 1, in 32 steps, and sweeping the input over the full scale for each of these steps. The plot elucidates a profile of the range of randomization of (N_1, N_2) across the analog input scale. For an arbitrary input K, the comparator trigger times, and hence the electromagnetic emissions in time, are randomized over the depicted time range. An attacker trying to lodge an estimation of K attack by electromagnetic sniffing and estimation of (N_1, N_2), is faced with an uncertainty in estimation proportional to the randomization range.

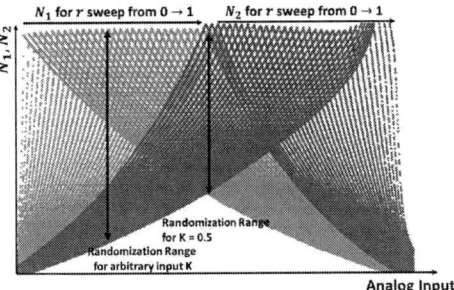

Fig.10. Set of 32 mapping function pairs generated on chip.

With reference to Fig.6(b), the attack detection sensitivity may be quantified by the probability of attack detection P_d and attack detection failure P_f, as in (4). For a particular kind of attack, a large P_d is a measure of sensitivity of the implementation towards detecting that attack, a large P_f is a measure against the suitability to detect that attack, a small P_d and P_f is an indication of the tolerance, of the implementation, to the attack. To demonstrate the physical abuse detection functionality of the proposed scheme, a hypothetical attack scenario was considered, in which the VDD of the chip was perturbed in 1% steps and the resulting (N_1, N_2) pairs for full sweep of analog input for a constant $r = 0.5$, was plotted at 0.1 LSB steps against ideal (no perturbation) and is as depicted in Fig.11(a). The probabilities P_d and P_f are plotted in Fig.11(b).

$$P_d = \frac{\begin{array}{c} No.\,of\,Points\,of\,Perturbed\,Curve \\ Falling\,Out\,of\,Ideal\,Curve \end{array}}{Total\,No.\,of\,Points\,on\,the\,Ideal\,Curve}$$

$$P_f = \frac{\begin{array}{c} No.\,of\,Points\,of\,Perturbed\,Curve\,on\,Ideal\,curve \\ for\,which\,digitized\,value\,is\,corrupted \end{array}}{Total\,No.\,of\,Points\,on\,the\,Ideal\,Curve}$$

(4)

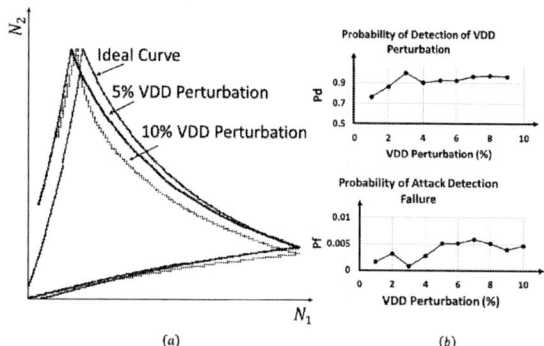

(a)   (b)

Fig.11. (a) Ideal and perturbed verification curves. (Ideal supply. 0 decision margin) (b) Plot of P_d and P_f against VDD perturbation percentage.

TABLE I.

1) THE DESIGNED AFE IS A PROOF-OF-CONCEPT AND NOT STATE-OF-THE-ART
2) THE CLOSEST APPROXIMATION IS TO TREAT THE CURRENT ADC AS SET OF 4 INDEPENDENT SINGLE SLOPE ADCs.

Chip Measurement Summary and Trade off Comparison with typical Slope Based ADCs			
Measurement Parameter	*This Work*	*[9]*	*[10]*
Technology	0.18-μm	0.18-μm	130-nm
ADC Operation Measurements			
Sample Rate (KS/s)	24	0.8	15
ENOB	7.64	6.49	7
DNL (LSB)	+0.6 / -0.6	+0.15/-0.11	0.6 (max)
INL (LSB)	+0.5/-0.4	+0.09/-0.79	0.95 (max)
SFDR (dB)	60.5	47.7	N/A
SNDR (dB)	47.7	40.8	43
Security Related Measurements			
EM Emission Randomization	Yes		
A/D Corruption Detection	Yes	N/A	
Detection Probability (VDD Attack)	> 0.9		
Tradeoff Comparison			
Area (mm^2)	1.475	0.054	0.06
Power Consumption (W)	5.38m	209.5n	22u
FoM (J/Conversion Step)	1124p	293p	14.3p

CONCLUSION AND DISCUSSION

An A/D conversion scheme with EM emission randomization and A/D corruption detection features was proposed. A solution was designed and an analog-front-end for an 8-bit ADC was realized in a 0.18um CMOS process, with ENOB = 7.64 bits, max DNL = 0.6 LSB, max INL = 0.9 LSB. The proposed scheme opens up a new model of authenticated-encryption. If point P $=(N_1, N_2)$ is randomly perturbed to P'= (N_{c1}, N_{c2}) using a shared secret, it acts as Cipher Text + MAC (Message Authentication Code). This scheme is Illustrated in Fig.12.

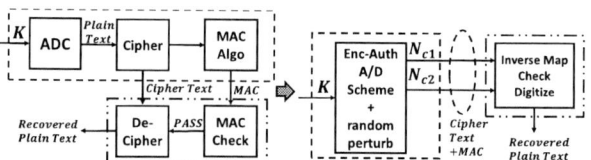

Fig.12. Sensor data security scheme using proposed ADC: The Digital Pair (N_{c1}, N_{c2}) can be used as (Cipher Text + MAC)

ACKNOWLEDGMENT

This work is supported by VLSI design and Education Center (VDEC), The University of Tokyo, in collaboration with Cadence Design Systems, Inc., Synopsys, Inc., Mentor Graphics, Inc., Rohm Corporation and Toppan Printing Corporation. The design work was also supported by Dr. T.Kikkawa, Dept. of EEIS, The University of Tokyo.

REFERENCES

[1] Gandolfi K. et al., (2001) Electromagnetic Analysis: Concrete Results. In: Koç Ç.K., Naccache D., Paar C. (eds) Cryptographic Hardware and Embedded Systems — CHES 2001. CHES 2001. Lecture Notes in Computer Science, vol 2162. Springer, Berlin, Heidelberg

[2] A. Dehbaoui et al., "Injection of transient faults using electromagnetic pulses Practical results on a cryptographic system," in Proc. of Workshop on Fault Diagnosis and Tolerance in Cryptography (FDTC 2012), 2012, pp. 7–15.

[3] M. Quilez et al., "Susceptibility of Dual-Slope ADCs to Electromagnetic Interference: An Experimental Analysis," *2007 IEEE Instrumentation & Measurement Technology Conference IMTC 2007*, Warsaw, 2007, pp. 1-4.

[4] F. Wan et al., "Effects of conducted electromagnetic interference on analogue-to-digital converter," in *Electronics Letters*, vol. 47, no. 1, pp. 23-25, 6 January 2011.

[5] Y. Zhang et al., "A Review of Compressive Sensing in Information Security Field," in IEEE Access, vol. 4, pp. 2507-2519, 2016.

[6] T. Bianchi et al., "On the security of random linear measurements," 2014 IEEE International Conference on Acoustics, Speech and Signal Processing (ICASSP), Florence, 2014, pp. 3992-3996.

[7] M. Verhelst et al., "Where Analog Meets Digital: Analog-to-Information Conversion and Beyond," in IEEE Solid-State Circuits Magazine, vol. 7, no. 3, pp. 67-80, Summer 2015.

[8] B. Murmann, "ADC Performance Survey 1997-2017," [Online]. Available: http://web.stanford.edu/~murmann/adcsurvey.html.

[9] Y. Osaki et al., "A low-power single-slope analog-to-digital converter with digital PVT calibration," 2012 19th IEEE International Conference on Electronics, Circuits, and Systems (ICECS 2012), Seville, 2012, pp. 613-616.

[10] H. Shan et al., "A low power programmable dual-slope ADC for single-chip RFID sensor nodes," 2017 IEEE 17th Topical Meeting on Silicon Monolithic Integrated Circuits in RF Systems (SiRF), Phoenix, AZ, 2017,

12-1 (8115)

IEEE Asian Solid-State Circuits Conference
November 5 - 7, 2018/Tainan, Taiwan

A 28nm 320Kb TCAM Macro with Sub-0.8ns Search Time and 3.5+x Improvement in Delay-Area-Energy Product using Split-Controlled Single-Load 14T Cell

Cheng-Xin Xue[1], Wei-Cheng Zhao[1], Tzu-Hsien Yang[1], Yi-Ju Chen[1], Hiroyuki Yamauchi[2] and Meng-Fan Chang[1]

[1]National Tsing Hua University, Hsinchu, Taiwan,
[2]Fukuoka Institute of Technology, Fukuoka, Japan
E-mail: mfchang@ee.nthu.edu.tw

Abstract—**This work proposes a Split-Controlled Single-Load (SCSL) 14T TCAM cell to achieve (1) compact cell area, (2) reduced search delays and energy, (3) less leakage current. A 320Kb 14T-TCAM macro was fabricated using a 28nm CMOS process and modified foundry compact-area 6T cell. The resulting macro lowered search delays to just 710ps and gained a 3.5+x improvement in the figure-of-merit (delay-area-energy product), compared to conventional 16T TCAMs.**

Keywords—TCAM、SRAM

I. INTRODUCTION

TCAM is used in networking and pattern-recognition chips to compare input data against all patterns stored in activated parallel rows within one access cycle [1-6]. Most TCAM cells (Fig. 1(a)(b)) comprise 16 transistors (16T), including two 6T SRAM cells and a 4T comparison circuits. Nodes Q0 and Q1 store pattern "1" (1, 0), "0" (0, 1) or don't-care (X, (0, 0)). The 4T comparison circuits compare input data on the search-line (SL) with pattern (Q0/Q1) stored in a TCAM cell, and generate a discharge current (I_{ML-MIS}) on the match-line (ML) for mismatch case. Typical TCAM faces a number of challenges: (1) large cell area due to the use of 16T; (2) high search power due to heavy parasitic load (2T per cell) and large voltage swings on each ML; and (3) considerable leakage current in 6T cells and 4T logics during standby and active modes.

This work proposes a SCSL-14T TCAM cell to achieve (1) compact cell area due to a reduced device count; (2) lower search delay and energy due to reduced loading on ML; (3) reduced current leakage in standby- and active-modes due to lower cell-VDD (CVDD) voltage (V_{CVDD}).

II. PROPOSED SPLIT-CONTROLLED SINGLE-LOAD 14T TCAM

SCSL-14T TCAM cell (Fig. 1(c)) comprises two 6T split-controlled CAM (SCAM) cells and 2T ML-driver (MLD). The two SCAM cells (cell-0 and cell-1) are placed vertically one above the other to store ternary data (1, 0, X), as in a 16T cell. The 6T SCAM cell has the same layout in devices (area) as typical compact 6T SRAM cells, but uses a different metal-layer layout and control scheme for the NMOS pass-gates (RPG and LPG). The gates of RPG and LPG are connected to the wordline (WL) and search-data line (SL or SLB),

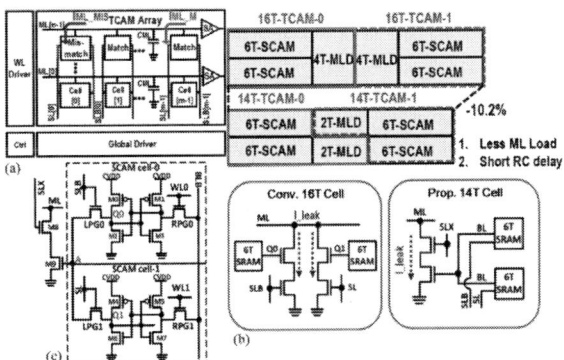

Fig. 1. (a) Typical TCAM structure (b) conventional 16T cell & SCSL-14T cell with L-shape layout (c) Proposed SCSL-14T cell

respectively. The sources of cross-coupled inverters are connected to cell-VDD (CVDD) and cell-VSS (CVSS). The drains of RPG in cell-0 and cell-1 are connected to the same bitline (BLB). The drains of LPGs in cell-0 and cell-1 are connected to node A, which is the gate of M9. The gate and drain of M8 are respectively connected to the search-data_don't_care line (SLX) and matchline (ML). Each ML is connected to an ML-header (MLH) and the Sense Amplifier. The MLH is powered by VDDML (V_{VDDML}). Only 1T is connected to the ML per 14T TCAM cell, rather than 2T per cell as in 16T TCAM cells.

An L-shape layout style (Fig. 1(b)) enables the 2T-MLD of two neighboring 14T cells to share the same extra cell-width beyond 6T cell. Thus, the cell area of a SCSL-14T cell is 10.2% smaller than that of a 16T TCAM with the same compact 6T-cell footprint. This also reduces the parasitic load and RC delay on ML to below that of a 16T TCAM.

In standby mode, the ML and all controls signals (WL, SL/SLB) are at 0V, SLX is held in its previous state, and MLH is disabled (SREN=0). The 2T-MLD does not consume standby current thanks to $V_{ML}=V_{CVSS}=0V$. The low V_{CVDD} suppresses the leakage current (I_{CELL_LEAK}) of TCAM cells.

Fig. 2 shows the SCSL14T cell's search operations, one SL/SLB in each column (IO bit) is raised to V_{SL}. SLX is high for valid input or 0V for Input=X. Note that SLX does not usually provide toggling across consecutive searches for most applications, unless there is change in the "X state" of an input

978-1-5386-6414-8/18 $31.00 © 2018 IEEE   127

12-1 (8115)

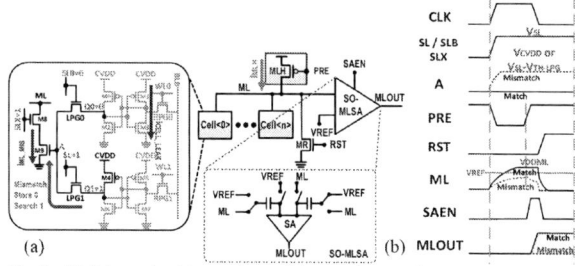

(a)

(b)

Fig. 2. (a) Schematic, (b) waveform of the search operation

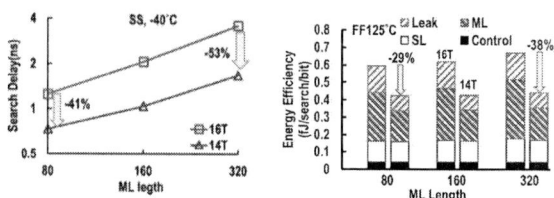

Fig. 3. (a) ML length v.s. search delay (b) ML length v.s. energy efficiency

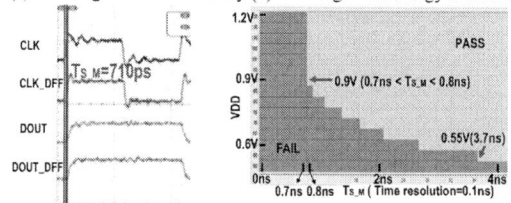

Fig. 4. (a) Testchip structure (b) captured waveform (c) measured shmoo plot of search operation

search bit. Then, MLH is turned on (SREN=1) to provide charging current (I_{MLH}) to the ML. The voltage-divider behavior between MLH and MLD of mismatched cells determines the ML voltage (V_{ML}).

In a matched cell (i.e., SL=1 and Q1=0), the LPG connected to "0" is turned on to discharge voltage at node A (V_A) to 0V. M9 is off and no ML discharge current (I_{ML-MIS}) is provided by the MLD of a matched cell. If all TCAM cells on a ML are match, then the ML is pulled to V_{VDDML} by MLH.

In a mismatch cell (i.e. SL=1 and Q1=1), the LPG connected to "1" are turned on to make VA= $V_{SL}-V_{TH-LPG}$ or V_{CVDD} (when $V_{SL} \geq V_{CVDD}+V_{TH}$). Then M9 is turned on to generate an I_{ML_MIS} and create the ML mismatch signal margin (V_{MSM}), $V_{ML}=VDD_{ML}-V_{MSM}$. For an ML with many mismatched cells, VML is low (large V_{MSM}). For an ML with few mismatched cells, V_{MSM} is high.

After a given ML developing time (T_{ML}), a self-timed signal turns off all MLH, and enable ML sense amplifier to detect V_{ML} and then outputs a digital signal (MLOUT) to indicate a match (MLOUT=1) or mismatch (MLOUT=0) state. Finally, the MLs are discharge to 0V. With a lower VCVDD, the I_{CELL_LEAK} and search energy (I_{ML-MIS}) is reduce, albeit at the expense of smaller V_{MSM}.

The write operation of the 14T-TCAM, in which BLB and the two WLs on RPG0 and RPG1 are used to write data. However, this pseudo single-ended write operation suffers less write margin as in [4]. Thus, we use column-wise data-aware CVDD-down and WL-boost schemes to assist write operations. Thanks to frequent-search and seldom-write applications, the energy-overhead due to write-assist scheme is not a main concern for many TCAM applications.

PERFORMANCE AND EXPERIMENT RESULTS

Fig.3 presents the performance of the proposed 14T TCAM macro with reduced parasitic load on the ML, the search delay (T_S) of the 14T TCAM was reduced by 41%. The lower V_{CVDD}(0.7V) reduces the active-mode standby current by 29%.

Fig. 5 shows the die photo and chip summary. A 28nm 320Kb 14T TCAM macro fabricated using a modified foundry compact-cell 6T cell. Fig. 4 is the measured search delay (T_{S_M}) was 710ps, excluding testchip search delay (T_{S_CHIP}) and the path-delay of the test board (T_{PATH}). TABLE I shows the proposed scheme achieved a FoM (delay-area-energy product) 3.5x~14x better than those in previous works.

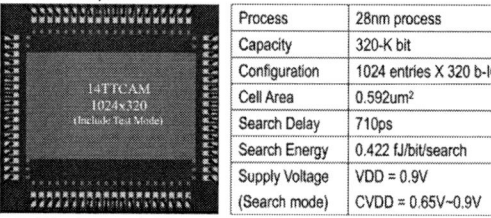

Process	28nm process
Capacity	320-K bit
Configuration	1024 entries X 320 b-IO (80X4)
Cell Area	0.592um²
Search Delay	710ps
Search Energy	0.422 fJ/bit/search
Supply Voltage (Search mode)	VDD = 0.9V CVDD = 0.65V~0.9V

Fig. 5. Die photo and chip summary

TABLE I. Comparison table with previous works

Scheme	Simulation	Measurement			
	16T_H2L VML=0.9	This Work	ISSCC 2017[1]	VLSI 2015[2]	ISSCC 2014[3]
Technology	28nm CMOS	28nm CMOS	14nm Din-FET	16nm Din-FET	28nm HKMG
VDD(V)	0.9	0.9	0.8	0.8	0.85
Normalized Cell Area, A⁺	1.114X	1X	1.114X	1.114X	1.114X
Energy Efficiency, E (fJ/bit/search)	1.394	0.422	0.316	0.85	1.6
Normalized Energy Efficiency, E⁺	3.3X	1X	1.9X	4.46X	4.25X
Search Rate, SR(GHz)	0.68	1.33	1.4	1.25	0.4
Normalized Search Rate, SR⁺	0.51X	1X	0.99X	0.6X	0.32X
FoM: (A⁺)x(E⁺)/(SR⁺)	7.2X	1X	3.57X	8.22X	14.87X

A⁺ = [1]-[3] normalize to 28nm 16T

E⁺ = E x (28/Technology) x (0.9/VDD)^2   SR⁺ = SR x (Technology/28) x (0.9/VDD)

ACKNOWLEDGMENT

The authors thanks UMC, and MOST-Taiwan for support.

REFERENCES

[1] I. Arsovski, et al., "1.4Gsearch/s 2Mb/mm2 TCAM Using Two-Phase-Precharge ML Sensing and Power-Grid Pre-Conditioning to Reduce Ldi/dt Power-Supply Noise by 50%", ISSCC, pp. 212-213, Feb. 2017.

[2] Y. Tsukamoto, et al., "1.8 Mbit/mm2 Ternary-CAM macro with 484 ps Search Access Time in 16 nm Fin-FET Bulk CMOS Technology", IEEE Symp. VLSI Circuits, pp. 274-275, June 2015.

[3] K. Nii, et al., "A 28nm 400MHz 4-parallel 1.6Gsearch/s 80Mb ternary CAM", ISSCC, pp. 240-241, Feb. 2014.

[4] M.-F. Chang et al., "A 28nm 256Kb 6T-SRAM with 280mV Improvement in VMIN Using a DualSplit-Control Assist Scheme" ISSCC, pp. 314-315, Feb. 2015.

[5] M.-F. Chang, et al., "A 3T1R Nonvolatile TCAM Using MLC ReRAM with Sub-1ns Search Time," ISSCC, pp. 318-319, Feb. 2015.

[6] C.-C. Lin, et al., "A 256b-Wordlength ReRAM-based TCAM with 1ns Search-Time and 14x Improvement in WordLength-EnergyEfficiency-Density Product using 2.5T1R cell," ISSCC, pp. 136-137, Feb. 2016

978-1-5386-6414-8/18 $31.00 © 2018 IEEE

A 6.8TOPS/W Energy Efficiency, 1.5μW Power Consumption, Pulse Width Modulation Neuromorphic Circuits for Near-Data Computing with SSD

Kota Tsurumi, Kenta Suzuki and Ken Takeuchi

Department of Electrical, Electronic, and Communication, Chuo University, Tokyo, Japan

Email: tsurumi@takeuchi-lab.org

Abstract— This paper proposes Pulse Width Modulation Neuromorphic (PWMNM) circuits, best suitable for near-data computing with SSD. The concept of neuromorphic computing is mimicking human brain with a number of mimicked-neurons of analog circuits. A mimicked-neuron in proposed PWMNM consists of 10 pairs of oscillators & clock counters, a micro controller and a charge pump. Proposed PWMNM achieves high energy efficiency of 6.8TOPS/W (Tera operations per second per Watt) and low power consumption of 1.5μW, thanks to matrix product operations are processed with number of output pulses as a function of the time instead of conventional current or voltage operation. Both proposed PWMNM and peripheral circuits of NAND flash memory are fabricated with a 180nm standard CMOS process and PWMNM can be embedded in SSD. Character recognition system is demonstrated by using proposed PWMNM, and the recognition accuracy of 93.3% is achieved.

Keywords —Neuromorphic; Pulse width modulation; Oscillator; Charge pump; Low power consumption; Near-data computing.

I. INTRODUCTION

Until now, miniaturization of transistors in LSIs improves computer performance in line with Moore's law. However, Moore's Law is coming to an end due to physical limits and cost problems of miniaturization. In addition, conventional von Neumann computing is based on sequential processing and its speed bottleneck is the large difference in access time between CPU and memory; this is called von Neumann bottleneck. In order to overcome the end of Moore's law and the von Neumann bottleneck, many novel accelerator circuits and computer architectures are under study, including a neuromorphic circuits [1, 2] and a near-data computing architecture [3].

A neuron consists of synapses (product operation), dendrites (sum operation), and an axon-hillock (spike generator). The neuromorphic circuits mimic neurons and computes a large number of matrix operations in parallel; thus, its power consumption and energy efficiency are essentially important. Besides, near-data computing is also a promising approach towards low energy consumption and high speed operation by integrating memories and processors in the same package [3].

This paper proposes low power and high energy efficiency Pulse Width Modulation Neuromorphic (PWMNM) circuits, which is suitable for near-data computing framework designed for a SSD. Both proposed PWMNM and peripheral circuits of NAND flash memory in the SSD are fabricated by the standard high voltage transistor (HV-Tr) CMOS process, so both can be configured on the SSD package together; this configuration is

called as near-data computing. A PWMNM consists of 10 pairs of oscillators & clock counters, a micro controller and a charge pump (Fig. 1). The PWMNM controls the pulse width using the oscillator. In other words, the time axis is used for a matrix product operation. In addition, in the oscillator, parallel computation is performed to improve computation efficiency. Therefore, power consumption of the PWMNM as an analog processor [1,2] is lower than that of digital processing [4,5] in low precision computation [6]. Moreover, conventional neuromorphic circuits [1, 2] use current or voltage to perform product and sum operations, their linearity is limited because of the output capability of analog devices used in these circuits and their power consumption. As a result, proposed PWMNM for near-data computing has lower power consumption and higher linearity than conventional voltage or current operation neuromorphic circuits.

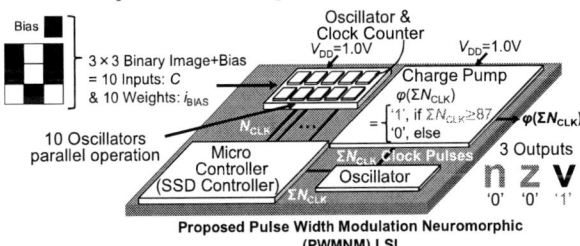

Fig. 1. Structure of proposed Pulse Width Modulation Neuromorphic (PWMNM) circuits.

II. NEAR-DATA COMPUTING ARCHITECTURE USING PWMNM WITH SSD

This section describes an architecture of near-data computing using proposed PWMNM. Figure 2(a) shows a schematic diagram of conventional von Neumann computing. In the von Neumann computing, the data movement across memory layers and CPU prevents improvement of computer performance. Moreover, improvement of computer performance due to miniaturization of transistors cannot be expected any more. Figure 2(b) shows the near-data computing architecture [3] with proposed PWMNM. In prior work [3], near-data computing is demonstrated and the data processing is 15× high-speed compared with conventional von Neumann computing. Proposed near-data computing with PWMNM is a promising concept to achieve high energy efficient product-sum operation due to its low data movement cost.

In a NAND flash memory, the peripheral circuits, such as the oscillator and a charge pump, are used to generate the program voltage (V_{PGM}) of 20V. The PWMNM also consists of the oscillator, a clock counter, a micro controller and the charge pump (Fig.1). Since both the peripheral circuits of NAND flash memory and PWMNM are fabricated with standard HV-Tr CMOS process, they can be integrated in a SSD with a negligible cost increase (Fig.2 (b)). In addition, the micro controller of PWMNM is replaced by SSD controller. In this way, computing data with in SSD, or near-data computing can be realized.

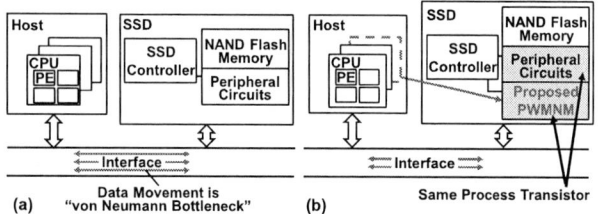

Fig. 2. Architecture of (a)conventional von Neumann comtuting, (b) near-data computing with proposed PWMNM.

III. PROPOSED PULSE MOUDLATION NEUROMORPHIC CIRCUITS AND SIMULATION RESULT

This section describes an analog neuromorphic circuits operation of PWMNM and its simulation results. Figure 3 shows the oscillator used in this work [7]. For neuromorphic circuits, linearity is important for product operation. In the proposed PWMNM, the oscillator in V_{DD} =1.0V controls the pulse width (T_{ON}) by the input (C) and the weight (i_{BIAS}) and performs the product operation. T_{ON} is given as follows:

$$T_{ON} = \frac{C_{ON}(V_{DD} - V_1)}{i_{REF}} = \frac{R \cdot C_{ON}}{V_{BIAS}} \quad (1)$$

T_{OFF} is adjusted to be equal to T_{ON} ($T_{ON} = T_{OFF}$). The reference current (i_{REF}) and the bias current (i_{BIAS}) are copied to nodes V_1 and V_2 using a current mirror. i_{REF} and bias voltage (V_{BIAS}) are given as follows:

$$i_{REF} = \frac{V_{DD} - V_{REF}}{R} \cdot \frac{V_{BIAS}}{V_{DD}} \quad (2)$$

$$V_{BIAS} = R \cdot i_{BIAS} + V_{REF} \quad (3)$$

C_{ON} and C_{OFF} are composed of two capacitors of 50fF and 200fF as shown in Figure 3. When both two capacitors are turned on by the controller, C_{ON} and C_{OFF} are set to high (250fF). On the contrary, C_{ON} and C_{OFF} are set to low (50fF) when only 50fF capacitor is turned on. Here, R and V_{REF} are fixed at 25kΩ and 500mV, respectively. When the connection strength represented as weight current (i_{BIAS}) changes from 1μA to 16μA, V_{BIAS} also changes from 525mV to 900mV according to Equation (3). Therefore, $T_{ON}/(T_{OFF})$ ranges 2.23ns to 33.8ns. In this way, T_{ON} (T_{OFF}), namely the switching frequency, is independently controlled by C_{ON} (C_{OFF}) and i_{BIAS}.

Figure 4 shows the simulated waveforms of the oscillator. The simulation settings are combinations of maximum or minimum values of C and i_{BIAS}. As a result, it can be confirmed that the pulse width changes according to C and i_{BIAS}. Figure 5 (a) shows T_{ON} ($=T_{OFF}$) when i_{BIAS} is changed from 1μA to 16μA with C of 50fF and 250fF, respectively. T_{ON} decreases as i_{BIAS} increases as shown in Figure 5 (a). Also, T_{ON} with C=250fF is larger than T_{ON} with C=50fF. Here, N_{CLK} means the number of output clock pulses per 100 ns counted by the clock counters in PWMNM. The expression of N_{CLK} is described as follows:

$$N_{CLK} = \frac{100ns}{(T_{ON} + T_{OFF})} = \frac{100ns}{2 \times T_{ON}} \quad (4)$$

Figure 5 (b) shows N_{CLK} with i_{BIAS} changing from 1μA to 16μA with C of 50fF and 250fF, respectively. As i_{BIAS} increases, N_{CLK} rises linearly. The linear approximation expression of N_{CLK} with C of 50fF and 250fF are derived as follows:

$$N_{CLK} = 1.34 \times i_{BIAS} \times 10^6 + 1.30 \quad (5)$$
$$(C=50fF)$$

$$N_{CLK} = 0.918 \times i_{BIAS} \times 10^6 + 1.41 \quad (6)$$
$$(C=250fF)$$

The value of N_{CLK} with C=250fF is larger than the value of N_{CLK} with C=50fF. Proposed PWMNM achieves linear characteristics by pulse width modulation using the oscillator.

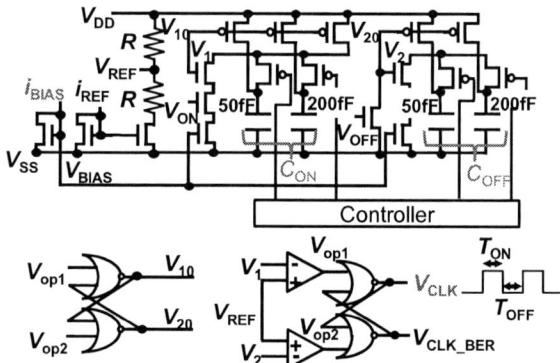

Fig. 3. Schematic of an oscillator [7]. T_{ON} ($=T_{OFF}$), is independently controlled by operation of by C ($=C_{ON}=C_{OFF}$), and i_{BIAS}.

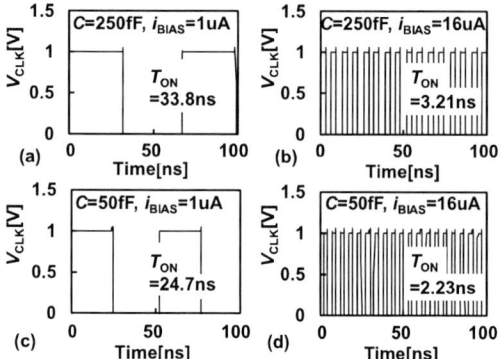

Fig. 4. Simulated V_{CLK} waveforms of an oscillator. (a)C=250fF, i_{BIAS}=1μA, (b) C=250fF, i_{BIAS}=16μA, (c)C=50fF, i_{BIAS}=1μA, (d) C=50fF, i_{BIAS}=16μA.

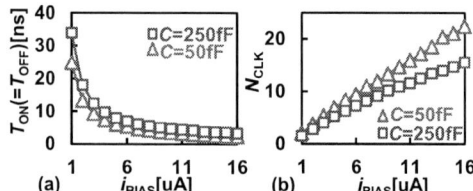

Fig. 5. (a) Simulated T_{ON} ($=T_{OFF}$) when i_{BIAS} is changed from 1μA to 16μA with C of 50fF and 250fF, respectively. (b) Calculated N_{CLK} when i_{BIAS} is changed from 1μA to 16μA using T_{ON}, with C of 50fF and 250fF, respectively.

Next, an operating mechanism of the charge pump is described (Fig. 6). The clock pulses of ΣN_{CLK} cycles ($T_{CPON}=T_{CPOFF}=12$ns) which are generated by the oscillator are applied to the charge pump to determine spiking. A cross-coupled charge pump [8] was used in PWMNM. Since the cross-coupled charge pump causes no threshold drop, high boost efficiency can be realized. In addition, overvoltage is not applied to the gate oxide film of the transistors in the circuits. The power supply voltage (V_{DD}) of the charge pump is 1.0V, and the output voltage of the charge pump (V_{CP}) rises as the ΣN_{CLK} increases. In this paper, the necessary number of ΣN_{CLK} is 87 due to the spiking threshold voltage (V_{TH}) is 1.1V. The spike waveform $\varphi(\Sigma N_{CLK})$ is generated by the comparator. V_{CP} is divided by 10/11 by two resistors, and compared with $V_{DD}=1.0$V by the comparator. The charge pump spikes if the divided voltage is higher than V_{DD}. Therefore, the charge pump spikes ($\varphi(\Sigma N_{CLK})$ is '1') when V_{CP} rises to 1.1V from 1.0V (ΣN_{CLK} is over 87).

Figure 7 shows the simulation results of the charge pump and the oscillator, (a) when ΣN_{CLK} is 50, and (b) when ΣN_{CLK} is 87. Since V_{CP} does not reach 1.1V, the step function $\varphi(\Sigma N_{CLK})$ is '0' in (a). On the other hand, $\varphi(\Sigma N_{CLK})$ is '1' in (b) because V_{CP} rise to 1.1V. This is how proposed PWMNM mimics neurons.

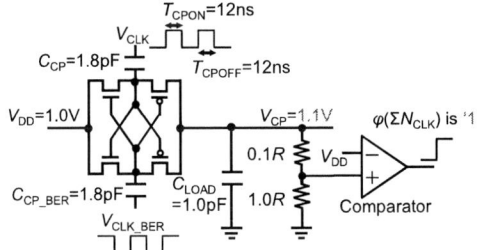

Fig. 6. Schematic of cross-coupled charge pump [8] and comparator. $\varphi(\Sigma N_{CLK})$ is '1' when ΣN_{CLK} is over 87.

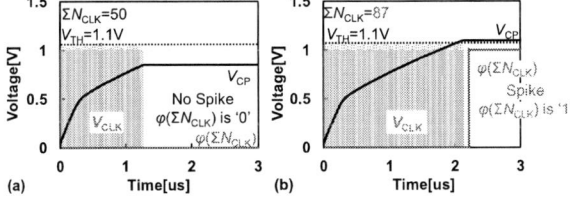

Fig. 7. Simulated V_{CP} and $\varphi(\Sigma N_{CLK})$ results of charge pump, (a) when ΣN_{CLK} is 50, and (b) when ΣN_{CLK} is 87.

IV. EXPERIMENT RESULT

This section describes character ('v', 'n', or 'z') recognition application using proposed PWMNM. A perceptron using PWMNM for the character recognition system as shown in Figure 8. Since this system recognizes 3×3-pixel monochrome images of 3 characters [9], the number of input node is 10 including bias and that of the output node is 3. The input (C) is binary precision corresponding to monochrome pixels, while the weight (i_{BIAS}) is 4bits precision. When the pixel is black, the Equation (5) is selected. On the other hand, the Equation (6) is selected, when the pixel is white. Figure 9 shows the flowchart during training phase in PWMNM. Neural network systems have inference phase to calculate the recognition rate and training phase to update the weights in order to improve the

recognition rate. In the inference phase, the product of the input (C) and the weight (i_{BIAS}) is calculated by using the oscillators for each pixel of training data and test data [9]. The number of the output clock pulse N_{CLK} is summed by the controller. Thereafter, the clock pulses of ΣN_{CLK} cycles are generated by the oscillator, and the charge pump judges whether to spike or not. When the charge pump judges to spike, $\varphi(\Sigma N_{CLK})$ is '1', otherwise $\varphi(\Sigma N_{CLK})$ is '0'. And the recognition rate is calculated by the controller. After inference, $\varphi(\Sigma N_{CLK})$ is compared with the correct answer data and the difference is calculated by the controller. Then, the weight (i_{BIAS}) is updated by the controller and the current generator. The answer data means the answer of the input image, for example, ('v', 'n', 'z') = ('1', '0', '0') for 'v'. When the difference is '1', the updating amount of the weight (Δi_{BIAS}) becomes large, while when the difference is '0', Δi_{BIAS} becomes small. Δi_{BIAS} is transmitted to the current generator to generate new i_{BIAS} to train the next image. Through this iteration, i_{BIAS} is updated, and the recognition rate is improved. The recognition rate of proposed PWMNM is calculated for both training data and test data using weight trained by training data.

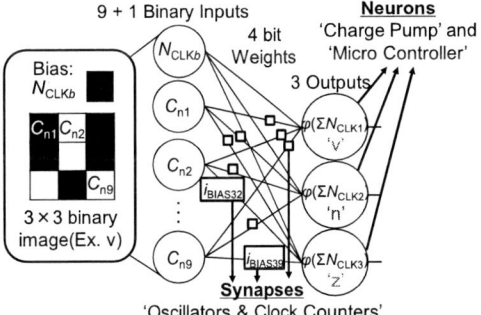

Fig. 8. Perceptron using proposed PWMNM for the character recognition system.

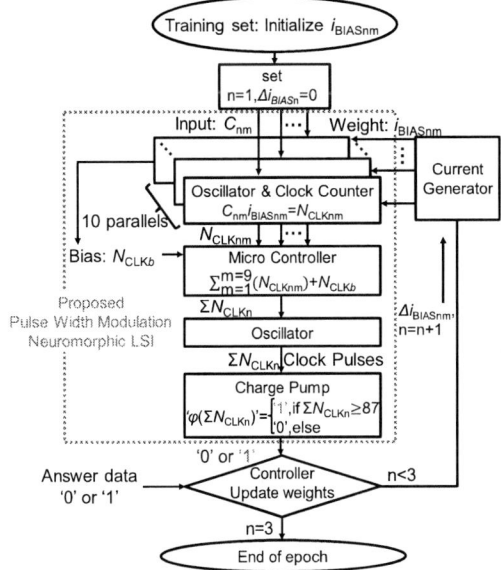

Fig. 9. Flowchart during training phase in proposed PWMNM.

Figure 10 (a) shows the recognition rate using training data and test data [9] in PWMNM, respectively. The maximum epoch is 100, and the recognition rate increases as the number of epochs increases. Figure 10 (b) shows the recognition rate at the 100th epoch in PWMNM and software. The recognition rate of PWMNM whose weight precision is 4-bits was 93.3% and 82.1% for training data and test data, respectively. When software recognition system was used, the recognition rate was 96.3% for training data and 86.2% for test data. Here, a simple perceptron coded in Python is used as software recognition system, whose weight precision is 32-bits. Thus, the recognition rate of software exceeds that of PWMNM. There are two reasons of lower recognition rate for test data. First reason is that the error of the test data is 2-bits and the error of the training data is 1-bit or no error. Second one is that the training data is used to update weights.

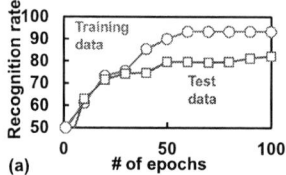

	Recognition rate	
	Software	Proposed PWMNM
Training data	96.3%	93.3%
Test data	86.2%	82.1%

Fig. 10. Perceptron using proposed PWMNM for the character recognition system.

V. MEASUREMENT RESULT

This section describes test chip and measurement result of proposed PWMNM. Figure 11 shows test chips of PWMNM. The oscillator (a) and the charge pump (b) are manufactured by using a 180nm standard CMOS process. The area of the oscillator is 8500um^2, and that of the charge pump is 45000um^2. The measured waveform of the oscillator (V_{DD}=1.0V) is shown in Figure 12. The measured conditions are similar to the simulation conditions (Fig. 4). The measured results of T_{ON} are 33.0ns for (a), 3.30ns for (b), 24.0ns for (c) and 2.20ns for (d) in Figure 12, then the simulated results of T_{ON} are 33.8ns for (a), 3.21ns for (b), 24.7ns for (c) and 2.23ns for (d) in Figure 4. Thus, the simulation reproduces the measurement faithfully.

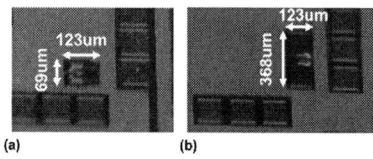

Fig. 11. Test chip of proposed PWMNM. (a)oscillator and (b) charge pump are fabricated with 180 nm standard CMOS process.

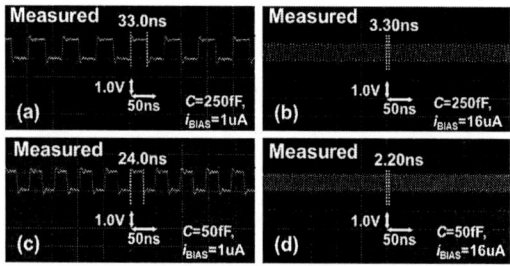

Fig. 12. Measured waveform V_{CLK} of oscillator.
(a)C=250fF, i_{BIAS}=1μA, (b) C=250fF, i_{BIAS}=16μA,
(c)C=50fF, i_{BIAS}=1μA, (d) C=50fF, i_{BIAS}=16μA.

VI. CONCLUSION

This paper proposes pulse width modulation neuromorphic (PWMNM) circuits for near-data computing. Proposed PWMNM can be fabricated by a standard CMOS process, which is compatible with the peripheral circuits of NAND flash memory. Table 1 shows the performance summary of proposed PWMNM test chip and comparison with other papers. Since proposed PWMNM controls the pulse width using oscillators, proposed PWMNM has lower power consumption and higher energy efficiency than conventional voltage or current analog operation and low precision digital operation. Concretely, proposed PWMNM can operate with low power consumption of 1.5μW and high energy efficiency of 6.8TOPS/W (Tera operations per second per Watt). The energy efficiency of proposed PWMNM are 6.8times better than analog implemented [1]. Furthermore, the energy efficiency of proposed PWMNM are 1.3times better than the digital implementation [4], whose weight precision is 4-bits. In addition, recognition rate of 3-character recognition using proposed PWMNM is 93.3%.

Table 1. Performance comparison.

	[1]	[2]	[4]	[5]	Proposed PWMNM
Process [nm]	130	130	65	40	180
Processing	Analog	Analog	Digital	Digital	Analog
Chip area [mm²]	0.36	0.11	19	120	0.14
Input [V]	3.0	1.2	1.2	1.1	1.0
Precision [bit]	8	8	4	4	4
Power [W]	11u	660n	4.0m	3.3	1.5u
Energy efficiency [TOPS/W]	1.0	0.060	5.1	0.59	6.8

ACKNOWLEDGMENT

The authors would like to thank S. Matsuda and S. Bando for their support and contributions.

REFERENCES

[1] J. Lu et al., "A 1 TOPS/W Analog Deep Machine-Learning Engine with Floating-Gate Storage in 0.13 μm CMOS," IEEE Journal of Solid-State Circuits (JSSC), vol. 50, no. 1, pp. 270-281, Jan. 2015.

[2] J. Zhang, Z. Wang and N. Verma, "A Matrix-Multiplying ADC Implementing a Machine-Learning Classifier Directly with Data Conversion," in IEEE International Solid-State Circuits Conference (ISSCC) Digest of Technical Papers, Feb. 2015, pp. 332-333.

[3] B. Gu et al., "Biscuit: A Framework for Near-Data Processing of Big Data Workloads," in ACM/IEEE Annual International Symposium on Computer Architecture (ISCA), June 2016, pp. 153-165.

[4] S. Yin et al., "A 1.06-to-5.09 TOPS/W Reconfigurable Hybrid-Neural-Network Processor for Deep Learning Applications," in IEEE Symposium on VLSI Circuits Digest of Technical Papers, Jun. 2017, pp. 26-27.

[5] K. Ueyoshi et al., "QUEST: A 7.49TOPS Multi-Purpose Log-Quantized DNN Inference Engine Stacked on 96MB 3D SRAM Using Inductive-Coupling Technology in 40nm CMOS", in IEEE International Solid-State Circuits Conference (ISSCC) Digest of Technical Papers, Feb. 2018, pp.216-217.

[6] R. Sarpeshkar, "Analog versus digital: Extrapolating from electronics to neurobiology," Neural Computing, vol. 10, no. 7, pp. 1601–1638, 1998.

[7] T. Tanzawa and T. Tanaka, "A stable programming pulse generator for single power supply flash memories," IEEE Journal of Solid-State Circuits (JSSC), vol. 32, iss.6, pp. 845-851, Jun 1997.

[8] M.-D. Ker, S.-L. Chen and C.-S. Tsai, "Design of charge pump circuit with consideration of gate-oxide reliability in low-voltage CMOS processes," IEEE Journal of Solid-State Circuits (JSSC), vol. 41, no. 5, pp. 1100-1107, May 2006.

[9] M. Prezioso et al., "Modeling and Implementation of Firing-Rate Neuromorphic-Network Classifiers with Bilayer Pt/Al2O3/TiO2-x/Pt Memristors", in IEEE International Electron Devices Meeting (IEDM) Technical Digest, Dec. 2015, pp. 455-458

12-3 (8228)

A 28nm FD-SOI 4KB Radiation-hardened 12T SRAM Macro with 0.6 ~ 1V Wide Dynamic Voltage Scaling for Space Applications

[1]Le Dinh Trang Dang, [1]Dongkyu Seo, [2]Jin-Woo Han, [1]Jinsang Kim and [1]Ik-Joon Chang
[1]Kyunghee University, Yongin-si, Korea, and [2]NASA Ames Research Center, Moffett Field, CA, USA

Abstract— We present a soft-error immune 12T SRAM cell for space applications, namely we-Quatro, and design a 4KB radiation-hardened macro of this SRAM in 28nm FD-SOI. Previously, 10T Quatro has been considered promising for this purpose, however suffers from extremely poor writability under parametric variations. Despite of two more transistors, we-Quatro delivers the same cell area as Quatro. Our simulations and measurements show that our 4KB macro provides robust read and write stabilities for wide supply voltage range of 0.6~1V. More critically, the measured soft-error resilience of we-Quatro is comparable to that of Quatro, significantly better than that of 6T SRAM.

Keywords—soft-error, single-event upset, SRAM

I. INTRODUCTION

Electrical circuit components for space applications such as satellite require radiation hardening since radiation effects such as single-event effect (SEE) or total ionizing dose (TID) significantly threaten the reliability of electrical circuits. Most of all, SRAM suffers from high soft-error rate due to SEE. Note that in modern VLSI architecture, the density of SRAM tends to increase for high performance, adversely increasing soft-error rate (SER). This motivates the necessity of soft-error resilient SRAM cells to provide robust operation even under severe radiation environment of space. 12T Dice [1] and 10T Quatro [2] are the most representative ones, which have interlocking structures to obtain high soft-error resilience. Their strong soft-error resilience has been proven through actual silicon measurements in previous works [3]. However, our study [4] showed that in scaled technologies, these SRAM cells are impractical due to large cell area (for Dice) or poor writability (for Quatro). To address these problems, we presented a new soft-error resilient 12T SRAM cell named as we-Quatro in [4]. The we-Quatro has the same cell area as Quatro despite two additional transistors. We design 4KB we-Quatro macro to support wide dynamic voltage scaling of 0.6~1V. We validate reliable operations over wide dynamic voltage range and strong soft-error resilience through actual silicon measurements.

II. DESIGN AND VALIDATION OF 4KB WE-QUATRO MACRO

A. we-Quatro SRAM Cell

As mentioned above, Dice and Quatro have interlocking structures. The interlocking structure contributes to significantly enhancing tolerance to soft-errors however, incurs considerable cell area penalty due to complex wiring of cell

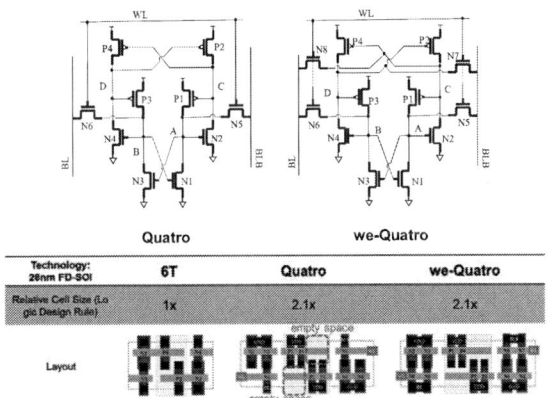

Fig. 1 Schematic and Layout of Quatro and we-Quatro

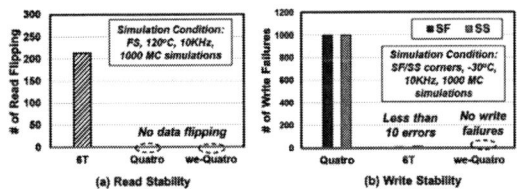

Fig. 2 Read and Write Stability MC Simulation Results at 0.6V V_{DD}

transistors. To the best of our knowledge, a thin-cell type layout of Dice does not have been reported yet. After our extensive effort, we conclude that it is difficult to draw a thin-cell type layout of Dice with moderate area overhead. For Quatro, we succeeded in implementing a thin-cell type layout in [4], where Quatro occupies 2.1x larger cell area compared to 6T SRAM under logic design rules of 28nm FD-SOI. However, our study of [4, 5] shows that Quatro cannot be applied under parametric variations due to weak writability.

We propose a new 12T SRAM cell, named as we-Quatro. Fig. 1 compares schematics and layouts of Quatro and we-Quatro. The we-Quatro have two more access transistors than Quatro however, these additional transistors are placed at empty space in the layout of Quatro. Hence, the cell area of we-Quatro is same as that of Quatro. We compare write and read stabilities of three SRAM cells: 6T, Quatro, and we-Quatro. We perform 1000 MC simulations at the worst process and temperature corners for write and read stabilities, shown in Fig. 2. At 0.6V supply power (V_{DD}), Quatro shows larger number of write failures while we-Quatro do not experience no write and read failures.

978-1-5386-6414-8/18 $31.00 © 2018 IEEE

B. 4KB Macro Design and Test-chip Implementation

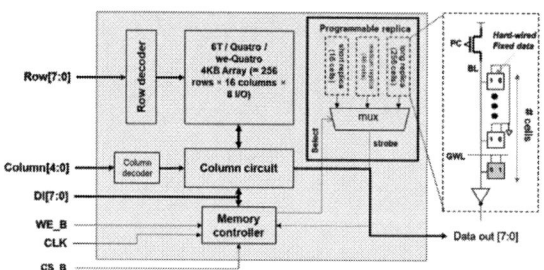

Fig. 3 4KB Macro Architecture

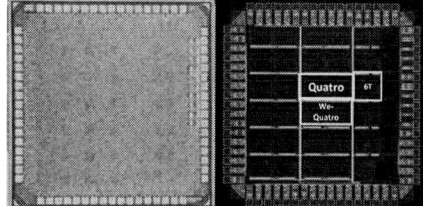

Fig. 4 Test-chip Die-photo and Layout

We design a 4KB macro of we-Quatro with the architecture of Fig. 3. We make the macro to support wide dynamic voltage scaling and radiation-hardening. It is challenging to properly generate strobe delays of sense amplifiers over wide supply voltage range. Further, TID effect due to radiation vary circuit delays in space. TID effect negatively shifts threshold voltages of NMOS and PMOS transistors [6]. This implies that after TID, drive current of NMOS becomes larger while that of PMOS decreases. This different variation trend of NMOS and PMOS makes it difficult to predict how TID effect affects the circuit to make strobe delays. To consider these problems, we generate strobe delays by employing column-replica circuits, which imitates column arrays of SRAM cells. Data patterns of these column-replica circuits is hard-wired to the worst one with respect to bit-line developing speed. The column-replica circuits occupy more area than simple inverter chains, however closely tracks bit-line developing speed varied by temperature variations or TID effect. To further provide robust sensing operation, we make programmable column-replica. The optimal one is selected through post-silicon or on-line validation. We fabricated a test-chip to verify the macro, whose photo and layout are shown in Fig. 4. We implement 4KB macros of 6T SRAM, Quatro, and we-Quatro in this test-chip. The macros of 6T SRAM and Quatro are used as our comparison target to verify the efficacy of our we-Quatro macro.

C. Measurment Results

We compare read and write stabilities, and soft-error rate of these three macros through actual silicon measurements. Fig. 5 shows the average number of read or write failures measured from five test-chips. At 0.6V, no read failures were observed for all macros. We make further voltage scaling to accelerate read failures. Then, at 0.4V V_{DD}, we found few read errors for 6T SRAM while Quatro and we-Quatro still provide robust

read operations. For write operations, 6T and we-Quatro do not show write failures for 0.6 ~ 1V V_{DD}. However, Quatro experiences some write failures even for the nominal V_{DD}, 1V. The measurement results demonstrate that Quatro suffer from extremely poor; we-Quatro delivers robust read and write operations for 0.6 ~ 1V V_{DD}.

We compare soft-error resilience of three SRAM macros by conducting alpha-foil tests for 23 hours (Fig. 6 (a)). Polonium-210 of 0.1μCi and 5407.5 keV is employed as the alpha source. For the macros of Quatro and we-Quatro, soft-errors are not observed up to 0.45V, as plotted in Fig. 6 (b). At 0.32V, very near to V_{min} of hold mode, Quatro and we-Quatro macros start to exhibit only 4 or 5 soft-errors. For 6T SRAM, considerable number of soft-errors are observed in even at 0.65V.

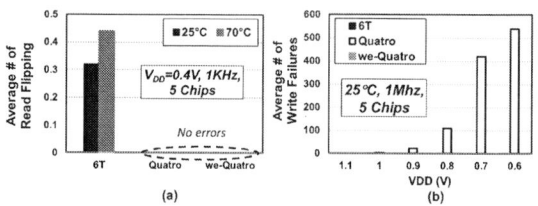

Fig. 5 (a) Measured # of Read Failures, (b) Measured # of Write Failures

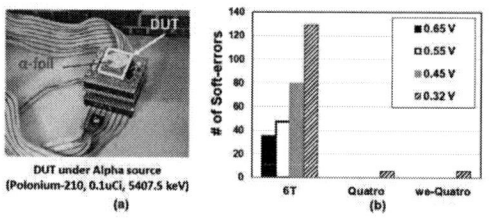

Fig. 6 (a) Photo of Alpha-testing (b) Measured # of Soft-errors

III. CONCLUSION

We design 4KB macros of 6T SRAM, Quatro and we-Quatro in 28nm FD-SOI technology. The silicon measurement results show that compared to 6T SRAM and Quatro, we-Quatro shows more robust operations under process variations. The soft-error resilience of we-Quatro is comparable to that of Quatro, more enhanced than that of 6T SRAM. These results well validate the efficacy of our we-Quatro.

ACKNOWLEDGMENT

This work is supported by National Research Foundation (NRF) of Korea under Grant NRF-2015M1A3A3A02010753 and Grant NRF-2017R1A2B4010828.

REFERENCES

[1] T. Calin et al., IEEE TNS, vol. 43, no. 6, pp. 2874–2878, 1996.

[2] S. M. Jahinuzzaman et al., IEEE TNS, vol. 56, no. 6, pp. 3768–3773, 2009.

[3] Q. Wu et al., IEEE TNS, vol. 62, no. 4, pp. 1898–1904, 2015.

[4] Le Dinh Trang Dang et al., IEEE TNS, vol. 64, no. 9. pp. 2489–2496, 2017

[5] Le Dinh Trang Dang et al., IEEE TNS, vol. 63, no. 4. pp. 2399-2401, 2016

[6] Daniel M. Fleetwood, IEEE TNS,Vol.60, No.3, pp. 1706-1730, 2013

Nonvolatile Crossbar 2D2R TCAM with Cell Size of 16.3 F² and K-means Clustering for Power Reduction

Keji Zhou[1+], Xiaoyong Xue[1+*], Jianguo Yang[1], Xiaoxin Xu[2], Hangbing Lv[2*], Mingyu Wang[1], Ming'e Jing[1], Wenjun Liu[1], Xiaoyang Zeng[1*], Steve S. Chung[3], Jing Li[4], Ming Liu[2]

[1]State Key Laboratory of ASIC and System, Department of Microelectronics, Fudan University, China;
[2]Key Laboratory of Microelectronics Devices and Integrated Technology, Institute of Microelectronics of the Chinese Academy of Sciences, China; [3]National Chiao Tung University, Taiwan; [4]University of Wisconsin-Madison, USA.
*Email: xuexiaoyong@ fudan.edu.cn, lvhangbing@ime.ac.cn, xyzeng@fudan.edu.cn

Abstract—A 28nm 16Kb nonvolatile ternary content addressable memory (nvTCAM) test chip based on logic-compatible resistive memory (ReRAM) is demonstrated with high density and low power for packet routing in network applications. The crossbar array exhibits a smallest cell size of 16.3F² by using 2-diode-2-ReRAM (2D2R) nvTCAM cell, useful for further 3-dimension (3D) stacking. K-means clustering is employed to allocate the storage of routing table entry for given bank count. Then the search of destination IP address can be classified and confined to a specific bank, reducing active banks for power saving. Evaluations show >3X improvement in cell density and >70% reduction in search energy with limited overhead in silicon area for bank count of four.

Keywords—TCAM, 2D2R, Crossbar, Machine Learning, K-means Clustering

I. INTRODUCTION

TCAM has been commercially used in high-throughput network routers to classify and forward internet protocol (IP) packets owing to its fast search capability. However, the speed of TCAM comes at the cost of increased silicon area and power consumption [1]. SRAM-based TCAMs are hardly acceptable due to poor density with 10T or 12T and prominent standby power [2]. The search power of TCAM is also highly remarkable with even several Watts for a single chip [3].

To address the density issue, emerging resistive memory such as MRAM, PCM and ReRAM have been widely investigated to design smaller nvTCAM cells. However, these cells still need transistors which generally reside on the silicon substrate, resulting in large cell area and excluding the possibility of 3D stacking [4-10]. Several cells depend on voltage division between loading transistor and storage resistor to realize search operation [3-5, 7]. Since loading transistor and storage resistor have different process-voltage-temperature-aging variability, the voltage division may suffer large unfavorable fluctuation, severely impairing the yield.

The power issue in TCAM comes from the fact that one search calls for all banks/rows being activated [1]. Fig. 1 shows that the conventional routing table allocates the entry storage by CIDR (classless inter-domain routing) prefix length. To perform search operation, all banks are searched simultaneously for the input destination IP address, causing large power consumption. Otherwise, the banks are searched in turns as needed, reducing power consumption but with large search delay overhead. Although some circuit or architectural

techniques such as split match line, ripple precharge, dual-supply voltage and statistical storage allocation help to alleviate the power issue, the overheads in search delay, control complexity, and throughput are still notable [11-14].

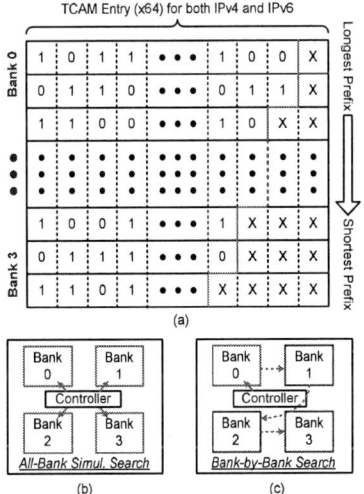

Fig. 1 Conventional entry storage allocation and search methods in routing table. (a) Entry storage is allocated by CIDR prefix length; (b) All banks are searched simultaneously for the input destination IP address, resulting in considerable power consumption; (c) The banks are searched in turns as needed, reducing power consumption but with large search delay overhead. Bank count of four is taken as an example for illustration.

In this work, two techniques are proposed to deal with the density and power issues in TCAM. Firstly, a novel 2D2R crossbar-based nvTCAM cell structure is proposed to improve the density by >3X. Because the diode and the ReRAM resistor can both be integrated in the back-end-of-line (BEOL), the 2D2R nvTCAM provides the potential of 3D stacking, meaning even higher density. Besides, the non-volatility of nvTCAM also saves the standby power. Secondly, the machine-learning concept is exploited in TCAM entry storage and search operations. By using K-means clustering in an unsupervised learning mode, the data characteristics in the TCAM entries are learned and the cluster centers are acquired for given bank count. Then the entry storage can be allocated to a specific bank based on the Euclidean distance from the learned cluster centers. The search of destination IP address can also be classified and targeted to a specific bank, avoiding activating all the banks.

978-1-5386-6414-8/18 $31.00 © 2018 IEEE

II. CROSSBAR 2D2R NVTCAM FOR HIGH DENSITY

A. 2D2R nvTCAM Cell

Fig. 2 (a) shows the proposed crossbar nvTCAM array architecture. Each nvTCAM cell consists of two 1D1R cells [15], which lie at the intersections of one match line, e.g., ML0, and two complementary search lines, e.g., SL0 and SLB0. The diode helps to restrain the disturbance of sneaking current to write and search operations and the ReRAM resistors feature unipolar set and reset operations. Fig. 1(b) shows the I-V curve of 1D1R cell. Fig. 1(c) shows the truth table for storing "0", "1" and "X" where LRS and HRS stand for low resistance state and high resistance state, respectively. Both diode selector and ReRAM resistor are integrated between adjacent metal layers, i.e. BEOL, supporting 3D stacking for higher density. Fig. 1(d) gives the layout of 2D2R cell measuring only $16.3F^2$, less than 1/3 of the state-of-the-art (Fig. 3). Owing to the configuration of two ReRAM resistors for storage, the 2D2R nvTCAM cell provides as sufficient sense margin against variations and leakage as 2T2R. Moreover, the two-bit encoding is also applicable to improve the sense margin [10].

B. Storage Operation

Fig. 4 shows the writing conditions for storage operation. The writing voltages are applied on MLs and SLs. The data is written into the TCAM array as routing table entry in two steps. Taking the writing of "0" as an example, the left ReRAM resistor is set to LRS in the first step and the right ReRAM resistor is reset to HRS in the second step. Vset (Vreset) is the set (reset) voltage of RRAM resistor and Vt is the threshold voltage of diode selector. Proper biasing voltages are also needed for the half-select and none-select SLs and MLs to restrain write disturbance and leakage current. The process for writing "1" or "X" is similar. Multiple ReRAM resistors in the same row can be written concurrently.

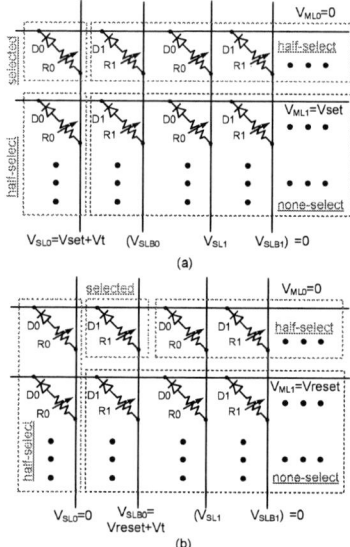

Fig. 4 Writing conditions for storing data "0": (a) first step to set the left ReRAM resistor; (2) second step to reset the right ReRAM resistor.

C. Search Operation

Fig. 5 shows the conditions and timing for search operation. The destination IP address is applied on SLs in complementary voltages. All MLs are first discharged to ground for equalization, then a mismatched ML is charged to a moderate voltage while a matched ML is only charged to a much lower voltage. Eliminating the precharge of highly capacitive match lines to a high voltage helps to save search power [4-10].

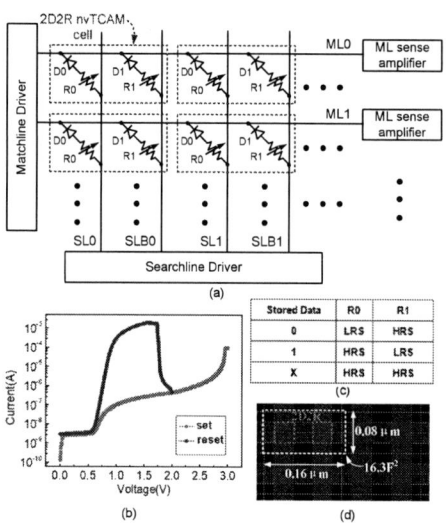

Fig. 2 Proposed crossbar-based 2D2R nvTCAM (a) array architecture, (b) IV curve of 1D1R, (b) truth table, and (c) cell layout.

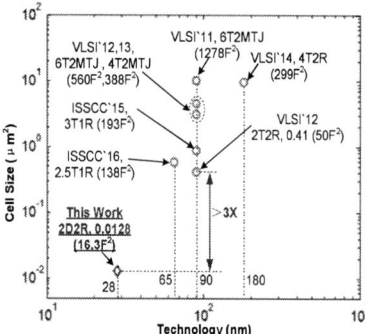

Fig. 3 Cell size comparison with the state-of-the-art. The cell size of 2D2R nvTCAM is at least 3X smaller.

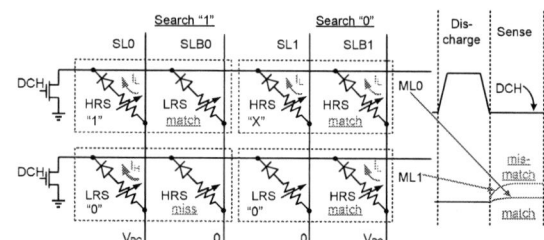

Fig. 5 Conditions and timing for search operation. All MLs are first discharged to ground, then a mismatched ML is charged to a moderate voltage while a matched ML is only charged to a low voltage for sense.

III. K-MEANS CLUSTERING FOR POWER REDUCTION

A. K-means Clustering for TCAM

Fig. 6 shows the proposed K-means clustering-assisted entry storage allocation and search methods. The CIDR prefix is first clustered using K-means algorithm to get the cluster centers, which are then registered in the TCAM controller. The clustering is performed periodically offline for updates. Online clustering is also possible using the TCAM controller with 2D2R array configured as working memory. From the latest BGP (Border Gateway Protocol) statistics shown in Fig. 7, it is known that almost all the CIDR prefix count for IPv4 and IPv6 exceeds 16 bits [16]. Therefore, the CIDR prefix (destination IP address) is classified by its first two numbers or bytes based on Euclidean distance in the controller before storage (search). A CIDR prefix (destination IP address) belongs to the same group with the cluster center from which the Euclidean distance is smallest. Different groups are

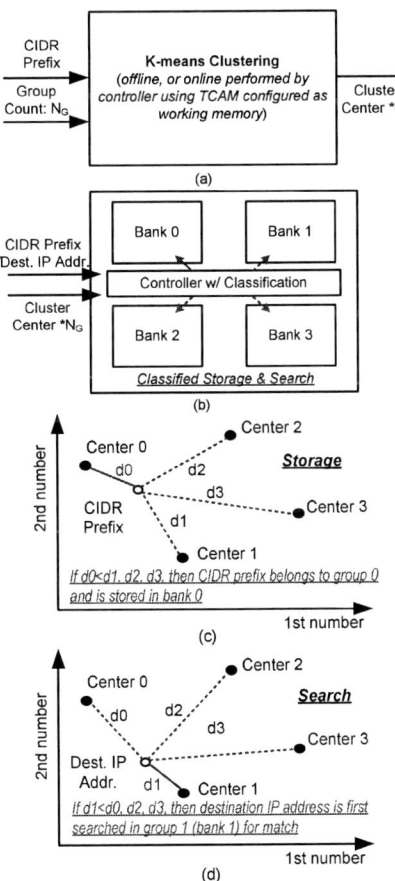

Fig. 6 Proposed K-means clustering-assisted entry storage allocation and search methods. (a) CIDR prefix is clustered using K-means algorithm to get the cluster centers, which are then registered in the TCAM controller; (b) CIDR prefix and destination IP address are classified based on Euclidean distance in the controller before storage and search respectively; (c) CIDR prefix storage allocation and (d) destination IP address search methods. Different groups are assigned to separate banks, and a miss at first search is followed by searching other banks.

assigned to separate banks. The CIDR prefix storage in the same bank still follows the conventional art and is allocated by its length. When a destination IP arrives, a miss at first search attempt is followed by searching other banks.

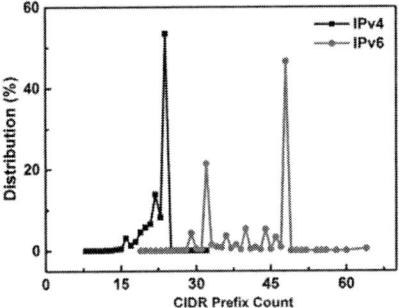

Fig. 7 Statistics for IPv4 and IPv6 CIDR prefix count from latest BGP reports

B. Simulation

Table I shows the clustering results for part of CIDR prefix and destination IP address for IPv6 in TCAM after using K-means clustering according to the distribution in Fig. 7. The case for IPv4 is similar. The storage allocation of CIDR prefix depends on not only its prefix length, but also its Euclidean distance from the cluster center of the specific bank. Because the miss rate for first search is low, as shown in Fig. 8(a), the average count of banks activated for each input destination IP address is almost always one. The clustering time increases remarkably with the group count, as shown in Fig. 8(b), impairing quick routing table update. Thus, the bank/group count of four is selected for implementation in this work. Fig. 9 shows that the average energy for search of each input destination IP address is reduced by >70% and >55% compared to conventional all-bank simultaneous search and bank-by-bank search, respectively.

TABLE I. CLUSTERING RESULTS OF CIDR PREFIX AND DESTINATION IP ADDRESS FOR IPv6 IN TCAM AFTER USING K-MEANS CLUSTERING

CIDR Prefix (IPv6, int8)								Bank	Destination IP Address (IPv6, int8)								Bank
140	183	110	86	118	59	0	0	1	183	110	86	118	59	129	26	140	1
248	250	205	185	139	39	0	0	1	250	205	185	139	39	84	139	248	1
239	159	43	144	132	230	0	0	1	159	43	144	132	230	189	206	239	1
140	255	245	171	190	180	0	0	1	255	245	171	190	180	131	187	140	1
142	144	213	72	128	0	0	0	1	144	213	72	130	186	138	142	142	1
205	220	175	133	0	0	0	0	1	220	175	133	116	75	123	12	205	1
198	163	44	153	0	0	0	0	1	163	44	153	1	241	246	35	198	1
195	183	32	228	0	0	0	0	1	183	32	228	204	231	212	196	195	1
211	242	40	72	0	0	0	0	1	242	40	74	165	200	14	72	211	1
177	222	117	144	0	0	0	0	1	222	117	146	163	208	106	15	177	1
63	209	92	32	184	78	0	0	2	209	92	32	184	78	133	25	63	2
40	233	24	32	242	228	0	0	2	233	24	32	242	228	132	244	40	2
36	198	163	89	153	216	0	0	2	198	163	89	153	216	172	141	36	2
104	255	74	227	33	32	0	0	2	255	74	227	33	47	225	71	104	2
125	172	200	19	105	64	0	0	2	172	200	19	105	67	2	138	125	2
42	170	31	148	40	0	0	0	2	170	31	148	44	221	215	126	42	2
114	177	52	231	200	0	0	0	2	177	52	231	207	85	125	92	114	2
121	192	253	233	0	0	0	0	2	192	253	233	33	119	114	46	121	2
81	251	117	119	0	0	0	0	2	251	117	119	192	96	69	131	81	2
97	137	233	210	0	0	0	0	2	137	233	211	252	47	87	176	97	2
238	76	248	98	116	85	0	0	3	76	248	98	116	85	57	80	238	3
158	61	115	37	199	149	0	0	3	61	115	37	199	149	142	201	158	3
133	118	77	109	74	114	0	0	3	118	77	109	74	114	125	28	133	3
138	68	77	47	123	94	0	0	3	68	77	47	123	94	221	74	138	3
180	18	105	198	77	24	0	0	3	18	105	198	77	24	190	11	180	3
220	8	220	166	219	0	0	0	3	8	220	166	219	89	121	72	220	3
147	70	66	105	150	0	0	0	3	70	66	105	150	3	80	172	147	3
145	51	4	44	116	0	0	0	3	51	4	44	116	44	136	23	145	3
249	62	226	12	32	0	0	0	3	62	226	12	37	111	197	23	249	3
175	23	219	60	0	0	0	0	3	23	219	60	202	249	50	216	175	3
115	23	110	96	213	255	0	0	4	23	110	96	213	255	122	224	115	4
51	63	34	222	83	205	0	0	4	63	34	222	83	205	150	11	51	4
26	88	46	175	248	145	0	0	4	88	46	175	248	145	231	83	26	4
30	53	83	237	210	106	0	0	4	53	83	237	210	106	212	178	30	4
34	13	183	198	102	46	0	0	4	13	183	198	102	46	89	243	34	4
53	88	188	62	77	70	0	0	4	88	188	62	77	70	136	27	53	4
38	37	127	237	165	38	0	0	4	37	127	237	165	38	95	213	38	4
17	41	60	154	193	46	0	0	4	41	60	154	193	46	165	207	17	4
127	27	150	91	64	0	0	0	4	27	150	91	71	154	225	146	127	4
122	11	223	32	0	0	0	0	4	11	223	32	234	31	197	137	122	4

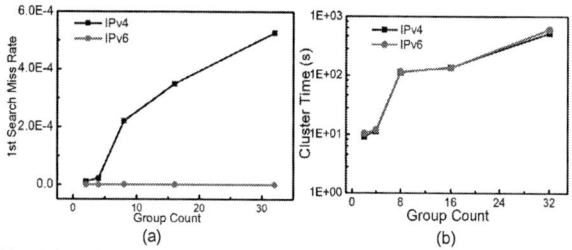

(a) (b)

Fig. 8 (a) Miss rate for first search is low for different group count, so the average banks activated for each input destination IP address is almost always one; (b) Clustering time increases remarkably with group count, impairing quick routing table update.

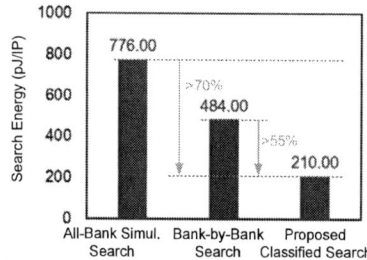

Fig. 9 Comparison of search energy for each input destination IP address with conventional methods.

IV. TEST CHIP

A 16 Kb nvTCAM test chip with proposed 2D2R nvTCAM cell and K-means clustering-assisted storage allocation and search has been demonstrated in 28 nm logic process. Fig. 10 shows the chip micrograph and key features. The area overhead by K-means clustering in the nvTCAM controller is ~2% and becomes neglected for larger TCAM capacity. The search delay is 2ns at nominal VDD, which is comparable with the state-of-the-art [4, 10], and reliable search can be performed for 45% reduction of VDD, as shown in Fig. 11.

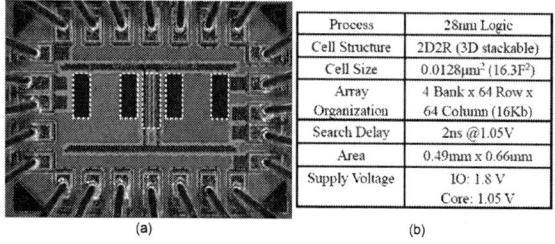

(a)

Process	28nm Logic
Cell Structure	2D2R (3D stackable)
Cell Size	0.0128μm² (16.3F²)
Array Organization	4 Bank x 64 Row x 64 Column (16Kb)
Search Delay	2ns @1.05V
Area	0.49mm x 0.66mm
Supply Voltage	IO: 1.8 V Core: 1.05 V

(b)

Fig. 10 Micrograph of fabricated nvTCAM test chip and its key features.

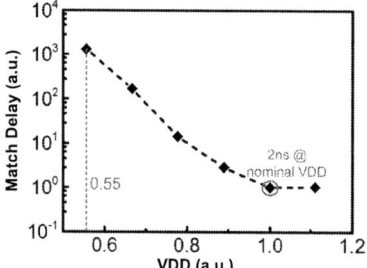

Fig. 11 Search delay at nominal VDD and its relation with VDD.

V. CONCLUSION

Crossbar-basesd 2D2R nvTCAM cell and K-means clustering-assisted storage allocation and search are proposed to deal with the density and power issues in TCAM. 2D2R cell achieves a record of smallest cell and provides 3D stackability. K-means clustering makes search operation more targeted, reducing the average count of active banks and saving energy.

ACKNOWLEDGMENT

This work was supported by STCSM (17ZR1446800), NSFC (61704029, 61874028, 61522408, 61521064) and the MOST of China under Grant (2016YFA0203800). [+]These authors contributed to this work equally.

REFERENCES

[1] K. Pagiamtzis and A. Sheikholeslami, "Content-addressable memory (CAM) circuits and architectures: a tutorial and survey," *IEEE Journal of Solid-State Circuits (JSSC)*, vol. 41, no. 3, pp. 712-727, 2006.

[2] C. Sungdae, *et al.*, "A 0.7-fJ/bit/search 2.2-ns search time hybrid-type TCAM architecture," *IEEE Journal of Solid-State Circuits (JSSC)*, vol. 40, no. 1, pp. 254-260, 2005.

[3] F. Shafai, *et al.*, "Fully parallel 30-MHz, 2.5-Mb CAM," *IEEE Journal of Solid-State Circuits (JSSC)*, vol. 33, no. 11, pp. 1690-1696, 1998.

[4] M. F. Chang, *et al.*, "A 3T1R nonvolatile TCAM using MLC ReRAM with Sub-1ns search time," in *IEEE International Solid-State Circuits Conference (ISSCC)*, 2015, pp. 318-319.

[5] C. C. Lin, *et al.*, "A 256b-wordlength ReRAM-based TCAM with 1ns search-time and 14X improvement in wordlength-energyefficiency-density product using 2.5T1R cell," in *IEEE International Solid-State Circuits Conference (ISSCC)*, 2016, pp. 136-137.

[6] S. Matsunaga, *et al.*, "A 3.14 um² 4T-2MTJ-cell fully parallel TCAM based on nonvolatile logic-in-memory architecture," in *Symposium on VLSI Circuits (VLSIC)*, 2012, pp. 44-45.

[7] S. Matsunaga, *et al.*, "Fully parallel 6T-2MTJ nonvolatile TCAM with single-transistor-based self match-line discharge control," in *Symposium on VLSI Circuits (VLSIC)*, 2011, pp. 298-299.

[8] H. Li-Yue, *et al.*, "ReRAM-based 4T2R nonvolatile TCAM with 7x NVM-stress reduction, and 4x improvement in speed-wordlength-capacity for normally-off instant-on filter-based search engines used in big-data processing," in *Symposium on VLSI Circuits (VLSIC)*, 2014, pp. 99-100.

[9] S. Matsunaga, *et al.*, "Fabrication of a 99%-energy-less nonvolatile multi-functional CAM chip using hierarchical power gating for a massively-parallel full-text-search engine," in *Symposium on VLSI Circuits (VLSIC)*, 2013, pp. 106-107.

[10] J. Li, *et al.*, "1Mb 0.41 μm² 2T-2R cell nonvolatile TCAM with two-bit encoding and clocked self-referenced sensing," in *Symposium on VLSI Circuits (VLSIC)*, 2013, pp. 104-105.

[11] S. Baeg, "Low-Power Ternary Content-Addressable Memory Design Using a Segmented Match Line," *IEEE Transactions on Circuits and Systems I: Regular Papers (TCAS I)*, vol. 55, no. 6, pp. 1485-1494, 2008.

[12] D. S. Vijayasarathi, *et al.*, "Ripple-precharge TCAM: a low-power solution for network search engines," in *International Conference on Computer Design*, 2005, pp. 243-248.

[13] T. S. Chen, *et al.*, "Filter-based dual-voltage architecture for low-power long-word TCAM design," in *International Conference on Intelligent Green Building and Smart Grid (IGBSG)*, 2016, pp. 1-5.

[14] F. Basci and T. Kocak, "Statistically partitioned, low power TCAM," in *IEEE Northeast Workshop on Circuits and Systems (NEWCAS)*. 2004, pp. 129-132.

[15] T. y. Liu, *et al.*, "A 130.7-mm² 2-Layer 32-Gb ReRAM Memory Device in 24-nm Technology," *IEEE Journal of Solid-State Circuits (JSSC)*, vol. 49, no. 1, pp. 140-153, 2014.

[16] BGP table statistics [Online]. Available: http://bgp.potaroo.net/

An Enhanced Built-off-Test Transceiver with Wide-range, Self-calibration Engine for 3.2 Gb/s/pin DDR4 SDRAM

Joung-Wook Moon[1], Hye-Sung Yoo[1], Hundai Choi[1], Il-Won Park[2], Seok-Yong Kang[1], Jun-Bae Kim[1], Haeyoung Chung[1], Kiho Kim[1], Dong-Hun Lee[1], Ki-Jae Song[2], Seok-Hun Hyun[1], Indal Song[1], Young-Soo Sohn[1], Yong-Ho Cho[1], Jung-Hwan Choi[1], Kwang-Il Park[1], and Seong-Jin Jang[1]

[1]Samsung Electronics, DRAM Design Team, Device Solution, Hwasung-si, Gyeonggi-do, Korea
[2]Samsung Electronics, Test Engineering Team, Test & Package Center, Asan-si, Chungcheongnam-do, Korea
austin.moon@samsung.com

Abstract— This paper presents a wide-frequency-range, self-calibrating, built-off-test (BOT) transceiver for a DDR4 SDRAM interface. The proposed BOT transceiver consists of a data transceiver, a phase-locked loop, a delay-locked loop, a self-timing calibration (STC) circuit with a 90-degree phase shifter, and a self-voltage calibration (SVC) circuit. In particular, with both the STC and SVC circuits, data channel skew is effectively compensated, and the voltage margin can be maximized while the DDR4 SDRAM is communicating with the test equipment. The BOT transceiver is realized in 20-nm DRAM CMOS technology. Using the fabricated BOT transceiver, we have successfully demonstrated 3.2-Gb/s/pin testing for DDR4 SDRAM with a 1.0-V supply voltage.

Keywords—Built-off-test (BOT), DDR4 SDRAM, self-timing calibration (STC), self-voltage calibration (SVC).

I. INTRODUCTION

There is a growing demand for high-performance DRAM to have a wider data bandwidth and larger data capacity. DDR4 SDRAM [1] operating up to 3.2 Gb/s/pin was developed to meet these requirements, and it is still expected to have a dominant market share in the semiconductor industry in the near future [2]. As the DRAM data bandwidth increases, the importance of high-performance device testing is also greatly increasing. Tests on high-performance devices traditionally require a considerable amount of facility investment mainly due to the increasing cost of automatic test equipment (ATE) [3]. In addition, it is becoming more difficult to generate the proper signal when communicating with DRAM over multiple I/O channels.

To overcome these problems, several studies on built-in self-test (BIST) circuits have been reported [4, 5]. With these circuits, the test burden for high-speed memory can be effectively reduced. However, the design-for-test circuitry increases the size and complexity of the device-under-test (DUT), and it produces results with relatively low accuracy between the actual high-speed I/O circuit and the test circuit of the DUT.

Recently, a new at-speed test solution with a built-off-test (BOT) chip that can relieve the ATE speed burden has been

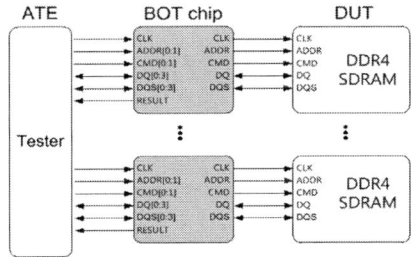

Fig. 1. Simplified block diagram of the proposed memory test system.

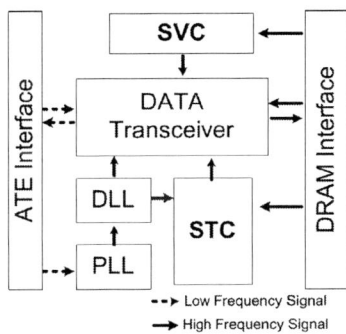

Fig. 2. Block diagram of the proposed built-off test (BOT) transceiver.

reported [6, 7]. The BOT chip functions as a bridge between an ATE and a DUT. It receives low-speed ATE signals and delivers high-speed signals to the DUT. However, there are limitations, such as data channel skew and clock data synchronization for various signal termination conditions.

In this paper, we report a BOT transceiver that includes a self-timing circuit and a self-voltage calibrated circuit. The proposed BOT transceiver minimizes data communication error between two different frequency sides and transmits seamless data. Thus, reliable test results between the ATE and DDR4 SDRAM can be achieved. The remainder of this paper is organized as follows. Section II presents the architecture of the proposed memory test solution and explains the component circuits in detail. Section III presents the measurement results. Finally, conclusions are given in Section IV.

978-1-5386-6414-8/18 $31.00 © 2018 IEEE

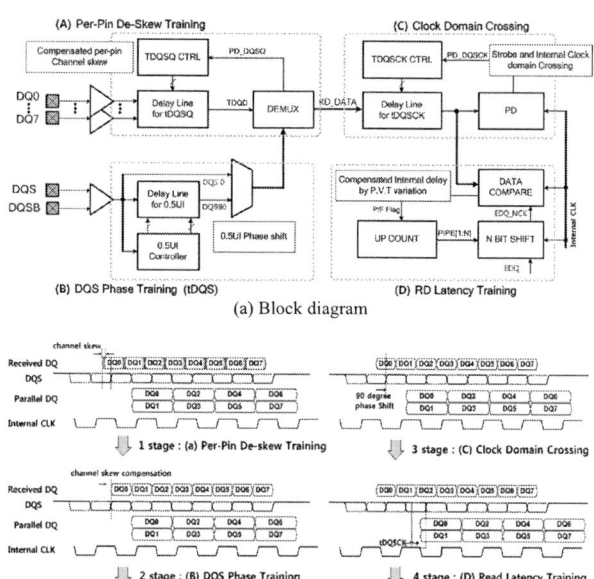

(a) Block diagram

(b) Timing diagram

Fig. 3. (a) Block diagram and (b) timing diagram of each calibration steps.

II. PROPOSED MEMORY TEST ARCHITECURE

Fig. 1 presents the proposed memory test system. It comprises ATE, BOT transceivers, and DUTs. The ATE generates low-speed test patterns, and the generated patterns are transmitted to the multiple BOT transceivers located between the ATE and DUTs. The low-speed test patterns are converted and aligned into high-speed test patterns inside a BOT transceiver to determine whether the DUT passes or fails. Fig. 2 shows a detailed block diagram of the proposed BOT transceiver architecture. It is composed of a phase-locked loop (PLL), a delay-locked loop (DLL), and a data transceiver. In particular, a self-timing calibration (STC) circuit and a self-voltage calibration (SVC) circuit are used to provide reliable interface communication.

Since the ATE operates at 50 to 400 MHz and the DDR4 SDRAM operates at 200 to 1600 MHz, we use an adaptive-bandwidth charge-pump PLL with an 8-bit controllable integer divider that is able to synthesize up to 3200 MHz clock signal using the reference clock from the ATE. The designed PLL tracks changes in the reference clock and then adjusts the loop-bandwidth of the PLL to achieve low-jitter performance for a wide operating range.

A. Self-Timing Calibration (STC) Circuit

Fig. 3(a) shows a block diagram of the STC circuit, and (b) shows a timing diagram of each calibration step. This calibration circuit has the following four distinct training states: (a) removal of time skew between data (DQ) channels, (b) 90-degree phase training between data strobes (DQS, DQS#) and DQ signals, (c) DQS to internal clock (CLK) training to prevent timing uncertainty of in-DRAM operation,

and (d) read-latency training to compensate latency difference between the BOT transceiver and the DRAM.

In the first self-calibration step, the STC circuit receives toggle-data patterns (01010101 or 00110011) from the DRAM using the "MPR Read" function. While DQ data is sent to the BOT receiver, each DQ has intrinsically unbalanced channel skew. Thus, calibration should be executed using the DLL, which iteratively delays the received signals by a small increment until the entire DQ window reaches the maximum timing margin.

Basically, read DQ data is transmitted with a DQS strobe, and the accurate 90-degree phase relationship between DQS and DQ is crucial to increase test speed. A 90-degree phase shifter with minimized quantization error is proposed for the second self-calibration step. Fig. 4(a) shows a block diagram of proposed 90-degree phase shifter, and Fig. 4(b) shows a schematic of sub-blocks.

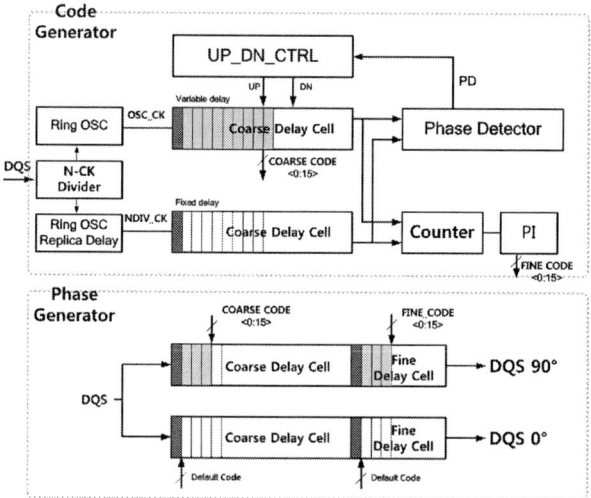

(a) 90-degree phase shifter block diagram

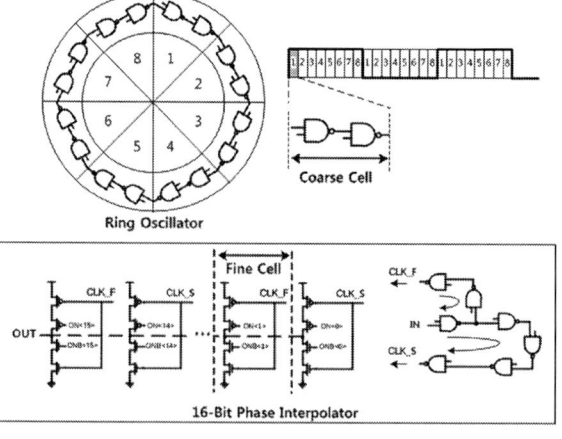

(b) Schematic of the phase shifter sub-blocks

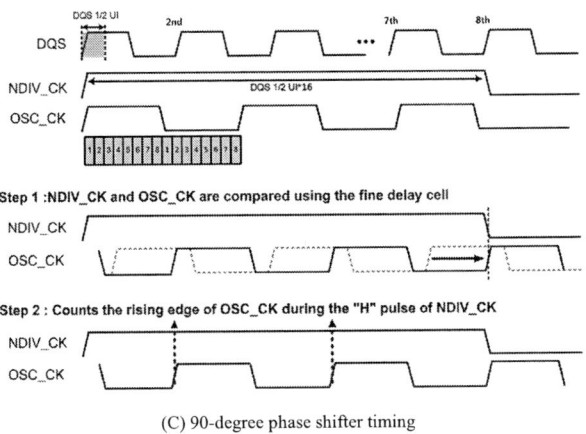

Step 1 :NDIV_CK and OSC_CK are compared using the fine delay cell

Step 2 : Counts the rising edge of OSC_CK during the "H" pulse of NDIV_CK

(C) 90-degree phase shifter timing

Fig. 4. (a) A block diagram, (b) schematic of sub-circuits and (c) timing diagram of proposed 90-degree phase shifter.

It consists of a ring-type oscillator having 8-unit delay cells, a divider, a counter, a phase detector, a phase interpolator, and two delay chains, consisting of coarse cells and fine cells. Fig. 4(c) shows a timing diagram of the circuit. An NDIV_CK pulse is generated that is enabled at the first rising edge of the DQS strobe and disabled at the eighth rising edge of the DQS strobe. This pulse is 16 times as wide as the 1/2UI of the DQS and is used as an indicator for the 90-degree shifter circuit. A ring-type oscillator is implemented, and the half frequency of the oscillator is divided into 8-unit delay cells. The total number of oscillator delay cells must be the same as the number of DQS rising edges counted to generate the NDIV_CK pulse. The 1-unit delay cell is defined as a coarse-delay cell. Each fine-delay cell is implemented as a phase interpolator (PI) with 16-bits. The PI circuit is made by combining the output passing through one coarse cell and two coarse cells, and dividing it into 16 phases. Therefore, the entire delay of 16 fine-delay cells is the same as that of one coarse-delay cell.

First, NDIV_CK and OSC_CK are compared using the fine delay cells. As the fine delay cells increase with increasing PI bit codes, the falling edge of NDIV_CK can detect the rising edge of OSC_CK, and the PI bit codes are stored as a fine-delay code (FINE_CODE). Second, the rising edges of OSC_CK during the "H" pulse of NDIV_CK are counted, and the number of rising edges is stored as a coarse-delay code (COARSE_CODE). Finally, a 0-degree DQS strobe is generated with a delay chain having a default code, and a 90-degree shifted DQS strobe is generated with the COARSE_CODE and the FINE_CODE. In this way, a 90-degree phase shifted DQS can be generated with a resolution of $1/16^2$; thus, the quantization noise can be significantly reduced.

Once the DQS phase training is completed, the read data is sampled and aligned with the DQS strobe. The sampled data is then de-serialized to the internal clock (CLK) after completion of the third self-calibration step, which is DQS to CLK training.

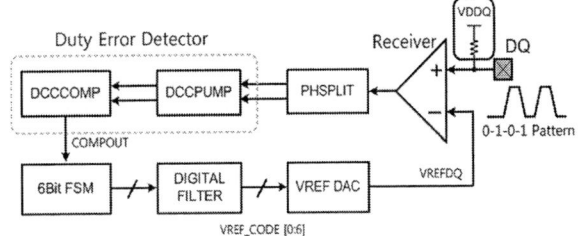

Fig. 5. Block diagram of self-voltage calibration circuit.

The DDR4 SDRAM has a "Read Latency" function defined by clock delay numbers from the read command to actual data output, depending on the clock rate. Thus, in the last self-calibration step, read latency training is performed between the BOT receiver and the DRAM.

B. Self-Voltage Calibration (SVT) Circuit

The DDR4 SDRAM uses a single-ended pseudo open-drain signal topology in which the data is terminated to a supply voltage, VDDQ, and it is driven by the single-ended input with a reference voltage, VrefDQ. The input signal and the reference voltage behave like a differential signal pair whose common-mode level, which is defined as $\Delta V_{cm} = \frac{1}{2}(V_+ - V_-)$, directly influences the output amplitude and duty. In particular, as the circuit operation speed increases, the output amplitude and duty can be easily distorted and further worsened due to the reduced bit-time, resulting in an increased data error rate of the receiver. To mitigate these problems, the SVC circuit is proposed as shown in Fig. 5. It is composed of a receiver, a phase splitter, a duty error detector, a VREF DAC, and a 6-bit finite-state machine (FSM). After the phase of the received data is split, a duty-error is estimated and represented by 6-bit error codes by the FSM. Referring to the 6-bit error codes, the VREF_DAC adequately controls the VrefDQ level of the receiver. Since the VrefDQ of the SVC is continuously controlled in a negative feedback manner, the BOT transceiver can receive the maximized valid data for both timing and voltage margin, regardless of the process, supply voltage, and temperature (PVT) variations.

III. EXPERIMENTAL RESULT

Test chip is fabricated in 20-nm DRAM CMOS technology. It has been packaged and mounted on FR-4 board for measurement. Fig. 6 shows a die microphotograph. The total area is 2.0 mm x 9.0 mm including bonding pads and ESD circuits. The chip has 18-I/O channels to communicate with an ATE and 22-I/O channels to communicate with a DRAM device.

Fig. 7 shows the measured PLL jitter characteristics at 800 MHz and 1600 MHz using 200 MHz and 400 MHz reference clock, respectively. The RMS jitter is 8.36 ps and 6.68 ps corresponding to 0.0067 UI and 0.01 UI, respectively.

Fig. 8 shows the measured results of (a) the DQ per-pin skew, and (b) the DQS to DQ training. All 3.2 Gbps data is properly

Fig.6. Die photograph.

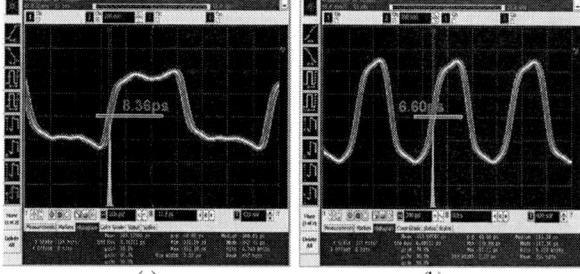

(a) (b)

Fig. 7. Measured PLL jitter performance at (a) 800 MHz and (b) 1600 MHz.

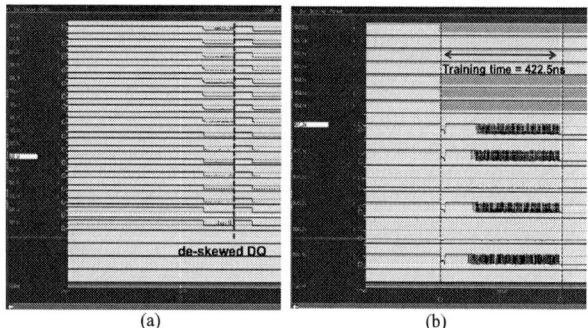

(a) (b)

Fig. 8. Measured results of (a) the DQ per-pin skew and (b) the DQS to DQ training.

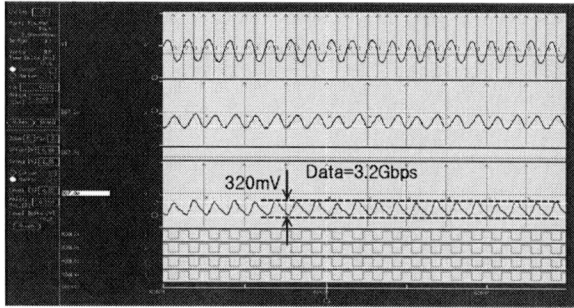

Fig. 9. Measured VrefDQ training result.

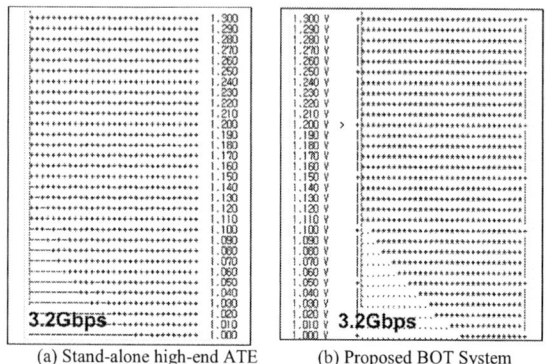

(a) Stand-alone high-end ATE (b) Proposed BOT System

Fig. 10. Comparison of the measured 2-dimensional Shmoo result. (x-axis: tCK, y-axis: VDD)

Shmoo plot with a stand-alone high-end ATE system and the proposed BOT system. This result indicates pass or fail of the DRAM operation while changing the clock rate of the X-axis and VDD of the Y-axis. It proves that the proposed transceiver is successfully testing 3.2 Gbps operating DDR4 with 1.0-V supply voltage.

IV. CONCLUSION

An enhanced BOT transceiver is demonstrated in 20-nm DRAM CMOS technology. With the proposed STC circuits, through several training sequences, data channel skew can be effectively compensated as well as 90-degree DQS strobe can be synchronized. The SVC circuits also maximizes the voltage margin on the receiver side. As a result, the DRAM test reliability can be further improved. We successfully achieve 3.2 Gbps DDR4 test under 1.0-V supply voltage, and the performance of the result is as equivalent as the test result of a stand-alone high-end ATE. Therefore, we believe that the test cost can be dramatically reduced by extending the use of the BOT transceiver and low-end ATE.

REFERENCES

[1] K. Sohn *et al.* "A 1.2 V 30 nm 3.2 Gb/s/pin 4 Gb DDR4 SDRAM with dual-error detection and PVT-tolerant data-fetch scheme", *IEEE J. Solid-State Circuits,* vol. 48, no. 1, pp. 168–177, Jan. 2013.

[2] JEDEC Server Memory Roadmap. [Online]. Available: https://www.jedec.org/sites/default/files/Ricki_Dee_Williams-Final_0.pdf

[3] R. Puhakka, "Cost of test - big driver in ATE," in *Proc. IEEE Int. Test Conf.*, pp. 1, Oct. 2006.

[4] M. Nelms, K. Gorman and D. Anand, "Generating at-speed array fail maps with low-speed ATE," *Proc. VLSI Test Symposium*, pp. 87–92, 2004.

[5] I. Schanstra *et al.*, "Semiconductor manufacturing process monitoring using built-in self-test for embedded memories," *Proc. International Test Conference*, pp. 872–881, 1998.

[6] S. Sunter and A. Roy, "A self-testing BOST for high-frequency PLLs, DLLs, and SerDes," in *Proc. IEEE International Test Conference*, pp. 1–8, 2007.

[7] J. Park *et al.*, "At-speed test of high-speed DUT using built-off test interface," *IEEE Asian Test Symposium*, pp. 269–274, 2010.

aligned and the DQS strobe is also trained in the center of DQ data within 422.5 ns of training time.

The measured VrefDQ training result using the SVC circuit is shown in Fig. 9. The 320-mV swing level, 3.2-Gbps input signals are successfully trained and converted to digital signals.

Fig. 10 shows the comparison test result of a VDD vs. Time

An Ultra-low Power 8T SRAM with Vertical Read Word Line and Data Aware Write Assist

Lu Lu, Taegeun Yoo, Le Van Loi, and Tony Tae-Hyoung Kim

School of Electrical and Electronic Engineering, Nanyang Technological University, Singapore

Email: LLU010@e.ntu.edu.sg

Abstract— **This paper presents an 8T SRAM macro with vertical read word line (RWL) and selective dual split power line techniques. The proposed vertical RWL reduces dynamic power consumption during read operation by charging and discharging only selected read bitlines (RBLs). The data-aware dual split power line enhances the write margin (WM) and the static noise margin (SNM) after combined with vertical write bitlines. The 16kb SRAM test chip in 65nm CMOS technology demonstrates the minimum energy consumption of 0.506 pJ at 0.4 V, and the minimum operating voltage of 0.26 V.**

Keywords— vertical read word line; data-aware; SRAM; low leakage; low power

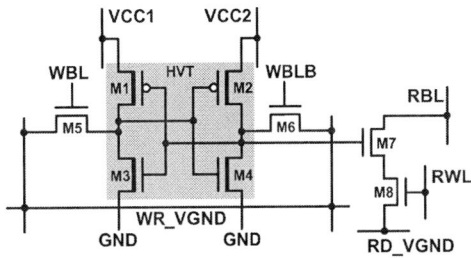

Fig. 1. Schematic of the proposed 8T SRAM cell.

I. INTRODUCTION

With the recent development of portable devices and wearable apparatus requiring long battery lifetime, ultra-low power consumption has been increasingly significant. As a crucial approach to meet ultra-low power and energy, supply voltage scaling has been commonly employed. In conventional 6T SRAMs, the minimum operating voltage is limited by the conflicting requirements from write ability and read stability [1]. In addition, SRAMs operating in the near- or sub-threshold have significantly degraded I_{on}-to-I_{off} ratios and exponentially increased variations, which deteriorates various SRAM design parameters [2]. In this work, two techniques are presented to reduce the power consumption and the leakage current of the proposed SRAM. They are: i) utilizing vertical RWL to reduce the dynamic power consumption during the read operation, and ii) a decoupled 8T SRAM cell with dual split control access and selective boosting cell voltage to improve WM and SNM compared to the conventional 6T or 8T cell. The proposed techniques are implemented on a 16kb SRAM test macro which was fabricated in 65 nm CMOS technology process.

II. PROPOSED SRAM DESIGN

A. Proposed 8T SRAM cell

Fig.1 shows the schematic of the proposed 8T SRAM respectively. The 8T SRAM consists of a cross-coupled latch and the read/write port. The RWL and read bitline (RBL) are controlled vertically (column-based), and the read virtual ground (RD_VGND) is controlled horizontally (row-based). In the write operation, the pass gate transistors are separately controlled by WBL and WBLB while the sources of pull-up PMOSs are biased dependently through the VCC1 and VCC2. The sources of two pass-gate transistors are connected to write virtual ground (WR_VGND) which is running horizontally.

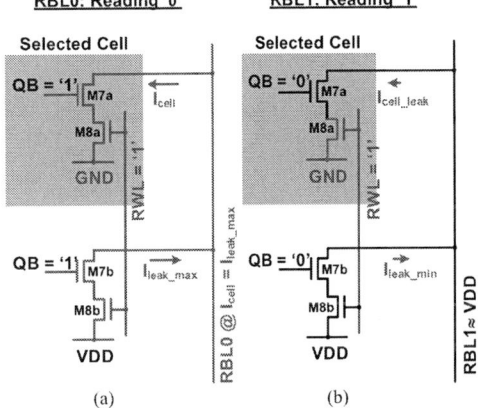

Fig. 2. Principle of the proposed vertical read word line scheme: (a) worst case reading '0' and (b) worst case reading '1'.

The data storage latch in the write port is implemented with HVT devices for leakage reduction while the access paths and read port employ RVT for performance.

B. Read operation with vertical RWL

For the proposed SRAM, the virtual ground keeps high potential in the hold status and helps to eliminate the sub-threshold leakage in transistors. Simultaneously, the dynamic power consumption is saved by reducing the number of RBLs being discharged. As shown in Fig.2 (a), when the reading data is '0' (QB = '1'), the selected cell RD_VGND is 0. RBL is discharged through M7a and M8a, while the half-selected cells in the same column are charging the RBL by the leakage current since their VGND is equal to VDD. If the QB in the half-selected cell is '1', only the leakage current flows through M7b even though the gate of M7b is held at high voltage potential. Because the RBL is always higher than VDD − V_{thn}

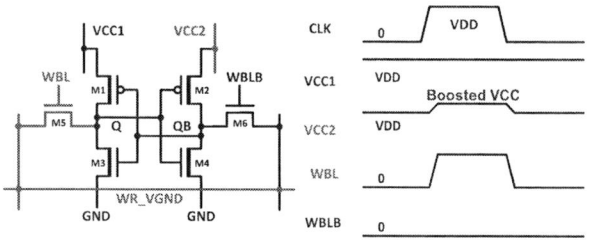

Fig. 3. Principle of the proposed dual split power line technique.

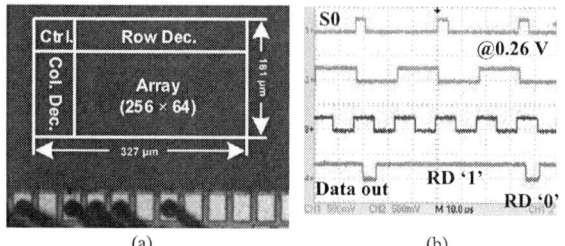

(a)   (b)

Fig. 4. (a) Die photo of the proposed 16 kb SRAM and (b) measured waveform at 0.26V voltage supply.

which results in the V_{gs} of M7b being smaller than V_{thn}. Therefore, only leakage current discharges RBL. When the selected cell data is '1', as shown in Fig.2 (b), only the leakage current is discharging the RBL. At the same time the RBL is charged by leakage currents from half-selected cells, which hold the RBL on the high potential.

The dynamic power is mainly consumed on the discharging and pre-charging of RBLs. During the discharge period, the dissipated dynamic power is close to the conventional one. In the pre-charge period, for the conventional one in the worst scenario, all of the RBLs are discharged and need to be pre-charged to a high potential. However, only the selected column is discharged in one CMUX (16 columns) for the proposed SRAM. In other words, the proposed SRAM saves majority dynamic power during the read period.

C. Write operation with selective dual split power line

The SRAM power consumption in the write operation relies on the minimum functional scaling voltage, which is determined by the stability of the half-selected cells and write ability in the selected cell. During the hold period, VCC1 and VCC2 maintain the potential of VDD, the pass gate transistors are disabled and all the WR_VGNDs are charged to VDD. During write operation, as shown in Fig.3, the selected WR_VGND is discharged to GND. If '0' is written, the WBL is raised, and node Q is discharged. A charge pump circuit is used to generate a boosted voltage on VCC2, which improves the write margin. If '1' is written, the WBLB is enabled, which raises VCC1 beyond VDD. For the half-selected cells, if Q is '0' when writing '0' in the selected cell, the value of Q will be raised. Because in the half-selected rows, the WR_VGND equals to VDD, the current flows into the Q. At the same time VCC2 is raised, QB is also driven up to consolidate the stability of data. If Q is '1', since the pull up transistor M2 is turned off, the rising VCC2 does not degrade the stability significantly.

III. MEASUREMENT RESULTS

The proposed 8T SRAM cell for the ultra-low power consumption application was fabricated in a 65nm CMOS process. The measurement result verified the proposed design works functionally at 0.26 V while the conventional SRAM failed to read '1' (in Fig.4). At 0.26 V, our fabricated chip achieved maximum frequency and power consumption of 60 kHz and 78 nW respectively. The leakage current is reduced significantly by using the virtual ground to eliminate the sub-threshold leakage in the read port. During the extremely low

voltage supply (under 0.35 V), the leakage current is similar to the total current. This means that when the voltage scaled down to sub-threshold level, the proposed design offers super low power consumption. The minimum experimental energy dissipation 0.506 pJ is achieved at 0.4V. As tabulated in Table I, in the ultra-low-voltage supply, the energy is considerably reduced since the active leakage power is similar to the total power consumption.

TABLE I. PERFORMANCE COMPARISON

	JSSC 2013[3]	JSSC 2014[4]	TCAS-I 2015[2]	JSSC 2017[5]	This Work
Technology	65nm	65nm	65nm	28nm	65nm
Density	2kb	128kb	16kb	256kb	16kb
Transistor count	9T	8T	9T	6T	8T
Cell area	$1.24x2.3\mu m^2$	N.A.	$2.63x0.72\mu m^2$	$0.1566\mu m^2$	$1.02x2.23\mu m^2$
VDDmin	0.28V	0.37V	0.26V	0.5V	0.26V
Frequency	330kHz (0.32V)	N.A.	N.A.	20MHz (0.5V)	60kHz (0.26V)
Leakage current	0.05μA (0.4V)	N.A.	1.4μA (0.1V)	120μA (0.5V)	0.23μA (0.26V)
Min. energy	0.57pJ	21.2pJ	2.07pJ	30pJ	0.506pJ (0.4V)
Normalized energy	278aJ/b	162aJ/b	126aJ/b	120aJ/b	31aJ/b

ACKNOWLEDGMENT

The author would like to thank Dr. Sultan Mohiuddin Siddiqui, Dr. Lei Wang and Alan Chang for their support

REFERENCES

[1] W. Choi, et al., "A Refresh-Less eDRAM Macro With Embedded Voltage reference and Selective Read for an Area and Power Efficient Viterbi Decoder," IEEE J. Solid-State Circuits, pp. 2451-2462.Oct. 2015

[2] B. Wang, et al., "Design of an Ultra-low Voltage 9T SRAM with Equalized Bitline Leakage and CAM-assisted Energy Efficiency Improvement," IEEE Transactions on Circuits and Systems-I, Vol. 62, pp. 441-448, 2015.

[3] S. Lutkemeier, "A 65 nm 32 b subthreshold processor with 9T multi-Vt SRAM and adaptive supply voltage control," IEEE J. Solid-State Circuits, vol. 48, no. 1, pp. 2438–2446, Jan. 2013.

[4] Y. Sinangil, et al., "A 128 Kbit SRAM With an Embedded Energy Monitoring Circuit and Sense-Amplifier Offset Compensation Using Body Biasing", IEEE J. Solid-State Circuits, pp. 172-179, January, 2014.

[5] S. L. Wu, et al., "A 0.5-V 28-nm 256-kb Mini-Array Based 6T SRAM With Vtrip-Tracking Write-Assist", IEEE J. Solid-State Circuits, pp. 1791–1802, 2017.

13-1 (8041)

A 0.46V-1.1V Transition-Detector with In-Situ Timing-Error Detection and Correction Based on Pulsed-Latch Design in AES Accelerator

Xinchao Shang[1], Weiwei Shan[1]*, Jiaming Xu[1], Minyi Lu[1], Yiming Xiang[2], Longxing Shi[1] and Jun Yang[1]

[1]Southeast University, Nanjing, Jiangsu, 210096, P. R. China
[2]Spreadtrum Communications, Tianjin, 300300, P. R. China
Email: wwshan@seu.edu.cn

Abstract—To overcome the minimum-delay constraint of latch based error detection and correction (EDAC) techniques, we propose a technique of using pulse latch and transition detector (TD). This method is also advantageous in no need of error recovery by time-borrowing characteristics of the latch. To detect timing violations and minimize the area overhead, we design a quick-response 15-transistor transition detector cover a wide-voltage range from near-threshold voltage (NTV) to Super-Vth. Test chips are fabricated in 28nm CMOS process. Silicon measurements demonstrate that the whole design has achieved up to 64.3% energy saving with 180mV additional voltage scaling, compared to the conventional worst-case design at the expense of 4.3% area overhead.

Keywords—Error detection and correction (EDAC); pulse latch; transition detector; time-borrowing

I. INTRODUCTION

In traditional digital integrated circuit designs, appropriate timing margins are reserved to ensure that the chip can work correctly under process, voltage and temperature (PVT) variations. Especially in near-threshold region, we need to remain larger timing margins due to severer PVT variations. Consequently, these margins result in losses in performance, energy, area and cost.

To address this issue, several adaptive voltage frequency scaling techniques have been proposed. Researchers have focused on the improvement of IC performance or reduce conservative timing margins. The in-situ timing monitoring techniques [1-9] such as Razor II [1], Transition-Detector (TD) based error detection and correction [2-3], Razor-lite [4], error detecting (ED) latch [5], Razor latch [6] and iRazor [8], have been proposed to reduce PVT margins. Razor is a classic EDAC monitor to control supply voltage through in-situ timing-error detection. However, the minimum-delay constraint exists in all Razor systems. To avoid hold-time violations, these approaches based on latch design need to insert a large number of buffers, incurring large area and power overhead.

In this paper, we propose a technique of using pulse latch and transition detector in no need of a recovery mechanism to eliminate worst-case safety margins. The key contributions of this paper are summarized as follows:

1) A quick-response error detection circuit that can operate at near-threshold voltage (NTV) with the detection delay of

This work is supported by the National Natural Science Foundation of China (61574033 and 61774038) and National High Technology Research and Development Program of China (863 Program) (2015AA016601) and the National Science and Technology Major Project (2014ZX01030-101).

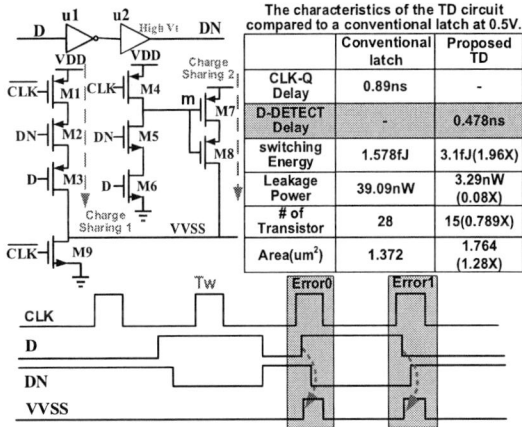

Fig. 1. The transition detector circuit schematic, the operational waveform and its characteristics.

only 0.53 times the CLK-Q delay at 0.5V.

2) Error-detection exploits a low-overhead TD and pulsed-latch based Razor Latch architecture to minimize hold buffer overhead. This method can reduce the number of hold buffers by 39× compared to the latch design (usually with 50% duty cycle clocking).

3) An efficient pulsed-latch utilization approach in physical design based on minimum-cost and maximum-flow is proposed to minimize the cost of pulsed latch.

II. TRANSITION DETECTOR CIRCUIT DESIGN

In this paper, a novel Transition-Detector with 15 transistors is proposed based on current sensing technique, which has a wide-voltage range from NTV to Super-Vth. TDs flag late-arriving transitions during negative phase of the clock as an error in response to the data transition, as shown in Fig .1.

Our 15-transistor TD has several advantages: (1) lower detection delay latency compared to other designs [4][6]. (2) A reliable wide operation range from 0.46V to 1.1V. (3) Low area and energy overheads of 10% and 28% respectively, which has an advantage over a conventional 19T latch with the reset signal.

A. Transition Detector Circuit and Its Working Principle

The detection mechanism is as follows: Because of the inverter (u1) and buffer (u2) delay, when input data D switches

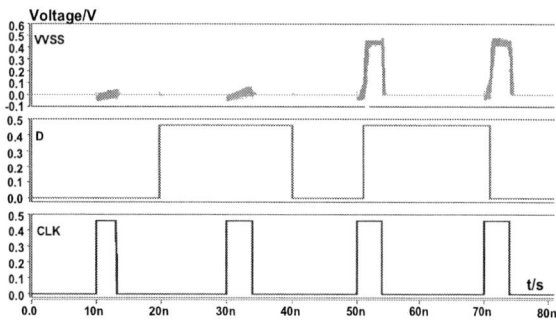

Fig. 2. The full Monte-Carlo SPICE simulation of the proposed transition detectors at 0.46V.

from 0 to 1 (1 to 0), the node DN need a certain delay when it transits from 1 to 0 (0 to 1). While the clock is high, the virtual rail VVSS floats: 1) when the input data D changes from 1 to 0, the node DN needs a certain delay when it transits from 0 to1. In this case, the transistors M1, M2, and M3 are turned on simultaneously in this certain delay. The node VVSS is charged through the PMOS (M1, M2, and M3). 2) When the input data D changes from 0 to 1, the transistors M5 and M6 are turned on simultaneously in a short period of time. The node m is discharged through the NMOS (M5 and M6) and the node VVSS will be charged through the PMOS (M7 and M8). When the clock turns low, M9 is always on and the node VVSS remains low. Thus, by monitoring the node VVSS, an error signal can be generated.

B. Data Race and Leakage Considerations

The TD circuit must consider the possibility of leakage-induced false positive error since the node VVSS is floating during the high clock phase. At the positive clock phase, When D stays low or high, the node m and VVSS are floating in this case. When the voltage of node m is reduced, there will be leakage current from the PMOS (M7 and M8) to the node VVSS. To ensure the correctness of the function, the stacked PMOS structure is adopted to minimize the leakage current. Note that, the VVSS node is refreshed every clock cycle and the pulse width is limited, which can relax restrictions on the leakage problem. To confirm immunity to this potential leakage issue, we performed 5k Monte-Carlo (MC) simulations. As shown in Fig. 2, no leakage-induced false errors were observed down to VDD=0.5V and CLK=50MHz.

C. Low Voltage Operation

This TD is designed to detect late transitions even at low voltage. Some errors may not be occasionally flagged at 0.6V and below, as the voltage of node VVSS lowers with the decrease of VDD. Therefore, the voltage of VVSS should be always larger than the dynamic OR gate threshold voltage to trigger an error. As shown in Fig. 2, MC simulations demonstrate the robustness of VVSS behavior for error detection at 0.46V.

III. Pulsed_Latch+TD Micro-architecture Design

In this design, as shown in Fig. 3, the critical path endpoints are replaced by Latch+TD and TD flags timing errors upon detection of excessive time-borrowing on circuit critical paths.

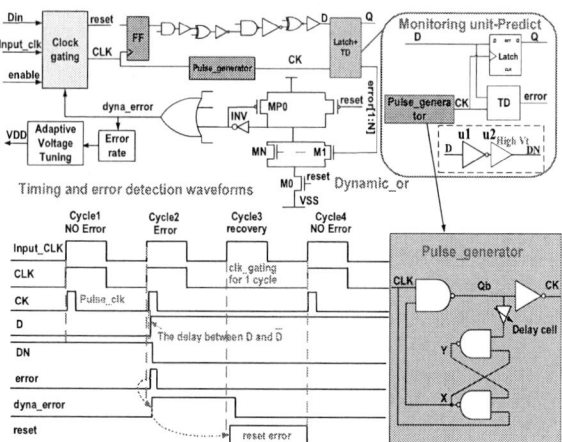

Fig. 3. The schematic, the operational waveform and the characteristics of the proposed transition detector circuit.

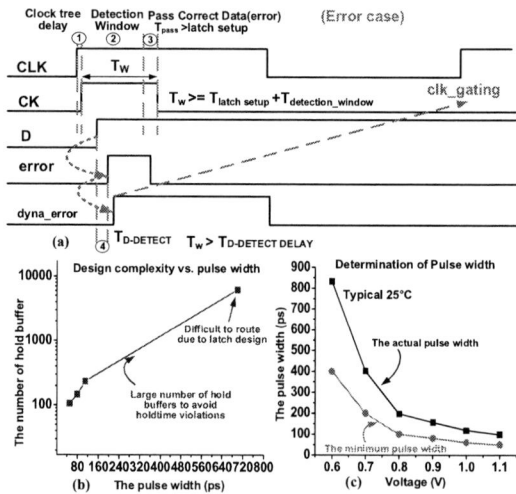

Fig. 4. (a) Timing constraints for correction scheme; (b) design complexity vs. pulse width; (c) determination of pulse width.

The timing diagrams that explain the principle of operation are also shown in Fig. 3. Under normal operation when no error occurs, data arrives before the rising clock edge and the error signal stays low. When an error occurs due to a late-arriving transition, an error signal can be generated as a positive pulse. To reduce propagation delay of the error signal, dynamic OR gates are employed for clustering the error signals from different paths. The error signal is captured by dynamic OR gates and used to control a clock-gating circuit. Thus, the timing error can be prevented through time-borrowing and dynamic clock gating.

A. Razor Latch

Local pulse generators (PG) are used to generate a pulse clock and the width of the pulse is determined by the delay of the pulse generator. To ensure correct error detection, as shown in Fig. 4(a), the pulse width (Tw) must meet a number of constraints: 1) allow valid transitions of D to pass through the

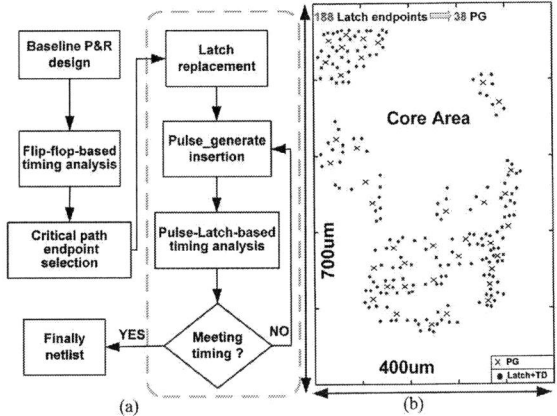

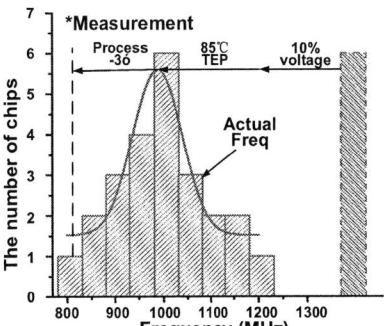

Fig. 7. Max frequencies and baseline test at 1.1V.

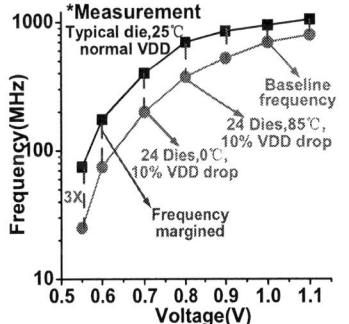

Fig. 8. Measured frequencies and baselines across 0.55-1.1V，where the x-axis is the margined VDD voltage.

(a)   (b)

Fig. 5. (a) The process of Pulsed-latch replacement and (b) the distribution of the shared pulse generator.

Technology node	28nm CMOS
Clock frequency	1.25GHz@1.1V
Die size	1 x 3.04 mm²
AVS supply voltage range	0.46-1.1V
Insertion rate	188/1856 (10.1%) (three 8bit-AES)
Area overhead	4.3%
Hold Buffer Area overhead	1.4%
Power gain (NTV@0.55V)	64.3% (max)

*Test on probecard in foundry

Fig. 6. Test chip platform and design details.

latch; 2) ensure valid error signal to be collected by dynamic OR gates. To assure that the input data is passed correctly, Tpass needs to be greater than the latch setup time. Thus, the minimum pulse width constraint is calculated as:

$$T_W \geq T_{latch\_setup} + T_{detection\_window} \tag{1}$$

$$T_W \geq T_{\text{D-DETECT}} \tag{2}$$

Time-borrowing can enable performance improvements. Additional time-borrowing is enabled by larger PG delay cell at the expense of increasing the minimum-delay constraint. To reduce hold-time constraints, the pulse width is set as small as possible while satisfying the minimum width constraint. Fig. 4(b) presents pulse width and design trade-offs. Plus, clock jitter and clock skew should be considered at the design time. The actual pulse width and the minimum pulse width are shown in Fig. 4(c). Overall, the pulse width can meet design requirements at all considered PVT corners because of the extra timing margin.

B. Pulsed-Latch-Aware Clustering

To reduce the area overhead, an efficient pulsed-latch utilization approach in physical design to minimize the cost of pulsed latch is proposed. As shown in Fig. 5(a), an automated flow is developed.

The design flow starts with a placed and routed baseline design. Then, flip-flops to be latched are selected, based on the tradeoff between the path coverage and the area overhead due

to hold buffers and timing requirement, which are required to guarantee the correctness of the function.

After latch insertion, placement of pulse-generators needs to be optimized. The tolerable load capacitance of a generator should be considered during the pulse-generator insertion. In summary, the following three issues should be considered during pulse-generator insertion: (1) maximization of pulse-generator utilization to drive pulse latches as many as possible; (2) the output load of pulse generators should be maintained at less than the maximum tolerable load capacitance; (3) the distance between a pulse generator and a pulsed latch should be less than the maximum distance.

In this design, one local pulse generator is shared by several nearby pulsed-latches. The distribution of the shared pulse generator is shown in Fig. 5(b). Finally, 38 local pulse generators are inserted.

IV. MEASUREMENT RESULTS

The proposed approach is applied to an 8bit-AES encryption circuit for IoT applications in 28nm CMOS. The system is mainly composed of the 8-bit AES8 circuit, an SPI interface, a PLL and an AVS control module. The test chip platform and implementation details are shown in Fig. 6. Out of a total of 1856 registers in this design, 188 TDs are inserted for timing-error detection. The area overhead due to TDs、PGs and control logic accounts for 4.3% of the total area.

Detailed measurement results are given in this session. To ensure correct operation across all dies and operating

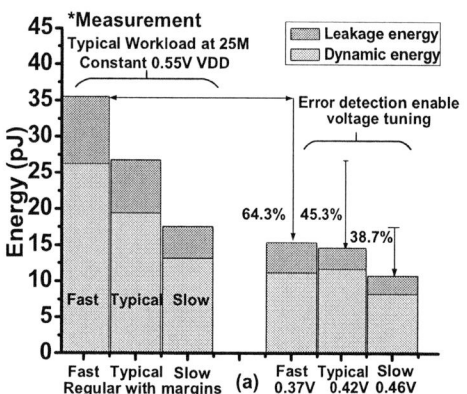

Fig. 9. Measured energy consumption and distribution of energy savings at 0.55V

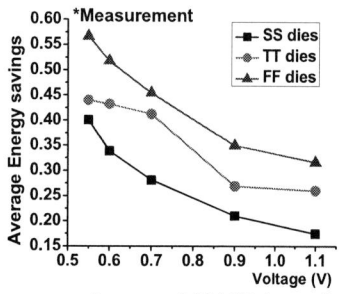

Fig. 10. Average energy savings across 0.55-1.1V, three wafers (SS, TT, FF), where the x-axis is the margined VDD voltage.

conditions, all chips need to operate at the worst-case voltage as in conventional baseline design. In the baseline test, as shown in Fig. 7, we assume a conventional worst case of 85°C (or -20°C for near-threshold voltage), 10% supply droop and 3σ process variation. The 3σ process variation is roughly estimated by choosing the worst die among the 24 total dies. Under these margined conditions, All 24 dies are measured across a wide voltage range under a probe-card in foundry directly and the baseline circuit operates at a maximum clock frequency of 800MHz at 1.1V.

As shown in Fig. 8, The maximum frequencies for correct operation of a typical die are measured and compared with baseline frequencies from 0.55V to 1.1V, indicating a maximum of 3× throughput improvement over the baseline design at 0.55V.

Energy savings compared to margined design for the worst-case condition at 0.55V and 25MHz are measured with results shown in Fig. 9. With error detection enabled voltage tuning, the proposed design improves the max energy consumption by 64.3% with 180mV additional voltage scaling. We also measure the energy saving across a wide voltage range (0.55~1.1V), as the average results of slow, typical and fast die shown in Fig. 10.

Compared with past EDAC approaches as shown in Table I, We use only 15 transistors for the transition detector, which leads to a small area cost with only 28% area overhead over a conventional latch. Plus, our proposed error detection technique is suitable for a wide range voltage from 0.46-1.1V.

TABLE I.   COMPARISON CHART OF PREVIOUS WORKS

	[3]	[5]	[8]	[9]	This work
TYPE	Latch	Latch	Latch	DFF	Latch
Extra # of Transistor	26	19	1.46	46	15
Extra Clock Loading	YES	NO	YES	YES	YES
ED area Overhead	Not reported	25%	4.3%	76.9%	28%
D-DETECT Delay	NO	1.17x (Latch)	1.11x (DFF)	NO	0.53X (Latch)
Vdd Range	0.6-1.0V	0.4-1.1V	0.6-1.1V	0.29-1.1V	0.46-1.1V
Technology	45nm	65nm	40nm	40nm	28nm
Insertion rate	12%	13%	8.6%	5.7%	10.1%
Area overhead	3.8%	9.5%	13.6%	7%	4.3%
Energy saving	22%	59% (max)	45%	26.4%	64.3% (max)

V.   CONCLUSION

In this paper, to address the minimum-delay constraint existed in all Razor systems, circuit-level timing-error detection is achieved using pulse latch and transition detector. To minimize the area overhead, a quick-response transition-detector and an efficient pulsed-latch utilization approach are proposed. These techniques offer a robust solution for error detection and correction by time-borrowing and dynamic clock gating. The test chips can detect timing errors and achieve 3× throughput improvement and a maximum of 64.3% higher energy efficiency at 0.55V over the margined baseline.

REFERENCES

[1] S. Das et al., "Razor II: In situ error detection and correction for PVT and SER tolerance," IEEE J. Solid-State Circuits, vol. 44, no. 1, pp.32–48, Jan. 2009.

[2] K. A. Bowman et al., "Energy-efficient and metastability-immune resilient circuits for dynamic variation tolerance," IEEE J. Solid-State Circuits, vol. 44, no. 1, pp. 49–63, Jan. 2009.

[3] K. A. Bowman et al., "A 45 nm resilient microprocessor core for dynamic variation tolerance," IEEE J. Solid-State Circuits, vol. 46, no. 1, pp. 194–208, Jan. 2011.

[4] I.Kwon et al., "Razor-lite: A light-weight register for error detection by observing virtual supply rails," IEEE J. Solid-State Circuits, vol. 49, no. 9, pp.2054–2066, Sep. 2014.

[5] Seongjong Kim et al., "Variation-Tolerant, Ultra-Low-Voltage Microprocessor with a Low-Overhead, Within-a-Cycle In-Situ Timing-Error Detection and Correction Technique," IEEE J. Solid-State Circuits, vol. 50, no. 6, pp. 1478–1490, June. 2015.

[6] P. N. Whatmough, et al. "A low-power 1-GHz razor FIR accelerator with time-borrow tracking pipeline and approximate error correction in 65-nm CMOS," IEEE J. Solid-State Circuits, vol. 49, no. 1, pp. 84–94, Jan. 2014.

[7] Xinchao Shang et al., "A 0.44V-1.1V 9-transistor transition-detector and half-path error detection technique for low power applications," in Proc. IEEE A-SSCC, Nov. 2017, 205–208.

[8] Y. Zhang et al., "iRazor: Current-based error detection and correction scheme for PVT variation in 40-nm ARM Cortex-R4 PROCESSOR," IEEE J. Solid-State Circuits, vol. 53, no. 2, pp. 619–631, Feb. 2018.

[9] Hans Reyserhove et al., "Margin Elimination Through Timing Error Detection in a Near-Threshold Enabled 32-bit Microcontroller in 40-nm CMOS," IEEE J. Solid-State Circuits, vol. 53, no. 7, pp. 2101–2113, July. 2018.

13-2 (8055)

Ultra-Lightweight 548 – 1080 Gate 166Gbps/W – 12.6Tbps/W SIMON 32/64 Cipher Accelerators for IoT in 14nm Tri-gate CMOS

Himanshu Kaul, Mark Anders, Sanu Mathew, Vikram Suresh,
Sudhir Satpathy, Amit Agarwal, Steven Hsu, Ram Krishnamurthy
Circuit Research Lab, Intel Corporation
Hillsboro, Oregon, USA
himanshu.kaul@intel.com

Abstract— A family of 32b data/64b key SIMON cipher accelerators, each reconfigurable for encrypt/decrypt modes and fabricated in 14nm tri-gate CMOS, is optimized for a range of area and performance targets: (i) a 1080-gate one round/cycle design occupying $136\mu m^2$ die area uses reconfigurable circuits for forward/reverse key generation, (ii) a 752-gate 16 cycles/round bit-serial design with a single round/key logic bit-slice lowers layout area by 38% to $85\mu m^2$, and (iii) latch-based key/text storage for the bit-serial circuit reduces area further by 22% to $66\mu m^2$ at 272 cycles/round for a 548-gate design with measured 11.4Mbps, $209\mu W$, 750mV operation. Ultra-low voltage circuit optimizations enable peak energy efficiency from 166Gbps/W at 380mV for the bit-serial latch-based design to 12.6Tbps/W at 260mV for the parallel, one round/cycle design.

Keywords—SIMON cipher; ultra-lightweight; encryption; IoT

I. INTRODUCTION

Ultra-lightweight cipher circuits are critical components for enabling energy-efficient security on small form factor IoT nodes, energy harvesting devices, and ultra-low-cost hashing accelerators [1-4]. Based on workload/platform constraints, a variety of light-weight cipher accelerators offer trade-offs in area, power, or performance [5-7], with gate counts scaling down to 2090 gates. The SIMON cipher [8] provides opportunities for circuit optimizations to further reduce area and energy cost, enabling sub-1000 gate implementations suited for energy and cost-constrained IoT platforms.

The SIMON block cipher uses bit-wise AND, XOR, and rotate operations, data/key sizes from 32/64 to 128/256, and 32-72 iterative rounds, where each round implements a Feistel stepping with a round-based key (Fig. 1a). The smallest SIMON 32/64 cipher, with 64b keys encrypting 32b data, offers a good trade-off for area/energy cost while providing 64b security to protect bursty IoT traffic with ephemeral keys. This cipher employs 32 rounds, 16b words, 4-round key history, and a 31b round-based constant sequence for generating subsequent 16b round keys (Fig. 1b). Prior SIMON implementations included only synthesis estimates without combined encrypt/decrypt support [8], or were implemented on FPGAs [9] with vastly different area-energy trade-offs.

This paper presents a family of 32b data/64b key SIMON cipher accelerators, each reconfigurable for encrypt/decrypt mode, and optimized for a range of area, power and performance targets. A 16b word-parallel design optimizes

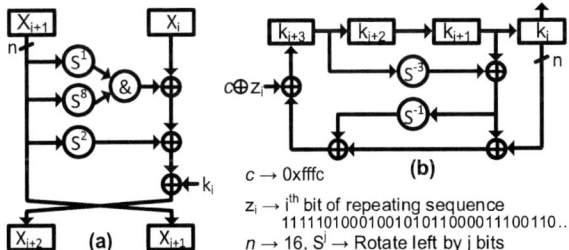

Fig. 1: SIMON 32/64 (a) round and (b) key generation functions [8].

round-based constant generation, key generation, and flip-flop circuits to reduce area and energy. A bit-serial design builds on these improvements by trading off latency to reduce key generation and round logic area. Further storage area optimizations realize a latch-based bit-serial SIMON accelerator. These designs progressively reduce area and performance, resulting in the smallest reported SIMON cipher silicon implementations on 14nm CMOS [10], with gate counts as low as 548 gates and measured sub-threshold operation down to 230mV.

II. WORD-PARALLEL SIMON ACCELERATOR

The word-parallel implementation computes the SIMON 32/64 cipher in 32 cycles (16b word or 1 round per cycle) using 16 bit-sliced key/round circuits, a 5b LFSR to generate the round-based constant sequence z_i, and 96 key/text storage registers (Fig. 2). Adding reconfiguration support for decrypt mode requires key generation in reverse order. In contrast to the conventional approach of reverse key traversal using separate inverse key generation logic, this design uses a unified reconfigurable forward/reverse key logic supporting encrypt/decrypt modes (*enc*) with only two additional multiplexers per key bit-slice (Fig. 3a). The sequence for z_i also needs to be reversed in decrypt mode. Instead of using a separate LFSR for this purpose, a single 5b LFSR with reconfigurable logic (Fig. 3b) generates z_i in the appropriate sequence. These optimizations enable the round keys *KeyNext* to be generated either in forward or backward direction for only 9% total area overhead. Round logic remains unchanged in decrypt mode, with input/output words reversed. Compared to the conventional approach of separate key logic to support encrypt/decrypt modes, this design results in 13% lower area. All storage registers are multi-bit (Fig. 3c), with shared local

978-1-5386-6414-8/18 $31.00 © 2018 IEEE

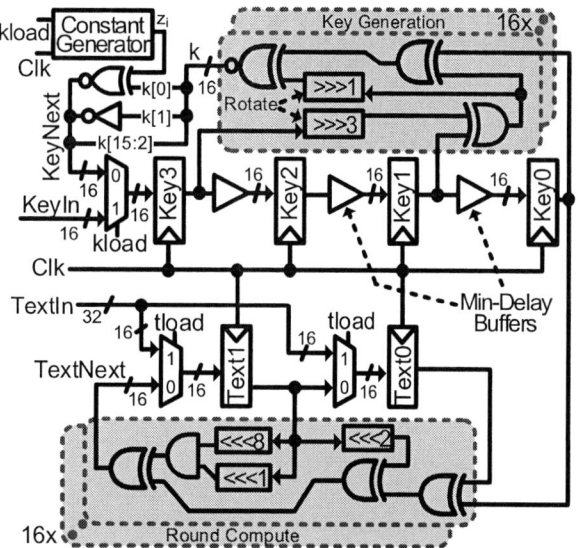

Fig. 2: Word-parallel SIMON accelerator with encryption only.

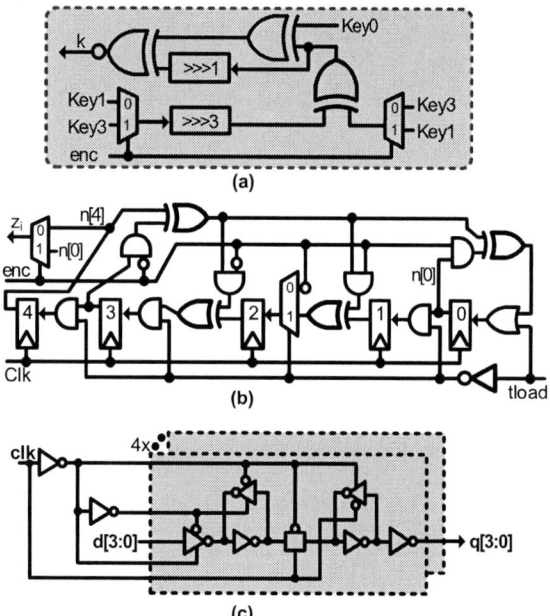

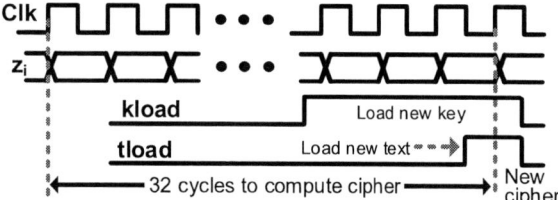

Fig. 3: SIMON circuit optimizations: (a) Reconfigurable key generation logic, (b) Reconfigurable round-based constant generator LFSR, (c) Multi-bit flip-flops.

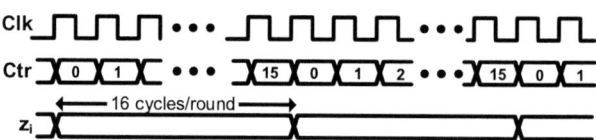

Fig. 4: Parallel SIMON accelerator timing.

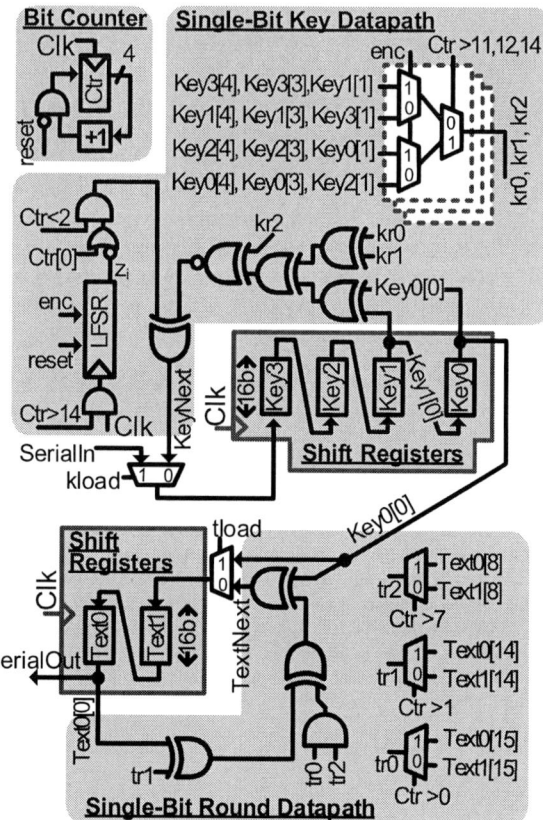

Fig. 5: Bit-serial flip-flop (FF) SIMON cipher accelerator.

Fig. 6: Bit-serial FF SIMON accelerator timing.

clock inverters across multiple flip-flops, to reduce clock power by 38%. Instead of loading the entire 64b key at the start of a new cipher, a 16b/cycle loading sequence for input keys overlaps new key load (*kload*) operation with the last 4 rounds of the current cipher compute (Fig. 4). This enables a reduction in multiplexers for 4% lower area, without impacting the 32-cycle throughput per text/key pair. All these optimizations result in an accelerator gate count of 1080 gates.

III. BIT-SERIAL SIMON ACCELERATOR

The bit-serial SIMON accelerator processes a bit every clock cycle, minimizing combinational gates by reducing key/round logic from 16 slices to a single bit-slice (Fig. 5). Key/text storage is organized as shift registers to enable bit-wise processing with a single logic slice. Loading of input operands is also serialized to reduce the associated multiplexers from 48 to 2. Bit-wise rotate operation on the intermediate 16b value within the key function is enabled by remapping it to the stored key words. Multiplexers select appropriate bits for rotations while accounting for both encrypt/decrypt modes and key/text movement within the shift registers. This selection is based on a 4b counter value (Ctr) tracking the current processed bit within the 16b word. Control circuits enabling bit-serial operation account for only 12% of total area, while key/text storage dominates area at 57%. Since z_i only affects one bit within a 16b word, the re-optimized LFSR is clocked once every 16 cycles for power reduction. Along with the bit counter state, it also controls cipher compute and key/text load operations. The latency for 32b block cipher computation is 512 cycles (Fig. 6). Compared to the 16b word-parallel design, the bit-serial SIMON accelerator reduces total gate count by 30% to a 752-gate implementation.

IV. LATCH-BASED SIMON ACCELERATOR

A latch-based implementation of the area-dominating key/text storage further reduces footprint of the bit-serial SIMON accelerator (Fig. 7). For every forward bit-shift operation, non-overlapping clocks traverse the ripple backwards through stored bits in a ripple fashion. The wavefront of this ripple has duplicated values in adjacent latches, requiring one extra latch for the entire storage. The bit-serial latch-based SIMON accelerator parallelizes the ripple shift at the 16b word level, with an extra latch for every 16b word and 17 clock cycles to process each bit. This ratio balances reduced storage vs. increased clock generation complexity and bit-shift latency. Multi-bit unbuffered latches result in 42% key/text storage reduction, while also eliminating all min-delay buffers. Seventeen serially-asserted local clocks ($lclk_i$) are derived

from the input clock using a 5b incrementer and a one-hot decode circuit to implement the ripple shift. Each local clock is shared by the same bit position in the four key and two round word latches. Compared to a conventional design where a 1 moves through shift registers to generate the serial one-hot clock enable signals, this design lowers local clock generator area by 26%. Even though the single-bit key/round computation remains unchanged from the flip-flop based bit-serial circuit, the datapath is re-optimized to operate in parallel with the multi-cycle bit-shift. Based on the latest arriving input to the key/round datapath during the multi-cycle ripple shift, and also exploiting latch-enabled time borrowing, this optimization enables a relaxed timing constraint of 4.5 cycles/1.5 cycles for key/round computations. The round-based constant LFSR and bit counter use appropriate local clocks $lclk_i$ to update Ctr every 17 cycles and z_i every 272 cycles, with 8704 cycles for cipher computation (Fig. 8). These optimizations further reduce total gate count to 548 gates, a 27% and 49% reduction compared to the bit-serial flip-flop and parallel designs, respectively.

V. 14NM MEASUREMENT RESULTS

All three SIMON 32/64 designs were implemented on a 14nm CMOS test-chip (Fig. 9). The 16b word-parallel design occupies a layout footprint of 136µm² (Fig. 10) with measured 750mV operation of 2.98GHz, 2.98Gbps and 1.35mW total power (Fig. 11). The bit-serial flip-flop implementation reduces layout area by 38% to 85µm², while the latch-based design reduces area by a further 22% to 66µm². Measured 750mV operation for the bit-serial flip-flop and latch-based designs are 3.18GHz, 187.1Mbps, 1.05mW and 3.31GHz, 11.4Mbps, 209µW, respectively. Larger area and lower latency implementations result in higher energy efficiency due to

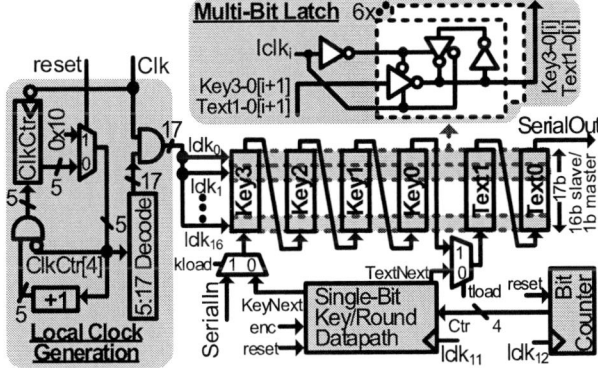

Fig. 7: Bit-serial latch-based SIMON cipher accelerator.

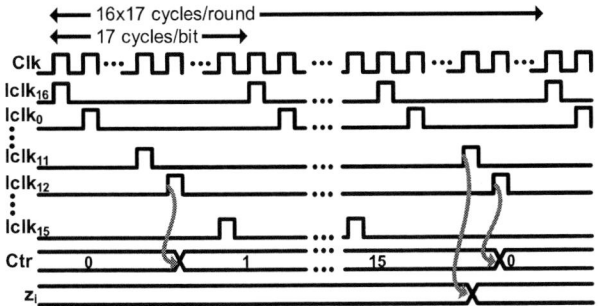

Fig. 8: Bit-serial latch-based SIMON accelerator timing.

Fig. 9: 14nm die micrograph.

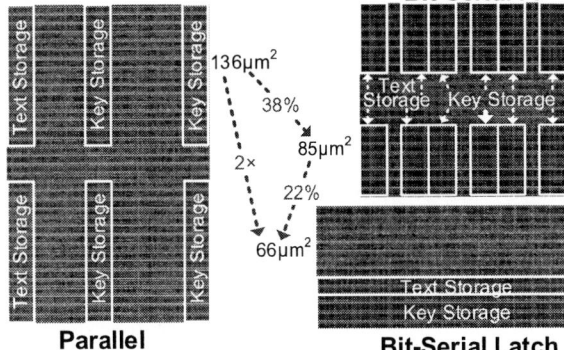

Fig. 10: SIMON 32/64 accelerator layout and area comparisons.

reduced clock switching activity per output bit, ranging from 54.7Gbps/W for the bit-serial latch-based design to 2.21Tbps/W for the parallel design (Fig. 11 and Table I). Ultra-low voltage circuit optimizations enable all 3 accelerators to operate over a wide supply voltage range from 900mV to a 230mV minimum for the bit-serial flip-flop design. Energy efficiency for all designs increases with scaled supply voltages, resulting in a peak of 12.6Tbps/W (5.7× improvement over nominal) at 260mV for the parallel SIMON accelerator. Compared to previously reported light-weight cipher silicon implementations, this work enables gate count reduction of >3.8× (Table II).

VI. SUMMARY

A family of encrypt/decrypt reconfigurable ultra-lightweight SIMON 32/64 accelerators is fabricated in 14nm tri-gate CMOS. The latch-based accelerator delivers smallest layout area, the parallel design enables highest performance, and the bit-serial flip-flop implementation provides an area-efficient trade-off. Across the family, these enable a range of 2× layout area, 261× nominal performance, and 40× nominal energy efficiency. These designs have the smallest reported area and gate count for light-weight cipher silicon implementations.

Table I: 14nm measurements summary at 25°C.

		Frequency, Throughput, Total Power, Energy Efficiency		
		Nominal (750mV)	Peak Performance (900mV)	Peak Energy Efficiency
Parallel		2.98GHz, 2.98Gbps, 1.35mW, 2.21Tbps/W	4.24GHz, 4.24Gbps, 2.82mW, 1.5Tbps/W	2.38MHz, 2.38Mbps, 188nW, 12.6Tbps/W at 260mV
Bit-Serial	FF	3.18GHz, 187.1Mbps, 1.05mW, 179Gbps/W	4.32GHz, 253.8Mbps, 2.08mW, 122.3Gbps/W	2.66MHz, 156kbps, 94nW, 1.67Tbps/W at 240mV
	Latch	3.31GHz, 11.4Mbps, 209µW, 54.7Gbps/W	4.69GHz, 16.2Mbps, 431µW, 37.7Gbps/W	74MHz, 256kbps, 1.54µW, 166.5Gbps/W at 380mV

Table II: Comparison to prior work.

				[5] 22nm AES-128		[6] 40nm AES-128	[7] 28nm PRINCE	
	This Work 14nm SIMON 32/64							
	Parallel	Bit-Serial					with De-glitch	without De-glitch
		FF	Latch					
Encrypt/ Decrypt	Both	Both	Both	Encrypt	Decrypt	Encrypt	Both	Both
Layout Area (µm²)	136	85	66	2200	2736	4290	7404	6927
Gate Count	1080	752	548	1947	2090	2228	N/A	N/A
Nominal Throughput (Mbps)	2982	187.1	11.4	432	671	494	25600	24620
Peak Energy Efficiency (Gbps/W)	12640	1669	166.5	186	289	446	9091	5556

REFERENCES

[1] J. Myers, A. Savanth, R. Gaddh, D. Howard, P. Prabhat, and D. Flynn, "A Subthreshold ARM Cortex-M0+ Subsystem in 65 nm CMOS for WSN Applications with 14 Power Domains, 10T SRAM, and Integrated Voltage Regulator," IEEE Journal of Solid State Circuits (JSSC), vol. 51, no. 1, pp. 31–44, Jan. 2016.

[2] U. Banerjee, C. Juvekar, A. Wright, Arvind, and A. Chandrakasan, "An Energy-Efficient Reconfigurable DTLS Cryptographic Engine for End-to-End Security in IoT Applications," IEEE International Solid State Circuits Conference (ISSCC), pp. 42-43, 2018.

[3] T. Karnik et al., "A cm-Scale Self-Powered Intelligent and Secure IoT Edge Mote Featuring an Ultra-Low-Power SoC in 14nm Tri-Gate CMOS," IEEE International Solid State Circuits Conference (ISSCC), pp. 46-47, 2018.

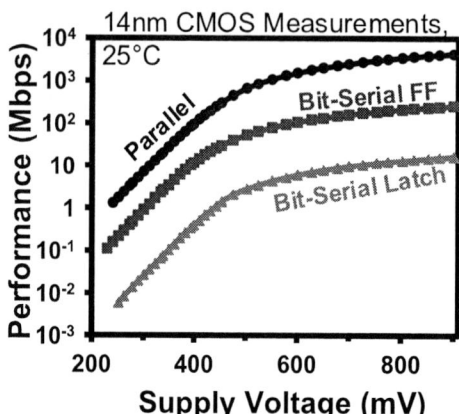

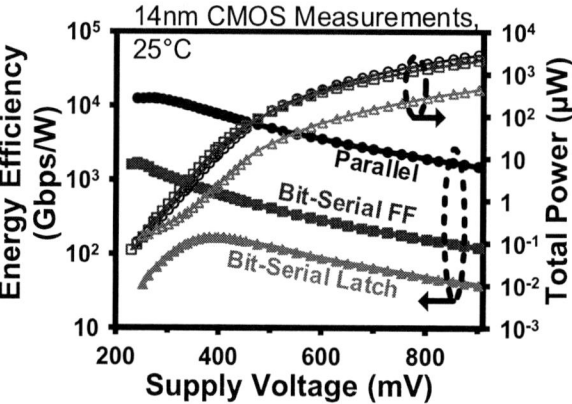

Fig. 11: Performance, power, and energy efficiency vs. supply voltage measurements.

[4] A. Bogdanov, G. Leander, C. Paar, A. Poschmann, M. Robshaw, and Y. Seurin, "Hash Functions and RFID Tags: Mind the Gap," IEEE Workshop on Cryptographic Hardware and Embedded Sytems (CHES) 2008. LNCS, vol. 5154, pp. 283–299. Springer, Heidelberg (2008)

[5] S. Mathew et al., "340 mV–1.1 V, 289 Gbps/W, 2090-Gate NanoAES Hardware Accelerator With Area-Optimized Encrypt/Decrypt GF(2⁴)² Polynomials in 22 nm Tri-Gate CMOS," IEEE Journal of Solid State Circuits (JSSC), vol. 50, no.4, pp. 1048-1058, Apr. 2015.

[6] Y. Zhang, K. Yang, M. Saligane, D. Blaauw, and D. Sylvester, "A Compact 446 Gbps/W AES accelerator for Mobile SoC and IoT in 40nm," IEEE Symposium on VLSI Circuits, pp. 246-247, June 2016.

[7] N. Miura et al., "A 2.5ns-Latency 0.39pJ/b 289µm²/Gb/s Ultra-Light-Weight PRINCE Cryptographic Processor," IEEE Symposium on VLSI Circuits, pp. 266-267, June 2017.

[8] R. Beaulieu, D. Shors, J. Smith, S. Treatman-Clark, B. Weeks, and L. Wingers, "The SIMON and SPECK families of lightweight block ciphers," Cryptology ePrint Archive, Report 2013/404, 2013, http://eprint.iacr.org/.

[9] A. Aysu, E. Gulcan, and P. Schaumont, "SIMON Says: Break Area Records of Block Ciphers on FPGAs," IEEE Embedded Systems Letters, vol. 6, no. 2, pp. 37–40, June 2014.

[10] C.-H. Jan et al., "A 14nm SoC Platform Technology Featuring 2nd Generation Tri-Gate Transistors, 70nm Gate Pitch, 52nm Metal Pitch, and 0.0499µm² SRAM cells, Optimized for Low Power, High Performance and High Density SoC Products," IEEE Symposium on VLSI Technology, pp. 12-13, 2015.

31.3 μs/Signature-Generation 256-bit $\mathbb{F}_p$ ECDSA Cryptoprocessor

Shotaro Sugiyama
Department of Electrical Engineering
and Information Systems
The University of Tokyo
Tokyo, Japan
Email: shotaro@silicon.u-tokyo.ac.jp

Hiromitsu Awano
VLSI Design and Education Center
The University of Tokyo
Tokyo, Japan
Email: awano@vdec.u-tokyo.ac.jp

Makoto Ikeda
Department of Electrical Engineering
and Information Systems
The University of Tokyo
Tokyo, Japan
Email: ikeda@silicon.u-tokyo.ac.jp

Abstract—A 256-bit $\mathbb{F}_p$ elliptic curve digital signature algorithm (ECDSA) cryptoprocessor featuring low latency, low energy consumption, and the ability to change the elliptic curve parameters is designed and fabricated in a 65-nm silicon on thin buried oxide (SOTB) CMOS process. We have demonstrated the lowest ever reported signature-generation time of 31.3 μs. Energy consumption is 3.28 μJ/signature-generation, which is same as the lowest reported to date.

I. Introduction

Public-key-based encryption and authentication are indispensable for many-to-many communications, such as vehicle-to-vehicle (V2V) or vehicle-to-infrastructure (V2I) communications. Elliptic-curve-based cryptography (ECC) [1], [2] is a promising alternative for RSA-based cryptography, as it can achieve the same security strength at 1/10 the key length [3]. With the increase in information exchange, high-speed encryption and authentication are desired and the necessity of cryptographic processors is therefore increasing. A good example can be found in V2V communications, which requires 1,000 authentications per second [4], or V2I communications, which requires even more. Furthermore, low-latency data transmission and authentication are essential for V2x communications. Moreover, it is an urgent issue for internet of things (IoT) devices to be equipped with data encryption, device authentication, and secure key-exchange capability; hence, lightweight ECDSA cryptoprocessors with lower energy consumption are in high demand. In this study, we mainly focus on realizing the fastest and most energy-efficient implementation of elliptic curve digital signature algorithm (ECDSA) by algorithm and coordinate selection and careful optimization of scheduling to fulfill the requirements of V2x communications.

II. ECDSA and Scalar Multiplication

Elliptic curve digital signature algorithm (ECDSA) [5] is a variant of the digital signature algorithm based on ECC, which can be used to assure that a message is sent from the right person and is unaltered. The signature-generation procedure is summarized in Alg. 1, where $[k]G$ is a scalar multiplication (SM) on an elliptic curve. The security of ECDSA depends on the difficulty of the discrete logarithm problem on the elliptic curve (ECDLP), i.e., given two points on an elliptic curve, G and $[k]G$, it is difficult to find scalar k. SM is decomposed into point additions and doublings on an elliptic curve. Given

---

Algorithm 1 ECDSA Signature Generation

Require: Base point $(G_X, G_Y, 1) \in E(\mathbb{F}_p)$, Order n s.t. Z-coordinate of $[n]G$ is 0, Digest $z = $ SHA256(Message), Secret key d, Random number k.

Ensure: Signature (r, s)

1: $(X_1, Y_1, Z_1) \Leftarrow [k]G$
2: $r \Leftarrow X_1 Z_1^{-1} \bmod n$
3: $s \Leftarrow k^{-1}(z + rd) \bmod n$
4: **return** (r, s)

---

two points $P_1 = (x_1, y_1)$ and $P_2 = (x_2, y_2)$ on an elliptic curve, the point addition $P_3 = P_1 + P_2$ is defined as

$$
\begin{aligned}
x_3 &= \lambda^2 - x_1 - x_2 \\
y_3 &= \lambda(x_1 - x_3) - y_1 \\
\lambda &= \frac{y_2 - y_1}{x_2 - x_1}
\end{aligned}
\tag{1}
$$

and the point doubling $P_3 = 2P_1$ is defined as

$$
\begin{aligned}
x_3 &= \lambda^2 - 2x_1 \\
y_3 &= \lambda(x_1 - x_3) - y_1 \\
\lambda &= \frac{3x_1^2 + a}{2y_1}.
\end{aligned}
\tag{2}
$$

For SM, we use the Montgomery ladder method [6] (Alg. 2), which has the unique characteristic that $P_2 - P_1 = G$ is always satisfied. Utilizing this characteristic, point addition can be computed without holding y, as follows [7]:

$$
x_3 \cdot g_x = \frac{(x_1 x_2 - a)^2 - 4b(x_1 + x_2)}{(x_1 - x_2)^2}.
\tag{3}
$$

The point doubling can also be simplified. To avoid an inverse operation over $\mathbb{F}_p$, the projective coordinate system is employed in which x and y in an affine coordinate system are converted to $x = X/Z$ and $y = Y/Z$, respectively. Point addition in the projective coordinate system can be computed as

$$
\begin{aligned}
X_3 &= (X_1 X_2 - aZ_1 Z_2)^2 - 4bZ_1 Z_2(X_1 Z_2 + X_2 Z_1) \\
Z_3 &= G_x(X_1 Z_2 - X_2 Z_1)^2.
\end{aligned}
\tag{4}
$$

With (3) and (4), we can reduce the data dependencies at the cost of increased number of multiplications on $\mathbb{F}_p$, as shown in Fig. 1. In our design, the increment of the operation cycles

Algorithm 2 Montgomery ladder (Scalar Multiplication)

Require: $P(g_x, g_y) \in E(\mathbb{F}_p)$, k
Ensure: $Q = [k]G$
1: $P_1 \Leftarrow \mathcal{O}$, $P_2 \Leftarrow G$
2: **for** $i = |k| - 1$ to 0 **do**
3:    **if** $k_i = 0$ **then**
4:       $P_1 \Leftarrow 2P_1$, $P_2 \Leftarrow P_1 + P_2$
5:    **else**
6:       $P_1 \Leftarrow P_1 + P_2$, $P_2 \Leftarrow 2P_2$
7:    **end if**
8: **end for**
9: **return** P_1

Algorithm 3 Montgomery Multiplication

Require: $X, Y \in \mathbb{F}_p$, $R = 2^{\lfloor \log_2 p \rfloor}$, $p' = -p^{-1} \bmod R$
Ensure: $Z = XYR^{-1} \bmod p$
1: $T \Leftarrow (XY \bmod R) \cdot p' \bmod R$
2: $Z \Leftarrow (XY + Tp)/R$
3: **if** $Z \geq p$ **then**
4:    $Z \Leftarrow Z - p$
5: **end if**
6: **return** Z

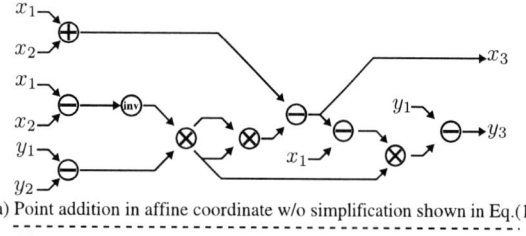

(a) Point addition in affine coordinate w/o simplification shown in Eq.(1).

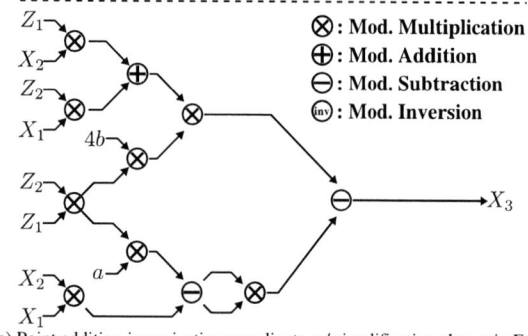

$\otimes$: Mod. Multiplication
$\oplus$: Mod. Addition
$\ominus$: Mod. Subtraction
(inv): Mod. Inversion

(b) Point addition in projective coordinate w/ simplification shown in Eq.(4).

Fig. 1. Formulae and Data Dependencies.

is mitigated by implementing additional hardware resources. Note here that even if the additional multipliers are implemented, our design still achieves the best area-delay product with the careful pipeline design and scheduling optimization, which are detailed in the next section.

III. PROPOSED ARCHITECTURE

A. Multiplier on $\mathbb{F}_p$

Fig. 2-(a) shows the 7-stage pipelined multiplier on $\mathbb{F}_p$ with a throughput of 1-output per cycle to minimize the number of operation cycles. Multiplications in different branches, as shown in Fig. 1-(b), can be performed in parallel by the pipelined multiplier. In order to make the curve parameters (a, b) variable, we use Montgomery multiplication [8](Alg. 3) instead of the reduction circuit specialized in a characteristic number. A 2-stage pipelined multiplier is adopted to equalize the critical-path length of each stage. Because the second multiplier discards the upper half of the result, its area is half that of the other multipliers and the total area of our Montgomery

multiplier unit is equalized to 2.5 multipliers. Compared to the Montgomery multiplier unit with a single multiplier in [9], whose throughput is 2 outputs per 8 cycles, our Montgomery multiplier unit achieves 4 times better throughput and 1.6 times better area efficiency.

B. Scheduling and Architecture

We optimized the scheduling of the operations within a loop of Alg. 2 to speed up ECDSA because the calculation time of SM is longer than any other operations in Alg. 1. The purpose of scheduling is to find a combination of sequences of 2-input modular arithmetic operations that minimize the number of clock cycles of Fig. 1-(b). The constraints of scheduling are the throughput and latency of arithmetic units and order dependencies. We used the job-shop problem solver for scheduling. Fig. 2-(b) shows the scheduling result of a point addition and point doubling in a Montgomery ladder loop. In order to enhance the SPA tolerance and minimize pipeline bubbling by data dependencies, point addition and doubling are optimized and scrambled. Fig. 2-(c) shows the block diagram of the designed ECDSA signature-generation cryptoprocessor. The inputs are the two parameters of the elliptic curve, a and b; random number k, hash digest of the message, z; base point of SM, G; and secret key, d. The outputs are the signatures r and s. SM and the inverse operation of k are conducted simultaneously, followed by inverse operation of Z. For the inverse operation, we used the Montgomery modular inverse operation [10], which is based on the extended Euclidean algorithm. The cycle count of the Montgomery modular inverse varies between 2 and 512, depending on the input. The total cycle count of signature generation varies from 6,945 to 7,455. Fig. 2-(d) shows the die photo of the architecture. The architecture is fabricated by using the 65-nm silicon on thin buried oxide (SOTB) CMOS process.

IV. PERFORMANCE

A. Measurement Results

Fig. 3 shows the measurement results. The maximum operating frequency is 238 MHz under the conditions of $V_{DD}=$ 1.4 V, $V_{BP}=$ 0.6 V, and $V_{BN}=$ 0.4 V, which is equivalent to a signature-generation time of 31.3 µs, and a minimum energy consumption of 3.28 µJ/signature-generation is achieved under the conditions of $V_{DD}=$ 0.45 V, $V_{BP}=$ 0.2 V, and $V_{BN}=$ 0.3 V.

Table I shows the performance comparison with the prior work. The number of signature-generation cycles of our architecture is 0.50 times that of [9] and the latency is 0.52 times that of [9]. (1/#Clk)/Area is an indicative value of

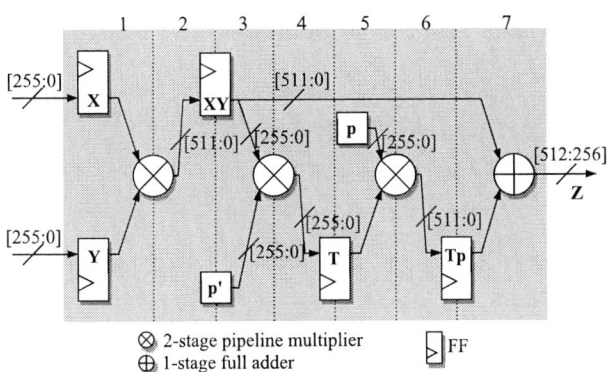

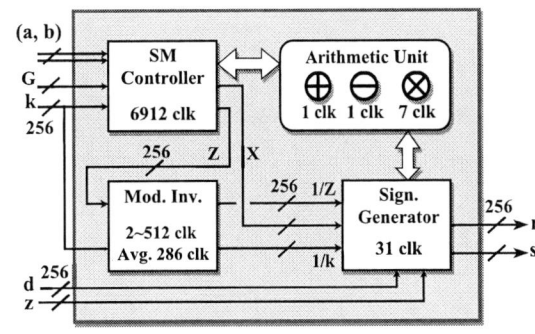

(a) 7-stage pipelined Montgomery multiplier. 2-stage pipeline multipliers are adopted to equalize critical-path length. Total area is about 2.5 2-stage multipliers since the second multiplier discards the upper half of the result.

(c) Block diagram of the processor. SM Controller generates control signals for calculating Alg. 2. Mod. Inv. converts the SM result in projective coordinate to affine coordinate. Sign. Generator outputs signatures according to Alg. 1.

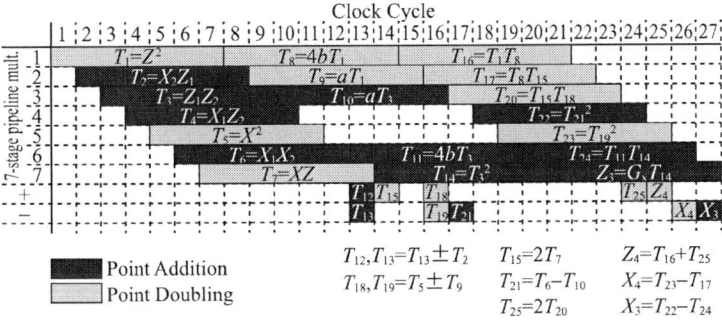

(b) Scheduling result of line 4 in Alg. 2. Line 6 in Alg. 2 is also calculated according to this scheduling.

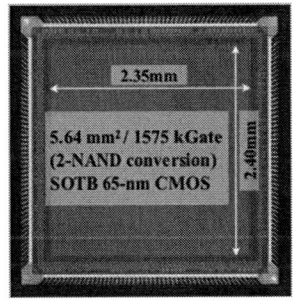

(d) Diephoto.

Fig. 2. Proposed Architecture, Scheduling, Data Flow Diagram, and Die Photo of the Processor.

the number of signature generations per cycle considering the increment of gate counts due to efforts to reduce operation cycles. The larger the value, the better the area efficiency. Our architecture is more efficient than that of any other studies from the perspective of both gate counts and physical area.

Table II shows the signature-generation delay normalized by the fan-out-of-4 delay of the inverter of each process. This value indicates design and scheduling optimality regardless of process differences. The normalized delay of our architecture is 0.73 times that of [9].

The process used for the most energy-efficient implementation [11] is SOTB LP, whereas the process of our architecture is SOTB LSTP. Table III shows the energy and oscillation frequency simulation results of a 5-stage ring oscillator, which indicates that the transistor of LSTP operates 0.46 times slower and consumes 1.98 times more energy per cycle compared to that of LP. Assume that our architecture is implemented with SOTB LP process, the energy consumption of our architecture is 1.66 μs per signature generation and The signature-

generation delay is 68.0 μs, which corresponds to 4.8 times improvement from [11].

V. CONCLUSION

In this study, we have proposed high-speed ECDSA signature-generation architecture with high area-cycle efficiency. The signature-generation delay is 31.3 μs, which is the fastest ever reported, and realizes more than 22,000 signatures per second, far exceeding the requirements of V2x communications. With voltage scaling, the same ECDSA processor realizes energy levels as low as 3.28 μJ per signature generation to fulfill the energy-consumption constraint for IoT devices. It is also possible to operate at 2 mW or less, which is approximately 3% of the power consumption of the wireless standard for IoT [14]. Therefore, the processor is useful for encryption of IoT communications.

ACKNOWLEDGMENT

This work is supported by the Council for Science, Technology and Innovation (CSTI), Cross-ministerial Strategic Inno-

TABLE I
PERFORMANCE COMPARISON OF ECDSA ON ANY CURVE OF 256-BIT $\mathbb{F}_p$.

	Platform	Area	V_{DD} [V]	Frequency [MHz]	#Clk [$\times 10^3$]	(1/#Clk)/Area [/kGE]	(1/#Clk)/Area [/mm^2]	$T_{Sig\text{-}Gen}$ [µs]	Power [mW]	$E_{Sig\text{-}Gen}$ [µJ]
This work	65 nm	1575 kGE (5.64 mm^2)	1.4 0.75 0.45	238 98.0 35.7	7.5	88,200	24.6	31.3 76.0 210	1,227 123 15.6	38.7 9.32 3.28
2016 [9]†	65 nm FDSOI	2493 kGE	1.0	236	15	26,700	NA	60	168	10.7
2016 [11]	65 nm FDSOI	1.92 mm^2	1.1 0.3	105 14	34.7	NA	15.0	325 2300	42.9 0.69	13.9 1.68
2012 [12]†	90 nm	540 kGE (2.72mm^2)	NA	131	22.3	83,000	16.5	170	NA	NA
2010 [13]	Stratix II (90 nm)	9,177 ALM 96 DSP	NA	157.2	106.9	NA	NA	320	NA	NA

† Post-synthesis results.

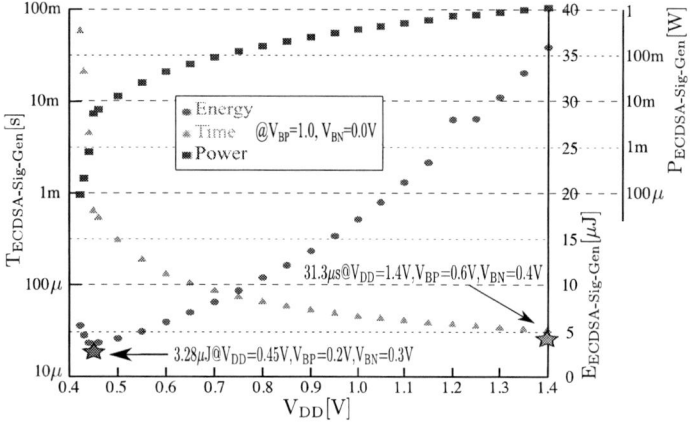

Fig. 3. Measurement results with conditions of V_{BP} = 1.0 V and V_{BN} = 0.0 V.

TABLE II
SIGNATURE-GENERATION DELAY NORMALIZED BY FAN-OUT-OF-4 DELAY OF INVERTER.

	V_{DD} [V]	T_{INV} [ps]	$T_{sig\text{-}gen}/T_{INV}$ [$\times 10^6$]
This work	1.4	24.6	1.27
	0.75	57.6	1.32
2016 [9]	1.0	33.8	1.78

TABLE III
ENERGY COMPARISON BETWEEN LSTP AND LP.

	Freq [MHz]	Energy/Cycle [fJ]
SOTB LSTP	431	8.21
SOTB LP	200	4.14

vation Promotion Program (SIP), "Cyber-Security for Critical Infrastructure" (funding agency: NEDO). The VLSI chip is designed and fabricated through VDEC, the University of Tokyo, in collaboration with Cadence, Synopsys, Mentor Graphics, and Renesas Electronics Corp. The authors also acknowledge Mr. D. Takahashi and Dr. M. Khanh for measurements.

REFERENCES

[1] N. Koblitz, "Elliptic curve cryptosystems." *Mathematics of computation*, vol. 48, no. 177, pp. 203–209, 1987.

[2] V. S. Miller, "Use of elliptic curves in cryptography," in *Conference on the theory and application of cryptographic techniques.* Springer, 1985, pp. 417–426.

[3] E. Barker, W. Barker, W. Burr, W. Polk, and M. Smid, "Recommendation for key management part 1: General (revision 3)," *NIST special publication*, vol. 800, no. 57, pp. 1–147, 2012.

[4] M. Knežević, V. Nikov, and P. Rombouts, "Low-latency ECDSA signature verification—a road toward safer traffic," *IEEE Transactions on Very Large Scale Integration (VLSI) Systems*, vol. 24, no. 11, pp. 3257–3267, 2016.

[5] D. Johnson, A. Menezes, and S. Vanstone, "The elliptic curve digital signature algorithm (ECDSA)," *International journal of information security*, vol. 1, no. 1, pp. 36–63, 2001.

[6] M. Joye and S.-M. Yen, "The montgomery powering ladder," in *International Workshop on Cryptographic Hardware and Embedded Systems.* Springer, 2002, pp. 291–302.

[7] T. Izu and T. Takagi, "A fast parallel elliptic curve multiplication resistant against side channel attacks," in *International Workshop on Public Key Cryptography.* Springer, 2002, pp. 280–296.

[8] P. L. Montgomery, "Modular multiplication without trial division," *Mathematics of computation*, vol. 44, no. 170, pp. 519–521, 1985.

[9] M. Tamura and M. Ikeda, "Montgomery multiplier design for ECDSA signature generation processor," *IEICE TRANSACTIONS on Fundamentals of Electronics, Communications and Computer Sciences*, vol. 99, no. 12, pp. 2444–2452, 2016.

[10] E. Savas and Ç. K. Koç, "The montgomery modular inverse-revisited," *IEEE Transactions on Computers*, vol. 49, no. 7, pp. 763–766, 2000.

[11] M. Tamura and M. Ikeda, "1.68 µj/signature-generation 256-bit ECDSA over GF (p) signature generator for IoT devices," in *Solid-State Circuits Conference (A-SSCC), 2016 IEEE Asian.* IEEE, 2016, pp. 341–344.

[12] S.-C. Chung, J.-W. Lee, H.-C. Chang, and C.-Y. Lee, "A high-performance elliptic curve cryptographic processor over GF (p) with SPA resistance," in *Circuits and Systems (ISCAS), 2012 IEEE International Symposium on.* IEEE, 2012, pp. 1456–1459.

[13] N. Guillermin, "A high speed coprocessor for elliptic curve scalar multiplications over $\mathbb{F}_p$," in *International Workshop on Cryptographic Hardware and Embedded Systems.* Springer, 2010, pp. 48–64.

[14] J.-S. Lee, Y.-W. Su, and C.-C. Shen, "A comparative study of wireless protocols: Bluetooth, UWB, ZigBee, and Wi-Fi," in *Industrial Electronics Society, 2007. IECON 2007. 33rd Annual Conference of the IEEE.* Ieee, 2007, pp. 46–51.

13-4 (8103)

A Physically Unclonable Function with 0% BER Using Soft Oxide Breakdown in 40nm CMOS

Kai-Hsin Chuang[*†], Erik Bury[†], Robin Degraeve[†], Ben Kaczer[†], Dimitri Linten[†] and Ingrid Verbauwhede[*]

*imec-COSIC, KU Leuven, Belgium

Email: kai.hsin.chuang@imec.be, ingrid.verbauwhede@esat.kuleuven.be

[†]imec, Belgium

Abstract—A physically unclonable function (PUF) utilizing the randomness of soft oxide breakdown (BD) locations in MOSFETs is presented. The so-called *soft-BD PUF* features a self-limiting mechanism to generate one single soft-BD spot in a pair of MOSFETs; the subsequent BD location is used as the source of entropy to generate a highly stable "0" or "1" bit with an equal probability of 0.5. The soft-BD PUF comprising all the essential periphery circuits are fabricated in a 40nm CMOS process. Experiments show that the PUF has no instability in most of the operating conditions using the proposed readout scheme. The native bit error rate remains zero from $V_{DD}=0.8V$ to 1.5V at room temperature and from -20°C to 120°C at nominal $V_{DD}=0.9V$. The throughput is shown to be at least 40 Mb/s and the PUF readout consumes only 51.8 fJ/bit. The randomness and uniqueness of the PUF are close to an ideal case, and no spatial correlation was observed.

I. INTRODUCTION

A physically unclonable function (PUF) in modern silicon technologies is an essential circuit primitive for on-chip security and cryptographic applications. As one of the most common applications, the PUF-based *key generation* procedure [1] harvests the entropy from uncontrollable process variations of a PUF, as illustrated in Fig. 1. Typically, the data obtained from a PUF, namely *raw* PUF data, do not have full-entropy and good stability. In order to generate a cryptographic key which meets certain standards, the raw data need to be post-processed, including entropy extraction and error correction [1]. An error correcting code (ECC), or called *helper data*, is stored in a non-volatile memory (NVM) to assist the post-processing circuits during the key-generation procedure.

A. Stability of PUF

In recent PUF works [2]–[7], the stability is considered as a major concern, since it usually takes more resource to be optimized. On the other hand, a PUF with an excellent *native* stability, i.e. the stability of raw PUF data, is beneficial, especially for an application in which no NVM is accessible. Since the ECC cannot be stored, it necessitates a bit-error-rate (BER) of 0%, which is considered as the *ideal* stability.

Temporary majority voting (TMV) [1] and dark-bit masking [2] methods are widely used to improve the bit stability of PUFs, but both cases have their own limitations. The TMV can reduce BER, but is insufficient for elimination. Using a dark-bit mask to filter out the unstable PUF bits was reported having 0% BER from the unmasked PUF bits [2], it however requires more testing time and/or additional circuitry to identify the

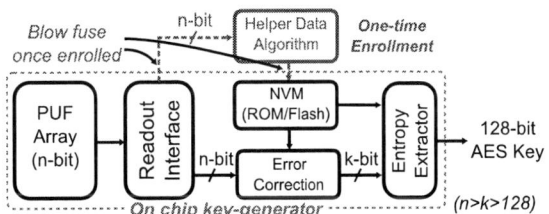

Fig. 1. An example block diagram of a PUF-based AES-128 key generator.

unstable bits. Moreover, once a NVM is required to store the masking information locally, there is not much distinction between the traditional error correction methods.

B. Approaches towards ideal stability

Recently, two major approaches were proposed to achieve an ideal native bit stability. The first one is to exploit the variability of emerging memory devices, such as the resistive-RAM (RRAM) based PUFs [8]. The bit stability in such case is usually good but is out of scope for this work, since it is linked to the specific device physics. The other approach is based on *aging* effects that can be electrically induced, such as biased-temperature instability (BTI) [3], hot-carrier injection (HCI) [4] and oxide breakdown (BD) [5]–[7]. In this paper, we will focus on the stability aspects of our proposed PUF using *soft* oxide breakdown (SBD), namely soft-BD PUF, with detailed experiments on the chips fabricated in 40nm.

II. CIRCUIT DESIGN AND OPERATION

The unit cell and array of the soft-BD PUF (Fig. 2), was first introduced in [6], without any periphery circuit. The three transistor (3T) unit cell consists of two minimum-sized NMOS transistors, which will be stressed to generate random soft-breakdown, and one PMOS transistor serves as the word line (WL) selector. The proposed PUF circuit in Fig. 2 is designed as a 32-by-32 array, with periphery circuits including sense amplifiers (SA) and other control logic. The operating concept of the PUF cell and array will be first introduced; followed by the readout scheme utilizing the proposed reference-free sense-amplifiers.

A. PUF cell and array

As illustrated in Fig. 2, a soft-breakdown path will occur in one of the two NMOS transistors, and the detailed procedure towarding this result can be found in [6]. Briefly, once a high

978-1-5386-6414-8/18 $31.00 © 2018 IEEE

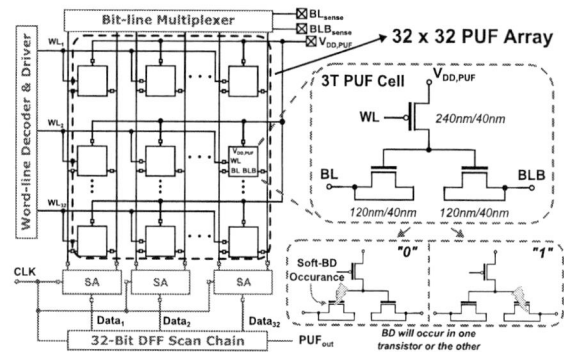

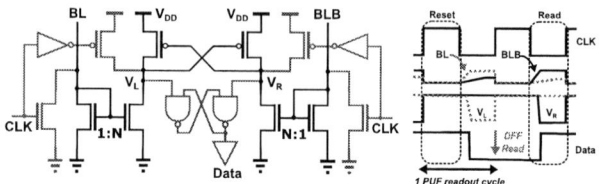

Fig. 4. The schematic of the proposed reference-free sense-amplifier and the corresponding timing diagram.

B. Reference-free sense-amplifier

The current flow through the soft breakdown spot exhibits approximately an exponential voltage dependence, as discussed in [6]. It leads to very low BD currents at low V_{DD}, in addition to a strong variation between devices (from 5nA to 200nA at V_{DD}=0.9V), as shown in Fig. 3. Consequently, the sense-amplifier has to sense the current in the nA range at nominal V_{DD} and below, which is not a typical current sensing range of a SA used for SRAMs. Note that in [7], a single-ended SA with reference voltage works well on sensing currents from the ruptured spots with equivalent resistance $< 100k\Omega$. This technique, however, cannot be directly adapted to fit the SBD spots, since it is rather difficult to define an optimized reference voltage or current.

In order to solve this issue, a dedicated reference-free SA is designed, as shown in Fig. 4, consisting of a current-mirror input stage and a cross-coupled pair second stage (in darker lines). As illustrated by the timing diagram, a readout cycle starts with resetting the SA when *CLK* is set to 1. Current sensing starts when *CLK* is set to 0, the soft-BD current from either *BL* or *BLB* will be amplified to cause a fast discharging on V_L or V_R. The second stage will be latched once sensing a certain difference between V_L and V_R. The state (*Data*) of the final SR-latch will be updated accordingly, which is ready for DFF registering at the next clock rising edge.

Each bit-line pair consists of 32 PUF cells is connected to an individual sense amplifier, in order to avoid the additional series resistance from a multiplexer, which degrades the sensing resolution. Having multiple SAs increases the overall throughput as well, since multiple bits can be readout in parallel. Note that only a small amount of area overhead is introduced by placing multiple SAs, as shown in Fig. 5.

Fig. 2. The schematic of the PUF array including the periphery circuits and the 3T PUF cell.

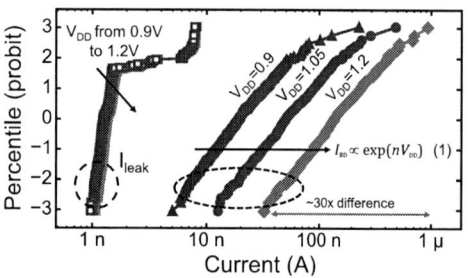

Fig. 3. The experimental statistics of the two current components from the soft-BD PUF cells. The BD current, which flows through the soft-BD spot is widely distributed and has an exponential voltage dependence, as described in (1), in which *n* is a scaling parameter. The leakage current, which flows through the unbroken gate oxides, is also widely distributed but has no strong voltage dependency. A more detailed explanation can be found in [6].

voltage stress is applied to $V_{DD, PUF}$ and a WL is enabled, a dielectric breakdown will eventually occur in a NMOS transistor, and the stress voltage and current will be both limited by the PMOS transistor as soon as this event occurs. The subsequent voltage stress will be too small to trigger another breakdown, and the breakdown path will not be able to grow towards the *"hard"* breakdown (HBD).

The main reason of aiming soft breakdown is to have less *visibility*, making it more secure against *invasive attacks*. Note that the HBD spots in nanoscale anti-fuse devices are not detectable by the *scanning electron microscope* (SEM) [9], so as the PUF in [7]. It can be, however, detected by the *transmission electron microscope* (TEM), as discussed in [10], while a SBD spot is much less obvious in this case. Even though using a powerful tool like TEM to attack a PUF is unrealistic, it is still reasonable to consider using SBD, as a precaution for the rapid advancing of attacking techniques.

The PUF cells are organized as a typical array with shared word lines and bit lines (BLs), the layout of the PUF cell and array are shown in Fig. 5. Note that for experiment and modeling purposes, the current flows through individual BLs can be directly multiplexed to sensing pads for external DC measurements (similar to [6]). The locations of BD spots obtained based on these DC measurements, exactly matches the PUF data generated by the SAs as expected (not shown).

III. EXPERIMENT DESCRIPTION AND RESULTS

For experimental characterization of the soft-BD PUF, the test chips are packaged and measured on a PCB; a packaged chip is shown in Fig. 5. An FPGA is being used to generate the control signals; it also receives and synchronizes the digital output signals from the PUF chips.

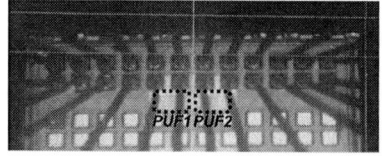 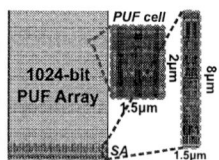

Fig. 5. Layout of the 1024-bit PUF array with sub-circuits and the die photo.

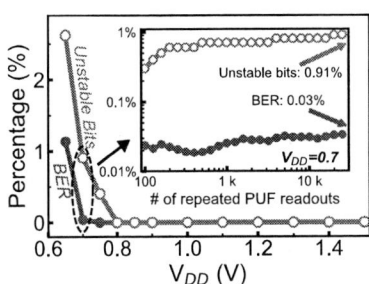

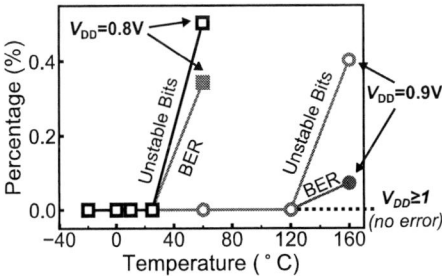

Fig. 6. The percentage of unstable bits and BER v.s. V_{DD} and v.s. repeating readout cycles at V_{DD}=0.7V (inset).

Fig. 7. The percentage of unstable bits and BER under different temperature, operating at V_{DD}=0.8V and 0.9V.

Using this setup, the maximum measured throughput is 40Mb/s (not the actual limit) for all tested conditions and the average energy consumption per PUF bit is 51.8fJ. The stability and the other properties are examined as follows.

A. Bit stability

As the voltage dependence is a major concern for the soft-BD PUF, we first examine the stability at different V_{DD}, as shown in Fig. 6 at the room temperature. The PUF is well functioning down to V_{DD}=0.7V; the bit-error-rate (BER) is below 0.1% at 0.7V and is 0% (ideal) for $V_{DD} \geq$0.8V. Note that the error-free region well covers the $\pm$10% of the nominal V_{DD} (0.9V for this 40nm technology).

The stability at reduced and elevated temperature has been tested using a temperature chamber. Note that an on-chip poly-heater, like the one characterized in [6], was used as an alternative heat source for 60 °C and above. The temperature inside the chamber cannot be further increased due to the temperature limit of the connecting cables.

As shown in Fig. 7, both the ratio of unstable bits and BER remain 0% at the room temperature and below. When operating at V_{DD}=0.9V, this excellent stability holds until 120°C. Once the V_{DD} is lowered to 0.8V, some of the PUF bits become unstable at 60°C. This result is in contrast to the temperature dependence of the current ratio observed in [6], where a better stability is expected at an elevated temperature. This effect can be attributed to the performance degradation of the SAs at higher temperature, and is verified by simulation, as shown in Fig. 8. The difference between V_L and V_R becomes smaller when the circuit is heated up, which makes it more sensitive to noise, and hence the output becomes less stable.

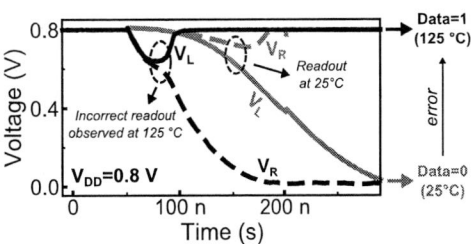

Fig. 8. Simulated voltage of SA nodes V_L and V_R (see Fig. 4) at V_{DD}=0.8V and 25°C/125°C. Here shows one of the error cycles observed at 125 °C.

B. Randomness and Uniqueness

The quality of PUF data is typically checked by three indices: randomness (bias), uniqueness (hamming distance) and spatial correlation (auto-correlation function). The normalized hamming weight (number of "1"s) distribution of 128-bit words from 20 measured PUF arrays shows no bias, as shown in Fig. 9 (a). The auto-correlation function (ACF) of PUF data, as the example shown in Fig. 9 (b), is also computed and plotted in Fig. 10. All the data sequences have passed the requirement of the auto-correlation test specified in AIS31 (T5) [11], which checks the bitwise correlation. As the result, we conclude that the PUF bits are uncorrelated, and hence no spatial correlation within PUF chips.

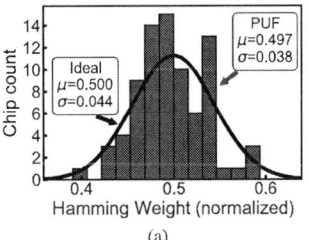 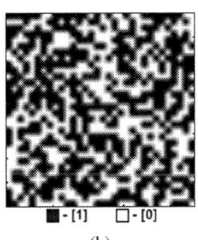

(a)   (b)

Fig. 9. (a) Normalized hamming weight distribution of the 128-bit PUF words generated from 20 PUF arrays and (b) an example PUF data.

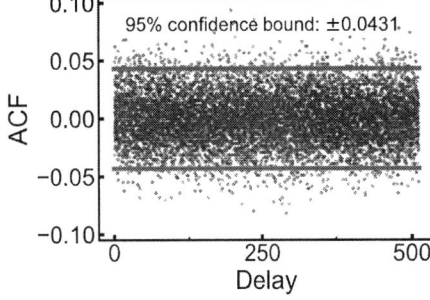

Fig. 10. The auto-correlation function (ACF) of the PUF arrays with the indication of 95% confidence bound, which shows no observable spatial correlation.

The resulting hamming distance from the 128-bit words are plotted in Fig. 11, which is almost identical to the ideal values. Moreover, the identifiability, which is defined as the ratio of the inter and intra hamming distance, is infinite at the nominal operating condition, as a benefit from the ideal stability.

978-1-5386-6414-8/18 $31.00 © 2018 IEEE

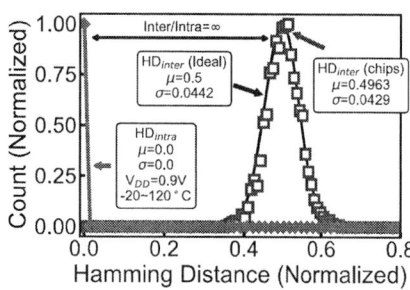

Fig. 11. The inter and intra hamming distance resulting from the 128-bit words, showing an uniqueness nearly indistinguishable from the ideal PUF.

IV. DISCUSSION AND COMPARISON

The comparison with prior works is shown in Table I. The proposed soft-BD PUF shows no downside among all the performance indices. The minimum V_{DD} to obtain 0% BER in the experiments is higher than [7], but the energy consumption is much lower. The relatively low resistance of the "hard" breakdown spot results in a higher current level comparing to the soft-BD current, especially at low V_{DD}. Consequently, the proposed soft-BD PUF is less robust against the extreme operating conditions comparing to [7], but it keeps several advantages, including *good energy efficiency* and *less visibility*. Moreover, we have noticed that the degrading stability at higher temperature is mainly originated from the custom designed sense-amplifier, i.e. not the PUF cell itself. In other words, the SA design can be further improved to obtain better temperature stability.

The SA-PUF with hot-carrier injection based stabilization technique [4] shows BER=0% as well, but it necessitates a relatively long burn-in period (25s) to achieve this target. The uniqueness is also worse as the average HD_{inter} is apart from 0.5, which may be attributed to the additional circuitry to enables this HCI burn-in procedure.

For the SRAM-like hybrid PUF [12], it utilize burn-in, TMV and dark-bit masking to reduce the instability while keeping an excellent energy efficiency, but it is insufficient to achieve a BER of 0%. The error correcting logic and NVM is therefore not removable for this case.

Besides all the advantages of the soft-BD PUF, it should be noted that the additional procedure to generate the oxide breakdown spots also occupies resources. It requires either an additional pin or an on-chip high-voltage generator and more testing time. Nevertheless, once considering the trade-off between an embedded NVM for error correction, the proposed soft-BD PUF stays competitive even in terms of cost.

V. CONCLUSION

A PUF using the randomness from the *soft* oxide breakdown mechanism in MOSFETs has been demonstrated. The reference-free sense amplifier can distinguish the current difference in the nA range, resulting in an excellent bit stability over a wide range of voltage and temperature variations. The soft-BD PUF also shows good randomness and uniqueness; it also consumes less energy per bit comparing to the designs

TABLE I
COMPARISON SUMMARY WITH PRIOR PUF WORK

	[4]	[5]	[7]	[12]	**This work**
Design	SA	Anti-fuse	Anti-fuse	Hybrid cell	**Soft-BD**
Technology	65nm	65nm	55nm	14nm	**40nm**
Stabilizing method	hot-carrier injection	native	native	delay-hardened	**native**
BER (%)	0	0	0	1.46	**0**
V_{DD} (V)	0.8–1.2	1.0–1.2	0.81–1.32	0.55–0.75	**0.9–1.5**
Temp. (°C)	-20–85	0–85	-40–150	25–110	**-20–120**
HD_{inter}	0.468	0.501	0.50	0.486	**0.496**
Energy/bit	N/A	340 fJ	5200 fJ *	4 fJ	**51.8 fJ**

* Including all peripheral blocks, in which the high-V source and BIST circuit are not implemented in this work. It was not clearly stated if these blocks are *active* during readout phase, hence it cannot be concluded whether this is a fair comparison or not.

using hard oxide breakdown. The experimental results prove that the soft breakdown mechanism is suitable for high quality PUF implementations, in particular for the on-chip stable key generation without error correction. The less-visible nature of the soft-BD spots also makes it a more viable candidate for the applications with higher security requirements.

ACKNOWLEDGMENT

This work was supported in part by the Research Council KU Leuven: C16/15/058. In addition, this work is supported in part by Cathedral ERC Advanced Grant 695305.

REFERENCES

[1] J. Delvaux, D. Gu, D. Schellekens, and I. Verbauwhede, "Helper data algorithms for puf-based key generation: Overview and analysis," *IEEE Transactions on Computer-Aided Design of Integrated Circuits and Systems*, vol. 34, pp. 889–902, 2015.

[2] M. Liu, C. Zhou, Q. Tang *et al.*, "A data remanence based approach to generate 100% stable keys from an sram physical unclonable function," in *2017 IEEE/ACM International Symposium on Low Power Electronics and Design (ISLPED)*, July, pp. 1–6.

[3] R. Maes and V. van der Leest, "Countering the effects of silicon aging on sram pufs," in *Hardware-Oriented Security and Trust (HOST), 2014*. IEEE, pp. 148–153.

[4] M. Bhargava and K. Mai, "A high reliability puf using hot carrier injection based response reinforcement," in *International Workshop on Cryptographic Hardware and Embedded Systems*. Springer, 2013, pp. 90–106.

[5] N. Liu, S. Hanson, D. Sylvester, and D. Blaauw, "Oxid: On-chip one-time random id generation using oxide breakdown," in *VLSI Circuits (VLSIC), 2010 IEEE Symposium on*. IEEE, pp. 231–232.

[6] K. H. Chuang, E. Bury, R. Degraeve *et al.*, "Physically unclonable function using cmos breakdown position," in *2017 IEEE International Reliability Physics Symposium (IRPS)*, April, pp. 4C–1.1–4C–1.7.

[7] M. Y. Wu, T. H. Yang, L. C. Chen *et al.*, "A puf scheme using competing oxide rupture with bit error rate approaching zero," in *2018 IEEE International Solid - State Circuits Conference - (ISSCC)*, pp. 130–132.

[8] R. Liu, H. Wu, Y. Pang *et al.*, "Experimental characterization of physical unclonable function based on 1 kb resistive random access memory arrays," *IEEE Electron Device Letters*, vol. 36, pp. 1380–1383, Dec 2015.

[9] N. Chen. (2016) The benefits of antifuse otp. [Online]. Available: http://semiengineering.com/the-benefits-of-antifuse-otp/

[10] K. L. Pey, C. H. Tung, M. K. Radhakrishnan *et al.*, "Dielectric breakdown induced epitaxy in ultrathin gate oxide - a reliability concern," in *Digest. International Electron Devices Meeting,*, Dec 2002, pp. 163–166.

[11] W. Killmann and W. Schindler, "A proposal for: Functionality classes for random number generators," 2011.

[12] S. Satpathy, S. K. Mathew, V. Suresh *et al.*, "A 4-fj/b delay-hardened physically unclonable function circuit with selective bit destabilization in 14-nm trigate cmos," *IEEE Journal of Solid-State Circuits*, vol. 52, pp. 940–949, April 2017.

978-1-5386-6414-8/18 $31.00 © 2018 IEEE

13-5 (8011)

A 373 F² 2D Power-Gated EE SRAM Physically Unclonable Function With Dark-Bit Detection Technique

Kunyang Liu, Yue Min, Xuan Yang, Hanfeng Sun and Hirofumi Shinohara

Graduate School of Information, Production and Systems
Waseda University
Kitakyushu, Japan
E-mail: konyo@fuji.waseda.jp

Abstract—This paper presents an Enhancement-Enhancement (EE) SRAM physically unclonable function (PUF) with a dark-bit detection technique based on an integrated V_{SS}-bias generator. The EE SRAM PUF cell improves native stability to 0.21% bit-error rate (BER). Bit cells that are potentially unstable due to environmental variations or aging are detected via the lightweight bias generator to ensure stability, and the effectiveness is verified with experimental results of dark-bit detection performed at room temperature. Measurement results of 10 chips in 130-nm CMOS show that after masking the detected dark bits, 1.3×10^{-6} BER is achieved across 0.8–1.4 V/–40–120 °C VT corners. The nMOS-only bit cell is also highly compact (i.e., 373 F²). Moreover, a 2D power-gating scheme is implemented for low operation energy, low standby power, and high attack tolerance.

Keywords—physically unclonable function (PUF); dark-bit masking; EE SRAM; IoT; hardware security

I. INTRODUCTION

Security is a growing concern due to the rapid development of the Internet-of-Things (IoT). To prevent an IoT device network from invasion, secret keys should be highly secure for safe authentication.

The physically unclonable function (PUF) is regarded as a reliable entropy source for secret-key generation. It gains entropy from intrinsic process variations and the volatile feature makes a PUF considered to be more secure than the conventional non-volatile memory-based key storage. Various types of circuits [1]–[10] have been utilized as PUF cells. However, existing SRAM-based PUFs [1]–[3] are generally vulnerable to thermal noise owing to their bi-stable feature. Hence, mono-stable PUF cells [4]–[7] were recently presented to address the bit-flipping problem of bi-stable cells to improve stability against thermal noise. PUF cells based on subthreshold leakage current [8] was also reported to improve mismatch-sensitivity for better anti-noise ability. However, the single-ended sensing scheme, which is utilized by mono-stable PUFs, has a concern of being attacked by the differential power analysis, and the subthreshold current-based PUF suffers from low operation speed, especially at low temperature. In [9], the hard oxide-breakdown scheme on anti-fuse is utilized for high stability but it can be vulnerable to reverse engineering [10].

This research is supported by ROHM Co., Ltd. and Kitakyushu Foundation for the Advancement of Industry, Science and Technology (FAIS).

IEEE Asian Solid-State Circuits Conference
November 5 - 7, 2018/Tainan, Taiwan

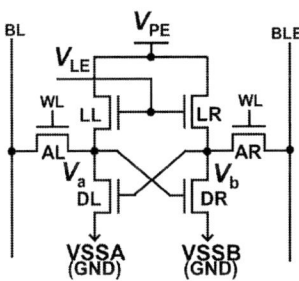

Fig. 1. Bit-cell schematic

Besides the thermal noise, supply voltage/temperature (VT) variations and long-term aging often cause large bit errors (such as 9% bit-error rate (BER) due to 25 °C variation in [2]) because they influence the characteristics of transistors. Conventional masking techniques [2], [5], [8] can screen out flipped bits but requires costly temperature sweeping [5], and it is difficult for these methods to reduce bit errors caused by long-term or unpredicted factors. Then, heavy error-correction codes (ECCs) are needed to correct the large bit errors [11], [12]. Therefore, reducing not only noise-induced bit errors but also potential bit errors that can be caused by VT variations and long-term usage is necessary. In addition, from the viewpoint of dependability, attack tolerance and margin checks are required for PUF designs.

In this work, an enhancement-enhancement (EE) SRAM PUF with a V_{SS}-bias generator-based dark-bit detection technique is proposed. This design has the following features: (a) high stability against thermal noise achieved with the novel EE SRAM structure; (b) high hidden dark-bit detectability using a V_{SS}-biasing technique; (c) sub-picowatt/bit standby power and improved operation energy efficiency due to a two-dimensionally (2D) power-gating scheme; (d) attack-tolerant designs of a differential bit-line structure and normally off bit cells; and (e) small bit-cell area due to the nMOS-only cell.

II. CIRCUIT AND OPERATING PRINCIPLES

A. PUF Cell Schematic and Butterfly Curve Analysis

The bit-cell schematic is shown in Fig. 1. V_{SSA} and V_{SSB} are normally connected to GND. The EE SRAM PUF cell is made

978-1-5386-6414-8/18 $31.00 © 2018 IEEE   161

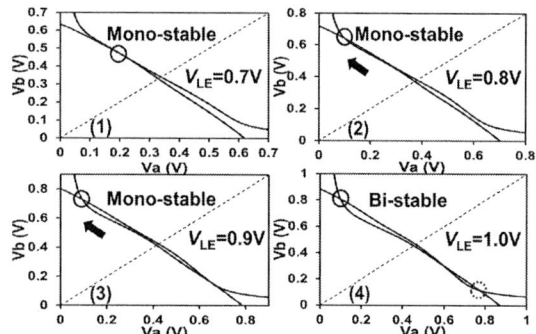

Fig. 2. Simulated transition of butterfly curves when powering up (20mV mismatch between driver transistors DL and DR)

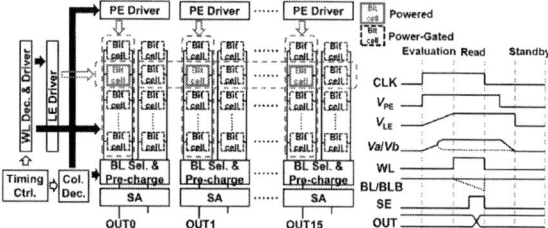

Fig. 3. Array architecture and waveforms

up of a pair of cross-coupled EE inverters. Generally, the gates of the load transistors can be connected to the drains. In this design, the gates and the drains are separated into V_{LE} and V_{PE} for the power gating. The V_{th} mismatch between the two EE inverters determines the power-up datum of a PUF cell.

On the basis of the long-channel model, the gain of the EE inverter (right side) is derived as follows (1):

$$\text{Gain} = \left| \frac{dV_b}{dV_a} \right| = \frac{1}{1 + \frac{\gamma}{2\sqrt{2\phi_F + V_b}}} \sqrt{\frac{\beta_D}{\beta_L}} \qquad (1)$$

where γ is the body-effect coefficient, $\beta = \mu C_{ox}(W/L)$, μ is the carrier mobility, C_{ox} is the gate capacitance of a unit area and W and L are the channel width and length, respectively. When the bit cell is powered up, V_b increases. Then, the gain of the EE inverter becomes larger because γ-related factor in (1) becomes smaller. When the gain becomes larger than one, a bit cell transfers from mono- to bi-stable state. The trip point at gain = 1 can be adjusted with β_D/β_L. The ratio in this design is set to four. As shown in Fig. 2, the mono-stable solution of a bit cell with a small V_{th} mismatch continuously moves away from $V_b = V_a$ line and finally advances into one of the bi-stable solutions (solid-circle). In the mono-stable state during power up, noises can hardly cause bit flipping and therefore, the PUF stability is improved.

B. 2D Power-Gated Array and Attack-Tolerant Schemes

Given that the EE SRAM PUF cell has DC current during evaluation, we propose a 2D power-gating scheme to reduce the power consumption. The attack tolerance is also enhanced by this technique. As shown in Fig. 3, power enable (PE) and load enable (LE) drivers are used to control the drain voltage (V_{PE}) and the gate voltage (V_{LE}) of the load transistors, respectively

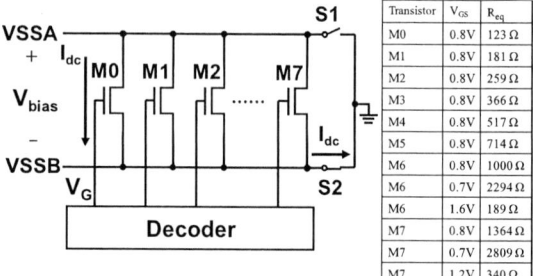

Transistor	V_{GS}	R_{eq}
M0	0.8V	123 Ω
M1	0.8V	181 Ω
M2	0.8V	259 Ω
M3	0.8V	366 Ω
M4	0.8V	517 Ω
M5	0.8V	714 Ω
M6	0.8V	1000 Ω
M6	0.7V	2294 Ω
M6	1.6V	189 Ω
M7	0.8V	1364 Ω
M7	0.7V	2809 Ω
M7	1.2V	340 Ω

Fig. 4. V_{SS}-bias generator and measured equivalent resistance of M0–M7 under different V_{GS} (polarity of bias voltage is changed by switching S1 and S2)

(Fig. 1). The PE and LE drivers are normally off except during evaluation and reading operations. During the operations, V_{PES} of the selected columns initially increases and then V_{LE} of the selected row turns on. The power-up speed of V_{LE} is set to be slower for a stable evaluation by using long-channel-length drivers. Then, the word line turns on and the evaluated cell data are read out. In this scheme, only one row and selected columns are activated to ensure that only the cross-point bit cells are powered up (Fig. 3). After reading, V_{PE} decreases before V_{LE} to clear the remanence charge of the PUF cells. In this way, bit cells are automatically powered off. Therefore, the standby power is significantly reduced. In summary, attack tolerance is enhanced in three ways. (a) The differential structure improves the robustness against potential differential power analysis attack [13]. (b) The remanence decay side-channel attack [14] is nullified by the remanence charge clearance scheme, as the simulation shows that the charge is cleared within 10 ns even under -40°C. (c) The normally powered-off bit cells increase the difficulty of reverse engineering.

C. V_{SS}-Bias Generator for Dark-Bit Detection

We propose an application of our V_{SS}-biasing technique that finds the distribution of mismatch [15] for efficiently detecting hidden unstable bit cells. A V_{SS}-bias generator (Fig. 4) is newly designed and is implemented on-chip. It artificially generates an unbalance on the differential bit cells so that a large number of hidden dark bits change their power-up data. Bit cells that change the power-up data with plus and minus V_{SS} biases are regarded as potentially unstable. Meanwhile, bit cells that do not change the data with large biases are regarded as highly stable. The address information of masked bits is stored outside the PUF (e.g., in the non-volatile memory), but no secret information is leaked.

The bias generator is connected to the two V_{SS} ports of the PUF cell array. During data evaluation phase, DC current of EE SRAM PUF cells (I_{dc}) flows through one of the M0–M7 transistors, and the IR drop works as a self-bias between V_{SSA} and V_{SSB}. The equivalent resistance can be tuned by selecting transistors M0–M7 with different sizes or changing V_{GS} of the transistors by tuning V_{DD} of the bias generator. The measured equivalent resistances of M0–M7 under various V_{GSS} are also shown in Fig. 4. Switches S1 and S2 control the polarity of the bias voltage.

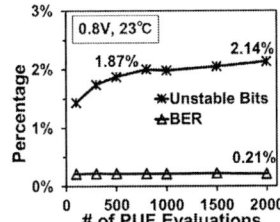

Fig. 5. Native BER and unstable bits

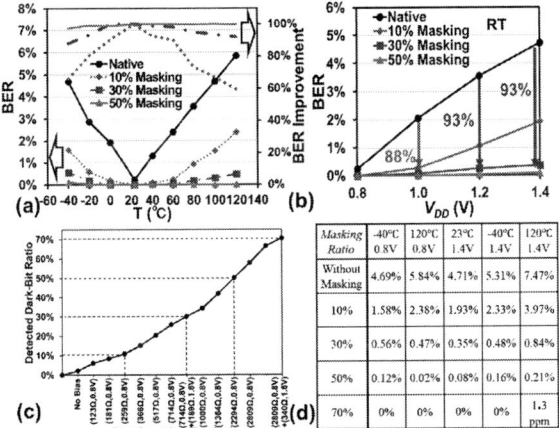

Fig. 6. (a) BER across T variation and improvement after the dark-bit masking (V_{DD}=0.8V); (b) BER across V_{DD} variation (room temperature); (c) masking ratio at various detection conditions; (d) BER at VT-corners

III. MEASUREMENT RESULTS

A. Stability at Nominal Condition

Stability is one of the most important metrics for PUFs, and it is evaluated with bit-error rate (BER) and the percentage of unstable bits. The stability at the nominal condition is affected by thermal noise [16]. Twenty 1k-bit (20k bits in total) PUFs from two wafers in 130-nm CMOS are measured at 0.8V/23 °C. Measurement results show that the percentage of native unstable bits is 2.14%, whereas the native BER is 0.21% (Fig. 5) at 2000 evaluations, which is more than 14× better than the 3.04% BER of the conventional CMOS SRAM-based PUFs [1].

B. Stability across VT Variations

VT variations are crucial factors for PUF stability. In VT-variation measurement, 10k bits (10 chips) are measured for 500 times in each condition. Black lines in Figs. 6(a) and 6(b) show the native BER induced by V_{DD} and T variations, respectively.

The detected dark-bit ratio of the proposed V_{SS}-biasing technique is shown in Fig. 6(c). The detections were performed under various bias conditions. The bit cells whose evaluation data never changed at biases in both polarities are regarded as stable and others are detected as dark bits. The detected dark-bit ratio is up to 70% in this work and can be higher with larger bias applied. The detection was performed only at the room temperature with the intention to confirm the effectiveness of the dark-bit detection method, although temperature sweeping can yield complete detection under the measured condition. At 30%

TABLE I. COMPARISON OF MASKING TECHNIQUES

(Room Temperature as Reference for BER Calculation)		Temperature					Dark Bits Detect?
		80° C	120° C				
This Work Vss Bias	Masking Ratio[a]	20%	10%	30%	50%	70%	Yes
	BER Improvement	88%	59.2%	92.0%	99.7%	100%	
ISSCC2016 [5] T Sweeps	Masking Ratio[b]	18.5%	–	–	–	–	No
	BER Improvement	90%[b] (@85°C)	–	–	–	–	
ISSCC2017 [6] Body Bias	Masking Ratio[a]	–	9%	–	–	–	Yes
	BER Improvement	–	60%	–	–	–	

[a.]Room temperature [b.]Temperature sweeps

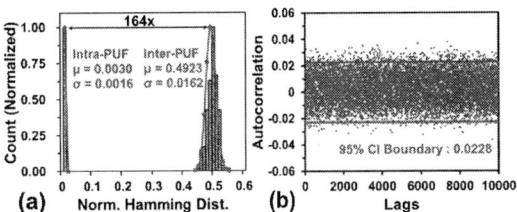

Fig. 7. (a) Normalized intra-/inter-PUF HD; (b) autocorrelation of bit cells

and 70% points in Fig. 6(c), the detections were performed at $V_{PE} = 0.8$ and 1.6V to efficiently locate the hidden dark bits, but data at $V_{PE} = 0.8$ and 1.6V were not compared.

Effects of masking the detected dark bits are demonstrated in Figs. 6(a), (b), and (c). Fig. 6(a) also shows the BERs across T variation after the masking. Masked bits are not regarded as the divisor for the BER calculation. A 10% masking reduces bit errors by more than 90% within T = 0–60 °C. For an extreme T variation range of T = -40–120 °C, masking 30% detected dark bits reduces bit errors by more than 88%, whereas a 50% masking reduces bit errors by up to 99.7% (<0.12% BER). Moreover, BERs are reduced to 0% at 70% masking. Fig. 6(b) shows that V_{DD} variation-induced bit errors are improved by 88% at 1.0 V after the 10% masking. With the 30% masking, a 93% BER improvement is achieved at both 1.2 and 1.4 V.

The BERs at VT corner conditions are measured to further investigate the effectiveness. Fig. 6(d) summarizes the results at five corners. By masking 30% and 50% detected dark bits, the BER is largely reduced to less than 0.84% and 0.21% at extreme VT corner conditions (1.4 V/-40 °C and 1.4 V/120 °C, respectively). When 70% dark bits are masked, only 1.3ppm bit errors are observed at 1.4 V/120 °C VT corner and no bit error appear at any other corners.

Notably, the proposed technique is effective at all of the corner conditions with the same masked bits (e.g., the same 30% bits). This fact suggests that the technique can also detect most of the dark bits caused by long-term aging.

These low BERs (e.g., <0.2% at 50% masking) enable the usage of lightweight ECCs with only 5-bit or 6-bit correction (e.g., BCH (63, 36, 11) or BCH (127, 85, 13)) to achieve <10^-7 key-error ratio (128-bit key length). By this combination, the overheads of additional PUF bits and logic circuits become much smaller than the conventional two-stage ECC, which requires ~10× larger PUF bits for the key generation [11], [12].

TABLE I shows a fair comparison with previous masking techniques. In comparison with [5], nearly the same BER

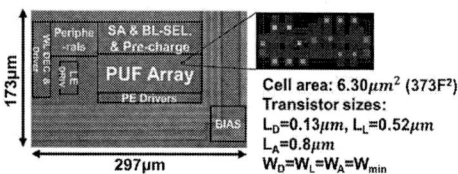

Cell area: 6.30μm^2 (373F^2)
Transistor sizes:
L_D=0.13μm, L_L=0.52μm
L_A=0.8μm
W_D=W_L=W_A=W_{min}

Fig. 8. Die micrograph, bit-cell layout, and transistor sizes.

TABLE II.   COMPARISON WITH PRIOR ARTS

	This work	ISSCC 2018 [9]	ISSCC 2018 [8]	ISSCC 2017 [6]	JSSC 2017 [3]	ISSCC 2016 [5]	JSSC 2008 [1]
Technology	130nm	55nm	180nm	180nm	14nm	45nm	130nm
Cell Area(F^2)	373	218	445	553	9388	2613	1092
Native BER (Nominal)	0.21%	-	0.72%	0.13%	-	0.10%^{b}	3.04%
Native Unstable Bits	2.14%	-	6.65%	1.67%	26.37%	-	-
Stabilizing Technique	V_{SS}-Bias Dark-Bit Detection	Oxide Rupture + Voltages Voting	Remap -ping	Body-Bias Masking	NBTI Burn-in + Masking	Glitch Detection + Masking^c	-
Measured V_{DD} (V)	0.8-1.4	0.75-1.35	1.2-1.8	0.8-1.8	0.55-0.75	-	0.9-1.2
Measured Temp (°C)	-40-120	-40-150	0-80	-40-120	25-110^d	-25-85	0-80
Stabilized BER @VT Corner	1.3ppm	0^{a,e}	-	1.28%^{e}	1.46%	0.21%^{c,e}	-
BER Improvement @VT corner	99.995%	100%^{e}	-	60%^{e}	51.3%	90%^{c,e}	-
Standby Power (pW/b)	0.46	-	-	26	2930 (70°C)	-	1260
Core Energy (fJ/b)	121	5200	9800	11.3	4	-	930

^{a}after hard oxide-breakdown processing ^{b}with 2-bit glitch detection
^{c}requires temperature sweeping ^{d}golden value set at 70°C ^{e}at T corner

improvement is achieved without costly temperature sweeping. Compared with [6], the proposed method can find out a much wider range of potential dark bits until BER approaches 0.

C. Uniqueness and Randomness

Uniqueness and randomness are basic requirements for a PUF. As shown in Fig. 7(a), the 164× separation between inter- and intra-PUF hamming distance (HD) shows the high uniqueness of the proposed PUF design. The randomness is verified with the close-to-zero autocorrelation (Fig. 7(b)). The proposed PUF also passed all of the applicable NIST randomness tests. After a 70% dark-bit masking, the average inter-PUF HD is 0.5025, showing no degradation on uniqueness.

D. Energy and Area Efficiency

The core operation power is 4.04 μW @ 33.33 Mbps throughput and the active energy is 121 fJ/bit. The total energy, including all of the peripheral circuits, is 243 fJ/bit. Due to the normally off V_{PE} and V_{LE} after the evaluation and reading, the PUF core static current is only 0.58 nA @ 0.8 V/23 °C and the standby power is 0.46 pW/bit. The die micrograph and bit-cell layout are shown in Fig. 8. The proposed bit-cell is more than 1.19× smaller than a recent compact PUF design [8]. The area overhead of the bias generator is 8%.

TABLE II shows the comparison with prior arts. The proposed design achieves a ~0% error across the extreme VT variations, a 99.995% BER improvement at the VT corner, and extremely low standby power. In addition, it has a high native stability that is comparable with the state-of-the-art mono-stable PUFs [5], [6] and a compact bit cell smaller than [1]–[8]. Compared with [9], this work implements several secure schemes for enhanced attack-tolerant ability.

IV. CONCLUSION

This paper has described an EE SRAM PUF with an integrated V_{SS}-bias generator for dark-bit detection. The proposed dark-bit detection can prevent potential bit errors with only 1.3×10^{-6} measured BER at the extreme VT corner. The EE structure of PUF cells introduces the mono-stable state into the SRAM power-up behavior to improve the stability against thermal noise. This design also achieves a compact cell area and ultra-low standby power. Moreover, the differential structure, remanence-charge clearance scheme, and normally powered-off bit cells provide better attack-tolerant ability.

ACKNOWLEDGMENT

The authors would like to thank ROHM Co., Ltd. and Kitakyushu FAIS for their supports in this research.

REFERENCES

[1] Y. Su et al., "A digital 1.6pJ/bit chip identification circuit using process variations," IEEE J. Solid-State Circuits, vol. 43, no. 1, pp. 69-77, Jan. 2008.

[2] S. K. Mathew et al., "A 0.19pJ/b PVT-variation-tolerent hybrid physically unclonable function circuit for 100% stable secure key generation in 22nm CMOS," in ISSCC Dig. Tech. Papers, 2014, pp. 278-279.

[3] S. Satpathy et al., "A 4-fJ/b delay-hardened physically unclonable function circuit with selective bit destabilization in 14-nm trigate CMOS," IEEE J. Solid-State Circuits, vol. 52, no. 4, pp. 940-949, Apr. 2017.

[4] A. B. Alvarez et al., "Static physically unclonable functions for secure chip identification with 1.9-5.8% native bit instability at 0.6-1.0V and 15fJ/bit in 65nm," IEEE J. Solid-State Circuits, vol. 51, no. 3, pp. 763-775, Mar. 2016.

[5] B. Karpinskyy et al., "Physically unclonable function for secure key generation with a key error rate of 2E-38 in 45nm smart-card chips," in ISSCC Dig. Tech. Papers, 2016, pp. 158-159.

[6] K. Yang et al., "A 553F^2 2-transistor amplifier-based physically unclonable function (PUF) with 1.67% native instability," in ISSCC Dig. Tech. Papers, 2017, pp. 146-147.

[7] S. Taneja et al., "A fully-synthesizable C-element based PUF featuring temperature variation compensation with native 2.8% BER, 1.02fJ/b at 0.8-1.0V in 40nm," in Proc. A-SSCC, 2017, pp. 301-304.

[8] J. Lee et al., "A 445F^2 leakage-based physically unclonable function with lossless stabilization through remapping for IoT security," in ISSCC Dig. Tech. Papers, 2018, pp. 132-133.

[9] M. -Y. Wu et al., "A PUF scheme using competing oxide rupture with bit error rate approaching zero," in ISSCC Dig. Tech. Papers, 2018, pp. 130-131.

[10] W. -C. Wang et al., "Implementation of stable PUFs using gate oxide breakdown," in Proc. AsianHOST, 2017, pp. 13-18.

[11] M. Hiller and A. G. Önalan, "Hiding secrecy leakage in leaky helper data," in Proc. CHES, 2017, pp. 601-619.

[12] R. Maes, Physically Unclonable Functions: Constructions, Properties and Applications. New York, NY, USA: Springer, 2013.

[13] T. Fujino et al., "The development of tamper resistant security hardware and demonstration for vehicle application," IEICE Trans. Fundamentals A, vol. J99-A, no. 2, pp. 83-93, Feb. 2016.

[14] S. Zeitouni et al., "Remanence decay side-channel: the PUF case," IEEE Trans. Inf. Forensics Security, vol. 11, no. 6, pp. 1106-1116, Jun. 2016.

[15] Z. Cui et al., "Measurement of mismatch factor and noise of SRAM PUF using small bias voltage," in Proc. ICMTS, 2017, pp. 1-4.

[16] H. Shinohara et al., "Analysis and reduction of SRAM PUF bit error rate," in Proc. VLSI-DAT, 2017, pp. D9-4.

978-1-5386-6414-8/18 $31.00 © 2018 IEEE

14-1 (8065)

An 82.1%-Power-Efficiency Single-Inductor Triple-Source Quad-Mode Energy Harvesting Interface with Automatic Source Selection and Reversely Polarized Energy Recycling

Chih-Lun Lo, Hao-Chung Cheng, Pei-Chun Liao, Yi-Lun Chen, Po-Hung Chen
Institute of Electronics
NCTU (National Chiao Tung University), Hsinchu, Taiwan
Email: hakko@nctu.edu.tw

Abstract—In this paper, a single-inductor triple-source quad-mode energy harvesting interface with automatic source selection is developed for multi-source energy harvesting. The interface employs buck-boost topology to convert the energy from photovoltaic (PV) cells and thermoelectric generator (TEG) to a regulated 1.2-V output. The proposed converter has four operating modes: harvesting mode (HM), recycling mode (RM), storage mode (SM), and backup mode (BM). The operating mode is selected automatically according to the input and load conditions to appropriately transfer the energy. The proposed reversely polarized energy recycling (RPER) technique extends the available output power range by 10x with 25.3% efficiency improvement at low input voltage. The measurement results demonstrates the proposed interface achieves the peak efficiency of 82.1%.

Keywords— energy harvesting, automatic source selection, reversely polarized energy harvesting

I. INTRODUCTION

Energy harvesting has been widely applied in various fields such as wireless sensors and the internet of things (IoT). The conventional single-source energy harvesting, however, may suffer from insufficient ambient energy. Therefore, the development of an energy harvesting system with multiple energy transducers is a promising solution for a self-sustaining system.

A previously published study [1] harvests energy from a photovoltaic (PV) cell or a thermoelectric generator (TEG); however, only a single transducer can be used. In [2], a single-inductor multiple-input-dual-output power converter, which manages multiple energy transducers, is demonstrated; however, the sources are manually selected. Recent work [5] demonstrated a multi-source energy harvesting interface whose input sources are selected automatically. However, the conversion efficiency at low input voltage is low and the startup circuit is not included.

This paper presents a single-inductor triple-source energy harvesting interface, which is capable of selecting an appropriate input source and four converter modes (HM, RM, SM, BM) automatically. The input voltage (V_{IN}) ranges from 0.1 V to 1.8 V to meet the requirement of indoor/outdoor photovoltaic (PV) cells and a thermoelectric generator (TEG). The converter

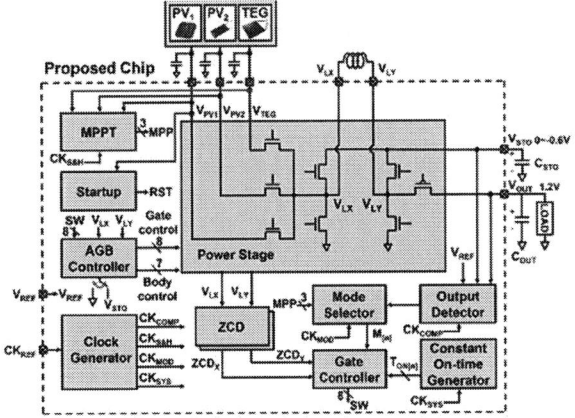

Fig. 1. Block diagram of proposed energy harvesting interface.

regulates the output (V_{OUT}) to 1.2 V and stores excessive energy in the storage element (C_{STO}). Moreover, the novel reversely polarized energy recycling (RPER) technique is proposed to enhance the conversion efficiency at low V_{IN}.

The paper is organized as follows. The system architecture and the operational sequences are introduced in Section II. Section III describes the controller circuits and the proposed RPER. The measurement results and a comparison of the state-of-the-art power converters are discussed in Section IV. Finally, conclusions are drawn in Section V.

II. SYSTEM ARCHITECTURE

The system block diagram of the proposed triple-source quad-mode energy harvesting interface is shown in Fig. 1. The power stage employs buck-boost converter topology with eight power switches to manage multiple energy sources and outputs. C_{STO} stores excessive energy in the SM as a negative voltage (V_{STO}) and recycles it in the RM to achieve both high conversion efficiency and wide output power range. Adaptive gate biasing (AGB) minimizes the drain and body leakage current of all the power transistors in each operating mode. A constant on-time generator determines the on time of the power transistors while two zero current detectors (ZCDs) optimize the off-time of the

978-1-5386-6414-8/18 $31.00 © 2018 IEEE

165

power transistors to minimize the reverse inductor current. The startup circuit allows the system to self-start without battery usage. Maximum power point tracking (MPPT) along with the mode detection circuit automatically determines the operating modes and extracts the maximum power from the energy transducers [6]. The MPPT circuit samples the open circuit voltage of input sources periodically and generates 0.8 times the PV V_{OC} (V_{PVMPP}) and half the TEG V_{OC} ($V_{TEG,MPP}$) via two on-chip switched-capacitor circuits with different conversion ratios.

Fig. 2 illustrates the operating principle of mode selection. The system automatically determines the operating mode by monitoring the V_{OUT} at the end of the previous cycle $T_{OFF}[n-1]$, V_{OUT} at the beginning of the current state $T_{ON}[n-1]$, and the input source conditions. The interface extracts the power from the transducer when the source voltage ($V_{PV1}/V_{PV2}/V_{TEG}$) is higher than its maximum power point voltage ($V_{PV1,MPP}/V_{PV2,MPP}/V_{TEG,MPP}$). The detailed operating mode selection is listed in table I. Fig. 3 shows the corresponding power flow in the power stage during the on-time/off-time period. It is worthy to note that the recycling mode combines the energy from the harvesters and the storage element to provide sufficient power to the load.

In HM, the converter extracts the power from the transducers to the output. In RM, the converter transfers energy from the transducers and the storage element to the output load in one switching cycle to increase the voltage difference across the inductor. The converter operates in SM when the harvested power is higher than load power. The converter stores the excessive energy from transducers to the storage element. If the harvested energy is insufficient, the converter operates in backup mode (BM). The converter transfers the power from the storage element to the output load.

TABLE. I: OPERATING MODE SELECTION

$T_{OFF}(N-1)$	$T_{ON}(N)$	$V_{PV1}<V_{MPP1}$ & $V_{PV2}<V_{MPP2}$ & $V_{PV3}<V_{MPP3}$	$V_{PV1}>V_{MPP1}$ / $V_{PV2}>V_{MPP2}$ / $V_{PV3}>V_{MPP3}$
$V_{OUT}>V_{REF}$	$V_{OUT}<V_{REF}$	Backup Mode	Harvesting Mode
$V_{OUT}<V_{REF}$			Recycling Mode
-	$V_{OUT}>V_{REF}$	- (Skip)	Storage Mode

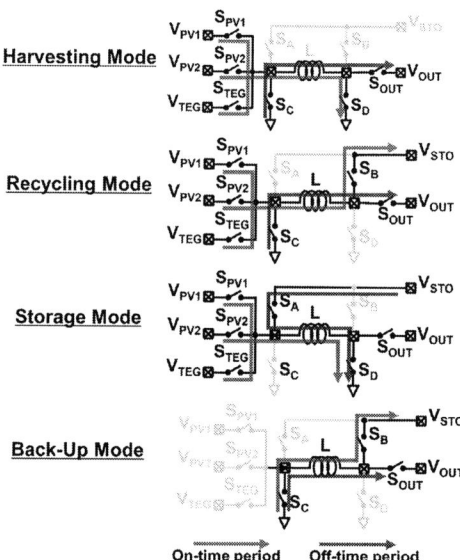

Fig. 3. Power flow under different operating modes.

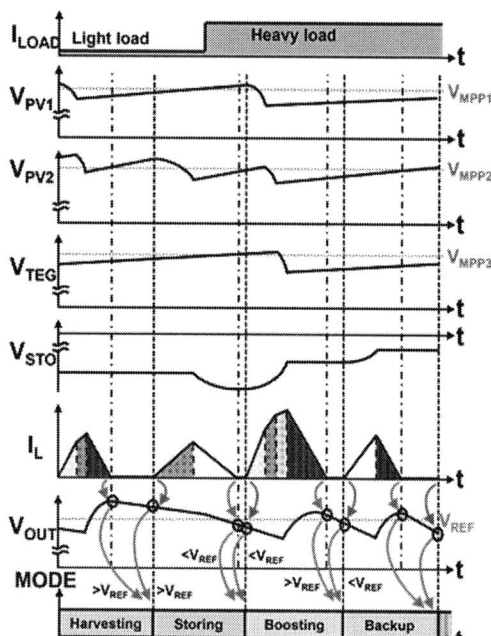

Fig. 2. Operating principle of the converter.

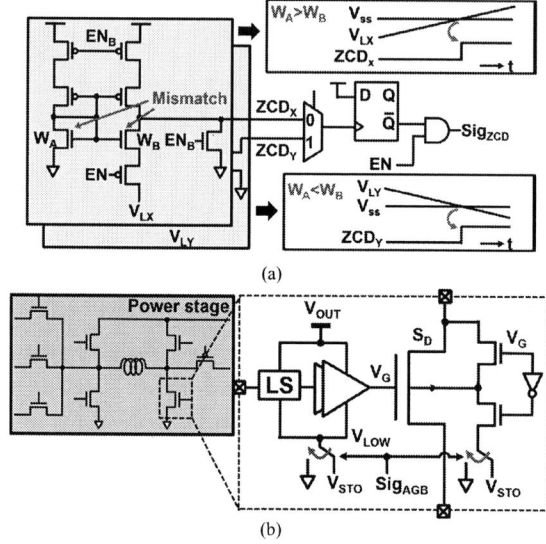

Fig. 4. Circuit schematic of proposed (a) ZCDs and (b) AGB controller.

978-1-5386-6414-8/18 $31.00 © 2018 IEEE

III. CONTROLLER CIRCUITS AND RPER

Fig. 4 shows the schematic of the ZCDs and AGB circuit. Two ZCD circuits determine the off-time period by comparing V_{LX}/V_{LY} with a pair of mismatched transistors to minimize the reverse inductor current (I_L) in each operating mode [7]. The AGB circuit dynamically switches the lower bound of V_G to V_{STO} when V_{LX}/V_{LY} is connected to V_{STO} to ensure the power transistors and the body diodes are fully turned off.

Fig. 5 compares the proposed RPER with the conventional approach. The proposed architecture is capable of combining the energy from the transducers and storage element in one switching cycle to deliver a higher power to the load. Since the slope of I_L is proportional to $V_{IN} - V_{STO}$ (V_{STO} is negative), the output power (P_{OUT}) increases for a fixed on time (T_{ON}). On the other hand, T_{ON} can be reduced to reach the required I_{PEAK} (P_{OUT}). Since the conduction losses can be reduced by the shortened T_{ON}, the proposed converter with RPER can achieve a higher conversion efficiency for a fixed P_{OUT}.

IV. MEASUREMENT RESUTLS

The proposed energy harvesting power converter is implemented in 180-nm CMOS process. Fig. 6 shows the chip micrograph and Fig. 7 shows measured efficiency under different operating modes. The proposed converter demonstrates a conversion efficiency higher than 82.1% with an output power range of 2.5 µW to 10 mW. Fig. 8 demonstrates the startup process and the MPPT function. The cold-startup voltage is as low as 370mV and the energy transducers achieve the maximum power point. Fig. 9 shows that the operating mode is automatically selected according to the status of the transducers and output load. Fig. 10 demonstrates that the proposed RPER improves the power efficiency and the output power range by 25.3% and 10x, respectively, at V_{IN}= 0.1V and V_{STO}= -0.4V.

Table II summarize the performance comparison with state-of-the-art energy harvesting dc-dc converters [1-5]. The proposed interface along with RPER is able to handle the energy from three energy transducers and automatically choose the energy source and operating mode cycle by cycle.

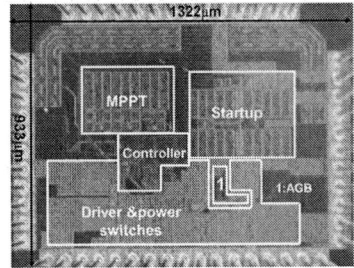

Fig. 6 Chip micrograph

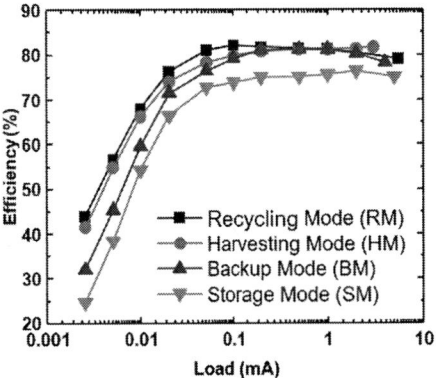

Fig. 7 Measured conversion efficiency under different operating modes.

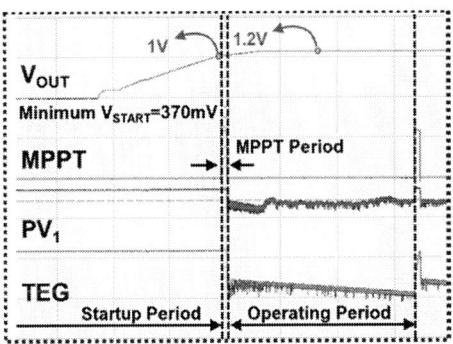

Fig. 8. Measured startup and MPPT waveforms of proposed energy harvesting interface.

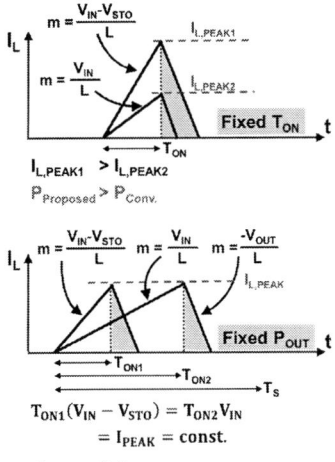

Fig. 5. Comparison between the proposed RPER and conventional approach.

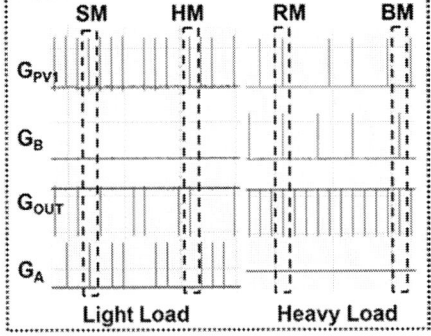

Fig. 9. Measured waveforms under different loading current.

978-1-5386-6414-8/18 $31.00 © 2018 IEEE

TABLE. II: PERFORMANCE COMPARISON WITH STATE-OF-THE-ART ENERGY HARVESTING POWER CONVERTERS

	[1] JSSC'15	[2] JSSC'12	[3] JSSC'15	[4] ISSCC'16	[5] JSSC'18	This work
Technology	130nm CMOS	350 nm CMOS	130nm CMOS	350 nm CMOS	180nm CMOS	180nm CMOS
Input voltage	0.01V-0.3V	0.02V-5V	-	0.03V-3.6V	0.05V-0.8V	0.1V-1.8V
Battery	No	No	Yes	Yes	No	No
Output Power	<22mW*	9μW-10mW	1μW-10mW	-	20μW-4mW	2.5μW-10mW
Peak eff.	83%	83%	83%	85%**	84%	82%
Inductor	10μH	22μH	10μH	22μH	4.7μH	4.7μH
Harvesting Source	PV/TEG	PV &TEG &PZT	PV	PV	PV &TEG	PV(outdoor) &PV(indoor) &TEG
Source Selection	-	-	-	-	Automatic	Automatic
No. of harvesting source in a cycle	1	1	1	1	2	3
Startup voltage	220mV /-14.5dBm(RF)	N/A	NA	300mV	N/A	370mV

* Minimum output power is not mentioned ** 2-stage power conversion

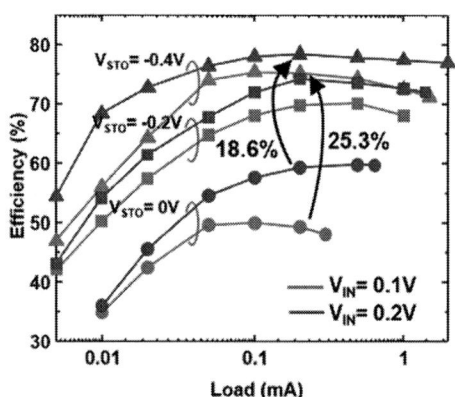

Fig. 10. Measured dependences of conversion efficiency on loading current under different storage voltage.

I. CONCLUSION

This paper presents a single-inductor triple-source quad-mode energy harvesting interface, capable of converting the energy from two PV cells and a TEG to a regulated 1.2V output. The proposed converter automatically operates in four operating modes (HM, RM, SM, BM) according to input and output conditions, to make the transducers operating at maximum power point while regulating the output voltage. The proposed RPER technique stores a negative voltage in the storage capacitor and reuses the energy in RM to improve the conversion efficiency and available output power at low V_{IN}. As a result, the proposed converter achieves a maximum conversion efficiency of 82.1% with output power ranging from 2.5μW to 10mW. The proposed (RPER) technique extends the available output power range by 10x with 25.3% efficiency improvement at low V_{IN}.

ACKNOWLEDGMENT

This work was partially supported by MOST, R.O.C. (107-2622-8-009-010-TA and 107-2221-E-009-133-MY2), and "Center for Neuromodulation Medical Electronics Systems" from The Featured Areas Research Center Program within the framework of the Higher Education Sprout Project by the Ministry of Education (MOE) in Taiwan. We would like to thank the National Chip Implementation Center (CIC) for chip fabrication.

REFERENCES

[1] A. Shrivastava, et al., "A 10 mV-input boost converter with inductor peak current control and zero detection for thermoelectric and solar energy harvesting with 220 mV cold-start and −14.5 dBm, 915 MHz RF kick-start," *IEEE Journal of Solid-State Circuits*, pp. 1820-1832, Apr. 2015

[2] S. Bandyopadhyay, et al, "Platform architecture for solar, thermal, and vibration energy combining with MPPT and single inductor," *IEEE Journal of Solid-State Circuits*, pp. 2199-215, Sep. 2012.

[3] G. Yu, et al., "A 400 nW single-inductor–tri-output dc–dc buck–boost converter with maximum power point tracking for indoor photovoltaic energy harvesting," *IEEE Journal of. Solid-State Circuits*, vol. 50, no. 11, pp. 2758-2772, Nov. 2015.

[4] Y. Lu, et al., "A 200nA single-inductor dual-input-triple-output (DITO) converter with two-stage charging and process-limit cold-start voltage for photovoltaic and thermoelectric energy harvesting," *IEEE International Solid-State Circuits Conference (ISSCC) Dig. Tech. Papers*, pp. 368-369, Feb. 2016..

[5] C. –W. Liu, et al., "Dual-Source Energy-Harvesting Interface with Cycle-by-Cycle Source Tracking and Adaptive Peak-Inductor-Current Control," *IEEE Journal of Solid-State Circuits*. (to be appear)

[6] K. Kadirvel et al., "A 330 nA energy-harvesting charger with battery management for solar and thermoelectric energy harvesting," *in IEEE Int. Solid-State Circuits Conf. (ISSCC) Dig. Tech. Papers*, Feb. 2012, pp. 106–108.

[7] X. Zhang et al., "A 0.6 V input CCM/DCM operating digital buck converter in 40 nm CMOS," *IEEE J. Solid-State Circuits*, vol. 49, no. 11, pp. 2377–2386, Nov. 2014.

A Transient-Enhanced Constant On-Time Buck Converter with Light-Load Efficiency Optimization

Mao-Ling Chiu, Tzu-Hsuan Yang, and Tsung-Hsien Lin

Graduate Institute of Electronics Engineering and Department of Electrical Engineering

National Taiwan University, Taipei, Taiwan

Abstract - Two key challenges in designing regulators of handheld devices are load transient time and light-load efficiency. A constant on-time (COT) buck converter is often adopted because it facilitates a fast response with reasonable light-load efficiency. For further improvement, two techniques are proposed in this work. One is the transient-enhanced technique which can reduce the settling time and minimize the voltage dips. The other one is the self-tuning technique to improve the light-load efficiency, termed as the near-optimal reverse current calibration (NORCC). The proposed COT buck converter is realized in 0.18-μm CMOS. Measurement results demonstrate settling time and undershoot voltage of 960 ns and 50 mV, respectively, during the load changes from 100 mA to 1.5 A. The efficiency is better than 79% from 10 mA to 1.7 A; the peak value is 93.6% at 300 mA load current.

I. INTRODUCTION

Power management (PM) plays an important role in portable devices. Its design is required to maintain high conversion efficiency over wide load range. In particular, the efficiency (η) at light-load condition tends to degrade, as depicted in Fig. 1(a), and is often an issue to be addressed. One reason for the decline in efficiency is the reverse inductor current which flows from the output back to ground. Conventionally, the zero-current detector (ZCD) is employed to avoid this situation. But, the delay of the detector and driver inevitably degrades the efficiency [1]. Hence, it still remained to be solved. In addition, the load transient performance is another critical issue for high-speed applications which demands fast transition between active mode and low-power idle mode. The system recovery speed also affects the output voltage dips, as shown in Fig. 1 (b).

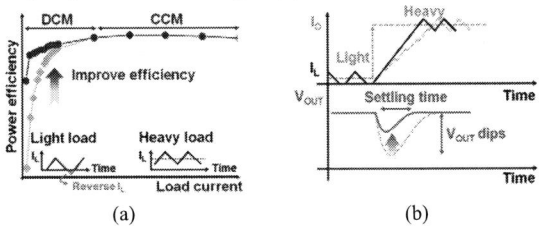

(a)   (b)

Fig. 1. Two key design challenges of a voltage converter, (a) efficiency; (b) settling time and voltage undershoot.

To address the aforementioned two design issues, the transient-enhanced (TE) technique and near-optimal reverse current calibration (NORCC) circuit are proposed in this work. This paper is organized as follows. In Section II, the proposed architecture and circuit implementation are introduced. Section III and IV present the experimental results and conclusion, respectively.

II. THE PROPSED ARCHITECTURE

The overall proposed COT buck converter with the TE technique and NORCC circuit are shown in Fig. 2.

A. Proposed TE Technique

There are two operation modes in the proposed COT converter. One is the reset-dominant mode, which operates as a conventional COT control. The other one is a set-dominant mode, termed as the transient-enhanced (TE) mode. In the TE mode, the reset signal (V_R) is ignored to keep the V_Q staying high when both V_S and V_R go high. The waveforms and schematics are shown in Fig. 3. In effect, the TE mode makes the high-side power switch (M_{P1}) remaining turned on for shortening the settling time and reducing the voltage dips.

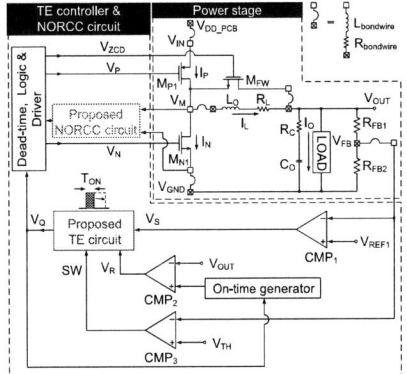

Fig. 2. The proposed TE buck converter.

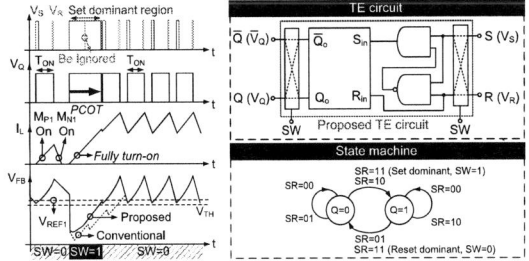

Fig. 3. Waveforms and schematics of the proposed technique.

B. Proposed NORCC Technique

In order to improve the light-load efficiency, the ZCD is usually employed to detect the reversed inductor current at light-load condition in a conventional synchronous buck converter. But the limited bandwidth of detectors and the long propagation delay of drivers degrade the converter performance. As a result, the inaccurate switching point causes

978-1-5386-6414-8/18 $31.00 © 2018 IEEE

reverse current and hurts the efficiency. It becomes more serious as the size of power inductors (L_O) decreases for higher switching frequency. In this work, the self-tuning algorithm based calibration is proposed to eliminate the time delay.

Fig. 4 shows the implementation of the proposed concept. The falling edge of signal V_N (=V_{Sample}) is used to close the Power NMOS (M_{N1}) and sample the switching node voltage V_M. After that, the sample-and-hold circuit adjusts I_{offset} based on the V_M situation to compensate the delay which is caused by the latency through the detector and drivers. After several cycles, the reverse current will be mitigated. The detailed circuit implementation is depicted in Fig. 5.

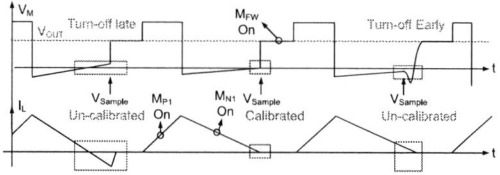

Fig. 4. Timing diagram of the proposed calibration.

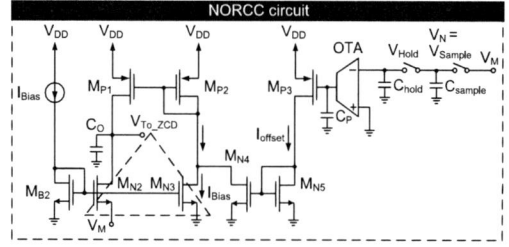

Fig. 5. Schematic of the proposed near-optimal reverse current calibration (NORCC) technique.

III. EXPERIMENTAL RESULTS

The proposed buck converter is fabricated in a TSMC 180-nm CMOS process. Fig. 6 shows the measured load transient waveforms for the conventional and the proposed converter, respectively. Compared to that of the conventional one, the undershoot voltage of the proposed technique is improved from 120 mV to 50 mV which realizes 58% enhancement and the settling time is shortened from 1.28 μs to 0.96 μs.

The measured calibration process of the self-tuning algorithms is shown in Fig. 7. In the CCM, the V_M is a negative value and the voltage of the capacitor C_O is low. According to the result of the V_M in the CCM, the NORCC will not add more current to the C_O in the first DCM cycle. It will result in inaccurate switching point. But after several cycles, this can be calibrated by the proposed circuit. The steady-state waveforms are also shown in Fig 7. Thus, the measured efficiency is better than 79% between 10 mA and 1.7 A which is optimized by the proposed technique, and the peak value is 93.6%, as shown in Fig. 8. Fig. 9 shows the chip micrograph that measures 2.8 mm² including the pads. Table I summarizes the measured performance and compares with other similar works.

IV. CONCLUSIONS

A transient-enhanced buck converter with optimized light-load efficiency is presented by using the proposed techniques. The measured settling time and the undershoot voltage are

improved to 960 ns and 50 mV, respectively. Moreover, the efficiency is optimized and its value is better than 79% for a load current range from 10 mA to 1.7 A. The peak efficiency is 93.6%. Thus, this work has shorter setting time and higher light-load efficiency under similar frequency.

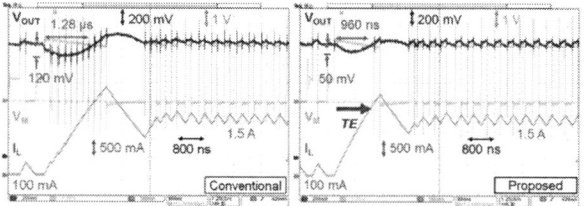

Fig. 6. Transient performances of the conventional and proposed control method.

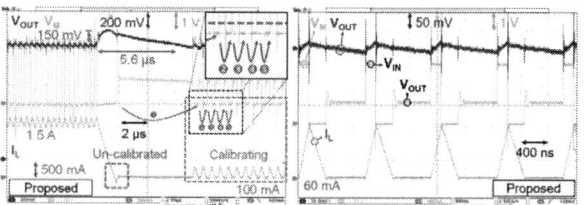

Fig. 7. Calibration process and steady-state waveforms of the proposed NORCC technique.

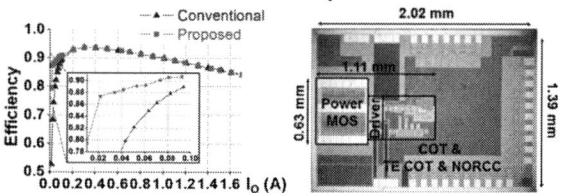

Fig. 8. Measured efficiency.   Fig. 9. Chip micrograph.

TABLE I. COMPARISON TABLE

	[2]	[3]	[4]	[5]	This work
Process (nm)	65	28	500	130	**180**
Topology	T-PID PWM/PFM	COT	CF-AOT V²	VM Hybrid LDO	**TE COT**
V_{IN} / V_{OUT} (V/V)	1.8 / 0.5	3.3 / 1.05	48 / 10	3.3 / 1.8	**3.3 / 1.2**
L (μH) / C (μF)	0.22 / 4.7	1 / 4.7	1.5 / 4.7	0.09 / 0.94	**1 / 4.7**
Sw. frequency (MHz)	10	~2.5	~2	30	**~2.5**
Settling time (μs) (L→H) (A)	20 (0.02-0.42)	8 (0.3-1.7)	20 ($\triangle I_O$=0.4)	0.125 ($\triangle I_O$=1.25)	**0.96 (0.1-1.5)**
Voltage drops (mV)	40	95	100	36	**50**
Peak efficiency (%) @ I_{LOAD} (A)	80 @ 0.2	89	90 @ 0.5	88	**93.6 @ 0.3**
Efficiency (%) @ 10 mA	75	< 75	< 75	< 73	**79.5**

ACKNOWLEDGMENT

The authors thank the National Chip Implementation Center for fabricating this chip. This work is supported in part by the MOST, Taiwan.

REFERENCES

[1] C.-L. Chen et al., *IEEE ISCAS*, pp. 2214-2217, May 2008.
[2] S. J. Kim et al., *IEEE CICC*, pp. 1-4, Apr. 2017.
[3] W. H. Yang et al., *IEEE TPEL*, pp. 1-27, Aug. 2017.
[4] J. Xue et al., *IEEE ISSCC*, pp. 226-227, Feb. 2016.
[5] L. Cheng et al., *IEEE ISSCC*, pp. 188-189, Feb. 2017.

A 99.2% Tracking Accuracy Single-Inductor Quadruple-Input-Quadruple-Output Buck-Boost Converter Topology with Periodical Interval Perturbation and Observation MPPT

Chao-Jen Huang[1,2], Yao-Sheng Ma[1], Wen-Hau Yang[1], Yen-Ting Lin[1], Chun-Chieh Kuo[1], Ke-Horng Chen[1], Hsiao-Jung Liu[2], Pei-Shan Yu[2], Fang-Chih Chu[2], Ching-Ju Lin[2], Hong-Wen Huang[2], Kuo-Chih Hung[2], Yuan-Hua Chu[2], Ying-Hsi Lin[3], Suhwan Kim[4], Krishnan Ravichandran[4]

[1]National Chiao Tung University, [2]Industrial Technology Research Institute, [3]Realtek Semiconductor Corp, Hsinchu, Taiwan
[4]Intel Corporation, Hillsboro, OR, US

Abstract — this paper provides a Single-Inductor Quadruple-Input-Quadruple-Output buck-boost converter to achieve maximum tracking efficiency of 99.2% due to continuous supply in one cycle. The Maximum Power Point (MPP) tracking mode and the Harvest Energy Delivering (HED) mode harvest each energy harvesting source, deliver energy to the battery and supply it to the loading system in one cycle. Each output voltage ranges from 600mV to 2.5V and the maximum output power is up to 60mW. The peak efficiency is 91.2%.

Keyword: Single-Inductor Quadruple-Input-Quadruple-Output (SI-QIQO), Maximum Power Point Tracking (MPPT), Harvest Energy Delivering (HED).

I. INTRODUCTION

Different energy harvesting sources such as photovoltaic (PV) modules, thermoelectric generators (TEGs), fuel cells (FCs) and other DC energy sources are used to provide sufficient energy to the system [1] - [5]. Thus, it is imperative to design maximum power point tracking (MPPT) techniques to accommodate each energy source, and to find a trade-off between tracking efficiency and energy delivery for loading system. In Fig. 1(a), the conventional energy harvesting converter for MPPT is disturbed by varying output voltages. If one of the outputs has a +/- ΔV_{OUT} disturbance as shown in Fig. 1(b), the falling slope has been distorted due to equation (1), and the MPPT point deviates from its peak value, decreasing the tracking efficiency. In addition, the energy delivered to the loading system is also affected by the acquisition and energy delivery sequences in different MPPT technologies [1] - [4]. If the energy source does not provide enough energy, the prior arts simply use additional storage source (e.g., Li-Ion battery, V_{BAT}) to supply energy to the loading system. Although multiple energy harvesting sources provide adequate energy, system loading and battery life may be limited to the weakest energy source. Therefore, this paper presents a Single-Inductor Quadruple-Input-Quadruple-Output (SI-QIQO) buck-boost (BB) converter with V_{BAT} for maximum tracking efficiency and continuous power supply to the loading system in one cycle. In other words, due to the single-inductor multi-output characteristics, the SI-QIQO converter uses a dual mode controller including MPP tracking mode and Harvesting Energy Delivering (HED) mode to make each energy harvesting source harvest efficiently as energy delivered to the V_{BAT}, and supply loading system at the same time if output requests.

$$I_{EH1.AVG} = \int_0^{T_{EH1}} I_L dt / T_{CP} \qquad (1)$$

II. PROPOSED SI-QIQO BB CONVERTER

SI-QIQO converter in Fig. 2 uses dual-mode controller including MPP tracking mode and harvesting energy delivering

(HED) mode to collect energy from each harvest source and deliver energy to V_{BAT} while providing loading system if each output requests. Compared with the prior arts, MPP tracking mode uses Path 1 and Path 2 to ensure MPP tracking stability because the battery is assumed to have a large storage capacitance and the collected energy is always stored in V_{BAT}. In other words, the voltage remains constant in MPP tracking mode to effectively suppress the variation +/- ΔV_{OUT} as shown in Fig. 1(b), and the time of Path2 is proportional to input energy harvesting power. On the other hand, the HED mode checks if energy from harvesting source is sufficient. As the energy is enough to provide loading outputs V_{OUT1}-V_{OUT3}, the HED mode works in sub-mode I, and control paths are in the Path1, Path4, and Path2 order.

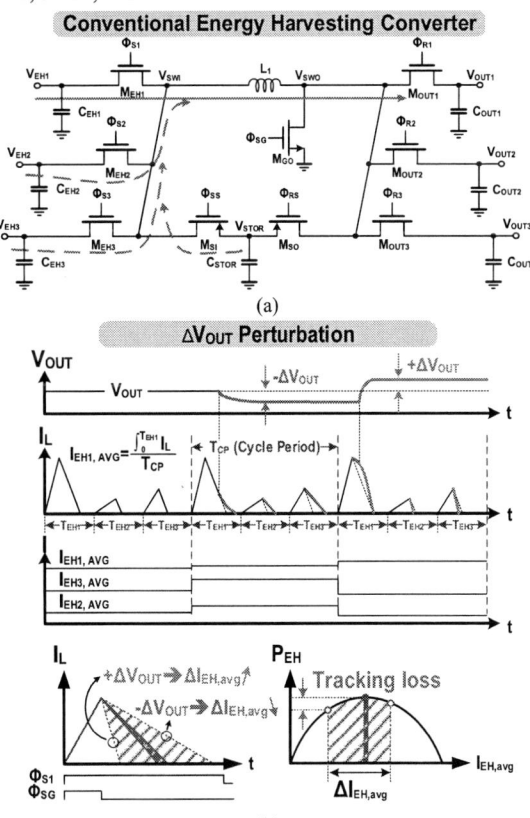

Fig. 1. (a) The conventional energy harvesting converter for MPPT. (b) The MPPT is disturbed by varying output voltage, and tracking efficiency is decreased.

In sub-mode I of the HED mode, the SI-QIQO BB converter harvests energy, regulates the outputs, and stores extra energy into V_{BAT} within one cycle. Conversely, if energy harvesting source has insufficient energy to provide outputs, the HED mode switches to sub-mode II to convert energy from the harvesting source and the V_{BAT}, and the control path sequence is Path 1, Path 4, and Path 3. Besides, V_{BAT} should avoid excessively discharging for battery protection. Owing to low power of the energy harvesting sources, MPPT mode is in low frequency operation to check the MPP point and prevent V_{OUT1}-V_{OUT3} from dropping extremely low.

III. CIRCUIT IMPLEMENTATION

A. PI-P&O MPPT controller

In Fig. 3(a), Periodical Interval Perturbation and Observation (PI-P&O) MPPT controller contains zero current detector, periodical power integrator and duty cycle generator. Without a high-speed counter in low power and high efficiency system, it is hard to have an accurate zero current signal. In addition, it is often affected by input noise, switching noise and crosstalk, so the tracking efficiency is severely affected. Therefore, this paper proposes a periodic power integrator (PPI) circuit to integrate the constant current controlled by the zero current detection (ZCD) signal, which is proportional to the input energy harvesting power and compared with a predefined threshold. After that, due to low-pass filter in charge pump module, the duty cycle generator (DCG) circuit derives duty cycle with high accuracy. Fig. 3(b) shows the MPPT controller operating in two modes. First, MPP tracking mode searches corresponding MPP values to each energy harvesting source. Take source 1 as example, $C_{IG,1}$ is charged and $V_{IG,1}$ increases every duty-off, after 26 duty cycles, $V_{IG,1}$ exceeds $V_{REF,1V}$ and the signal "UP" increases duty-on by charging node V_{ERR} from the charge pump.

The counter counts the value to 26. In case of sufficient source energy, $C_{IG,1}$ is charged more in each cycle, thus $V_{IG,1}$ exceeds $V_{REF,1V}$ just after 17 duty cycles and duty-on is increased again; if MPP of harvesting source is exceeded, counter will decrease next time. By this means, the MPP is obtained by searching minimum number, and mode operation switches to HED sub-mode I to regulate multiple outputs, following the flow chart in Fig. 3(b). Regardless of MPPT or HED sub-mode I, the corresponding duties are exported to the Power Stage Dispenser (PSD) in Fig. 3(c). If the energy harvesting source is insufficient, operation switches to HED sub-mode II.

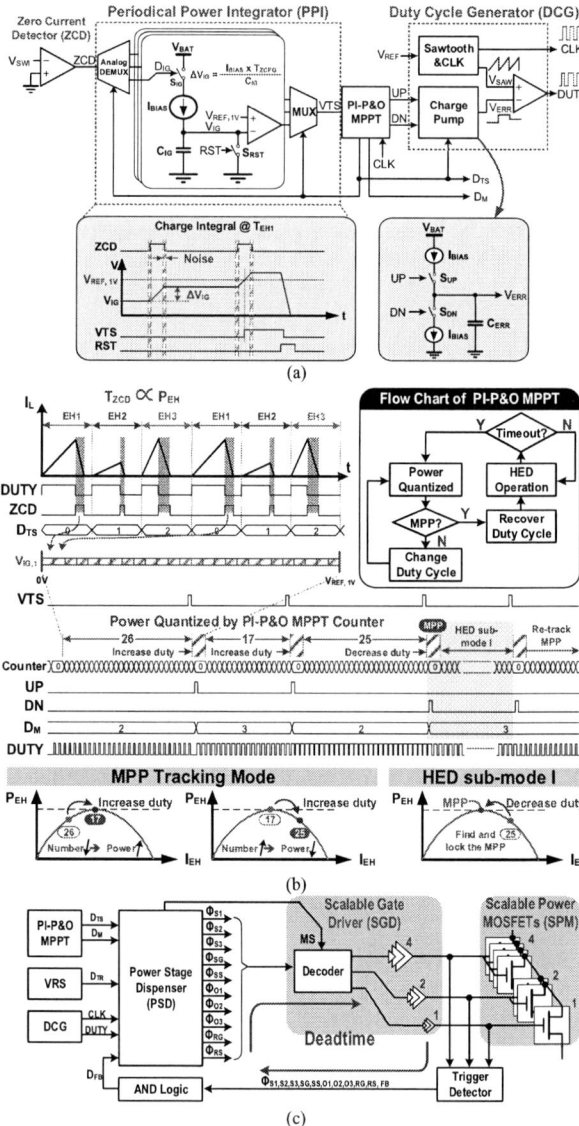

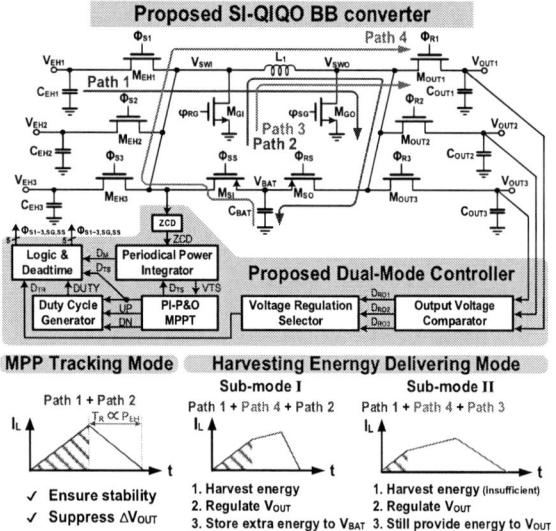

Fig. 2. The proposed SI-QIQO BB converter contains the MPP tracking mode and the HED mode to extend tracking efficiency and regulate multiple outputs including the V_{BAT}.

Fig. 3. (a) The proposed Periodical Interval Perturbation and Observation (PI-P&O) MPPT controller. (b) The MPPT controller operation. (c) Circuit implementation of the PSD to control energy delivery and optimize converter efficiency with the SGD and SPM.

978-1-5386-6414-8/18 $31.00 © 2018 IEEE

B. Voltage Regulation Selector (VRS) circuit

The Voltage Regulation Selector (VRS) circuit in Fig. 4(a) determines which outputs requires energy and sequences all output requests to the PSD, then the PSD outputs a combination of on/off control signals to each power switch. D_{RO1}, D_{RO2}, and D_{RO3} are sampled by the VRS circuit from the output voltage comparator on the falling edge of clock. On the rising edge of clock, the PI-P&O MPPT determines harvesting signal "D_{TS}" which inputs harvests, and driven signal "D_M" to optimize conversion loss. If D_{RO1}-D_{RO3} are both high, the converter charges V_{BAT}. When signal "Duty" is from low to high, the PSD selects signal "D_{TS}" from PI-P&O MPPT to control energy harvesting. Conversely, when "Duty" goes from high to low, the PSD selects signal "D_{TR}" from VRS and transfers energy to selected output. At clock rising edge, the PSD samples signal "D_M" then outputs control signal "MS" in one cycle. The signals "Φ" and "MS" are inputs of scalable gate drivers (SGDs) which drives scalable power MOSFETs (SPMs). The SGDs and SPMs are divided into 1: 2: 4 ratios. When "Φ" turns on/off, the trigger detector gathers all gate driver output signals and returns feedback signals "Φ_{FB}". The SI-QIQO BB conversion efficiency is improved because PI-P&O MPPT has power quantization information to select appropriate size of SGDs and SPMs for energy loss reduction, shown in Fig. 4(b).

IV. EXPERIMENTAL RESULTS

In Fig. 5(a) measured waveforms, the enable signal activates MPPT, and the first and second energy harvesting sources are 2-cell PV modules and TEGs. In the initial state, the task is set to an initial value, which triggers the HED mode to obtain energy from the three energy harvesting sources. After HED mode, MPPT mode starts. In the PI-P&O circuit operation, if open-circuit voltage $V_{PV,OC}$ is 1.17V, the derived MPP voltage $V_{PV,MPP}$ of 2-cell PVs is 937mV. Similarly, if open circuit voltage $V_{TEG,OC}$ is 0.8V, the MPP voltage $V_{TEG,MPP}$ is 414mV.

In Fig. 5(b), the conduction loss in MPPT mode is analyzed based on inductor current and power dissipation caused by body diode conduction. Conduction loss and switching loss are 4.78% and 4.05%, respectively, and overall power efficiency is 91.17%. Fig. 5(c) shows the tracking efficiency of PV and TEG, and the highest power tracking accuracy is 99.2% of TEG source.

In Table I, the comparison table shows that the SI-QIQO BB converter has the largest number of inputs and outputs, and the highest power tracking accuracy of 99.2%. Each output voltage can be controlled between 600mV and 2.5V. The maximum output power is up to 60mW. If the acquisition source voltage is below 1.2V and V_{BAT} is 4.2V, the peak efficiency is as high as 91.2% due to scalable power MOSFET size. The test chip in 0.18μm process occupies silicon area 1.9505mm² in Fig. 6. To achieve a compact layout, this paper presents a modular MOSFET unit (MMU) layout. Empty channels above the MMU's SGDs are sufficient to have automatic place and route (APR). The technology easily minimizes and integrates SI-MIMO converter layouts with vertical M4-M5 layers and horizontal M6 layer.

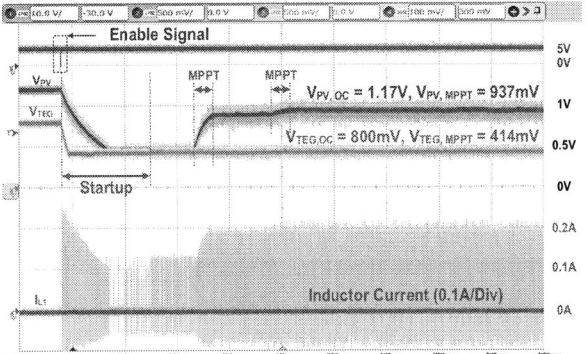

(a)

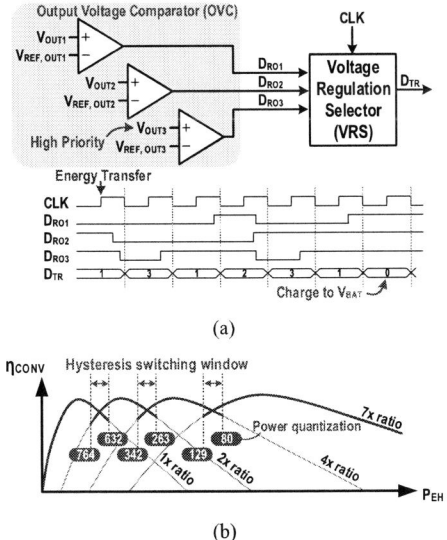

(a)

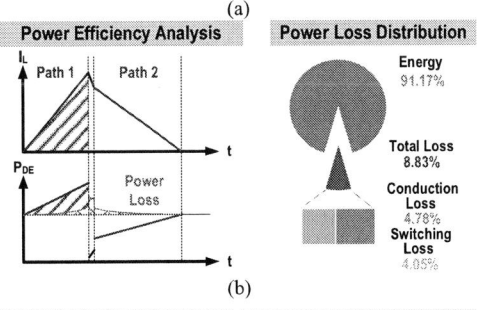

(b)

Fig. 4. (a) Output voltage regulation selection control. (b) Conversion efficiency with SGD and SPM sizing

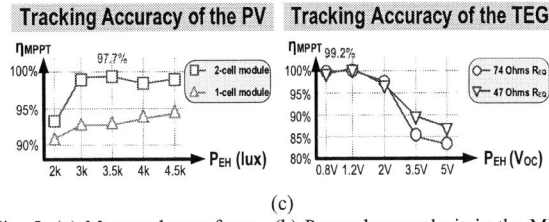

(c)

Fig. 5. (a) Measured waveforms. (b) Power loss analysis in the MPP tracking mode. (c) Tracking accuracy of the PV and the TEG.

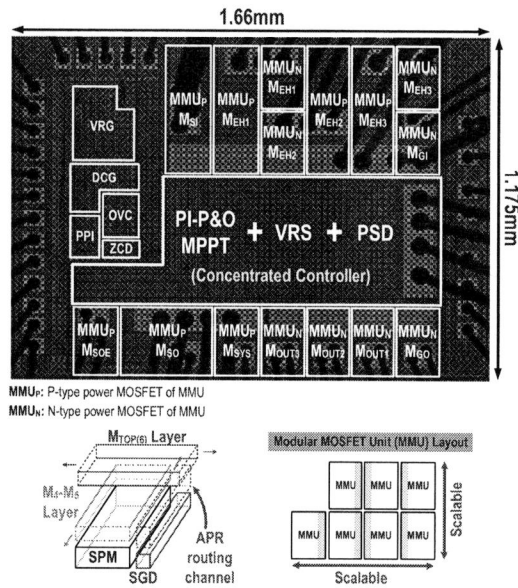

MMUₚ: P-type power MOSFET of MMU
MMUₙ: N-type power MOSFET of MMU

Fig. 6. Die micrograph and the modular MOSFET unit (MMU) layout.

V. CONCLUSION

The proposed SI-QIQO BB converter with dual-mode controllers including MPP tracking mode and harvesting energy transfer (HED) mode SMD technique effectively extend the tracking efficiency and the regulation of multiple outputs

including the VBAT, and the highest power tracking accuracy of 99.2%. Moreover, with power quantization information from the PI-P&O MPPT controller, the overall peak efficiency is as high as 91.2% by sizing scalable power MOSFET.

REFERENCES

[1] K. W. R. Chew, Z. Sun, H. Tang, and L. Siek, "A 400nW Single-Inductor Dual-Input-Tri-Output DC-DC Buck-Boost Converter with Maximum Power Point Tracking for Indoor Photovoltaic Energy Harvesting", *ISSCC Dig. Tech. Papers*, pp. 68-69, Feb. 2013.

[2] S. Kim, and G. A. Rincoón-Mora, "Dual-Source Single-Inductor 0.18μm CMOS Charger-Supply with Nested Hysteretic and Adaptive On-Time PWM Control", *ISSCC Dig. Tech. Papers*, pp. 400-401, Feb. 2014.

[3] H. J. Chen, Y. H. Wang, P. C. Huang, and T. H. Kuo, "An Energy-Recycling Three-Switch Single-Inductor Dual-Input Buck/Boost DC-DC Converter with 93% Peak Conversion Efficiency and 0.5mm2 Active Area for Light Energy Harvesting", *ISSCC Dig. Tech. Papers*, pp. 374-375, Feb. 2015.

[4] Y. Lu, S. Yao, B. Shao, and P. Brokaw, "A 200nA Single-Inductor Dual-Input-Triple-Output (DITO) Converter with Two-Stage Charging and Process-Limit Cold-Start Voltage for Photovoltaic and Thermoelectric Energy Harvesting", *ISSCC Dig. Tech. Papers*, pp. 368-369, Feb. 2016.

[5] S. Uprety, and H. Lee, "A 93%-Power-Efficiency Photovoltaic Energy Harvester with Irradiance-Aware Auto-Reconfigurable MPPT Scheme Achieve > 95% MPPT Efficiency across 650μW to 1W and 2.9ms FOCV MPPT Transient Time", *ISSCC Dig. Tech. Papers*, pp. 378-379, Feb. 2017.

Table I: Performance summary and the comparison with state-of-the-arts.

	ISSCC 2013 [1]	ISSCC 2014 [2]	ISSCC 2015 [3]	ISSCC 2016 [4]	This Work
Process	0.18μm	0.18μm	0.5μm	0.35μm	0.18μm BCD
Topology	SI-DITO	SI-DIDO	SI-DISO	SI-DITO	SI-QIQO
Harvester Type	PV	N.A.	PV	PV / TE	PV / TE / FC
Peak Efficiency	83%	83%	93%	85%	91.2%*
Cold Startup	N.A.	N.A.	N.A.	300mV	1.4V
Input Voltage Range	N.A.	1.1V ~1.3V	N.A.	30mV ~ 3.6V	300mV ~ 5V
Output Voltage	1.0V / 1.8V	0.8V	1.0V ~3.3V	-	0.6V ~ 2.5V@ Vₒᵤₜᵧ** 4.2V@ V_BAT
Maximum Output Power	10mW	6.4mW	15mW	N.A.	60mW @ Vₒᵤₜᵧ** and V_BAT
Power of Current Consumption	400nW	N.A.	1uA @ 4V	200nA	313 μA @5V
Tracking Accuracy	N.A.	-	N.A.	N.A.	99.2%
MPPT Method	OCV	-	F-OCV	P&O	PI-P&O
Converter Architecture	Buck-Boost	Buck-Boost	Buck / Boost	Boost	Buck-Boost
Die Size	4.6225mm²	N.A.	0.79mm²	4mm²	1.9505mm²

*Converter efficiency is a calculation result for this work.
**y of output voltage Vₒᵤₜᵧ means 1, 2 or 3.

14-4 (8186)

A Digital Multiphase Converter with Sensor-less Current and Thermal Balance Mechanism

Kai-Yu Hu, Yu-Sin Chen, and Chien-Hung Tsai

Department of Electrical Engineering
National Cheng-Kung University
Tainan, Taiwan
s972882@gmail.com, stanley1459@yahoo.com.tw, chtsai@ee.ncku.edu.tw

Abstract—This paper presents a sensor-less approach to achieve current balance and thermal balance in a digital voltage mode controlled four-phase buck converter. Rather than utilize current and temperature sensors to gather the current and thermal information as most of previous researches, the sensor-less equivalent resistance (R_{eq}) ratio estimation is used in this work to obtain the current and thermal information implicitly. Based on the estimative results, the duty offset is further calculated and compensated for each phase to carry out either equal current sharing or uniform thermal distribution. The digital controller was manufactured by TSMC 0.18-μm 1P6M standard CMOS process. The experimental results are proved the practicability of the proposed balancing mechanism. The current balance scheme improves the current sharing error from 35% to 6.3%, while the thermal balance technique narrows down the peak temperature difference from 6.6°C to 1.8°C.

Keywords—Multiphase buck converter, digitally-controlled, current balance, thermal balance, duty disturbance.

I. INTRODUCTION

Recently, artificial intelligence (AI) has been widely used in various fields and the utilization of powerful processors for computing becomes a trend. The high-performance processors need high current density voltage regulators (VRs) to power them on. A single-phase buck converter is not suited for high current density use due to low efficiency at high load current condition. Instead, a multiphase topology which combines several single converters in parallel is commonly adopted to supply high current demand applications. Nevertheless, in practical use, multiphase converters suffer from phase current unbalance [1]-[8] because of mismatches in parasitic resistances of power-stage components among all phases. Phase current unbalance may cause thermal hotspots, inductor saturations, and switching transistor damages which lead to deterioration in efficiency, stability, and reliability of multiphase converters [1]. There are generally two kinds of balancing methods in previous literatures: current balance [1]-[5] and thermal balance [6]-[8].

To realize current balance or thermal balance, most of the previous works require a current sensor [1]-[3] or a temperature sensor [6]-[7] in each phase as shown in Fig. 1. The controller converts the sensed current or thermal data into reasonable voltage values which are fed back to the current balancer and thermal balancer respectively. The balancer receives the compensator output and current or thermal

information to adjust the duty of each phase to attain the desired balance scheme. However, the higher number of phase counts, the more sensing circuitries are needed, and in turn increase extra costs and layout complexities.

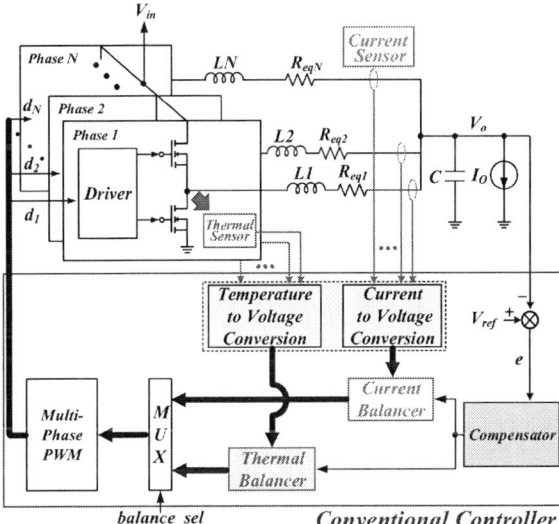

Fig. 1 Conventional controller with current and thermal sensors and conversion blocks to achieve current or thermal balance scheme

For a digitally-controlled switching converter, it is possible to perform an estimative technique to realize current balance or thermal balance without analog sensors [4]-[5], [8]. In [4]-[5], the resistance estimation is conducted on-line and off-line respectively. Once the estimation is done, duty in each phase can be calculated and compensated accordingly. In [8], the even thermal distribution between phases is implemented by calculating conduction losses of all phases. In this work, considering both phase current and thermal unbalance are relevant to the parasitic resistances, a sensor-less equivalent resistance (R_{eq}) ratio estimation is employed to eliminate the need of external current/thermal sensors, and the corresponding conversion blocks in Fig. 1 can be cancelled out. Therefore, the implementation of this work can further reduce power consumption and die area of current/thermal sensors.

This paper is organized as follows. Section II introduces the proposed digital sensor-less current and thermal balance algorithm and explains its implementation. Section III presents

978-1-5386-6414-8/18 $31.00 © 2018 IEEE

175

experimental results to prove the effectiveness of the proposed balance scheme, and finally summarize this work in Section IV.

II. THE PROPOSED DIGITAL FOUR-PHASE BUCK CONVERTER

The proposed digital sensor-less current and thermal balance mechanism is demonstrated with a four-phase digital buck converter. Fig. 2 shows the system block diagram of this work. The power stage consists of four powertrain modules with input and output tied together respectively. The controller consists of an encoder, a digital compensator, a sensor-less current and thermal balancer (SCTB) and a multiphase digital pulse-width modulator (DPWM). The output voltage of power stage is sampled by a analog to digital converter (ADC) to the digital controller. The digital compensator stabilizes the control loop and generates a reference duty command to the SCTB. The SCTB consists of a R_{eq} ratio estimator, a current balancer and a thermal balancer. In order to minimize the hardware implementation, the current balance mechanism and thermal balance mechanism share the result of R_{eq} ratio estimator to perform the mechanism accordingly. Multiphase DPWM generates the duty signal from the SCTB with 90° (360°/4) phase shift to the gate driver of each phase.

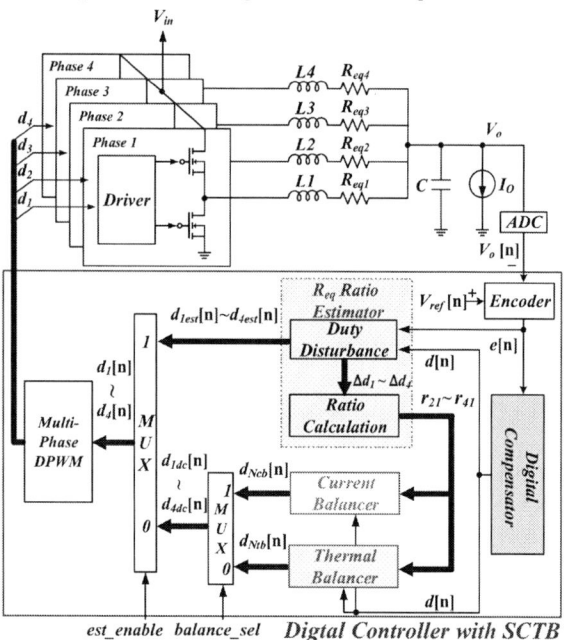

Fig. 2. The proposed system block diagram

Fig. 3 illustrates the DC model of a N-phase multiphase buck converter. The parasitic resistance of the Nth phase is modeled as R_{eqN} which is composed of all the resistive parasitic components along the power train, including the resistance of a inductor, on resistance of the power transistor and trace resistances of the printed circuit board (PCB). The power switches is modeled as the product of duty (d_N) and input voltage (V_{in}). The current of the Nth phase (I_{LN}) can be derived from the voltage difference between R_{eqN}. Considering

a traditional multiphase converter which duty of every phase is the same, the phase current among each phase will be different if R_{eq} are asymmetrical. On the other hand, the power dissipation of each phase can be derived from I_{LN} and R_{eqN} as shown in Fig. 3. The thermal stress between phases will mismatch if the power dissipation in each phase is unequal. As the result, the asymmetric parasitic resistances among phases leads to current and thermal unbalance in multiphase converters. Moreover, the estimation for the relationship of the parasitic resistances between phases is possible to replace the need of sensing circuits and the detail implementations of this work will be explained in the following sessions.

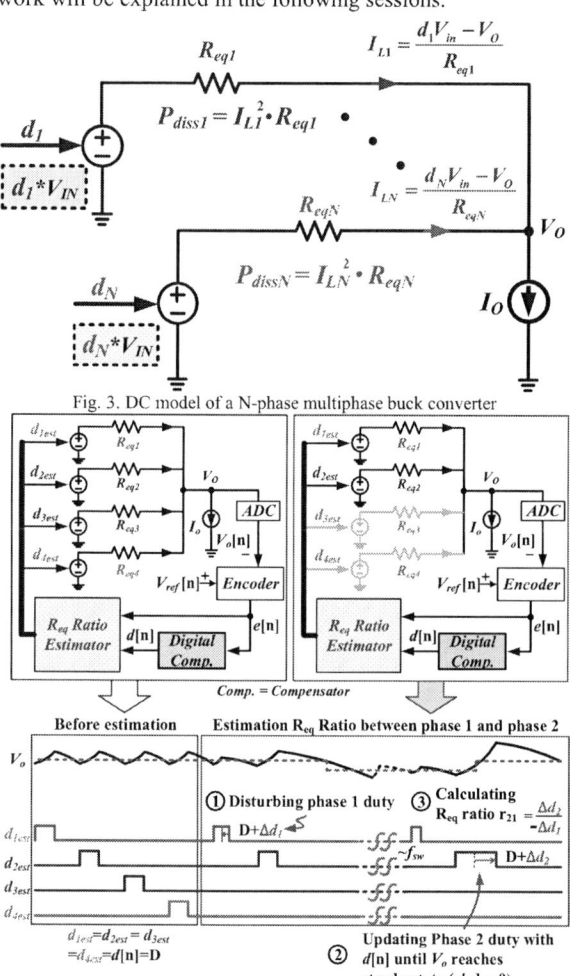

Fig. 3. DC model of a N-phase multiphase buck converter

Comp. = Compensator

Fig. 4 R_{eq} Ratio Estimation operation for phase 1 and phase 2

A. Senor-less R_{eq} Ratio Estimator

The R_{eq} ratio estimator gets the R_{eq} ratio of the Nth phase to the first phase. For a four-phase buck converter, three estimation rounds should be taken to obtain complete R_{eq} ratio information. Fig. 4 shows the operation of ratio estimation between the first two phases. Before estimation, every phase is active and control with the same steady-state duty (D). When

978-1-5386-6414-8/18 $31.00 © 2018 IEEE

the estimation is enabled, the R_{eq} ratio estimation block is activated with three steps for every round. In the first step, a duty disturbance Δd_1 is added into the 1st phase and let the 2nd phase duty directly connects to the output of the digital compensator, while disabling 3rd phase and 4th phase. During the disturbance, the output voltage (V_o) will drop (or rise) according to negative (or positive) Δd_1. The second step is to wait until that output voltage is regulated to the nominal output voltage ($e[n] = 0$). In the steady state, the output voltage and load current remain the same, the duty of the 2nd phase changes (Δd_2) due to the duty disturbance of the 1st phase (Δd_1) which is used to estimate R_{eq} ratio [4]. The 3rd step is to compute the R_{eq} ratio of the 1st phase to the 2nd phase (r_{21}) based on equation (1), where r_{N1} stands for R_{eq} ratio of the Nth phase to the 1st phase.

$$r_{N1} = \frac{R_{eqN}}{R_{eq1}} = -\frac{\Delta d_N}{\Delta d_1} \tag{1}$$

For a 4-phase converter, N in (1) ranges from 2 to 4. The r_{31} and r_{41} estimations can be carried out by the same steps with disabling the 2nd & 4th phase and 2nd & 3rd phase respectively.

B. Current Balance Mechanism

By using internal duty command and the previous R_{eq} ratio estimation, it is able to achieve equal current sharing without external current sensors. The phase current of the Nth phase is shown in (2), where R_{eqN} is the equivalent resistance of the Nth phase.

$$I_{LN} = \frac{d_N \cdot V_{in} - V_o}{R_{eqN}} \tag{2}$$

In order to balance all phase current, every phase current is normalized to the current of the first phase I_{L1}. The duty command to achieve current balance can be obtained by equating I_{L1} and I_{LN}, as shown in equation (3).

$$d_N = \frac{R_{eqN}}{R_{eq1}} \cdot d_1 + (1 - \frac{R_{eqN}}{R_{eq1}}) \cdot \frac{V_o}{V_{in}} \tag{3}$$

To minimize the utilization of multipliers, the duty of the Nth phase can be realized by summing the duty of first phase and the duty difference to the first phase as shown in equation (4).

$$\Delta d_{cbN1} = d_N - d_1 = (\frac{R_{eqN}}{R_{eq1}} - 1) \cdot (d_1 - \frac{V_o}{V_{in}}) \tag{4}$$

The resistance ratio in (4) can be accessed by the R_{eq} ratio estimator. After the duty adjustment Δd_{cbN1} of each phase is implemented, the phase currents of multiphase converters will be equalized. The hardware implementation of the sensor-less current balance mechanism is shown in Fig. 5.

C. Thermal Balance Mechanism

The thermal unbalance among phases is mainly from the power dissipation mismatches, especially from the conduction loss mismatches. Therefore, it is possible to achieve thermal balance by making every phase conduction power loss the same as shown in equation (5).

$$I_{L1}^2 R_{eq1} = I_{LN}^2 R_{eqN} \tag{5}$$

Substitute the equation (2) into (5), the duty command of the Nth phase to achieve thermal balance can be derived as (6).

$$d_N = \sqrt{\frac{R_{eqN}}{R_{eq1}}} \cdot d_1 + (1 - \sqrt{\frac{R_{eqN}}{R_{eq1}}}) \cdot \frac{V_o}{V_{in}} \tag{6}$$

To simplify the hardware usage, the duty of the Nth phase can be realized by summing the duty of the first phase and the duty difference to the first phase as shown in equation (7).

$$\Delta d_{tbN1} = d_N - d_1 = (\sqrt{\frac{R_{eqN}}{R_{eq1}}} - 1) \cdot (d_1 - \frac{V_o}{V_{in}}) \tag{7}$$

The resistance ratio in (7) can be calculated by the R_{eq} ratio estimator. The square root is implemented with a lookup table in this work for reducing area and calculation time. The hardware implementation of the sensor-less thermal balance mechanism is shown in Fig. 5.

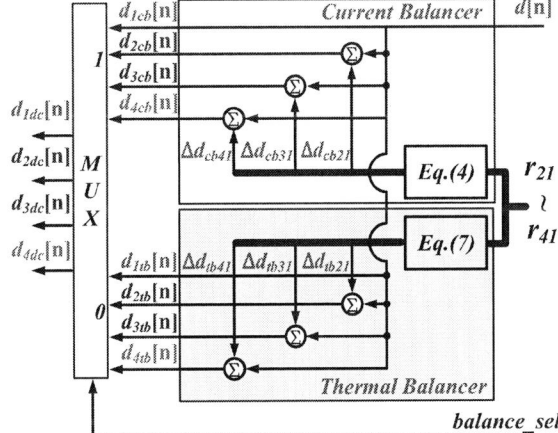

Fig. 5 The hardware implementation of current balancer and thermal balancer mechanism

III. EXPERIMENTAL RESULTS

The experimental prototype of the digitally-controlled four-phase buck converter is presented in Fig. 6. The complete system includes four phase buck powertrain modules, a voltage ADC, and a digital controller which was fabricated by TSMC 0.18-μm CMOS technology. The core cell and i/o pad are supplied via external 1.8V and 3.3V voltage source respectively. To verify the effectiveness of the proposed SCTB controller, the PCB is intentionally added with different resistances in each phase to create mismatches. The system specifications are summarized in Table I. The measurement is tested under steady state with 12A load. Fig. 7 shows the current waveforms of each phase with the Current Balancer enabled or not. Without current balance mechanism the current sharing error is around 35%, while the error is reduced to 6.3% by enabling the Current Balancer. The thermal image is used to observe the thermal distribution of the four-phase buck converter with the Thermal Balancer is enabled or not. In Fig.8 with marked M1 to M4 are located at the highest temperature of the 1st phase to the 4th phase respectively. Without activating the thermal balance technique, the highest temperature is at the 1st phase due to carrying most of the load current while the lowest one is at the 4th phase. The maximum temperature difference is 6.6°C. When enabling the Thermal Balancer, the peak thermal difference is reduced to 1.8°C.

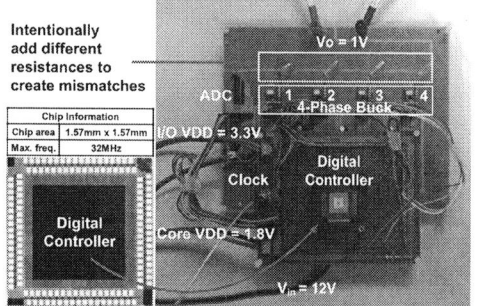

Fig. 6 Prototype of the proposed digitally controlled 4-phase buck converter

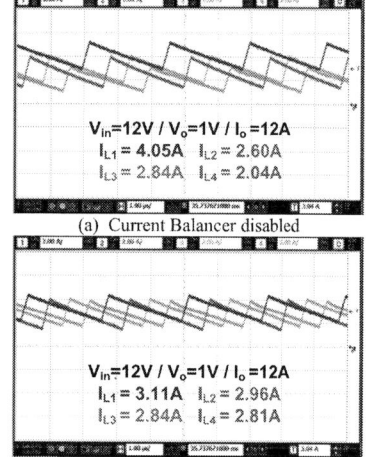

(a) Current Balancer disabled

(b) Current Balancer enabled

Fig. 7 Current waveforms of the 4-Phase Buck Converter with the Current Balancer disabled (a) and enabled (b)

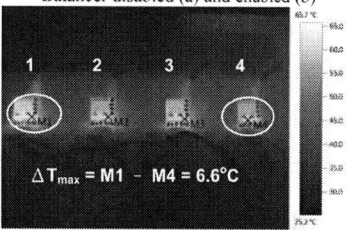

(a) Thermal Balancer disabled

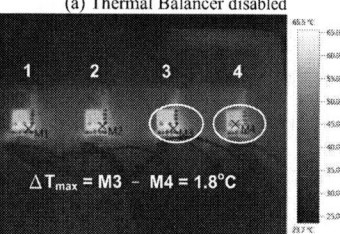

(b) Thermal Balancer enabled

Fig. 8 Thermal Images of the 4-Phase Buck Converter with the Thermal Balancer disabled (a) and enabled (b)

IV. CONCLUSION

This paper demonstrates both current balance and thermal balance mechanisms without using current and temperature

sensors in a digitally-controlled four-phase buck converter. The measured current sharing error and thermal distribution error is 6.3% and 1.8°C respectively. The performance comparison is listed in TABLE I. Even though without external sensors, the performances on current distribution and thermal distribution are still competitive to the previous works with sensors.

TABLE I. Performance comparison table

Publication	[2]	[3]	[7]	This work
Control Mode	Peak Current	Voltage	Peak Current	Voltage
Technology	0.18μm BCD	FPGA	FPGA	0.18μm CMOS
Current Balance	√	√	√	√
No. of current sensors	4	3	4	0
Current Sharing Error	N.A.	6 %	N.A.	6.3%
Thermal Balance	×	×	√	√
No. of thermal sensors	0	0	4	0
Peak thermal difference	×	×	1.9°C	1.8°C
Phase Count	4	3	4	4
V_{in} (V)	2.7-5	12	12	12
V_o (V)	0.8-V_{in}	5	1	1
$I_{o(max)}$ (A)	4.5	15	50	12
f_{sw} (kHz)	2000 (×4)	195 (×3)	250 (×4)	500 (×4)

ACKNOWLEDGMENT

This work was supported by M.O.S.T. project. The authors would like to thank National Chip Implementation Center (CIC) for chip fabrication support.

REFERENCES

[1] L. Wang, S. K. Khatamifard, O. A. Uzun, U. R. Karpuzcu and S. Köse, "Efficiency, Stability, and Reliability Implications of Unbalanced Current Sharing Among Distributed On-Chip Voltage Regulators," in *IEEE Transactions on Very Large Scale Integration (VLSI) Systems*, vol. 25, no. 11, pp. 3019-3032, Nov. 2017.

[2] Y. Ahn, I. Jeon and J. Roh, "A Multiphase Buck Converter With a Rotating Phase-Shedding Scheme For Efficient Light-Load Control," in *IEEE Journal of Solid-State Circuits*, vol. 49, no. 11, pp. 2673-2683, Nov. 2014.

[3] K. I. Hwu and Y. H. Chen, "Current Sharing Control Strategy Based on Phase Link," in *IEEE Transactions on Industrial Electronics*, vol. 59, no. 2, pp. 701-713, Feb. 2012.

[4] X. Zhang, L. Corradini and D. Maksimovic, "Sensorless Current Sharing in Digitally Controlled Two-Phase Buck DC-DC Converters," *2009 Twenty-Fourth Annual IEEE Applied Power Electronics Conference and Exposition*, Washington, DC, 2009, pp. 70-76.

[5] J. Gordillo and C. Aguilar, "A Simple Sensorless Current Sharing Technique for Multiphase DC–DC Buck Converters," in *IEEE Transactions on Power Electronics*, vol. 32, no. 5, pp. 3480-3489, May 2017.

[6] A. Elbanhawy, "Current Sharing in Multiphase Converters Using Temperature Equalization," *2005 IEEE 36th Power Electronics Specialists Conference*, Recife, 2005, pp. 1464-1468.

[7] P. Cao, W. T. Ng, and O. Trescases, "Thermal management for multiphase current mode buck converters," in *Proc. IEEE Appl. Power Electron. Conf.*, 2011, pp. 1124-1129.

[8] Z. Lukić, S. M. Ahsanuzzaman, Z. Zhao, and A. Prodic, "Sensorless selftuning digital CPM controller with multiple parameter estimation and thermal stress equalization," *IEEE Trans. Power Electron.*, vol. 26, no. 12, pp. 3948-3963, Dec. 2011.

A Fully-integrated LC-Oscillator Based Buck Regulator with Autonomous Resonant Switching for Low-Power Applications

Tianyu Jia, and Jie Gu

Department of Electrical Engineering and Computer Science
Northwestern University
Evanston, IL, 60208, USA

Abstract— In this paper, a fully integrated, *LC*-oscillator based buck regulator is presented for low power applications. The proposed design uses a self-resonator as both high-speed clock source and a resonant switch driver for the regulator, which not only achieves significantly improved energy efficiency at 2GHz switching frequency but also delivered a high-quality clock source eliminating requirement of the external high-speed clocks. Measurement on a 65nm testchip shows a wide tuning range from 0.35V to 0.82V with an input voltage 1.1V. The proposed design achieves a peak efficiency of 70.3% and an autonomous clock-less operation. The regulator core area is only 0.079mm².

Keywords— *Buck regulator, fully integrated, input-clock free, resonant switching, low power applications.*

I. INTRODUCTION

Energy-efficient near-threshold computing has been proposed to increase energy efficiency across a wide range of applications [1]. Recent developments on low power wearable electronics and IoT devices bring new design requirements to power regulators. For example, achieving high power efficiency performance at low output voltages/current is one of the keys for near or subthreshold operation. The conventional buck regulator design normally utilizes large off-chip inductors (>1uH) with slow switching frequency (<5MHz) [2-3]. The fully integrated regulator solutions using on-chip inductors are more suitable for integration with the microprocessors and attracts more attentions recently [4-7]. However, most of the previous fully-integrated buck regulators are designed for high output power with hundreds of milliamps output current and show poor efficiency at low voltage regime, e.g. only 40% at 0.5V [5]. For near-threshold computing, the nominal operation voltage is normally around 0.4-0.6V, with only a few to tens of milliamps load current [1]. Therefore, the design strategy for buck regulator needs to be adjusted to optimize the regulator's performance at low power low voltage conditions.

For a conventional buck regulator, its step-down voltage ΔV is determined by the inductance value L and the changing rate of inductor current di_L/dt, as expressed by the equation (1):

$$\Delta V = L \cdot \frac{di_L}{dt} = L \cdot \frac{\Delta i_L}{D} \cdot f_{sw} \qquad (1)$$

in which D is the duty cycle of the switching clock and f_{sw} is the regulator switching frequency. Δi_L is the current ripple of inductor L. For the fully-integrated buck regulator designs, the small on-chip inductance value L causes the increasing of the inductor current ripple Δi_L, which potentially leads to reverse

current and severely degraded efficiency assuming simple continuous-conduction mode (CCM) operation is maintained. As a result, to maintain small inductor current ripple Δi_L and small size of on-chip inductor L, the switching frequency f_{sw} of fully integrated buck regulator has to been typically pushed much higher to 200~500MHz comparing with the off-chip counterpart [5-6]. For low power applications where the current is in the order of only milliamp to tens of milliamp, the switching frequency has been further increased to 2GHz to maintain a small size of the inductor [7].

Fig.1 shows that inductance value and power loss trend with the scaling of switching frequency f_{sw}. As shown in Fig. 1(a), the on-chip inductance value, which translates to the inductor area, scales inversely proportional with switching frequency. Fig. 1(b) shows the simulated power loss components, i.e. inductor, clock, power switches as a function of switching frequency. The power loss from inductor, in fact, drops with higher switching frequency due to the reduced peak-to-peak ripple Δi_L at higher switching frequency, which leads to 2-4X less inductor loss at 2GHz compared with 500MHz. Fig. 1(c) and (d) shows the current and voltage ripple at switching frequency of 500MHz and 2GHz. The output voltage ripple is improved with higher frequency, which also leads to the benefits of reduction of required output decoupling capacitors (decap), e.g. from a few of nF [4-6] to 120pF in this work.

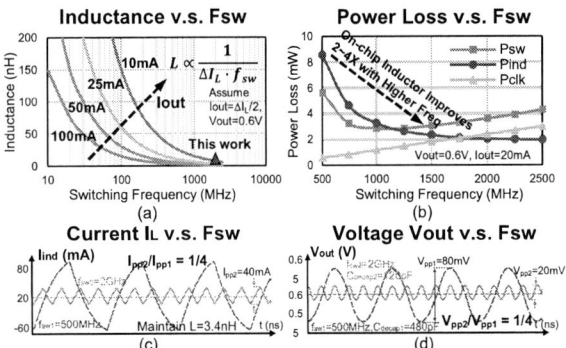

Fig. 1. (a) Inductance value, (b) power loss breakdown, (c) inductor current, and (d) output voltage ripples scaling with switching frequency.

As shown in Fig. 2, the conventional buck regulator normally needs two clocks, in which one provide the switching frequency for the power switches and the other one provides the loop sampling frequency. These two frequencies can be designed to be same frequency value or may choose two different frequency values [4-5]. In the conventional buck regulator operating with slow switching frequency, i.e. a few MHz, the clock source power is negligible. However, for fully integrated buck regulator running at 500MHz or even higher frequency, the power consumption from the clock sources could become significant. Unfortunately, none of the previous work addresses the issue of the generation of clock source [5-7].

In this work, we proposed a LC-oscillator based buck regulator design, in which the oscillator serves both as the switching clock source and the power switch gate driver. A resonant switching operation is naturally formed between the LC oscillator and the gate capacitance of power switches, which eliminating the requirement of resonant drivers [7]. Our simulation shows a 7% to 20% efficiency improvement is observed due to the use of resonant switching, especially at low voltages, e.g. 0.3V~0.5V, where the clock power loss dominates, as shown in Fig. 2(b). In addition, the loop frequency is provided by a divided clock from the LC oscillator. As a result, no external clock generator is needed.

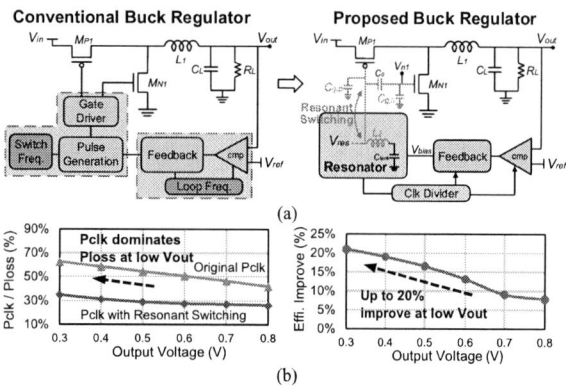

Fig. 2. (a) Concept of the buck regulator with autonomous switching. (b) Power saving benefit from the resonant swithcing.

II. PROPOSED BUCK REGULATOR DESIGN

A. Overview

Fig.3 shows the overview of the proposed buck regulator. The conventional multi-stage non-overlapping PMOS/NMOS switch drivers are replaced by a LC oscillator topology self-resonator, which naturally creates a resonance between the inductor L_2 and capacitance $C_{g,sw}$ from the gates of power switches M_{P1} and M_{N1}. The resonant frequency is determined by the total capacitance of $C_{g,sw}$ and a tunable capacitor C_{tune} with the oscillator inductor L_2. In this work, the self-resonator oscillation frequency, i.e. the switch frequency, is set around 2GHz, which is determined by the size limit of inductors and the low output current according to equation (1).

A resonant switching operation is formed by the self-resonator, which saves the switching clock power. When the power switches are fully turned on/off, the energy within the gate capacitance $C_{g,p}$ and $C_{g,n}$ is gradually charged or discharged into the oscillator inductor L_2. During the transition of the power switch, the stored energy is released to assist the power switch transitions. To utilize the resonant switching, the gate of the power switches M_{P1} and M_{N1} are jointly switched by the self-resonator. A coupling capacitor C_1 is added between the

PMOS/NMOS gate with a high impedance DC biasing voltage V_b to shift the switching voltage of NMOS and suppress the short-circuit current. To control output voltages, a duty cycle distortion is deliberately introduced through the feedback control of the Gm transistor in the self-resonator. Feedback is realized through hysteresis comparators with reference voltages and converted into Gm tuning voltage through a R-2R DAC. The oscillation frequency is tunable from 1.6GHz to 2.4GHz through tunable capacitors C_{tune}, and is divided down by 64 times to provide baseband clock for feedback control. Therefore, there is no external clock generator required.

In the proposed design, two on-chip inductors are utilized, which is the L_1 of 3.4nH for the buck regulator and L_2 of 11.6nH for the LC self-resonator. The inductor L_1 is designed with wide width and less number of turns to reduce the conduction resistance. The current that flow through inductor L_2 is the charging and discharging current for the gate capacitance of power switch $C_{g,sw}$, which is much smaller than the load current. Therefore, there is less stringent Q requirement for the inductor L_2 leading to larger inductance value but similar size as L_1. Programmable static on-chip resistance load is utilized for the efficiency measurement, with an on-chip scan chain control the testing configurations.

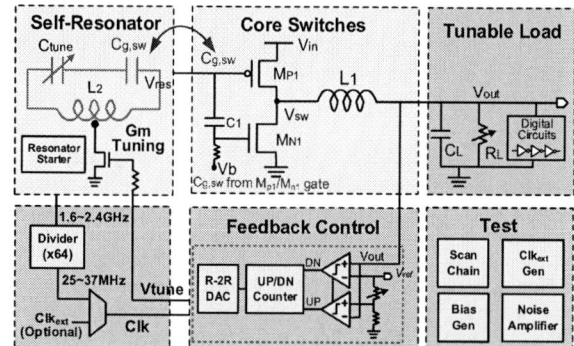

Fig. 3. Overview of the proposed buck regulator design.

B. Self-Resonantor Design

Fig. 4 shows the detailed implementation of the LC oscillator based self-resonator. The oscillation tank is formed from an inductor and the capacitance contributed from gate capacitance $C_{g,sw}$ of both power switches, M_{P1} and M_{N1} and additional tunable capacitors C_{tune}.

There are three special design techniques used to control duty cycle and boost slew rate of oscillation. (1) Different from the conventional self-biased inverter for generating negative resistance for oscillation, as shown in Fig. 4, the proposed design uses a separate gate bias V_{tune} for Gm transistor enabling tuning of duty cycle. Simulation shows a duty cycle from 32%~76% can be introduced at the output node of the self-resonator leading a wide range of output voltage. (2) Coupling capacitor C_3 couples the oscillation node V_{OSC} to PMOS driver M_{P2} to boost swing at V_{RES} signal and improve its slew rate. (3) Coupling capacitor C_4 is utilized to couple the square wave V_{SW} to M_{P2} to further boost the slew rate of V_{RES}. As a result, a rail-to-rail swing at the self-resonator output V_{RES} is achieved and the slew rate is improved, which leading to a significant reduction of short-circuit current and power switch loss reduction.

978-1-5386-6414-8/18 $31.00 © 2018 IEEE

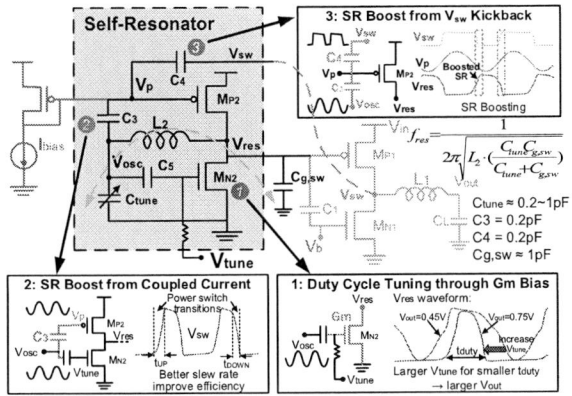

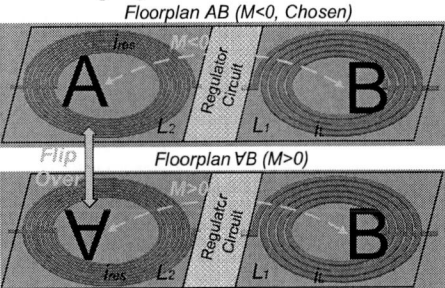

Fig. 4. Design of the self-resonator integrated with power switches and the description of duty cycle control and slew rate enhancement.

The special bias condition used here does not guarantee startup condition for the oscillation. Hence, a special startup sequence with kick-start is used to establish the oscillation, as shown in Fig. 5. Before the oscillation, a startup PMOS header is enabled to provide a constant current $i_{startup}$. The Gm transistor is pre-biased by voltage V_{tune} into the saturation region before a kick-start is applied. After completing all these initialization setup, the $V_{startup}$ signal is released and kick the oscillation. After the oscillation is built-up, the feedback control loop is enabled to generate desired voltages through duty cycle control. As the measurement waveform shown in Fig. 5, the self-resonant startup takes about 25ns to generate a 0.55V output voltage from 0V.

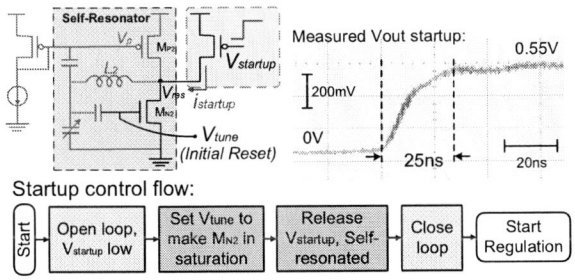

Fig. 5. The startup control flow for the self-resonator.

C. EM Simulations for Inductor Floorplan

The electromagnetic (EM) analysis and mutual coupling simulations are conducted with the actual inductor sizes and distance. Fig. 6 shows the consideration of inductor placement. As the resonant tank is placed right next to the main power inductor and consumes much less current, extra care is given to avoid pulling or glitches to the oscillator from the magnetic field of the main inductor. The effective inductance of L_1 and L_2 can derived as equation (2) and (3).

$$L_{eff,1} = L_1 + M \frac{di_{res}/dt}{di_L/dt} \approx L_1 \qquad (2)$$

$$L_{eff,2} = L_2 + M \frac{di_L/dt}{di_{res}/dt} \qquad (3)$$

As the inductor current i_L is much larger than the resonant inductor current i_{res}, the mutual coupling between two inductors

has negligible impact to the buck regulator inductor value L_1. Meanwhile, the effective resonant inductance $L_{eff,2}$ becomes a little larger/smaller with positive/negative mutual coupling. As shown in the Fig. 6, the current rotation direction determines the polarity of the mutual coupling. The mutual coupling EM simulation results are shown in Fig. 9(d). The simulation results show that the negative mutual coupling M achieves 10~30dB better isolation than positive M at the second and higher harmonic frequency. Therefore, the negative mutual coupling inductor floorplan is chosen in this design. In addition, only a 30MHz center frequency shift of the self-resonator is observed due to mutual coupling.

Fig. 6. Inductor placements based on EM simulations, with i_{res} rotating at reverse direction.

III. MEASUREMENT RESULTS

The proposed buck regulator is fabricated in a 65nm CMOS process with 1.1V input voltage. The resonant inductor L_2 has the same area as inductor L_1. The regulator core area including two inductors, power switches, self-resonator and the feedback logics only takes 0.079mm². The overall area including input/output decaps is 0.137mm². The phase noise of the self-resonator was measured from the test chip at the output of the resonant tank, as shown in Fig. 7. The measured self-resonator phase noise achieves -105dBc/Hz at 1MHz frequency offset when resonating at 2GHz, which is significantly better than a typical ring oscillator and indicating a potential high-quality clock source as a by-product. The strong supply noise and lack of filtering of bias current during measurement is suspected to contribute a 6dB degradation compared with simulation results.

Fig. 8 shows the measurement results of transient tracking. The transient reference tracking waveform shows a response speed of 100ns for the voltage change of 140mV (with a single step of adjustment of 40mV at 22ns). With the switching frequency at 2GHz, a 38mV ripple is observed with only 120pF output decap. Under a transient load current change of 10mA (~50% change), the undershoot/ overshoot is around 100mV due to small decap in use and the settling time is 74~78ns. By utilizing larger decap, the undershoot/overshoot will be smaller.

Fig 9(a) and Fig. 9(b) shows the measured power efficiency versus different output voltage level and switching frequency. With the input voltage of 1.1V, a wide output ranges from 0.35V to 0.82V with output power from 3mW to 33mW is achieved. The peak efficiency is 70.3% and remains above 55% for voltage down to 0.45V. The optimal switching frequency to achieve peak efficiency is observed around 2GHz. The power loss breakdown at 0.6V and 2GHz is shown in Fig. 9(c), in which the power loss is dominated by the conduction loss from the power switch M_{P1} and the on-chip inductor L_1.

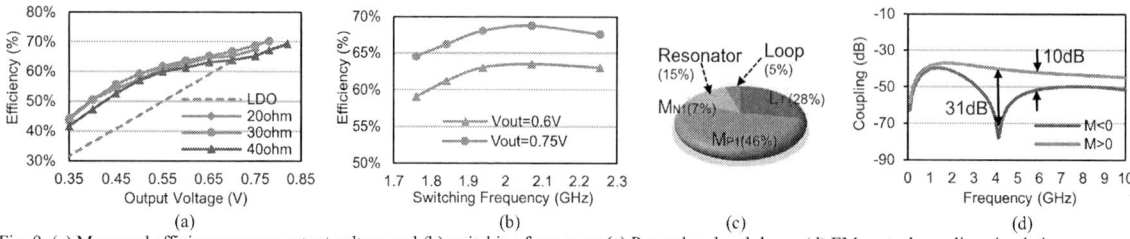

Fig. 9. (a) Measured efficiency versus output voltage and (b) switching frequency. (c) Power loss breakdown. (d) EM mutual coupling simulation.

Table. 1 compares the performance of the proposed buck regulator with previous state-of-the-art fully integrated buck regulators. For low voltage performance, the proposed design achieves ~10% better efficiency at ultra-low voltage range of 0.3~0.6V with substantially smaller area [4-5]. The peak efficiency is on part with the previous 73% with the on-chip inductor [7] and 71% with a wirebond inductance [6]. However, the use of the proposed self-resonator eliminates the need of external clock sources and complicated clock drivers. If taking into account the power saving by eliminating the high-speed clock source, e.g. a typical ring oscillator, an equivalent extra efficiency improvement of around 5~9% is obtained. Fig. 10 shows the die micrograph.

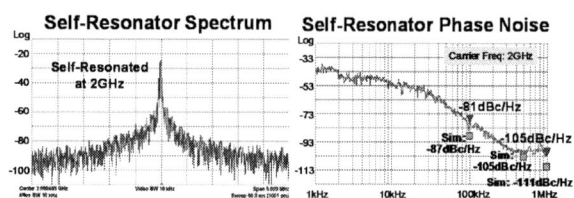

Fig. 7. Self-resonator spectrum and phase noise measurement at 2GHz.

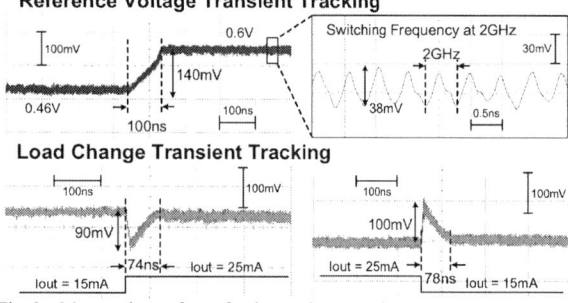

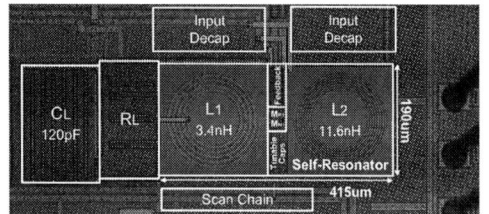

Fig. 8. Measured waveforms for the regulator transient response.

Fig. 10. Chip micrograph.

IV. CONCLUSION

This paper presents a fully integrated *LC*-oscillator based buck regulator which uses a self-resonator as both clock source and a resonant switch driver for low power applications. The proposed self-resonator design serves as both a high-speed clock source and a resonant switch driver for the regulator, which significantly improves the energy efficiency at high switching frequency and eliminates the external clocks. Measurement on a 65nm testchip shows the proposed design achieves a peak efficiency of 70.3% and an autonomous clock-less operation, with the regulator core area only 0.079mm^2.

TABLE I.   PERFORMANCE SUMMAYR AND COMPARISON

	[4] 14' VLSI	[5] 17' ISSCC	[6] 17' JSSC	[7] 17' VLSI	This work
Process (nm)	22	14	130	65	65
Clk source	External	External	External	External	Self-Resonator
Fsw (MHz)	500	100	125	2000	2000
Iout (mA)	250	90-330	70	10-40	10-50
L (nH)	1.5	1.5	11.8	3	3.4
C$_L$ (nF)	10	5	3.2	0.12	0.12
Vin (V)	1.5	1.5	1.2	1.1	1.1
Vout (V)	0.7-1.2	0.4-1.15	0.45-1.05	0.3-0.86	0.35-0.82
Ripple (mV)	43	>50	84	32	38
Response (ns)	100	700	80	5	20
Peak Eff. (%)	68	84	71	73	70.3
Eff. @0.5Vin (%)	--	<50	~55	65	63
Core Area (mm^2)	1.5	0.4	0.5	0.073	0.079

ACKNOWLEDGMENT

This work is supported in part by NSF grants CCF-1618065. We thank Integrand Software, Inc. for supporting EMX simulation.

REFERENCES

[1] N. Pinckney, D. Blaauw, D. Sylvester, "Low-power near-threshold design: techniques to improve energy," *IEEE Solid-State Circuits Magazine*, vol. 7, no. 2, pp. 49–57, Jun. 2015.

[2] A. Paidimarri, and A. Chandrakasan, "A buck converter with 240pW quiescent power, 92% peak efficiency and a 2×106 dynamic range," in *IEEE Int. Solid-State Circuits Conf. (ISSCC)*, Feb. 2017, pp. 192–193.

[3] P. Chen, C. Wu, and K. Lin, "A 50 nW-to-10 mW output power tri-mode digital buck converter with self-tracking zero current detection for photovoltaic energy harvesting," *IEEE J. Solid-State Circuits (JSSC)*, vol. 51, no. 2, pp. 523–532, Feb. 2016.

[4] H. Krishnamurthy *et al.*, "A 500MHz, 68% efficient, fully on-die digitally controlled buck voltage regulator on 22nm tri-gate CMOS," in *Proc. IEEE Symp. VLSI Circuits (VLSI)*, Jun. 2014.

[5] H. Krishnamurthy *et al.*, "A digitally controlled fully integrated voltage regulator with on-die solenoid inductor with planar magnetic core in 14nm tri-gate CMOS," in *IEEE Int. Solid-State Circuits Conf. (ISSCC)*, Feb. 2017, pp. 336–337.

[6] M. Kar *et al.*, "An all-digital fully integrated inductive buck regulator with a 250-MHz multi-sampled compensator and a lightweight auto-tuner in 130-nm CMOS," *IEEE J. Solid-State Circuits (JSSC)*, vol. 52, no. 7, pp. 1825–1835, Jan. 2017.

[7] T. Jia, and J. Gu, "A 0.3-0.86V fully integrated buck regulator with 2GHz resonant switching for ultra-low power applications," in *Proc. IEEE Symp. VLSI Circuits (VLSI)*, Jun. 2017.

A bulk 65nm Cortex-M0+ SoC with All-Digital Forward Body Bias for 4.3X Subthreshold Speedup

Pranay Prabhat[1], Graham Knight[1], Supreet Jeloka[2], Sheng Yang[1], James Myers[1]

[1]Arm Ltd, Cambridge, UK; [2]Arm Inc., Austin, US
first.last@arm.com

Abstract—**IoT devices demand ultra-low power operation while still achieving the performance demanded by application constraints. Dynamic forward body biasing can help to achieve this by providing a speed-up during active operation without incurring a leakage penalty during standby periods. While body biasing has been fully explored in FD-SoI technology, bulk CMOS can also benefit from efficient forward body biasing. At subthreshold voltage levels, the Low Voltage Swapped Body (LVSB) technique, in which n-well and p-well are driven to VSS and VDD respectively, helps to realize a significant speedup without incurring analog bias generation overheads. This work presents key advances to leverage LVSB, proven on a bulk 65nm subthreshold Arm Cortex-M0+ system. The system achieves a 4.3X speedup on the ULPBench benchmark at a cost of only 11% average power and 10.4% area, while showing that LVSB can be usefully applied up to 0.50V.**

Keywords— Arm; Cortex-M0+; LVSB; subthreshold; IoT; FBB

I. INTRODUCTION

The energy efficiency demands of IoT devices are driving renewed interest in subthreshold systems [1-3]. However, reduced performance is a major barrier to adoption. Fine-grained VDD control or analog body bias are commonly used to improve performance at the expense of area, energy and complexity. Low Voltage Swapped Body (LVSB) is a performance boosting technique introduced in [5], in which n-well and p-well may be "swapped" to power and ground rails respectively as long as the device is operating below the substrate latch-up voltage (Fig. 1). This provides a large forward body bias to both n and p devices, which translates to a significant speedup. Prior work [4-6] applied LVSB primarily to standalone logic blocks and reported speed increases of up to 3.7×.

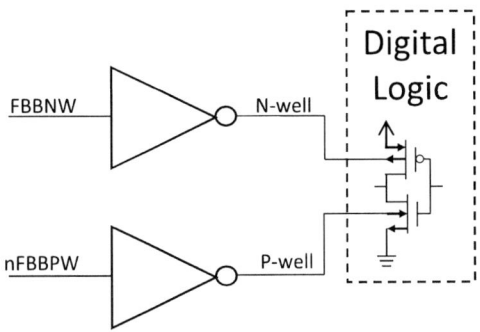

Fig. 1: LVSB operating principle ($V_{DD} < 0.6V$) [4-6].

This work proposes: 1. A no-trim, all-digital, fully synthesized LVSB implementation with µs-order mode switching. 2. Fine-grained independent bias control per power domain. 3. Split bias for SRAM array and periphery. 4. Adaptive clock generation to track system performance. A bulk 65nm Arm Cortex-M0+ LVSB SoC using these techniques achieves a 4.3× subthreshold speedup with only 11% power penalty for an industry standard duty-cycled workload.

This paper is organized as follows. Section II details the digital logic, memory, clock generation and power control sequencing considerations with block-level silicon results. Section III presents full system silicon results and section IV presents conclusions.

II. LVSB CIRCUITS

A. Logic and well drivers

Careful implementation is required to integrate dynamic LVSB into synthesized logic with multiple power domains. The CPU, bus and each peripheral are implemented as separate power domains for fine-grained bias control (Fig. 2). A physically separate deep n-well under each power domain isolates p-wells from the substrate at a marginal area cost. This enables per-domain well control to minimize leakage by forward biasing only active domains. Special tie cells are used to disconnect wells from the supply rails, allowing an industry-standard router to route wires as signals from well driver buffer outputs to well ties. Custom well driver buffer cells (Fig. 2, zoomed inset) are supplied by the same low-voltage power rail as the rest of the logic. The well driver cells incorporate a deep n-well hole, to prevent their internal n-well from shorting to the driven n-well and creating unwanted feedback. A pair of p- and n-well drivers is combined into a single cell to reduce DRC keep-out overhead. To accommodate these constraints in a library cell footprint, the buffer cell is 3 rows high with a jogged power rail. Well drivers are kept on the always-on ground rail so that wells do not float when power-gating the rest of the logic, as this would forward-bias power-gated logic and increase sleep power. When enabled, power-gating footers are forward biased along with the rest of the logic to supply the increased current demand from forward-biased operation.

Buffer density and sizing is determined from SPICE simulation including extracted well diodes, to meet an LVSB transition target of 16 cycles for all domains (Fig. 2, bottom right inset). Place and route is fully automated, with 4.3% area and 1.4% (simulated) active leakage overhead.

978-1-5386-6414-8/18 $31.00 © 2018 IEEE

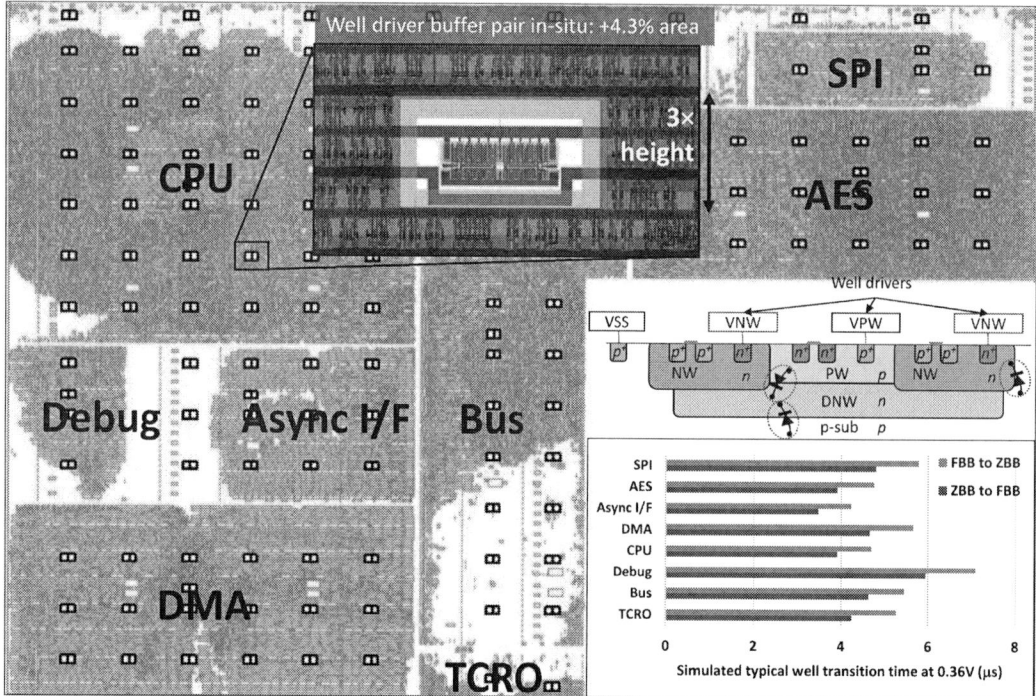

Fig. 2: Logic floorplan showing well buffer layout, extracted well diodes and simulated well transition time.

Logic LVSB is tested with a 128b AES encrypt/decrypt accelerator driven by on-chip LBIST, showing a 6.2× speedup on silicon at a Minimum Energy Point (MEP) of 0.30V with only 6% energy penalty (Fig. 6).

B. Memory

LVSB logic alone is not sufficient for a functional and efficient LVSB system. If the entire SRAM is at Zero Body Bias (ZBB), clock buffers inside it will increase clock skew in system Forward Body Bias (FBB) mode, causing hold time violations. Also, SRAM access can be up to 50% of system cycle time, reducing potential FBB speedup and degrading system energy efficiency. However, the SRAM array has a low activity factor – a simulated FBB array drives up the MEP, increasing system energy by 16% with only 16% performance improvement.

To solve these issues, SRAM array and periphery are split into separate deep n-wells (Fig. 3). The array is kept ZBB while the periphery incorporates LVSB. Well drivers are integrated into the macro to drive periphery wells. Tested on silicon with on-chip MBIST, the 4KB 10T SRAM macro with LVSB shows a 3× speedup near the individual MEP of 0.40V with 19% energy penalty. The SRAM is fully functional down to 0.29V in both FBB and ZBB modes.

C. Clock generation and control

A Tuned Clock Ring Oscillator (TCRO) [7] is used to generate a voltage and temperature-tracking system clock. The delay chain is physically spread out (Fig. 4) to capture representative wire load and gate delay effects. It also needs to track well bias,

considering SRAM and logic performance scaling. To mimic simulated system behavior, inverter chains representing 26% of the maximum TCRO delay are kept at ZBB and the rest may be FBB to match the system critical path in both bias modes. TCRO voltage and temperature tracking is intrinsic and tuning bits are only used to trim chip-to-chip variation margin. Fig. 4 shows excellent matching within 25% of system frequency across more than two orders of magnitude frequency variation due to voltage, temperature and bias conditions – without any runtime re-trimming.

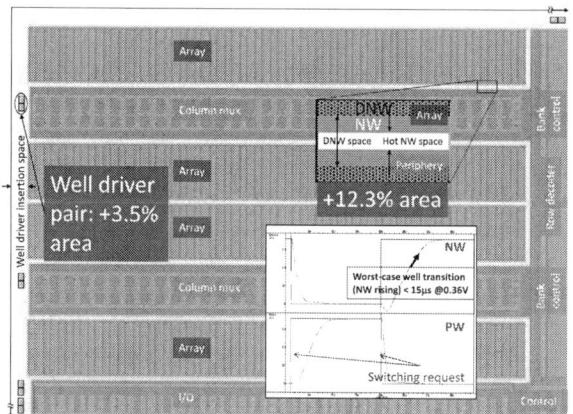

Fig. 3: SRAM layout section showing well split and simulated waveforms. Total SRAM area penalty is 16%.

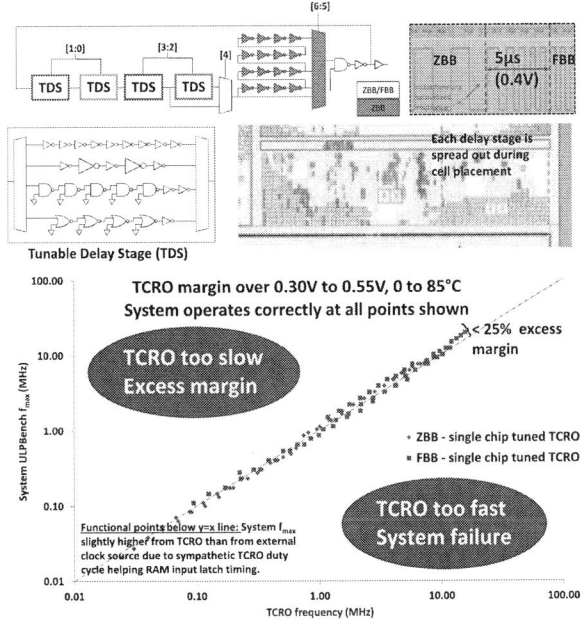

Tunable Delay Stage (TDS)

Fig. 4: TCRO schematic, layout, clock capture during bias transition and timing margin across voltage, temperature and bias mode.

D. Power and bias sequencing

An on-chip Power Control State Machine (PCSM) safely sequences LVSB transitions and generates per-domain control signals (Fig. 5) in response to power-on reset, wake from sleep, or SW power/performance mode requests. When switching bias mode, integrated clock gates are disabled while wells are switched. No state is lost during well transitions.

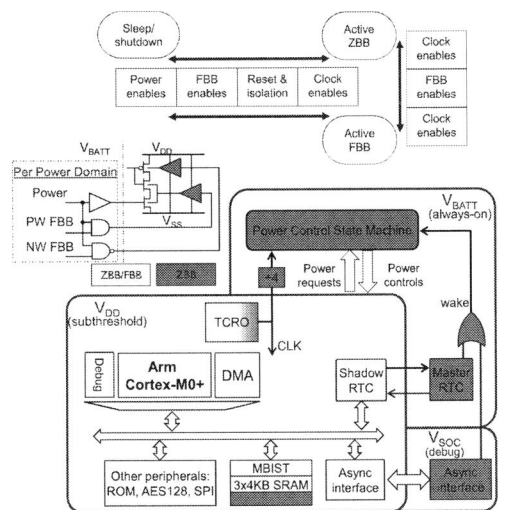

Fig. 5: System block diagram and well switching control sequence. Power gate footers are 2.5V I/O devices driven by 1.2V VBATT control.

FBB is turned off during sleep – well switching is incorporated into the sleep and wakeup sequence. Most mode transitions count 16 TCRO cycles, including enabling/disabling FBB, which required careful simulation. The TCRO output clock was measured to settle within this time frame as expected, without any glitches. As the PCSM is always-on and supplied by 1.2V VBATT it has significant timing slack and does not require any body bias.

III. SYSTEM RESULTS

Measured results are shown in Fig. 6. The change in FBB/ZBB relative leakage power and frequency with VDD causes a leakage energy penalty for logic above 0.25V. As logic and SRAM scaling under FBB are quite different, the system efficiency is quite sensitive to the amount of SRAM used. When the system runs with 8KB SRAM enabled, this results in a total energy penalty of 24% with a 4.3× speedup at 0.4V. The moderate energy penalty shows that LVSB offers a viable fast dynamic system operating point from a single supply voltage. For a duty-cycled or intermittent workload, the system can go into ZBB retention resulting in an average power penalty of only 11% for EEMBC's ULPBench at 0.40V. Without the LVSB capability, Fig. 6 shows that equivalent performance would need 0.47V operation and incur a 78% power penalty. Fig. 7 shows annotated logic analyzer capture for these scenarios. LVSB can also compensate for temperature inversion: at 0°C the 2× reduction from 25°C performance becomes a 2.6× increase at only 11% energy cost if FBB is asserted.

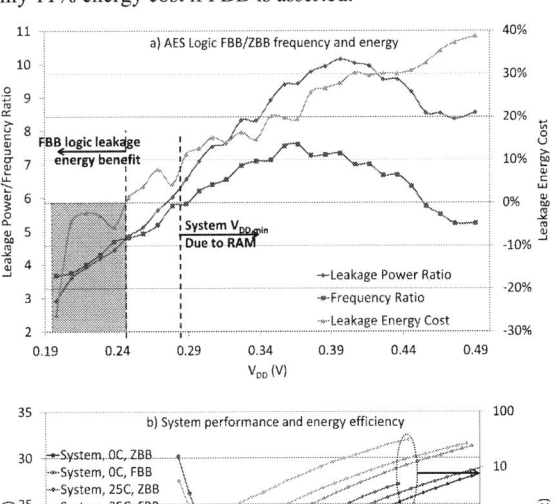

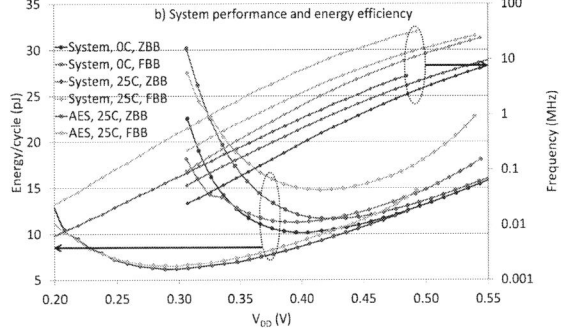

Fig. 6. (a) Measured AES (logic only) FBB/ZBB frequency and leakage.
(b) Measured system FBB/ZBB frequency and energy efficiency.

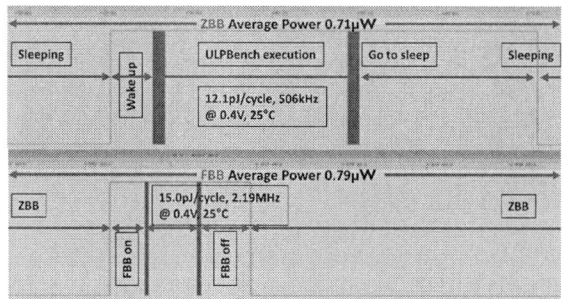
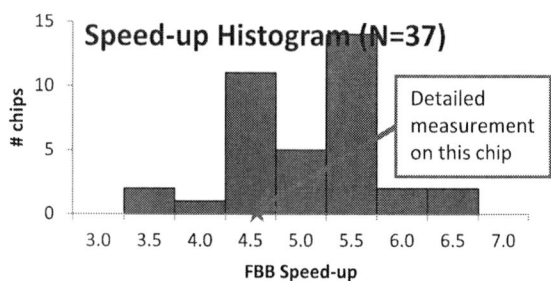

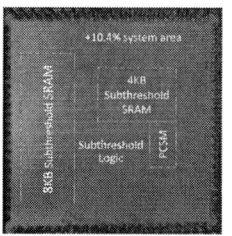

Figure 7: Measured behaviour of ULPBench running at 0.4V with and without active mode body bias.

IV. CONCLUSION

As far as the authors are aware, this work is the first application of LVSB to a complete digital SoC including SRAM and clock generator. Careful system design through fine-grained bias control, a voltage and temperature tracking clock, and a core/periphery partitioned RAM macro ensures that the excess leakage power due to large forward bias is effectively mitigated, demonstrating only 11% power overhead during system operation. Densely distributed well buffers ensure that the forward bias can be switched off in a few cycles when entering sleep mode, resulting in a negligible sleep power penalty.

Table 1 shows a comparison with LVSB prior art. This work achieves the best speedup (other than a ring oscillator in [5]) as well as the highest level of integration. Fig. 8 shows an FBB system speedup histogram, and annotated chip photo.

LVSB shows promise for system and circuit co-optimization of cost-sensitive ultra-efficient IoT systems that are torn between the energy efficiency of sub-threshold operation and the performance demands of existing software. While sophisticated back-bias schemes have been proposed on FD-SoI [8], this work demonstrates a low-overhead, fully synthesizable zero-trim all-digital biasing scheme to get a low-voltage performance boost on widely available, cost-effective and mature bulk CMOS processes.

Figure 8: Speedup histogram and annotated chip photo.

ACKNOWLEDGMENTS

The authors thank An Nguyen and Dave Ondricek of Arm Inc. in San Jose for extensive layout support, and Anand Savanth of Arm Ltd in Cambridge for invaluable test and PCB assistance.

REFERENCES

[1] M. Fojtik *et al.*, "A Millimeter-Scale Energy-Autonomous Sensor System With Stacked Battery and Solar Cells," in *IEEE Journal of Solid-State Circuits*, vol. 48, no. 3, pp. 801-813, March 2013.

[2] S. Clerc *et al.*, "8.4 A 0.33V/-40°C process/temperature closed-loop compensation SoC embedding all-digital clock multiplier and DC-DC converter exploiting FDSOI 28nm back-gate biasing," *2015 IEEE International Solid-State Circuits Conference - (ISSCC) Digest of Technical Papers*, San Francisco, CA, 2015, pp. 1-3.

[3] S. Paul *et al.*, "An energy harvesting wireless sensor node for IoT systems featuring a near-threshold voltage IA-32 microcontroller in 14nm tri-gate CMOS," *2016 IEEE Symposium on VLSI Circuits (VLSI-Circuits)*, Honolulu, HI, 2016, pp. 1-2.

[4] W. Zhao, Y. Ha and M. Alioto, "Novel Self-Body-Biasing and Statistical Design for Near-Threshold Circuits With Ultra Energy-Efficient AES as Case Study," in *IEEE Transactions on Very Large Scale Integration (VLSI) Systems*, vol. 23, no. 8, pp. 1390-1401, Aug. 2015.

[5] S. Narendra *et al.*, "Ultra-low voltage circuits and processor in 180nm to 90nm technologies with a swapped-body biasing technique," *2004 IEEE International Solid-State Circuits Conference (IEEE Cat. No.04CH37519)*, 2004, pp. 156-518 Vol.1.

[6] J. S. Wang, J. S. Chen, Y. M. Wang and C. Yeh, "A 230mV-to-500mV 375KHz-to-16MHz 32b RISC Core in 0.18μm CMOS," *2007 IEEE International Solid-State Circuits Conference. Digest of Technical Papers*, San Francisco, CA, 2007, pp. 294-604.

[7] J. Myers *et al.*, "A 12.4pJ/cycle sub-threshold, 16pJ/cycle near-threshold ARM Cortex-M0+ MCU with autonomous SRPG/DVFS and temperature tracking clocks," *2017 Symposium on VLSI Circuits*, Kyoto, 2017, pp. C332-C333.

[8] A. Quelen, G. Pillonnet, P. Flatresse and E. Beigné, "A 2.5μW 0.0067mm² automatic back-biasing compensation unit achieving 50% leakage reduction in FDSOI 28nm over 0.35-to-1V VDD range," *2018 IEEE International Solid - State Circuits Conference - (ISSCC)*, San Francisco, CA, 2018, pp. 304-306.

TABLE I. COMPARISON TO LVSB PRIOR ART.

	[4]	[5]	[6]	This work
Technology	65nm LL	180, 130, 90nm	180nm	65nm LP
Design	AES (8b datapath)	Ring oscillators, TCP core	32b CPU	32b Cortex-M0+ SoC
Circuits	Logic	Logic	Logic, power gates, control	Logic, SRAM, clock, power gates, control
Bias Scheme	LVSB, ZBB, SFBB	LVSB, ZBB	LVSB/D-NPSBB	LVSB, ZBB
On-chip Well Drivers	No	P-well only	Yes	Yes
Voltage Range	0.50 to 1.20V	0.48 to 2.00V	0.23 to 0.50V	0.28 to 1.20V
Bias Speedup	3.7× (LVSB) 1.6× (SFBB)	1.6× (90nm TCP) 4.4× (130nm RO)	3× (vs. 250nm design)	4.3×

A 2.1 pJ/bit, 8 Gb/s Ultra-Low Power In-Package Serial Link Featuring a Time-based Front-end and a Digital Equalizer

Po-Wei Chiu, Muqing Liu, Qianying Tang and Chris H. Kim

Department of Electrical and Computer Engineering
University of Minnesota, Minneapolis, MN, USA
Email: Chiux148@umn.edu

Abstract— **An 8 Gb/s time-to-digital converter (TDC) based receiver with a time-based front-end in 65nm CMOS is specifically designed for in-package serial link applications. The proposed receiver converts the channel signal to a corresponding time delay which is amplified by a novel delay line based time amplifier. Next, a time-to-digital converter generates a 4-bit code which is used for digital equalization. The proposed design is digital intensive and hence highly resilient to voltage headroom and/or PVT issues. A bathtub curve and time domain eye-diagram were measured by an in-situ bit-error-rate (BER) monitor circuit. An energy-efficiency of 2.1 pJ/b was achieved at 8 Gb/s for a 7 mm link. The receiver area is 240×120μm².**

Keywords— *Time-based, digital equalization, time-to-digital converter (TDC), system-in-package (SiP), digital intensive, inverter-based.*

I. INTRODUCTION

Emerging packaging technologies such as system-in-package (SiP), 2.5D integration, through-silicon-via based 3D ICs, and silicon interposers enable ultra-small form factors while allowing heterogeneous technologies to be integrated into the same chip package [1]-[2]. Serial links for such in-package applications must be more compact, energy-efficient, and digital friendly compared to their chip-to-chip counterparts. As shown in Fig. 1, analog front-ends (AFE) usually contain analog-intensive circuits such as continuous linear equalizer (CTLE), variable gain amplifier (VGA), and current-mode-logic (CML) based decision feedback equalizer (DFE). These circuits are not suitable for in-package links as they suffer from voltage headroom and PVT variation issues, and are typically powered by a separate high power supply. Recently, ADC-based receivers have been drawing attention for off-chip links where RX equalization is performed entirely in the digital domain by a DSP unit [3]-[6]. While ADC-based receivers can operate at a lower voltage than their analog counterparts, and hence take full advantage of technology scaling, they still rely on analog-intensive circuits for signal conditioning. This paper presents a time-to-digital converter (TDC) based receiver with a digital-intensive time-based front-end (TBFE) for in-package link applications. The proposed TBFE consists of a voltage-to-time converter (VTC), a delay line based time amplifier (TA) and a 4-bit TDC. The VTC converts the channel signal to a time delay which is then

amplified by a delay line based TA. A Vernier-line based TDC converts the amplified delay difference to a 4-bit digital code which is fed to the digital equalization block. The proposed TBFE obviates the need for a sample and hold (S/H) circuit as the VTC converts the instantaneous voltage seen by the passing signal edge.

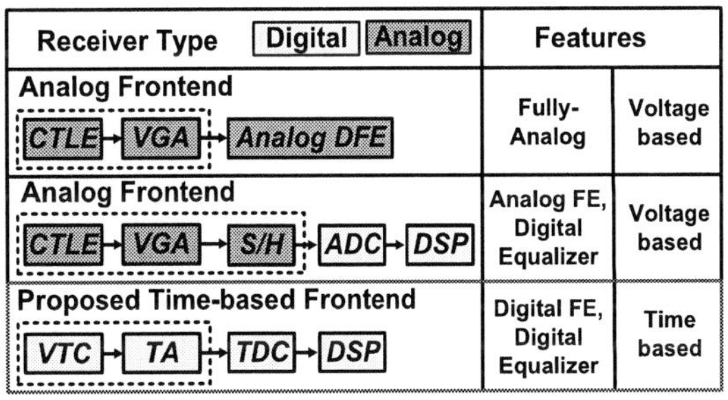

Receiver Type	Features	
Analog Frontend CTLE → VGA → Analog DFE	Fully-Analog	Voltage based
Analog Frontend CTLE → VGA → S/H → ADC → DSP	Analog FE, Digital Equalizer	Voltage based
Proposed Time-based Frontend VTC → TA → TDC → DSP	Digital FE, Digital Equalizer	Time based

Fig. 1. Comparison between the proposed time-based front-end design with conventional RX designs.

II. TIME-BASED TRANSCEIVER IMPLEMENTATION

The block diagram of the proposed transceiver system, including a half rate voltage mode transmitter, a 1/4 rate time-based receiver and an in-situ bit error rate (BER) monitoring circuit, is shown in Fig. 2. The transmitter is based on a 3-tap half-rate feedforward equalizer (FFE) with an inverter-based driver. The receiver contains four lanes of TBFE+TDC circuits, followed by a DSP for digital equalization. Each lane operates at 2Gb/s. A 2^{15}-1 pseudo random bit generator (PRBS) and an in-situ BER monitor are implemented to characterize the circuit performance. The circuit implementation and operation of the VTC are shown in Fig. 3 [7]. A 2GHz clock enters two identical inverter-based VTCs, generating the reference and RX clock signals. The time delays of the VTC circuits are determined by the channel voltage V_{RX} and reference voltage V_{REF}, respectively. A low channel voltage (=data '0') induces a larger delay difference, and vice versa. The reference delay path is fixed to the longest delay to ensure it is always slower than the RX path delay. Inter-symbol-interference noise corrupts the delay of the RX clock path which is filtered out later by the digital DFE.

978-1-5386-6414-8/18 $31.00 © 2018 IEEE

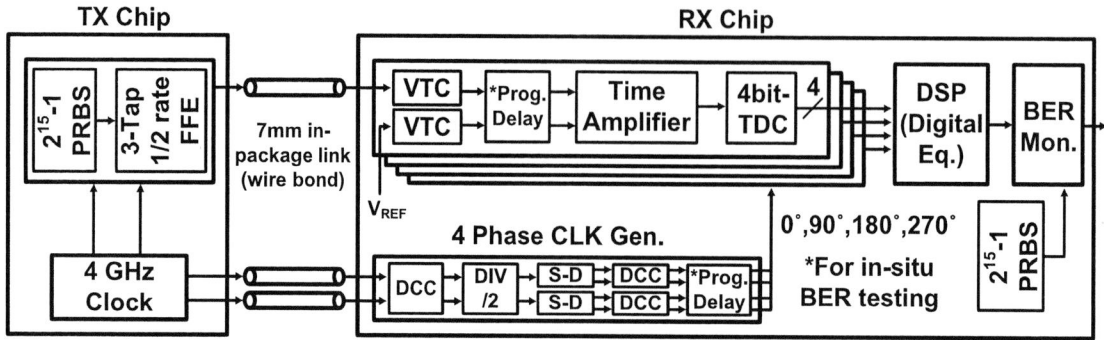

Fig. 2. Block diagram of the proposed digital-intensive time-based transceiver.

III. PROPOSED INVERTER-BASED TIME AMPLIFIER

The delay difference generated by the VTC block is amplified by the fully-digital delay line based TA shown in Fig. 4. The TA circuit consists of two identical tri-state inverter based delay lines. Each stage is driven by two parallel tri-state inverters with 1X and NX sizing, respectively. Initially (i.e. STARTi=0, STOPi=0), the enable signal EN is high which activates all the NX tri-state inverters. Since a total of (N+1)X tri-state inverters are driving the output, the rising edge of STARTi experiences a short propagation delay. Once the STARTi signal arrives, EN is set to low after a fixed delay which disables the parallel NX tri-state inverters. The delay line is now driven only by the 1X tri-state inverters, resulting in a longer propagation delay seen by the STOPi signal. Since the STARTi edge travels faster than the STOPi edge, an (N+1) times longer delay difference appears at the end of the delay line. The timing diagram is shown in Fig. 5. A ring-oscillator based TA using a similar concept was reported in [8]. However, in the previous design, the performance was limited by the ring oscillator frequency. Furthermore, a NAND gate based implementation was required to ensure circuit oscillation. To increase the TA operating speed, in this work, we proposed an open-loop delay line configuration, and utilized tri-state inverters. Post-layout simulation results of the proposed TA with size N=1 and 4 in Fig. 6 confirm high linearity between the input delay and output delay. The proposed delay line based VTC and TA implementation can reduce the circuit complexity while canceling out voltage and temperature induced delay shifts in the delay lines.

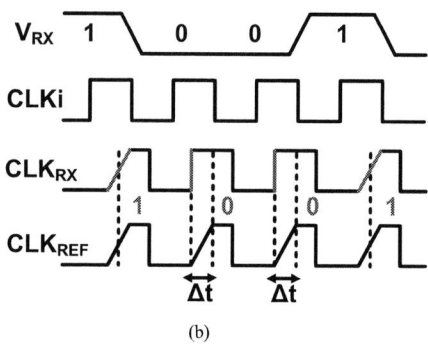

Fig. 3. (a) Voltage to time converter (VTC) circuit implementation and (b) timing diagram.

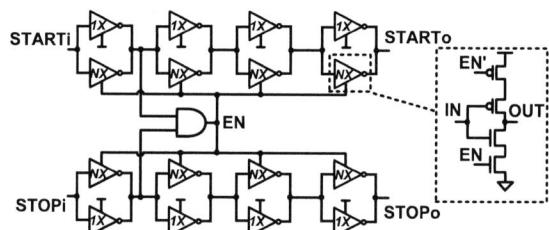

Fig. 4. Schematic of delay line based time amplifier (TA) with open-loop configuration.

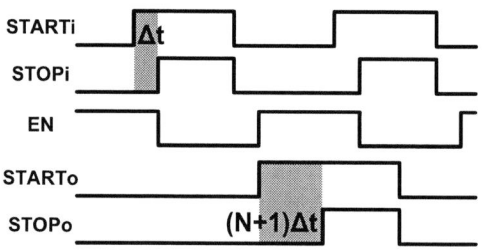

Fig. 5. Timing diagram of the proposed TA.

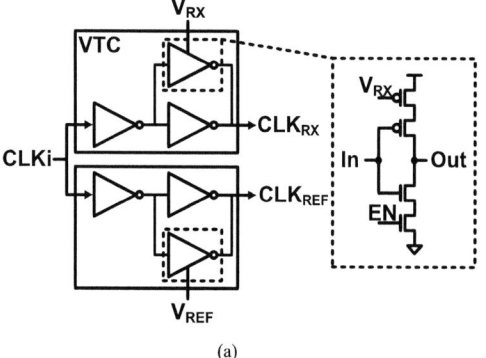

(a)

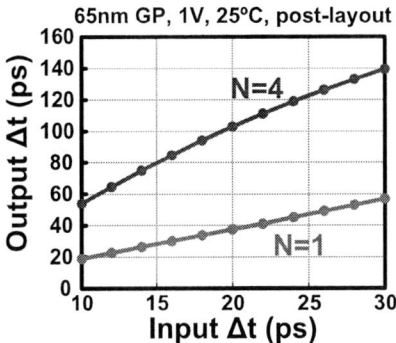

Fig. 6.   TA gain simulation results for N=1 and 4 (Gain=2 and 5).

IV.   TDC, DIGITAL EQUALIZER, AND IN-SITU BER MONITOR

Fig. 7 shows the implementation of the TDC. The Vernier-line based TDC consists of four cascaded delay units with each unit having four delay buffers and four arbiters. A 16-bit thermometer code generated by the Vernier-line is converted to a 4-bit binary code using a thermometer-to-binary (T2B) decoder. The 4-bit output from the TDC is fed to the DSP for digital equalization, as shown in Fig. 8. A bank of 4-bit digital comparators in the DSP compares the new TDC output with predetermined weights w0000, w0001, etc. The correct result from the comparator is selected based on the previous decision results D1-D4. A 16:1 digital MUX outputs the final RX data.

To verify the performance, the in-situ monitor circuit is adopted as shown in Fig. 9. The PRBS in the RX chip, identical to the one in the TX chip, is clocked using a delayed clock to generate the ground-truth data needed for BER measurements. An 11-bit BER counter increments whenever an error is detected. The error count is serially read out using a scan chain. To measure the BER eye-diagram using the in-situ circuit, we swept the two programmable delays denoted as phase delay in red box and time offset in blue box. The two programmable delays correspond to the x and y axes of the BER eye-diagram.

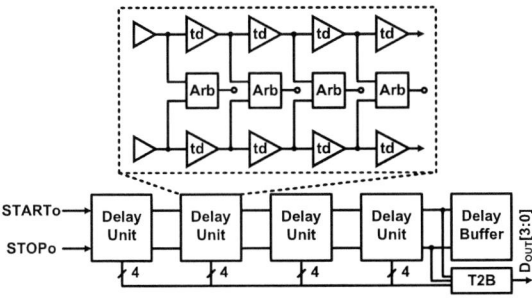

Fig. 7.   Implementation of the 4-bit Vernier-line based TDC.

V.   65 NM TEST CHIP

The chip photo and SiP prototype building with both TX and RX are shown in Fig. 10. Two dies were integrated into a single package to mimic the link behavior of an SiP system. A single-ended data signal and a differential clock were transferred from the TX chip to the RX chip. To test the proposed circuits for different channel conditions, we built packages with varying link distances from 1mm to 7mm. The bathtub data in Fig. 11 shows an eye width of 0.12 UI for a BER of 10^{-12} for 7mm link. The time-domain eye-diagram measured using in-situ circuits is shown in Fig. 12. To save the measurement time, BER down to 10^{-11} is reported. Lower BER values such as 10^{-12} or 10^{-13} can be measured using the same setup. Fig. 13 shows the comparison with previous link designs. When operating at a data rate of 8 Gb/s, the proposed system achieves an energy-efficiency of 2.1 pJ/b (including TX, RX, and DSP power) at 1V. The circuit areas of the time-based front-end (including TDC) and DSP are 0.0192 mm^2 and 0.0096 mm^2, respectively.

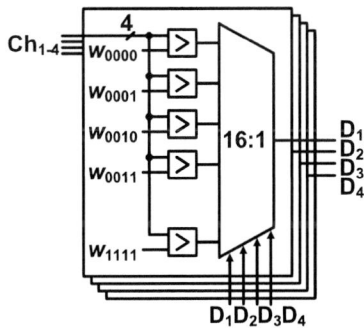

Fig. 8.   DSP for digital equalization.

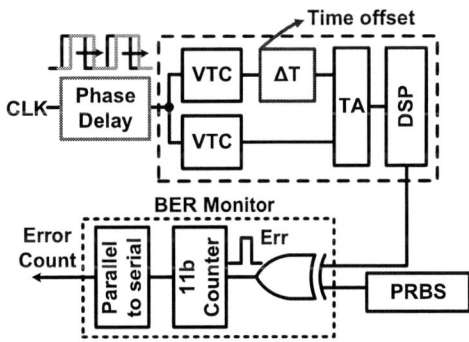

Fig. 9.   In-situ bathtub and BER eye diagram measurement circuits.

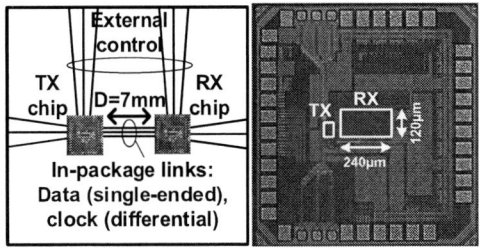

Fig. 10. 65nm chip microphotograph and test package with TX and RX chips for in-package link demo.

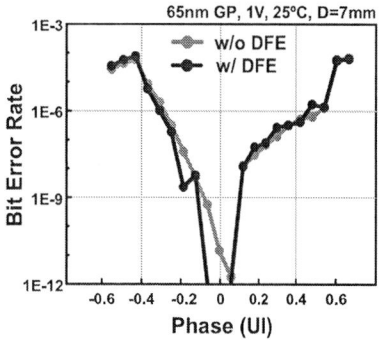

Fig. 11. Measured BER bathtub.

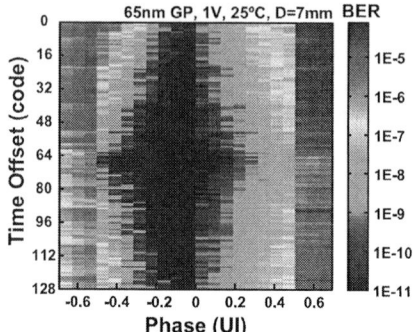

Fig. 12. Measured BER eye diagram.

	JSSC'12 [3]	JSSC'13 [4]	JSSC'15 [5]	JSSC'16 [6]	This work
Application	Off Chip	Off Chip	Off Chip	Off Chip	SiP
RX Architecture	4x Flash ADC	4x Flash ADC	4x Flash ADC	32x SAR ADC	4x TDC
Front-end Type	Voltage-Based (CTLE +VGA)	Voltage-Based (VGA)	Voltage-Based (VGA)	Voltage-Based (Analog FFE)	Time-Based (VTC+TA)
Data Rate	10 Gb/s	10.3125 Gb/s	8.5-11.5 Gb/s	10 Gb/s	8 Gb/s
Technology	65nm	40nm	40nm	65nm	65nm
Voltage	1.1V	0.9V	1V	1V	1V
Resolution	4 bit	6 bit	6 bit	6 bit	4 bit
BER	<1E-9	<1E-12	<1E-12	<1E-10	<1E-12
RX Area (w/o DSP)	0.288 mm^2	0.27 mm^2	0.82 mm^2	0.38 mm^2	0.0192 mm^2
Power Efficiency (pJ/b)	8.1 (RX only)	15.1 (RX only)	18.9 (RX, includes Clock)	7.9 (RX only)	2.1 (TX+RX, includes DSP power)

Fig. 13. Comparison with state-of-the-art link designs.

VI. CONCLUSION

In this paper, a TDC based receiver with TBFE is demonstrated on in-package serial link in 65nm GP process. To our best knowledge, this is the first TDC based receiver with TBFE. A highly linear TA is proposed to amplify the small time difference generated from VTC. A BER less than 10^{-12} is verified using the in-situ measurement circuits. Our proposed TBFE is highly digitalized, low voltage operation and has good compatibility with post digital circuit. The compact size and high energy efficiency show that the proposed time-based receiver is promising for SiP applications.

REFERENCES

[1] C. Y. Ho, H. H. Cheng, P. C. Pan, C. C. Wang and C. P. Hung, "Dielectric Characterization of Ultra-Thin Low-Loss Build-Up Substrate for System-in-Package (SiP) Modules," in *IEEE Trans. Microw. Theory Techn.*, vol. 63, no. 9, pp. 2923-2930, Sept. 2015.

[2] D. Greenhill *et al.*, "A 14nm 1GHz FPGA with 2.5D transceiver integration," *IEEE Int. Solid-State Circuits Conf. Dig. Tech. Papers*, Feb. 2017, pp. 54-55.

[3] E. H. Chen, R. Yousry and C. K. K. Yang, "Power Optimized ADC-Based Serial Link Receiver," in *IEEE J. Solid-State Circuits*, vol. 47, no. 4, pp. 938-951, April. 2012.

[4] A. Varzaghani *et al.*, "A 10.3-GS/s, 6-Bit Flash ADC for 10G Ethernet Applications," in *IEEE J. Solid-State Circuits*, vol. 48, no. 12, pp. 3038-3048, Dec. 2013.

[5] B. Zhang *et al.*, "A 40 nm CMOS 195 mW/55 mW Dual-Path Receiver AFE for Multi-Standard 8.5–11.5 Gb/s Serial Links," in *IEEE J. Solid-State Circuits*, vol. 50, no. 2, pp. 426-439, Feb. 2015.

[6] A. Shafik, E. Z. Tabasy, S. Cai, K. Lee, S. Hoyos and S. Palermo, "A 10Gb/s hybrid ADC-based receiver with embedded 3-tap analog FFE and dynamically-enabled digital equalization in 65nm CMOS," in *IEEE J. Solid-State Circuits*, vol. 51, no. 3, pp. 671-685, April. 2016.

[7] P. W. Chiu, S. Kundu, Q. Tang and C. H. Kim, "A 65-nm 10-Gb/s 10-mm On-Chip Serial Link Featuring a Digital-Intensive Time-Based Decision Feedback Equalizer," in *IEEE J. Solid-State Circuits*, vol. 53, no. 4, pp. 1203-1213, April 2018.

[8] B. Kim, H. Kim, and C. H. Kim, "An 8bit, 2.6ps two-step TDC in 65nm CMOS employing a switched ring-oscillator based time amplifier," in *Proc. IEEE Custom Integr. Circuits Conf. (CICC)*. Sep. 2015, pp. 1-4.

15-3 (8102)

A 2.69 Mbps/mW 1.09 Mbps/kGE Conjugate Gradient-based MMSE Detector for 64-QAM 128×8 Massive MIMO Systems

Guiqiang Peng, Leibo Liu*, Qiushi Wei, Yao Wang, Shouyi Yin, and Shaojun Wei
Institute of Microelectronics, Tsinghua University, Beijing, China
Email: liulb@tsinghua.edu.cn

Abstract—This paper proposes a very-large-scale-integration (VLSI) architecture for an MMSE detector in an uplink 128×8 64-QAM massive MIMO system to achieve high energy and area efficiencies. A soft-output recursion conjugate gradient (RCG)-based minimum mean square error (MMSE) detector is fabricated onto a 3.5 mm^2 silicon with TSMC 65 nm CMOS technology for a 64-QAM 128×8 massive MIMO system. A processing element (PE) array and a user-level pipeline are introduced for this architecture to achieve high energy and area efficiencies. The chip achieves a 1.5 Gbps throughput under a 500 MHz working frequency while dissipating 557 mW at 1.2 V. The energy efficiency (throughput/power) and area efficiency (throughput/area) are 2.69 Mbps/mW and 1.09 Mbps/kGE, which are 2.39-to-2.47× and 1.15-to-8.81× those of the normalized state-of-the-art designs, respectively.

I. INTRODUCTION

Massive MIMO scales up antennas by orders of magnitude while serving tens of users. It is generally accepted that this technology will be applied in future wireless communication techniques such as 5G and beyond. However, the complicated signal detection for uplink processing in a base station (BS) makes it difficult to be efficiently implemented in circuits [1]. Maximum likelihood (ML) detection is an optimal detection algorithm. Nonetheless, the computing load increases markedly with the number of users and modulation orders, thus preventing the practical application of the ML algorithm. Other non-linear algorithms [2]–[4], such as K-best, sphere decoding (SD), expectation-propagation detection (EPD), and message-passing detector (MPD), are able to achieve near-optimal ML detection performance. However, these non-linear architectures are of low throughput and require abundant area and power. To reduce the computational complexity, various linear detection algorithms have been proposed [5], [6], which can be employed in a massive MIMO system with a large but finite number of antennas and a comparatively small number of users. However, considering the hardware architecture, MMSE detection involves complicated matrix inversions and multiplications as well as low parallelism, which produces difficulties for hardware implementation with increasing numbers of users. Consequently, numerous methods have been proposed to further reduce the computational complexity, improve the parallelism of MMSE, and optimize hardware architectures

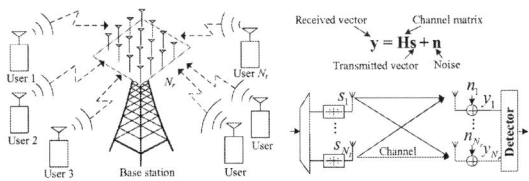

Fig. 1. Uplink massive MIMO system.

[1], [5]–[7]. However, the throughput is limited, and the architectures require large amounts of hardware resources.

This paper proposes a VLSI architecture for signal detection to achieve higher energy and area efficiencies (i.e., throughput/power and throughput/area) in an uplink massive MIMO system. Given that the results of a Gram matrix are required in subsequent detection, a parallel processing element (PE) array with single-sided input is designed. This parallel PE array substantially reduces latency using limited hardware, and the architecture operates in a deeply pipelined manner. In addition, a user-level pipeline architecture is proposed to achieve detection based on the recursion conjugate gradient (RCG) method, which is proposed to scale down the high computational complexity and improve the parallelism of the MMSE detection. The proposed VLSI architecture is verified on an FPGA and fabricated onto a chip with TSMC 65 nm CMOS technology in an uplink massive MIMO system. The measurement results show that this architecture achieves substantial improvements in energy and area efficiencies compared with other state-of-the-art architectures.

This paper is organized as follows. Section II presents the proposed RCG massive MIMO detection algorithm, and Section III describes the architecture of the RCG detector. Then, the silicon implementation results and the algorithm simulation results are shown in Section IV. Finally, a conclusion is given in section V.

II. MASSIVE MIMO DETECTION BY RECURSION CONJUGATE GRADIENT

In an $N_r \times N_t$ MIMO system with N_t transmitters on the user side and N_r antennas on the BS side (predominantly $N_r \gg N_t$), the uplink system can be modeled as $\mathbf{y} = \mathbf{Hs} + \mathbf{n}$,

978-1-5386-6414-8/18 $31.00 © 2018 IEEE

where $\mathbf{H} \in \mathbb{C}^{N_r \times N_t}$ represents a Rayleigh flat-fading channel matrix; $\mathbf{s} \in \mathbb{C}^{N_t \times 1}$ denotes the transmitted signal vector, which is based on the 64-QAM modulation constellation set Ω in this paper; $\mathbf{n} \in \mathbb{C}^{N_r \times 1}$ is an additive white Gaussian noise vector with zero mean and variance σ^2; and $\mathbf{y} \in \mathbb{C}^{N_r \times 1}$ is the received signal vector at the BS. The uplink massive MIMO system is shown in Fig. 1. In classical MMSE detection, the estimation of the transmitted signal can be expressed as

$$\mathbf{s} = \left(\mathbf{H}^H\mathbf{H} + N_0 E_s^{-1}\mathbf{I}_{N_t}\right)^{-1}\mathbf{H}^H\mathbf{y} = \mathbf{W}^{-1}\mathbf{y}^{\mathrm{MF}}, \quad (1)$$

where $\mathbf{W} = \mathbf{H}^H\mathbf{H} + N_0 E_s^{-1}\mathbf{I}_M$ is the MMSE filtering matrix with a noise power spectral density of N_0 and a signal power spectral density of E_s, and $\mathbf{y}^{\mathrm{MF}} = \mathbf{H}^H\mathbf{y}$ denotes the matched-filter vector.

The RCG iteration can be expressed by

$$\mathbf{s}^{(k+1)} = \mathbf{s}^{(k)} + \boldsymbol{\alpha}^{(k)}\mathbf{p}^{(k)}, \quad (2)$$

where $\mathbf{p}^{(k)}$ is an orthogonal basis [8], k is the iteration number, and the parameter $\boldsymbol{\alpha}^{(k)}$ can be calculated by

$$\boldsymbol{\alpha}^{(k)} = \frac{\left(\mathbf{z}^{(k)}, \mathbf{z}^{(k)}\right)}{\left(\mathbf{W}\mathbf{p}^{(k)}, \mathbf{z}^{(k)}\right)}. \quad (3)$$

In (3), $\mathbf{z}^{(k)}$ represents the residual vector, which can be expressed as

$$\mathbf{z}^{(k)} = \mathbf{y}^{\mathrm{MF}} - \mathbf{W}\mathbf{s}^{(k)}. \quad (4)$$

In addition, according to the Lanczos orthogonalization algorithm, the residual vector $\mathbf{z}^{(k+1)}$ can be expressed as [8]

$$\mathbf{z}^{(k+1)} = \rho^{(k)}\left(\mathbf{z}^{(k)} - \gamma^{(k)}\boldsymbol{\eta}^{(k)}\right) + (1 - \rho^{(k)})\mathbf{z}^{(k-1)}, \quad (5)$$

where $\rho^{(k)}$, $\gamma^{(k)}$, and $\boldsymbol{\eta}^{(k)} = \mathbf{W}\mathbf{z}^{(k)}$ are iterative parameters. Combining (4) and (5), the estimation of the transmitted vector can be expressed as

$$\mathbf{s}^{(k+1)} = \rho^{(k)}\left(\mathbf{s}^{(k)} + \gamma^{(k)}\mathbf{z}^{(k)}\right) + \left(1 - \rho^{(k)}\right)\mathbf{s}^{(k-1)}. \quad (6)$$

Next, the iterative parameters are computed. Because the vectors $\mathbf{z}^{(k+1)}$, $\mathbf{z}^{(k)}$, and $\mathbf{z}^{(k-1)}$ are mutually orthotropic [8], there are

$$\left(\mathbf{z}^{(k+1)}, \mathbf{z}^{(k)}\right) = \left(\mathbf{z}^{(k-1)}, \mathbf{z}^{(k)}\right) = \left(\mathbf{z}^{(k-1)}, \mathbf{z}^{(k+1)}\right) = 0. \quad (7)$$

Hence, combining (5) and (7), the parameters can be calculated as

$$\gamma^{(k)} = \frac{\xi^{(k)}}{\phi^{(k)}}; \quad \rho^{(k)} = \frac{\xi^{(k-1)}}{\xi^{(k-1)} + \gamma^{(k)}\left(\boldsymbol{\eta}^{(k)}, \mathbf{z}^{(k-1)}\right)}, \quad (8)$$

where $\xi^{(k)} = \left(\mathbf{z}^{(k)}, \mathbf{z}^{(k)}\right)$, and $\phi^{(k)} = \left(\boldsymbol{\eta}^{(k)}, \mathbf{z}^{(k)}\right)$. In addition, according to (5) and (7), there are

$$\left(\boldsymbol{\eta}^{(k)}, \mathbf{z}^{(k-1)}\right) = \left(\mathbf{z}^{(k)}, \boldsymbol{\eta}^{(k-1)}\right),$$
$$\boldsymbol{\eta}^{(k-1)} = -\frac{\mathbf{z}^{(k)}}{\rho^{(k-1)}\gamma^{(k-1)}} + \frac{\mathbf{z}^{(k-1)}}{\gamma^{(k-1)}} + \frac{\left(1 - \rho^{(k-1)}\right)\mathbf{z}^{(k-2)}}{\rho^{(k-1)}\gamma^{(k-1)}}. \quad (9)$$

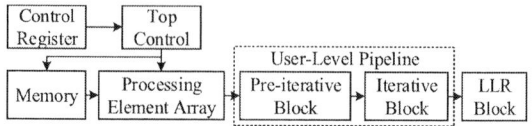

Fig. 2. Top-level block diagram for the proposed massive MIMO detector.

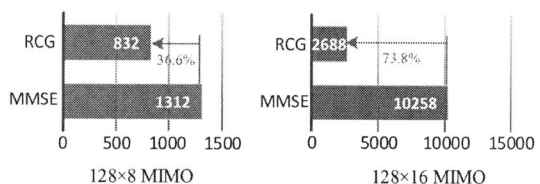

Fig. 3. Number of real-valued multiplications.

Therefore, the parameter $\rho^{(k)}$ can be computed as

$$\rho^{(k)} = \left[1 - \frac{\gamma^{(k)}}{\gamma^{(k-1)}}\frac{\xi^{(k)}}{\xi^{(k-1)}}\frac{1}{\rho^{(k-1)}}\right]^{-1}. \quad (10)$$

In this proposed algorithm, to perform the iteration and calculate the parameters, there are several required initial settings such as $\rho^{(0)} = 1$, $\mathbf{z}^{(-1)} = \mathbf{z}^{(0)}$, and $\mathbf{s}^{(-1)} = \mathbf{s}^{(0)}$.

III. VLSI Architecture

In this section, a VLSI architecture is designed to achieve massive MIMO detection based on the RCG detection algorithm. The architecture was designed for a 64-QAM, 128×8 massive MIMO system case study. Fig. 2 shows the top-level block diagram for the proposed massive MIMO detector. To achieve a higher throughput with limited hardware resources, the top-level architecture is fully pipelined. There are three computation blocks in the detector. The first PE array is used to compute the matched-filter $\mathbf{y}^{\mathrm{MF}}$ and matrix $\mathbf{W}$. The outputs are used to compute the initial solution of the estimated vector in the pre-iterative block. Next, a user-level parallelism-based full pipeline iteratively applies the RCG method. The high computational complexity of the MMSE arises from a series of matrix multiplication and matrix inversion operations. A modified RCG-based MMSE detector is proposed to reduce the number of real-valued multiplications (RMULs) from $O(N_t^3)$ to $O(N_t^2)$. For example, as shown in Fig. 3, the RMULs are reduced by 36.6% and 73.8% when compared to the traditional MMSE detector for a 128×8 and 128×16 MIMO system, respectively. In addition, the proposed detector enhances the parallelism of each iteration. Finally, the soft outputs are computed in a log-likelihood-ratio (LLR) block. The two key blocks (processing element array and user-level pipeline) will be detailed in the following subsections.

A. Processing Element Array

The architecture of the proposed PE array multiplies the vector $\mathbf{y}^{\mathrm{MF}}$ and matrix $\mathbf{W}$, as shown in Fig. 4. In the array, there are two types of PEs: N_t PE-As and $\frac{N_t^2 + N_t}{2}$ PE-Bs. The PE-As are used to compute the diagonal elements of

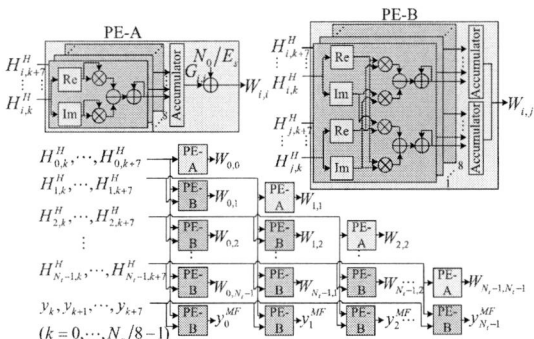

Fig. 4. Architecture of the processing element array.

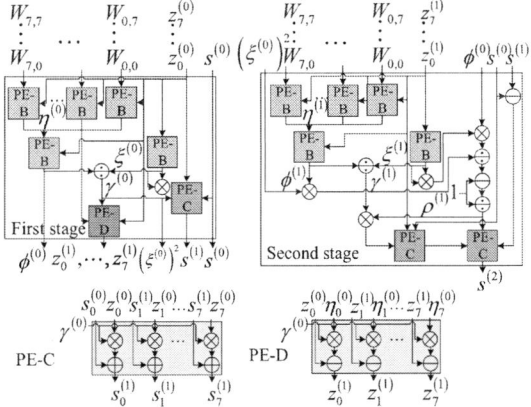

Fig. 5. Architecture of the iterative block in the user-level pipeline.

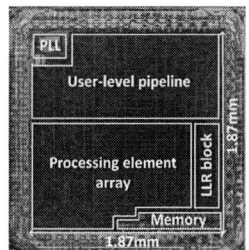

Fig. 6. Die micrograph.

TABLE I
COMPARISON OF ASIC IMPLEMENTATION RESULTS

Algorithm	This work[a]		[4]	[5]	[7]	[6]
	RCG		EPD	CHD	WeJi	LUD
MIMO ($N_r \times N_t$)	128×8		128×16	128×8	128×8	4×4
Modulation	64		256	256	64	256
Silicon proof	Yes		Yes	Yes	Yes	Yes
Technology [nm]	65		28	28	65	65
Voltage [V]	1.2		1.0	0.9	1.0	1.0
Frequency [MHz]	500		569	300	680	517
Throughput [Gbps]	1.5		1.8	0.3	1.02	1.379
Preprocessing	No	Yes	No	No	Yes	Yes
Area (Gate count) [kGE]	596	1372	3607	148	1070	347
Power [mW]	120	557	127	18	650	26.5
Normalized[bc] energy efficiency[d] [Mbps/mW]	12.5	2.693	1.826	1.740	1.090	1.129
Normalized[bc] area efficiency[d] [Mbps/kGE]	2.517	1.093	0.215	0.873	0.953	0.124

[a] According to the simulation results, the number of iterations assumed is $K = 2$, which achieves near-optimal performance under the RCG method.

[b] Technology normalized to 65 nm technology assuming the following: $f_{clk} \sim s$ and $P_{dyn} \sim (1/s)(V_{dd}/V'_{dd})^2$.

[c] MIMO size normalized to 128×8. According to [2], throughput is increased by $\frac{8}{N_t} \times \frac{\log_2 N_t}{\log_2 8}$, and the power and area are increased by $\frac{128}{N_r} \times \frac{8}{N_t} \times \frac{\log_2 N_t}{\log_2 8}$.

[d] The energy and area efficiencies are defined as throughput/power and throughput/area, respectively.

the matrix $\mathbf{W}$; these elements are real valued. For a 128×8 massive MIMO system, there are 8 PE-As and 36 PE-Bs. The PE-Bs are used to compute the off-diagonal elements of the matrix $\mathbf{W}$ (28 PE-Bs) and the vector $\mathbf{y}^{MF}$ (8 PE-Bs). In one clock cycle, 8 elements of $\mathbf{H}^H$ and $\mathbf{y}$ are input for computation, and the PEs exhibit echelonment with high processing speed because the Gram matrix $\mathbf{G} = \mathbf{H}^H \mathbf{H}$ is an asymmetric matrix. As shown in Fig. 4, each PE-A includes eight groups of the same arithmetic logical units (ALUs), one accumulator, and an adder for diagonal elements. The ALU is used to compute the part of the diagonal element in the Gram matrix. Then, the results of all ALUs are accumulated to obtain the value of $G_{i,i}$. Fig. 4 also shows the details of the PE-B, which performs the computation of the complex-valued multiplications and accumulations. In addition, the hardware utilizations of both PE-A and PE-B approaches are high. Therefore, this PE array achieves high throughput and high energy and area efficiencies.

B. User-level Pipeline

The user-level pipeline (Fig. 5) is designed to achieve the RCG iteration, which includes two blocks: the pre-iterative block and the iterative block. The initial solution and some parameters are computed in the pre-iterative block. Instead of

zero, each element of the initial solution can be chosen with a certain point of one quadrant according to the located quadrant of the $\mathbf{y}^{MF}$. In addition, there is no extra computational cost. The iterative block estimates and updates the signal $\mathbf{s}^{(K)}$ and residual $\mathbf{z}^{(K)}$ according to the channel matrix $\mathbf{H}$. The iterative block includes two stages in total. The first stage calculates $\gamma^{(0)}$, which is required to update the signal $\mathbf{s}^{(1)}$ and the residual $\mathbf{z}^{(1)}$ according to the initial solution of $\mathbf{s}^{(0)}$; then, it completes the first update of $\mathbf{s}^{(1)}$ and $\mathbf{z}^{(1)}$. The second stage calculates $\rho^{(1)}$ and $\gamma^{(1)}$; then, it updates the signal $\mathbf{s}^{(2)}$ according to outputs of the first stage. The calculations of these two stages were deeply pipelined based on the user level, resulting in high parallelism. The hardware utilizations of both PE-B, PE-C and PE-D are high.

IV. MEASUREMENT RESULTS

The proposed MMSE detector was implemented onto a 1.87×1.87 mm^2 silicon using TSMC 65 nm CMOS technology. Fig. 6 shows the die micrograph of the chip. This chip achieved a 1.5 Gbps data rate at a 500 MHz working frequency while dissipating 557 mW. Table I lists the implementation results of the proposed detector and the state-of-the-art designs in [4]–[7]; these designs were those with results closest to the

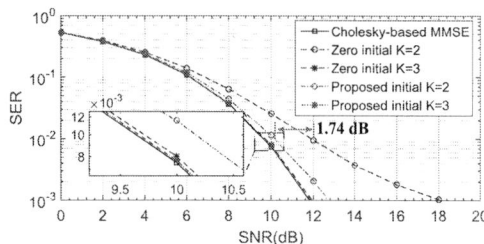

Fig. 7. SER performance comparison between the proposed initial solution and the traditional zero-vector initial solution.

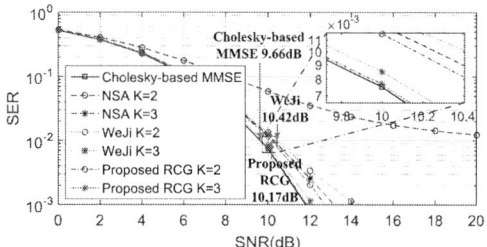

Fig. 8. SER performance comparisons between the proposed algorithm and other algorithms.

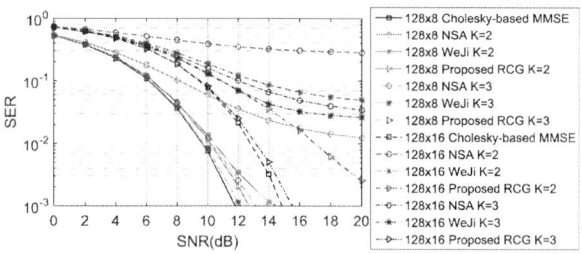

Fig. 9. SER performance comparisons at different MIMO dimensions.

results of the proposed detector. To achieve a fair comparison, the energy and area efficiencies are normalized to the same MIMO dimension and process (full-scaling approach); the same normalization method is also used in [2]. The proposed architecture achieves $2.47\times$ ($1.15\times$) and $2.39\times$ ($8.81\times$) normalized energy (area) efficiencies compared with that of [7] and [6], respectively. Despite the pre-iterative processing in the detector (only considering post pre-iterative processing design), the proposed architecture achieves 12.5 Mbps/mW energy efficiency and 2.52 Mbps/kGE area efficiency, which are $6.85\times$ ($11.71\times$) and $7.18\times$ ($2.88\times$) those of the designs in [4] and [5], respectively.

To evaluate the performance of the proposed RCG method, simulated symbol-error-rate (SER) results for RCG are compared with state-of-the-art methods such as NSA and WeJi. The SER performance of exact MMSE detection based on the CHD method is also provided for comparison. In these simulations, the settings are adopted as a 64-QAM modulation scheme and a rate-1/2 industry standard convolutional code with a $[133_o \ 171_o]$ polynomial along with a random

interleaver. In addition, the coding is performed over 120 symbols, and the number of frames is 10,000. The channels were assumed to exhibit i.i.d. Rayleigh fading across the coded symbols. At the receiver, the LLRs are used as the soft input for Viterbi decoding. In addition, as in [4]–[7], the signal-to-noise ratio (SNR) is defined at the receiver. Fig. 7 shows a comparison between the proposed initial solution and the traditional zero-vector initial solution. According to Fig. 7, to achieve the same SER (10^{-2}), the proposed initial solution has a 1.74 dB gain when compared with the traditional zero-vector initial solution. Fig. 8 shows the simulated SER results of the proposed RCG method and different detection algorithms in a 64-QAM 128×8 MIMO system. When $K = 2$, the SNR required to achieve an SER of 10^{-2} was 10.17 dB, which was close to the performance of the Cholesky-based MMSE detection (9.66 dB) [5]. In contrast, the required SNRs of the WeJi and NSA methods mentioned in [7] and [1] are 10.42 dB and >20 dB, respectively. Fig. 9 shows that to achieve the same SER, the SNR required by the proposed algorithm is also smaller than that by the NSA and WeJi methods, proving that the proposed algorithm can maintain its advantages at different MIMO dimensions.

V. CONCLUSION

This paper proposes an ASIC implementation of a massive MIMO detector based on a recursion conjugate gradient method, therein achieving near-optimal performance for massive MIMO systems. This architecture achieves high energy and area efficiencies when compared with state-of-the-art designs. Thus, we believe that this detector makes an important contribution to next-generation massive MIMO communication systems.

REFERENCES

[1] M. Wu, B. Yin, G. Wang, C. Dick, J. R. Cavallaro, and C. Studer, "Large-scale MIMO detection for 3GPP LTE: algorithms and FPGA implementations," *IEEE J. Sel. Topics Signal Process.*, vol. 8, no. 5, pp. 916–929, Oct. 2014.

[2] Y. T. Chen, C. C. Cheng, T. L. Tsai, W. C. Sun, Y. L. Ueng, and C. H. Yang, "A 501mW 7.6lGb/s integrated message-passing detector and decoder for polar-coded massive MIMO systems," in *Proc. IEEE Symp. VLSI Circuits*, Kyoto, Japan, Jun. 2017, pp. 330–331.

[3] W. Tang, C.-H. Chen, and Z. Zhang, "A 0.58 mm² 2.76 Gb/s 79.8 pJ/b 256-QAM massive MIMO message-passing detector," in *Proc. IEEE Symp. VLSI Circuits*, Honolulu, USA, Jun. 2016, pp. 250–251.

[4] W. Tang, H. Prabhu, L. Liu, V. Öwall, and Z. Zhang, "A 1.8Gb/s 70.6pJ/b 128×16 link-adaptive near-optimal massive MIMO detector in 28nm UTBB-FDSOI," in *Proc. IEEE Int. Solid-State Circuits Conf. Dig.*, San Francisco, CA, Feb. 2018, pp. 224–225.

[5] H. Prabhu, J. N. Rodrigues, L. Liu, and O. Edfors, "A 60pJ/b 300Mb/s 128×8 Massive MIMO precoder-detector in 28nm FD-SOI," in *Proc. IEEE Int. Solid-State Circuits Conf. Dig.*, San Francisco, CA, Feb. 2017, pp. 60–61.

[6] C. Chen, W. Tang, and Z. Zhang, "A 2.4mm² 130mW MMSE-nonbinary-LDPC iterative detector-decoder for 4×4 256-QAM MIMO in 65nm CMOS," in *Proc. IEEE Int. Solid-State Circuits Conf. Dig.*, San Francisco, CA, Feb. 2015, pp. 338–339.

[7] G. Peng, L. Liu, S. Zhou, S. Yin, and S. Wei, "A 1.58 Gbps/W 0.40 Gbps/mm² ASIC Implementation of MMSE Detection for 128 × 8 64-QAM Massive MIMO in 65 nm CMOS," *IEEE Trans. Circuits Syst. I. Reg. Papers*, vol. 65, no. 5, pp. 1717–1730, May 2018.

[8] D. R. Kincaid and E. W. Cheney, *Numerical analysis: mathematics of scientific computing*, 3rd ed. Belmont, CA: Wadsworth, 2002.

A Fully Standard-Cell Based On-Chip BTI and HCI Monitor with 6.2x BTI sensitivity and 3.6x HCI sensitivity at 7 nm Fin-FET Process

Mitsuhiko Igarashi, Yuuki Uchida, Yoshio Takazawa, Yasumasa Tsukamoto, Koji Shibutani and Koji Nii

Design Platform Business Dep., Renesas Electronics Corporation, Tokyo, Japan
Mitsuhiko.igarashi.xv@renesas.com

In autonomous driving era, it is inevitable to pursue higher performance for automotive LSIs such as advanced driver-assistance systems (ADAS) and so on. This should be achieved under limited power budget at high temperature condition of the car. Therefore, requirement of higher performance and higher energy efficiency are one of the key challenges and motivations to apply cutting-edge process technologies such as 7 nm Fin-FET process [1] and so on. On the other hand, as process technology is scaled down, device reliability such as bias temperature instability (BTI) and hot carrier injection (HCI) are becoming more important because of decreasing design margin with supply voltage (Vdd) scaling [2]. Moreover, an annual mileage of the car is expected to be increased in autonomous driving era since paradigm shift occurs in terms of the car usage. As a result, reliability design becomes more severe by both process scaling and car usage change. In general, LSI designers take guard-band (GB) into account for aging such as BTI and HCI to prevent delay failures at the end of the products lifetime. Therefore, the aging monitor system, which is easy to implement in products with features of high sensitivity and NBTI/PBTI/HCI separations, could be a solution to optimize required GB and detect outlier of aging at testing [3, 5].

Fig. 1 and 2 show comparison of ring-oscillator (RO) topology and BTI sensitivity under DC stress at 7 nm Fin-FET process. DC stress means non-oscillated at stress period. The NBTI and PBTI sensitivity mean the ratios of delay degradation (ΔTpd) over threshold voltage degradation (ΔVth) for PMOS and NMOS, respectively. The standard configuration of RO is inverter (INV) based (INV-RO) [6]. As shown in Fig. 1 (a2) and (a3), NAND or NOR based RO control output signal of each cell as high or low by control pin, result in applying NBTI or PBTI selectively at DC stress [4] although its NBTI or PBTI sensitivity is almost the same as that of INV-RO. As shown in Fig. 1 (a4), the combination of NAND and NOR enhances NBTI and PBTI sensitivity [4] although it has almost the same NBTI and PBTI sensitivity. In contrast, NBTI-RO shown in Fig. 1 (b1) has 4.2x NBTI sensitivity as described in ref. [5]. Furthermore, ΔTpd becomes negative for R-N/PBTI-RO which is reverse configuration of N/PBTI-RO to invert input voltage of each NAND and NOR cell under DC stress. It is because when parallel PMOS is degraded by NBTI and stacked NMOS is not in NOR cell of R-NBTI-RO, rise input delay of NOR is improved by decreasing PMOS drivability, compared with fresh one. This unique characteristic enables to increase and decrease NBTI and PBTI sensitivity respectively by taking difference of operation frequency degradation of NBTI-RO (ΔF1) and R-NBTI-RO (ΔF2). Fig. 3 shows proposed NBTI monitor configuration as a representative. NBTI monitor has 6.2x NBTI sensitivity and negligibly small PBTI impact shown in Fig. 4(a).

We also correct the influence of temperature variation on ΔF to multiply ΔF by temperature dependent table (T-table), which is determined by information collected from on-chip temperature sensor. As shown in Fig. 4(b), simulation result shows error rate of this correction based on typical condition is less than 4% in various process and voltage conditions.

In logic circuit, the operation waveform of small input and large output transition is the worst case for HCI degradation in the advanced process generations [3]. The small drive strength cells of ACHCI-RO have the operation waveform described above. Fig. 5 shows transition time per cycle time of product design from statistical timing analysis, upper limit of design constraint and ACHCI-RO. The transition time of ACHCI-RO is set slightly beyond the product design. In this way, ACHCI-RO simulates the worst HCI degradation operating waveform of the logic circuit. We also changed the cell type used in ACHCI-RO from NAND [3] to INV because stacked NMOS of NAND decrease the drain-source voltage at transient time, resulting in reduction of NMOS HCI degradation.

We have fabricated a test chip in a 7 nm Fin-FET process to confirm the basic characteristics of each RO. Fig. 6 shows die photo and RO configuration of the test chip. A high voltage and high temperature stress is applied. First, we estimated ΔVth of NBTI/PBTI by multiplying the simulated BTI sensitivity of NOR-RO and NAND-RO by the measured delay degradation ΔTpd of them, respectively. It is because NOR-RO and NAND-RO can be selectively applied by NBTI or PBTI stress, respectively. Fig. 7 shows estimated ΔVth for both 7 nm and 16nm Fin-FET case [7]. NBTI is the dominant factor for both 7 nm and 16 nm. The main difference between 16 nm and 7 nm is Vth type dependency; In 7 nm case, lower Vth device obviously provides the worst NBTI, which is similar result with ref. [8].

Fig. 8 shows comparison between measured and estimated ΔTpd of each RO caused by BTI under DC stress. We estimate ΔTpd by using estimated ΔVth shown in Fig. 7 and simulated NBTI/PBTI sensitivity of each RO. We successfully confirm that measured and estimated ΔTpd of all RO types are well correlated and maximum error rate, defined as ratio of measured data per estimated data, of NBTI monitor is less than 6%. Fig. 9(a) shows normalized measurement and simulation result of INV-RO and two ACHCI-ROs with different stage number under AC stress. AC stress means RO is oscillating during stress period. HCI degradation is only considered in this simulation. HCI degradation is increased by reproducing the worst-case waveform of logic circuit like the difference between INV-RO and ACHCI-RO at the same stages number. It is also increased with decreasing stage number of RO because a ratio of switching operation time per unit time is increased, resulting in 3.6x sensitivity compared with INV-RO. Measurement results of ΔTpd are well explained by simulation results. Furthermore, as shown in Fig. 9(b), the recovery effect was not observed as with the characteristics of HCI even if the recovery time, which is the total value of the power-off time, increased. In this way, we confirm that the dominant degradation factor of ACHCI-RO is HCI and has 3.6x sensitivity compared with INV-RO.

CONCLUSION

We propose an on-chip NBTI, PBTI and HCI monitor by using standard cell based unbalanced RO at 7 nm Fin-FET process. The NBTI monitor consists of two ROs; one is NBTI-RO and the other is R-NBTI-RO with reversed cell order of NBTI-RO. R-NBTI-RO gets fast after NBTI stress where as other ROs is degraded. As a result, 6.2x NBTI sensitivity compared with normal INV-RO and negligibly small PBTI sensitivity are achieved. PBTI monitor is achieved in a similar manner. In HCI monitor, HCI degradation is 3.6x emphasized by simulating worst-case waveform of logic circuit by using unbalanced drive strength configuration of INV cell. The measurement result of our test chip fabricated in 7 nm Fin-FET process shows that measured result of each RO is well matched to the simulation one. These high sensitive NBTI/PBTI/HCI monitor can be a solution to optimize required GB in a field, detect variations and outliers of aging at time of testing to achieve both high performance and high reliability of autonomous driving LSI.

REFERENCES

[1] S-Y. Wu et.al., "A 7nm CMOS Platform Technology Featuring 4th Generation FinFET Transistors with a 0.027um2 High Density 6-T SRAM cell for Mobile SoC Application", IEDM, pp. 298-301, 2016

[2] R. Huang et. al., "Variability- and Reliability-Aware Design for 16/14nm and Beyond Technology," IEDM, pp. 298-301, 2017

[3] M. Igarashi et. al., "An On-die Digital Aging Monitor against HCI and xBTI in 16 nm Fin-FET Bulk CMOS Technology," ESSCIRC, pp. 112-115, 2015

[4] M. Chen et. al., "Aging Sensors for Workload Centric Guardbanding in Dynamic Voltage Scaling Applications," IPRS 2013.

[5] M. Igarashi et. al., "NBTI/PBTI separated BTI monitor with 4.2x Sensitivity by Standard Cell Based Unbalanced Ring Oscillator", A-SSCC, pp. 201-204, 2017

[6] T. H. Kim et. al., "Silicon Odometer: An On-Chip Reliability Monitor for Measuring Frequency Degradation of Digital Circuits," IEEE J. Solid-State Circuits, vol. 43, no. 4, pp. 874-880, March 2008

[7] Shien-Yang et. al., "A 16nm FinFET CMOS Technology for Mobile SoC and Computing Applications," IEDM, pp. 224-227, 2013

[8] Wen Liu, et. al., "Cap Layer and Multi-Work-Function Tuning Impact on TDDB / BTI in SOI FinFET Devices," IRPS, pp. 6D.2-1-6F.2-5, 2018

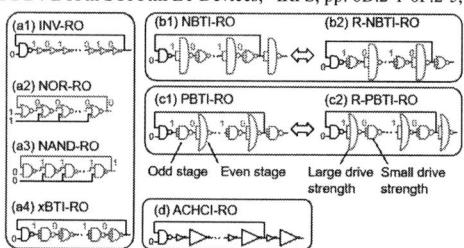

Fig. 1. Comparison between topology of proposed and conventional ROs

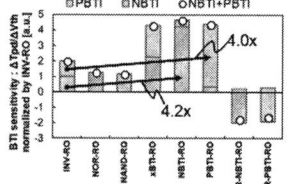

Fig. 2. Simulated NBTI and PBTI sensitivity at 7 nm Fin-FET process

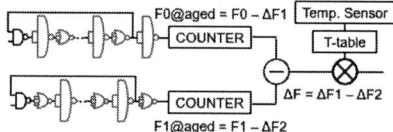

Fig. 3. Proposed NBTI-monitor configuration

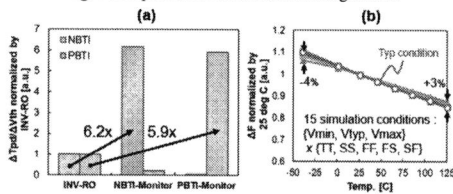

Fig. 4. Characteristics of proposed NBTI/PBTI-monitor. (a) NBTI/PBTI-Monitor sensitivity, (b) Temperatute dependency

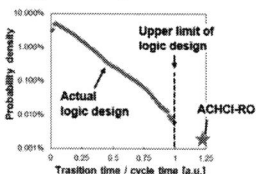

Fig. 5. Transition time of ACHCI-RO and product design

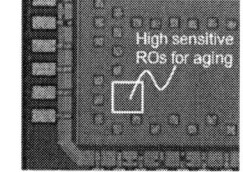

Fig. 6. Die photo of a test chip in a 7 nm Fin-FET process

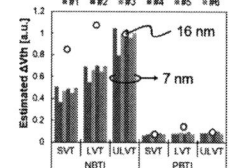

Fig. 7. ΔVth estimation by NOR-RO and NAND-RO measurement result

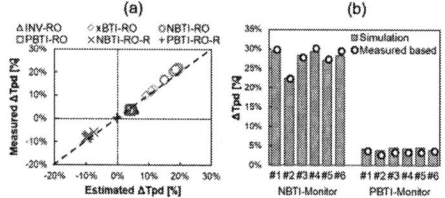

Fig. 8. Measurement and estimated result under DC stress. (a) Measured result of each RO, (b) NBTI/PBTI-monitor characteristics

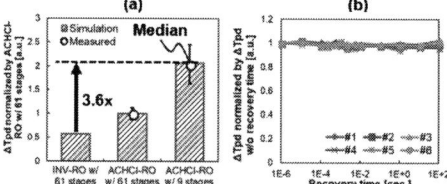

Fig. 9. Measurement result of ACHCI-RO under AC stress. (a) Measured and HCI simulation result, (b) Recovery effect of ACHCI-RO

978-1-5386-6414-8/18 $31.00 © 2018 IEEE

15-5 (8006)

A 140 nW, 32.768 kHz, 1.9 ppm/°C Leakage-Based Digitally Relocked Clock Reference with 0.1 ppm Long-Term Stability in 28nm FD-SOI

Guénolé Lallement[*†], Fady Abouzeid[*], Thierry Di Gilio[*], Philippe Roche[*] and Jean-Luc Autran[†]

[*]STMicroelectronics, 850 Rue Jean Monnet, F-38926 Crolles Cedex, France.
[†]IM2NP, Aix-Marseille University and CNRS, UMR7334, Marseille, France.
Email: guenole.lallement@st.com

Abstract—The design of Ultra-Low Power clock reference systems with highly energy-efficient operations is a key concept to achieve autonomous Internet-of-Things applications. In this work, a System-on-Chip is presented, embedding an area-efficient ultra-low voltage clock reference generator built on a digitally controlled leakage-based Ring Oscillator. Through a relocking scheme using a 2^{22} Hz external quartz reference, an associated digital compensation circuit ensures a stable output frequency of 32.768 kHz over inherent Process, Voltage and Temperature variations. The whole design has been fabricated in 28 nm FD-SOI technology and operates at a fixed supply voltage V_{dd} of 0.5 V. By combining Ultra-Low Power techniques, a 15 nW power consumption is achieved for the Oscillator and 125 nW for the digital compensation. The circuit area of the proposed clock source is 56.2 μm x 29.1 μm. A 90 ppm/V voltage accuracy has been measured over 10 packaged dies for V_{dd} ± 8%. A temperature accuracy of 1.9 ppm/°C is also reported from 0 °C to 50 °C. Lastly, the long-term frequency stability is characterized by an Allan deviation floor of 0.1 ppm.

I. INTRODUCTION

With billions of devices expected in the next decade, the Internet-of-Things (IoT) ecosystem will require a wide variety of low cost, battery-operated and inter-connected applications. From implantable biomedical sensors to industrial machine diagnosis devices or personnal IoT applications [1], a clock source is always required as a substantial building block of digital Systems-on-Chip (SoCs). Actually, a low frequency clock from a few Hz to 10's of kHz is used as a Real Time Clock (RTC) or applied as an input reference for further embedded clock multipliers (*e.g.* PLLs). More aggressive power reduction strategies are using this reference to clock digital circuits during sleep modes. In the specific context of Ultra-Low Power (ULP) applications the proposed clocking solution must also ensures insensitivity to PVT variations as well as inherent low power consumption.

In this work a leakage-based approach has been developed combined with a fully digital controller. The system has been implemented in 28 nm Fully Depleted Silicon-On-Insulator (FD-SOI) technology with an operating supply voltage of 0.5 V. The ULP leakage-based Ring Oscillator (LRO) is coupled with a digital compensation block encapsulated in a Control Logic Unit (CLU) to offer an accurate and stable 32.768 kHz output frequency with 90 ppm/V voltage, 1.9 ppm/°C temperature accuracy, and 0.1 ppm long-term frequency stability. A 2^{22} Hz external reference is used for LRO

calibration, yet periodically deactivated to offer energy savings in comparison to a XTAL approach.

In Section II, the implementation challenges and chosen system architecture are presented. Thereafter, Section III focus onto the clock generation using a programmable leakage-based oscillator, whereas Section IV gives an overview of the digital compensation developed. Lastly, the silicon measurements and results are given in Section V.

II. SYSTEM ARCHITECTURE

In the context of IoT-oriented systems with ∼100 μW power budget [2], the timer power must be drastically reduced. Indeed, as the time reference remains ON for the entire lifetime utilization of the SoC, the regular power consumption of the device may easily dominate the energy budget.

As shown in Fig. 1 a complete system has been implemented. The whole design is composed of three building blocks:

- A Full-Custom leakage-based Ring-Oscillator (LRO) producing an output clock CLK_{LRO}, and digitally controlled using 10 control bits and an enable signal;
- A Control Logic Unit (CLU) embedding a digital compensation circuit to calibrate the LRO with the help of an external known reference CLK_{REF};
- A SPI controller and configuration interface to easily communicate with the IP and guarantee a proper configuration of the device.

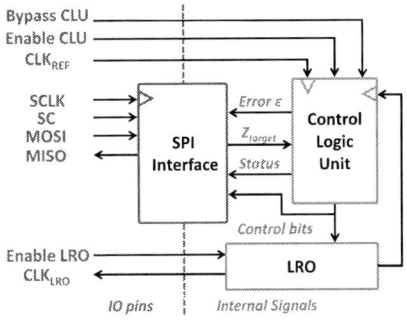

Fig. 1: Overview of the clock generator architecture.

978-1-5386-6414-8/18 $31.00 © 2018 IEEE

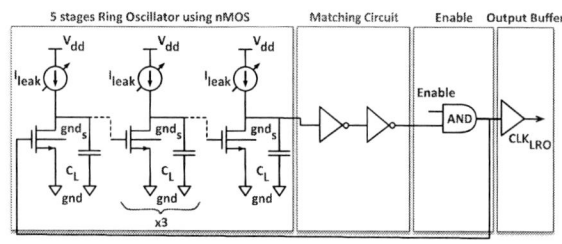
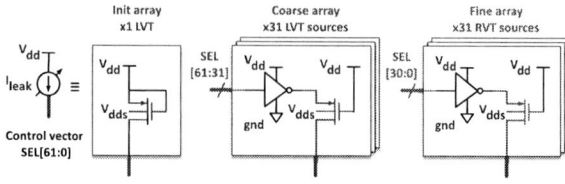

(a) Schematic of the Ring Oscillator.

(b) Schematic of the digitally controlled leakage sources.

Fig. 2: Leakage-based Ring Oscillator Architecture.

III. Clock Generation using a Leakage-Based Ring Oscillator

A 5 stages current controlled Ring Oscillator architecture has been designed (see Fig. 2a). This topology offers a very low power consumption as explained in [3]. nMOS devices are used to drive a current source I_{leak} to load an output capacitor C_L. Thus, the output frequency is directly given by the current as in:

$$f_{CLK} \propto \frac{I_{leak}}{C_L.V_{TH}}$$

with V_{TH} the threshold voltage of the nMOS transistors.

A matching circuit is attached to the last transistor of the ring. It helps to maintain the dynamic swing of the signal between V_{dd} and gnd to drive the following stage composed of standard cells. Lastly, an AND gate is added into the feedback loop of the RO to act as an enable whereas a final output buffer ensures the correct driving of the output stages. The full system is operating at a fixed $V_{dd} = 0.5$ V.

Digitally controlled leakage-based current sources have been implemented as shown in Fig. 2b. A set of three different arrays of transistors is used to generate the output current I_{leak}. By selecting X number of Low-Voltage Threshold (LVT) pMOs and Y number of Regular-Voltage Threshold (RVT) pMOS, a given amount of current is produced, leading to a modulation of the LRO output frequency. The array based on LVT pMOS helps to produce coarse steps whereas RVT pMOS are used for fine steps. Besides, the Init array is added to guarantee oscillations when no other sources are activated and the RO enabled. The current sources in Fig. 2b are replicated 5 times to compose the 5 RO stages.

To ensure high linearity when adding one more digitally controlled leakage source, a thermometer approach is chosen. Thus, a conversion stage converts the binary vector CTRL to a thermometer code SEL dispatched to the leakage sources.

To compensate the PVTs variations, the LRO must offer adequate current to always ensure correct operation at the 32.768 kHz target frequency. Simulations were performed around five corner cases; FastFast (FF), FastSlow (FS), Slow-Fast (SF), SlowSlow (SS) and Typical (TT), from 0 °C to 50 °C, and 0.46 V to 0.54 V voltage ranges.

IV. Digital Relocking Scheme

In order to guarantee absolute frequency stability of the across PVT variations calibration methods are required. Hence, two approaches could be developed; reducing the sensitivity trough analog circuit techniques or compensating the variations using an external digital feedback loop. In advance technology nodes, high noise immunity can be achieved using simple (i.e. negligible power) digital logic, whereas decreasing the sensitivity of the RO implies a more complex analog part. A digital approach using a 2^{22} MHz XTAL reference has therefore been chosen.

As circuits using Successive Approximation Registers tend to have high jitter sensitivity [4], a Proportional-Integral (PI) corrector solution has been selected. It ensures a fast settling time of the compensation loop, a good accuracy and a simple implementation.

Fig. 3 describes the Control Logic Unit implementation. As a first step, a synchronization and counting stage produces the digital word $Z_{LRO/REF} = f_{REF}/f_{LRO}$. This frequency ratio is then fed to the compensation stage where it is compared with $Z_{target} = 2^{22}/2^{15} = 128$ to produce the error signal ϵ. Subsequently, ϵ is used into the PI stage presenting a programmable integral gain K_i ($\leqslant 0$) to generate the binary control signal CTRL. However, to maximize the accuracy of the system two modules have been added.

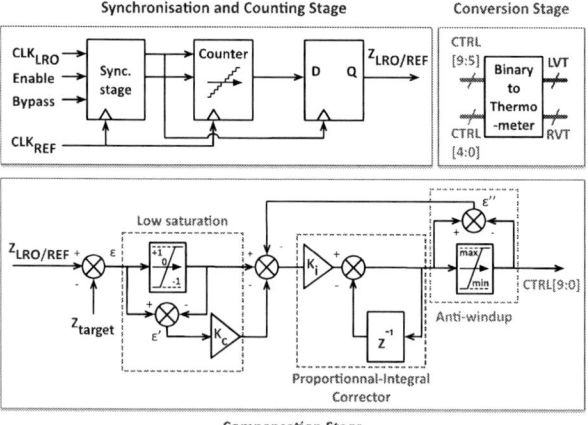

Fig. 3: Control Logic Unit.

First, a low saturation module avoids oscillations due to the discrete error. On the one side, when a large frequency error is detected a large gain ensures fast convergence. On the other side, when $\epsilon \to 0$, a smaller feedback gain is required to set the tuning word and target the correct output frequency. By detecting this low saturation a second error signal ϵ' is produced to automatically tune the effective gain of the corrector and achieve accuracy with high jitter tolerance.

Secondly, an anti-windup stage avoids overshooting and continuous increasing of the accumulated error. Indeed, if ϵ remains positive (resp. negative) for a certain period of time, the control signal saturates at a min. value (resp. a max value) due to the limited number of leakage sources available. Yet, if the error stays positive (or negative) after saturation, the integrator continues to accumulate an error that will be difficult to compensate in a reasonable amount of time. This can lead to a significant error on the output or system instability. Therefore, a loop is added that uses a third error signal ϵ'' defined as the difference between the PI output and the effective output CTRL. Finally, a conversion stage converts the binary code into a thermometer code to address the LVT and RVT array (see Section III).

The CLU is using CLK_{REF} as the clocking element. As the error is updated every period of CLK_{LRO}, this solution helps relax the design timing constraints by allowing the K_i multiplication to be done over multiple CLK_{REF} periods. Moreover, when the relocking is done, disabling the reference clock leads to removing the dynamic power of the digital block and the XTAL power.

V. SILICON MEASUREMENTS

The full system has been integrated into an ULP SoC [2] and fabricated in 28 nm FD-SOI technology. The oscillator provides a 32.768 kHz reference for a frequency multiplier or enables direct clocking of the Always-On domain of an ARM-based microprocessor during deepsleep operations. In Fig. 4, an overview of the LRO layout is given as well as a micrograph of the testchip. The CLU and SPI module have been integrated directly into the Always-On logic.

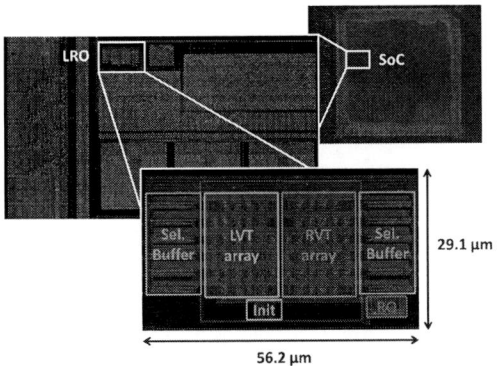

Fig. 4: Testchip micrograph and view of LRO layout. LRO area is $1635\,\mu m^2$.

Thereupon, 42 dice were packaged and measured using a custom development board and a Kintex®-7 FPGA from Xilinx. The power consumption of the LRO is 15 nW and 125 nW for the CLU, both measured at 0.5 V/25 °C with an output frequency of 32.768 kHz and a reference at 2^{22} Hz.

A. LRO Free Oscillations and Jitter

The output frequency of the oscillator according to the input control code has been measured for 42 dice and reported in Fig. 5. The grey area represents the whole set of measured values. For each sample the target frequency can be reached showing proper process compensation capabilities. Moreover, the measured mean frequency is compared to the simulated value in TT, showing matching between CAD models and silicon results. From these results, 10 dice were randomly selected for further analysis.

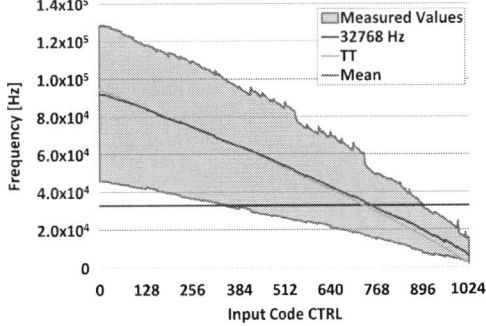

Fig. 5: Comparison of simulated and measured LRO frequency ranges according to the input code for 42 dice at 0.5 V/25 °C.

B. Digital Compensation Evaluation

A measurement of a typical locking cycle is shown in Fig. 6 with the correction word CTRL and the error ϵ in blue and red respectively. Starting from the fastest configuration (all sources activated), the system is able to lock in 0.342 ms showing a fast settling time.

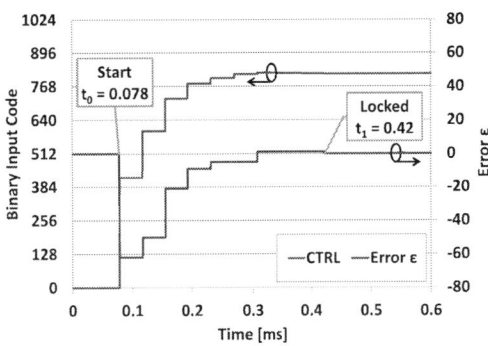

Fig. 6: Evaluation of the locking scheme during a typical initial calibration at 0.5 V/25 °C.

The CLU compensation capabilities across voltage and temperature were measured for 10 dice when the system has locked. At the nominal operating point 0.5 V, a median output frequency of 32 768.3687 Hz is calculated leading to a relative variation of 11 ppm. Across the [0.46 V-0.54 V] voltage range (*i.e.* $V_{dd} \pm 8\%$), an average median frequency of 32 768.2870 Hz is measured leading to a ~90 ppm/V voltage stability (Fig. 7). From 0 °C to 50 °C, an average median frequency of 32 771.0670 Hz is obtained leading a ~1.9 ppm/°C temperature stability.

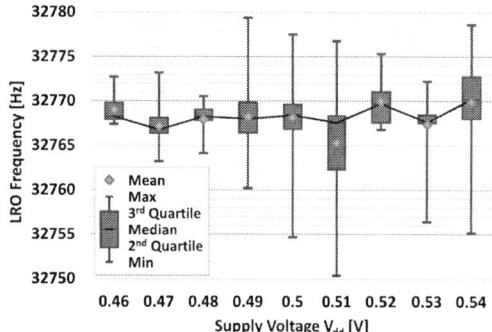

Fig. 7: Measured output frequency when relocked versus supply voltage for 10 dice at 25 °C.

Long-term stability is often described using the two-sample deviation $\sigma_y(\tau)$, also called Allan deviation. Timer long-term stability is mandatory when used for SoC's sleep modes. In Fig. 8, the Allan deviation is calculated for averaging periods τ up to 100 s with the max and min confidence interval. For intervals up to 20 s, $\sigma_y(\tau)$ is limited by white noise. Then, the Allan deviation is bounded by 1/f noise, which is reduced in advanced FD-SOI nodes [9]. This helps to achieve a 0.4 ppm Allan deviation floor after 30 s.

The system performances are compared in Table I with the latest State of the Art CMOS-based oscillators operating around 32.768 kHz.

VI. CONCLUSIONS

This research presents a fully-integrated Ultra-Low Power oscillator and an associated compensation unit in 28 nm FD-SOI both operating at 0.5 V. It offers high power efficiency combined with very low voltage and temperature variations. In the end, by demonstrating an efficient low-cost clock, this work offers a versatile time reference and clock source for standard digital systems.

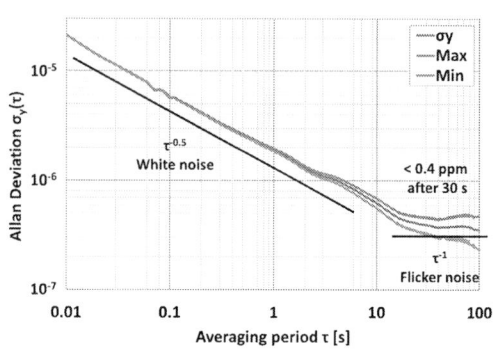

Fig. 8: Measured Allan deviation with min and max confidence intervals at 0.5 V/25 °C.

REFERENCES

[1] J. A. Stankovic, "Research directions for the internet of things," *IEEE Internet of Things Journal*, pp. 3–9, 2014.

[2] G. Lallement *et al.*, "A 2.7 pJ/cycle 16 MHz, 0.7 µW Deep Sleep Power ARM Cortex-M0+ Core SoC in 28 nm FD-SOI," *IEEE Journal of Solid-State Circuits*, pp. 1–13, 2018.

[3] B. Razavi, *Design of Analog CMOS Integrated Circuits.* McGraw-Hill Education, 2001.

[4] M. Scholl *et al.*, "A 80 nW , 32 kHz Charge-Pump based Ultra Low Power Oscillator with Temperature Compensation," *IEEE European Solid-State Circuits Conference (ESSCIRC)*, pp. 343–346, 2016.

[5] D. Kamakshi *et al.*, "A 36 nW, 7 ppm/°Con-Chip Clock Source Platform for Near-Human-Body Temperature Applications," *Journal of Low Power Electronics and Applications (JLPEA)*, 2016.

[6] D. Griffith *et al.*, "A 190nW 33kHz RC oscillator with ±0.21% temperature stability and 4ppm long-term stability," *IEEE International Solid-State Circuits Conference (ISSC)*, pp. 300–301, 2014.

[7] K. Tsubaki *et al.*, "A 32.55-kHz, 472-nW, 120ppm/°C, fully on-chip, variation tolerant CMOS relaxation oscillator for a real-time clock application," *IEEE European Solid-State Circuits Conference (ESSCIRC)*, pp. 315–318, 2013.

[8] K.-J. Hsiao, "A 32.4 ppm/°C 3.2-1.6V Self-chopped Relaxation Oscillator with Adaptive Supply Generation," *IEEE Symposium on VLSI Circuits (VLSI)*, pp. 14–15, 2012.

[9] E. G. Ioannidis *et al.*, "Low frequency noise variability in high-k/metal gate stack 28nm bulk and FD-SOI CMOS transistors," *International Electron Devices Meeting, IEDM*, pp. 449–452, 2011.

TABLE I: SUMMARY OF THE ACHIEVED PERFORMANCES AND STATE OF THE ART COMPARISON

Feature	This work	JLPEA'16 [5]	ESSCIRC'16 [4]	ISSC'14 [6]	ESSCIRC'13 [7]	VLSI'12 [8]
Technology Architecture Osc. Area [mm²] Frequency [kHz]	**28 nm FD-SOI** **Ring Oscillator** **0.001635** **32.768**	130 nm CMOS Ring Oscillator 0.269 12-150	130 nm CMOS Charge-Pump 0.014 32	65 nm CMOS RC Oscillator 0.015 33	180 nm CMOS Relax. Oscillator 0.105 32.55	60 nm CMOS Relax. Oscillator 0.048 32.768
Power	**15 nW (Osc.)** **125 nW (Digital)**	20 nW (Osc.) 12 nW (Digital)	80 nW (Osc.) 260 nW (Bandgap)	190 nW	472 nW	2.8 µW
Voltage Stability [ppm/V]	**90** **[0.46 – 0.54] V**	10000 [0.65 – 0.75] V	N/A	900 [1.15 – 1.45] V	11000 [1.0 – 1.8] V	6250 [1.6 – 3.2] V
Temperature Stability [ppm/°C]	**1.9** **[0 – 50] °C**	7 [20 – 40] °C	10 [10 – 100] °C	38 [-20 – 90] °C	120 [-40 – 100] °C	32.4 [-20 – 100] °C
Allan Deviation floor	**0.4 ppm** **τ = 100 s**	N/A	60 ppm τ = 10 s	4 ppm τ = 100 s	N/A	N/A

978-1-5386-6414-8/18 $31.00 © 2018 IEEE

16-1 (8027)

A 2× Blind Oversampling FSE Receiver with Combined Adaptive Equalization and Infinite-Range Timing Recovery

Seuk Son[1], Hwanseok Yeo[2], Sigang Ryu[1] and Jaeha Kim[1]

[1]Department of Electrical and Computer Engineering and Inter-university Semiconductor Research Center, Seoul National University, Seoul, Korea, [2]Samsung Electronics, Hwasung, Korea.
seuk@mics.snu.ac.kr and jaeha@snu.ac.kr

Abstract— A 2× blind-oversampling, fractionally-spaced equalizer (FSE) receiver is presented as an effective way to combine adaptive equalization and infinite-range timing recovery. A FSE can perform equalization as well as timing adjustment via data-interpolation and the presented work demonstrates an infinite-range timing recovery using a set of two half-UI-spaced, 4-tap FSEs that seamlessly switch across the UI boundaries. A current-integrating summer and multi-input regenerative latch help the 4-tap FSEs and 4-tap DFEs achieve low power dissipation, respectively. A prototype receiver fabricated in a 28nm CMOS consumes 3.5pJ/bit and 0.10mm² at 9Gb/s while compensating a 22-dB channel loss and a 100ppm frequency offset between the transmitted data and blind sampling clocks.

Keywords— *wireline, receiver, CDR, adaptive equalizer, fractionally-spaced-equalizer, decision-feedback-equalizer*

I. INTRODUCTION

To receive high-speed wireline data sent across lossy channels with low bit-error rates (BERs), equalization and timing-recovery are essential. A previous work [1] demonstrated that a fractionally-spaced equalizer (FSE) can combine adaptive equalization and timing recovery in a mesochronous system. The 0.5-UI-spaced FIR-based FSE shown in Fig. 1(a) could not only perform feedforward equalization but also timing adjustment across a 2-UI range. Instead of adjusting the timing of the sampling clocks, the FSE adjusts the FIR tap coefficients, which also serve as weights interpolating the blindly-oversampled data values, as illustrated in Fig. 2(a) [1,2]. However, the previous work was not suitable for a plesiochronous system where the timing offset between the data and sampling clock would drift indefinitely due to a small frequency offset.

This paper presents a 2× blind-oversampling FSE receiver that extends the previous one to support infinite-range timing recovery for plesiochronous systems. As shown in Fig. 1(b). The infinite-range timing recovery is achieved by using a set of two 0.5-UI-spaced, 4-tap FSEs, which seamlessly alternate whenever the timing offset drifts by 0.5-UI and the main-cursor tap position shifts. The tap coefficients of the FSE are adapted by a sign-sign least-mean-square (SS-LMS) algorithm, which serves the roles of both equalizer adaptation and timing calibration. The proposed receiver was fabricated in a 28nm CMOS and successfully operates at 8.3~9.1Gb/s with a 100ppm frequency offset between the data and blind sampling clocks. The receiver consumes 3.5pJ/bit of power and 0.10mm² of area, and can compensate up to 22-dB of channel loss at 4.5-GHz with 0.2-UI_pp high-frequency jitter tolerance (JTOL) for BER less than <10^{-12}.

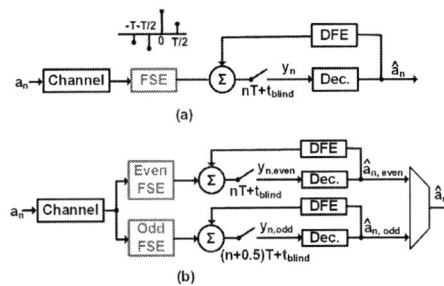

(a)

(b)

Fig. 1. (a) The previous FSE receiver [1] and (b) the proposed 2× blind oversampling FSE receiver with infinite timing recovery range.

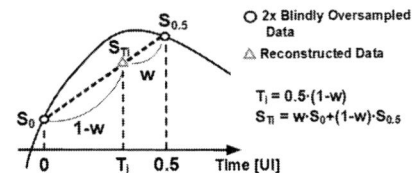

Fig. 2. The timing recovery with FSE via data interpolation.

II. ARCHITECTURE

The overall architecture of the proposed FSE receiver is illustrated in Fig. 3. First, a quarter-rate sample-and-hold (S/H) circuit samples and deserializes the incoming 9-Gb/s data stream into a set of eight, 0.5-UI-spaced data samples (S_{1-8}) to establish a delayed version of input for each FSE tap [4]. Using a S/H circuit instead of an analog delay line saves power and area. The 8-phase, quarter-rate blind clocks (Φ_{1-8}) used by the S/H circuit are generated by frequency-dividing the 9-GHz differential full-rate clocks from a phase-locked loop (PLL). While the receiver in this work demonstrates a full-rate architecture in this paper, an extension to a double-data-rate (DDR) architecture is also possible by using the half-rate quadrature clocks.

To support an infinite range of timing recovery, the receiver employs two FSE/DFE front-ends (called *even* and *odd* front-ends) that process the data samples from the S/H circuit. The even front-end processes the data samples (S_{2-5}) of which timings are 0.5-UI shifted from those of the odd front-end (S_{1-4}). Each front-end computes the weighted sum of the four input data samples for the 0.5-UI-spaced, 4-tap FFE and data interpolation and recovers the digital bits (De_{1-4} or Do_{1-4}) assisted by a 1-UI-spaced, 3-tap decision-feedback equalization (DFE).

978-1-5386-6414-8/18 $31.00 © 2018 IEEE   201

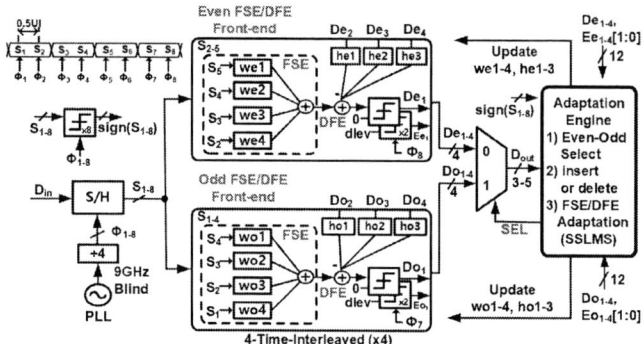

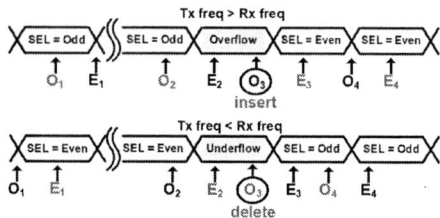

Fig. 3. The overall architecture of the proposed FSE receiver.

Fig. 4. Illustration of the data overflow/underflow cases when there is a frequency offset between the transmit data and receive clock.

Finally, a multiplexer stage selects one 4-bit output (D_{out}) between those of the even front-end (D_{even}) and odd front-end (D_{odd}) as instructed by the adaptation engine via the selection signal (*SEL*). This multiplexer stage also inserts or deletes 1-bit data when a data underflow or overflow occurs due to the frequency offset between the transmitted data and receiver sampling clock [3]. As illustrated in Fig. 4, when the transmit frequency is higher than the receive frequency, a data overflow may occur while the multiplexer selection is switched from odd to even. In this case, the multiplexer adds the lost bit (O_3) back to the output data stream. On the other hand, when the transmit frequency is lower, the same data bit can be selected twice while the selection is switched from even to odd. In this case, the multiplexer stage deletes the duplicate bit from the output data stream.

III. ADAPTATION AND SELECTION ALGORITHM

Adapting the FSE and DFE tap coefficients serve two purposes simultaneously: equalizer adaptation and timing calibration. The adaptation is done via a sign-sign least-mean-square (SS-LMS) algorithm [5], which requires only the polarity information of the error signals (Ee_{1-4} and Eo_{1-4}) and the input signals scaled by the FSE/DFE taps (S_{1-8}, De_{1-4} and Do_{1-4}).

Fig. 5 shows the block diagram of SS-LMS adaptation engine. The proposed receiver adds two error slicers that produce the outputs $Ee_1[0]$ and $Ee_1[1]$ comparing the equalized signal with the reference data levels (+*dlev* and -*dlev*). The first-stage multiplexers then select the proper error signals according to the current data and the subsequent XNOR gates compute the update directions $up_w[3:0]$ and $up_h[3:0]$ for the FSE and DFE taps according to the following SS-LMS formula:

$$w_{n+1} = w_n - sign(e) \cdot sign(s),$$

where w is the FSE/DFE tap coefficient, e is the error signal, s is the input signal scaled by w. To lower the operating frequency of the synthesized digital adaptation engine and reduce possible dithering of the tap coefficient values, a set of 32:1 decimation filters aggregate 32 update directions before updating each 6-bit FSE/DFE tap coefficient.

The decision whether the final output Dout should be selected from the even or odd front-end is made based on the current values of the even/odd FSE tap coefficients. Fig.6 (a) shows that each front-end is designed to cover a different half-UI range of

the timing offset between the data and blind-sampling clock. To let the FSE suppress the pre-cursor ISI while the DFE cancels the post-cursor ISI, the even/odd selection is made to position the main-cursor tap of the FSE at its third tap (*we3*, *wo3*). In other words, the selection is switched to the other front-end when the third tap coefficient of the currently-selected front-end no longer has the largest tap coefficient value. This is when the timing offset is about ±0.25-UI, which is the ideal point to make a switch between the two front-ends while maximizing the overall timing margin.

Fig. 6(b) shows the flowcharts of the proposed front-end selection algorithm. When an even or odd front-end is first selected, its third tap coefficient is initialized to the maximum

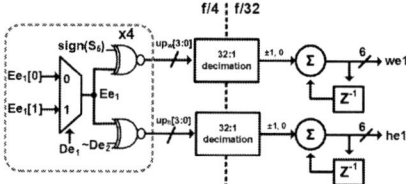

Fig. 5. The block diagram of the SS-LMS adaptation engine for the proposed FSE receiver.

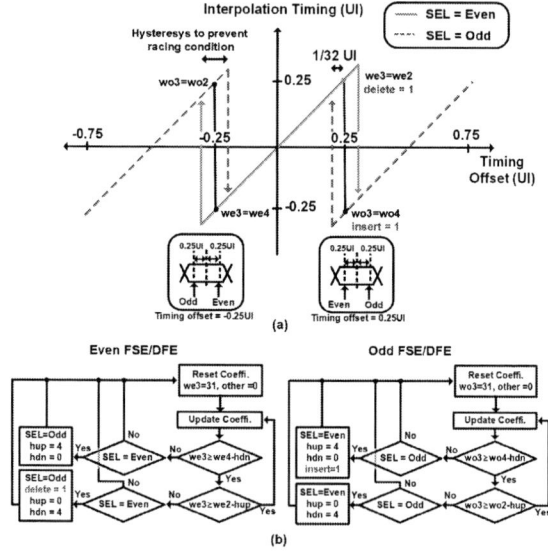

Fig. 6. (a) The selection of the even/odd front-ends versus the timing offset between the data and blind-sampling, and (b) the flowchart illustrating the selection algorithm of the proposed FSE receiver.

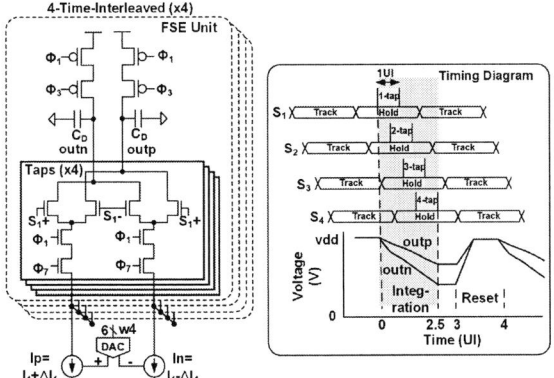

Fig. 7. The 4-tap FSE front-end circuit with its timing diagram.

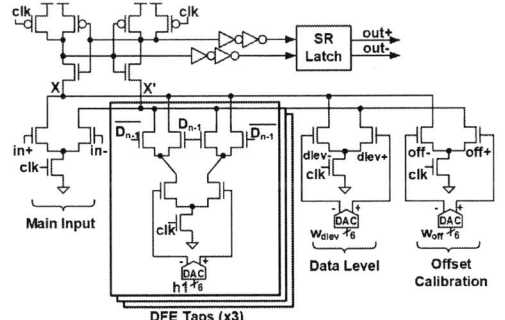

Fig. 8. The multi-input regenerative latch circuit for 3-tap DFE operation.

value of 31 while all others are reset to 0, and the SS-LMS adaptation begins. The algorithm keeps the current selection as long as the third tap coefficient has the largest value. When it becomes smaller than one of the other tap coefficients due to the timing drift, the algorithm prepares the other front-end with the main-cursor position reset to the third tap and makes a switch. As mentioned earlier, the algorithm also inserts or deletes 1-bit

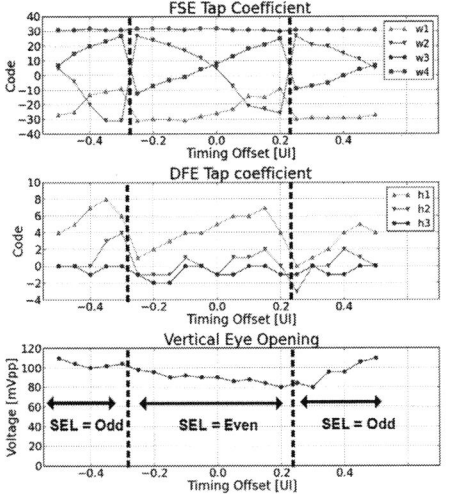

Fig. 9. The measured trajectories of the FSE/DFE tap coefficients and the voltage margin of the proposed receiver at each timing offset position.

data as necessary to address data underflows or overflows. And to avoid redundant switching when the timing offset is near ±0.25-UI, a hysteresis of 4 in tap coefficient values is applied, which corresponds to a hysteresis of 1/32-UI in timing offsets.

IV. CIRCUIT IMPLEMENTATION

The weighted summation required by the 4-tap FSE front-end is done by a low-power, current-integrating summer circuit [4]. Fig. 7 illustrates one of the FSE front-ends including four time-interleaved FSE units and its timing diagram. Each input (S_1–S_4) drives a set of two differential pairs that differentially discharge the outputs (*outp* and *outn*) depending on the difference between their bias currents (Ip and In). Since these bias currents are also controlled differentially by a 6-bit weight code via a current-steering DAC, the overall circuit can compute the weighted sum of 4 inputs. The operation is sequenced by the multi-phase clocks (Φ_1–Φ_8). First, in the reset phase, the output voltages are pre-charged to the supply voltage. Then in the integration phase, each FSE input sequentially discharges the output as weighted by the 6-bit digital code. Since each FSE unit integrates a given input contribution at different intervals, the 4 FSE units in each even/odd front-end can share the same current source device, avoiding the mismatch issue.

Fig. 8 describes a multi-input regenerative latch circuit [6], which performs the 3-tap DFE operation. This circuit is basically a StrongARM latch which performs the static-current-free signal summation during the sampling aperture period. By performing the summation within the latch, the circuit offers fast

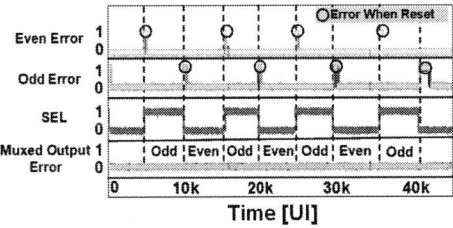

Fig. 10. The measured waveforms of the bit-error and SEL signals with 100ppm frequency offset, demonstrating the infinite-range operation of the proposed receiver.

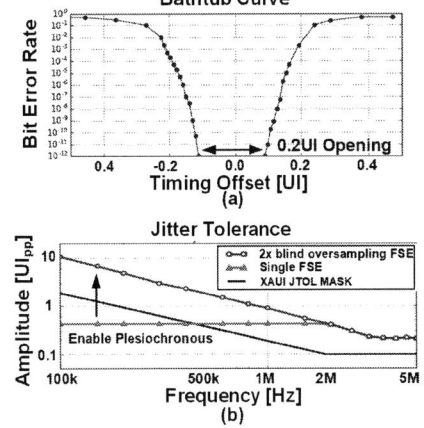

Fig. 11. The measured (a) BER bathtub curve and (b) jitter tolerance.

TABLE I.   COMPARISON WITH PREVIOUS BLIND-OVERSAMPLING
RECEIVERS.

	This Work	[1]	[2]	[3]	ESSCIRC 2014
Technology	28-nm	90-nm	28-nm	22-nm	28-nm
Data Rate (Gbps)	9	6.25	32	5	8
Loss@Nyquist (dB)	22	N/A	22.3	8	N/A
CDR Type	FSE	FSE	Data-interpolation	Blind-Oversampling	Blind-oversampling
Plesiochronous	Yes	No	Yes	Yes	Yes
EQ Type	4-tap FSE 3-tap DFE	2-tap FSE 1-tap DFE	CTLE 2-tap DFE	N/A	N/A
Area (mm²)	0.10	0.03	0.24	0.041	0.16
Power (mW)	31.5	22.5	308.4	10	56.8
Power Efficiency (pJ/bit)	3.5	3.6	9.6	2	7.1
FOM (pJ/bit/dB)*	0.16	N/A	0.43	0.25	N/A

* FOM = (Power Efficiency)/(Compensated Channel Loss @ Nyquist)

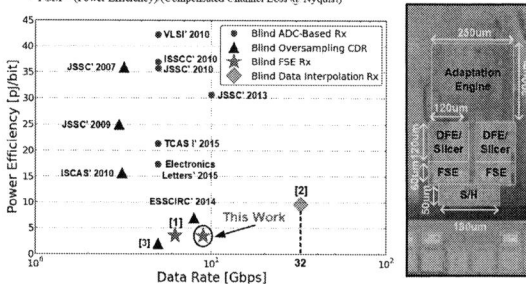

Fig. 12. Die Photo

timing to close the DFE feedback without loop-unrolling and low power by eliminating a separate summation stage. The slicer has a total of six input pairs driven by one main input (*in*), three DFE feedback inputs (*h1–3*), one data-level reference input (*dlev*), and one offset input for calibration (*off*). When the clock (*clk*) rises, the current differences steered by the input pairs are linearly added onto the pre-charged node X and X', and the final decision is made by the two cross-coupled inverters via regeneration. The DFE tap coefficient inputs, data-level input, and offset input are each generated by a DAC controlled by a 6-bit digital code.

V. MEASUREMENT RESULT

The prototype chip was fabricated in a 28-nm CMOS technology, and its die photograph is shown in Fig. 12. The receiver is tested by feeding a 9Gb/s, 2^{31}-1 PRBS pattern through a channel with a 22dB loss at 4.5GHz and measuring the BER with an on-chip PRBS checker. The receiver including the S/H circuit, two FSE/DFE front-ends, and digitally synthesized adaptation engine consumes the power of 31.5mW and active area of 0.10mm², operating at a 0.85V supply. The PLL supplying the blind-oversampling clocks consumes 17mW.

Fig. 9 plots the measured FSE/DFE tap coefficients and achieved vertical eye opening for BER of 10^{-12} while sweeping the timing offset between the blind sampling clock and input data. The receiver achieves the worst-case voltage margin of $80mV_{pp}$ across all timing offsets. Fig. 10 demonstrates the receiver can operate in a plesiochronous mode by supporting the infinite range of time adjustment. It shows the measured transient waveforms of the bit-error signals of the even/odd front-ends and the selection signal (*SEL*) when the receiver operates with a 100_{ppm} frequency offset between the data and

blind-sampling clock. As expected, the receiver periodically alternates its selection between the even and odd front-ends. While the unselected front-end may produce momentary bit errors, the final multiplexed output from the selected front-ends is free of bit errors.

Fig. 11 shows the measured BER bathtub curve and jitter tolerance (JTOL) curve of the prototype receiver. The JTOL curve satisfies JTOL mask with the high-frequency JTOL of $0.2\text{-}UI_{pp}$ and the corner frequency of 3-MHz. Note that the previous FSE receivers could not satisfy this JTOL mask at low frequencies because of their limited timing recovery range. The measured BER bathtub curve shows the 0.2-UI horizontal eye opening with BER of 10^{-12}. Table 1 compares the proposed receiver with the previously-reported blind-oversampling receivers. The proposed FSE receiver has a competitive power-efficiency and best figure-of-merit (FOM) while supporting the plesiochronous timing recovery.

VI. CONCLUSION

This work demonstrated that a 2× blind-oversampling FSE receiver can support plesiochronous links with infinite timing-recovery range. The infinite-range timing recovery is realized by having a set of two FSE/DFE front-ends each covering a different half-UI period and seamlessly switching the selection between the two. The adaptation and selection algorithm are developed to update FSE/DFE coefficients and produce the proper selection signal. The FSE/DFE implementation based on the current-integrating summers and multi-input regenerative latches helps realize a power-efficient design. The measurement results show that the proposed FSE receiver architecture is an efficient way to build a power-efficient plesiochronous blind-oversampling CDR with combined adaptive equalization.

ACKNOWLEDGMENT

The authors would like to thank Jongshin Shin, Taeik Kim, and Jaehong Park for their technical advice and Samsung Electronics, Co., Ltd, for its funding and IC fabrication support.

REFERENCES

[1] S. Song and V. Stojanovic, "A 6.25 Gb/s voltage–time conversion based fractionally spaced linear receive equalizer for mesochronous high-speed links," *IEEE J. Solid-State Circuits*, vol. 46, no. 5, pp. 1183–1197, May 2011.

[2] Y. Doi *et al.*, "A 32 Gb/s data-interpolator receiver with two-tap DFE fabricated with 28-nm CMOS process," *IEEE J. Solid-State Circuits*, vol. 48, no. 12, pp. 3258–3267, Dec. 2013.

[3] S. Shekhar *et al.*, "A 1.2–5Gb/s 1.4–2pJ/b serial link in 22nm CMOS with a direct data-sequencing blind oversampling CDR," in *Proc. Symp. VLSI Circuits*, Jun. 2015.

[4] A. Agrawal *et al.*, "A 19-Gb/s Serial Link Receiver With Both 4-Tap FFE and 5-Tap DFE Functions in 45-nm SOI CMOS," *IEEE J. Solid-State Circuits*, vol. 47, no. 12, pp. 3220–3231, Dec. 2012.

[5] J. Kim *et al.*, "A Four-Channel 3.125-Gb/s/ch CMOS Serial-Link Transceiver With a Mixed-Mode Adaptive Equalizer," *IEEE J. Solid-State Circuits*, vol. 40, no. 2, pp. 462–471, Jan. 2005.

[6] S. Son *et al.*, "A 2.3-mW, 5-Gb/s low-power decision-feedback equalizer receiver front-end and its two-step, minimum bit-error-rate adaptation algorithm," *IEEE J. Solid-State Circuits*, vol. 48, no. 11, pp. 2693–2704, Nov. 2013.

A Bimodal (NRZ/PAM-4) ISI Tolerant Timing Recovery with Adaptive DDJ Equalization

Masum Hossain[1], Aurangozeb[1], Nhat Nguyen[2]
[1]University of Alberta, Edmonton, Canada, [2]Rambus Inc, Sunnyvale
masum@ualberta.ca

Abstract— This paper describes low latency bimodal NRZ/PAM-4 timing recovery. This scheme reduces latency and power consumption by eliminating the need for data equalization in the timing recovery path for inter-symbol-interference limited channels. Rather it directly equalizes the data dependent jitter by adaptively shifting the ISI effected zero crossings. The implemented prototype in 65nm CMOS supports both 10 Gb/s NRZ and 20 Gb/s PAM-4 consuming only 23 mW. The CDR achieves more than $f_{baud}/500$ peaking free tracking bandwidth and adapts to optimized jitter tolerance for both PAM-4 and NRZ for the given input eye.

Keywords— Clock and Data Recovery; DDJ Equalization; Timing adaptation;Digital receiver timing recovery.

I. INTRODUCTION

Increased bandwidth demand in data center has motivated circuits, architecture and signal integrity innovation in wireline link design. But channel losses are not improving at the expected rate and that has motivated more spectrum efficient signaling such as PAM-4. Along with the higher order modulation receiver architectures have also evolved significantly. Although analog mixed signal receivers are preferred for lower power and complexity, their usage is limited to short reach (SR) channels only [1,2]. Medium reach (MR) and longer reach (LR) channels require more sophisticated equalization that is only realizable in digital domain [3]. Therefore, ADC based receiver architecture as shown in Fig. 1 are becoming more popular for solutions for PAM-4 signalling.

In ADC based receiver the analog frontend (AFE) provides partial equalization of the signal, more significant part of the equalization is performed in the digital domain after the ADC. Although this arrangement allows for more sophisticated equalization, timing recovery becomes more challenging for two reasons. First, the eye at the AFE output has significant ISI that causes data dependent jitter (DDJ). In the presence of such large amount of DDJ, timing recovery becomes very challenging and the CDR may fail to lock. Second, an ISI free eye is available at the digital equalizer output but only data samples are available since digitizing both data and edge would lead to prohibitive power consumption. Therefore, we can only implement baud rate timing recovery such as muller-muller (MM-CDR) that is known to be less robust compared to data and edge sampling CDR. In addition to that loop latency is a major concern since the digital equalizer runs at a much slower clock rate – usually $f_{baud}/32$[3]. An example breakdown of the loop latency is shown in Fig. 1 that suggests 250 UI+ loop latency that translates to peaking free bandwidth less than MHz. With such latency constrain it becomes more challenging

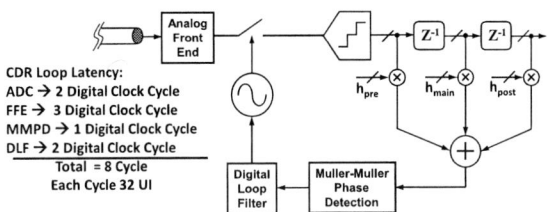

Fig. 1. timing Recovery for Conventional Digital receiver with loop latency breakdown.

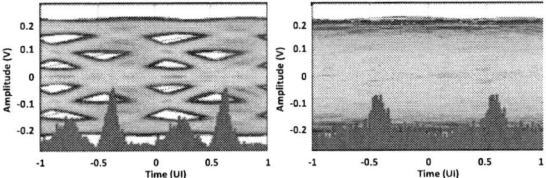

Fig. 2. ISI limited NRZ and PAM-4 eye with zero crossing distribution.

to meet traditional jitter tolerance corner frequency of $f_{baud}/1667$. One approach is to reduce the jitter tolerance specification as seen in 56G OIF standard that has relaxed the jitter tracking bandwidth of 56Gb/s link to 4 MHz. However, when backward compatibility to 28 Gb/s NRZ link is considered, we would need to support 16+ MHz bandwidth and that would be challenging to accommodate in conventional digital receivers when the loop latency is considered.

II. CDR ARCHITECTURE

Conventional implementation of PAM-4 phase detection requires separate detection of edges from three eyes compared to a single eye detection in NRZ signaling. In addition to that in PAM-4 signalling we need to filter the asymmetric edges [4]. Therefore, hardware complexity of a PAM-4 PD is 3x compared to NRZ. However, when we consider the ISI impacted case, distinct signal levels are smeared by ISI in similar manner. This is because post cursors when combined with the main creates multiple levels similar to the multilevel signaling case as shown in Fig. 2. In both cases zero crossing distribution shows multiple peaks that will degrade the timing recovery performance.

Timing recovery loop proposed in this work is motivated by following objectives: first goal is to eliminate the need for data equalization for timing recovery. Therefore, we can remove the digital FFE and save significant power latency in the timing recovery path. Second, objective is to enable zero-crossing-based timing recovery for PAM-4 ADC based receivers. MMPD extracts timing information from signal levels. Since for *N*-bit modulation translate to 2^N levels,

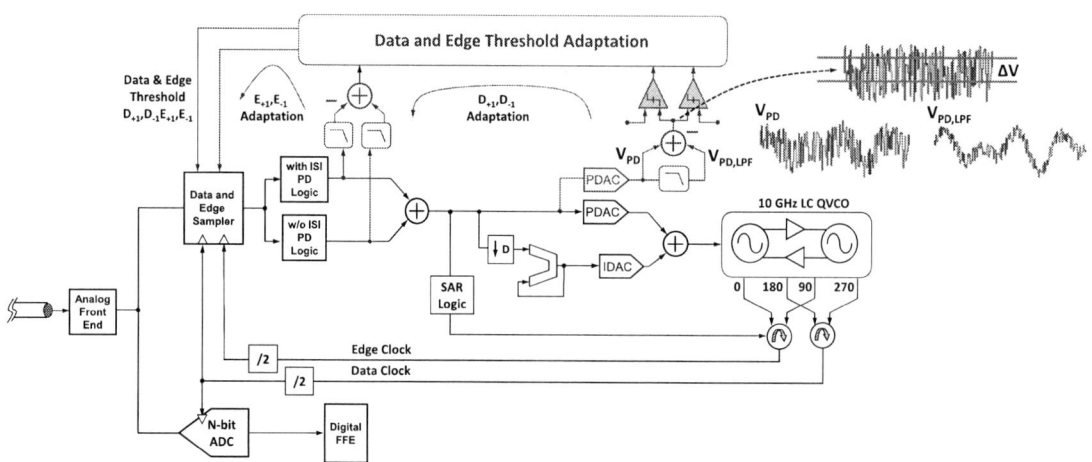

Fig. 3. Proposed timing recovery with adaptive DDJ Equalization. Adaptation loop is shown blue.

corresponding hardware also increases exponentially. Compared to that zero-crossing contains most of the transition information especially in the presence of ISI and that allows re-use of the same edge comparators NRZ/PAM-4 and robust phase tracking. Third objective is to enable jitter adaptation where the loop bandwidth would be adapted to meet target jitter performance in mission mode. The proposed receiver architecture shown in Fig. 3 is designed to provide the above functionalities.

A challenge for next generation SerDes is to provide compatibility to different modulation format and still maintain energy efficiency over range of data rates. Since the equalization and linearity requirements are different, it is difficult to make analog mixed signal solutions to be optimal for both NRZ and PAM-4. An ADC based solution in this case allows more flexibility with number of taps and bits to make the digital equalization more energy efficient. The functionality of the PD described in Fig. 4 requires 3 data and 3 edge comparators. Although in this implementation we have separate data comparators for flexibility, in ADC based receiver we can re-use the ADC outputs eliminating the data comparators by selecting three from 2^N for phase detection. When it comes to edge, we use three edge comparators with adjustable threshold for zero crossing detection. With three selectable data samples and three edge samples we can extract phase error information for both PAM-4/NRZ using the same hardware without any additional overhead. The detail truth table for early late detection is given in Fig. 4.

A. Contribution:

Existing literature has extensively addressed the characteristics of bang-bang PD and it's impact on the loop dynamics of the CDR. However, in most cases input is considered equalized i.e. DDJ is low. In this work we explain and demonstrate how timing loop can recover clock even in the presence of large data dependent jitter. ISI filter technique introduced in [5], considers only ISI free edges for timing recovery therefore negatively impacts the CDR bandwidth due reduced PD gain. Instead in this work we use both ISI free and ISI affected edges for timing recovery but ISI affected zero crossing are corrected to eliminate DDJ impact on CDR

performance. More importantly, ISI filter thresholds are adapted to maximize the loop bandwidth and minimize the DDJ impact at the same time. Mission mode adaptation of CDR loop bandwidth is recently reported in [6], but it is only applicable to latency induced periodic jitter. However, in most cases DDJ is the main limiting factor. This work addresses DDJ in two ways – first a novel DDJ equalizer is used to reduce to reduce the DDJ impact and second loop gain is adapted to reduce the recovered clock DDJ to improve timing margin and high frequency jitter tolerance.

III. DATA DEPENDENT JITTER EQUALIZATION

Traditional 2x CDR uses a single edge comparator to detect the zero crossing. However, in the presence of ISI the zero crossings are poorly distributed and makes timing recovery challenging. Proposed DDJ equalization can use the offset edge comparators to improve the effective distribution. To provide bimodal functionality with maximum hardware reuse, we only use the middle eye of PAM-4 for timing recovery. For simplicity, we describe the technique for NRZ case in Fig. 4. Here edge distribution of an ISI limited channel is plotted with patterns of interest. From the decomposed distributions it becomes clear that 1010 and 1101 patterns are mainly responsible for bimodal distribution, therefore should be filtered. Zero crossing of 1100 or 0011 patterns are more suitable for timing recovery since pre and post cursors negate each other at zero crossing. We consider these patterns to be ISI free. These zero crossings are distinguished using data comparators D-1, D0 and D+1 as described in Fig. 4. To make best use of these ISI free edges – this zero crossing is compared to three reference levels from E-1 to E+1 which translate to 2 bit phase error resolution. When it comes to ISI affected patterns such as 1010 or 1101, to avoid bi-modality appropriate threshold levels E+1 and E-1 are used instead of E0. Note that if E+1 and E-1 threshold levels are appropriately chosen, then ISI affected zero crossing point would coincide with the ISI free zero crossing. Note that when the zero crossings are misaligned, over all PD transfer curve also shows poor gain or even false lock point as shown by the dotted line in Fig. 4. Once appropriate edge threshold levels are used, both ISI free and ISI affected Phase detector would have the same lock

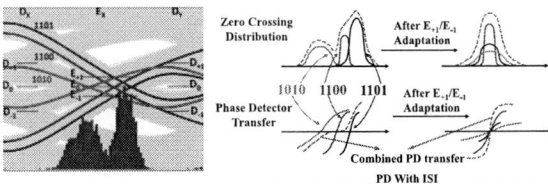

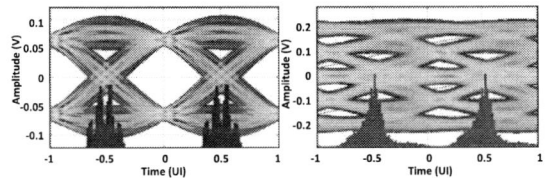

PD Without ISI			
D_X	D_Y	E_X	Early/Late
$D_X > D_{-1}$	$D_Y < D_{-1}$	$E_X > E_{+1}$	Earlier
		$E_{+1} > E_X > E_0$	Early
		$E_0 > E_X > E_{-1}$	Late
		$E_{-1} > E_X$	Later
$D_X < D_{-1}$	$D_Y > D_{-1}$	$E_X < E_{-1}$	Earlier
		$E_{-1} < E_X < E_0$	Early
		$E_0 < E_X < E_{+1}$	Late
		$E_{+1} < E_X$	Later
$D_X > D_0$	$D_Y > D_0$	----	No Change

PD With ISI			
D_X	D_Y	E_X	Early/Late
$D_X > D_{+1}$	$D_0 > D_Y > D_{-1}$	$E_X > E_{-1}$	Early
		$E_X < E_{-1}$	Late
$D_X < D_{-1}$	$D_0 < D_Y < D_{-1}$	$E_X > E_{-1}$	Early
		$E_X < E_{-1}$	Late
$D_0 < D_X < D_{-1}$	$D_0 > D_Y > D_{-1}$	$E_X > E_{-1}$	Early
		$E_X < E_{-1}$	Late
$D_0 > D_X > D_{-1}$	$D_0 < D_Y < D_{+1}$	$E_X > E_{-1}$	Early
		$E_X < E_{-1}$	Late
$D_X > D_0$	$D_Y > D_0$	----	(000)No Change

Fig. 4. DDJ Equalization with adaptive edge threshold. Here zero crossing distribution is decomposed into zero crossings from specific patterns 1010, 1100 and 1101 etc.

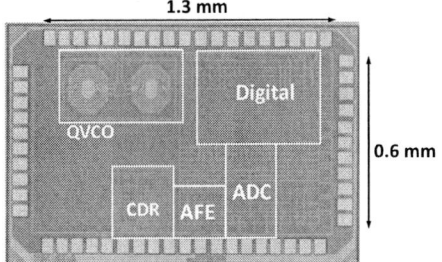

Fig. 6. Implemented digital receiver prototype with timing recovery and ADC in 65nm CMOS

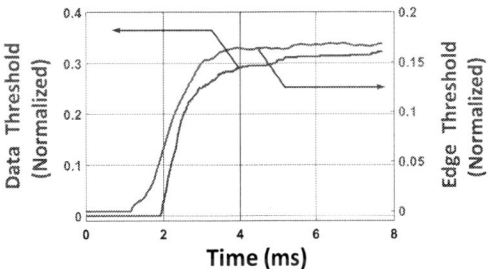

Fig. 5. Comparison of DDJ Equalization using (a) conventional FFE (b) Proposed zero-crossing-based equalization.

Fig. 7. Data and edge threshold adaptation convergence

point, but ISI affected zero crossings may have a poor distribution. To further improve, we combine these phase detector outputs with appropriate weighting factors to shape the combined effective edge distribution.

It is constructive to compare the resultant edge distribution to an equalized NRZ eye in Fig. 5. Since the goal is to provide timing recovery, the proposed technique does not need to improve the data eye, but the edge distribution has significantly improved – it is comparable or even better than a fully equalized eye. This DDJ equalization can be implemented with much lower latency and power consumption without needing the DSP FFE.

IV. JITTER ADAPTATION

The CDR needs both data and edge threshold voltages that must be adapted for different channel responses. As mentioned in the previous section, the edge comparators should be placed such that ISI impacted edge distribution's average to match the ISI free edge distribution. To ensure that, the ISI free and ISI affected edges are separately processed by the phase detector with the logic described in Fig. 4. To align these two phase detectors locking point their average values are compared and accordingly the edge thresholds are updated. The loop converges where both PD outputs are aligned on average.

Adapting the data threshold level requires further consideration. In general, higher threshold would result in tighter edge distribution but reduces the CDR bandwidth since very few edges will be used. On the contrary, lower threshold would cause higher DDJ induced jitter in the recovered clock.

Ultimately, our goal is to optimize the Jitter tolerance both at low and high frequency. Unfortunately, during mission mode there is no simple way to measure the jitter tolerance. But recovered clock jitter is often an indicator of high frequency jitter but their measurement is also challenging. Instead we measure the dithering in the control voltage that directly indicates the VCO output jitter in locked condition. This can be done using a replica of the proportional DAC that drives the VCO. Ideally, by integrating the filtered DAC output we can estimate CDR's recovered phase and therefore, the recovered clock jitter can be estimated by the deviation from the average DAC output. Therefore, to achieve target jitter performance control voltage fluctuation should be limited to a target value. In this case we are using the control voltage fluctuation to adapt the data threshold voltages.

Note that CDR is often required to track low frequency sinusoidal jitter (SJ) and thereby meets the jitter tolerance. Fortunately, the DDJ frequency content is at higher frequency therefore, can be separated from the low frequency SJ. Therefore, the lowpass pole frequency in the V_{CNTL} path is chosen such that we can filter the DDJ without effecting the low frequency SJ tracking as shown in Fig. 3

V. IMPLEMENTATION AND MEASUREMENT

The proposed timing recovery is implemented as part of ADC based receiver in 65nm CMOS targeting 10Gb/s NRZ and 20Gb/s PAM-4 signalling (Fig. 6). In a bang-bang PD, dithering jitter reduces timing margin significantly. Therefore, SAR TDC based PD is adopted to linearize loop dynamics and reduce in-band jitter. 10 GHz quadrature LC VCO is used to interpolate and generate data and edge clock with arbitrary skew setting. Required half-rate clock (5 GHz) is generated for the ADC and CDR from the 10 GHz using divide-by-2

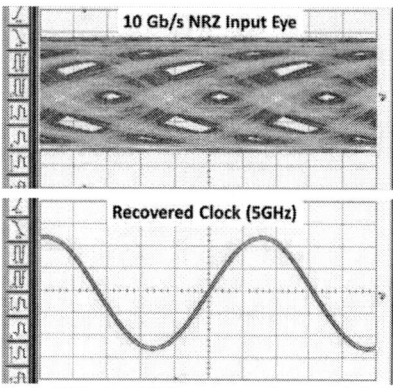

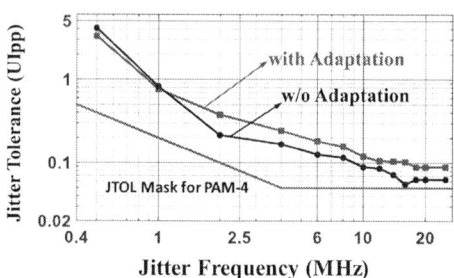

Fig. 10. Jitter Tolerance with PAM-4 input with BER lower than 10^{-10}

Fig. 8. ISI limited NRZ input eye and recovered clock

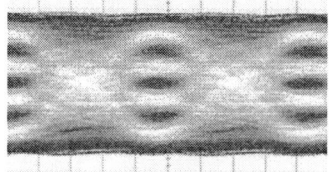

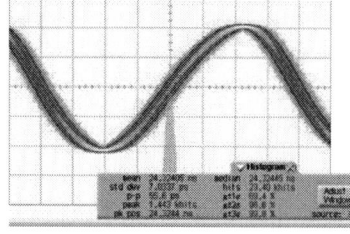

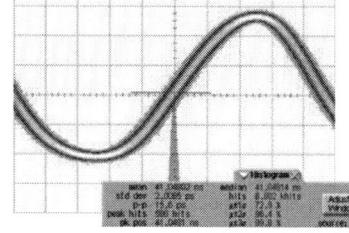

Fig. 9. ISI limited PAM-4 input eye and recovered clock with and without DDJ adaptation.

Table I: State-of-art comparison

	[6] ISSCC 2017	[2] JSSC 2015	This work
Data Rate	56 Gb/s	28 Gb/s	20 Gb/s
Technology	40 nm	28 nm	65 nm
PAM-4/NRZ	Both	NRZ only	both
DDJ Equalization	No, relies on front end EQ	No, relies on front end EQ	Yes
Jitter Adaptation	No	Yes	Yes
Jitter Tolerance	---	0.25 UI (at high frequency) (NRZ)	0.1 UI at high frequency (PAM-4)
Power Consumption	7.5 pJ/bit (PAM-4) 3.9 pJ/bit (NRZ)	3.7 pJ/bit (NRZ)	1.15 pJ/bit (CDR) 3.75 pJ/bit (Rx)

is also improved. Compared to existing solutions, the CDR can recover clock from an un-equalized eye consuming only 23 mW. Including the ADC power, receiver consumes 73 mW. Although the proposed timing recovery concept is implemented for ADC based receiver in this work, it is possible to implement this concept for analog-mixed signal solution. Note that ISI impacts are less in mixed signal receiver. But as we are moving towards loop unrolled architectures, CDR input eye will have more ISI.

ACKNOWLEDGEMENT

The authors would like to thank CMC Microsystems for CAD tools. This research has been funded in part by the Computing Hardware for Emerging Intelligent Sensory Applications (COHESA) project financed by National Sciences and Engineering Research Council of Canada (NSERC) Strategic Networks grant number NETGP485577-15.

REFERENCES

[1] Jay. Im *et al.*, "A 40-to-56Gb/s PAM-4 receiver with 10-tap direct decision-feedback equalization in 16nm FinFET," *2017 IEEE International Solid-State Circuits Conference (ISSCC)*, 2017

[2] J. Lee, P. C. Chiang, P. J. Peng, L. Y. Chen, and C. C. Weng, "Design of 56 Gb/s NRZ and PAM4 SerDes Transceivers in CMOS Technologies," *IEEE Journal of Solid-State Circuits*, vol. 50, no. 9, pp. 2061–2073, Sep. 2015.

[3] Y. Frans *et al.*, "A 56Gb/s PAM4 wireline transceiver using a 32-way time-interleaved SAR ADC in 16nm FinFET," in *2016 IEEE Symposium on VLSI Circuits (VLSI-Circuits)*, 2016, pp. 1–2.

[4] V. Stojanovic *et al.*, "Autonomous dual-mode (PAM2/4) serial link transceiver with adaptive equalization and data recovery," *IEEE Journal of Solid-State Circuits*, vol. 40, no. 4, pp. 1012–1026, Apr. 2005.

[5] Aurangozeb *et al.*, "Channel adaptive ADC and TDC for 28 Gb/s PAM-4 digital receiver," *IEEE Journal of Solid-State Circuits*, vol. 53, no. 3, pp. 772–788, March. 2018

[6] Joshua. liang *et al.*, "A 28Gb/s digital CDR with adaptive loop gain for optimum jitter tolerance," *2017 IEEE International Solid-State Circuits Conference (ISSCC)*, 2017

eliminating the need for DCD correction. Fig. 7 shows the adaptation loop convergence. Initially, Data threshold are set to a minimal value and these thresholds are gradually increased till the control voltage dithering reaches the dither target and jitter distributions are all aligned. Fig. 8 and 9 shows the bimodal functionality with NRZ and PAM-4 excitation. Effectiveness of the DDJ filtering is evident from the CDR's ability to recover clock from a completely closed eye. In addition, adaptation loop improves the recovered clock jitter from 7 ps to 2 ps and as a result high frequency jitter tolerance

16-3 (8022)

A 12-Gb/s AC-Coupled FFE TX With Adaptive Relaxed Impedance Matching Achieving Adaptation Range of 35-75Ω Z_0 and 30-550Ω R_{RX}

Minsoo Choi, Myungguk Lee, and Byungsub Kim

Dept. of Electrical Engineering, Pohang University of Science and Technology (POSTECH)
Pohang, South Korea,
byungsub@postech.ac.kr

Abstract—A 12-Gb/s feed forward equalizing transmitter which automatically adapts to arbitrary impedances of AC-coupled channels and receivers is proposed for the first time. The proposed transmitter detects impedances of the channel and the receiver within 8% and 5% errors, respectively, and adaptively relaxes its impedance matching to maximize eye size at cost of a negligible penalty in signal integrity. In experiment, the proposed transmitter adapted to any combination of a channel impedance of 35-75Ω and a receiver impedance of 30-550Ω while achieving eye size larger than the conventional transmitter with impedance matching. At most, a completely closed eye was increased to 136mV by the proposed transmitter.

Keywords—*AC-coupled interconnects; feed forward equalization; impedance adaptaion; impedance matching; relaxed impedance matching*

I. INTRODUCTION

For decades, the legacy of 50Ω standard has set various limitations on high-speed links [1]. Because 50Ω standard was originally developed for high-voltage coaxial cables by compromising channel loss for high breakdown voltage [2], characteristic impedance Z_0 of 50Ω is not optimal for modern high-speed links where high voltage is no longer used. In addition, 50Ω termination makes transmitters (TXs) and receivers (RXs) consume unnecessarily large power, preventing advancement of low-power links; if termination impedances could be optimized without the restriction of the 50Ω standard, power efficiency would be greatly improved. For better power efficiency, other impedance standards were customized for specific applications [3], but they could not be widely used due to problems of compatibility with the 50Ω standard and among themselves; multiple standards cause huge inconvenience for users and large development cost for companies. Furthermore, impedance matching for strict zero-reflection is an over-constraint, which can be relaxed with a negligible penalty in signal integrity [1], [4].

To overcome these limitations, a feed-forward equalizing (FFE) TX that automatically adapts to arbitrary Z_0 and RX impedances R_{RX} by optimally relaxing impedance matching

This work was supported in part by National Research Foundation (NRF) of Korea (No. 2018R1A2A2A16022248), in part by Ministry of Science and ICT (MSIT) of Korea under the "ICT Consilience Creative Program" (IITP-2018-2011-1-00783) supervised by the Institute for Information & Communications Technology Promotion (IITP), and in part by the Samsung POSTECH Research Center (SPRC) funded by Samsung Electronics.

IEEE Asian Solid-State Circuits Conference
November 5 - 7, 2018/Tainan, Taiwan

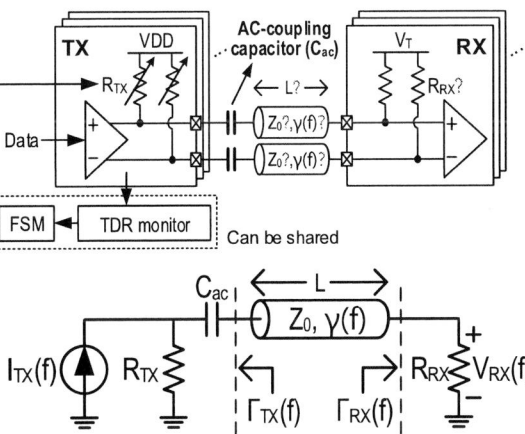

Fig. 1. The concept of adaptive relaxed impedance matching.

was recently proposed for DC-coupled interconnects [1], [4]: this FFE will be referred as a DC-coupled FFE TX with adaptive relaxed impedance matching (RIM). However, this technique cannot be applied to AC-coupled interconnects, such as USB, SATA, HDMI, and PCIe cables, which are much more widely used than DC-coupled interconnects, because Z_0 and R_{RX} cannot be detected by the previous method [1], [4] when DC is blocked. Furthermore, the formulas [1], [4] used to calculate Z_0 and R_{RX} in DC-coupled interconnect are complex requiring large hardware cost. This paper proposes an AC-coupled FFE TX with adaptive RIM. The proposed FFE TX can automatically adapt to any combination of Z_0=35-75Ω and R_{RX}=30-550Ω, and achieves better performance and power efficiency than the conventional TX with impedance matching. In addition, the formulas for Z_0 and R_{RX} calculation in the proposed technique is simpler than in the previous DC-coupled FFE TX with adaptive RIM [1], [4].

II. RELAXED IMPEDANCE MATCHING

Fig. 1 describes the proposed concept of adaptive relaxed impedance matching (RIM). When a current mode TX with an impedance R_{TX} is connected to an RX with an unknown R_{RX} and a channel with unknown parameters (length L, Z_0, and propagation constant $\gamma(f)$) through AC-coupling capacitors C_{ac},

978-1-5386-6414-8/18 $31.00 © 2018 IEEE   209

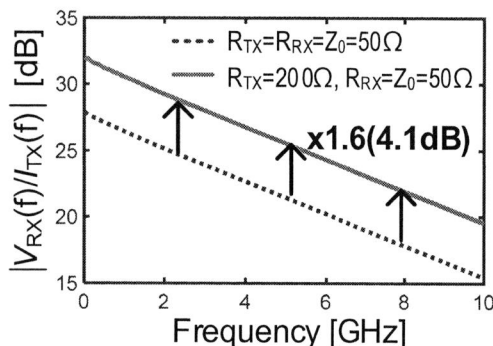

Fig. 2. $|V_{RX}(f)/I_{TX}(f)|$ of an example interconnect in Fig. 1 with 1) $R_{TX}=R_{RX}=Z_0=50\Omega$ and 2) $R_{TX}=200\Omega$ and $R_{RX}=Z_0=50\Omega$.

the transfer function can be rigorously expressed as (1) [5].

$$\frac{V_{RX}(f)}{I_{TX}(f)} = \frac{R_{TX}Z_0}{R_{TX}+Z_0} 2e^{-L\gamma(f)} \frac{R_{RX}}{R_{RX}+Z_0}\left(\frac{1}{1-\Gamma_{TX}(f)\Gamma_{RX}(f)e^{-2L\gamma(f)}}\right)$$

(1)

In (1), $\Gamma_{TX}(f)$ and $\Gamma_{RX}(f)$ are the reflection coefficients of TX and RX, respectively [4]. If all impedances are matched ($R_{TX}=R_{RX}=Z_0$), the spectral shape of (1) is proportional to $e^{-L\gamma(f)}$ because $\Gamma_{TX}(f)\Gamma_{RX}(f)e^{-2L\gamma(f)}=0$ in (1). Even though impedance matching is violated ($R_{TX}\neq Z_0$ and $R_{RX}\neq Z_0$), the spectral shape can be preserved as long as the impedances satisfy (2).

$$\left|\Gamma_{TX}(f)\Gamma_{RX}(f)e^{-2L\gamma(f)}\right| \ll 1$$

(2)

Satisfying (2) allows approximation $\Gamma_{TX}(f)\Gamma_{RX}(f)e^{-2L\gamma(f)}\approx 0$ in (1). Therefore, TXs, channels, and RXs with different impedances can be connected without sacrificing signal integrity if (2) is satisfied; in this sense, we will refer to (2) as relaxed impedance matching (RIM). Moreover, clever selection of such impedances can increase transmission gain. For instance, $R_{TX}=200\Omega$ and $R_{RX}=Z_0=50\Omega$ improve transmission gain by 60% compared with the matched one ($R_{TX}=R_{RX}=Z_0$) while preserving the same spectral shape (Fig. 2). Exploiting this principle, the proposed TX automatically configures R_{TX} for arbitrarily connected channels and RXs to maximize transmission gain while achieving good signal integrity. To this end, the TX calculates the largest R_{TX} value satisfying $|\Gamma_{TX}(f)\Gamma_{RX}(f)|<0.03$, a sufficient condition of (2), by detecting Z_0 and R_{RX} with a time-domain reflectometer (TDR) monitor.

III. TRANSMITTER ARCHITECTURE

The proposed concept is implemented in a 4-tap FFE TX with a TDR monitor (Fig. 3). The TX is designed in half-rate architecture exploiting a CML driver with resistor banks for termination. The input data of taps can be flipped for tap-polarity control, or overridden for Z_0 and R_{RX} detection. The monitor consists of a slicer, a 96-bit snapshot, and a timer. The snapshot is a fast shift register running at half-rate clock [6]. During detection, the sliced TX output is fed to the snapshot which is enabled by the transition of the input data and disabled by the timer after a programmed delay. When the snapshot is disabled, 96 consecutive sliced bits are held and read by a finite state machine (FSM) at low speed. The 96 bits are processed to

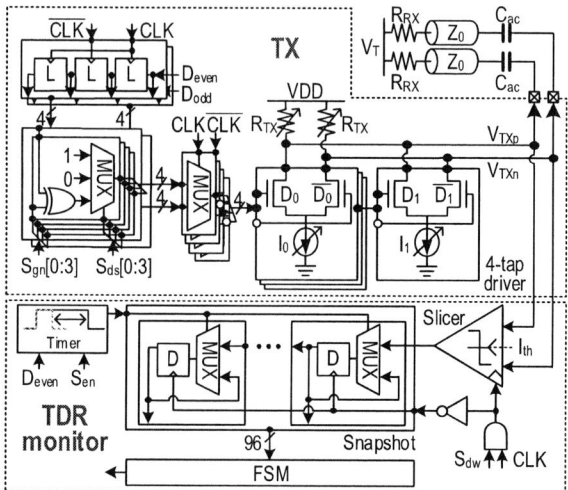

Fig. 3. A simplified block diagram of the proposed TX.

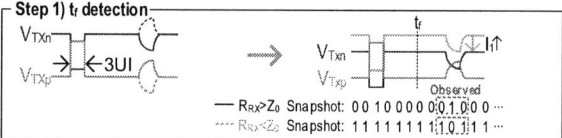

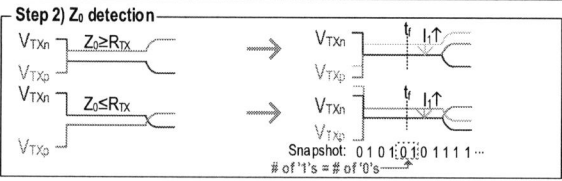

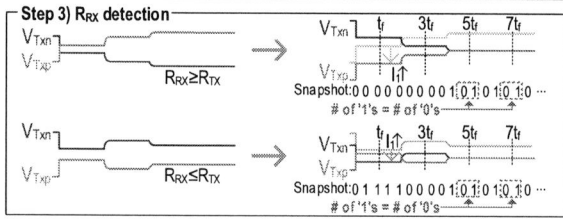

Fig. 4. Detection procedures for t_f, Z_0, and R_{RX}.

optimally configure R_{TX}. The monitor can be totally shutdown to save power after one-time configuration, and can be shared by many TXs to amortize hardware cost.

IV. IMPEDANCE DETECTION/ADAPTATION

The previous method [1], [4] used to detect R_{RX} and Z_0 in DC-coupled interconnects cannot be used for AC-coupled interconnects. Since TX's output voltage level at DC is affected by R_{RX} in DC-coupled interconnect, R_{RX} can be detected by observing the DC voltages of the TX output in various TX configurations in DC-coupled links [1], [4]. However, R_{RX} cannot be detected by the same method [1], [4] in AC-coupling due to DC-blocking. Likewise, DC-blocking requires modification of Z_0 detection in AC-coupled interconnect. In DC-coupled interconnect [1], [4], Z_0 is detected by observing the TX's output voltage after

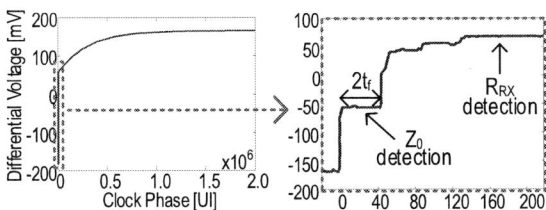

Fig. 5. TX output voltage while transmitting a step pulse.

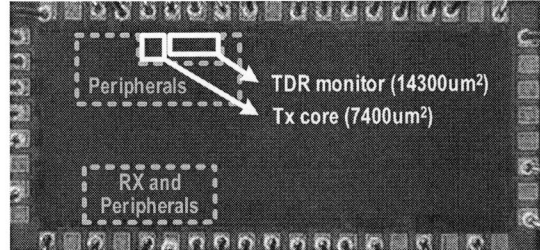

Fig. 6. Chip micrograph.

transmitting a step pulse. With DC-blocking, this voltage level is no longer affected by R_{RX}.

For R_{TX} adaptation, flight time t_f, Z_0, and R_{RX} are detected in order by algorithms described in Fig. 4. In all these algorithms, the 1st tap I_0 is used to transmit pulses while the 2nd tap I_1 is used to adjust DC offset V_{TXoff} of the TX output.

In Step 1, t_f is detected by the same procedure used for DC-coupled interconnects [1], [4]. A 3UI-wide pulse is transmitted and the monitor searches for the position of '010' or '101' in the snapshot bits. The position corresponds to $2t_f$. If t_f is not detected, the TX uses the maximum $R_{TX}=200\Omega$ and the rest detection steps are ignored.

In Step 2, Z_0 is detected by observing the TX voltage at around t_f after transmitting a step pulse. The monitor searches for V_{TXoff} which makes $V_{TXp}=V_{TXn}$ at around t_f by examining whether the snapshot bits at around t_f contain equal number of '1's and '0's. Because the TX voltage at t_f is only determined by R_{TX} and Z_0 in an AC-coupled interconnect, Z_0 can be calculated by (3) or (4) depending on the polarity of the searched V_{TXoff}; it is '+' if $Z_0 \geq R_{TX}$, or '−' otherwise.

$$Z_0 = R_{TX}\left(\frac{I_0+I_1}{I_0-I_1}\right) \quad \text{if} \quad Z_0 \geq R_{TX} \qquad (3)$$

$$Z_0 = R_{TX}\left(\frac{I_0-I_1}{I_0+I_1}\right) \quad \text{if} \quad Z_0 \leq R_{TX} \qquad (4)$$

Equations (3) and (4) are very different from and much simpler than the ones for DC-coupled interconnect [1], [4] because (3) and (4) are independent on R_{RX} due to DC blocking.

In Step 3, R_{RX} is detected by observing the saturated TX voltage after transmitting a step pulse. The monitor searches for V_{TXoff} which makes $V_{TXp}=V_{TXn}$ at around every $(2n-1)t_f$, where n is an integer, by examining the numbers of '1's and '0's as in Z_0 detection. If the same values of V_{TXoff} are consecutively searched at $(2n-1)t_f$ and $(2n+1)t_f$, the TX voltage

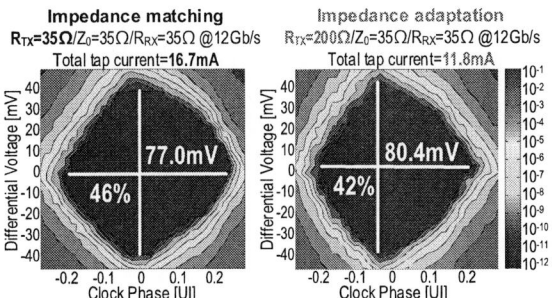

Fig. 7. Eye diagrams with impedance matching and with impedance adaptation measured at 12Gb/s.

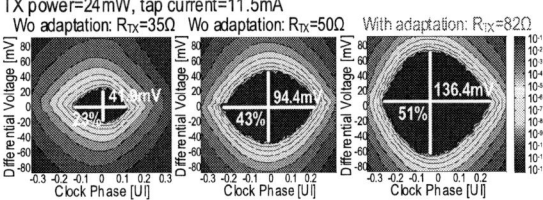

Fig. 8 Measured eye diagrams with/without impedance adaptation.

is considered saturated. Since the saturated TX voltage is determined by R_{TX} and R_{RX} in an AC-coupled interconnect, R_{RX} can be calculated by (5) or (6) depending on the polarity of the searched V_{TXoff}; it is '+' if $R_{RX} \geq R_{TX}$, or '−' otherwise.

$$R_{RX} = R_{TX}\left(\frac{I_0+I_1}{I_0-I_1}\right) \quad \text{if} \quad R_{RX} \geq R_{TX} \qquad (5)$$

$$R_{RX} = R_{TX}\left(\frac{I_0-I_1}{I_0+I_1}\right) \quad \text{if} \quad R_{RX} \leq R_{TX} \qquad (6)$$

Although the saturated TX voltage level increases very slowly due to C_{ac}, this effect is negligible because a typical cut-off frequency by C_{ac} is much lower than operating frequency (Fig. 5). This R_{RX} detection method is completely different from the method [1], [4] relying on observation of the TX's DC voltage level in DC-coupled interconnect.

After Z_0 and R_{RX} detection, R_{TX} is configured to the largest value satisfying $|\Gamma_{TX}(f)\Gamma_{RX}(f)|<0.03$ to maximize the TX's output swing and the eye opening at the RX for a negligible signal integrity penalty.

V. MEASUREMENT RESULTS

The proposed FFE TX was fabricated in 65nm CMOS technology (Fig. 6). The TX core and monitor occupy about

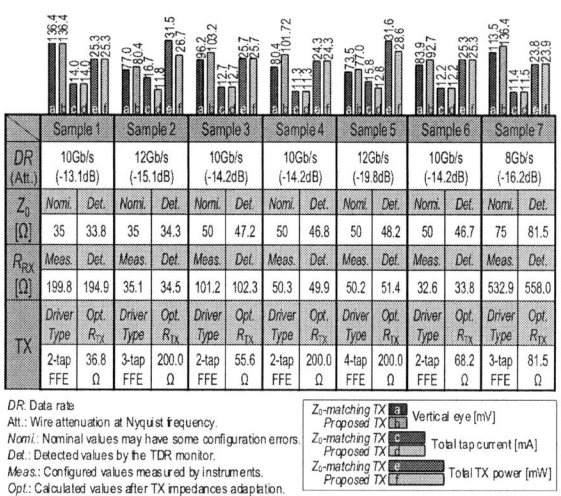

	Sample 1		Sample 2		Sample 3		Sample 4		Sample 5		Sample 6		Sample 7	
DR (Att.)	10Gb/s (-13.1dB)		12Gb/s (-15.1dB)		10Gb/s (-14.2dB)		10Gb/s (-14.2dB)		12Gb/s (-19.8dB)		10Gb/s (-14.2dB)		8Gb/s (-16.2dB)	
Z_0 [Ω]	Nomi.	Det.	Nomi.	Det.	Nomi.	Det.	Nomi.	Det.	Nomi.	Det.	Nomi.	Det.	Nomi.	Det.
	35	33.8	35	34.3	50	47.2	50	46.8	50	48.2	50	46.7	75	81.5
R_{RX} [Ω]	Meas.	Det.	Meas.	Det.	Meas.	Det.	Meas.	Det.	Meas.	Det.	Meas.	Det.	Meas.	Det.
	199.8	194.9	35.1	34.5	101.2	102.3	50.3	49.9	50.2	51.4	32.6	33.8	532.9	558.0
TX	Driver Type	Opt. R_{TX}	Driver Type	Opt. R_{TX}	Driver Type	Opt. R_{TX}	Driver Type	Opt. R_{TX}	Driver Type	Opt. R_{TX}	Driver Type	Opt. R_{TX}	Driver Type	Opt. R_{TX}
	2-tap FFE	36.8 Ω	3-tap FFE	200.0 Ω	2-tap FFE	55.6 Ω	2-tap FFE	200.0 Ω	4-tap FFE	200.0 Ω	2-tap FFE	68.2 Ω	3-tap FFE	81.5 Ω

DR: Data rate
Att.: Wire attenuation at Nyquist frequency.
Nomi.: Nominal values may have some configuration errors.
Det.: Detected values by the TDR monitor.
Meas.: Configured values measured by instruments.
Opt.: Calculated values after TX impedances adaptation.

Z_0-matching TX / Proposed TX : Vertical eye [mV]
Z_0-matching TX / Proposed TX : Total tap current [mA]
Z_0-matching TX / Proposed TX : Total TX power [mW]

Fig. 9. Summary of TX's adaptation ability test.

TABLE I. PERFORMANCE SUMMARY AND COMPARISON

		Conventional TX	This work	
Technology		65nm CMOS		
Architecture		4-tap FFE TX		
Target interconnect applications		AC-coupled interconnects		
Adaptation to Channel impedance		X	O	
			Range	35Ω~75Ω
			Error	< 8.0%
Adaptation to RX impedance		X	O	
			Range	30Ω~550Ω
			Error	< 4.5%
PRBS23, Data rate=10Gb/s, Wire attenuation= -13.1dB, Z_0=35Ω, R_{RX}=200Ω	TX termination	75Ω	36.8Ω (Automatically adapted)	
	Tap current	6.7mA	14mA	
	TX power	18.0mW	25.3mW	
	Vertical eye	Closed eye	136.41mV	
	Horizontal eye	Closed eye	54%	
PRBS23, Data rate=12Gb/s, Wire attenuation= -15.1dB, Z_0=35Ω, R_{RX}=35Ω	TX termination	35Ω	200Ω (Automatically adapted)	
	Tap current	16.7mA	11.8mA (1.4x)	
	TX power	31.5mW	26.7mW (1.2x)	
	Vertical eye	77.0mV	80.4mV	
	Horizontal eye	46%	42%	

$7400\mu m^2$ and $14300\mu m^2$, respectively. Fig. 7 shows eye diagrams measured at 12Gb/s with the conventional impedance matching [7] and with the proposed impedance adaptation. The latter consumes only 71% tap current of the former for the same eye size. The proposed TX's capability of adaptation to arbitrary channels and RXs was tested at 10Gb/s, and compared with standard TXs with the same open-circuit voltage (Fig. 8). When 50Ω-/75Ω-standard TXs are blindly connected to a 35Ω-channel and a 200Ω-RX through AC-coupling capacitors, the eye size of the former is reduced to 66.4mV (Fig. 8) from 136.4mV (Z_0-matching, Fig. 9 sample 1), and the eye of the latter is completely closed (Fig. 8) due to increased reflection. When the proposed TX is connected, however, 136.4mV is achieved by adaptively configuring R_{TX} to the optimal value: 37Ω. In summary, the proposed TX improves the eye size by 2.1x compared with the 50Ω-standard TX, and increases the closed-eye of the 75Ω-standard TX to 136.4mV, verifying the superiority in R_{TX} adaptation. The proposed TX's adaptation ability was also tested at 8Gb/s with a 75Ω-channel and a 550Ω-RX, and compared with 35Ω-/50Ω-standard TXs dissipating the same power (Fig. 8). By adaptively configuring R_{TX} to 82Ω, the proposed TX improves the eye size by 3.3x and 1.5x compared with the 35Ω- and 50Ω-standard TXs, respectively. The adaption ability of the proposed TX was also tested with various channels and RXs, and compared with Z_0-matching (Fig. 9). The proposed TX successfully adapted to any combination of 35-75Ω channels and 30-550Ω RXs, achieving the same or better eye size and power efficiency compared with Z_0-matching TX in AC-coupled interconnects. In all tests, the Z_0 and R_{RX} were detected within 8% and 5% errors, respectively. The highlight results are summarized in Table I.

VI. CONCLUSION

We proposed the first FFE TX that can automatically adapt to arbitrary impedances of the channel and RX for AC-coupled interconnects. The proposed TX can accurately detect channel and RX impedances, and automatically and adaptively relax the impedance matching of the TX to maximize the eye size at

cost of a negligible penalty in signal integrity. The proposed FFE TX was implemented in 65-nm CMOS technology. In experiment, the proposed TX adapted to any combination of a channel impedance of 35-75Ω and a RX impedance of 30-550Ω while achieving eye size larger than the conventional TX with impedance matching. When blindly connected to an arbitrary channel and an arbitrary RX through AC-coupling capacitors, the proposed TX increased a completely closed eye to 136mV. In addition, the proposed TX consumed only 71% tap current of the conventional TX with impedance matching for the same eye size at the maximum data rate of 12Gb/s.

ACKNOWLEDGMENT

The authors would like to thank IDEC for CAD support.

REFERENCES

[1] M. Choi, S. Lee, M. Lee, J.-H. Lee, J.-Y. Sim, H.-J. Park, and B. Kim, "An FFE TX with 3.8x eye improvement by automatic impedance adaptation for universal compatibility with arbitrary channel and RX impedances," in *IEEE Symp. VLSI Circuits Dig. Tech. Papers*, Jun. 2017, pp. 58–59.

[2] G. Breed, "There's nothing magic about 50 ohms," *High Frequency Electronics*, vol. 6, no. 6, pp. 6-7, Jun. 2007.

[3] R. A. Aroca and S. P. Voinigescu, "A large swing, 40-Gb/s SiGe BiCMOS driver with adjustable pre-emphasis for data transmission over 75 Ω coaxial cable," *IEEE J. Solid-State Circuits*, vol. 43, no. 10, pp. 2177–2186, Oct. 2008.

[4] M. Choi, S. Lee, M. Lee, J.-H. Lee, J.-Y. Sim, H.-J. Park, and B. Kim, "An FFE transmitter which automatically and adaptively relaxes impedance matching," *IEEE J. Solid-State Circuits*, vol. 53, no. 6, pp. 1780–1792, Jun. 2018.

[5] M. Choi, J.-Y. Sim, H.-J. Park, and B. Kim, "An approximate closed-form channel model for diverse interconnect applications," *IEEE Trans. Circuits Syst. I, Reg. Papers*, vol. 61, no. 10, pp. 3034–3043, Oct. 2014.

[6] M. Lee, S. Han, J.-Y. Sim, H.-J. Park, and B. Kim, "A 10-GHz multi-purpose reconfigurable built-in self-test circuit for high-speed links," in *IEEE Asian Solid-State Circuits Dig. Tech. Papers*, Nov. 2017, pp.73-76.

[7] V. Stojanović, A. Ho, B. Garlepp, F. Chen, J. Weil, E. Alan, C. Werner, J. Zerbe, and M. A. Horowitz, "Adaptive equalization and data recovery in a dual-mode (PAM2/4) serial link transceiver," in *IEEE Symp. VLSI Circuits Dig. Tech. Papers*, Jun. 2004, pp. 348-351.

16-4 (8114)

IEEE Asian Solid-State Circuits Conference
November 5 - 7, 2018/Tainan, Taiwan

A 40 Gb/s PAM-4 Receiver with 2-Tap DFE Based on Automatically Non-Even Level Tracking

Chia-Tse Hung
Institute of Electronics
National Chiao Tung University
Hsinchu, Taiwan, 30010

Yu-Ping Huang
Institute of Electronics
National Chiao Tung University
Hsinchu, Taiwan, 30010

Wei-Zen Chen
Institute of Electronics
National Chiao Tung University
Hsinchu, Taiwan, 30010

Abstract—A 40 Gb/s PAM-4 receiver comprised of a continuous-time linear equalizer (CTLE) and 2-tap decision-feedback equalizers (DFE) based on a novel level tracking circuit (ANLT) is proposed. A sign-sign LMS engine is embedded for the DFE and ANLT coefficients adaptation to accommodate different channel loss. The ANLT is capable of automatically tracking a non-evenly spaced PAM-4 signal, allowing the receiver to demodulate a distorted input with 2-bit flash ADCs. Fabricated in a TSMC 40nm CMOS technology, the whole receiver consumes 241.8 mW at 40 Gb/s operation. Core area is 0.274mm².

Keywords—*Pulse-amplitude modulaiton (PAM), continuous-time linear equalizer (CTLE), decision-feedback equalizer (DFE), automatically non-even level tracking (ANLT), sign-sign least mean square (SS-LMS).*

I. INTRODUCTION

The data rates of high speed I/Os keep increasing as are driven by versatile new applications, such as VR, AR, AI machines, and data centers for cloud services. In contrast to NRZ data transmissions, pulse-amplitude modulation (PAM) signaling provides better spectral efficiency to relax the bandwidth requirement of data channels. As the data rate is approaching 50 Gb/s and beyond, skin effect and dielectric losses in electrical channels become more and more pronounced, which lead to serious channel loss at higher frequencies. Compared to their NRZ counterparts, a PAM transceiver can reduce the equalizer efforts at both the transmitter and receiver side, so as to improve the overall energy efficiency for high speed data links.

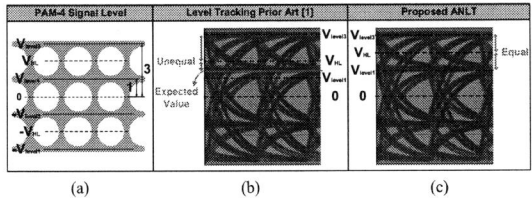

(a)   (b)   (c)

Fig. 1 (a) PAM-4 Signal Level (b) Level Tracking Prior Art [1]

(c) Proposed ANLT

In a typical PAM-4 transceiver, FFE and CTLE are commonly used to compensate part of the high frequency signal attenuation. Due to their gain nonlinearities, it will lead to an unevenly spaced amplitude modulated signal before being fed into the PAM-demodulator. Conventional level tracking circuit detects the peak amplitude of the PAM signal and generates the multi-thresholds of data slicers based on an

evenly spaced amplitude distribution assumption, as is shown in Fig. 1(a). It can hardly be held considering the nonlinearity of the transceiver. Thus the PAM demodulator might encounter severe SNR degradation, as is shown in Fig. 1(b). This issue can be remedied by employing more quantizers and then recover the data in the digital domain. But it also imposes significant hardware and power overhead, which degrades the energy efficiency. This paper proposes a PAM-4 receiver with 2-tap DFE based on a novel automatically non-even level tracking (ANLT) circuit. It is capable of recovering the thresholds of an unevenly spaced PAM-4 signal, as is shown in Fig. 1(c). Compared to the prior art in [1][2], it can cope with the amplitude distortion issues while maintaining the maximum SNR with minimum hardware overhead.

II. PROPOSED ARCHITECTURE

Fig. 2 describes the proposed receiver architecture, which is composed of a CTLE, followed by time-interleaved 2-tap DFEs, an ANLT engine that generates the threshold voltages for the data slicers, a SS-LMS engine for the adaptation of DFE and ANLT concurrently, and a clock generator that provides multiphase clock signals for the digital circuitries. Finally, the input signal is demodulated at quarter rate through a Gray-code decoder.

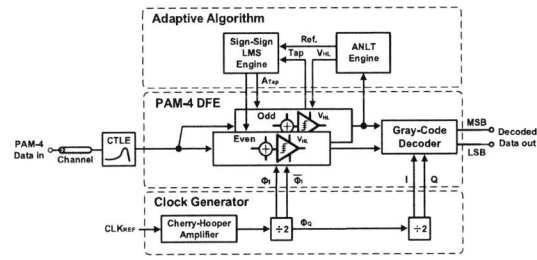

Fig. 2 PAM-4 Receiver Architecture

In a PAM demodulator, level tracking is essential for the DFE coefficients adaptation as well as providing a proper threshold to the data slicer for a maximum SNR. [1] generates the threshold voltage as $2/3V_{pp}$ (V_{pp} : average peak-to-peak swing). Similarly, [2] provides the threshold voltage as $2V_{min}$ (V_{min} : minimum pulse amplitude). Both rely on the fidelity of an evenly spaced PAM-4 signal. Considering the amplitude distortion in the transceiver, an improper level estimation not only misleads the coefficients adaptation in DFE, but also degrades the SNR of the demodulator.

978-1-5386-6414-8/18 $31.00 © 2018 IEEE

213

Fig. 3(a) illustrates the proposed ANLT circuit, which is embedded in the DFE engine. A pattern filter is incorporated in the ANLT, which steers tracking of V_{level1} and V_{level3} of the PAM-4 input signal independently. To circumvent the speed bottleneck of the critical path, speculative error quantizers are also implemented in the ANLT. To avoid the ANLT being falsely locked onto the 0 amplitude state, V_{level3} and V_{level1} are preset to a high and low voltage levels respectively in the initial state. The threshold levels of the PAM-4 signal are tracked by averaging V_{level1} and V_{level3} through a level divider. The DFE, ANLT, and SS-LMS engine share most of the hardware, thus the area overhead can be minimized. Meanwhile, the DFE and ANLT can be converged concurrently irrespective of the non-evenly spaced PAM-4 input signal, as is shown in Fig. 3(b).

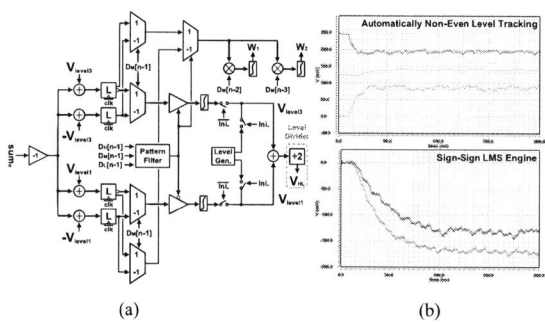

(a)   (b)

Fig. 3 (a) ANLT Engine (b) Adaptation of ANLT and DFE Coefficients

III. EXPERIMENTAL RESULTS

The experimental prototype chip is measured by applying a 2^7-1, 40 Gb/s (20 GBd) PAM-4 data through a FR-4 PCB trace whose channel loss is 10 dB at the Nyquist frequency. Fig. 4(a) shows the data eye at the receiver input, which is almost totally closed. The 2-way time-interleaved receiver demodulates the input data through half rate DFEs followed by quarter rate Gray-code decoders. Fig 4(b) shows the measured eye diagram at the 5 Gb/s output. The measured bit error rate is less than 10^{-7} with a sampling timing margin of 0.4 UI.

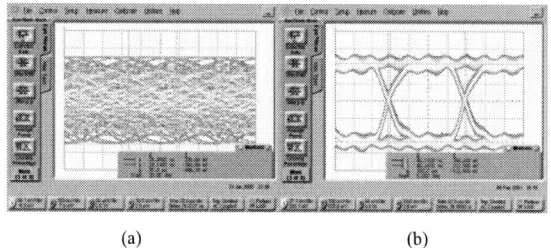

(a)   (b)

Fig. 4 (a) 40 Gb/s (20 GBd) Input Data (b) 5 Gb/s Digital Output

Fig. 5 shows the chip micrograph. Implemented in a TSMC 40nm CMOS process, the core area is about 0.274 mm². The CTLE and ClockGen consume 26.8 mW and 20.4 mW respectively from a 1.2V supply, and the DFE consumes 194.6 mW from a 1V supply.

Table I shows the performance benchmark compared to the prior art. The proposed PAM-4 receiver demonstrates the first non-evenly spaced level tracking capability for the PAM-4

demodulator. Integrating CTLE, DFE, ANLT, SS-LMS engine, and 1-to-4 de-multiplexer on a single chip, it manifests a higher level of integration compared to the previous works with an energy efficiency of 6.05 pJ/bit.

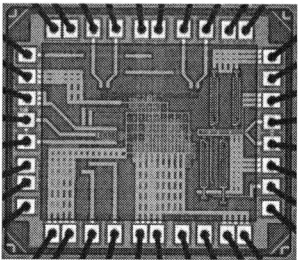

Fig. 5 Chip Micrograph

TABLE I.   PERFORMANCE COMPARISON

	This Work	[1] 2017 ISSCC	[2] 2018 ISSCC	[4] 2017 JSSC
Technology	40nm CMOS	40nm CMOS	65nm CMOS	65nm CMOS
Supply	1V/1.2V	1V	1.2V	1V
Data Rate	40Gb/s PAM-4	28Gb/s NRZ 56Gb/s PAM-4	32Gb/s PAM-4	16Gb/s NRZ 32Gb/s PAM-4
Architecture	CTLE 2-tap DFE	CTLE 3-tap DFE CDR	CTLE 1-tap DFE CDR	3-tap DFE
Ch. Loss	10 dB	24 dB	23 dB	13.5 dB
Adaptation	Yes	Yes	Yes**	No
Area	0.274 mm²	1.26 mm²*	0.16 mm²	0.074 mm²
Power	241.8 mW	382 mW	80 mW	176.3 mW
Power Efficiency	6.05 pJ/bit	7.64 pJ/bit	2.5 pJ/bit	5.51 pJ/bit
Non-Even Level Detection	Yes	No	No	No

* Chip area.

** Only partial adaptation.

ACKNOWLEDGMENT

This work was supported by the Center for mmWave Smart Radar Systems and Technologies, Ministry of Education (MOE) under contract MOST 107-3017-F-009-001 and National Science Council under contract MOST 106-2622-8-009-017-and MOST 107-2218-E-009-041 in Taiwan. The authors would like to thank TSMC University Shuttle Program for chip fabrication.

REFERENCES

[1] P.-J. Peng, J.-F. Li, L.-Y. Chen, and J. Lee, "A 56Gb/s PAM-4/NRZ Transceiver in 40nm CMOS," *ISSCC*, pp. 110-111, Feb. 2017.

[2] L. Tang, W. Gai, L. Shi, X. Xiang, K. Sheng, and A. He, "A 32Gb/s 133mW PAM-4 Transceiver with DFE Based on Adaptive Clock Phase and Threshold Voltage in 65nm CMOS," *ISSCC*, pp. 114-115, Feb. 2018.

[3] Z.-H. Hong, Y.-C. Liu, and W.-Z. Chen, "A 3.12 pJ/bit, 19-27 Gbps Receiver with 2-Tap DFE Embedded Clock and Data Recovery," *IEEE J. Solid-State Circuits*, vol. 50, no. 11, pp. 2625-2634, Nov. 2015.

[4] A. Roshan-Zamir, O. Elhadidy, H.-W. Yang, and S. Palermo, "A Reconfigurable 16/32 Gb/s Dual-Mode NRZ/PAM4 SerDes in 65-nm CMOS," *IEEE J. Solid-State Circuits*, vol. 52, no. 9, pp. 2430-2447, Sep. 2017.

A 1.6-GHz 3.3-mW 1.5-MHz Wide Bandwidth ΔΣ Fractional-N PLL with a Single Path FIR Phase Noise Filtering

Jingcheng Tao and Chun-Huat Heng

National University of Singapore, Kent Ridge, Singapore

Email: jingcheng_tao@u.nus.edu

Abstract—This paper describes a novel ΔΣ quantization noise filtering method for wideband ΔΣ fractional-N PLL. A single path finite impulse response (FIR) filtering technique is realized through two-step phase interpolation. A prototype 1.6GHz ΔΣ fractional-N PLL with 1.5MHz wide loop bandwidth is implemented in 130nm CMOS process. Measurement results show that the proposed technique effectively reduces the high frequency quantization noise by 12dB and achieves an in-band phase noise of -101dBc/Hz at 400kHz and 2.14ps integrated jitter, while consuming only 3.3mW power and an area of 0.24mm^2.

Keywords—wideband; fractional-N PLL; FIR; single path; phase noise filtering

I. INTRODUCTION

ΔΣ fractional-N PLLs have played a key role in modern wireless and wireline applications, offering very fine frequency resolution without being limited by the reference frequency. ΔΣ fractional-N PLL can also enable direct digital phase modulation, which is critical for polar transmitter. However, wide loop filter bandwidth is needed to meet the bandwidth requirement of the modulated data. For PLLs with smaller loop bandwidth, the issue is overcome by either applying pre-emphasis on the transmitted data, or employing 2-points modulation to achieve all-pass characteristic [1]. Both techniques require sophisticated calibration to characterize the PLL loop filter and VCO characteristics. Wider loop filter bandwidth also allows faster frequency hopping. Nevertheless, the ΔΣ fractional-N PLL loop bandwidth is usually limited by the shaped ΔΣ quantization noise at high frequencies.

Various techniques have been reported to suppress the high frequency shaped phase noise for fractional-N PLL to achieve wider PLL loop bandwidth. One approach is to adopt a digital-to-analog converter (DAC) to compensate for the accumulated phase errors generated by the ΔΣ modulator [2]. However, complicated calibration is often needed to mitigate the analog mismatches and non-linearity. This incurs larger power and area penalty. Hybrid FIR filtering technique has been proposed to filter out the high frequency shaped phase noise due to the ΔΣ modulator [3][4]. However, it requires multiples parallel circuit paths, with each path containing their own charge pump (CP), phase frequency detector (PFD) and divider. In [5], a high frequency phase blender and an interpolator are utilized to reduce the number of dividers and PFDs, but multiple high frequency circuit paths are still required. These duplicated circuit paths incur significant area and power penalty. In [6], a

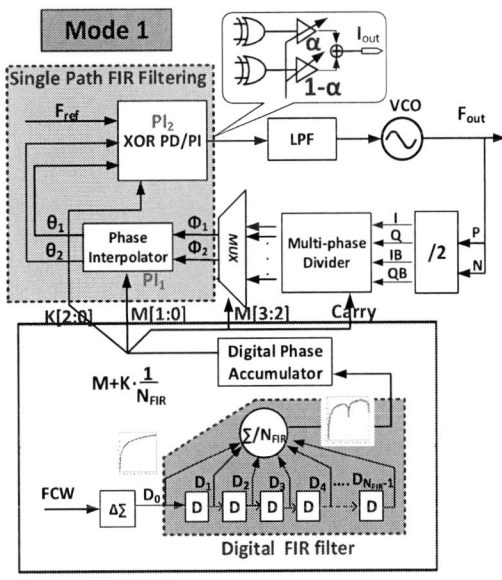

(a)

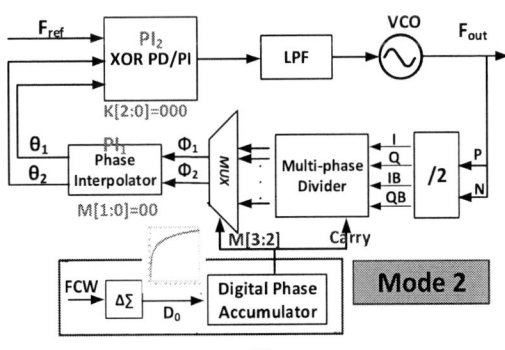

(b)

Fig. 1: Block diagram of the proposed ΔΣ fractional-N PLL(a) Mode 1with FIR filtering (b) Mode 2 without filtering

high over sampling ratio (OSR) is achieved by using cascaded PLLs to obtain high reference frequency of 800MHz. However, the circuit suffers from limited frequency resolution and higher modulator power consumption.

In this paper, we propose a novel FIR phase noise filtering technique with only a single circuit path, which leads to significant circuit area and power saving. The proposed technique is demonstrated with a 1.6 GHz ΔΣ fractional-N PLL.

II. SINGLE PATH FIR FILTERING ARCHITECTURE

The single path FIR noise filtering technique is illustrated in Fig. 1(a). The phase rotation mechanism of the proposed fractional-N PLL is controlled by two sets of control words, i.e. 4-bit M[3:0] and 3-bit K[2:0]. M[3:2] controls the multi-phase divider output while M[1:0] determines the 1st phase interpolator (PI$_1$). K[2:0] will adjust the 2nd phase interpolator (PI$_2$) output. A divide-by-2 stage generates four equally spaced phases denoted as I, Q, IB, QB at half the VCO frequency. Then the four phases are fed into a multi-phase divider to generate $P_0, P_{90}, P_{180}, P_{270}, P_{360}$ with an evenly spaced phase of half VCO cycle. QB is then fed to a dual modulus divider that is controlled by the carry bit of the digital phase accumulator.

The PLL has two working modes: mode 1 (with single path FIR filtering, Fig. 1(a)) and mode 2 (without FIR filtering, Fig. 1(b)). In mode 2, a 2nd order ΔΣ modulator is used to generate fractional frequency control output (D$_0$) with a high-pass noise profile. D$_0$ is then directly sent to the digital phase accumulator. The M[3:2] and carry bit from the accumulator are then used to control multi-phase divider and the dual modulus divider. Through M[3:2], the MUX will select the two neighbouring phases (Φ_1, Φ_2) from $P_0 \sim P_{360}$ to feed into XOR PD. In mode 2, the single-path FIR filtering is disabled by keeping M[1:0]=00 and K[2:0]=000. This essentially results in a phase-rotator (PR) based fractional-N PLL with a phase resolution of 1/2 VCO cycle.

In mode 1, the modulator output (D$_0$) will first go through a digital FIR filter to suppress the high frequency shaped phase noise. This effectively generates an average of N$_{FIR}$ delayed ΔΣ modulator outputs (N$_{FIR}^{th}$-order FIR, in this design, N$_{FIR}$=8). As a result, the high frequency shaped noise is suppressed by the low pass FIR filter but the digital filtered output will have larger bit-width. The digitally filtered output is then passed to the digital phase accumulator. The accumulator will in turn generate carry bit, M[3:0] and K[2:0]. M[3:2] and the carry bit function similarly as in mode 2. To provide finer phase control that corresponds to the larger digital output bit-width, a two-stage phase interpolator PI$_1$ controlled by M[1:0] is inserted to further interpolate the MUX outputs into 4 finer phases with a phase resolution of 1/8 VCO cycle. Hence, M[3:0] now controls a total of 16 phase rotation steps. Although more phase interpolations can be incorporated into PI$_1$, the inherent non-linearity of voltage-to-phase conversion limits the number of interpolated phases in PI$_1$.

It should be pointed out that the 3-bit K[2:0] from accumulator is used to represent the fractional phase rotation in addition to the integer part M[3:0]. To correctly represent this finer digital fractional phase rotation, we need to create their corresponding higher resolution analog phase representation to avoid the truncation error that will otherwise worsen the overall phase noise. In the hybrid FIR approach, this is achieved in analog domain by summing the phase errors from

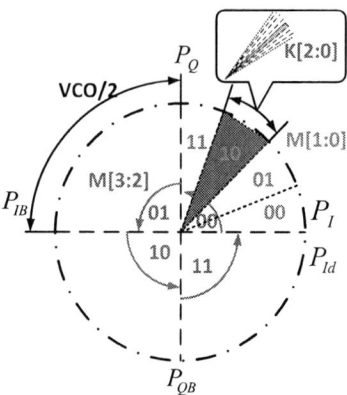

Fig. 2: Phase rotation diagram

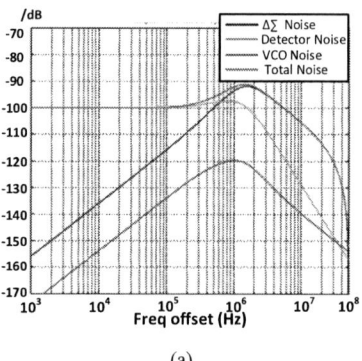

(a)

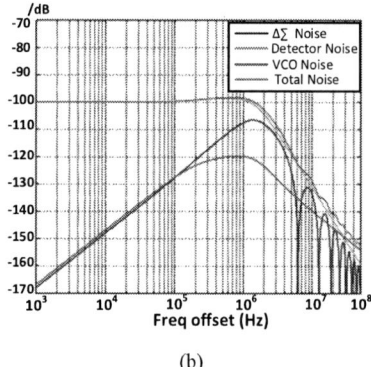

(b)

Fig. 3: (a) Mode 2 : w/o noise filtering
(b) Mode 1: w/ single path FIR noise filtering

multiple delayed circuit paths at the CP. In the proposed single path FIR approach, higher resolution analog phase control is achieved by embedding PI$_2$ inside the XOR phase detector/interpolator (the XOR PD/PI). The PI$_2$ operation is described as follows: First, the two interpolated finer adjacent phases (θ_1, θ_2) from PI$_1$ described earlier are fed into the XOR PD/PI. The phase detector part consists of two XOR gates (XOR_PD1 and XOR_PD2) that compare (θ_1, θ_2) with the reference signal. Finer phase interpolation is achieved by weighting the output currents (α and 1-α) of the two phase detectors based on K[2:0] and then summing them to generate the desired CP output current. Much better linearity and resolution can be achieved in PI_2 because the phase interpolation is performed in current domain. The total CP current ($N_{cp}I_0$) is set by the desired PLL loop bandwidth. With this approach, fractional phase resolution can be achieved without extra power consumption. Fig. 2 illustrates the phase rotation details of the proposed technique. As an example, assume the initial phase is 0°, and the FIR output is 2.375: M=0010, K=011. M rotates the phase to the red region; K will then further interpolate the phase within this region as shown in the zoom-out sub-figure. Fig. 3(a) and (b) compares the phase noise performance in mode 1 and mode 2 when the loop bandwidth is 1.5 MHz. As shown, the $\Delta\Sigma$ quantization noise (blue line) is significantly suppressed in mode 1.

III. CIRCUIT DESIGN

A. Multi-phase Divider

The multi-phase divider is shown in Fig. 4. The divide-by-2 quadrature phase generator employs two D-latches in a master-slave configuration [8]. The full swing output of this structure allows easier interfacing with the subsequent frequency divider. The dual-modulus divider output is delayed by a retiming circuit. The amount of delay in the retiming circuit is tuned automatically by detecting output phase sequence to make sure that the sequence of P_0- P_{360} is always correct.

B. Pipelined Phase Interpolator (PI₁)

PI_1 is an inverter-based two-stage pipelined phase interpolator as shown in Fig. 5(a). The stack inverters are used to minimize the overshooting effect. Small inverter buffers between each stage slow down the rise/fall time to achieve better interpolation at the output. All these help to improve the PI linearity as depicted in Fig. 5(b).

C. XOR PI/PD

The schematic of the XOR PD/PI is shown in Fig. 6 [7]. The charge pump function is embedded within this block. Four XOR PDs are used here to steer the current into or away from the actual charge pump path. XOR_PD 1 and XOR_PD 2 are employed to detect the phase difference between the reference signal and the divider feedback signal. Two additional phase detectors, XOR-PD 3 and XOR-PD 4, are used to steer the tail current when XOR-PD 1 and XOR-PD 2 are both turned off. A unity gain buffer is added to maintain the same voltage for both the replica node and output node. The phase interpolation is done by scaling the output currents of the two phase detectors by varying the 3-bit tail current DAC.

IV. MEASUREMENT RESULTS

This PLL is fabricated in 130-nm CMOS process as shown in Fig. 7. Its active die area is only 0.24 mm². The designed loop bandwidth is 1.5 MHz. The 50 MHz reference signal using external crystal is generated on chip. The measured phase noise of free-running VCO is -103 dBc/Hz at 200 KHz offset and 120 dBc/Hz at 1 MHz offset. Fig. 6 shows the measured phase noise performance. When the single path FIR filtering is enabled, the phase noise due to $\Delta\Sigma$ modulator quantization error is reduced by 12 dB. Compared to phase noise of integer-N mode, it is clear that the proposed technique

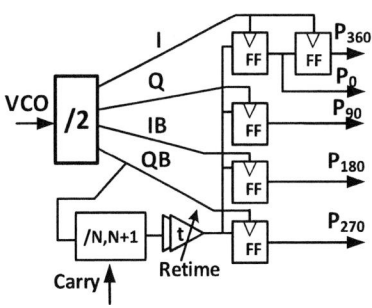

Fig. 4: Multi-phase divider

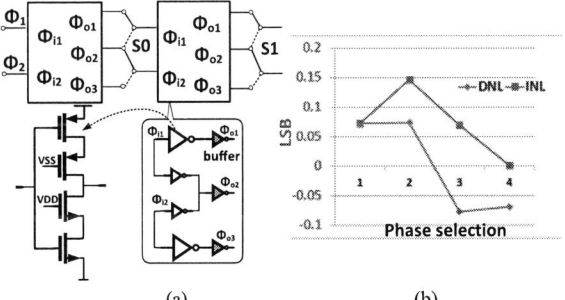

(a) (b)

Fig. 5: (a) Pipelined phase interpolator (b) Simulated PI linearity

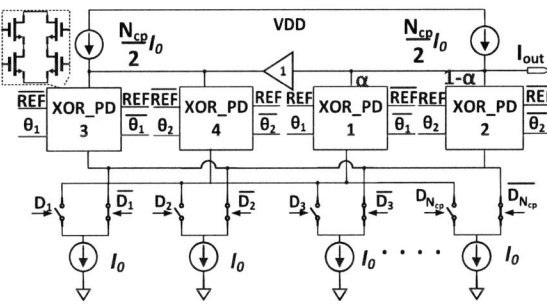

Fig. 6: XOR PD/PI

effectively filters out the out-of-band $\Delta\Sigma$ quantization noise. The in-band phase noise is about -101 dBc/Hz at 400 kHz and the measured integrated RMS jitter from 40 kHz to 40 MHz is 2.14 ps. The reference spur as shown in Fig. 8(a) is -57.6 dBc. The total power consumption is only 2.7 mA under 1.2 V DC supply and Fig. 8(b) shows the system power breakdown.

Our work is benchmarked with recent works on $\Delta\Sigma$ quantization noise cancellation in Table I. Thanks to the proposed single path FIR filtering technique, our work demonstrates the lowest power (3.3 mW) and smallest circuit area (0.24 mm²) compared to other $\Delta\Sigma$ quantization noise suppression techniques, while achieving excellent and comparable phase noise filtering performance.

TABLE I

PERFORMANCE COMPARISON WITH PREVIOUS REPORTED FRACTIONAL-N PLLS

	This work	[2]TMTT'2016	[3]JSSC'2009	[4]JSSC'2009	[5]JSSC'2013	[6]JSSC'2012
Output frequency	1.6 GHz	1.8 GHz	1 GHz	2 GHz	1 GHz	2.4 GHz
BW	1.5 MHz	1 MHz	1 MHz	200 kHz	3.2 kHz	2 MHz
In-band phase noise	-101 dBc/Hz (400 kHz)	-101 dBc/Hz (100 kHz)	-85 dBc/Hz (10 kHz)	-92 dBc/Hz (200 kHz)	-106 dBc/Hz (100 kHz)	-102 dBc/Hz (100 kHz)
Out-of-band phase noise	-124 dBc/Hz (6 MHz)	-129 dBc/Hz (3 MHz)	N.A.	-128 dBc/Hz (3.5 MHz)	-107 dBc/Hz (6 MHz)	130 dBc/Hz (10 MHz)
Noise cancelling method	Single Path FIR	DAC	FIR	FIR	FIR embedded	High-OSR
Fractional spur level	-57 dBc	-68 dBc	N.A.	-63 dBc	-66 dBc	-54.7 dBc
Core power	3.3 mW	8.3 mW	6.1 mW	17.2 mW	16.8 mW	9.6 mW
Core area	0.24 mm²	0.33 mm²	0.5 mm²	1.5 mm²	0.31 mm²	0.46 mm²
Technology	130 nm	130 nm	180 nm	180 nm	130 nm	130 nm
VCO type	LC	LC	Ring	LC	Ring	LC

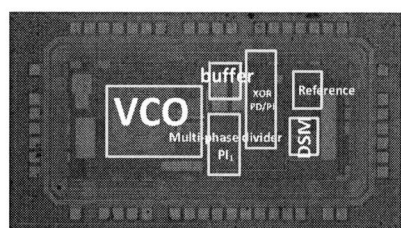

Fig. 7: PLL die photo

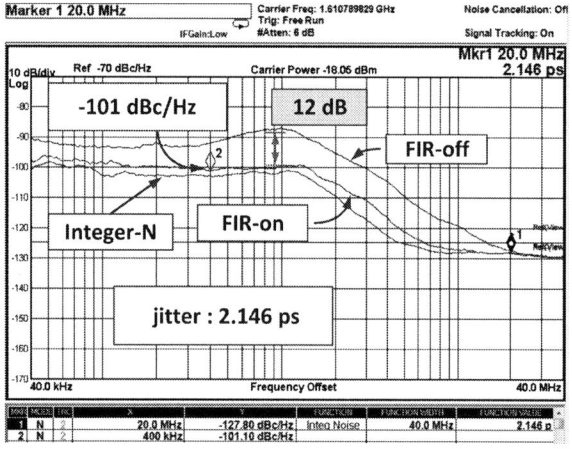

Fig. 8: Phase noise measurement

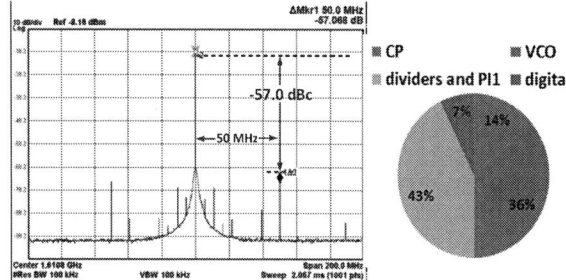

Fig. 9: (a) Reference spur measurement   (b) Power breakdown

V. CONCLUSION

In this paper, we demonstrate an architecture that can suppress the shaped $\Delta\Sigma$ phase noise at higher frequencies by 12 dB. It has the smallest area of 0.24 mm² and power of 3.3 mW compared to other high frequency shaped phase noise suppression techniques. This is made feasible by avoiding multiple feedback analog circuit paths or high frequency phase blender. The proposed 1.6-GHz PLL exhibits 1.5-MHz loop bandwidth, and achieves -101 dBc/Hz in-band phase noise and 2.14 ps RMS jitter. This allows direct data modulation without the need of pre-emphasis or sophisticated calibration.

ACKNOWLEDGMENT

This work is funded by MOE Tier 2 grant MOE2016-T2-1-123.

REFERENCES

[1] D. Cherniak, C. Samori, R. Nonis and S. Levantino, "PLL-Based Wideband Frequency Modulator: Two-Point Injection Versus Pre-Emphasis Technique," in *IEEE Transactions on Circuits and Systems I: Regular Papers*, vol. 65, no. 3, pp. 914-924, Mar. 2018.

[2] Y. Zhang et al., "A wideband fractional-N synthesizer with low effort adaptive phase noise cancellation for low-power short-range standards," *2015 IEEE Radio Frequency Integrated Circuits Symposium (RFIC)*, Phoenix, AZ, 2015, pp. 71-74.

[3] X. Yu et al., "A fractional-N synthesizer with customized noise shaping for WCDMA/HSDPA applications," *IEEE J. Solid-State Circuits*, vol. 44, no. 8, pp. 2193–2200, Aug. 2009.

[4] X. Yu, Y. Sun, L. Zhang, W. Rhee, and Z. Wang, "An FIR-embedded noise filtering method for fractional-N PLL clock generators," *IEEE J. Solid-State Circuits*, vol. 44, no. 9, pp. 2426–2436, Sep. 2009.

[5] D. W. Jee, Y. Suh, B. Kim, H. J. Park and J. Y. Sim, "A FIR-embedded phase interpolator based noise filtering for wide-Bandwidth fractional-N PLL," *IEEE J. Solid-State Circuits*, vol. 48, no. 11, pp. 2795-2804, Nov. 2013.

[6] P. Park, D. Park and S. Cho, "A 2.4 GHz Fractional-N Frequency Synthesizer With High-OSR $\Delta\Sigma$ Modulator and Nested PLL," in *IEEE Journal of Solid-State Circuits*, vol. 47, no. 10, pp. 2433-2443, Oct. 2012.

[7] R. K. Nandwana et al., "A Calibration-Free Fractional-N Ring PLL Using Hybrid Phase/Current-Mode Phase Interpolation Method," in *IEEE Journal of Solid-State Circuits*, vol. 50, no. 4, pp. 882-895, Apr. 2015.

[8] B. Razavi, K. F. Lee and R. H. Yan, "Design of high-speed, low-power frequency dividers and phase-locked loops in deep submicron CMOS," in *IEEE Journal of Solid-State Circuits*, vol. 30, no. 2, pp. 101-109, Feb. 1995.

17-2 (8203)

A 37-GHz-Input Divide-by-36 Injection-Locked Frequency Divider with 1.6-GHz Lock Range

Sangyeop Lee, Kyoya Takano, Ruibing Dong, Shuhei Amakawa, Takeshi Yoshida, and Minoru Fujishima

Graduate School of Advanced Sciences of Matter, Hiroshima University
Higashihiroshima, Japan
Email: sangyeop@ieee.org

Abstract—**This paper presents a divide-by-36 injection-locked frequency divider (ILFD). It locks onto a 37-GHz input over a locking range of about 1.6 GHz (4.3%) and outputs a frequency of approximately 1 GHz, which is sufficiently low for further division by a programmable divider. The high division ratio is realized by time-gating the superharmonic input signal and injecting it into a 9-stage ring VCO at 9 feeding points. Nine gating signals are generated by the ring VCO itself by logic operation such that the input signal is injected only during the correct time slice for each of the feeding points. The ILFD was fabricated using a 55-nm deeply depleted channel (DDC) CMOS process, occupies an area of 0.020 mm^2, and consumes a power of 9.1 mW.**

Index Terms—**Injection-locked frequency divider (ILFD), lock range, ring VCO, CMOS.**

I. INTRODUCTION

Studies on high-speed data communications using millimeter-wave (mmW) and sub-THz frequencies are becoming ever more active. Phase-locked loops (PLLs), shown in Fig. 1, are commonly used for frequency synthesis even at such high frequencies (> 30 GHz). Their role is to generate high-quality signals from a fixed, low-frequency external reference.

One critical issue in designing a mmW/sub-THz PLL is the prescaler design. Since both the input and output frequencies of a prescaler are relatively high owing to its fixed and typically low division ratio N, extremely high speed operation is required for the following programmable divider (Fig. 1). If a single frequency divider cannot provide a sufficiently low frequency for the programmable divider, multiple dividers might need to be cascaded to obtain the required value of $N = f_{in}/f_{out}$, increasing the total power consumption and size of the prescaler. In addition, it is difficult to design high-speed prescaler based on conventional flip flops.

Injection-locked frequency dividers (ILFDs) are often used as a fixed high-speed prescaler [1], [2]. Usually, divide-by-2 or -3 LC-VCO-based ILFDs are employed because LC VCOs can have sufficiently high oscillation frequencies required for that purpose. However, they occupy a large area, and their low division ratio tends to necessitate a chain of ILFDs to achieve the required division ratio. In this paper, we demonstrate a ring-VCO-based ILFD with a high division ratio of $N = 36$. A wide lock range of 1.59 GHz is realized with self-time-gated multiphase pulse injection. Such an ILFD can be used as a compact alternative to a chain of large ILFDs and can also

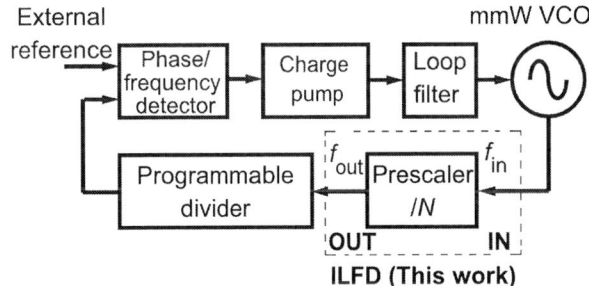

Fig. 1: Millimeter-wave phase-locked loop (PLL).

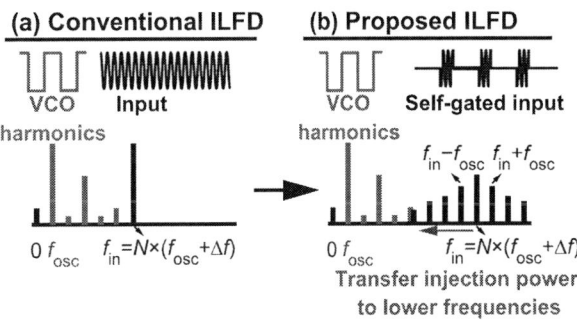

Fig. 2: Relationship between the VCO oscillation frequency f_{osc} and the input frequency f_{in} in (a) a conventional ILFD and (b) the proposed ILFD.

reduce the overall power of a PLL by allowing the use of a programmable divider (Fig. 1) operating at a lower frequency.

II. ILFD WITH HIGH DIVISION RATIO

A superharmonically injection-locked VCO, oscillating at a fundamental frequency f_{osc}, locks onto an injected signal at a frequency of $f_{in} = N(f_{osc} + \Delta f)$ if $N\Delta f$ is within the lock range, in which case $\Delta f \to 0$ as locking is established. It is generally difficult to realize a high-division-ratio ILFD because, as shown in Fig. 2(a), the harmonic output power of the VCO at Nf_{osc} becomes very low as N becomes large. For locking to take place, N cannot be very high.

One possible way of increasing N is to transfer some of the power from f_{in} to lower frequencies, to which the VCO output harmonics can lock on more easily (Fig. 2(b)). This can be

978-1-5386-6414-8/18 $31.00 © 2018 IEEE

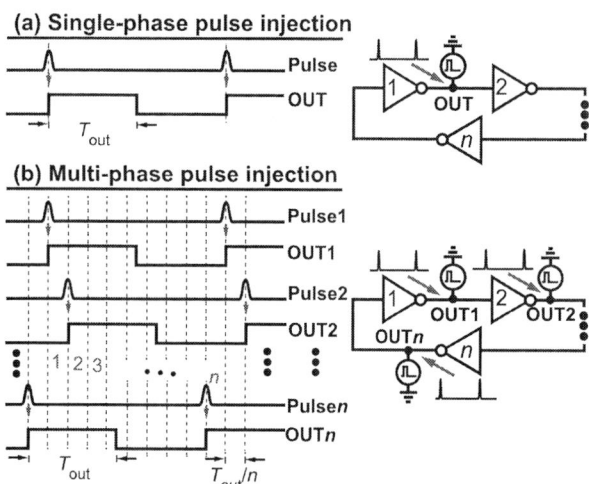

(a) Single-phase pulse injection

Pulse
OUT
T_{out}

(b) Multi-phase pulse injection

Pulse1
OUT1
Pulse2
OUT2
1 2 3 ··· n
Pulse$n$
OUTn
T_{out} T_{out}/n

Fig. 3: (a) Single-phase pulse injection into a ring VCO. (b) Multi-phase pulse injection.

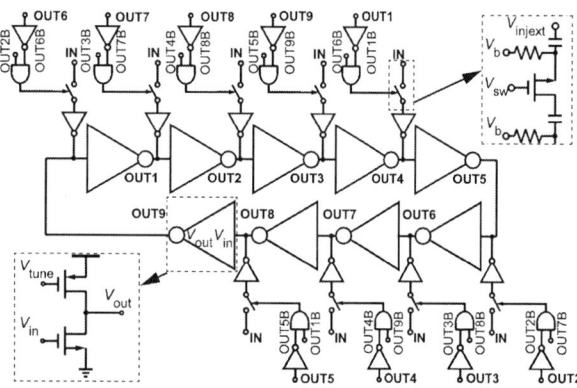

Fig. 4: Proposed ILFD including a 9-stage ring VCO and gating circuits.

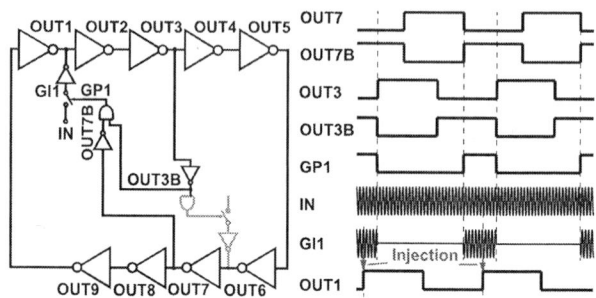

Fig. 5: Self-gated input generation in time domain.

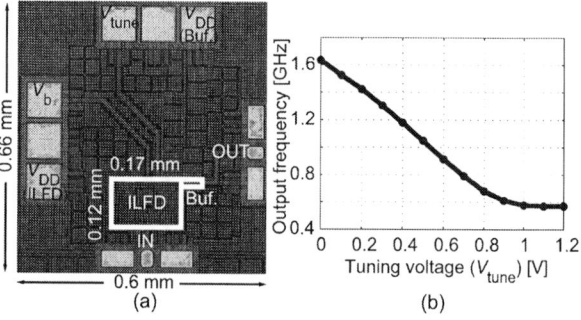

Fig. 6: (a) Chip micrograph of the ILFD. (b) Tuning range of the ring VCO.

accomplished by time-gating (or modulating) the input signal with the frequency f_{osc}. A similar idea was used to realize an injection-locked frequency multiplier (ILFM) [3]. An n-phase ring VCO is required to increase the output power of the nth harmonic (nf_{osc}) and its integer multiples. With an n-phase ring VCO, a division ratio of $N = m \times n$ ($m = 1, 2, \cdots$) can be realized easily.

In the time domain, injection locking occurs more efficiently if the input superharmonic signal at f_{in} is injected into the VCO as pulses at appropriate moments, rather than as a continuous wave. It is usually best to inject the pulses at zero-crossing points of the VCO oscillation waveform [4]–[6]. Fig. 3(a) shows single-phase pulse injection into one inverting stage of a ring VCO. In Fig. 3(a), fundamental pulse injection locking can be achieved because the pulse and the VCO output have the same frequency, f_{osc} [6], [7]. However, the lock range is limited because the injected power is

dramatically reduced compared with that when injecting the original sinusoidal signal. Therefore, multi-phase pulse injection, shown in Fig. 3(b), is desirable to widen the lock range [8], [9].

In the proposed ring-VCO-based ILFD, 9-phase self-gated inputs are injected at 9 feeding points. A circuit schematic is shown in Fig. 4. The lower inset of Fig. 4 shows the topology of the delay cell, which consists of an nMOS-based common source amplifier with a pMOS resistive load. The pMOS resistive load enables the VCO output frequency to be tuned. Gated input signals are generated by a gate-pulse window generator circuit and a switch circuit. Fig. 5 shows how a gating pulse, applied to the node GP1, is generated from OUT7 and OUT3 by logic operation. The duty ratio is about 30%. In total, nine phase-shifted self-gated input signals are generated similarly. Owing to the CMOS deeply-depleted-channel (DDC) process, which provides a high maximum oscillation frequency (f_{max}) and low threshold voltage [10], [11], the switch circuit can operate at input frequencies higher than 30 GHz.

III. MEASUREMENT RESULTS

Fig. 6(a) shows a chip micrograph of the ILFD, which was fabricated using a 55 nm CMOS DDC process. The chip area is $0.020\,\mathrm{mm}^2$. The supply voltages were $V_{DD} = 1.2\,\mathrm{V}$ and $V_b = 0.4\,\mathrm{V}$. The output frequency ranged from 0.58 GHz ($V_{tune} = 1.2\,\mathrm{V}$) to 1.6 GHz ($V_{tune} = 0\,\mathrm{V}$) as shown in Fig. 6(b).

978-1-5386-6414-8/18 $31.00 © 2018 IEEE

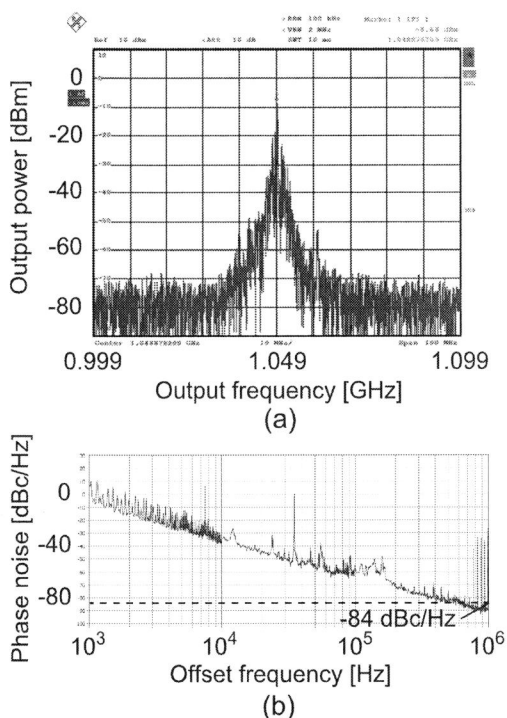

(a)

(b)

Fig. 7: Measured (a) spectrum and (b) phase noise of the free-running VCO. Spurs are shown in red.

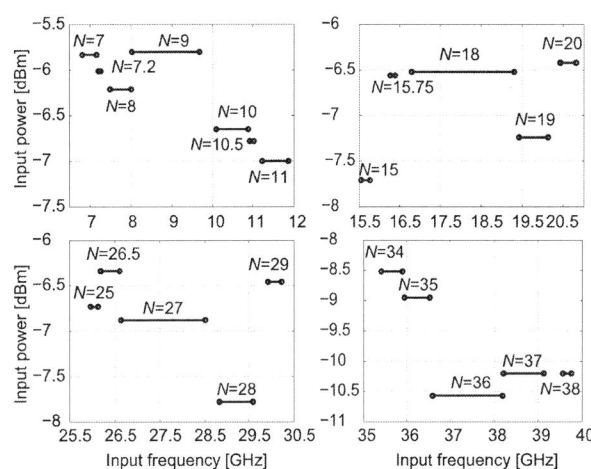

Fig. 8: Measured lock ranges when the output power of the signal generator was 0 dBm.

Fig. 7(a) shows the output power spectrum of the free-running ring VCO at $f_{osc} = 1.05\,\mathrm{GHz}$ under $V_{tune} = 0.5\,\mathrm{V}$, which was measured using a Rohde & Schwarz FSU spectrum analyzer. Fig. 7(b) shows the corresponding phase noise characteristics measured by an Aeroflex PN9000 phase noise analyzer. The 1-MHz-offset phase noise was $-84\,\mathrm{dBc/Hz}$. Some spurs were observed presumably due to the measurement environment. Under these supply and bias conditions, the power consumption (P_{cons}) of the ILFD was 8.9 mW, not including the output-buffer power consumption of 7.2 mW.

The input signal was supplied by a Keysight E8247C signal generator. The lock range was measured while sweeping the input frequency, f_{in}, downward from 40 to 6 GHz. Fig. 8 shows the measured lock ranges. The input power was calibrated using probe- and cable-loss data from 6 to 40 GHz, while the output power of the signal generator was 0 dBm throughout. Since the 9-stage ring VCO generates stronger harmonics at $9mf_{osc}$ ($m = 1, 2, \cdots$) than at other harmonic frequencies, it is expected to show wider lock ranges at $N = 9, 18, 27, 36, \cdots$, as indeed shown in Fig. 8. In addition, the ILFD also showed some fractional division ratios such as $N = 7.2$ and 26.5, as was reported in [4], [5]. $N = 7.2$, for example, can be understood as the VCO output at $36f_{osc}$ locking onto the 5th harmonic of $f_{in} = 7.2f_{osc}$ ($5f_{in} = 36f_{osc}$) owing to frequency mixing caused by the

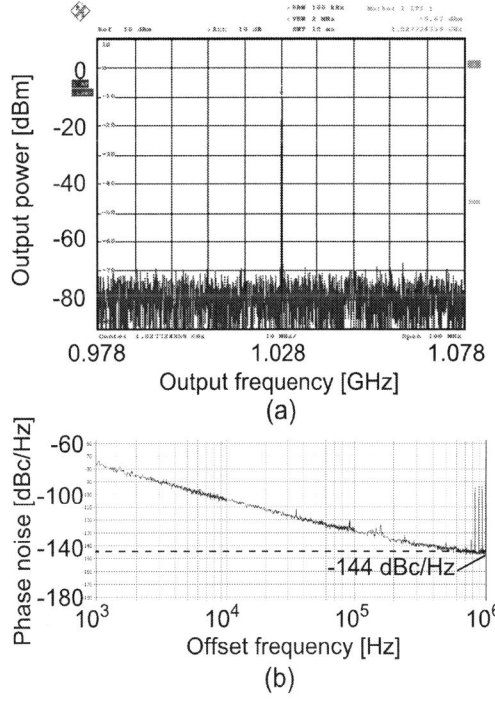

(a)

(b)

Fig. 9: Measured (a) spectrum and (b) phase noise of 1.03-GHz VCO output injection-locked onto 37-GHz input ($N = 36$).

TABLE I: Performance summary and comparison with state-of-the-art ILFDs.

	This work	[12]	[13]	[14]	[15]	[16]
CMOS tech. [nm]	55	90	65	180	180	180
V_{DD} [V]	1.2	2	1.2	1.8	1.8	N.A.
Division ratio	9, 18, 27, **36**, etc.	10	4, 8	8	7	3, 6, 9
Lock range [GHz] (W/o tuning)	1.65 ($N = 9$, 18.7%)	4.6 ($N = 10$, 7.6%)	4 ($N = 4$, 32%)	3.1 ($N = 8$, 47%)	3 ($N = 7$, 125%)	3.2 ($N = 3$, 32.4%)
	2.50 ($N = 18$, 13.9%)					1.3 ($N = 6$, 6.5%)
	1.88 ($N = 27$, 6.8%)		2.9 ($N = 8$, 17%)			0.26 ($N = 9$, 0.8%)
	1.59 ($N = 36$, 4.3%)					
Input power [dBm]	≤ -5.8	0	N.A.	N.A.	-4	≤ 0.25
P_{cons} [mW]	9.1	12.9	7.1	3.6	0.86	12.5 ($N = 9$)
Core area [mm^2]	0.020	0.50 (Chip area)	0.006	0.001	0.011	0.0048

nonlinearity of the MOSFETs including the switches. In the case of $N = 26.5$, the VCO output at $54 f_{osc}$ has locked onto a mixing product of $2 f_{in} = 53 f_{osc}$ and f_{osc}.

The frequency spectrum and phase noise characteristics under an injection-locked condition are respectively shown in Figs. 9(a) and 9(b), which were measured at $f_{osc} = 1.03\,\text{GHz}$ ($N = 36$, $f_{in} = 37\,\text{GHz}$). The power consumption of the ILFD increased slightly from 8.9 to 9.1 mW when an input signal was injected. The ILFD showed a low 1-MHz-offset phase noise of $-144\,\text{dBc/Hz}$ (Fig. 9(b)) owing to injection locking.

Table I gives a performance summary and a comparison with other ILFDs having division ratios higher than 6. The proposed ILFD has the highest division ratio of $N = 36$ despite having the smallest input power.

IV. CONCLUSION

We demonstrated a divide-by-36 ILFD that can be used in a mmW PLL as a prescaler. The output frequency (f_{osc}) at around 1 GHz is suitably low for further division by a programmable frequency divider. A wide lock range of 1.59 GHz at around 37 GHz was achieved. This was made possible by (i) the high harmonic output power of the 9-stage ring VCO at $9m f_{osc}$ ($m = 1$, 2, $\cdots$), (ii) the self-time-gated signal injection only during the correct time slice, thereby reducing the phase error, and (iii) 9-phase injection that increases the injected power. In addition, the gating also causes frequency mixing and generates a comblike frequency spectrum desirable for locking at large/multiple values of the division ratio, N.

We experimentally verified the operation of the ILFD fabricated using a 55-nm CMOS DDC process. The ILFD consists of a simple 9-stage ring VCO and gating circuits. It occupies an area of $0.020\,\text{mm}^2$ and consumes a power of 9.1 mW. Such a compact high-division-ratio ILFD may be a good alternative to a chain of low-division-ratio LC-VCO-based ILFDs in mmW frequency synthesizers.

ACKNOWLEDGMENT

This work was supported in part by JSPS KAKENHI Grant Number JP18H03781, Mie Fujitsu Semiconductor Limited, and VLSI Design and Education Center (VDEC), the University of Tokyo in collaboration with Cadence Design Systems and Mentor Graphics, Inc.

REFERENCES

[1] V. Szortyka et al., "A 42mW 230fs-jitter sub-sampling 60GHz PLL in 40nm CMOS," ISSCC, pp. 366–367, Feb. 2014.

[2] Y. Zhao et al., "A 0.56 THz phase-locked frequency synthesizer in 65 nm CMOS technology," IEEE J. Solid-State Circuits, vol. 51, no. 12, pp. 3005–3019, Dec. 2016.

[3] S. Lee et al., "A ring-VCO-based injection-locked frequency multiplier with novel pulse generation technique in 65 nm CMOS," IEICE Trans. Electron., vol. E95-C, no. 10, pp. 1589–1597, Oct. 2010.

[4] S. Hara et al., "10 MHz to 7 GHz quadrature signal generation using a divide-by-4/3, -3/2, -5/3, -2, -5/2, -3, -4, and -5 injection-locked frequency divider," IEEE Symp. VLSI Circuits, pp. 51–52, Jun. 2010.

[5] W. Deng et al., "A compact and low-power fractionally injection-locked quadrature frequency synthesizer using a self-synchronized gating injection technique for software-defined radios," IEEE J. Solid-State Circuits, vol. 49, no. 9, pp. 1984–1994, Sep. 2014.

[6] S. Lee et al., "An inductorless injection-locked PLL with 1/2- and 1/4-integral subharmonic locking in 90 nm CMOS." IEEE RFIC Symp., pp. 189–192, Jun. 2012.

[7] K. Takano et al., "4.8GHz CMOS frequency multiplier with subharmonic pulse-injection locking," IEEE A-SSCC, pp. 336–339, Nov. 2007.

[8] A. Mirzaei et al., "Multi-phase injection widens lock range of ring-oscillator-based frequency dividers," IEEE J. Solid-State Circuits, vol. 43, no. 3, pp. 656–671, Mar. 2008.

[9] J. Chien and L. Lu, "Analysis and design of wideband injection-locked ring oscillators with multiple-input injection," IEEE J. Solid-State Circuits, vol. 42, no. 9, pp. 1906–1915, Sep. 2007.

[10] K. Fujita et al., "Advanced channel engineering achieving aggressive reduction of VT variation for ultra-low-power applications," IEEE IEDM, pp. 32.3.1–32.3.4, Dec. 2011.

[11] K. Katayama et al., "An 80–106 GHz CMOS amplifier with 0.5 V supply voltage," IEEE RFIC Symp., pp. 308–311, Jun. 2017.

[12] S. Li et al., "A V-band 90-nm CMOS divide-by-10 injection-locked frequency divider using current-reused topology," IEEE Microw. Compon. Lett., vol. 28, no. 1, pp. 76–78, Jan. 2018.

[13] A. Musa et al., "Progressive mixing technique to widen the locking range of high division-ratio injection-locked frequency dividers," IEEE Trans. Microw. Theory Tech., vol. 61, no. 3, pp. 1161–1173, Mar. 2013.

[14] S. Cheng et al., "A fully differential low-power divide-by-8 injection-locked frequency divider up to 18 GHz," IEEE J. Solid-State Circuits, vol. 42, no. 3, pp. 583–591, Mar. 2007.

[15] Y. Lo et al., "A 1.8V, sub-mW, over 100% Locking Range, divide-by-3 and 7 complementary-injection-locked 4 GHz frequency divider," IEEE CICC, pp. 259–262, Sep. 2009.

[16] S. Sim et al., "A CMOS direct injection-locked frequency divider with high division ratios," IEEE Microw. Compon. Lett., vol. 19, no. 5, pp. 314–316, May 2009.

A 37.5-45.1GHz Superharmonic-Coupled QVCO with Tunable Phase Accuracy in 28nm Bulk CMOS

Luya Zhang[1], Ali Ameri[1], Yi-An Li[1], Nai-Chung Kuo[1], Mekhail Anwar[2] and Ali M. Niknejad[1]
[1]Berkeley Wireless Research Center, University of California, Berkeley, CA 94709 USA
[2]University of California, San Francisco, CA 94143 USA

Abstract—This paper presents a new superharmonic coupled millimeter-wave (mmW) quadrature voltage-controlled oscillator (QVCO) with high phase accuracy. The mechanism of using superharmonic coupling to ensure quadrature locking and its capability of rejecting phase error is analyzed. Based on the analysis, a new superharmonic coupling structure is proposed, which can adjust quadrature error against mismatch to satisfy practical system requirements. For verification, a 40GHz QVCO is implemented in 28nm bulk CMOS. The measurement shows a frequency coverage from 37.5GHz to 45.1GHz with 8.4mW power consumption and 0.18° quadrature error.

I. INTRODUCTION

Quadrature signal generation at mm-Wave frequencies is critical as it enables the implementation of direct down-conversion architecture, phased-array transceivers, and half-rate clock and data recovery (CDR) circuits. Popular methods of generating quadrature signals include using poly-phase filters or quadrature hybrid, divide-by-2 frequency dividers, and coupled VCOs. For mm-Wave applications, quadrature coupled oscillators (QVCO) are attractive due to their superior power and area efficiency. The original QVCO proposed in [1] used a parallel coupling scheme which suffered from poor phase noise performance and severe trade-off between quadrature accuracy and phase noise due to off-resonance injection locking. Series coupling [2] showed an improved phase noise and phase accuracy performance at the expense of increased voltage headroom. Moreover, coupling networks that are directly connected to the IQ differential ports are extensively studied to achieve better phase noise and phase accuracy through in-phase-injection-locking [3], [4]. However, these coupling devices add extra parasitics and loss to the LC tank which impairs both tuning range and tank quality factor.

Instead of coupling through fundamental components, quadrature locking can also be achieved by enforcing a 180° phase difference between the two 2nd harmonics extracted from IQ oscillator common-mode nodes, i.e., by super-harmonic (S-H) coupling. For example, [5] used an inverting transformer to resonate out the tail capacitor at $2\omega_0$ and to ensure quadrature lock at ω_0 (Fig. 1(b)), and [6] used a capacitive-cross-coupled pair to generate phase quadrature (Fig. 1(c)). Theoretically speaking, no significant phase noise penalty should occur using superharmonic coupling as it is inherently able to inject in-phase components at the fundamental frequency. Moreover, locking range and tank Q remain intact. Despite these benefits, superharmonic coupling is predominantly used in sub-10GHz CMOS QVCOs with only

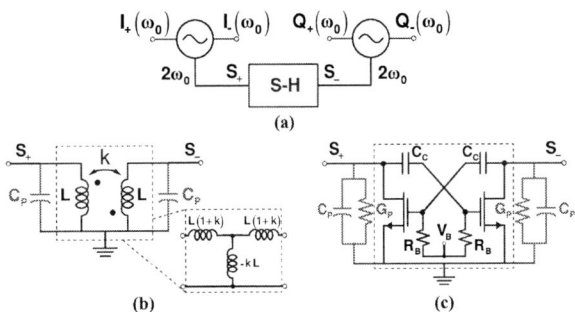

Fig. 1. (a)Two identical VCOs locked through superharmonic (S-H) coupling network, (b) a transformer based and (c) a capacitive-cross-coupled pair based S-H coupling network.

moderate phase accuracy. In addition, the mechanism of how a superharmonic network ensures quadrature lock (i.e., ensures a 180° phase difference between the coupled 2nd harmonics) has not been well identified yet.

This paper points out that the superharmonic coupling network shapes the oscillator loop gain based upon the phase offset between the two coupled oscillators. Quadrature locking is achieved by maximizing the loop gain at 90° phase offset only. A concept of effective negative Gm considering 2nd harmonic is developed to show that the loop gain shaping is realized by stimulating an oscillator-phase-offset-dependent impedance value at the two coupled nodes ($S_{+,-}$ in Fig. 1). Based on the analysis, a new superharmonic coupling network, suitable for mm-Wave QVCOs, is proposed with low phase error and no extra headroom requirement. A 40GHz QVCO utilizing the proposed network is implemented achieving 0.18° phase accuracy. More importantly, the proposed coupling network can adjust the degree of quadrature accuracy against oscillator mismatch to satisfy practical system requirements.

II. SUPERHARMONIC COUPLED QVCO

A. Effective Negative Gm

Consider the half circuit of a tail-coupled superharmonic QVCO shown in Fig. 2. Z_s represents the total tail impedance, including parasitic capacitance from the oscillator itself and the impedance presented by the coupling network. The oscillator is operated in the voltage-limited regime for optimal phase noise performance, therefore in the steady-state, the voltage minima at the tail node aligns with the the differential output

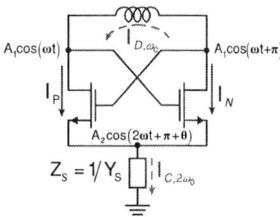

Fig. 2. Half circuit of a S-H coupled QVCO showing source impedance.

extrema (maxima and minima) with a fixed phase offset θ ($-\pi/2 < \theta < \pi/2$) caused by the node net reactance [7]. Assume the differential output voltage to be $A_1 cos(\omega_0 t)$, the tail node can be approximated as a sinusoid with a frequency of $2\omega_0$: $A_2 cos(2\omega_0 t + \pi + \theta)$. Considering the 2nd harmonic at the tail node, the current flowing through each transistor can be modeled by:

$$I_{p,n} = K(V_{OV0} \mp A_1 cos(\omega_0 t) - A_2 cos(2\omega_0 t + \pi + \theta))^2 \quad (1)$$

where V_{OV0} is the DC overdrive voltage. From Eq. (1), the differential current going into the tank at ω_0 is:

$$I_{D,\omega_0} = -K(2V_{OV0}A_1 cos(\omega_0 t) + A_1 A_2 cos(\omega_0 t + \theta)) \quad (2)$$

The second term in the current expression originates from the 2nd harmonic at the tail node which is down-converted by the cross-coupled pair. The phasor expression of effective negative Gm can then be derived as:

$$G_m = -K(2V_{OV0} + A_2 e^{j\theta}) \quad (3)$$

Two conclusions can be drawn from Eq. (3) regarding the 2nd harmonic voltage at the tail node. First, increasing its magnitude A_2 increases the effective negative Gm and thereby the oscillator loop gain. Second, reducing the absolute value of the phase angle θ caused by tail reactance also helps increase effective Gm and loop gain. Moreover, in-phase injection locking happens at $\theta = 0°$. Since the 2nd harmonic tail current $I_{C,2\omega_0} \approx K A_1^2 cos(2\omega_0 t)$ is mostly set by A_1, the same conclusions can be directly applied to the tail impedance Z_s.

So far, the relationship between effective Gm (thereby loop gain) and tail impedance Z_s has been established. As will be explained in the following section, in the case of two superharmonic coupled oscillators with a phase offset of ϕ (i.e., $A_1 e^{j\phi/2}$ and $A_1 e^{-j\phi/2}$), the tail impedance Z_s of each oscillator is a function of ϕ, and the magnitude of $Z_s(\phi)$ reaches its maximum under the condition of quadrature locking (i.e., when $\phi = 90°$). As maximizing $\|Z_s(\phi)\|$ (thereby Gm) is equivalent to maximizing oscillator loop gain, it makes the quadrature mode prevail over all the other modes during the oscillator start up, and thereby ensures quadrature locking.

B. Tail Impedance Analysis

This section derives the analytical expression of the phase-dependent tail impedance $Z_s(\phi)$ for transformer based [5] and capacitive-cross-coupled pair based [6] S-H coupling network.

For simplicity, two extreme situations (i) in-phase lock ($\phi = 0°$) and (ii) quadrature lock ($\phi = 90°$) are considered without loss of generality. Based on the derived expressions, it will be shown that these two coupling schemes can produce even larger quadrature errors when used in mm-Wave QVCOs.

In the case of transformer based S-H coupling network (Fig. 1(b)), the tail parasitic capacitance C_P is relatively constant, therefore only transformer inductance value is calculated and compared. Under quadrature lock ($\phi = 90°$), the voltages at the coupled tail nodes $S_{+,-}$ are 180° out of phase and the coupling network is in differential-mode. On the other hand, when the two oscillators are locked in-phase ($\phi = 0°$), $S_+ = S_-$ and the coupling network is in the common-mode. Using the transformer T-model, the effective inductance at $S_{+/-}$ in two modes are

$$\begin{aligned} L_{DM,\phi=90°} &= L(1+k) \\ L_{CM,\phi=0°} &= L(1-k) \end{aligned} \quad (4)$$

The transformer is sized such that at $2\omega_0$ the differential inductance $L_{DM,\phi=90°}$ is in resonance with C_P. Therefore the tail impedance, as well as the effective Gm and loop gain, are maximized at $\phi = 90°$, which makes the two oscillators run in quadrature. Since $L_{CM,\phi=0°} - C_P$ cancellation causes in-phase locking, the ratio $L_{DM,\phi=90°}/L_{CM,\phi=0°} = \frac{1+k}{1-k}$ reflects to what extent the quadrature mode is preferred over other modes. Apparently as ω_0 increases, the transformer inductance L has to drop, which makes it challenging to achieve a high coupling coefficient k. k degradation ($\frac{1+k}{1-k} \to 1$) leads to a reduced preference for quadrature mode, and thereby an increased phase error under oscillator mismatch.

In the case of using a capacitive-cross-coupled pair as coupling network (Fig. 1(c)), for now assume low frequency range where the parasitic capacitance C_P is negligible compared with the transconductance of the cross-coupled pair g_m. The tail admittance at $\phi = 90°$ and $\phi = 0°$ can be expressed as:

$$\begin{aligned} Y_{DM,\phi=90°} &= -g_m + G_P \\ Y_{CM,\phi=0°} &= g_m + G_P \end{aligned} \quad (5)$$

where G_P represents the parasitic conductance at the tail node. Note that $Y_{DM,\phi=90°}$ is non-negative because of the transistor non-linearity. The quadrature mode is selected for the same reason. However, when scaling to mm-Wave frequencies, $\omega_0 C_P$ becomes comparable with g_m and eventually erases the impedance contrast established through g_m sign inversion: $Y_{DM,\phi=90°}/Y_{CM,\phi=0°} \approx \frac{\omega_0 C_P}{\omega_0 C_P + g_m} \approx 1$. The oscillators may even lock in phase. Increasing g_m helps marginally but at the cost of extra voltage headroom or more capacitive parasitics.

C. Proposed Superharmonic Coupling Network

Fig. 3(a) shows the schematic of the implemented QVCO with proposed superharmonic coupling network. A third inductor L_S is employed to resonate out the parasitic tail capacitance at $2\omega_0$. It's worth mentioning that this additional inductor is very small compared with the tank inductors in the main oscillators (see Fig. 3(b)), because its working frequency is doubled and the tail capacitance is usually high in the

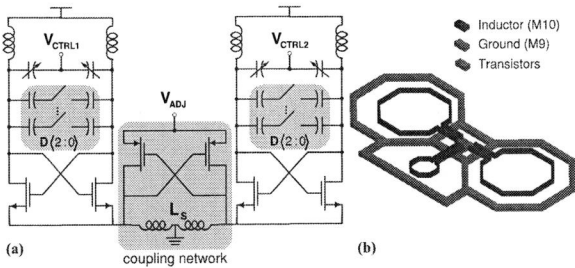
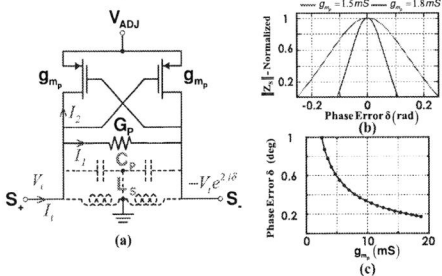

Fig. 3. (a) Proposed QVCO schematic and (b) Inductor Layout (not to scale).

Fig. 4. (a) $Z_{+/-}(\delta)$ nodal analysis under small mismatch, (b) calculated $\|Z(\delta)/Z(0)\|$ assume $G_p = 1mS$ and $g_{m_p} = 1.5mS, 1.8mS$ respectively and (c) simulated quadrature error with 0.5% tank mismatch.

practical layout. After the capacitance is canceled, a PMOS cross-coupled pair is used to ensure quadrature locking in the same manner as depicted in Eq. (5), without consuming any oscillator voltage headroom. No high-pass biasing network ($R_B - C_C$ in Fig. 1) is required either.

It is well known that quadrature error occurs in the presence of oscillator mismatch. A nice feature of the proposed coupling network is that the quadrature accuracy under given tank mismatch can be adjusted with a control voltage V_{ADJ} without affecting the main oscillators. To prove this, assume small mismatch and hence an approximate quadrature lock with a phase error equal to δ ($\delta \ll 1$). The voltage at the two coupled nodes (see Fig. 4(a)) can be written as $S_+ = A_2 e^{-j(\delta+\pi/2)}$ and $S_- = A_2 e^{j(\delta+\pi/2)}$. Since the mismatch is small, reactance cancellation (L_S and C_P) still holds roughly. By applying KCL, the admittance at nodes $S_{+/-}$ is derived as following:

$$Y_{+/-}(\delta) = (-g_{m_p} + G_p)e^{\pm 2j\delta} + G_p \qquad (6)$$

Note that $Y_+(\delta)$ and $Y_-(\delta)$ have same magnitude but opposite phases. Fig. 4(b) plots the normalized magnitude of tail impedance $\|Z(\delta)/Z(0)\|$ in the vicinity of $\delta = 0$ for two hypothetical g_{m_p} values. As shown in the plot, with a larger g_{m_p} value (higher V_{ADJ}), the tail impedance (and thereby the loop gain) sees a steeper downfall when deviating from perfect quadrature lock. This implies that as g_{m_p} increases, a higher level of mismatch is required to drag the two oscillators to a given phase error. In other words, the quadrature accuracy can be improved by increasing g_{m_p}. Fig. 4(c) plots the simulated QVCO phase error under 0.5% tank mismatch with different g_{m_p}. The error decreases monotonically as g_{m_p} increases.

III. PROTOTYPE AND MEASUREMENT RESULTS

A 40GHz QVCO was designed and fabricated in 28nm CMOS (no ultra-thick metal option). The die photo is shown in Fig. 5. The QVCO occupies 0.068mm² and consumes 8.4mW under 0.75V supply.

Tuning range and phase noise measurements are performed through on-wafer probing. As ground-signal-ground (GSG) pads are not available on chip, a Cascade infinity GS probe was landed in between the I/Q inductors to capture the near-field signal. The probe output is amplified with 65dB gain before feeding into the spectrum analyzer (Agilent N9030A) for phase noise measurement. The QVCO can be tuned from

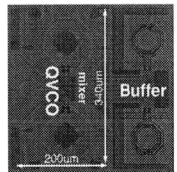

Fig. 5. Chip micrograph of the proposed QVCO.

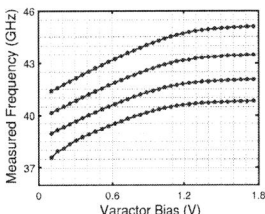

Fig. 6. Measured frequency tuning curves.

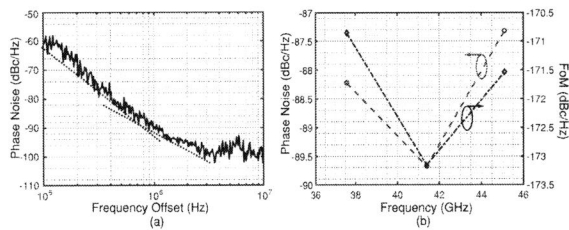

Fig. 7. Measured (a) phase noise at 41GHz and (b) phase noise/FoM across tuning range.

37.5GHz to 45.1GHz, indicating a tuning range of 18.4%, as shown in Fig. 6. The QVCO exhibits a relatively constant phase noise level across the entire frequency tuning range, from -87.3dBc/Hz to -89.7dBc/Hz at 1MHz offset, corresponding to a FoM from -171dBc/Hz to -173dBc/Hz (see Fig. 7).

Quadrature accuracy is characterized through the DC product of I/Q self mixing (i.e., $cos(\omega t - \delta/2)sin(\omega t + \delta/2)|_{DC} \rightarrow sin(\delta)$). To do this, a double-balanced voltage commutative passive mixer followed by a VGA (variable gain amplifier) is integrated on chip. Chopping is employed to remove DC

TABLE I
PERFORMANCE SUMMARY AND COMPARISON

Reference	Coupling Method	CMOS Tech.	Freq. (GHz)	TR (%)	DC Power (mW)	Phase Error (deg)	PN @1MHz (dBc/Hz)	FoM* (dBc/Hz)	FoM$_T$** (dBc/Hz)	Adjustable Phase Accuracy
ISSCC'14 [8]	Superharmonic	40 nm	58	16.2	30	2	-92.5	-173	-177	No
JSSC'14 [9]	NMOS diode	65 nm	63	16.6	11.4	0.7	-94.2	-180	-184	No
TMTT'15 [3]	Bidirectional diode	65 nm	26	15.4	11.8	0.36	-99	-176	-180	No
TMTT'16 [4]	Transformer	65 nm	54	9.1	24	2	-95.5	-180	-179	No
This Work	Superharmonic	28 nm	41	18.4	8.4	0.18	-89.7	-173	-178	Yes

*FoM=PN-20log$_{10}$($\frac{f_0}{\Delta f}$)+10log$_{10}$($\frac{P_{DC}}{1mW}$), **FoM$_T$=FoM-20log$_{10}$($\frac{TR[\%]}{10}$)

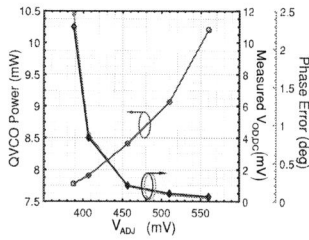

Fig. 8. Quadrature accuracy and DC power measurement.

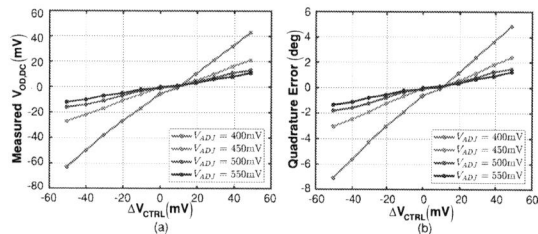

Fig. 9. (a) Measured I/Q self-mixing differential DC output and (b) back-calculated phase error decrease as V_{ADJ} increases ($V_{CTRL,CM}$=1.2V).

offset and flicker noise. Fig. 8 shows the measured differential output $V_{OD,DC}$ as a function of V_{ADJ} and total QVCO power consumption. Quadrature error is back-calculated using the mixer phase-to-voltage conversion gain obtained from post-layout simulation, and plotted in the same figure. A phase error as low as 0.18° is achieved with V_{ADJ}=0.45V. Taking into account the extraction and model uncertainty, the actual phase error is believed to be less than 0.4°. To validate that the proposed coupling network can improve phase accuracy using V_{ADJ} in the presence of mismatch, artificial offsets are created between the free-running frequencies of I/Q oscillators using individually-accessible varactor control voltages. Fig. 9(a) shows the measured differential output as a function of ΔV_{CTRL} under different V_{ADJ}. Fig. 9(b) plots the back-calculated phase error. It can be seen that as V_{ADJ} increases from 400mV to 450mV, 500mV and 550mV, respectively, a 2×, 3.3× and 4× improvement in quadrature accuracy is obtained. Table I summarizes and compares our work with recent publications on mm-Wave QVCOs. A good tuning range and very low quadrature error is achieved with very

low power consumption. A new mechanism of adjusting phase error against mismatch is provided.

IV. CONCLUSION

A new superharmonic coupling scheme is proposed for mm-Wave QVCO to improve phase accuracy without consuming voltage headroom or loading the tank. In the presence of mismatch, the quadrature error can be adjusted by the proposed coupling network to satisfy practical system requirement without affecting main oscillators. A 40GHz prototype is fabricated in 28nm CMOS with 18.4% frequency tuning range. The QVCO achieves as low as 0.18° phase error with only 8.4mW power consumption.

ACKNOWLEDGMENT

The authors wish to acknowledge the TSMC University Shuttle Program for chip fabrication.

REFERENCES

[1] A. Rofougaran, J. Rael, M. Rofougaran, and A. Abidi, "A 900 MHz CMOS LC-oscillator with quadrature outputs," in *1996 IEEE International Solid-State Circuits Conference. Digest of Technical Papers, ISSCC*, Feb. 1996, pp. 392–393.

[2] P. Andreani and X. Wang, "On the phase-noise and phase-error performances of multiphase LC CMOS VCOs," *IEEE Journal of Solid-State Circuits*, vol. 39, no. 11, pp. 1883–1893, Nov. 2004.

[3] N. C. Kuo, J. C. Chien, and A. M. Niknejad, "Design and analysis on bidirectionally and passively coupled QVCO with nonlinear coupler," *IEEE Transactions on Microwave Theory and Techniques*, vol. 63, no. 4, pp. 1130–1141, Apr. 2015.

[4] T. Xi, S. Guo, P. Gui, D. Huang, Y. Fan, and M. Morgan, "Low-phase-noise 54-GHz transformer-coupled quadrature VCO and 76-/90-GHz VCOs in 65-nm CMOS," *IEEE Transactions on Microwave Theory and Techniques*, vol. 64, no. 7, pp. 2091–2103, Jul. 2016.

[5] S. L. J. Gierkink, S. Levantino, R. C. Frye, C. Samori, and V. Boccuzzi, "A low-phase-noise 5-GHz CMOS quadrature VCO using superharmonic coupling," *IEEE Journal of Solid-State Circuits*, vol. 38, no. 7, pp. 1148–1154, Jul. 2003.

[6] P. Tortori, D. Guermandi, M. Guermandi, E. Franchi, and A. Gnudi, "Quadrature VCOs based on direct second harmonic locking: theoretical analysis and experimental validation," *Int. J. Circuit Theory Appl.*, vol. 38, no. 10, pp. 1063–1086, Dec. 2010.

[7] A. Hajimiri and T. Lee, "Design issues in CMOS differential LC oscillators," *IEEE Journal of Solid-State Circuits*, vol. 34, no. 5, pp. 717–724, May 1999.

[8] V. Szortyka, Q. Shi, K. Raczkowski, B. Parvais, M. Kuijk, and P. Wambacq, "A 42mW 230fs-jitter sub-sampling 60GHz PLL in 40nm CMOS," in *2014 IEEE International Solid-State Circuits Conference Digest of Technical Papers (ISSCC)*, Feb. 2014, pp. 366–367.

[9] X. Yi, C. C. Boon, H. Liu, J. F. Lin, and W. M. Lim, "A 57.9-to-68.3 GHz 24.6 mW frequency synthesizer with in-phase injection-coupled QVCO in 65 nm CMOS technology," *IEEE Journal of Solid-State Circuits*, vol. 49, no. 2, pp. 347–359, Feb. 2014.

17-4 (8133)

A Fast Auto-Frequency Calibration Technique for Wideband PLL with Wide Reference Frequency Range

Zhao Zhang, Jincheng Yang, Liyuan Liu[*], Nan Qi, Peng Feng, Jian Liu, and Nanjian Wu[*]

State Key Laboratory of Superlattice and Microstructures
Institute of Semiconductors, Chinese Academy of Sciences,
University of Chinese Academy of Sciences, Beijing, China
Email: liuly@semi.ac.cn, nanjian@red.semi.ac.cn

Abstract—This paper proposes a fast auto frequency calibration (AFC) technique for wideband PLL with wide reference frequency range. The AFC circuit block adopts a proposed clock controller and a current-mode logic (CML) divider-by-2 divider to accelerate the AFC process without the penalty of AFC resolution. It also adopts the adjustable AFC counting period technique to speed up the AFC process at low reference frequency and to reduce AFC time variation within wide frequency range of reference clock. A 0.1~5 GHz $\Delta\Sigma$ fractional-N PLL with this AFC technique is designed and implemented in 65-nm CMOS process. The measurement results show that in the reference frequency range from 15 to 50 MHz, the AFC time varies only from 1.25 to 1.86 μs with AFC resolution range from 3 to 5 MHz.

Keywords—*Fast AFC; AFC resolution; PLL; wide reference freqeuncy range.*

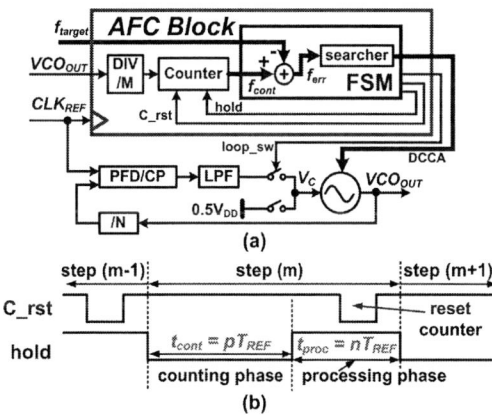

Fig. 1. Conceptual block diagram of the AFC in the PLL and the timing diagram of each AFC searching step.

I. INTRODUCTION

The wideband fractional-N PLL [1-2] is the critical building block for software-defined radio (SDR) transceiver. In such PLLs, the fast auto frequency calibration (AFC) technique [2-6] is necessary not only to set the optimum control word for digital controlled VCO capacitor array (DCCA) but also to achieve fast locking. In addition, such PLLs should operate in wide reference frequency range to be flexible for multiple applications [7]. With the support of wide reference frequency range, the PLL can share one crystal oscillator with other building blocks or systems and thus reduce the cost of external crystal [8].

Several AFC techniques [2-6] have been proposed to speed up the AFC process. However, the time of the AFC process [2-6] is proportional to the period of reference clock. So the AFC time becomes longer if using lower frequency reference clock. The longer AFC time results in a slower settling process of PLLs, which is not suitable for some applications.

In this paper, a fast auto frequency calibration (AFC) technique is proposed. It can not only get faster AFC process than prior works, but also reduce the variation of AFC time within a wide reference frequency range. It adopts a proposed AFC clock controller and a current-mode logic (CML) divider-by-2 divider to shorten the AFC time. We use an adjustable AFC counting period technique not only to speed up the AFC process at low reference frequency but also to reduce the variation of AFC time within a wide frequency range of reference clock. The proposed AFC technique is demonstrated by the implementation of a $\Delta\Sigma$ fractional-N PLL with a wide reference frequency range from 15 to 50 MHz. This paper is organized as follows. Section II explains the limitation of prior AFC techniques. Section III introduces the proposed fast AFC. Section IV presents the measurement result. The conclusions are drawn in Section V.

II. LIMITATION OF PRIOR AFC TECHNIQUES

In this section, we breifly analyze the limitation of prior AFCs to get the overall guidance to speed up the AFC process without AFC resolution degrading while reduce the AFC time variation over wide reference frequency range.

Fig. 1(a) shows the conceptual block diagram of prior AFCs [2-6] in the PLL. The AFC part mainly consists of a counter and a finite-state machine (FSM). The counter estimates the frequency of the VCO. The FSM compares the counter output f_{cont} with the target frequency f_{target} first, and then search for the optimal DCCA word for VCO. To get the optimal DCCA code, the AFC block needs k searching steps, where k is the number of bits of DCCA in the VCO. Fig. 1(b) shows the timing diagram of each searching steps. Each step includes counting phase and processing phase. In the counting phase, the counter counts the output clock to get the value of VCO frequency f_{cont}. In the processing phase, the FSM gets the frequency error f_{err} between f_{cont} and f_{target} first, and then the searcher in the FSM find the frequency band code and update it to the VCO for the operation of next AFC searching step. After k AFC searching steps, the FSM selects the DCCA code

978-1-5386-6414-8/18 $31.00 © 2018 IEEE

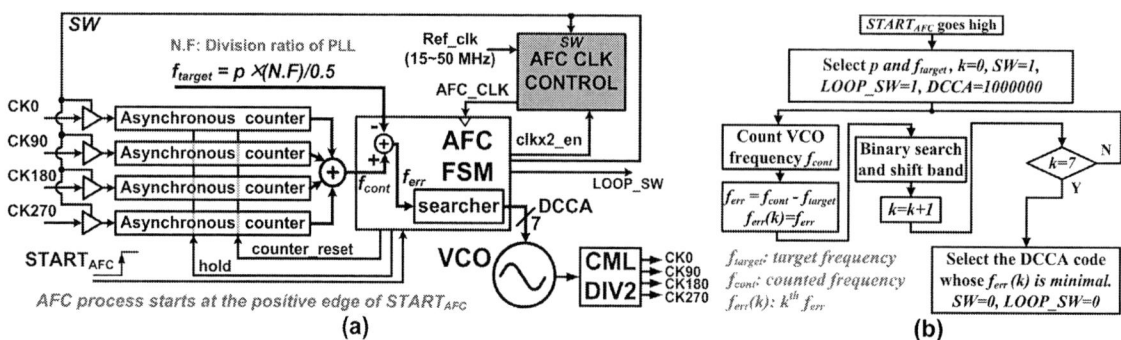

Fig. 2. (a) Block diagram of proposed AFC, (b) flowchart.

whose f_{err} is minimal as optimal.

As indicated in Fig. 1(b), the total time of AFC process can be approximately expressed as:

$$t_{AFC} = (p+n) \times k \times T_{REF} \quad (1)$$

where p is the periods number of the counting phase, n is the periods number of the processing phase and T_{REF} is the period of reference clock. The calibration resolution of AFC [3], $f_{resolution}$, is given by

$$f_{resolution} = \frac{M \times f_{REF}}{p} \quad (2)$$

where M is the division of pre-divider between VCO and AFC counters (Fig. 1(a)), f_{REF} is the reference frequency of PLL, and p is the periods number of the counting phase. The lower the $f_{resolution}$, the better the calibration resolution of AFC will be.

As indicated in (1), t_{AFC} is proportional to the period of reference clock, which causes long AFC time as well as long locking time of PLL at low f_{REF}. To accelerate the AFC process without AFC resolution degrading while reduce t_{AFC} variation over wide range of f_{REF}, our proposed AFC takes the following methods: 1) reduce M and p to reduce the counting phase time without AFC resolution degrading; 2) reduce n to shorten the processing phase time; 3) set lower p for low f_{REF} and larger p for high f_{REF} to reduce variation of t_{AFC} over wide range of f_{REF} without AFC resolution degrading. The details of our AFC are shown in Section III.

III. PROPOSED FAST AFC

The block diagram of the proposed fast AFC circuit is shown in Fig. 2(a). It mainly consists of a proposed AFC clock controller, four tri-state clock buffers, four counters, a CML DIV2 and an AFC FSM. Fig. 2(b) shows the flowchart of the proposed AFC. The asynchronous structure is adopted for the four counters to get fast counting speed. The binary search algorithm [3] is used for the searcher (Fig. 2(a)) in the AFC FSM to get the optimal DCCA word for VCO. The AFC process starts at the rising edge of the signal START$_{AFC}$. During the AFC process, the signal SW is set 1 to turn on tri-state buffers and the AFC clock controller. After the AFC process is finished, the SW goes to 0 and then the tri-state buffers, the AFC clock controller and the counters turn off to save power consumption of the whole PLL. The number of bits of DCCA in the VCO is 7 in this design. The LOOP_SW is set

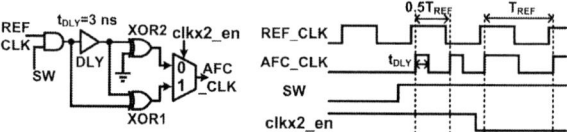

Fig. 3. Proposed AFC clock controller.

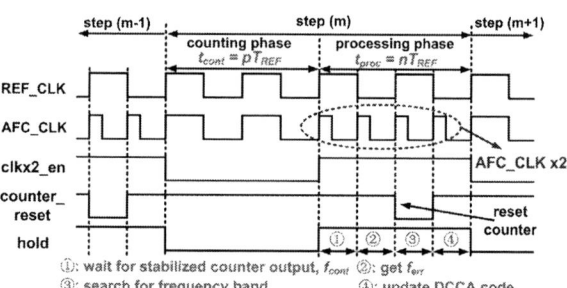

①: wait for stabilized counter output, f_{cont} ②: get f_{err}
③: search for frequency band ④: update DCCA code

Fig. 4. Timing diagram of each AFC searching step. (p=2 for example).

1 to open the PLL loop during the AFC process and set 0 after the AFC process is finished.

Equation (1) and (2) indicate that if p as well as M decrease and the ratio between them keeps constant, the t_{AFC} can be shortened due to the smaller p without the penalty of AFC resolution. Since the interval of each adjacent phase of the 4-phase clock from CML DIV2 equals to the half of the period of VCO, the sum of the four counters' output, f_{cont}, represents the double value of VCO frequency. So the M (Fig. 1(a)) here equals to 0.5 and the calibration resolution of AFC can be rewritten as

$$f_{resolution} = \frac{0.5 \times f_{REF}}{p} \quad (3)$$

Equation (3) indicates that with the same p, the frequency resolution of this AFC can be once higher than that in [3-5], or more than once higher than that in [2] and [6]. In other words, the same resolution can be kept by reducing the value of p and thus the counting period can be shortened to reduce the t_{AFC}. So the proposed AFC has shorter t_{AFC} than prior AFC techniques [2-6] with the comparable AFC calibration resolution.

Equation (1) also indicates that the processing phase in each step of the AFC process spends a large portion of the whole t_{AFC}, especially at low frequency reference clock. So it is necessary to shorten the time in such phase not only to

978-1-5386-6414-8/18 $31.00 © 2018 IEEE

TABLE I. VALUE OF P AT DIFFERENT REFERENCE FREQUENCIES

value of p	Reference frequency f_{REF} (MHz)
2	15~20
3	20~30
5	30~40
6	40~50

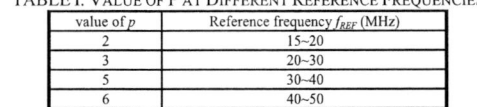

Fig. 5. (a) Calculated AFC resolution and (b) calculated AFC time.

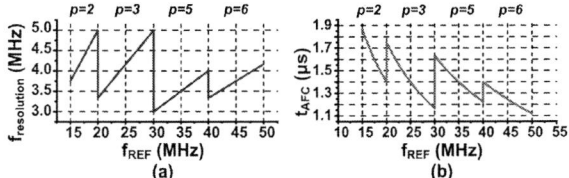

Fig. 6. Block diagram of the wideband PLL with proposed AFC technique.

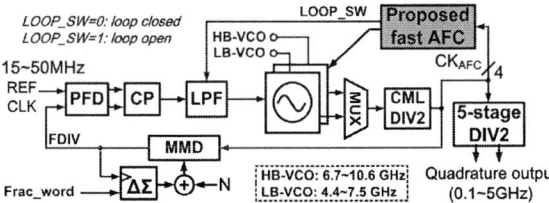

Fig. 7. Chip micrograph.

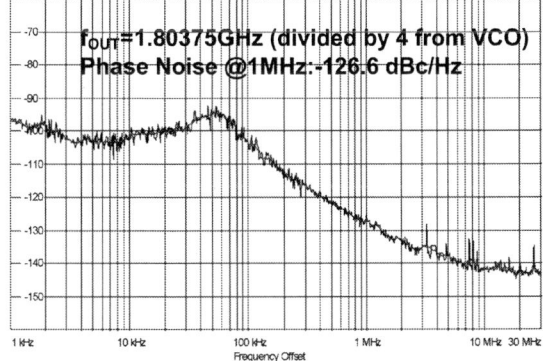

Fig. 8. Measured phase noise at 1.80375 GHz.

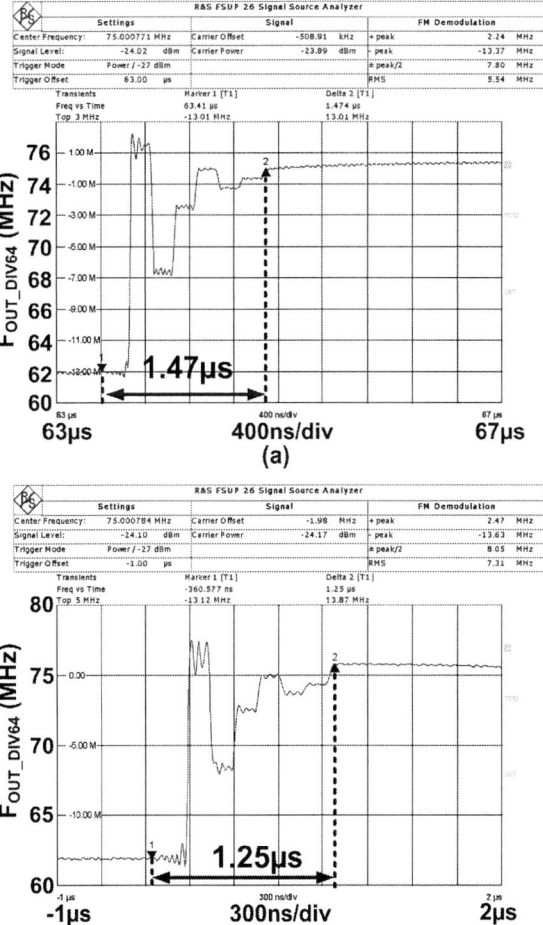

Fig. 9. AFC process at (a) 20 MHz and (b) 50 MHz reference frequency, respectively.

accelerate the AFC process, especially at low f_{REF}, but also to reduce the t_{AFC} variation over wide range of f_{REF}. In this AFC, we propose an AFC controller (Fig. 3) to double the frequency of AFC clock during the comparing and setting phases and thus reduce the t_{AFC}. The AFC controller mainly includes two XOR gates and one MUX. The frequency of AFC clock AFC_CLK can be doubled by selecting the control signal clkx2_en to be high. As the timing diagram of the proposed AFC shown in Fig. 4, the n can be reduced from 4 to 2. So with the help of the AFC clock controller, not only the AFC process can be further sped up, especially when using the low frequency reference

clock, but also the t_{AFC} variation at difference reference frequencies can be reduced.

In order to further reduce the variation of t_{AFC} over wide range of f_{REF}, p can be set to be programmable. Table I shows the value of p at different reference frequencies. The calculated AFC resolution and t_{AFC} is shown in Fig. 5(a) and (b), respectively. As can be seen, the t_{AFC} varies only from 1.12 to 1.87 μs, which indicates that the t_{AFC} doesn't vary much over covered range of f_{REF}. What's more, the resolution can be kept no more than 5 MHz.

IV. IMPLEMENTATION AND MEASUREMENT RESULTS

The wideband $\Delta\Sigma$ fractional-N PLL to demonstrate the proposed AFC technique was designed and implemented in 65 nm CMOS process, as shown in Fig. 6. The details of other building blocks except the proposed AFC are similar to our prior work [9]. The core area is 0.675 mm^2, as shown in Fig. 7. The measurement results show the PLL can generate frequency range from 0.1 to 5 GHz. The PLL can operate in wide reference range from 15 to 50 MHz. The power consumption

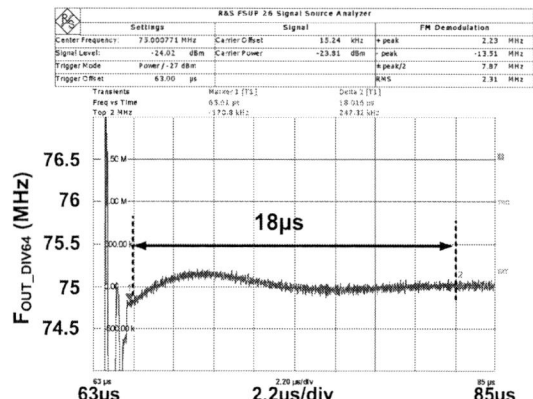

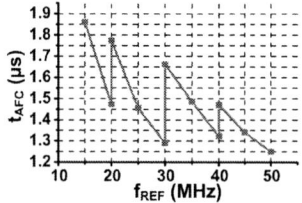

Fig. 10. Phase tracking process.

Fig. 11. AFC time at reference frequency range from 15MHz to 50MHz.

TABLE II. PERFORMANCE SUMMARY AND COMPARISON

	This work	[2]	[3]	[4]	[5]	[6]
Process	65nm	130nm	130nm	130nm	180nm	65nm
Freq.(GHz)	0.1~5	1.8~6	2.34~3.94	1.9~3.8	0.975~1.96	0.96~2.06
Power (mW)	14~20.5	35.6~52.62	19	15	25.2	NA
Supply (V)	1.2	1.2	1.2	1.2	1.8	1.2
Phase Noise (dBc/Hz @1MHz)	-126.6 (1.80375 GHz)	-115 (5.18 GHz)	-124.1 (1.492 GHz)	-117.57 (1.9 GHz)	-126 (1.339 GHz)	NA
Ref. Freq.(MHz)	15~50	40	19.2	40	25	40
DCCA bit	7	7	6	7	8	5
Total AFC time (µs)	1.25~1.86	6.4	2.03	1.925	6.4	4.03
AFC resolution (MHz)	3~5	2.5*	4.8	5	3.125**	NA
AFC time per bit (µs)	0.179~0.266	0.914	0.338	0.275	0.8	0.8
PLL Area (mm²)	0.675	1.86	NA	0.651	1.58	NA

*The AFC resolution: 4*f_{ref}/p, p=64

**The AFC resolution: f_{ref}/p, p=8, p is the periods number of the counting phase

proposed and demonstrated by a 0.1~5 GHz ΔΣ fractional-N PLL in 65 nm CMOS process. By using proposed technology, this AFC can achieve fast settling process over wide reference range. The measurement shows the total AFC time varies from 1.25 to 1.86 µs and the AFC time per bit is no more than 0.266 µs with frequency resolution from 3 to 5MHz. The reference frequency range is from 15 to 50 MHz.

ACKNOWLEDGMENT

This work is financially supported by Natural Science Foundations of China No. 61474108, 61331003 and 61306027, and National Science and Technology Major Projects of the Ministry Science and Technology of China Under Grant No. 2016ZX03001002-002.

REFERENCES

[1] J. Zhou et al., "A 0.4-6 GHz Frequency Synthesizer Using Dual-Mode VCO for Software-Defined Radio," *IEEE Trans. on Microw. Theory and Tech.*, vol. 61, No. 2, pp. 848-859, Feb. 2013.

[2] D. Huang et al., "A Frequency Synthesizer With Optimally Coupled QVCO and Harmonic-Rejection SSBmixer for Multi-Standard Wireless Receiver," *IEEE J. Solid-State Circuits*, vol. 46, no. 6, pp. 1307-1320, Apr. 2011.

[3] J. Shin and H. Shin, "A Fast and High-Precision VCO Frequency Calibration Technique for Wideband Fractional-N Frequency Synthesizers," *IEEE Trans. on Circuits and Syst. I: Regular Papers*, vol. 57, no. 7, pp. 1573-1582, July 2010

[4] J. Shin and H. Shin, "A 1.9–3.8 GHz Fractional-N PLL Frequency Synthesizer with Fast Auto-calibration of Loop Bandwidth and VCO Frequency," *IEEE J. Solid-State Circuits*, vol. 47, no. 3, pp. 665-675, Mar. 2012.

[5] L. Lu et al., "A 975-to-1960MHz Fast-Locking Fractional-N Synthesizer with Adaptive Bandwidth Control and 4/4.5 Prescaler for Digital TV Tuners," in *ISSCC Dig. Tech. Papers*, Feb. 2009, pp. 396-397.

[6] H. Ryu et al, "Fast Automatic Frequency Calibrator Using an Adaptive Frequency Search Algorithm," *IEEE Trans. on Very Large Scale Integ. (VLSI) Syst.*, vol. 25, No. 4, pp. 1490-1496, Apr. 2017.

[7] M. Ferriss et al., "A 13.1-to-28 GHz Fractional-N PLL in 32nm SOI CMOS with a ΔΣ Noise-Cancellation Scheme," in *ISSCC Dig. Tech. Papers*, Feb. 2016, pp. 192-193.

[8] S. Ye et al., "A Multiple-Crystal Interface PLL With VCO Realignment to Reduce Phase Noise," *IEEE J. Solid-State Circuits*, vol. 37, no. 12, pp. 1795-1803, Mar. 2002.

[9] Z. Zhang et al., "A 0.1-to-5 GHz Wideband ΔΣ Fractional-N Frequency Synthesizer for Software-Defined Radio Application", in *IEEE Int. Conf. on Solid-State and Integ. Circ. Tech. (ICSICT)*, Oct. 2016, pp. 1570-1572

of the PLL varies from 14 to 20.5 mW exclude the output buffer for testing purpose.

The measured phase noise at carrier frequency of 1.80375 GHz (divided by 4 from VCO output) is shown in Fig. 8 with f_{REF} of 20 MHz. The phase noise at 1 MHz offset frequency is -126.6 dBc/Hz.

The measured AFC time at f_{REF} of 20 MHz and 50 MHz is 1.47 µs and 1.25 µs, respectively, as shown in Fig. 9(a) and (b). Fig. 10 shows that the PLL phase tracking time after the AFC process is 18 µs. The output for AFC testing and the phase tracking testing is divided by 64 from VCO output due to the limited tracking bandwidth of the signal source analyzer. The measured AFC time within the range of f_{REF} from 15 to 50 MHz is presented in Fig. 11. It indicates that the AFC time varies from 1.25 to 1.86 µs. The variation of the AFC time is 0.62 µs, which is a negligible value compared with the whole PLL locking process that includes the AFC process and the phase tracking process. The target frequency of the VCO output in this measurement is 4.8 GHz.

Table II presents the performance summary of the proposed AFC and the ΔΣ fractional-N PLL. Compared with the prior fast AFC techniques, the proposed AFC technique has the shortest AFC time and the shortest AFC time per bit, even at the lowest reference frequency (15 MHz). The AFC resolution of this work is comparable with other works'. In addition, this work has the widest suitable reference frequency range.

V. CONCLUSION

A fast auto frequency calibration technique suitable for wideband PLL with wide reference frequency range was

17-5 (8020)

IEEE Asian Solid-State Circuits Conference
November 5 - 7, 2018/Tainan, Taiwan

A Sub-Picosecond Hybrid DLL for Large-Scale Phased Array Synchronization

Matan Gal-Katziri and Ali Hajimiri
Department of Electrical Engineering
California Institute of Technology
Pasadena, CA 91125, USA
Email: mgal@caltech.edu

Abstract—**A large-scale timing synchronization scheme for scalable phased arrays is presented. This approach utilizes a DLL co-designed with a subsequent 2.5GHz PLL. The DLL employs a low noise, fine/coarse delay tuning to reduce the in-band rms jitter to 323fs, an order of magnitude improvement over previous works at similar frequencies. The DLL was fabricated in a 65nm bulk CMOS process and was characterized from 27MHz to 270MHz. It consumes up to 3.3mW from a 1V power supply and has a small footprint of 0.036mm².**

Keywords—CMOS integrated circuits, phased-arrays, radio frequency, tracking loops, delay-lines, phase locked loops, phase noise.

I. INTRODUCTION

Phased arrays are extensively used in radar, sensing, and communication systems due to their electronic beam steering capabilities combined with the added directivity and enhanced SNR/SIR which scale with number of array elements [1]. Example applications are 5G networks, which are currently targeting hundreds to thousands of elements, and very large-scale arrays, sometimes referred to as million-element arrays [2]. The practical implementation of such systems necessitates a broad range of architectural and technological innovations, such as scalable structures and highly-integrated silicon-based RFICs [2-5]. In this scalable array transceiver architecture, a single low-frequency reference clock is distributed to identical blocks (tiles), where high-frequency signals are synthesized (often using an integrated PLL-based frequency synthesizer) and used for coherent RF signal generation and reception in concert with the other array elements, as shown in Fig. 1.

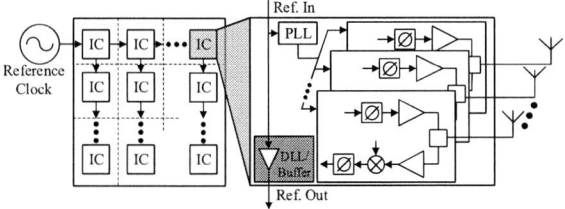

Fig. 1. Clock distribution to CMOS driven phased array

One major challenge with this architecture is maintaining the timing accuracy of the reference signal in the distribution process. A central star or H-tree distribution is impractical in the case of a large scale array as the number of traces and the

This work was sponsored by Caltech's Space Solar Power Project (SSPP).

electrical load of all the driven elements become prohibitively large. On the other hand, sequential buffering of the reference suffers from large accumulated timing deviations due to variations in the supply, temperature, and the driven load. These challenges can be mitigated utilizing a delay-locked loop (DLL) in the repeater buffer. While fundamentally sound, this approach presents new challenges since the low reference frequency, usually a few tens of MHz, necessitates a relatively large delay which can lead to unacceptable timing jitter. We propose a hybrid DLL architecture that utilizes several noise reduction techniques as well as a novel semi-digital loop control scheme with a single phase detection path. Moreover, the co-design of the DLL with the subsequent PLL-based synthesizer is exploited to further reduce the overall timing jitter by proper alignment of the two phase noise transfer functions; one loop provides rejection over the frequencies where the other has a large noise contribution. This approach opens the design space, leading to superior overall performance.

II. HYBRID DLL

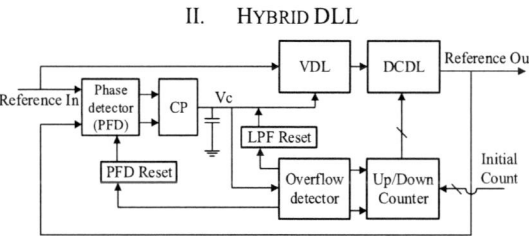

Fig. 2. Hybrid DLL block diagram

In a DLL, the output signal must practically be delayed by at least half a clock period compared to the reference in order to correct both negative and positive timing errors. A standard implementation does so with a single continuous delay line, which is usually the main noise contributor due to the large delay range it needs to cover. A hybrid DLL can solve this problem by using two different sets of delay elements, as shown in Fig. 2. A digitally controlled delay line (DCDL) composed of low-noise fixed-delay elements is used for coarse delay tuning, while a short, continuously variable delay line (VDL) is used to fine tune within the digital segments.

In order to achieve delay lock, we use an analog DLL architecture and continuously monitor its charge pump (CP) output control voltage (Vc) to adjust the required DCDL value.

978-1-5386-6414-8/18 $31.00 © 2018 IEEE   231

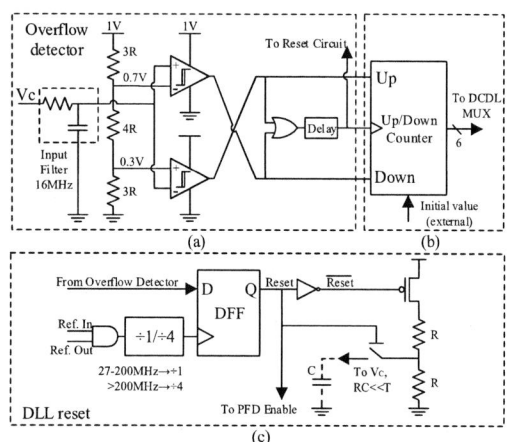

(a)　　　　　　　　　　(b)

(c)

Fig. 3. (a) Overflow detect/actuate circuit (b) DCDL MUX set (c) reset circuit for hybrid operation

Initially, the up/down counter of Fig. 3b is set to fix the DCDL state, and the DLL loop of Fig. 2 continuously controls the VDL. If an unattainable VDL tuning value is required, the control voltage Vc will rail, crossing some lower or upper threshold along the way. This activates the overflow detector of Fig. 3a to pause the continuous control loop, initiate a single increase/decrease of a DCDL cell, and restart VDL tracking. Unlike [6], we are not changing the continuous delay range by flipping a state machine to set discrete phase states but are instead adding or removing fixed amount of low noise delay as required. This significantly improves the noise performance. In addition, we are tracking the same edge in a monotonous, continuous, and overlapping manner—which, when combined with the fact that the DLL is a first-order control loop, guarantees its stability. The reset circuitry in Fig. 3c is crucial to temporarily disable the phase detector and force Vc to mid-supply when a DCDL shift occurs and is synchronized such that the phase detector starts at a consistent state once the VDL tracking restarts. The noise-optimized, pseudo-differential delay elements of Fig. 4 also allow tracking of the falling edge of the output clock, which effectively reduces the minimum delay required by *T/2* and enables usage of the same delay line at lower reference frequencies.

This architecture offers enhanced robustness because (1) it necessitates neither lock detect indication nor dual phase detection circuitry as in [7][8], (2) the small signal gain is identical for all DCDL values, and (3) the DCDL state changes in single up/down steps. The latter indicates that subsequent VDL tracking starts from a well-defined, nearby position, unlike a digital controller with automatic delay step adjustment. Our implementation favors clock distribution applications where lock time is not a major consideration. If necessary, fast lock is achievable with an *a priori* estimate of the DCDL delay step values and external programming of the up/down counter state.

III. PLL DESIGN

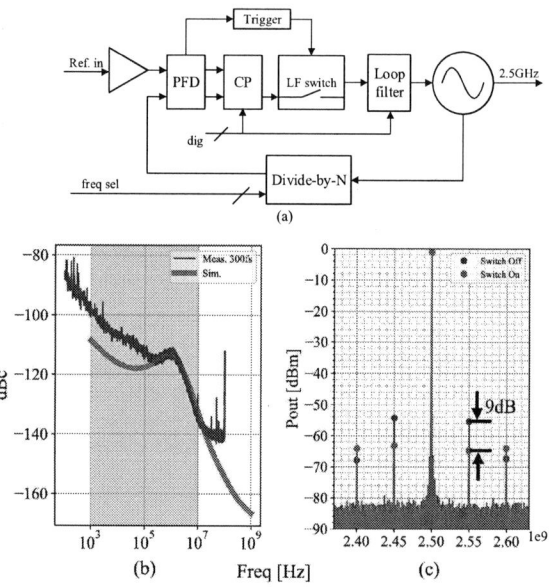

Fig. 5. (a) Application PLL (b) PLL phase noise and rms jitter at 1kHz-10MHz (c) reference spurs with and without reduction

The hybrid DLL was co-designed with its intended load PLL (Fig. 5) for a 50MHz clock distribution of an existing RF phased array application. The PLL itself is fully integrated and operates at an output frequency of 2.5GHz with a loop bandwidth of 1MHz. It contains a mechanism similar to [9] to reduce its reference spurs, which, when present at the output of a large-scale transmitter array, might become a significant spectral disturbance. In order to minimize the DLL in-band noise, its loop filter bandwidth was optimized to be around 1MHz in order to sufficiently reject the delay line noise while maintaining a relatively flat noise shape around the PLL loop filter knee frequency.

IV. MEASUREMENT RESULTS

Both the DLL and the PLL were fabricated in a 65nm bulk CMOS process (Fig. 6). They occupy 0.036mm² and 0.4mm² of active area, respectively, and their joint operation was characterized at an output frequency of 2.5GHZ with the input reference ranging from 27MHz-270MHz.

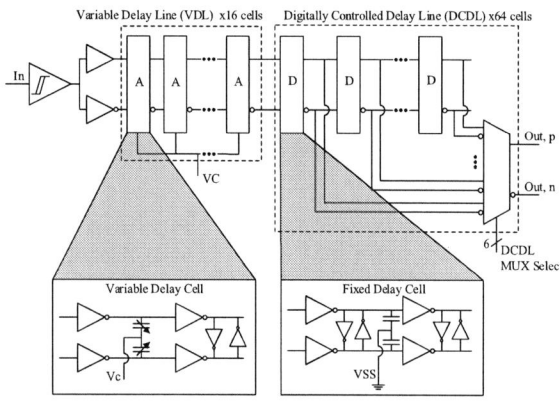

Fig. 4. Hybrid delay line structure

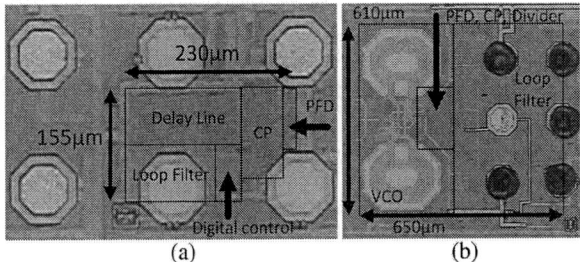

Fig. 6. Die micrographs of the (a) DLL and (b) the driven PLL

Fig. 7 shows the delay locking mechanism while the DLL drives either 50Ω or 10pF loads. The control voltage Vc in Fig. 7a overflows and resets until it reaches the necessary DCDL value, while fine-tuning persists indefinitely. The delay between the output and reference signals (Fig. 7b) was calculated from the waveforms' zero-crossing points, emphasizing how proper sizing of the overlapping DCDL step size and VDL range allow for proper operation of the circuit.

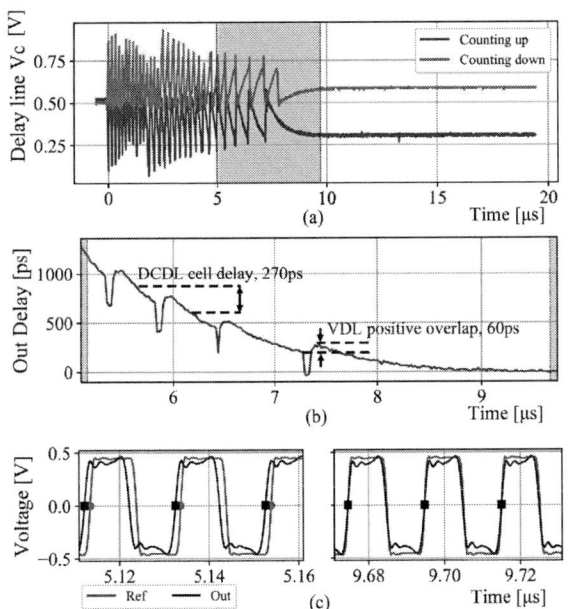

Fig. 7. Hybrid lock process at different time scales. (a) Loop filter control voltage, (b) time delay between reference and output clocks, and (c) time domain waveforms (adjusted)

In our phased-array application, the expected temperature fluctuation is less than 10°C in steady state, and the measured closed-loop control voltage tracks the temperature at a rate of 2.4mV/°C. The nominal control voltages for locking are 340mV and 660mV when counting up and down, respectively, and the overflow detector has a nominal hysteresis of 30mV. Therefore, temperature variations are not expected to toggle the digital counter and add additional, unaccounted noise. In our clock distribution scheme, static buffer phase offset is programmatically removed when the array is calibrated and therefore not of a major concern.

Fig. 8 shows how the DLL degrades the noise performance of a reference clock source by examining the phase noise spectral density profile of the cascaded application blocks, measured using a Keysight PXA N9030B signal analyzer. Notably, the frequency band of interest is above 1kHz, where phase errors are presumably correctable by external phased array adjustment algorithms, and below 10MHz, far away from the load PLL loop filter knee frequency.

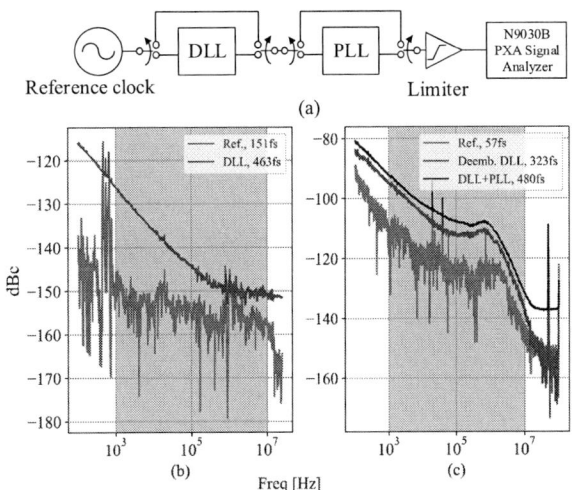

Fig. 8. (a) Phase noise test setup. Blocks are taken off when not measured. (b) 50MHz DLL phase noise and rms jitter (c) 2.5GHz post PLL phase noise and rms jitter. The red curves are the rms measurement uncertainty

Figs. 8b and 8c clearly show how the PLL loop filer rejects most of the DLL noise and thus brings it to contribute as little as 323fs rms jitter in the relevant frequency band.

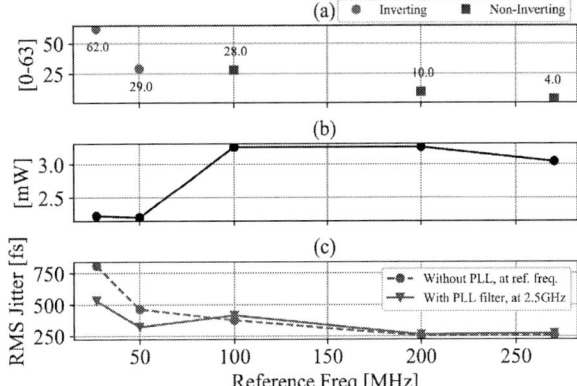

Fig. 9. Performance vs. frequency. (a) Participating DCDL cell count, (b) power consumption of participating DLL blocks, and (c) rms jitter within the 1kHz - 10MHz band

These measurements were repeated at different frequencies between 27MHz and 270MHz, and a summary is shown in Fig. 9. The lower and upper frequency ranges are limited by the maximum DCDL delay and overflow actuation timing accuracy, respectively. Fig. 9b demonstrates how this DLL is advantageous in that an increase in the frequency of operation decreases the number of DCDL elements that participate in the delay chain, and thus the power consumption remains roughly constant. Fig 9c emphasizes how the system is optimized for 50MHz operation. At lower frequencies, the high DCDL count adds more noise to the output, while at higher frequencies the subsequent PLL loop filter has little effect on noise rejection.

Because the end goal is the phased array reference distribution scheme of Fig. 1, noise performance was characterized for several, cascaded DLLs. If the noise of each stage is uncorrelated with the others, the total noise measured at the output of an N DLL cascade is expected to be:

$$n^2_{total} = n^2_{ref} + n^2_{meas} + N \cdot n^2_{DLL} \qquad (1)$$

where n_{meas} is the measuring instrument noise, n_{ref} is the reference noise, and the single device noise can be estimated from the slope of the linear fit. Fig. 10 shows the linear behaviour of the DLL cascade jitter variance at different frequencies and the resulting rms jitter is summarized in Table I, showing good agreement with the single device measurements of Figs. 8b and 9c.

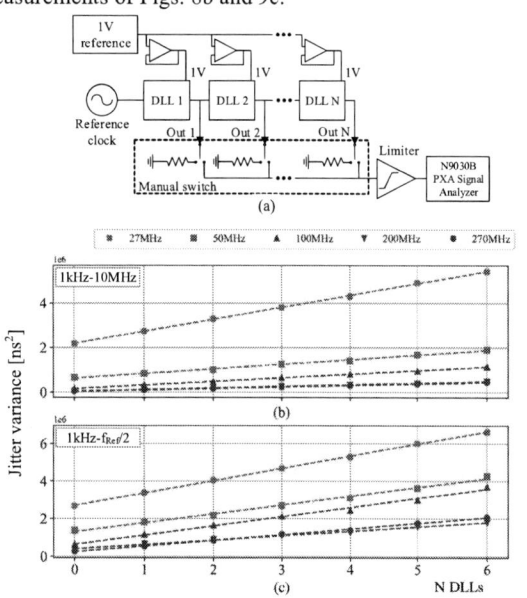

Fig. 10. Cascaded DLL jitter (a) test setup, (b) 1kHz - 10MHz measurement, and (c) 1kHz-$f_{Ref}/2$ measurement. Red and blue curves indicate locking to inverted/non-inverted output, respectively.

DLL NOISE PERFORMANCE, BASED ON FIG. 10

Reference Frequency [MHz]	27	50	100	200	270
RMS Jitter, 1kHz-10MHz [fs]	733	456	402	268	261
RMS Jitter, 1kHz-$f_{Ref}/2$ [fs]	809	685	698	481	549

V. CONCLUSIONS

The task of distributing a low noise reference to very large-scale phased arrays is challenging because it does not enjoy the shorter period times of GHz range clocks. Table II shows a performance comparison of the hybrid DLL/PLL scheme with prior art at similar frequency ranges, and demonstrates how combining new circuit architectures with application-aware design can result in an order-of-magnitude improvement over the state-of-the-art.

REFERENCES

[1] W. L. Stutzman and G. A. Thiele, Antenna Theory and Design, 2nd ed. New York: Wiley, 1998.

[2] S. Jeon et al., "A Scalable 6-to-18 GHz Concurrent Dual-Band Quad-Beam Phased-Array Receiver in CMOS," in IEEE JSSC, vol. 43, no. 12, pp. 2660-2673, Dec. 2008.

[3] Xiang Guan, H. Hashemi and A. Hajimiri, "A fully integrated 24-GHz eight-element phased-array receiver in silicon," in IEEE JSSC, vol. 39, no. 12, pp. 2311-2320, Dec. 2004.

[4] A. Natarajan, A. Komijani and A. Hajimiri, "A fully integrated 24-GHz phased-array transmitter in CMOS," in IEEE JSSC, vol. 40, no. 12, pp. 2502-2514, Dec. 2005.

[5] F. Bohn, B. Abiri and A. Hajimiri, "Fully integrated CMOS X-Band power amplifier quad with current reuse and dynamic digital feedback (DDF) capabilities," 2017 IEEE RFIC 2017, pp. 208-211.

[6] Byung-Guk Kim and Lee-Sup Kim, "A 250-MHz-2-GHz wide-range delay-locked loop," in IEEE JSSC, vol. 40, no. 6, pp. 1310-1321, June 2005.

[7] Yeon-Jae Jung et al., "A dual-loop delay-locked loop using multiple voltage-controlled delay lines," in IEEE JSSC, vol. 36, no. 5, pp. 784-791, May 2001.

[8] K. H. Cheng and Y. L. Lo, "A Fast-Lock Wide-Range Delay-Locked Loop Using Frequency-Range Selector for Multiphase Clock Generator" in IEEE TCAS-II, vol. 54, no. 7, pp. 561-565, July 2007.

[9] B. Zhang, P. E. Allen and J. M. Huard, "A fast switching PLL frequency synthesizer with an on-chip passive discrete-time loop filter in 0.25-μm CMOS," in IEEE JSSC, vol. 38, no. 6, pp. 855-865, June 2003.

[10] C. N. Chuang and S. I. Liu, "A 20-MHz to 3-GHz Wide-Range Multiphase Delay-Locked Loop," in IEEE TCAS-II, vol. 56, no. 11, pp. 850-854, Nov. 2009.

[11] D. Zhang et al., "A fast-locking digital DLL with a high resolution time-to-digital converter," Proceedings of the IEEE 2013 CICC, San Jose, CA, 2013, pp. 1-4.

TABLE II
PERFORMANCE SUMMARY AND COMPARISON TO PRIOR WORK

	This work		JSSC 05' [6]	TCASII 07' [8]	TCASII 09' [10]	CICC 13' [11]
Frequency range	27-270MHz		0.25-2GHz	32-320MHz	0.02-3GHz	80-450MHz
Comparison frequency	50MHz	270MHz	250MHz	200MHz	50MHz	180MHz
RMS jitter [ps]	0.685	0.55	5.25	4.44	7 (approx.)	2.3
In band RMS jitter	0.33ps	0.26ps	NA	NA	NA	NA
Power consumption [mW]	2.25	3	1.2	15 (320MHz)	0.4-3.6	26
Supply voltage [V]	1		1.8	2.5	1	1.5
Technology process [CMOS]	65nm		180nm	250nm	90nm	130nm
Die area [mm²]	0.036		0.046	0.07	0.005	0.08

A 7b 2 GS/s Time-Interleaved SAR ADC with Time Skew Calibration Based on Current Integrating Sampler

Wenning Jiang, Yan Zhu, Chi-Hang Chan, Boris Murmann[2], Seng-Pan U[3,4], Rui Paulo Martins[3,5]

State Key Laboratory of Analog and Mixed-Signal VLSI, University of Macau, Macao, China

2 – Stanford University, USA

3 – Also with Department of ECE, Faculty of Science of Technology, University of Macau, Macao, China

4 – Also with Synopsys Macau Ltd.

5 – On leave from Instituto Superior Técnico/Universidade de Lisboa, Portugal

email: ivorchan@ieee.org

Abstract - This paper presents a time-interleaved (TI) SAR ADC that utilizes the characteristic of the current integrating (CI) sampler for sampling time skew background calibration, while it also provides buffering and anti-aliasing filtering functions, simultaneously. The inter-sample interaction in the CI sampler enables the mapping of the time domain information to the amplitude domain. Time skew errors can therefore be extracted by comparing the output-code variance among channels without requiring a reference path. A 2-channel 2 GS/s 7b TI-SAR prototype realized in 28-nm CMOS achieves a 36.4 dB SNDR at Nyquist with >2.6 GHz ERBW after calibration. The ADC with CI sampler consumes 7.62 mW, leading to a Walden FoM of 70.8 fJ/conversion-step.

Keywords— Time-interleaved ADC, current integrating sampler, background calibration, timing skew.

I. INTRODUCTION

The time-interleaved (TI) topologies have been widely explored in high-speed, moderate-resolution ADCs [1]-[9]. However, nonidealities from the TI sampling front-end, e.g., due to kick-back and sampling time skew, are often limiting the performance of such converters. A buffer can be placed before the sample-and-hold (S/H) circuit to reduce its kick-back [1][2][8], but the added power is a costly investment, given that the buffer acts merely as a unity-gain element. The issue of sampling time skew is a similarly costly issue, since its detection is significantly more difficult than measuring gain and offset mismatches among TI channels [1]. While it is possible to infer the time skew using a reference-free autocorrelation approach [3][4], this solution is limited to scenarios where the input closely resembles wide-sense stationarity. Alternatives that alleviate this issue use a window detector [5] or a reference channel [6]. However, the added circuitry is prone to injecting asynchronous input disturbances [4].

In this paper, we adopt a current integrating sampler [10] to address the above-described challenges. In addition to input buffering, the CI sampler provides anti-alias filtering and simultaneously enables reference-free time skew detection. Unlike other self-referenced schemes [3][4], the proposed calibration approach works under various input conditions and

does not require wide-sense stationarity. Our prototype design with CI sampler front-end achieves ERBW > 2.6 GHz and exhibits 36.4 dB SNDR at Nyquist input, consuming only 7.62 mW total power with the CI sampler power included. The Walden FoM is 70.8 fJ/conversion-step based on the SNDR at Nyquist.

II. TI-ADC ARCHITECTURE AND CI SAMPLER OVERVIEW

A. TI-ADC Architecture

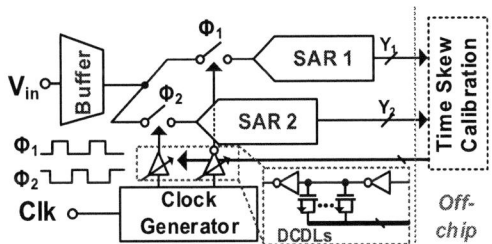

Fig. 1. Block diagram of a two-way TI-SAR ADC.

Fig. 1 depicts the two-way TI-SAR architecture used in our design. It consists of the CI-based input buffer, clock generator, two SAR ADCs that were re-used from [9], digitally controlled delay lines (DCDLs) and a time skew calibration block. The time skew information is extracted off-chip with the proposed calibration algorithm based on the ADC output codes. The input buffer alleviates the kick-back effect from the switched-capacitor sampler and provides a time-invariant input impedance. Conventionally, the buffer implemented with a source follower (SF) [1][8] tends to be power hungry given the stringent bandwidth requirement dictated by short sampling windows. In this work, we adopt the CI-based buffer to relax such power tradeoff [10], and simultaneously benefit from its inherent anti-alias filtering and time-to-voltage conversion for skew detection.

B. Overview of CI Sampler and its Characteristics

Fig. 2 illustrates the block diagram of the CI sampler with its corresponding timing strategy. First, let's consider the model without the capacitor (C_I). The CI sampler circuit consists of a Gm cell followed by the integration-and-hold path which comprises two switches with non-overlapping phases (Φ_S and Φ_r). Different from the voltage sampler, the sampling voltage

This work was financially supported by the NSFC Grant 61604180 and the Macao Science & Technology Development Fund (FDCT) with Ref no: 117/2016/A3

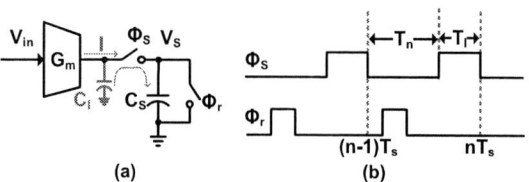

Fig. 2. (a) CI sampler model. (b) Timing strategy.

(V_S) must be fully reset before the next integration cycle to eliminate inter-symbol-interference (ISI). In the CI sampler structure, V_S not only depends on the input signal (V_{in}) and the transconductance (G_m) but also the integration time (T_I) and sampling capacitor (C_S). The sampled voltage can be expressed as:

$$V_S(n) = \frac{G_m}{C_S} \int_{nT_S - T_I}^{nT_S} V_{in}(t)\, dt, \quad (1)$$

where T_S is the sampling clock period. The impulse response of the integration process in (1) is a periodic rectangular window function whose window width and height are T_I and G_m/C_S [7], respectively. Accordingly, the frequency response is:

$$\left| H(f) \right| = \frac{G_m \cdot T_I}{C_S} \left| \frac{\sin(\pi \cdot T_I \cdot f)}{\pi \cdot T_I \cdot f} \right|. \quad (2)$$

When considering C_I in the sampling process, unlike the C_S, the C_I is still charged by the Gm cell during $\Phi_S = 0$, time slot that we call idle time (T_n) as shown in Fig. 2(b). The charge stored at C_I during the idle time will be shared to C_S at the next sample, thus leading (1) to:

$$V_S(n) = \frac{G_m}{C_I + C_S} \int_{nT_S - T_I}^{nT_S} V_{in}(t)\, dt + \frac{G_m}{C_I + C_S} \int_{(n-1)T_S}^{(n-1)T_S + T_n} V_{in}(t)\, dt + \frac{C_I}{C_I + C_S} V_S(n-1). \quad (3)$$

The first order FIR filter of the CI sampler provides an anti-aliasing function and cascades with an IIR filter (the $1^{st} + 2^{nd}$ term in (3)) introduced by the charge-sharing process between C_I and C_S [7]. The 3rd term in (3) is the unwanted ISI while its effect is not significant and can be compensated with little power overhead in the Gm cell of our design. It can be recognized from (3) that V_S is not only dependent on T_I but also T_n. Such characteristic enables the proposed time skew calibration (detailed next). Assuming r_o is the output resistance of the G_m cell, the transfer function of the CI sampler becomes:

$$H(s) = \frac{G_m}{(C_I + C_S)} \frac{1 - e^{-(s+1/\tau)T_I}}{(s + 1/\tau)}, \quad (4)$$

where $\tau = r_o(C_S + C_I)$ is the time constant. The depth of the nulls determines the attenuation ability of the FIR filter:

$$\frac{\left| H(m/T_I) \right|}{\left| H(0) \right|} \approx \frac{1}{m\pi} \left(1 + \frac{T_I}{2\tau} \right) \frac{T_I}{2\tau}, \quad (5)$$

where m is an integer that denotes the location of the mth null. In this work, C_I is about 60 fF due to the parasitics of the G_m cell and routing, leading to an extra ~15 dB attenuation at the 1st null.

III. TIMNG SKEW CALIBRATION

A. Time Skew in TI-CI Sampler

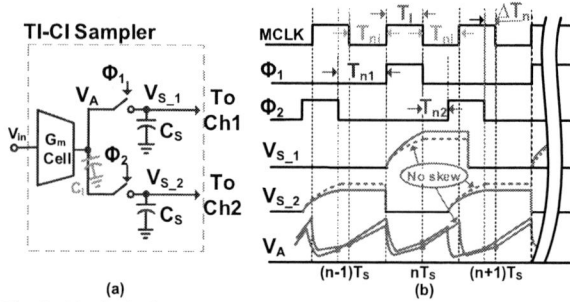

Fig. 3. (a) Block diagram of the TI-CI sampler. (b) Time diagram of the TI-CI sampler.

Fig. 3(a) displays the block diagram of the TI-CI sampler front-end, which consists of a G_m cell and two integration-and-hold channels followed by two sub-ADCs. V_{S_1} and V_{S_2} are the TI-CI sampler voltages from channel 1 and 2, respectively. Fig. 3(b) shows its timing diagram and the time skew detection concept. The master-clock (MCLK) is the sampling clock and T_S is its clock period. T_{n1} and T_{n2} are the idle times of the two channels in which none of them are in the integration mode and they are identical without time skew error. Due to the characteristic of the CI sampler as denoted in (3), the integrated charge on C_I (Q_I) also depends on the duration of the idle time. When the interleaved sampling clock suffers from time skew, the idle time of each channel deviates from the ideal value (T_{ni}) that affects Q_I. Furthermore, such charge eventually passes to C_S (from C_I) at the next sample, therefore altering the sampling voltage and mapping the time skew into the amplitude domain.

As exemplified in Fig. 3(b), channel 2 suffers from time skew with a leading time ΔT_n that reduces the idle time of channel 2 (T_{n2}) from T_{ni} to $T_{n2} = T_{ni} - \Delta T_n$, and the idle time of channel 1 increases to $T_{n1} = T_{ni} + \Delta T_n$. Fig. 3 (b) shows the voltage on C_I (V_A) with different idle times. V_A tends to become larger with longer T_{n1} and smaller with shorter T_{n2}. Then, the error of V_A resulting from the time skew transfers to (charge sharing between C_I and C_S) V_{S_1} and V_{S_2} during Φ_1 and Φ_2, respectively. The sample of channel 1 and 2 $(V_{S\_1}$ and $V_{S\_2})$ with time skew can be expressed in a form similar to (3). It is worth observing that the time skew only affects the second term in (3), and the sampling voltage of the two channels can be rewritten as (6) and (7), respectively.

$$V_{S\_1}(n) = \frac{G_m}{C_I + C_S} \left(\int_{(n-1)T_S}^{nT_S} V_{in}(t)\, dt + \int_{(n-1)T_S - \Delta T_n}^{(n-1)T_S} V_{in}(t)\, dt \right) + \frac{C_I}{C_I + C_S} V_{S\_2}(n-1) \quad , (6)$$

$$\approx \frac{G_m}{C_I + C_S} \int_{(n-1)T_S}^{nT_S} V_{in}(t)\, dt + \frac{G_m \cdot V_{in}((n-1)T_S)}{C_I + C_S} \Delta T_n + \frac{C_I}{C_I + C_S} V_{S\_2}(n-1)$$

$$V_{S\_2}(n) = \frac{G_m}{C_I + C_S} \left(\int_{nT_S}^{(n+1)T_S} V_{in}(t)\, dt - \int_{(n+1)T_S - \Delta T_n}^{(n+1)T_S} V_{in}(t)\, dt \right) + \frac{C_I}{C_I + C_S} V_{S\_1}(n) \quad , (7)$$

$$\approx \frac{G_m}{C_I + C_S} \int_{nT_S}^{(n+1)T_S} V_{in}(t)\, dt - \frac{G_m \cdot V_{in}((n+1)T_S)}{C_I + C_S} \Delta T_n + \frac{C_I}{C_I + C_S} V_{S\_1}(n)$$

The second terms of (6) and (7) are the sampling voltage errors with time skew. From (6) and (7), it can be seen that $|V_{S\_1}|$ and $|V_{S\_2}|$ are proportional (with reversed polarity in channel 1 and 2) to the amount of time skew ΔT_n. As depicted in the example in Fig. 3(b), with a smaller T_{n2} and larger T_{n1}, $|V_{s\_1}|$ increases and $|V_{s\_2}|$ decreases. Consequently, the skew information is inherently transferred into the amplitude domain.

B. Proposed Time Skew Calibration Algorithm

From (6) and (7), the CI sampler front-end allows the time skew to be observed in the amplitude domain. On the other hand, a simple amplitude detection scheme at the ADC outputs (Y_1 and Y_2 in Fig. 1), such as max/min function, fails to provide sufficient accuracy as it can be easily affected by both quantization and thermal noises. To mitigate such impacts, we use the sample variances of Y_1 and Y_2 as skew estimators. These quantities are robust to ADC noise due to the inherent averaging in computing the sample statistics. Correspondingly, the convergence condition of the proposed calibration becomes,

$$Var(Y_1) - Var(Y_2) \approx \frac{8\sigma_0^2 \left[C_I / (C_I + C_S) \right]}{1 + \left[C_I / (C_I + C_S) \right]^2} \cdot \frac{\Delta T_n}{T_S} \quad , \qquad (8)$$

where σ_0^2 is the digitized output variance without time skew. Then, it can be affirmed that the variance difference among the channels is proportional to the time skew. We calculate the variance between channels 1 and 2 from the sub-ADCs' outputs with the digitally controlled delay lines managed to minimize the difference among the variances.

As the proposed scheme extracts time mismatch from the amplitude domain, it exhibits certain different characteristics relative to its time-domain counterparts [3]-[6]. First, it does not require an extra reference channel or window detector information that minimizes the analog circuit modification overhead. Second, the proposed calibration ability does not rely on specific input properties while correlation-based schemes [3][4] need to approximate a wide-sense stationary input. Nevertheless, the input characteristic affects the convergence time of our calibration. Unlike the time-domain approach where the largest voltage error occurs at the zero-crossing region with the steepest slope, the largest error of the proposed scheme happens when the largest amount of charge accumulates on C_S during ΔT_n. Therefore, the detection of ΔT_n is more sensitive with a low input frequency where the integration area under ΔT_n interval is relatively large, requiring less samples to converge for the same accuracy. Fig. 4 plots the behavioral simulated result of the convergence at different input frequencies with similar quantization and thermal noise in this design. At high input frequency, the same calibration accuracy requires the double number of samples.

IV. Circuit Implementation

Fig. 5 depicts the schematic of the CI sampler front-end circuit. The main design considerations of the G_m cell are the transconductance (G_m), parasitic capacitance (C_I), linearity and noise. We design G_m to match the desired gain in conjunction with T_I as well as C_S. The r_o and parasitic capacitance must be designed cautiously since they lead to a deterioration in the bandwidth and gain of the CI sampler. To minimize these

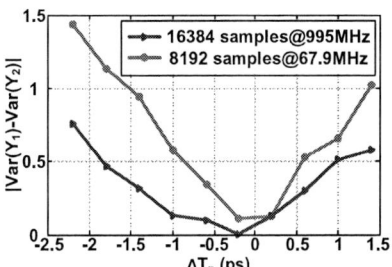

Fig. 4. Simulated convergence of variance versus time skew.

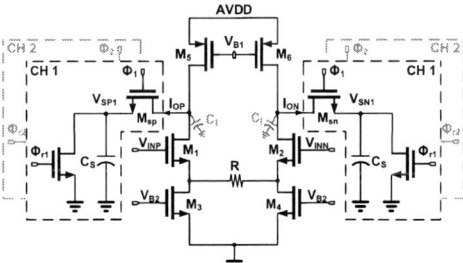

Fig. 5. CI sampler front-end circuit.

impacts we adopt a single-stage class-A architecture to implement the G_m cell, whose circuit structure determines the linearity of the CI sampler front-end; therefore, we add a source degeneration resistor (R) between the sources of M_1-M_2. In this work, C_S is 64 fF with a full swing of ~ 0.8 V peak-to-peak.

V. Measurement Results

Fig. 6 exhibits the chip micrograph of the 7-bit TI-ADC prototype fabricated in 28 nm CMOS. Its active area is 80 μm × 103 μm which includes two ADC channels with background offset calibration [9], a G_m cell buffer and a clock generator with tunable delay.

Fig. 7 shows the measured variance difference among channels (16384 samples/channel) at different input rates against the DCDL control code. When the variance difference approaches zero, it minimizes the time skew. Fig. 8 displays the measured frequency response of the CI sampler. The first notch of the CI sampler at 5.5 GHz reaches -20 dB to reject aliasing and the achieved bandwidth is around 2.25 GHz. With a T_I of 180 ps in this design, the location of the first null should be ideally at 5.55 GHz, but experiences a slight deviation due to process and mismatch effects. Fig. 9 plots the measured SNDR/SFDR versus input frequency and we obtain an ERBW>2.6 GHz. Fig. 10 depicts the ADC output spectrum at Nyquist before and after the proposed calibration. The spur from time skew reduces by more than 20 dB, and the SNDR improvement is around 8.5 dB. The maximum |DNL| and |INL| are < 1 LSB and 1.5 LSB, respectively. The power of the ADC core and clock generation is 5.58 mW, with 0.9 V power supply. The G_m cell consumes 2.04 mW from a 1.2 V supply. Table I summarizes and compares the proposed design with state-of-the-art ADCs that have an input buffer and similar

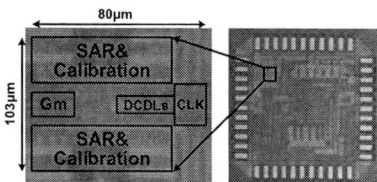

Fig. 6. Chip microphotograph.

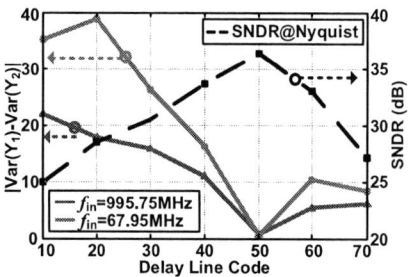

Fig. 7. Measured SNDR and variance difference versus skew calibration code.

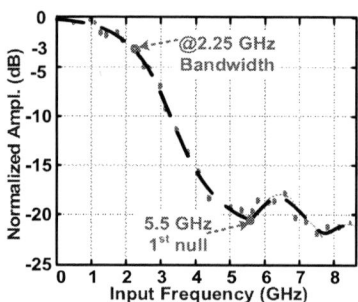

Fig. 8. Measured CI sampler frequency response.

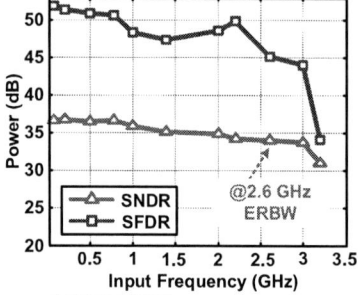

Fig. 9. Measured SFDR/SNDR versus input frequency.

specifications. Our design reaches the best energy efficiency with additional anti-aliasing ability.

VI. CONCLUSIONS

A 7b 2 GS/s 2-way TI-SAR ADC with CI sampler front-end and background time skew calibration has been presented. The CI sampler acts as an input buffer and simultaneously enables anti-aliasing and time skew calibration ability. The proposed calibration does not require an auxiliary channel and is

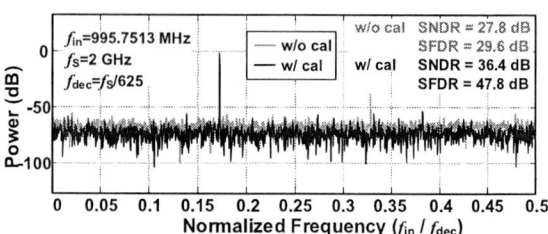

Fig. 10. Measured FFT before and after gain, offset and time skew calibration (8192 samples after decimation by 625).

TABLE I. COMPARISON WITH STATE-OF-THE-ART DESIGNS

	CICC 2014[1]	JSSCC 2017[2]	JSSCC 2012[8]	This work
Architecture	TI-SAR	TI-SAR	TI-Subranging	TI-SAR
Technology (nm)	40	40	65	28
f_s (GS/s)	2.64	2	2.2	2
Resolution	8	8	7	7
Supply voltage (V)	1.2	2.5/1.1	1	1.2/0.9
SNDR (dB) @Nyquist	38	39.4	38	36.4
Power (mW)	39	48.5	38	7.62
FoM (fJ/conv.step)	230	318	269.9	70.8
Area (mm^2)	0.18	0.54	0.3	0.0082
Anti-aliasing filter	No	No	No	Yes

insensitive to the input signal statistics. The ADC achieves Walden FoM of 70.8 fJ/conv-step with input buffer at Nyquist, which is superior to state-of-the-art designs.

REFERENCES

[1] S. Kundu, et al., "A 1.2 V 2.64 GS/s 8bit 39 mW skew-tolerant time-interleaved SAR ADC in 40 nm digital LP CMOS for 60 GHz WLAN," in Proc. *IEEE CICC*, pp. 1-4, Sep. 2014.

[2] T. Miki, T. Ozeki, and J. Naka, "A 2-GS/s 8-bit Time-Interleaved SAR ADC for Millimeter-Wave Pulsed Radar Baseband SoC," in *IEEE JSSC*, vol. 52, no. 10, pp. 2712-2720, Oct. 2017.

[3] A. Haftbaradaran, and K. W. Martin, "A Sample-Time Error Compensation Technique for Time-Interleaved ADC Systems," in Proc. *IEEE CICC*, pp. 341–344, Sep. 2007.

[4] B. Razavi, "Problem of timing mismatch in interleaved ADCs," in Proc. *IEEE CICC*, pp. 1-8, Sep. 2012.

[5] J. Song, K. Ragab, X. Tang, and N. Sun, "A 10-b 800MS/s time-interleaved SAR ADC with fast time-skew calibration," in Proc. *IEEE A-SSCC*, pp. 73-76, Nov. 2016.

[6] M. El-Chammas, and B. Murmann, "A 12-GS/s 81-mW 5-bit TimeInterleaved Flash ADC With Background Time Skew Calibration," in *IEEE JSSC*, vol. 46, no. 4, pp. 838-847, Apr. 2011.

[7] A. Mirzaei, S. Chehrazi, R. Bagheri and A. A. Abidi, "Analysis of first-order anti-aliasing integration sampler," in *IEEE TCAS. I*, vol. 55, no. 10, pp. 2994-3005, Nov. 2008.

[8] I. N. Ku, Z. Xu, Y. C. Kuan, Y. H. Wang and M. C. F. Chang, "A 40-mW 7-bit 2.2-GS/s Time-Interleaved Subranging CMOS ADC for Low-Power Gigabit Wireless Communications," in *IEEE JSSC*, vol. 47, no. 8, pp. 1854-1865, Aug. 2012.

[9] C. H. Chan, Y. Zhu, W. H. Zhang, S. P. U and R. P. Martins, "A Two-Way Interleaved 7-b 2.4-GS/s 1-Then-2 b/Cycle SAR ADC With Background Offset Calibration," in *IEEE JSSC*, vol. 53, no. 3, pp. 850-860, Mar. 2018.

[10] B. Malki, T. Yamamoto, B. Verbruggen, P. Wambacq and J. Craninckx, "A 70dB DR 10b 0-to-80MS/s current-integrating SAR ADC with adaptive dynamic range," in *IEEE ISSCC*, pp.470-472, Feb. 2012.

A 15.1-mW 6-GS/s 6-bit Flash ADC with Selectively Activated 8x Time-Domain Interpolation

Il-Min Yi, Naoki Miura, Hiroyuki Fukuyama, and Hideyuki Nosaka

NTT Device Technology Labs, NTT Corporation, Atsugi, Japan

E-mail: yilmin.yi.zc@hco.ntt.co.jp

Abstract—A selectively activated 8x time-domain interpolation is proposed for a low-power high-speed 6-bit flash ADC. By improving the linearity of the voltage-to-time conversion gain, a 3-bit resolution is achieved in time-to-digital conversion. Hence, the number of the dynamic comparators is reduced from conventional 63 to 10. Also, unlike other time-domain interpolation schemes, time-to-digital converters are selectively activated to reduce the power consumption of the time-to-digital conversion. The flash ADC fabricated in a 1-V 65-nm CMOS process achieves 6 GS/s with 15.1-mW power consumption. It shows a 31.18-dB SNDR and an 85 fJ/conv.-step FoM with a Nyquist frequency input.

I. INTRODUCTION

In single channel ADCs, the flash architecture achieves the fastest conversion speed among various ADC architectures. However, the large number of dynamic comparators (CMPs) in the flash architecture not only results in large power consumption, but also limits operating speed with large input capacitance. To reduce the number of the dynamic CMPs, a time-domain interpolation scheme was proposed [1]-[3]. With the time-domain interpolation scheme, the lower bits of the digital output code are generated by using time-to-digital converters (TDCs) and the time differences between the CMPs' outputs [Fig. 1(a) and (b)]. Due to the TDCs, the burden of the dynamic CMPs reduces, and hence the overall ADC power consumption reduces. Moreover, the time-domain interpolation scheme helps to increase the conversion speed by reducing the input capacitance. The higher resolution in the time-to-digital conversion further reduces the power consumption. However, the nonlinearity of the voltage-to-time conversion (VTC) gain based on the comparator clock-to-Q delay limits the resolution of the time-to-digital conversion to 1 or 2 bits [1]-[3]. Therefore, the nonlinearity of the VTC gain should be improved to increase the resolution.

In this work, a selectively activated 8x time-domain interpolation is proposed for a low-power high-speed 6-bit flash ADC. Compared to the conventional interpolation, the voltage-to-time conversion in the proposed interpolation uses a 4x larger voltage as input, thereby increasing the linear region of the VTC gain by 4 times. This enables a 3-bit time-to-digital conversion. With the proposed interpolation, the number of the dynamic CMPs is reduced from 63 to 10, where 63 is the number of a conventional 6-bit flash ADC. In addition, unlike other time-domain interpolation schemes, only selected TDCs operate to reduce the power consumption of the time-to-digital conversion. The proposed 6-bit flash

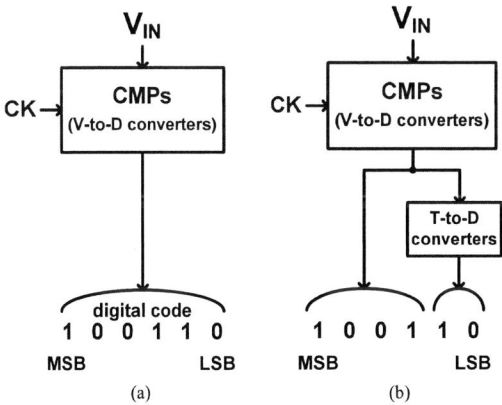

Fig. 1. Conventional flash ADC architectures: (a) without and (b) with time-domain interpolation.

ADC achieves a 6-GS/s sampling frequency, a 15.1-mW power consumption, a 31.18-dB SNDR, and an 85-fJ/conversion-step FoM. Compared to other 65-nm single-channel 6-bit ADCs, it shows the best conversion speed and energy efficiency.

II. TIME-DOMAIN INTERPOLATION IN FLASH ADC

The dynamic CMPs compare the ADC input voltage (V_{IN}) with voltage references (VRs) and generate digital outputs after clock-to-Q delays (TCQs). Because a TCQ varies with the voltage difference between the input and the VR, there are time differences between the digital outputs. The time differences are converted to digital codes by using TDCs. A TDC consists of time-domain comparators (TD CMPs), and a TD CMP has a time reference (TR). The TD CMPs compare the time difference with several TRs to generate digital outputs. Conventionally, two adjacent dynamic CMPs and two adjacent VRs are used for time-domain interpolation [1], [2]. In the 4x time-domain interpolation of Fig. 2(a), three TRs (TR[1]~TR[3]) are compared with the time difference of the outputs of the two CMPs (C[n] and C[n+1]), which use VR[n] and VR[n+1] of a 4-LSB difference [2]. The 4x interpolation reduces the number of the dynamic CMPs from conventional 63 to 17 with using 48 TD CMPs for a 2-bit time-to-digital conversion. The resolution in time-to-digital conversion is limited by the nonlinearity of the VTC gain. As $|V_{IN}-VR[n]|$ reduces, the VTC gain changes drastically [Fig. 2(b)]. This makes the dynamic CMP output pulse narrow, which limits the operation speed and the resolution.

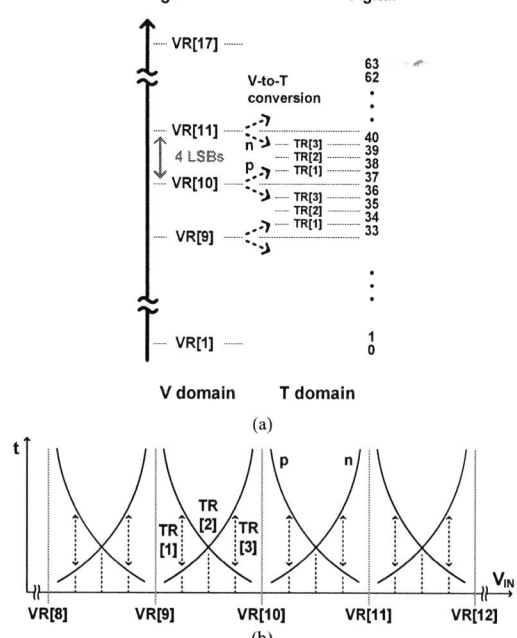

Fig. 2. Conventional 4x time-domain interpolation: (a) concept, (b) voltage-to-time conversion.

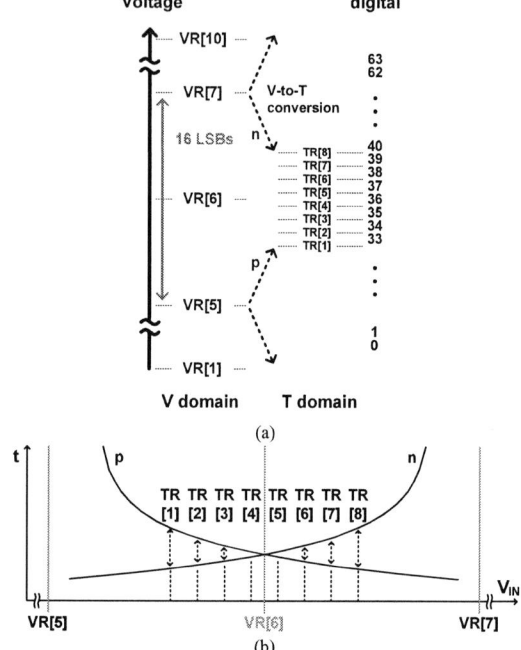

Fig. 3. Proposed 8x time-domain interpolation: (a) concept, (b) voltage-to-time conversion.

III. PROPOSED ADC ARCHITECTURE

The proposed 8x time-domain interpolation further reduces the number of the dynamic CMPs to 10. Unlike other time-domain interpolation schemes, the proposed approach uses two comparators of C[n] and C[n+2] for interpolation. The two VRs of C[n] and C[n+2], VR[n] and VR[n+2], have a voltage difference of 16 LSBs. Around the center of the two VRs, the 8-LSB linear region is used as the input of the VTC for a 3-bit time-to-digital conversion [Fig. 3(a) and (b)]. Namely, the proposed time-domain interpolation with two outputs of C[n] and C[n+2] is valid only when V_{IN} is in the range between $VRC - 4 \cdot VLSB$ and $VRC + 4 \cdot VLSB$, where VRC is $(VR[n + 2] + VR[n])/2$ and $VLSB$ is $(VR[n + 2] - VR[n])/16$. By using 8 TD CMPs and 8 TRs in a TDC, the time difference between the two outputs is converted to a 3-bit digital code. The proposed time-domain interpolation alleviates the sensitivity requirement of the dynamic CMPs because their input voltage difference is large enough in the valid region. This helps in achieving a low-power and high-speed flash ADC.

To reduce the power consumption of the time-to-digital conversion, 8 switches are used following the dynamic CMPs. Depending on the results of the dynamic CMPs, only 2 among 8 switches are turned on, and the turned-on switch transmits its input to its 8 TD CMPs (Fig. 4). Therefore, 16 among the 64 TD CMPs are activated. Deactivated TD CMPs consume almost zero current. As a result, only 26 comparators (10 dynamic CMPs + 16 TD CMPs) operate in a cycle, so the number of the operating CMPs is reduced by 58% compared to the conventional flash ADC.

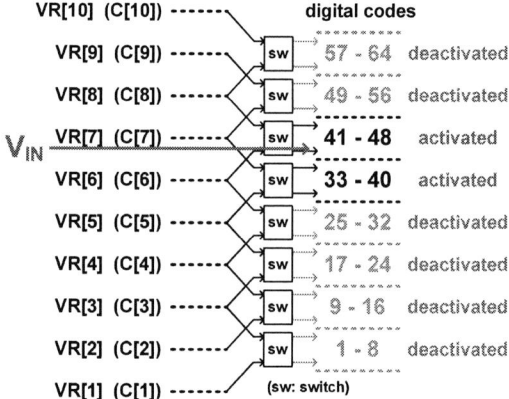

Fig. 4. Selectively activated time-domain interpolation.

IV. CIRCUIT IMPLEMENTATION

The proposed ADC consists of a track-and-hold (T/H) circuit, 10 dynamic CMPs (C[1]~C[10]), 10 SR latches, 8 time-to-digital stages (TDSs), and a 6-bit encoder [Fig. 5(a)]. The dynamic CMPs are latch-based dynamic ones. In the dynamic CMP, a built-in offset is used as a VR instead of using a resistor ladder [1]. With the outputs of the dynamic CMPs, the SR latches generate 10 thermometer codes. The 8 TDSs generate 8 sets of a carry bit and a 3-bit binary code by using the proposed 8x time-domain interpolation. In the 6-b encoder, the 10 thermometer codes and the 8 carry bits

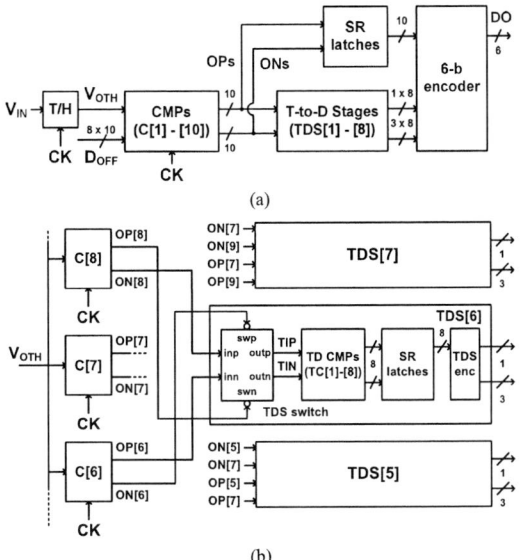

(a)

(b)

Fig. 5. (a) Proposed ADC block diagram. (b) Proposed connection between dynamic CMPs and time-to-digital stages.

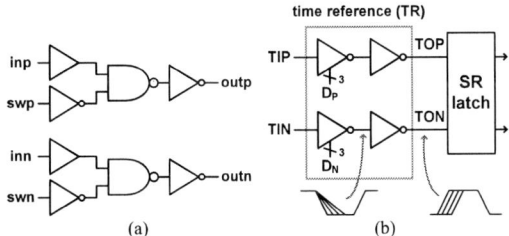

(a)   (b)

Fig. 6. (a) TDS switch circuit. (b) TD CMP circuit.

generate the upper 3 bits among 6 ADC output bits (DO). They also select one set among 8 TDS output sets as the lower 3 bits. A TDS consists of a TDS switch, 8 TD CMPs, 8 SR latches, and a TDS encoder (TDS enc) [Fig. 5(b)]. The 4 outputs of two dynamic CMPs (OP[n], ON[n], OP[n+2], and ON[n+2]) are connected to TDS[n] as shown in Fig. 5(b). With TDS[n], ON[n+2] and OP[n] are used for interpolation and ON[n] and OP[n+2] are used to control the TDS switch.

With C[n], both OP[n] and ON[n] are low (0) while CK is low. When CK transitions to high, OP[n] or ON[n] transitions to high (1) according to V_{IN}-VR[n] after a clock-to-Q delay. When V_{IN}>VR[n], OP[n] becomes 1. When V_{IN}<VR[n], ON[n] becomes 1. With the TDS switch in Fig. 6(a), inp is transmitted to outp when swp is 0. When swp is 1, outp becomes 0 regardless of inp. Namely, OP[n] and ON[n+2] are transmitted to the TD CMPs in TDS[n] only when V_{OTH} is between VR[n] and VR[n+2] (Fig. 7). On the other hand, when V_{OTH} is smaller than VR[n] or larger than VR[n+2], both the TIP and the TIN remain 0, and hence the TD CMPs of TDS[n] become deactivated. A high-frequency glitch noise, which might occur due to the logic operation, is suppressed by the bandwidth limit of the following logics. A TD CMP consists of a time reference (TR) block and a NAND-based SR latch [2] [Fig. 6(b)]. The 8 TD CMPs (TC[1]~TC[8])

compare the time difference input (T_P-T_N) with 8 different TRs (TR[1] ~ TR[8]). In TC[n], T_P-T_N becomes T_P-T_N-TR[n] after the TR block, where TR[n] is generated by the propagation delay difference between two buffers in the TR block. Then the SR latch makes a decision by comparing the rising edges of the TR block outputs [2]. If T_P-T_N-TR[n]<0, the rising edge of TOP comes before TON and the SR latch generates 0. The TDS encoder accepts the 8 thermometer codes and generates 1 carry bit and a 3-bit binary output.

In this work, the simulated VTC gain is 0.26 ps/mV with a sampling frequency of 6 GS/s, and it can be translated to a voltage gain of 6.5 with the slew rate of 500 mV/20 ps. The time resolution of the TDS is approximately 2.8 ps. To improve the sensitivity, the comparator offset and the time reference offset are calibrated with two 4-bit binary codes per a dynamic CMP and two 3-bit binary codes per a TR block.

V. MEASUREMENT RESULTS

The proposed 6-bit ADC chip was fabricated in a 65-nm CMOS process with an area of 0.023 mm^2 (Fig. 8). The full-scale input range of the ADC is 0.7 V_{pp} differential. Without offset calibration, the measured DNL and INL of the ADC chip are +1.75/-1 and +3.22/-4.48 LSB, respectively. They are reduced to +0.42/-0.46 and +0.48/-0.55 LSB with offset calibration [Fig. 9(a) and (b)]. Fig. 10(a) shows the measured SNDR and SFDR at 6 GS/s with several input frequencies. With a low frequency input, the measured SNDR and SFDR are 33.6 dB and 44.25 dB, respectively. With a Nyquist frequency input, the measured SNDR and SFDR are 31.18 dB and 41 dB, respectively, and the measured spectrum is shown in Fig. 11. Fig. 10(b) shows the measured SNDR and SFDR at several sampling frequencies with a 10-MHz input. The power consumption is 15.1 mW at 6 GS/s with a 1-V supply. With a Nyquist frequency input, the ADC chip shows an FoM of 85 fJ/conversion-step. Table I and Fig. 12 show the performance comparison with other ADCs. The proposed flash ADC chip achieves the best conversion speed and energy efficiency among 65-nm single-channel 6-bit ADC chips.

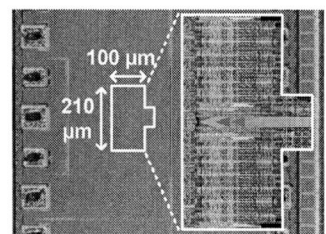

Fig. 8. Chip micrograph and layout.

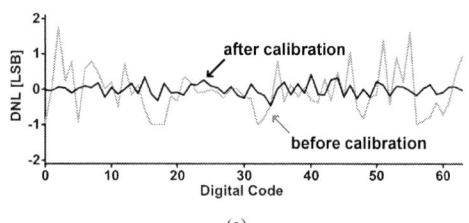

(a)

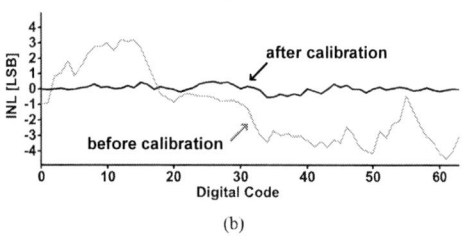

(b)

Fig. 9. Measured (a) DNL and (b) INL.

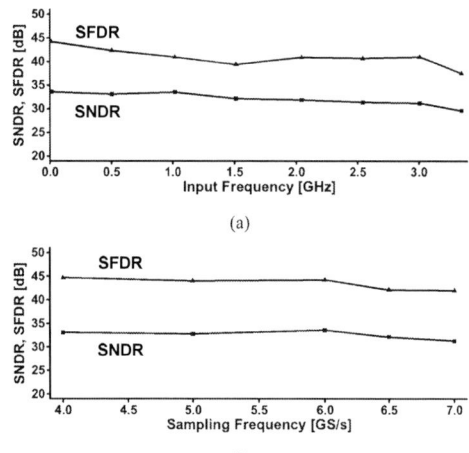

(a)

(b)

Fig. 10. Measured SNDR and SFDR vs. (a) input frequency and (b) sampling frequency.

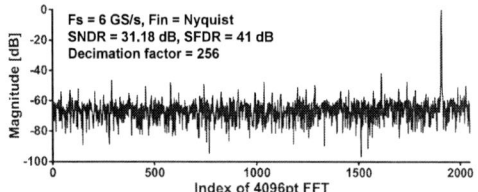

Fig. 11. Measured spectrum with a 3010.9-MHz input.

Table I. Performance Comparison of Flash ADCs

	VLSI 2012 [1]	VLSI 2013 [4]	JSSC 2015 [3]	ASSCC 2015 [2]	TCS-I 2017 [5]	This work
Architecture	Flash					
Technology (nm)	40	32 SOI	65	65	65	65
fs (GS/s)	3	5	2	3.4	5	6
Resolution (bit)	6	6	7	6	5	6
SNDR @ Nyq.	33.1	30.9	38.12	34.2	26.19	31.18
Supply (V)	1.1	0.85	1.2	1	1	1
Power (mW)	11	8.5	20.7	12.6	7.8	15.1
FoM (fJ/conv.step)	100	59.4	157	89	94.6	85

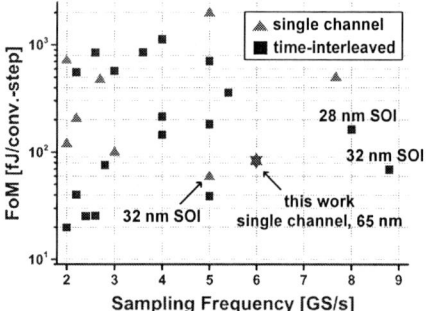

Fig. 12. ISSCC and VLSI 10-year ADC survey and this work.

VI. CONCLUSION

An 8x time-domain interpolation is proposed to reduce the number of the dynamic CMPs for a low-power high-speed 6-bit flash ADC. By using two dynamic CMPs which have two voltage references of a 16-LSB difference, the clock-to-Q delays of the two CMPs are compared. The linear region around the center of the two voltage references is used in the interpolation for a 3-bit time-to-digital conversion. With the proposed interpolation, the number of the dynamic CMPs is reduced from conventional 63 to 10. In addition, only 16 among 64 time-domain comparators are selectively activated for low-power operation. The proposed 6-bit flash ADC was fabricated in a 1-V 65-nm CMOS process. The measured DNL and INL are +0.42/-0.46 and +0.48/-0.55 LSB, respectively. At 6 GS/s, the ADC chip consumes 15.1 mW and achieves an SNDR of 31.18 dB and an FoM of 85 fJ/conversion-step.

REFERENCES

[1] Y.-S. Shu, "A 6b 3GS/s 11mW Fully Dynamic ADC in 40nm CMOS with Reduced Number of Comparators," in *Dig. Symp. VLSI Circuits*, 2012, pp. 26-27.

[2] J. Liu, *et al.*, "A 89fJ-FOM 6-bit 3.4GS/s Flash ADC with 4x Time-Domain Interpolation," in *Proc. IEEE A-SSCC*, 2015, pp. 1-4.

[3] J.-I. Kim, *et al.*, "A 65 nm CMOS 7b 2 GS/s 20.7 mW Flash ADC With Cascaded Latch Interpolation," *IEEE J. Solid-State Circuits*, vol. 50, no. 10, pp. 2319–2330, Oct. 2015.

[4] V. H.-C. Chen and L. Pileggi, "An 8.5mW 5GS/s 6b Flash ADC with Dynamic Offset Calibration in 32nm CMOS SOI," in *Dig. Symp. VLSI Circuits*, 2013, pp. 264-265.

[5] C.-H. Chan, *et al.*, "A 7.8-mW 5-b 5-GS/s Dual-Edges-Triggered Time-Based Flash ADC," *IEEE Trans. Circuits Syst. I, Reg. Papers*, vol. 64, no. 8, pp. 1966–1976, Aug. 2017.

18-3 (8106)

A 38-mW 7-bit 5-GS/s Time-Interleaved SAR ADC with Background Skew Calibration

Yung-Hui Chung[1], Chia-Yi Hu[1,2], and Che-We Chang[1]

[1]National Taiwan University of Science and Technology, Taipei, Taiwan
[2]Realtek Semiconductor Cooperation, Hsin-Chu, Taiwan
Email: yhchung@mail.ntust.edu.tw

Abstract—This paper presents a 7-bit 5-GS/s time-interleaved SAR ADC with background timing skew calibration. The two-step approaching skew calibration was proposed to reduce the tuning range of the digital control delay circuit, thus suppress the additional clock jitter. The ping-pong domino-SAR ADC architecture was proposed to speed up channel-ADCs. The prototype ADC consumes a total power of 38 mW from a 1.2V supply and occupies an active area of 0.69 mm^2 in a 55 nm low-power CMOS technology. For 10 MHz input, the measured SNDR and SFDR are 42.7 and 65 dB, respectively. The ENOB is 6.8 bits, equivalent to the peak FOM of 69 fJ/conversion-step. At the Nyquist rate, this ADC achieves 35.9 dB SNDR and 45 dB SFDR. The ENOB is 5.7 bits, equivalent to the Nyquist FOM of 150 fJ/conversion-step.

Keywords—ADC; DAC; SAR ADC; skew calibration; time-interleaved ADC

I. INTRODUCTION

Wireless systems operating in the 60GHz band and millimeter-wave pulsed radar systems require low-power low to medium resolution over-GHz sampling rate ADCs. In this decade, the successive-approximation-register (SAR) analog-to-digital converter (ADC) architecture presents its strength of excellent energy-efficiency and low cost implementation [1]. But, its operating speed is still limited by the ADC resolution. To improve this, the time-interleaved (TI) architecture is the best candidate to be integrated with SAR ADCs. Therefore, different from a prior flash ADC [2], TI-SAR ADCs can be implemented without reaching the speed limit of CMOS technologies [3]–[7]. However, the interchannel mismatches, including offset, gain, and timing skew, must be well corrected. The offset and gain mismatches can be compensated in the digital domain by using long-term averaging and correction schemes [3]–[5]. But, the timing mismatch needs more efforts to correct in the foreground [6] or background [7].

This paper presents a 7-bit 5-GS/s TI-SAR ADC in 55nm low-power (LP) CMOS. In this study, a background timing skew calibration scheme with two-step approaching (TSA) is proposed to reduce one half skew tuning range, thus get lower additional clock jitter. To achieve better energy-efficiency, the channel-ADC uses the ping-pong domino-SAR ADC architecture [8]. This paper is organized as follows. Section II presents the proposed skew calibration scheme. Section III depicts the proposed ADC architecture and circuit implementation of key blocks. The experimental results are shown in Section IV. Section V concludes this work.

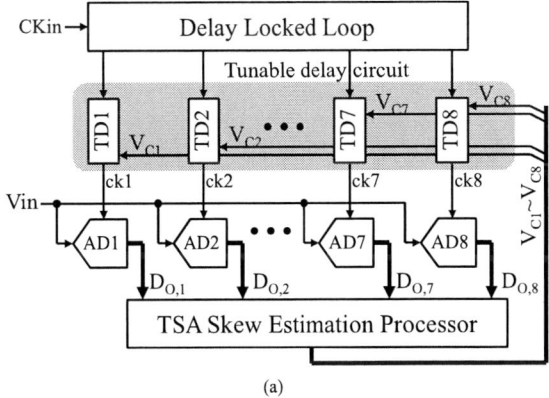

(a)

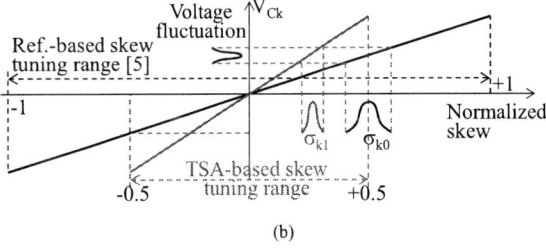

(b)

Fig. 1. (a) The proposed timing skew calibration loop and (b) the transfer curve between the in-out delay and control voltage (V_{Ck}) of the k-th tunable delay circuit (TDk).

II. PROPOSED SKEW CALIBRATION SCHEME

Fig. 1(a) presents the proposed skew calibration including the skew estimation processor and tunable delay (TD) circuit. There are eight clock signals generated from a delay-locked loop (DLL) with the input frequency of 625 MHz. These clock signals are passed to eight channel-ADCs through their individual routing traces. Ideally, the phase difference between adjacent clock signals is 45° to perform an equivalent sampling rate of 5 GHz. However, considering routing trace mismatches and the timing skew in the DLL, the timing deviation from the ideal sampling instant is contributed to degrade the ADC performance. To compensate the timing skew for each channel-ADC, the TD circuit can be inserted into the routing trace to adjust each sampling clock to have the equal phase delay of 45°. Although the TD circuit can compensate the channel clock

978-1-5386-6414-8/18 $31.00 © 2018 IEEE

skew, it also introduces additional clock jitter to worsen the signal-to-noise ratio (SNR) for high input frequencies. For example, with certain supply fluctuations, the resultant clock jitter is also generated. Therefore, how to reduce the jitter contribution from the TD circuit becomes one of the key issues in TI ADCs.

Fig. 1(b) presents the relevance between the clock jitter and the skew tuning range. The larger skew tuning range will introduce more clock jitter. In prior skew compensation schemes, one channel clock is used as the reference to adjust other channel clocks to align with the reference channel. For example, with certain supply fluctuations, the resultant clock jitter (σ_{j0}) is generated while using the reference-based skew calibration [7]. If the skew tuning range can be cut down by half, the clock jitter (σ_{j1}) is also reduced accordingly.

Fig. 2(a) depicts the basic concept of the proposed TSA skew calibration including the variance convergence tuning (VCT) and reference convergence tuning (RCT) schemes. In the beginning, eight channel skews are distributed in a wide range and the first channel has the largest timing skew (Δt_1). In [7], all other channel skews approach Δt_1 to have a final skew error (Δt_{f0}). The skew tuning range must cover the maximum skew difference among all channel clocks. Different from the reference-based skew calibration, the proposed TSA skew calibration utilizes the VCT scheme firstly to reduce the total skew range. After all channel skews are close to each other (around the average of all skews), the RCT scheme chooses one of them as the reference skew (Δt_k) and drives other skews approaching Δt_k to have a final skew error (Δt_{f1}). In the worst case, the TSA-based skew tuning range is only one half of that in [7]. Furthermore, smaller Δt_{f0} and Δt_{f1} can be obtained by using finer skew tuning steps which may require higher cost for the circuit implementation.

The TSA skew calibration concept is described as follows. The i-th channel skew detection (Δt_i) is realized using a correlation-based estimator [7]. Δt_i is represented as

$$\Delta t_i = K \cdot \left(E\left[|D_i - D_{i-1}| \right] - E\left[|D_{i+1} - D_i| \right] \right), \tag{1}$$

where K is a constant and $E[x]$ is an average of x. As shown in Fig. 2(b), taking four ADC channels as an example, the TSA algorithm updates each channel skew according to the following criterion,

$$\Delta t_{i,next} = \begin{cases} \Delta t_i + sign(\Delta t_i) \times t_{step}, & i=1\text{-}8, \text{ at VCT mode} \\ \Delta t_i + sign(\Delta t_1 - \Delta t_i) \times t_{step}, & i=2\text{-}8, \text{ at RCT mode} \end{cases}, \tag{2}$$

where t_{step} is the tuning step and $sign(x)$ is the polarity of x. Fig. 2(b) presents a small skew error if the initial skews are the integers of t_{step}. However, if the initial skews are not integers of t_{step}, there exists a possible fluctuation which is shown in Fig. 2(c). A large skew error of $3 \times t_{step}$ is presented. Generally, such a large skew fluctuation can be suppressed by using a much smaller tuning step. But, it may be hard to implement at low cost. In this work, to solve this fluctuation, the RCT scheme is applied to use one channel clock as the reference. Then, it drives other channel skews to approach the reference channel. As shown in eq.(2), the first channel is chosen as the reference

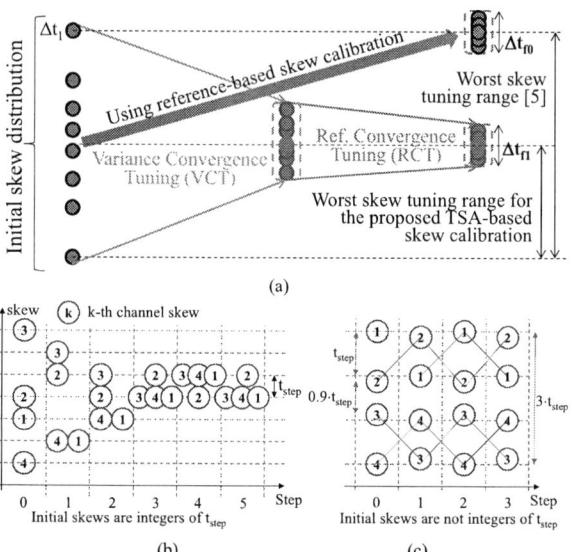

Fig. 2. (a) the concept of the TSA skew calibration and two kinds of skew convergence at the VCT mode: (b) a small skew range and (c) a large skew fluctuation range.

channel. The RCT scheme behaves like the skew calibration in [7]. Differently, the proposed TSA scheme avoids a large skew tuning range since all channel skews are close to each other after applying the VCT scheme. In comparison with the RCT-only calibration scheme, the TSA scheme saves at most one half convergent time.

III. CIRCUIT IMPLEMENTATION

A. ADC Architecture

Fig. 3 shows an overall block diagram of the proposed 7-bit TI-SAR ADC. There are eight channel-ADCs (ADC1–8) to achieve a sampling rate of 5GHz. To maintain higher input tracking bandwidth, the distributed input buffer (IB) is applied individually for each channel-ADC. The digital outputs ($DX<7:0>$) are carried off-chip for interchannel gain and offset correction and timing skew estimation according to the proposed TSA skew calibration scheme. The estimation result ($CSC<1:8>$) travels back to the chip on a parallel bus and adjusts each channel clock phase to reduce the interchannel timing skew by increasing or decreasing one delay step. To simplify the main clock generation circuit, a delay-locked-loop (DLL) is used to generate eight clock signals according to the input clock ($CKin$) with the frequency of 625 MHz.

The digital control TD circuit is shown in Fig. 4. It is divided into the coarse ($CC<6:0>$) and fine ($FC<3:0>$) delay control circuits. The coarse delay circuit mainly covers a wide tuning range for the skew contributed from the DLL and clock routing traces. The fine delay is applied to have a small delay step to suppress the skew error. Once the coarse delay is determined, the TD controller enters the fine control mode. By simulation results, the coarse delay range and step are 60 ps and 450 fs, respectively. The fine delay range and step are 2 ps

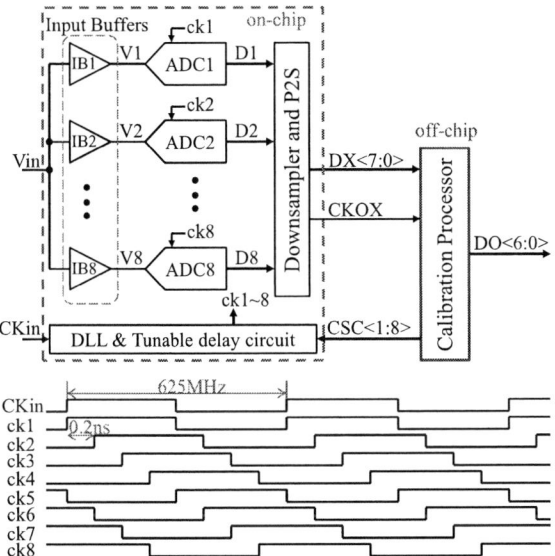

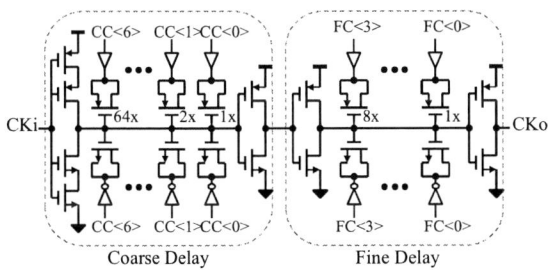

Fig. 3. Overall 8x TI SAR ADC architecture and its timing chart.

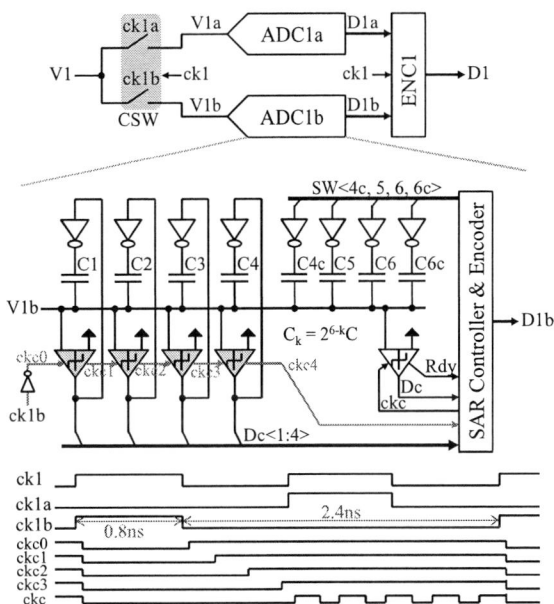

Fig. 5. Proposed channel ADC schematic.

Coarse Delay   Fine Delay

Fig. 4. Proposed digital tunable delay schematic.

and 150 fs, respectively. The additional clock jitter is less than 100 fs contributed from the TD circuit.

B. Channel-ADC

In 55nm LP CMOS, the ping-pong SAR ADC architecture was implemented to relax the conversion time [8]. The ping-pong operation gives a longer conversion time of 2.4 ns to implement a 312.5-MS/s subchannel SAR ADC. As shown in Fig. 5, taking the channel-ADC1 as an example, *ck1a* and *ck1b* are used to clock ADC1a and ADC1b, respectively. Both *ck1a* and *ck1b* control the channel switch (CSW) with the master clocking scheme [9]. It uses the main channel clock *CK1* to determine the sampling instant for both subchannel-ADCs. Furthermore, the ping-pong channel-ADC can reduce the complexity of the skew calibration circuit. The number of ADC channels is reduced from 16 to 8. This can reduce the hardware implementation cost while using the skew calibration in TI ADCs. Moreover, the additional clock jitter caused by a large skew tuning range of excessive ADC channels is also reduced.

Fig. 5 depicts the channel-ADC using the ping-pong domino-SAR ADC architecture. Traditionally, the SAR ADC needs to go through the SAR controller and then controls each

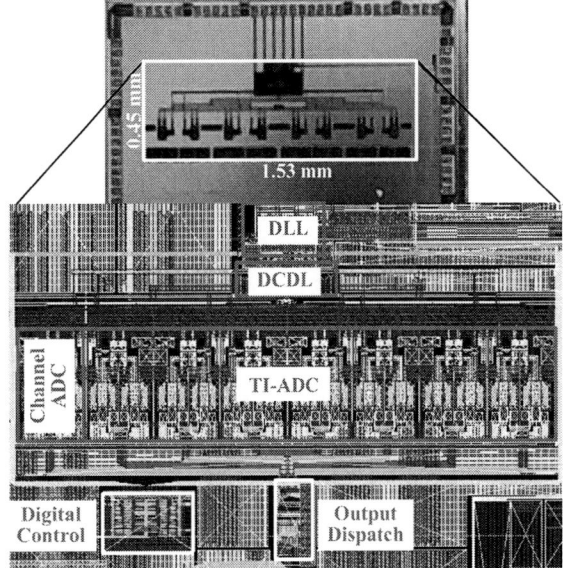

Fig. 6. Chip micrograph and layout.

DAC driver sequentially from MSB to LSB conversion cycles. Using the domino operation, the domino comparator output can directly control the individual DAC driver without going through the SAR controller. Thus, the delay time of the SAR controller is saved. Actually, the domino comparator acts like both a quantizer and a data latch. After the domino operation being done, the SAR operation is followed to complete the last five conversion cycles. In the capacitor array, two redundant capacitors, C4c and C6c, are used for the error tolerance. Using more-than-one comparators in SAR ADCs, the comparator

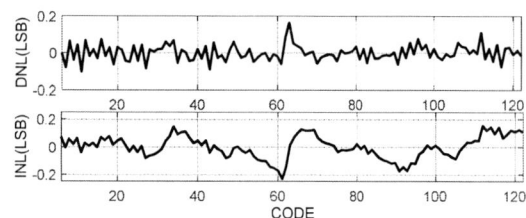

Fig. 7 Measured DNL/INL plots.

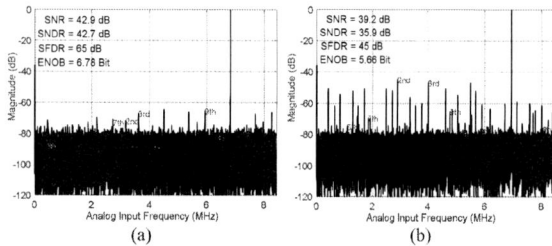

Fig. 8. Measured spectra at 5-GS/s for (a) 10 MHz input and (b) 2.4 GHz input (32768-pt FFT, the digital output is decimated by 297x).

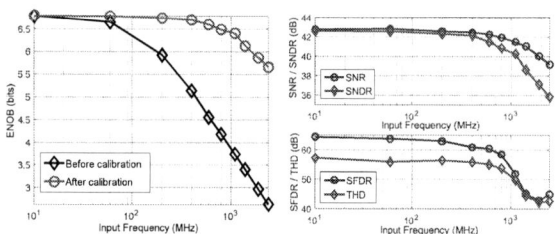

Fig. 9. Dynamic performance before and after calibration.

offset deviation must be concerned. In this work, the offset deviation between domino-comparators and main comparator can be tolerated by the digital redundancy.

IV. EXPERIMENTAL RESULTS

The prototype ADC was fabricated in a 55nm LP CMOS technology and occupies an active area of 0.69 mm². Fig. 6 shows the chip micrograph and layout. Operating at 5-GS/s, the ADC consumes 28 mW from 1.2V supplies and 10 mW from one 1.5V supply for the input buffer array. The output is decimated by 297x to reduce the measurement cost. Fig. 7 presents the measured DNL/INL plots. The achieved DNL and INL are -0.1/+0.16 LSB and -0.23/+0.16 LSB, respectively. Fig. 8 shows the measured spectra for 10 MHz and 2.4 GHz input. For the input frequency of 10 MHz, the measured SNDR and SFDR after using skew calibration are 42.7 dB and 65 dB, respectively. It leads to a peak FOM of 69 fJ/c.-s. With a 2.4 GHz input, the measured SNDR and SFDR are 35.9 dB and 45 dB, respectively. Fig. 9 presents the measured dynamic performance versus input frequencies before and after using the skew calibration. For the 1.1GHz input, the ENOB is improved from 3.75 to 6.4 bits. The performance degradation at high input frequencies are mainly caused by the underestimated input clock jitter. The excessive clock jitter may be caused by the on-chip clock buffer. Table I shows the performance summary and comparison with recently reported 7–8b TI-SAR

TABLE I. Performance summary and comparison.

Specification	This work	[4]	[6]	[7]
Process (nm)	55 LP	40 GP	40	65
Architecture	8x TI	8x TI	16x TI	4x TI
Resolution (bit)	7	7	8	8
Speed (GS/s)	5	3.5	2	4
Supply (V)	1.2/1.5	0.9	NA	1.2/1.4
Power (mW)	38	6.2	54.2	120
SNDR* (dB)	42.7/35.9	38/34.7	45/39.4	46/44
SFDR* (dB)	65/45	50/39	59/55	55
FOM* (fJ/c.-s.)	69/150	27/40	180/355	184/219
Area (mm²)	0.69	0.139	0.54	1.35

* x/y: DC/Nyquist

ADCs. In comparison with the best FOM ADC [4], this work achieves better SNDR and SFDR.

V. CONCLUSIONS

A 7-bit 5-GS/s TI SAR ADC was implemented in 55nm LP CMOS. The ping-pong domino-SAR ADC architecture further reduces the number of ADC channels to reduce the hardware cost for the skew calibration processor. The proposed TSA skew calibration incorporating VCT and RCT schemes effectively compensate the channel skew with lower additional clock jitter. Using the proposed skew calibration, the Nyquist ENOB can be improved from 2.7 to 5.7 bits. The measurement results prove the TSA skew calibration scheme.

ACKNOWLEDGMENT

The authors would like to thank the UMC University Shuttle Program for supporting this work.

REFERENCES

[1] K. D. Choo, J. Bell, and M. P Flynn, "Area-Efficient 1GS/s 6b SAR ADC with Charge-Injection-Cell-Based DAC," in *Proc. IEEE Int. Solid-State Circuits Conf. Dig. Tech.Papers*, Feb. 2016, pp. 460–461.

[2] Y. Nakajima, N. Kato, A. Sakaguchi, T. Ohkido, and T. Miki, "A 7-bit 1.4 GS/s ADC with offset drift suppression techniques for one-time calibration," *IEEE Trans. Circuits Syst. I. Reg.Papers*, vol. 60, no. 8, pp. 1979–1990, Aug. 2013.

[3] C.-H. Chan, *et al*, "A 5mW 7b 2.4GS/s 1-then2b/cycle SAR ADC with Background Offset Calibration," in *Proc. IEEE Int. Solid-State Circuits Conf. Dig. Tech.Papers*, Feb. 2017, pp.282–283..

[4] A. Spagnolo, *et al*, "A 6.2mW 7b 3.5GS/s Time Interleaved 2-Stage Pipelined ADC in 40nm CMOS," in *IEEE ESSCIRC Dig. Tech. Papers*, Sep. 2014, pp.75–78.

[5] S. Kundu, *et al*, "A 1.2 V 2.64 GS/s 8bit 39 mW Skew-Tolerant Time-interleaved SAR ADC in 40 nm Digital LP CMOS for 60 GHz WLAN," in *Proc. IEEE Custom Integrated Circuits Conf*, Sep. 2014, pp.1–4.

[6] T. Miki, *et al*, "A 2GS/s 8b Time-Interleaved SAR ADC for Millimeter-Wvae Pulsed Radar Baseband SoC," in *Proc. IEEE Asian Solid-State Circuits Conf. Dig. Tech.Papers*, Feb. 2016, pp.5–8.

[7] H. Wei, et al, "An 8-Bit 4-GS/s 120-mW CMOS ADC," in *Proc. IEEE Custom Integrated Circuits Conf*, Sep. 2013, pp.1–4.

[8] Yung-Hui Chung, Wei-Shu Rih, and Che-Wei Chang, 'A 6-bit 1.3-GS/s Ping-Pong Domino-SAR ADC in 55nm CMOS,' *IEEE Trans. on Circuits and Systems II: Briefs*, vol. 65, no. 8, pp. 999–1003, Aug. 2018.

[9] M. Waltari and K. Halonen, "Timing skew insensitive switching for double sampled circuits,"in *Proc. International Symposium on Circuits and Systems*, June 1999, pp. 61–64.

18-4 (8220)

IEEE Asian Solid-State Circuits Conference
November 5 - 7, 2018/Tainan, Taiwan

A 0.6-to-1V 10k-to-100kHz BW 11.7b-ENOB Noise-Shaping SAR ADC for IoT sensor applications in 28-nm CMOS

Young-Ha Hwang, Yoonho Song, Jun-Eun Park and Deog-Kyoon Jeong

Department of Electrical and Computer Engineering

Seoul National University, Seoul, Korea

Email : yhhwang@isdl.snu.ac.kr, dkjeong@snu.ac.kr

Abstract— **This paper presents a noise-shaping SAR ADC for IoT sensor applications. The ADC exploits a 2nd-order passive noise-shaping loop without a quiescent current. To reduce harmonic distortion induced by mismatches between MSBs, thermometer-coded 3-bit MSBs are implemented with a simple shift register-based dynamic element matching (DEM) technique. Furthermore, a programmable majority-voting (PMV) technique for LSB decision is applied in order to relax noise requirement of a comparator. With the DEM and PMV, SFDR, SNDR and SNR are enhanced by 9.8, 4.7 and 1.9 dB at a 1.0 V supply, respectively. For 50 kHz BW, the modulator dissipates 74.5 μW from a 1.0 V supply and achieves a peak SNDR of 72 dB, a peak SNR of 72.2 dB and a DR of 73.8 dB. The prototype modulator is fabricated in 28 nm CMOS technology, occupying an area of 0.0575 mm².**

Keywords— *SAR ADC, noise shaping, dynamic element matching, majority voting, scalable bandwidth, IoT sensor.*

I. INTRODUCTION

Recently, noise-shaping SAR ADCs with a low oversampling ratio (OSR) have been presented to replace a conventional delta-sigma modulator [1, 2]. For energy-efficient operation, the ADC is required to have a scalable bandwidth and power consumption with supply voltage scaling for the IoT sensor nodes, which is difficult to implement with an operational transconductance amplifier (OTA)-based design [3]. In this work, a passive noise-shaping SAR ADC for the IoT sensor applications is presented. The SAR ADC features a 2nd-order passive noise shaping with a low OSR of 16 and a scalable BW of 10 k–100 kHz at a supply voltage of 0.6–1 V with a dynamic element matching (DEM) and a programmable majority voting (PMV) technique to enhance SFDR and SNDR.

II. PROPOSED NOISE-SHAPING SAR ADC

A. Proposed Architecture

Fig. 1 shows the proposed noise-shaping SAR ADC. The SAR ADC consists of a pair of bootstrap switches, 10-bit capacitive DACs (CDACs), and a 2nd-order passive noise shaping (NS) loop based on switched capacitors [1], a 3-input dynamic comparator, a 3-bit thermometer decoder with a DEM technique, a SAR with a digital error correction (DEC) logic for redundant bits, and a PMV block. After sampling the input signal and 11-bit conversion, the residue of the CDAC are transferred to the NS loop by charge sharing.

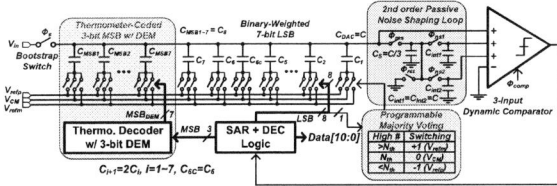

Fig. 1. Block diagram of proposed noise-shaping SAR ADC.

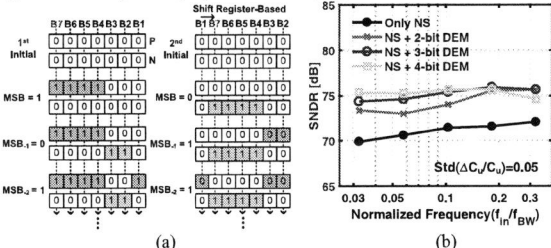

Fig. 2. (a) Operation of 3-bit DEM based on shift registers. (b) Simulated SNDR with NS and DEM.

B. 10-bit CDAC with DEM of 3-bit MSBs

The 10-bit capacitive DAC consists of the thermometer-coded 3-bit MSBs from C_{MSB7} to C_{MSB1}, and the binary-weighted 7-bit LSBs from C_7 to C_1, as shown in Fig. 1. Fig. 2(a) shows that the 3-bit DEM is based on simple shift registers with a low overhead in terms of power consumption and area. Fig. 2(b) shows the simulated average SNDR with the NS and shift register-based DEM for MSBs. The 3-bit DEM technique is chosen to average mismatches between MSBs in order to reduce in-band harmonic distortion induced by mismatches between MSBs. The thermometer decoder converts the 3-bit MSB to 7-bit MSB_{DEM}. There is a DEC logic with a redundant capacitor C_{6C} to correct wrong decision among the 5-bit MSB conversion by the comparator noise.

C. Supply-Insesntive Second-order Noise-Shaping Loop

To reduce quantization noise with maintaining a power efficiency of the SAR ADC, the passive integrators for the 2nd-order noise shaping loop are implemented by three capacitors, C_S, C_{int1}, and C_{int2} [1]. The 3-input dynamic comparator is designed to compensate gain losses of the passive integrators, maintaining -30 dB suppression of in-band quantization noise over a supply voltage of 0.6–1 V, as shown in Fig. 3(a). The

978-1-5386-6414-8/18 $31.00 © 2018 IEEE   247

$$NTF = \frac{[1-(1-c)z^{-1}]^2}{1+(1-c)(\frac{g_{m3}}{g_{m1}}c^2 + \frac{g_{m2}}{g_{m1}}c-2)z^{-1}+(1-c)^2(1-\frac{g_{m2}}{g_{m1}}c)z^{-2}}, c = \frac{C_S}{C_S + C_{DAC}}$$

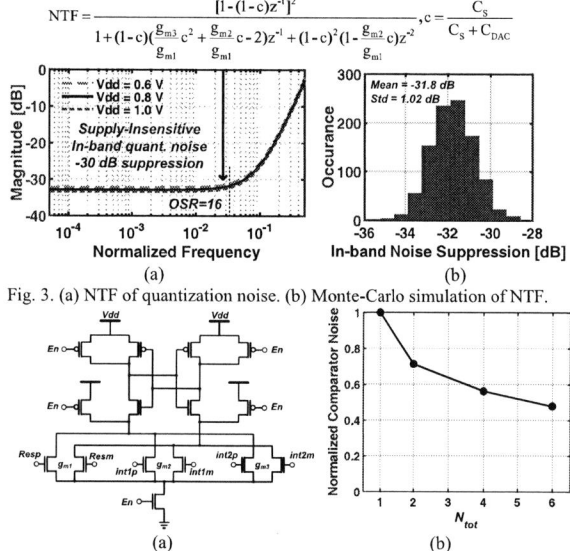

Fig. 3. (a) NTF of quantization noise. (b) Monte-Carlo simulation of NTF.

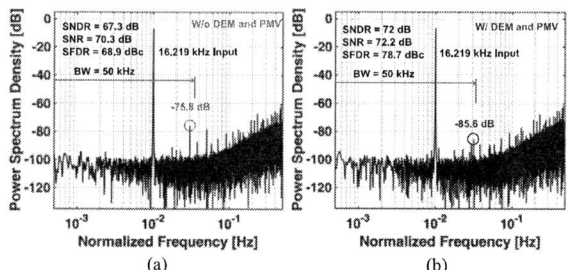

Fig. 6. Measured power spectrum density (a) without and (b) with DEM and PMV.

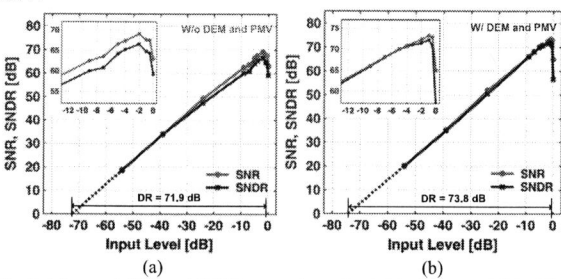

Fig. 7. Measured SNR and SNDR versus input level (a) without and (b) with DEM and PMV.

Fig. 4. (a) Schematic of 3-input comparator. (b) Simulated equivalent RMS input-referred noise of comparator with respect to N_{tot}.

comparator is laid out with common-centroid configuration. Fig. 3(b) shows Monte-Carlo simulation results of the in-band noise suppression with mean of -31.8 dB for std(g_m) of 0.15 and std(c) of 0.05.

D. LSB Decision with PMV

Fig. 4(a) shows the 3-input comparator. The PMV technique reduces the input-referred noise of the comparator by using consecutive comparisons with the controllable number of LSB comparison N_{tot}, as shown in Fig. 4(b). The PMV for upper LSBs can further improve SNR of the ADC.

III. MEASUREMENT RESULTS

The prototype noise-shaping SAR ADC has been fabricated in a 28 nm CMOS process with an area of 0.0575 mm², as shown in Fig. 5. Fig. 6 shows the measured power spectrum density. For a 50 kHz BW at a supply voltage of 1 V, the measured SNDR, SNR and SFDR are 67.3, 70.3 and 68.9 dB, respectively, without the DEM and PMV. With the DEM and PMV, the measured SNDR, SNR and SFDR are 72, 72.2 and 78.7 dB, respectively. The measured SNDR, SNR and SFDR are enhanced by 4.7, 1.9 and 9.8 dB, respectively. Fig. 5 shows that the measured SNR and SNDR are also improved with the DEM and PMV, as well as dynamic range increases by 1.9 dB.

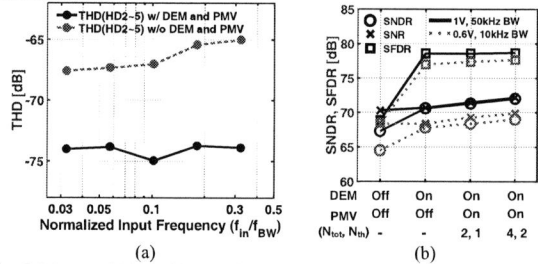

Fig. 8. Measured (a) total harmonic distortion (THD) and (b) SNDR, SNR and SFDR with DEM and PMV.

Fig. 8 shows the measured performances enhanced by the DEM and PMV. For 50 kHz BW, the DEM improves SFDR and SNDR by 9.7 and 2.7 dB, and the PMV improves SNR and SNDR by 1.6 and 1.4 dB, respectively.

REFERENCES

[1] W. Guo, *et al.*, *Symp. VLSI Circuits Dig.*, Jun. 2017, pp. C236–C237.

[2] M. Miyahara, *et al.*, Apr. 2017, *CICC*, pp. 1–4.

[3] J.-E. Park, *et al.*, *TCAS-II*, vol. 64, no. 12, pp. 1417–1421, Dec. 2017.

[4] A. Hussain, *et al.*, *TVLSI*, pp. 364–374, Jan. 2017.

[5] Y. Yoon, *et al.*, *JSSC*, pp. 2342–2352, Oct. 2015.

[6] A. F. yeknami, *et al.*, *TCAS-I*, pp. 358–370, Feb. 2014.

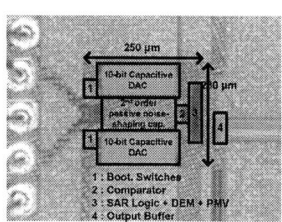

Fig. 5. Chip microphotograph.

TABLE I. PERFORMANCE SUMMARY AND COMPARISON

	This Work				TCAS'17 [3]	TVLSI'17 [4]			CICC'17 [2]			JSSC'15 [5]	TCAS'14 [6]
Process[nm]	28				65	65			65			130	65
Architecture	NS SAR				DT DSM	DT DSM			NS SAR			DT DSM	DT DSM
Noise Shaping (NS) Type	Passive				Two-Step Hybrid Integrator-Based	Active + Passive SC-Based			Open-Loop Integrator-Based			Hybrid SR + SC-Based	Passive
NS Order	2				3	3			3			3	2
OSR	16				50				260			160	250
Supply[V]	0.6	0.7	1.0	1.0	0.4	0.5	1		1			0.4	0.7
BW[kHz]	10	20	50	100	7.5	20	400		25	62.5	250	625	20	0.5
SNDR[dB]	69	70.7	72	69.3	60.5	60.8	61.1		88.2	82.3	83.4	80.4	76.1	65
ENOB[bit]	11.2	11.5	11.7	11.2	9.8	9.8	9.9		14.4	13.4	13.6	13.1	12.3	10.5
Power[μW]	6.9	15.6	60	118	12.7	43.4	948		73.6	66.3	258	630	63	0.43
FoM$_W$[fJ/conv]	147	139	180	251	980	1210	1280		70	49.8	42.6	58.9	310	290
Area[mm²]	0.0575				0.38	0.38			0.1			0.33	0.125

A Calibration-Free 0.7-V 13-bit 10-MS/s Full-Analog SAR ADC with Continuous-Time Feedforward Cascaded (CTFC) Op-Amps

Kwuang-Han Chang and Chih-Cheng Hsieh

Department of Electronic and Electrical Engineering
National Tsing Hua University, Hsinchu, Taiwan
Email: kh.chang.roger@gmail.com, cchsieh@nthu.ee.edu.tw

Abstract—A calibration-free 13-bit 10-MS/s full-analog SAR ADC integrates the functions of comparator, SAR logic, and DAC switches into multiple inverter-based regenerative amplifiers (IRAs) to have a double timing budget for settling and relax the bandwidth requirement of analog circuits compared to the conventional SAR ADC. The continuous-time feedforward cascaded (CTFC) Op-amps are proposed to enhance the residue SNR using open-loop low gain-bandwidth amplifiers instead of closed-loop high-precision amplifiers. The prototype in 40nm CMOS occupies 0.013 mm² and achieves 67.6 dB SNDR, 77.2 dB SFDR, 3.2 fJ/conv.-step FoM$_W$, and 176.6 dB FoM$_S$ without any calibration.

Keywords— SAR ADC, high-resolution, calibration-free, full-analog, continuous-time, feedforward cascaded, stability.

I. INTRODUCTION

Recently, SAR ADC has become the most prevailing candidate in the region of medium speed (10 - 100 MS/s) and medium resolution (10 - 14 bit) [1]. The dynamic-operated architecture without Op-amps makes SAR ADC simpler and more power efficient than the Op-amp based ADCs (pipelined, cyclic, multi-step ADCs). However, in the high-resolution applications (> 12 bit), SAR ADC encounters the difficulty of degraded residue SNR (SNR$_{res}$) which gradually diminishes from MSB to LSB conversions. Hence, the comparator requires a considerable low-noise level to achieve a high-precision quantization at the expense of speed and power.

The alternative solution is to reconfigure as pipelined-SAR architecture by inserting a residue amplifier (RA) between two adjacent sub-ADCs to enhance the SNR$_{res}$ for the subsequent sub-ADCs (fine ADC) as in Fig. 1. High-precision RA is required for weights matching between the coarse (V$_{R1}$) and fine (V$_{R2}$) ADCs, which is usually implemented in a closed-loop configuration with high gain-bandwidth (GBW) Op-amp at the expense of large power. For power efficiency, the RA could be replaced with a low GBW open-loop Op-amp or a G$_m$/C amplifier, but the additional calibration or digital post-processing is required to compensate the non-linearity and gain variation in open-loop configuration.

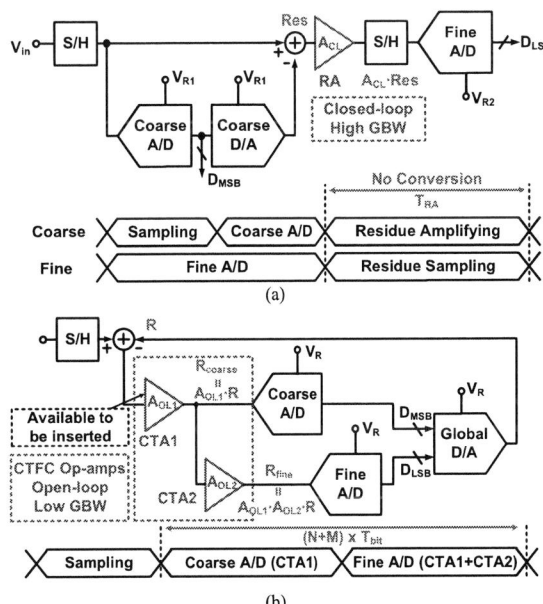

Fig. 1. The architectures and timing diagrams of the (a) pipelined-SAR ADC and (b) CTFC-SAR ADC.

TABLE I

COMPARISON TABLE OF THE DESIGN PARAMETERS

	Pipelined-SAR	CTFC-SAR
	N-bit Coarse ADC + M-bit Fine ADC	
DC gain	2^{N+M}	2^N
f_t	$2^N \cdot M \cdot \ln 2 / T_{RA}$	$2^N / T_{bit}$
Loading	$C_{DAC} + C_F$	C_{Gate}

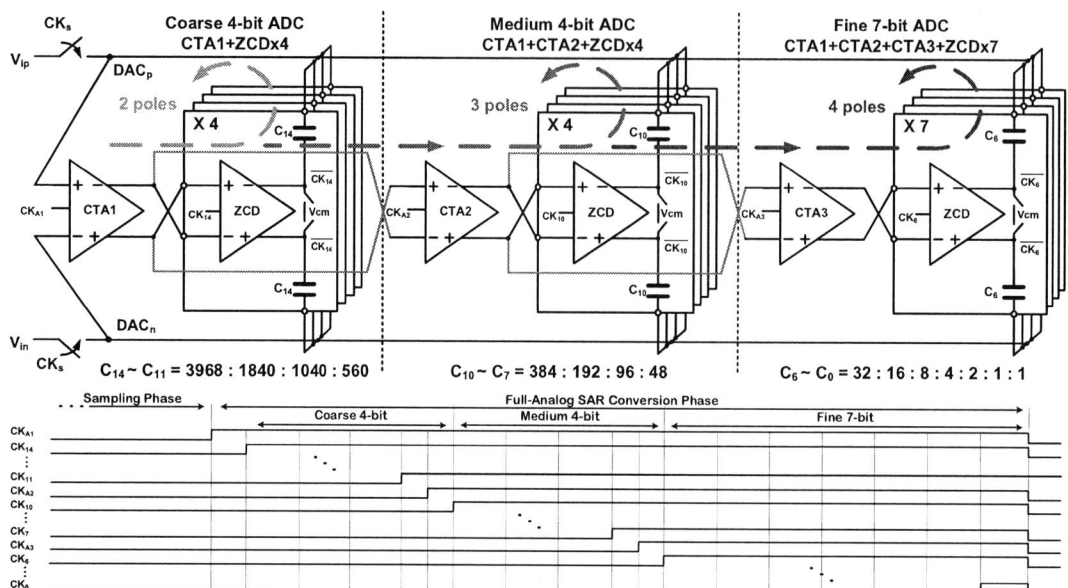

Fig. 2. The architecture and timing diagram of the proposed calibration-free full-analog CTFC-SAR ADC.

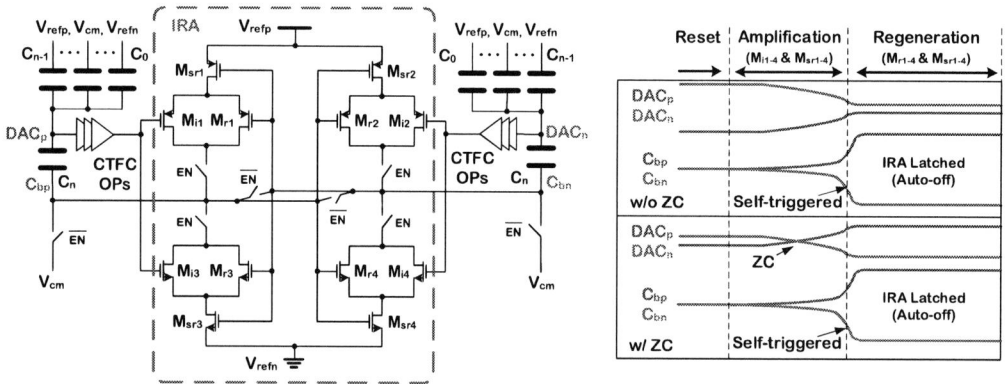

Fig. 3. The proposed ZCD (IRA) to achieve full-analog SAR operation.

II. OVERALL ARCHITECTURE

To achieve both SNR_{res} enhancement and power efficiency without the additional calibration or post processing, this work proposes a calibration-free full-analog SAR ADC with continuous-time feedforward cascaded (CTFC) Op-amps as in Fig. 1. By sharing the identical global DAC for all sub-ADCs, the weights among sub-ADCs are implicitly matched. By incorporating multiple 1-bit quantizers (zero-crossing detector, ZCD) in sub-ADCs for serial bit-conversions, no quantization threshold is required. Since only the polarity of residue is detected, the absolute magnitudes of amplified residues (R_{coarse} and R_{fine}) are inessential even saturated. Therefore, instead of requiring a high-precision RA using closed-loop amplifier in pipelined-SAR ADC, CTFC-SAR ADC uses the power-efficient and fast open-loop amplifier for residue amplification ahead of

the sub-ADCs. Moreover, the feedforward cascaded gain stages are realized to achieve a high feedforward gain for SNR_{res} enhancement of the subsequent sub-ADCs.

The comparison of Op-amp's requirements for (N+M)-bit pipelined-SAR and CTFC-SAR ADCs are depicted in Fig. 1. The pipelined-SAR ADC needs an extra phase (T_{RA}) for RA to settle to the required precision of fine ADC with the loading of DAC and feedback capacitor ($C_{DAC} + C_F$). On the contrary, the CTFC Op-amps only drive a smaller loading (MOS gates of the ZCD, C_{Gate}) along with the bit-conversion loops. As in Table I, to achieve the same SNR_{res} enhancement, the CTFC Op-amp requires only $1/2^M$ of DC gain (ADC) and M-times lower unit-gain bandwidth (f_T) with a smaller output loading (C_L). Moreover, proper redundancy is also implemented by the sub-radix DAC and the additional 1-bit LSB-conversion for error tolerance.

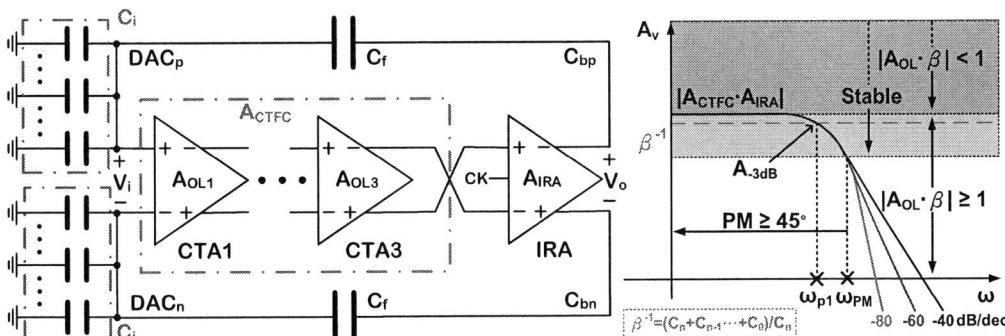

Fig. 4. Bode plot and stability analysis of CTFC Op-amps.

III. CIRCUIT IMPLEMENTATION

Fig. 2 shows the proposed CTFC-SAR ADC consisting of 3 sub-ADCs: coarse 4-bit, medium 4-bit, and fine 7-bit ADCs. Each sub-ADC contains an amplifier ahead of multiple ZCDs for SNR_{res} enhancement. By cascading the sub-ADCs, the subsequent sub-ADCs obtain the increasing residue gains to compensate the diminishing SNR_{res}. To improve the power efficiency, the inverter-based regenerative amplifier (IRA) in Fig. 3 is employed as a ZCD to perform the integrated functions of comparator, SAR logic and DAC switches by directly driving the corresponding capacitors (weights) in negative feedback configuration. Each IRA realizes a full-analog 1-bit SAR A-to-D conversion with the self-triggered amplification and regeneration, which fully utilizes the whole bit-conversion time (t_{bit}) with a relaxed bandwidth requirement. Moreover, since there is no reset phase (50% duty cycle) in the full-analog operation, the gain stages (CTAs) ahead of sub-ADCs are implemented in continuous-time to utilize the double timing budget for settling and without the need of high frequency clock compared to the dynamic comparator.

Each bit-conversion can be regarded as a switched-capacitor C_i-C_f feedback amplifier with cascaded feedforward gain stages as depicted in Fig. 4, where C_f is the corresponding weight capacitor to drive and C_i is the total capacitance of reset capacitors excluding the C_f. The loop stability should be noticed since there are 2-, 3-, and 4-cascade stages (poles) in the closed-loop configuration as in Fig. 2 for the coarse, medium, and fine ADCs, respectively. To ensure the stability in a multi-pole system, the frequency (ω_{PM}) of unity loop gain ($|A_{OL}\cdot\beta| = 1$) should be placed smaller than the second pole (ω_{p2}) to ensure the stability (PM > 45°). The $|A_{OL}\cdot\beta|$ of each conversion depends on the corresponding feedforward gain ($A_{OL} = A_{CTFC}\cdot A_{IRA}$) and capacitor ratio ($\beta^{-1} = (C_i+C_f)/C_f$). The coarse ADC is implicitly stable since ω_{p2} is greater than the unity-gain frequency (ω_t) due to the low A_{OL} and small ω_{p1} (large output capacitor). In the medium and fine ADCs, the stability is ensured by the proper DAC radix arrangement which either makes β^{-1} closed to $A_{OL,-3dB}$ (at ω_{p1}, PM = 135°)

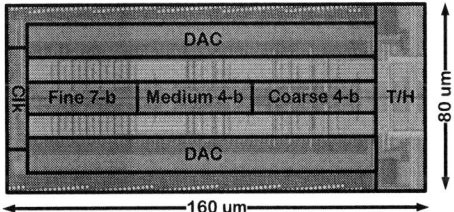

Fig. 5. Chip micrograph.

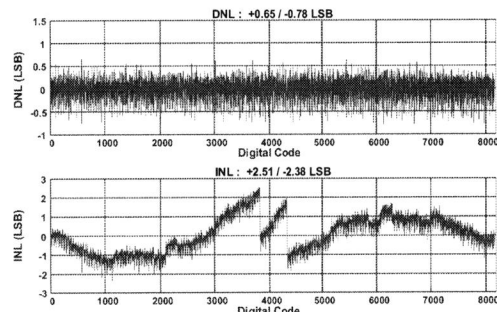

Fig. 6. Measured static linearity.

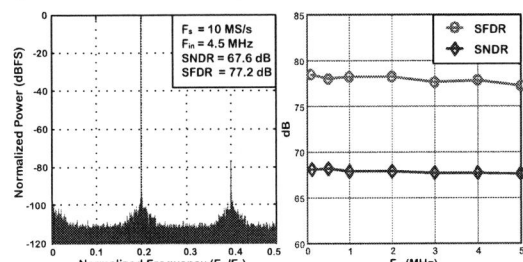

Fig. 7. Measured dynamic linearity.

in the preceding conversions or makes β^{-1} greater than A_{OL} ($|A_{OL}\cdot\beta| < 1$, stable region) in the subsequent conversions.

TABLE II

COMPARISON TABLE OF THE STATE-OF-THE-ART ADCS

	[3] Ding, JSSC 2017	[4] Shen, VLSI 2017	[5] Kim, JSSC 2016	[6] Liu, TCAS-I 2016	This work
Architecture	SAR	Flash-SAR	SATI SAR (2× CHs)	SAR	Full-analog CTFC-SAR
Technology	40 nm CMOS	55 nm CMOS	65 nm CMOS	180 nm CMOS	40 nm CMOS
Supply Voltage	1.0 V	3.3 / 1.2 V	0.6 V	1.8 V	0.7 V
Resolution	13 bit	16 bit	12 bit	12 bit	13 bit
Sampling Rate	6.4 MS/s	16 MS/s	10 MS/s	10 MS/s	10 MS/s
Sampling Capacitance (pF)	1.3	1	1.2	15.8	3.2
SNDR @ Nyq	64.1 dB	78.0 dB	64.3 dB	66.9 dB	67.6 dB
SFDR @ Nyq	81.9 dB	98.0 dB	83.8 dB	75.8 dB	77.2 dB
DNL (LSB)	1.08	N / A	+0.25 / -0.26	+0.36 / -0.33	+0.65 / -0.78
INL (LSB)	3.79	+2.30 / -1.90	+0.90 / -0.91	+0.55 / -0.60	+2.51 / -2.38
Power	46 µW	1630 µW	83 µW	820 µW	62.5 µW
FoM$_W$ @ Nyq (fJ/conversion-step)	5.5	156.9	6.2	44.2	3.2
FoM$_S$ @ Nyq (dB)	172.5	165	172.1	164.8	176.6
Core Area	0.068 mm^2	0.55 mm^2	0.1 mm^2	0.36 mm^2	0.013 mm^2
Calibration	Yes	No	Yes	No	No

The input-referred noise (IRN) is dominated by the CTA1 instead of the ZCDs since the noise of subsequent stage is suppressed by the preceding CTFC Op-amps. Similarly, the input-referred offset mismatch (IRO$_{mis}$) between two adjacent ZCDs is also suppressed by the preceding CTFC Op-amps. Therefore, the IRN and IRO$_{mis}$ are suppressed by the CTFC Op-amps and covered by the redundancies without any calibration. In the coarse and medium ADCs, sub-radix DAC arrangement based on [2] provides the speed-optimized and error-tolerated redundancies. In the fine ADC, the additional 1-bit LSB-conversion provides an additional ±0.5 LSB redundancy in each conversion. As a result, this work achieves a calibration-free full-analog CTFC-SAR ADC in a 13-bit resolution.

IV. MEASUREMENT RESULTS

Fig. 5 shows the ADC micrograph fabricated in 1P9M 40nm CMOS with an active area of 0.013 mm^2. Fig. 6 and Fig. 7 illustrate the measured static and dynamic performance (64x down-sampled), respectively. At 10MS/s under a 0.7 V supply with 3.2 pF input capacitance, the total power consumption is 62.5 µW where CTFC-SAR and clock generator consume 58.3 µW (93%) and 4.2 µW (7%), respectively. The measured SNDR and SFDR at Nyquist rate are 67.6 dB and 77.2 dB, respectively. The measured DNL and INL are +0.67/-0.75 LSB and +2.35/-2.26 LSB, respectively. Table II summarizes the performance comparison. The realized FoM$_W$ and FoM$_S$ at Nyquist rate are 3.2 fJ/conv.-step and 176.6 dB, respectively.

ACKNOWLEDGMENT

The authors acknowledge the support of National Chip Implementation Center (CIC) Taiwan, and Signal Sensing and Application Laboratory (SiSAL), National Tsing Hua University (NTHU). This research was supported by NOVATEK Fellowship and Ministry of Science and Technology, Taiwan under contract number MOST 104-2221-E-007-103- MY3.

REFERENCES

[1] B. Murmann, "ADC Performance Survey 1997-2017," [Online]. Available: http://web.stanford.edu/~murmann/adcsurvey.html.

[2] K.-H. Chang, and C.-C. Hsieh "A Hybrid Analog-to-Digital Conversion Algorithm with Sub-radix and Multiple Quantization Thresholds," in IEEE Tran. Circuits Syst. I (TCAS-I), vol. 64, no. 6, pp. 1400-1408, Jun. 2017.

[3] M. Ding, P. Harpe, Y.-H. Liu, B. Busze, K. Philips, and H. de Groot, "A 46 µW 13 b 6.4 MS/s SAR ADC with Background Mismatch and Offset Calibration," in IEEE J. Solid-State Circuits, vol. 52, no. 2, pp. 423-432, Feb. 2017.

[4] J. Shen, A. Shikata, L. Fernando, N. Guthrie, B. Chen, M. Maddox, N. Macarenhas, R. Kapusta, and M. Coln, "A 16-bit 16MS/s SAR ADC with on-chip calibration in 55nm CMOS," in IEEE J. Solid-State Circuits, vol. 53, no. 4, pp. 1149-1160, Aug. 2018.

[5] W. Kim, H.-K Hong, T.-J. Roh, H.-W Kang, S.-I. Hwang, D.-S. Jo, D.-J. Chang, M.-J. Sco, and S.-T. Ryu, "A 0.6 V 12 b 10 MS/s Low-Noise Asynchronous SAR-Assisted Time-Interleaved SAR (SATI-SAR) ADC," in IEEE J. Solid-State Circuits, vol. 51, no. 8, pp. 1826-1839, Aug. 2016.

[6] S. Liu, Y. Shen, and Z. Zhu, "A 12-Bit 10 MS/s SAR ADC with High Linearity and Energy Efficient Switching," in IEEE Tran. Circuits Syst. I (TCAS-I), vol. 63, no. 10, pp. 1616-1627, Oct. 2016.

978-1-5386-6414-8/18 $31.00 © 2018 IEEE

18-6 (8030)

IEEE Asian Solid-State Circuits Conference
November 5 - 7, 2018/Tainan, Taiwan

An 89.55dB-SFDR 179.6dB-FoM$_S$ 12-bit 1MS/s SAR-Assisted SAR ADC with Weight-Split Compensation Calibration

Yao-Sheng Hu, Jhao-Huei Lin, Ding-Guo Lin, Kai-Yue Lin and Hsin-Shu Chen

Department of Electrical Engineering and Graduate Institute of Electronics Engineering, National Taiwan University
Taipei, Taiwan 10617, R. O. C.
hschen@ntu.edu.tw

Abstract - This paper presents an 89.55dB-SFDR 2.55µW 12-bit 1MS/s SAR-assisted SAR ADC in 40nm CMOS at a 0.7V supply. The proposed weight-split compensation provides an accurate mapping between capacitor mismatch and digital weight to take advantages of both low-power skipping switching method and robust digital calibration. The reconfigurable redundancy region with tracking bits is used to speed up the calibration time to only 112 clock cycles. The SFDR is improved by 19.45dB with unit capacitors of 0.25fF for power-saving. The prototype ADC achieves an SNDR of 69.1dB at Nyquist rate. It results in an FoM$_S$ of 179.6dB and an FoM$_W$ of 1.43fJ/c.-s.

Index Terms - *low-power, SAR ADC, SAR-assisted, on-chip capacitor mismatch calibration.*

I. INTRODUCTION

High-linearity low-power SAR ADC array is widely utilized in numerous mobile sensor applications such as touchscreen/ fingerprint sensor, biomedical wireless sensor network and Internet of Everything (IoE). To prolong the battery life, shrinking the size of unit capacitor (C_u) [1] with the skipping switching method [2], [3] is an effective approach. But its linearity performance would be limited by the process variation. In addition, its non-switching feature prevents the skipping switching method from the usage of the digital capacitor mismatch calibration [4], [5], which can enhance the linearity. This work proposes a weight-split compensation (WSC) calibration combining the compact digital calibration with the skipping switching method, which reduces switching energy and inherently get $\sqrt{2}\times$ linearity improvement. The proposed WSC calibration provides accurate mapping from the capacitor mismatch of the skipping switching method to the corresponding digital weight. Also, the reconfigurable redundancy region and the shift registers with the tracking bits [6] tolerate more mismatch during the calibration mode and save power consumption during the conversion mode. Moreover, it not only accelerates the total calibration time for the low-cost circuit probing test or fast start-up applications, but also increases the calibration accuracy. As a result, this work accomplishes an SFDR of as high as 89.55dB with a C_u of only 0.25fF. At a 0.7V supply, its FoM$_S$ and FoM$_W$ reach 179.6dB and 1.43fJ/c.-s., respectively.

II. PROPOSED WEIGHT-SPLIT COMPENSATION CALIBRATION

The DACs of the conventional non-skipping switching methods, such as monotonic or merged capacitor [1], [4]-[6], switch the corresponding capacitors, $C_i + \Delta C_i$, in every

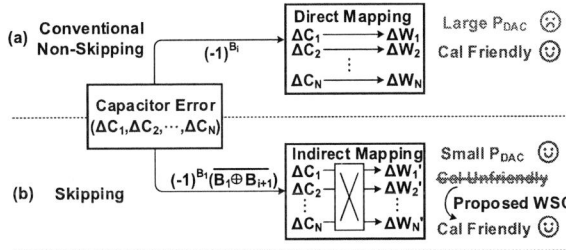

Fig. 1. The direct and indirect mapping equations from ΔC_i to ΔW_i according to the (a) non-skipping and (b) skipping switching method.

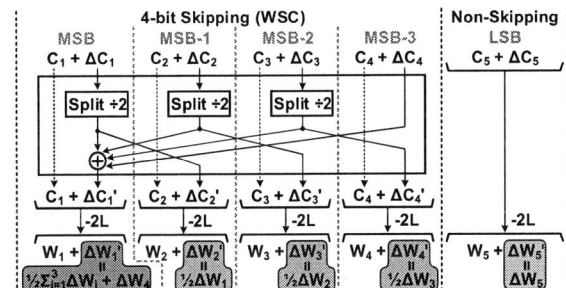

Fig. 2. The 5-bit mapping example of the proposed WSC calibration

conversion, where $i = 1, 2, ..., N$ and N is the number of bits. Its switching voltage, V_{SW_NOSK}, is written as

$$V_{SW\_NOSK} = \frac{V_{ref}}{C_{tot}} \sum_{i=1}^{N} (-1)^{B_i} (C_i + \Delta C_i), \quad (1)$$

where B_i is the digital code of the ith bit conversion. C_{tot} and V_{ref} are the total value of the capacitor array and the reference voltage, respectively. V_{SW_NOSK} is proportional to the inner product of $(-1)^{B_i}$ and $C_i + \Delta C_i$. Equation (1) can be simplified as

$$V_{SW\_NOSK} = V_{ref} - \frac{2V_{ref}}{C_{tot}} \sum_{i=1}^{N} (B_i C'_i), \quad (2)$$

where $C'_i = C_i + \Delta C_i$. Assume that $C'_i = LW'_i = L(W_i + \Delta W_i)$, where L is the mapping coefficient from C_i to W_i. Then, V_{SW_NOSK} is expressed as

$$V_{SW\_NOSK} = V_{ref} - \frac{2LV_{ref}}{C_{tot}} \sum_{i=1}^{N} (B_i W'_i)$$
$$= V_{ref} - \frac{2LV_{ref}}{C_{tot}} D_{Out}. \quad (3)$$

978-1-5386-6414-8/18 $31.00 © 2018 IEEE

253

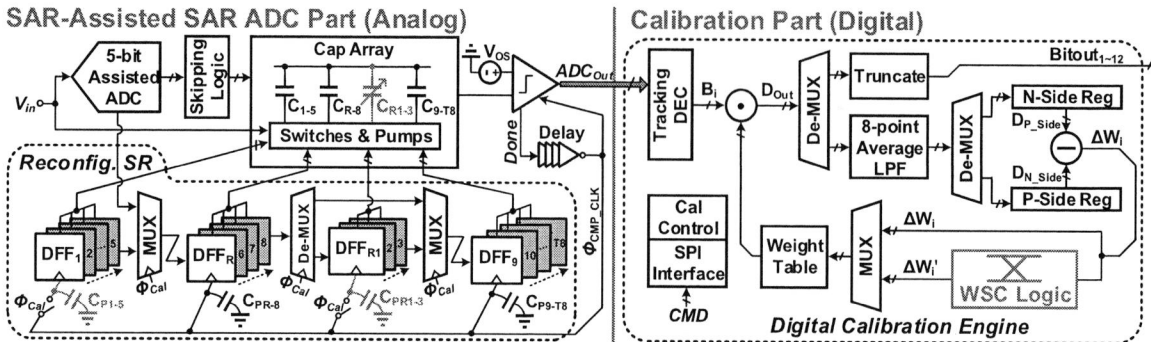

Fig. 3. Simplified block diagram of the proposed WSC calibration system.

After this transformation, V_{SW_NOSK} is transformed to be proportional to the inner product of B_i and $W_i + \Delta W_i$. Fig. 1(a) shows that ΔC_i is directly mapped to ΔW_i. That is, if the ratio of ΔC_i and C_{tot} is given, the corresponding ΔW_i can be directly mapped. According to (3), ΔW_i can be used to compensate D_{Out}. The mapping from the capacitor mismatch error ΔC_i to the digital weight W_i' results in the feasible implementation of digital calibration circuit with the non-skipping switching methods.

On the other hand, the skipping switching method [2] possesses non-switching states for switching energy reduction. As illustrated in Fig. 1(b), its $(-1)^{B_1}\overline{(B_1 \oplus B_{i+1})}$ judgement criterion leads ΔC_i indirectly proportional to $\Delta W_i'$. Therefore, it is necessary for the proposed WSC to map its switching voltage, V_{SW_SK}, to the equivalent threshold. Note that if the comparable format of (3) can be derived, the mapping from the capacitor mismatch error ΔC_i to the digital weight W_i' will enable the implementation of digital calibration with the low-power skipping switching method. Its switching voltage, V_{SW_SK}, can be computed as

$$V_{SW\_SK} = \frac{V_{ref}}{C_{tot}} \left\{ \begin{matrix} \left[\sum_{i=1}^{N-1} (-1)^{B_1} \overline{(B_1 \oplus B_{i+1})} (C_i + \Delta C_i) \right] \\ + (-1)^{B_1}(C_N + \Delta C_N) \end{matrix} \right\}. \quad (4)$$

By substituting $1 - (B_1 + B_{i+1})$ into $(-1)^{B_1}\overline{(B_1 \oplus B_{i+1})}$, V_{SW_SK} is written as

$$V_{SW\_SK} =$$
$$\frac{V_{ref}}{C_{tot}} \left\{ \begin{matrix} C_{tot} \\ - \left[B_1 \left(\sum_{i=1}^{N-1} C_i + \sum_{i=1}^{N-1} \Delta C_i + 2(C_N + \Delta C_N) \right) \right] \\ + \sum_{i=1}^{N-1} B_{i+1}(C_i + \Delta C_i) \end{matrix} \right\}. \quad (5)$$

A binary weighted capacitor array, $C_i = 2C_{i+1}$, is utilized to further simplify (5) as

$$V_{SW\_SK} = \frac{V_{ref}}{C_{tot}} \left\{ \begin{matrix} C_{tot} \\ -2 \left[\begin{matrix} B_1 \left(C_1 + \Delta C_N + \frac{1}{2} \sum_{i=1}^{N-1} \Delta C_i \right) \\ + \sum_{i=1}^{N-1} B_{i+1}(C_{i+1} + \frac{1}{2}\Delta C_i) \end{matrix} \right] \end{matrix} \right\}. \quad (6)$$

Let $C_i = LW_i$, $\Delta C_i = L\Delta W_i$ and substitute them into (6).

The switching direction of every $C_i + \Delta C_i$ in the skipping switching method depends on B_1 and B_{i-1}, so its $\Delta W_i'$ has component of every ΔW_i and the other $\Delta W_i'$ has the component of previous weight mismatch, ΔW_{i-1}, where i is from 2 to N. Consequently, V_{SW_SK} gets the comparable format of (3), which is

$$V_{SW\_SK} = V_{ref} - \frac{2LV_{ref}}{C_{tot}} \sum_{i=1}^{N} B_i (W_i + \Delta W_i'), \quad (7)$$

where

$$\Delta W_1' = \frac{1}{2} \left(\sum_{i=1}^{N-1} \Delta W_i \right) + \Delta W_N \quad (8a)$$

$$\Delta W_i' = \frac{1}{2} \Delta W_{i-1}, \quad i = 2 \sim N. \quad (8b)$$

By self-detecting the ratio of ΔC_i, the accurate $\Delta W_i'$ can be calculated for every B_i. The proposed WSC can calibrate the capacitor mismatch error in the skipping switching method.

Fig. 2 clearly visualizes (8) in a 5-bit mapping example for the WSC calibration with the first 4 bits of skipping switching method and the last bit of non-skipping switching method. The binary capacitor array, C_i, and the digital weight, W_i, are used to set a split ratio as 2 for the simplicity of digital logic. According to the V_{SW_SK} indirect mapping results of (8), ΔC_i of the first 3 bits are split by 2. The half is added to $\Delta C_i'$ and the other becomes $\Delta C_{i+1}'$. The switching direction of the last skipping capacitor, C_4, only depends on B_1, so all of ΔC_4 is added to $\Delta C_1'$. Without the skipping state, the switching direction of C_5 only depends on $(-1)^{B_5}$. ΔC_5 maps directly to $\Delta C_5'$. Hence, every $\Delta W_i'$ can be compensated by the value of ΔC_i. By using the proposed WSC calibration, this work attains $\sqrt{2}\times$ linearity, power and accuracy improvement from both the skipping switching method and the digital calibration.

III. ARCHITECTURE AND CIRCUIT IMPLEMENTATION

Fig.3 shows the simplified block diagram of this work. It consists of a SAR-assisted SAR ADC part and a digital calibration part. The ADC part is composed of several analog blocks, such as a 5-bit assisted ADC, skipping logic, a main ADC and its reconfigurable shift register. The digital calibration part is fully on-chip logic including tracking digital error correction (DEC) logic, WSC logic, weight look-up table, calibration control logic and SPI interface. The 5-bit

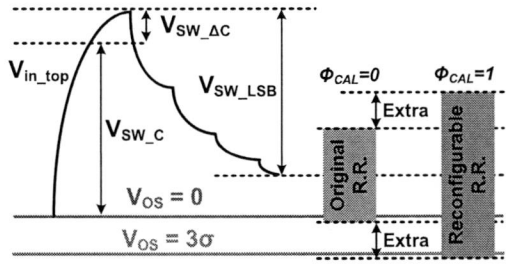

*R.R. means Redundancy Region.

Fig. 4. Top-plate voltage of capacitor array during Φ_{CAL}=0 and the concept of reconfigurable redundancy region.

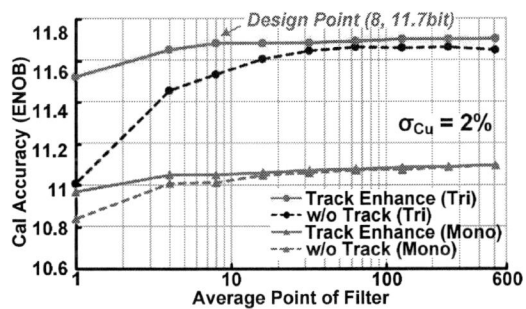

Fig. 5. The calibration accuracy simulation of different average point of filter.

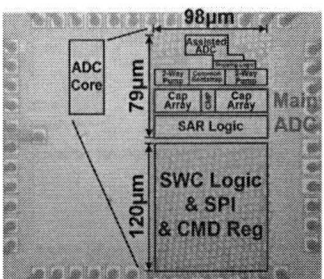

Fig. 6. Chip micrograph.

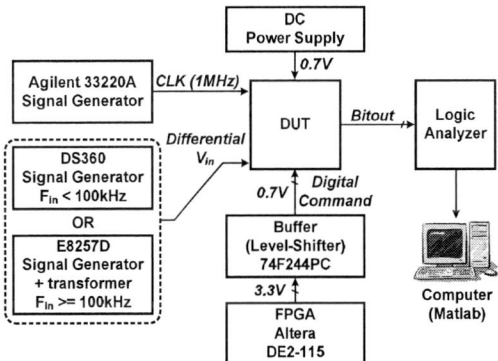

Fig. 7. Measurement setup.

assisted ADC and the main ADC perform the subranging operation during the conversion mode, Φ_{CAL}=0. A pump system is also used to increase sampling linearity and settling accuracy at a low-voltage supply. The common bootstrap circuit is shared by both ADCs to sample the same input signal, V_{in}, with high linearity.

Fig.4 depicts the top plate voltage of the main capacitor array, V_{in_top}, during the calibration mode, Φ_{CAL}=1. The switched voltage of the corresponding main ADC capacitors, $V_{SW\_C} + V_{SW\_\Delta C}$, and ΔC_i can be quantized by the switched voltage of the rest LSB capacitors, V_{SW_LSB}. Then, the corresponding ΔW_i can be calculated by the WSC logic and compensate the capacitor mismatch to W_i'. The WSC system calibrates from MSB-6 to MSB capacitors, C_{1-7}. Both p-side and n-side of ADC$_{Out}$ execute inner product with the weight look-up table, $W_i + \Delta W_i$. Then, the WSC logic calculates $\Delta W_i'$ from $D_{P\_Side} - D_{N\_Side}$ and corrects the weight look-up table according to (8). Hence, ΔC_i of the skipping switching method can match the equivalent digital thresholds during the conversion mode.

Although the comparator offset, V_{OS}, can be removed by $D_{P\_Side} - D_{N\_Side}$, it still needs to be covered by a sufficient redundancy region during the calibration mode. Insufficient redundancy region causes quantization error in ΔW_i, which makes $\Delta W_i'$ inaccurate. Thus, 3 tunable cover capacitors, C_{R1-3}, provide extra +/-112LSB red reconfigurable redundancy region than the gray original cover region during

the calibration mode, Φ_{CAL}=1, as shown in Fig. 4. It makes sure the 3σ of V_{OS} and C_u mismatch are in the cover region. This reconfigurable redundancy made by C_{R1-3} and the shift registers of the first 5 MSBs are disconnected during Φ_{CAL}=0. As a result, their parasitic capacitors, C_{P1-5} and C_{PR1-3}, are gated to save digital power and speed up Φ_{CMP_CLK}.

Fig. 5 illustrates the calibration accuracy of the different filter average points. The longer filter average time is, the longer the calibration time is. The standard deviation of C_u mismatch, σ$_{Cu}$, is 2% with a unit capacitance of 0.25fF. It is an extrapolated result of the fabrication process datasheet. To achieve 11.7 bit of ENOB, this work chooses an 8-point average filter to reduce the noise disturbance of D$_{Out}$. The single-end switching feature of the conventional monotonic method switching limits the calibration accuracy. Instead, using additional track bits with the tri-level DAC switching method not only enhances the calibration accuracy, but also reduces the total calibration time. Therefore, this work only takes 112 clock cycles to calibrate the first 7 MSB capacitors.

IV. MEASUREMENT RESULTS

The prototype ADC fabricated in 40nm CMOS. Fig. 6 displays the chip micrograph. The core circuit occupies an area of 0.0198mm^2 (202μm×98μm), including the ADC part of 0.0077mm^2 (79μm×98μm) and the digital calibration part of 0.0118mm^2 (120μm×98μm). The gate counts of the digital circuit is 4.3k, including 3.1k of the WSC arithmetic logic and 1.2k of the SPI interface and command registers for the prototype testing. Fig. 7 shows the measurement setup of this work. The FPGA sends the command functions through the SPI interface to the test chip. F_{in} higher than 100kHz is measured by transformers with the associated band-pass filter (BPF), while F_{in} lower than 100kHz is measured by a differential signal generator with the associated passive low-pass filter (LPF).

Fig. 8 illustrates the static performance at the conversion rate of 1MS/s. Thanks for the proposed WSC calibration, the effective value of C_u is down to 0.25fF, which is close to the limitation of KT/C noise. Before calibration, the maximal DNL and INL are 0.66 and 2.63LSB, respectively. After calibration, the DNL and INL are down to +0.61/-0.57 and +0.93/-0.92LSB. Fig. 9 and Fig. 10 show the Nyquist rate FFT and the dynamic performance, respectively. Before calibration, the measured SFDR is 70.1dB and SNDR is 61dB at the Nyquist rate. After calibration, the measured SNDR, SFDR, SNR, and THD achieve 66.54dB, 89.55dB, 67.09dB, and 75.68dB, respectively. The SFDR performance is improved by 19.45dB.

The measured power dissipation is 2.47µW with a 0.7V supply. It can be broken down as follows: the DACs use only 4.8% of the power; the analog circuits use 51.4%; the digital circuits use 43.8%. Because of the noise requirement, the comparator (CMP) of the main ADC dominates the power distribution of this work. The start-up power of the calibration mode is about 4.5µW for 120µs, but it is shut down during the conversion mode. The FoMs and FoMw achieves 179.6dB and 1.43fJ/c.-s., respectively. Table I lists the performance summary of the prior-art ADCs in recent years.

ACKNOWLEDGEMENT

This work is supported by the Donation Grant FD105012.

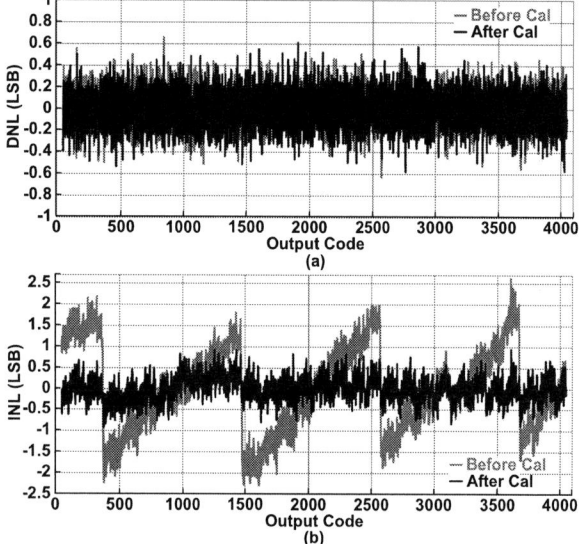

Fig. 8. Measured static performance of (a) DNL and (b) INL.

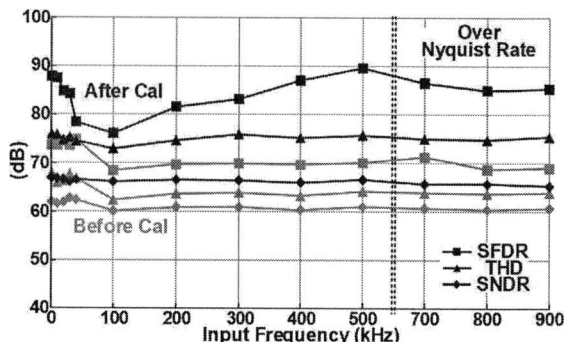

Fig. 9. Measured dynamic performance.

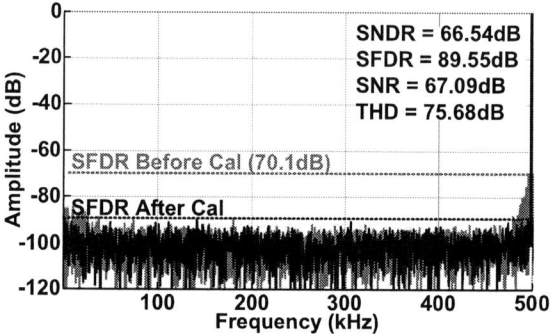

Fig. 10. Measured Nyquist rate FFT plot.

Table I. Performance summary and comparison.

	ISSCC'13 P. Harpe	ISSCC'14 P. Harpe	ISSCC'15 Y. Lim	ISSCC'15 M. Ding	VLSI'16 S.-E Hsieh	This Work
Technology (nm)	65	65	65	40	90	40
Resolution (bits)	12	14	13	13	11	12
Active Area (mm^2)	0.076	0.18	0.054	0.0675	0.035	0.0198
Supply (V)	0.6	0.8	1.2	1	0.3	0.7
F_S (MS/s)	0.04	0.032	50	6.4	0.6	1
SFDR$_{Nyquist}$	68.8	78.5	84.6	81.9	73.4	89.55
DNL$_{max}$ (LSB)	0.97	1.75	0.58	1.08	0.63	0.61
INL$_{max}$ (LSB)	1.9	3.5	0.96	3.79	0.72	0.93
Power (µW)	0.097	1.367	1000	46	0.187	2.47
ENOB$_{Nyquist}$	10.1	11.3	11.48	10.36	9.46	10.76
FoMs (dB)	175.7	176.3	174.9	172.5	180.8	179.6
FoMw (fJ/c.-s.)	2.2	8.2	6.9	5.5	0.44	1.43

REFERENCES

[1] P. Harpe, *et al.*, "An Oversampled 12/14b SAR ADC with Noise Reduction and Linearity Enhancements Achieving up to 79.1dB SNDR," *IEEE ISSCC Dig. Tech. Papers*, pp. 194-195, Feb. 2014.

[2] H.-Y. Tai, *et al.*, "A 0.85fJ/conversion-step 10b 200kS/s Subranging SAR ADC in 40nm CMOS," *IEEE ISSCC Dig. Tech. Papers*, pp. 196-197, Feb. 2014.

[3] Y. Lim, *et al.*, "A 1mW 71.5dB SNDR 50MS/s 13b Fully Differential Ring Amplifier Based SAR-Assisted Pipeline ADC," *IEEE ISSCC Dig. Tech. Papers*, pp. 458-459, Feb. 2015.

[4] Y. Zhou, *et al.*, "A 12 bit 160 MS/s Two-Step SAR ADC with Background Bit-Weight Calibration Using a Time-Domain Proximity Detector," *IEEE J. Solid-State Circuits*, pp. 920-931, Apr. 2015.

[5] W.-H Tseng, *et al.*, "A 12-bit 104 MS/s SAR ADC in 28 nm CMOS for Digitally-Assisted Wireless Transmitters," *IEEE J. Solid-State Circuits*, pp. 2222-2231, Oct. 2016.

[6] T. Miki, *et al.*, "A 4.2mW 50MS/s 13bit CMOS SAR ADC with SNR and SFDR Enhancement Techniques," *IEEE J. Solid-State Circuits*, pp.1372-1381, Jun. 2015.

An Image Recognition Processor with Time-domain Accelerators using Efficient Time Encoding and Non-linear Logic Operation

Zhengyu Chen and Jie Gu

Department of Electrical Engineering and Computer Science, Northwestern University
2145 Sheridan Road, Evanston, IL, 60208, USA
zhengyuchen2015@u.northwestern.edu, jgu@northwestern.edu

Abstract— This paper presents novel time-domain circuit techniques including double encoding strategy, shared time generator (TG) and bit-scalable design which significantly improve the performance of time-domain signal processing (TDSP) and error tolerance. A feature-extraction and vector-quantization processor accelerated by TDSP has been developed for real-time image recognition. A 55nm prototype chip shows 72 fps/core (@1.33 GHz) operation with significant enhancement from time-domain techniques compared with conventional digital implementation.

Keywords—time-domain signal processing; image recognition; bit-scalable design; double-encoding scheme.

I. INTRODUCTION

Time-domain signal processing (TDSP) design has been attracting attention recently on a variety of applications, e.g. neural network [1, 2, 3], due to the fact that it combines the benefits from both analog and digital computing. Time-based design offers several attractive features: (1) it provides inherent error resiliency for applications like image processing because the error probability decreases exponentially with higher position of bit; (2) Many non-linear operations from neural network and image processing applications can be implemented more efficiently; (3) It is more energy efficient in term of encoding since multiple bits can be encoded in the pulse width within one pair of rise and fall transition. In addition, the TDSP circuits are implemented in a fully digital method using standard cells, which makes it scales well with technology and supply.

Time-based designs have been recently proposed on a variety of applications, e.g. neural network and low-density parity check (LDPC) [1, 2, 3, 4]. However, most of existing work suffers from the following issues: (1) the existing design utilized a low efficient multiple-gate time encoding (TE) circuit limiting the advantages of the technique [4]; (2) existing design only contains low bits precision, e.g. 3 bits in [4], and does not address the process variation impact to the design; (3) The strong capability of TDSP in various nonlinear operations, e.g. MAX, MIN operation, has not been well explored leaving limited improvement from the techniques [1, 2, 3].

In order to overcome these drawbacks, this work proposes a series of circuit techniques with a demonstration on image recognition processor design. The contributions of this paper are highlighted as below: (1) new energy efficient circuits techniques such as shared time generator (TG) design, double-encoding strategy and bit-scalable design, are proposed to significantly improve the efficiency of TDSP; (2) Time-based median filter and winner-take-all algorithms and circuits are proposed to demonstrate the energy, area and performance improvement comes from TDSP; (3) A test chip on image recognition application is provided achieving state-of-art energy and computing efficiency.

II. TIME DOMAIN SIGNAL PROCESSING

A. System Overview

In TDSP, the digital binary inputs are encoded and processed in time domain and reconverted back into digital domain through time decoder. Fig. 1(a) shows the overall system view of the TDSP which consists of key building blocks of (1) time encoder; (2) Time logic for signal processing. (3) time decoder. Compared with conventional CMOS logic design, many operations can be performed much more efficiently. For example, the MAX, MIN and Compare (CMP) operations are realized by a single or two logic gates leading to tremendous saving from conventional CMOS operation. Although the information is processed in time domain, the information carriers are still binary digital signals processed by conventional logic circuits, such as inverters or NOR gates, making the technique suitable for advanced CMOS technology compared with the existing analog signal processing.

B. Proposed Highly Efficient Circuit Techniques

Fig. 1(b) shows the proposed circuit techniques used to enhance the efficiency of time domain processing and tolerance to process variation including: (a) shared time generator, (b) double-edge operation scheme, (c) bit-split technique and (d) high efficient time-domain operations.

1) Shared time generator

We propose shared time generator (TG) which uses a common invertor chain to generate timing signals relieving the variation impact. Compared with previous individual time encoder (TE) design whose mismatch in a n-bit TE is proportional to $\sqrt{2^n + n}$ [4], the shared TG has mismatch only proportional to $\sqrt{n}$ leading to 3~4X less mismatch rendering shortened single bit delay and smaller cell size.

2) Double-edge operation

We propose a double-edge operation, where logic operation is processed at both rising and falling transitions [5]. Because for any single gate, energy is only consumed during the rising edge of the

978-1-5386-6414-8/18 $31.00 © 2018 IEEE

output, the energy efficiency from the double-encoding is improved by 2X compared with previous work [4]. Area consumption is also reduced by around 30% because the buffer stage is shared for both rising and falling transitions.

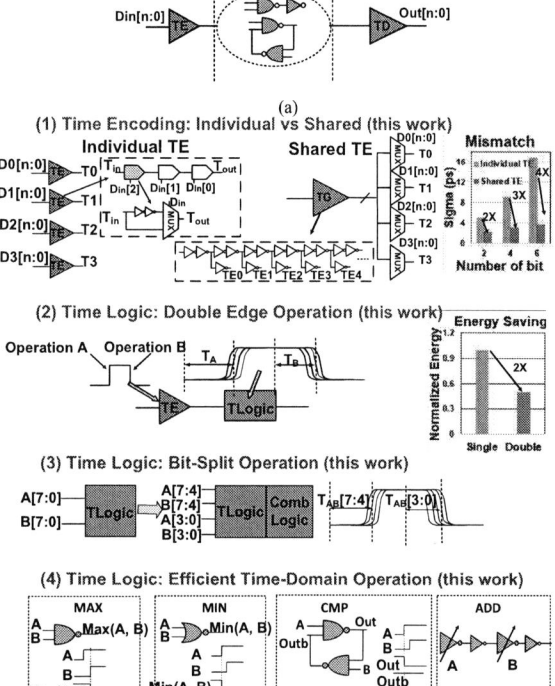

Fig. 1. Concept of Time-domain signal process: (a) system overview of TDSP, (b) proposed shared time encoder scheme, double-edge operation, bit-split techniques and efficient time-domain operations.

3) Bit-split technique

We propose a bit-split technique that splits an input vector into smaller bit groups leading to a scalable high-resolution encoder without exponentially increasing the delay. This work encodes 4-bit MSBs operation in falling-edge and 4-bit LSBs in rising-edge rending 16X reduction of delay and 2-4X reduction on energy. This technique also makes TDSP designs scalable with number of bits since large number of bits can be split into small groups.

4) Efficient time-domain operations

While it takes significant effort for conventional digital design to perform MAX/MIN/CMP operation, it only takes a single or two logic gates for TDSP to perform the same operation. For example, the determination of the winner in winner-take-all (WTA) module can be easily done by using a CMP operation while the passing of winner to the next stage can be simply realized by a NAND/NOR gate. For comparison, the digital counterpart takes entire 6-bit ADD/SUB operation in conventional digital design to realize the above operation.

III. DESIGN IMPLEMENTATION DETAILS

To demonstrate the proposed circuit techniques, we adopt a basic image recognition algorithm as shown in Fig.2 into a hybrid ASIC design with time-domain accelerators [6].

A. Implemented Image Processing Algorithm

As shown in Fig.2, the operations of the image recognition algorithm involve three main steps: (1) feature extraction which detects edges in four directions: horizontal, vertical, +45° and -45°. In order to determine the threshold value for edge detection, all the absolute-value differences between each two neighboring pixels are calculated in the 3×3 kernel and the median detection of the 12 difference-value is adopted as the threshold; (2) Vector formation where edge flags in all directions are counted and the spatial distribution of edge flags is represented by a vector of 64 elements; (3) Classification: the generated feature vector is then classified by a winner-take-all (WTA) classifier. The heavily used nonlinear computations such as comparison (CMP), MIN/MAX function, are expensive for CPU/GPU based design or even state-of-art ASIC design. In particular, the MF consumes 70% of total CPU cycles due to its enormous amount of CMP and swapping operation [6]. Previously, analog signal processing, have been proposed to accelerate the computation [6,7]. In this work, TDSP based accelerators are used to remove the bottlenecks of the algorithm, i.e. with significant speedup as will be shown later.

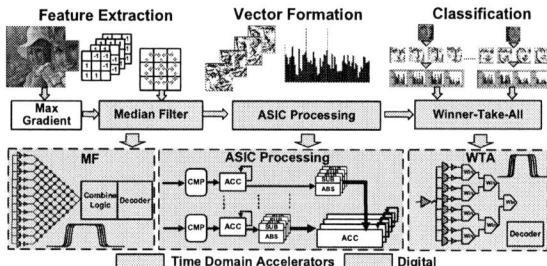

Fig. 2. Overview of image recognition algorithm used in this work.

B. High Efficient Time-Domain Median Filter

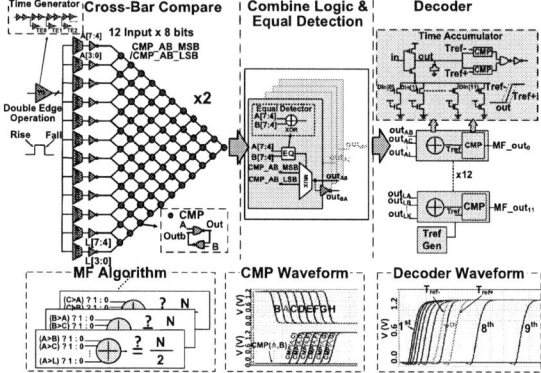

Fig. 3. Time-domain MF implementation and circuit diagram.

Fig.3 shows the algorithm and detailed implementation of the proposed 12-input 8-bit time-domain median filter design. The digital inputs are first converted into time-domain by the proposed shared TEs. Each pair of input is compared parallelly in time-domain with double-edge and bit-split design, with overall 66 comparisons for both MSBs and LSBs for all 12 input vectors. All the comparison results are then processed by the following digital logic for purpose of equal detection and MSB/LSB grouping to obtain the comparison results for all input vectors. To decode the comparison results into final median value, a special time accumulator design is implemented where all 11 digital comparison results from each input are summed in time-domain and compared with a reference median time-domain signal. The final area of proposed time-based MF is improved by 24% compared to conventional digital implementation.

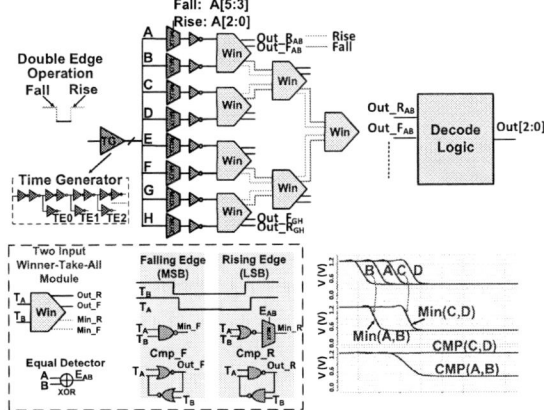

Fig. 4. Topology and implementation of WTA in time-domain.

C. High Efficient Time-Domain Winner-Take-All

Fig.4 shows the design of the proposed 8-input 6-bit winner-take-all accelerator. The algorithm of WTA is based on binary comparison tree. The winners or MIN value from each branch in each stage are selected in parallel and propagated to the subsequent stage to be compared again. After converting the input digital value into time-domain, the comparison can be simply made by using time-domain CMP. The MIN function which is built by a single NAND/NOR gate directly propagates the winner to next stage without intermediate restoration or regeneration. As a result, a massive parallel operation with mostly NAND or NOR gates, is realized in TDSP. All comparison results are finally decoded in digital domain to find the final winner. Shared TG, double-edge operation and split-bit techniques are also utilized. The area of proposed time-based WTA is improved by 42% compare to digital implementation.

D. Test Chip Implementation

Fig.5 shows the test chip implementation of the proposed image recognition processor in a 55nm low power CMOS process at 1.2V. Scan chains are used to fetch image data to on-chip register files and read out all internal register/comparator values for test verification. A special timing test module is built to exam the linearity and robustness of the proposed shared TE design. A Vernier-delay-chain based TDC with ~5ps bit-resolution is used to characterize the timing variation of TE. The TE used throughout

this work is implemented with a ~25ps single-bit resolution which can be tuned from 13ps to 35ps for further evaluation.

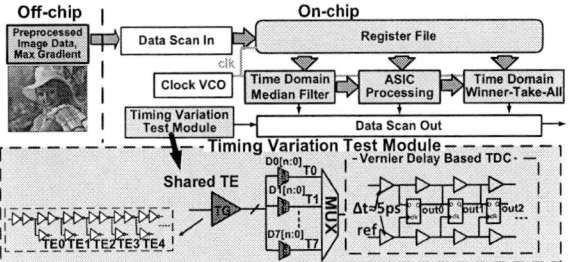

Fig. 5. Top level implementation of the proposed test chip.

IV. MEASUREMENT RESULT

Fig. 6 shows measurement results. Robustness of the design was verified across 10 chips. By default, no error was observed at the design target speed of 1.33GHz. As shown in Fig. 6 (a), when pushing the TE resolution beyond 22ps, small error was observed at the MF's output at LSBs while no error was observed at the final WTA output. The error rate from MF reached 0.6% when reducing the resolution to 13ps which led to an operating speed of 1.5GHz, a 13% boost of performance without observing error at the final output. This also shows the strength of TDSP where small errors may be generated at LSBs at stringent timing condition but does not lead to significant error at final output. The linearity of TE was also measured for all eight inputs across 10 chips and supply voltages from 1.1V to 1.4V (Fig. 6 (b)). Only single-bit deviation (~8 ps) from ideal value was observed across all the measurement leading to an integral-nonlinearity (INL) of less than 0.3 LSB. The measured (blue histogram) vs. the simulated variation from SPICE Monte Carlo simulation of TE across chips are shown in Fig. b (c). The design is compared with conventional ASIC design in the same process with standard synthesis and place and route implementation (Fig. 6 (d)). A 24% to 42% area saving is observed in MF and WTA accelerators compared with ASIC implementation. A 1.7X speedup and a 20% to 23% power saving are also observed using TDSP. The overall image recognition processor operates at 1.33GHz with a throughput of 72 frames per second (fps). Fig. 7 shows the die micrograph and the detailed design specifications. As the focus of this work is on robust and efficient techniques for time-based design, a direct comparison with prior work is difficult. We made comparison in two aspects:

As shown in Table I and Fig. 8 (a), compared with prior time-based work [1, 4, 10], this work achieved (1) the fastest operation speed with a single-bit delay which is 2X~4X shorter, (2) the largest number of bits by the bit-split technique, (3) the lowest mismatch/variation which is 3X smaller compared with [1, 4, 10], (4) the least encoding effort with lowest transistor count.

As shown in Table I and Fig. 8 (b), compared with image recognition processors with similar algorithms, e.g. feature vector based, this work achieved (1) the highest throughput per core which is 5X more that prior designs, (2) the highest energy efficiency for single processor core with more than 9X improvement. However, it is notable that prior work involves more configurations and numbers of processing units [6, 8, 9].

enhancement from time-domain techniques compared with conventional digital implementation.

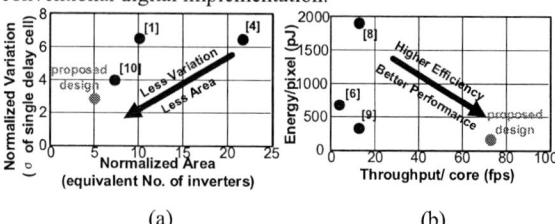

(a)   (b)

Fig. 8. (a) Efficiency vs. variation of timing encoding circuits from prior work [4, 1, 10] and our proposed work; (b)Performance vs. energy for prior image processing designs [6, 8, 9].

TABLE I. PERFORMANCE COMPARISON

		[6] JSSC 2007	[8] JSSC 2014	[9] JSSC 2017	[4] JSSC 2014	[1] CICC 2017	[10] ASSCC 2016	This work
Image Recognition	Technology	180nm	180nm	40nm	65nm	65nm	55nm	55nm
	Voltage (V)	1.8	1.8	0.6	1.2	1.2	1.2	1.2
	Area (mm²)	33.64	82.3	5.9	0.063	0.24	3.61	0.64
	Power (mW)	85	630	23	4	0.3	-	75
	Frequency (GHz)	0.1	0.05	0.1	-	0.1	-	1.33
	Throughput/core (fps) *	6.1	16	14.7	-	-	-	72
	Energy/pixel (pJ)	756	2126	84	-	-	-	54
Time-based	Single bit delay (ps)	-	-	-	100	50	-	25
	Maximum No. of bit encoded	-	-	-	3	3	5	8
	Timing Mismatch**	-	-	-	6.5	6.5	4	2.8
	No. of equiv. inverters to generate 1-bit delay (4-bit TE)	-	-	-	22	10	7	5

* Based on 256×256 image with 64×64 scan window.

** Based on 4-bit time-encoding. Normalized to sigma (σ) of single delay cell.

REFERENCES

[1] Muqing Liu, Luke R. Everson, Chris H. Kim, " A Scalable Time-based Integrate-and-Fire Neuromorphic Core with Brain-Inspired Leak and Local Lateral Inhibition Capabilities", *CICC*, 2017.

[2] Anvesha Amravati, Saad Bin Nasir, Sivaram Thangadurai, Insik Yoon, Arijit Raychowdhury," A 55nm Time-Domain Mixed-Signal Neuromorphic Accelerator with Stochastic Synapses and Embedded Reinforcement Learning for Autonomous Micro-Robots", ISSCC, 2018.

[3] Daisuke Miyashita, et al., Toshiba Corporation, Kawasaki, Japan, "Time-Domain Neural Network: A 48.5 TSOp/s/W Neuromorphic Chip Optimized for Deep Learning and CMOS Technology", ASSCC, 2016.

[4] Daisuke Miyashita, et al.," An LDPC Decoder With Time-Domain Analog and Digital Mixed-Signal Processing", JSSC, vol. 49, no. 1, pp. 73-83, 2014.

[5] Zhengyu Chen, Jie Gu, "Analysis and Design of Energy Efficient Time Domain Signal Processing", ISLPED, 2016.

[6] Hideo Yamasaki, Tadashi Shibata, " A Real-Time Image-Feature-Extraction and Vector-Generation VLSI Employing Arrayed-Shift-Register Architecture", JSSC, vol. 42, no. 9, pp. 2046-2053, 2007.

[7] Chia-Heng Wu, et al., "Low-Latency Voltage-Racing Winner-Take-All (VRWTA) Circuit for Acceleration of Learning Engine",VLSI-DAT, Apr. 2017.

[8] Cong Shi, Jie Yang, Ye Han, Zhongxiang Cao, Qi Qin, Liyuan Liu, " A 1000 fps Vision Chip Based on a Dynamically Reconfigurable Hybrid Architecture Comprising a PE Array Processor and Self-Organizing Map Neural Network", JSSC, VOL. 49, NO. 9, 2014.

[9] Dongsuk Jeon, et al., "A 23-mW Face Recognition Processor with Mostly-Read 5T Memory in 40-nm CMOS JSSC", VOL. 52, NO. 6, 2017.

[10] Daisuke Miyashita, Shouhei Kousai, Tomoya Suzuki, Jun Deguchi, Toshiba Corporation, Kawasaki, Japan, "Time-Domain Neural Network: A 48.5 TSOp/s/W Neuromorphic Chip Optimized for Deep Learning and CMOS Technology", IEEE ASSCC, 2016.

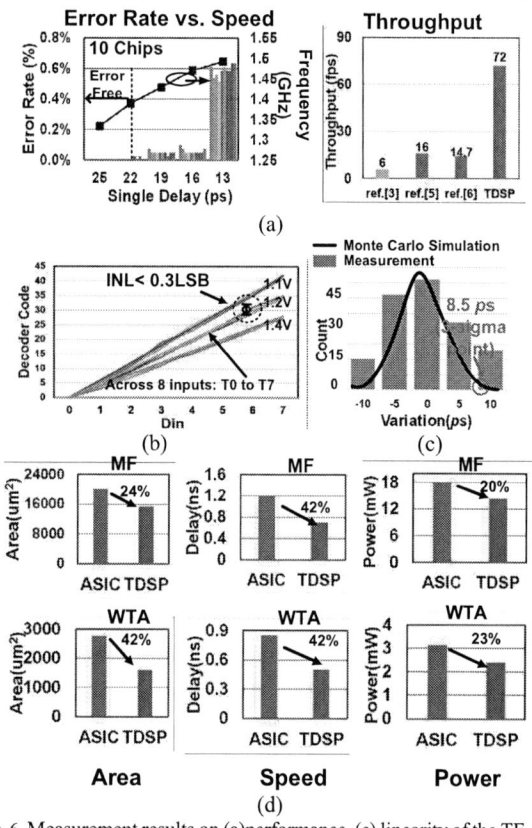

Fig. 6. Measurement results on (a)performance, (c) linearity of the TE, (b) measured (blue histogram) vs. simulated variation through chips, (d) area, speed and power compared with ASIC.

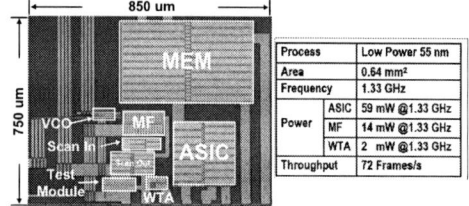

Fig. 7. Die micrograph and specifications.

V. CONCLUSION

This paper proposed a series of highly efficient time-domain signal processing techniques including: shared time generator, double-edge operation scheme, bit-split technique and high efficient time-domain operations. In our approach, the use of TDSP accelerates the pipeline operation bottleneck by 40% due to the limitation of MF and WTA operations. The test chip on image recognition processor is fabricated in 55-nm low power CMOS showing 72 fps/core (@1.33 GHz) operation with significant

A 95pJ/label Wide-Range Depth-Estimation Processor for Full-HD Light-Field Applications on FPGA

Li-De Chen, Yu-Ta Lu, Yu-Ling Hiao, Bo-Hsiang Yang, Wei-Chi Chen, and Chao-Tsung Huang
Department of Electrical Engineering
National Tsing Hua University, Hsinchu, Taiwan
Email: ldchen@gapp.nthu.edu.tw

Abstract—**High-resolution and wide-range depth maps are the key to enable novel light-field applications, such as digital refocusing, view synthesis, and 3D reconstruction. In this paper, we present an energy-efficient depth-estimation processor on FPGA to meet this purpose. There are two major contributions. First, image-guided depth inference and upsampling is adopted and implemented to provide accurate depth maps while lowering the working frequency from 215MHz to 54MHz. Second, octave search range sampling is proposed to efficiently allocate depth labels for wide-depth-range scenes to save computation and maintain accuracy. Finally, the implementation result on Xilinx ZC706 shows ASIC-comparable energy efficiency—95pJ/label—for Full-HD five-view light fields at 30fps.**

Keywords—*light field, depth estimation, FPGA, Full HD*

I. INTRODUCTION

Multi-view stereo matching can provide high-quality depth maps for high-resolution light-field applications. However, its demanding computation complexity could introduce high implementation cost and high power consumption and thus defy its usage on embedded devices. For example, [1] requires 1.5M gates in 40nm CMOS for Full-HD five-view depth estimation and consumes 611mW, i.e. 153pJ per depth label. In addition, as image resolution goes up, the corresponding depth range will become wider. Accordingly, conventional designs will need to increase the number of depth labels and therefore demand more hardware resources, such as computation logics, on-chip SRAM size, and off-chip DRAM bandwidth.

In this paper, we address these issues mainly by two novel design approaches. One is image-guided depth inference and upsampling, which allows us to perform stereo matching using half-resolution views to save computation and then recover the depth resolution back by referencing the full-resolution center view. The other one is octave search range sampling, which smartly allocates the limited number of depth labels for wide-range scenes to save hardware resources. In the following, we will present the details of their designs and also the final implementation on Xilinx ZC706 FPGA.

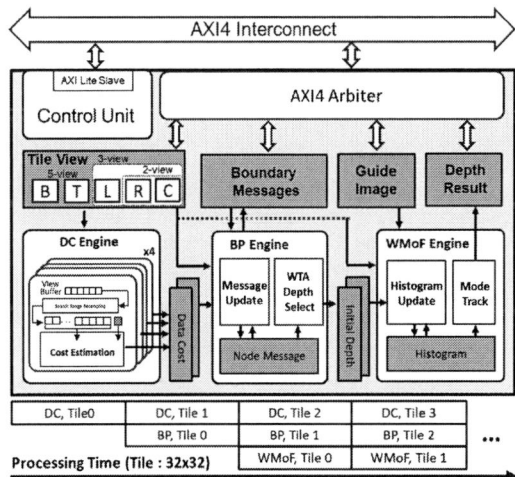

Fig. 1. System block diagram and tile-based pipelining.

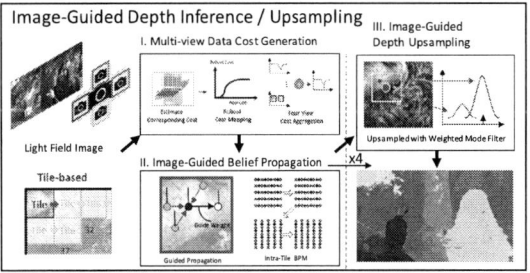

Fig. 2. Computation flow of the image-guided inference and upsampling.

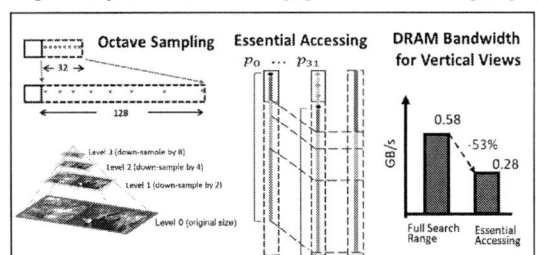

Fig. 3. Octave search range sampling and its bandwidth-efficient accessing.

978-1-5386-6414-8/18 $31.00 © 2018 IEEE

II. DEPTH ESTIMATION PROCESSOR

Fig.1 shows the system diagram of the processor. It follows a 32x32 tile pipeline and consists of three major engines: data cost (DC), belief propagation (BP), and weighted mode filter (WMoF). The DC estimates and aggregates data costs from four half-resolution surrounding views for each depth label. The BP performs intra-tile belief propagation and estimates a half-resolution depth map based on minimum costs. Finally, the WMoF upsamples and refines the depth map back to the full resolution using the high-quality center view.

A. Energy-Efficient Design with Image-Guided Upsampling

Fig. 2 shows the computation flow of the image-guided depth inference and upsampling for the three engines. To preserve object boundaries in full resolution, we adopt WMoF for depth upsampling. In contrast to the universal 4x upsampling in [2], we implemented 2x upsampling (in side length) in this processor to preserve more texture details. As a result, we can use the same amounts of processing elements in DC and BP as [1] while reducing the working frequency to 25% for saving power consumption. To increase the half-resolution depth quality, we also apply image-guided BP, i.e. conditional random fields, to enhance depth sharpness, and the hardware overhead is negligible. In addition, the DC engine is also optimized by only using the luma channel to save 78% of gates.

B. Memory-Efficient Design for Octave Sampling

Conventional stereo matching samples depth labels based on equally spaced disparities; however, this will overly sample near regions with unnecessarily high depth accuracy and thus waste lots of computation power. To address this issue, we proposed the octave search range sampling. It samples the depth labels with equally spaced depth (inverse disparity) to enable wide-depth-range stereo matching.

However, the memory access, including on-chip SRAM and off-chip DRAM, for such wide depth range becomes a problem for multi-view stereo matching. To resolve this memory bottleneck, we adopted the image pyramid and devised an essential accessing method as shown in Fig. 3. Combined with an on-the-fly downsampling unit, only the essential regions of the search range are required to be stored on SRAM. As a result, we save 53% of on-chip SRAM size and external DRAM bandwidth for accessing surrounding views.

III. IMPLEMENTATION RESULTS AND CONCLUSION

The proposed design is implemented on Xilinx ZC706 platform. It occupies 167K LUTs, 67K Registers, and 329K bytes of BRAM. Table I shows the resource utilization of the whole demo system and the depth estimation core. The depth estimation processor operates at 54 MHz and consumes 384 mW based on Xilinx Power Estimator. It equivalently achieves 95pJ per depth label for Full-HD 30fps.

A depth display system for the proposed processor is also implemented for verification and demonstration purpose.

Table I. Utilization Report

	LUT	LUTRAM	FF	BRAM	DSP
Available	277400	108200	554800	755	2020
Whole System	175423 (63.2%)	710 (0.7%)	74376 (13.4%)	240.5 (31.9%)	274 (13.6%)
Depth-Estimation Core	167206	48	67653	240.5	274

Table II. Comparison

	ISSCC 2015[1]	VLSI2017[3]	TCSVT2015[4]	This work
Technology / FPGA	40nm	28nm	Altera 5SGSMD5K2	Xilinx ZC706
Clock	215 MHz	300 MHz	180 MHz	54 MHz
Logic or [LUT, Registers]	1.5M gates	3.2M gates	[222K, 149K]	[167K, 67K]
Memory	352K bytes	582.5K bytes	2M bytes	329K bytes
Algorithm	MRFI	SGM	SGM	MRFI + Upsampling
Stereo type	5	3	2	5
Depth label / Range	64 / 32	128 / 128	128	64 / 128
Throughput	1920 x 1080, 30fps	2048 x 1080, 32fps	1600 x 1200, 42fps	1920 x 1088, 30fps
Power	611 mW	380 mW	-	384 mW*
Energy Efficiency**	153 pJ	42 pJ	-	95 pJ

* The Power result is estimated from Xilinx Power Estimator
** Energy Efficiency = Core Power / (Frame Rate * Depth Image Resolution * number of depth labels)

Integrated with ARM CPU and other peripheral ICs on ZC706 platform, we provide a demo of depth estimation for real-world light-field scenes. The tested light field is captured by Lytro Illum.

Table II shows the performance comparison between the proposed work with previous works [1,3,4]. The proposed processor improves resource and BRAM utilization by 63% and 84% respectively compared to [4]. It also achieves comparable energy efficiency on FPGA to the ASIC implementation [1].

In conclusion, an energy-efficient and cost-effective depth estimation processor is proposed. It is implemented on FPGA with real-world light-field demo. The proposed framework using image-guided depth inference and upsampling enables Full-HD depth throughput at merely 54MHz. The proposed octave search range sampling enables wide-range depth estimation for high-resolution light fields and also alleviates hardware cost for multi-view stereo matching. This work demonstrates ASIC-comparable energy efficiency, which facilitates the use of multi-view Full-HD depth estimation on FPGA and also ASIC.

ACKNOWLEDGEMENT

This work was supported by Novatek Microelectronics Corp.

REFERENCES

[1] H. Chen, et al., "A 1920x1080 30fps 611mW five-view depthestimation processor for light-field applications", in IEEE ISSCC Dig. Tech. Papers, pp. 422-423, 2015.

[2] B.-H. Yang, et al.,"A 320M Pixel/s VLSI Architecture Design of Weighted Mode Filter for 4K Ultra-HD Depth Upsampling", in IEEE International Conference on Acoustics, Speech and Signal Processing (ICASSP), 2018.

[3] Lee, Jinsu, et al. "A 31.2 pJ/disparity· pixel stereo matching processor with stereo SRAM for mobile UI application." VLSI Circuits, 2017 Symposium on. IEEE, 2017.

[4] Wang, Wenqiang, et al. "Real-time high-quality stereo vision system in FPGA." IEEE Transactions on Circuits and Systems for Video Technology 25.10 (2015): 1696-1708.

A 280mV 3.1pJ/code Huffman Decoder for DEFLATE Decompression Featuring Opportunistic Code Skip and 3-way Symbol Generation in 14nm Tri-gate CMOS

Sudhir Satpathy, Sanu Mathew, Vikram Suresh, Vinodh Gopal, James Guilford, Mark Anders, Himanshu Kaul, Amit Agarwal, Steven Hsu, Ram Krishnamurthy

Intel Corporation, Hillsboro, USA

Abstract— A 10,790μm² dynamic Huffman decoder targeted for DEFLATE header decompression in area and energy constrained IoT platforms is fabricated in 14nm Tri-gate CMOS. Ternary CAM (TCAM) assisted concurrent literal-length and distance tree generation, opportunistic code-skipping with register file tagging to leverage symbol sparsity, and 3-way forward-reverse-parallel Huffman decoding results in 300 cycle header processing latency, 2.4× faster than conventional serial decoding approach. Absence of custom circuits enables a fully synthesizable design operating over a wide supply range of 210-900mV with 895M codes/s throughput measured at 750mV, 25°C. Near-threshold voltage operation and clock power reduction with vector latch insertion provides peak energy-efficiency of 3.1pJ/code (5.9× higher than nominal) at 280mV, 15MHz operation with 32μW and 6μW total and leakage power consumption.

Keywords—DEFLATE, dynamic Huffman Decoder, TCAM

I. INTRODUCTION

Lossless data compression is a key enabler of energy-efficient IoT networks because of its promise to reduce communication bandwidth by upto 50% resulting in significant IO power savings over wired/wireless channels [1]. Compressed payloads also reduce storage requirements and access latency making them an integral component of energy and area constrained SoCs in the IoT era [2,3]. The widely used compression algorithms that include GZIP, ZLIB, 7-ZIP, PUTTY and PNG. are based off the DEFLATE algorithm, that uses a combination of LZ77 and Huffman entropy coding to optimally compress data [4,5,6]. LZ77 eliminates duplicated byte sequences in raw literal stream by replacing them with references comprising of a pair of length and distance symbols. Huffman coding further reduces data footprint by converting fixed 8b LZ77 symbols to tokens of variablelengths that are inversely proportional to their corresponding frequency of occurrences. Huffman code tables for eachpayload block are prepended to the compressed packet as header metadata in the form of literal-length (lit-len) and distance trees that must be recovered prior to decompression (Fig. 1). These trees appear as a stream of code-lengths that undergo canonical Huffman decoding to extract the actual codes. To reduce packet size a standard specified set of Huffman codes (CLEN codes) with special symbols for run-length encoding is used to further compress the metadata. Such double encoding scheme with non-deterministic symbol widths introduce a serial-dependency between conjugate symbols necessitating specialized arithmetic and CAM units for fast Huffman

tree recovery, rendering on-die hardware acceleration prohibitively expensive for cost/energy constrained platforms [7,8,9].

Light-weight implementations of DEFLATE hardware accelerators use static Huffman encoding that leverages standard specified pre-determined Huffman trees for entropy coding while trading off compression efficiency for area savings. Data transfer between IoT nodes in home network originate from a wide variety of devices comprising of smart appliances, monitoring/surveillance infrastructure, personal electronics, and sensor hubs, with dramatically different symbol statistics. This motivates the use of dynamic Huffman coding that generates a fresh pair of trees for every payload block with respect to its unique symbol distribution for optimal file compression. Pre-processing the compressed header at the gateway via hardware acceleration enables fast recovery of Huffman trees reducing decompression latency without sacrificing compression efficiency while isolating the impact of cost and design overhead from area constrained IoT edge nodes. Conventional Huffman decoders serially process the header code-lengths stream in a standard specified order, and require complete recovery of all prior symbols before attempting to decode a new symbol limiting throughput [4]. This paper presents a fully DEFLATE compliant dynamic Huffman decoder fabricated in 14nm Tri-gate CMOS that leverages symbol sparsity in compressed IoT payloads to opportunistically skip redundant codes and concurrently enable 3-way code generation for 52% speed-up over conventional designs resulting in 300 cycle overall tree-recovery latency while achieving 895 Million codes/s throughput.

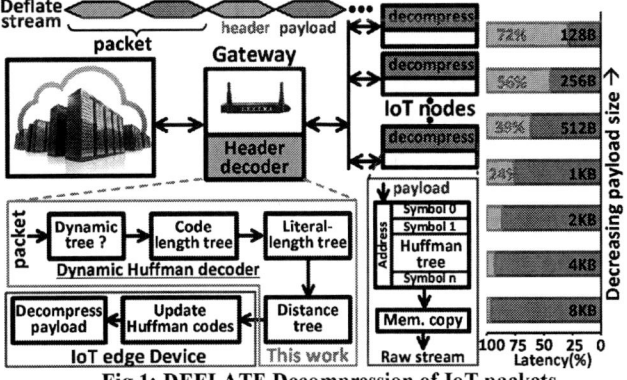

Fig.1: DEFLATE Decompression of IoT packets

The rest of this paper is organized as follows: Section II provides an overview of the accelerator datapath and key building blocks; TCAM circuit assisted simultaneous lit-len and distance tree decode

978-1-5386-6414-8/18 $31.00 © 2018 IEEE

is presented in section III; section IV describes register-file tagging circuits and optimal array partitioning for opportunistic code skipping; in section V the 3-way symbol generation approach that includes concurrent forward-reverse and check-pointed parallel decode is discussed; measurement results from 14nm CMOS test-chip are presented in section VI, and finally we summarize this paper in section VII.

II. DYNAMIC HUFFMAN DECODER OVERVIEW

Unlike cloud and server networks that stream multiple gigabits of data between communicating nodes per transaction, typical IoT payloads tend to span < 1KB. Performance analysis for such small packet transfers indicate that header processing accounts for 24%-72% of total decompression latency for 1KB-128B compressed payloads respectively, necessitating specialized circuits for fast recovery of Huffman trees to reduce overall packet delivery latency and improve energy-efficiency (Fig. 1). Compressed packets undergo in-transit header pre-processing at the gateway where the proposed dynamic Huffman decoder recovers lit-len and distance Huffman trees before dispatch to corresponding IoT edges. Storing these codes in register-file or CAM macros at the edge node allows accurate decompression of the payload via simple table look-up operations. Subsequent LZ77 reconstruction is accomplished with memory copy using load/store instructions, eliminating the need for any additional specialized logic at the IoT edge.

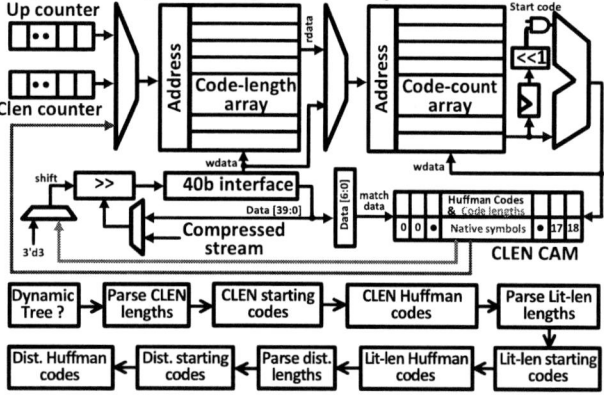

Fig.2: Dynamic Huffman decoder architecture

The Huffman decoder is organized around a pair of register-file arrays and a CAM unit to sequentially recover the CLEN tree of 19 symbols and subsequently use them to generate a pair of DEFLATE compliant trees comprising of upto 286 lit-len and 30 distance codes respectively (Fig. 2). Each tree generation process involves 3 steps: i) parsing the code-lengths sequence and creation of the code-count histogram; ii) updating the code-count histogram to derive starting Huffman codes; iii) revisiting the code-lengths in a second parse to compute the entire tree of Huffman codes. A 286-entry 4b register file array caches all code-lengths sequentially during initial parse, while another 15-entry 15b array keeps track of histogram count for all code-lengths from 1b to 15b. Caching the code-lengths locally within the accelerator enables a streaming datapath eliminating the need for header rewind to gather code-lengths for second parse. This simplifies the interface to a 40b shift-register that reads 32bits of compressed data at a time and appropriately progresses the header in accordance with the number of bits consumed in each cycle. The

total number of codes for each of the 3 trees (HCLEN, HLIT, HDIST) read from header configures the datapath to identify termination of various steps in each tree generation process (Fig. 3). CLEN symbol lengths are encoded using fixed 3b codes, while lit-len and distance symbols use variable length codes spanning 1b-7b. Besides, CLEN codes appear in a standard specified shuffled order (16,17,18,0,…1,15) that is tracked by a 4b "clen counter", while a 9b up-counter is used for the other trees that pack symbols in serial order (Fig. 2).

III. CONCURRENT LIT-LEN AND DISTANCE DECODE

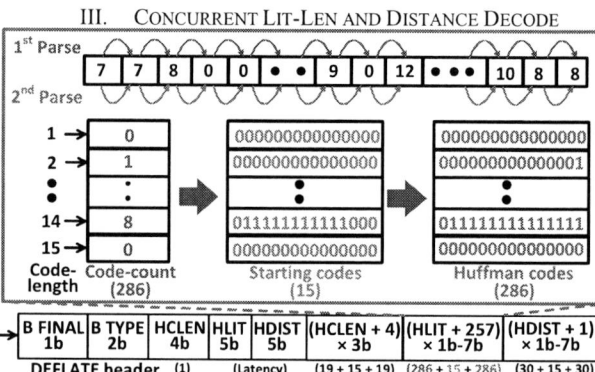

Fig.3: Conventional Huffman decoding approach

The canonical nature of Huffman trees enable recovery of variable length codes from code-lengths using shift/add operations (Fig. 3). After completion of first parse and initialization of starting 1b code to 0, the code-counts are serially added to create cumulative counts that undergo left-shift to generate starting codes for longer lengths. The actual Huffman codes are then sequentially generated during 2nd parse by accessing the array of starting codes and selectively incrementing the entry for the corresponding code-length. This serial scheme takes 2*N + 15 cycles to completely recover a tree of N symbols, where N cycles are incurred for each parse with 15 additional cycles accounting for initialization of starting codes.

A 19-entry 7b-match 7b-data TCAM stores Huffman codes computed after recovery of the first tree (Fig. 4).Afterwards, the codes for 19 CLEN symbols are matched against leading 7bits of header payload to identify the native symbol (0 to 19) and its corresponding length while processing the subsequent trees. A binary to thermometer conversion circuit translates code-length into don't care tags to configure TCAM contents for smaller code-lengths (1b-6b). TCAM sequentials get programmed during CLEN tree recovery and later accessed for HLIT and HDIST tree processing. Absence of simultaneous read/write operations enables a latch based fully synthesizable TCAM implementation resulting in 30% area and 42% power savings over flip-flop based design. Latches storing content and data are grouped to maximize quad and dual latch insertion enabling additional 6% area and 8% clock power reduction. Following CLEN tree generation, the header interface switches from 3b shift to 1-7b variable shift per cycle in accordance with the number of bits that the decoded symbol spans.

Although lit-len and distance tree generation process involves 3 steps, only the 1st parse requires TCAM access since raw code-lengths are locally stored following initial decode. To maximize TCAM utilization, and reduce decode latency the accelerator

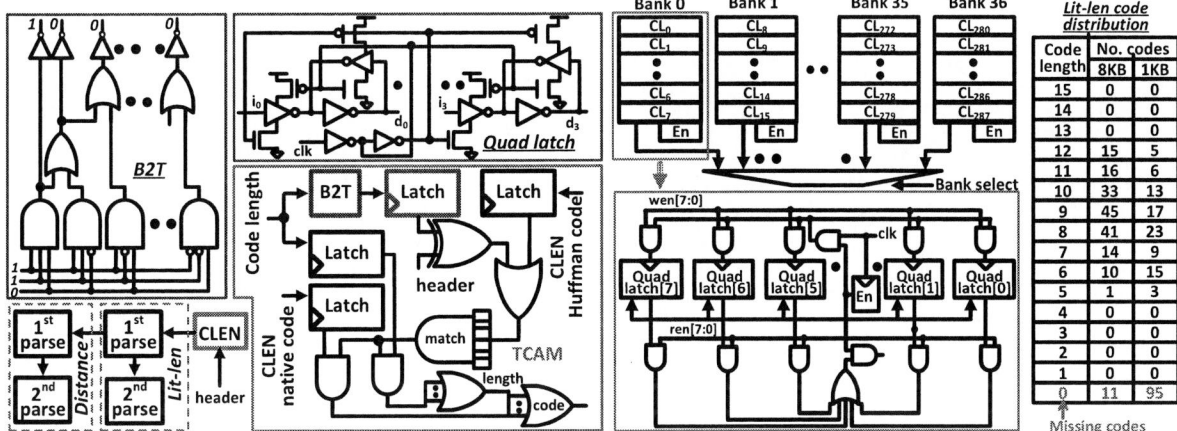

			Lit-len code distribution	
			Code length	No. codes
				8KB	1KB
			15	0	0
			14	0	0
			13	0	0
			12	15	5
			11	16	6
			10	33	13
			9	45	17
			8	41	23
			7	14	9
			6	10	15
			5	1	3
			4	0	0
			3	0	0
			2	0	0
			1	0	0
			0	11	95

Fig.4: TCAM based concurrent lit-len & distance decode

Fig.5: Register file tagging for sparse Huffman trees

progresses to distance code-length parsing after lit-len tree 1st parse without waiting for complete recovery of all lit-len symbols (Fig. 4). An extra pair of code-length (30entry 4bit) and code-count (15entry 15bit) arrays enable completion of distance tree computation alongside lit-len tree resulting in 10% overall latency reduction at 8%. additional area. The asymmetry between the two trees (286 vs 30 symbols) results in elimination of distance tree from critical timing path, allowing the remaining optimizations to be applied only to lit-len Huffman decode for further performance improvement.

IV. OPPORTUNISTIC CODE-SKIPPING

Although DEFLATE compliant Huffman trees can comprise upto 286 lit-len symbols, the limited size of IoT payloads render a lot of the symbols unused. Missing symbols in the header stream are represented by zero code-lengths. Analysis of symbol distribution histogram for two compressed files with 8KB and 1KB payloads shows that 33% of symbols remain unused for typical IoT packets (Fig. 5). Unlike conventional designs that operate at a fixed processing rate of one symbol per cycle, the proposed accelerator opportunistically skips non-existing symbols for early decode completion. Special CLEN symbols (code 17 and 18) that allow encoding upto 138 sequences of zero code-lengths, trigger the datapath to advance multiple symbols in a cycle during 1st parse. Besides, keeping track of these redundant symbols also allows skipping them during 2nd parse. This is accomplished using a register file tagging scheme where an enable flag identifies a bank as filled with zero lengths without requiring to explicitly clear all entries in the corresponding bank (Fig. 5).

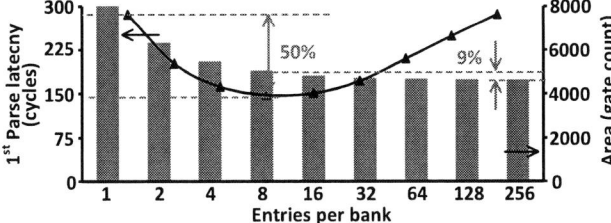

Fig.6: Optimal register file partitioning for code-skip

Sequences of zero code-lengths are intermittently broken with genuine symbols. Extreme coarse grain banking (> 64 entries per bank) renders this scheme ineffective because of reduced likelihood

of existence of such long sequences in compressed header, though it offers the opportunity to skip more symbols per cycle (Fig. 6). Similarly, ultra-fine grain banking (< 4 entries per bank) improves the likelihood of code-skipping at the cost of increased area overhead owing to increasing number of flag bits and limited performance gains because of smaller skips. An intermediate approach of 8entries/bank maximizes performance improvement resulting in 50% 1st parse (22% overall) latency reduction at 9% additional area over a conventional untagged register-file array.

V. 3-WAY SYMBOL GENERATION

Following initialization of starting codes, conventional decoding scheme serially generates Huffman codes for all 286 lit-len symbols by incrementing the initial code values for corresponding code-lengths during 2nd parse. The sequential nature of Huffman codes limits throughput to 1code/cycle for such one-way forward decoding approach from symbol 0 towards 285. The proposed design mitigates this performance bottleneck by using an additional 15-entry 9b array to keep track of the total count of left-over codes. For every symbol that the forward thread generates, the count corresponding to its code-length is decremented in this new array. The sum of original Huffman codes and left-over code counts indicates the last Huffman code for respective code-lengths (Fig. 7). A reverse thread leverages this technique to decode symbols backwards from 285 towards 0, thus breaking the serial-dependency to generate 2 codes/cycle. The code generation process completes when the forward and reverse decoders converge at a code mid-way between the starting and ending indices, reducing 2nd parse latency by 50%, improving overall decode performance by 28%.

To further improve accelerator throughput, the engine launches an additional check-pointed thread to enable Huffman code generation at any arbitrary index (Fig. 7). An additional 15entry 15b array checkpoints a copy of the partial code-count histogram at a chosen symbol index during 1st parse. The sum of this array elements with corresponding starting codes of forward decode thread generates a new set of values that are used as initial codes for a parallel thread starting at the selected index during 2nd parse. This parallel thread operates concurrently alongside forward and reverse decoders enabling 3-way code generation. The forward thread completes on reaching the check-pointed index, while the reverse and parallel threads continue to decode until convergence at same

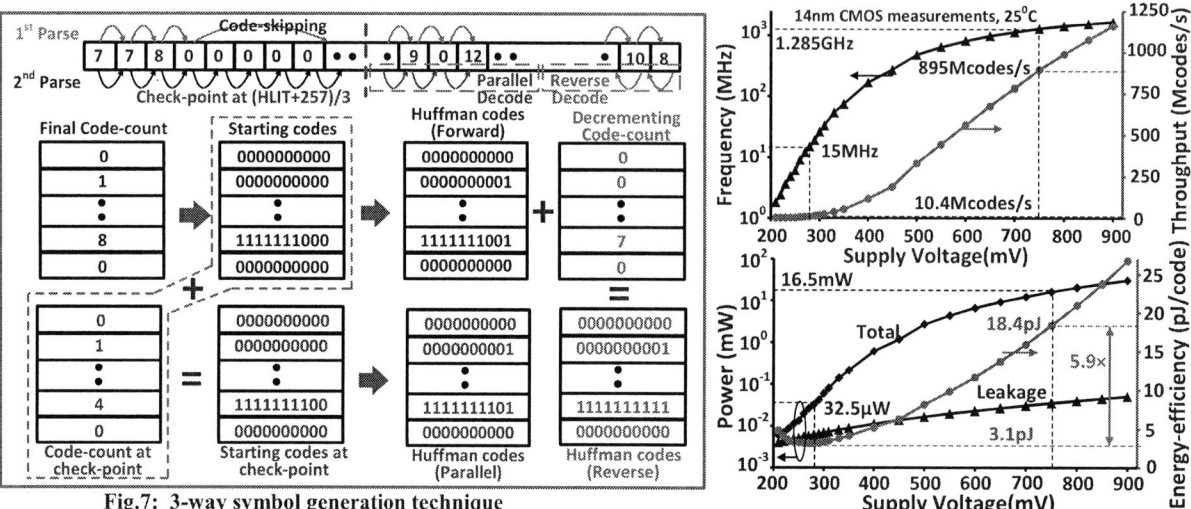

Fig.7: 3-way symbol generation technique

Fig.8: Power and performance measurements

index. Though the forward and reverse decoders start from symbol 1 and 285 respectively, the index for parallel thread is uniquely set midway for each header after accounting for missing symbols to minimize idle cycles because of early termination of one of the threads. Analysis of 1KB DEFLATE streams on maximum compression test-suite shows that the 3-way adaptive decoding scheme provides an additional 17% higher code generation throughput.

VI. MEASUREMENT RESULTS

The dynamic Huffman decoder operates at a maximum frequency of 1.285GHz at nominal 0.75V, 25°C resulting in 895Million Huffman codes/s average throughput measured for 1KB files from maximum compression standard test suite. At nominal conditions, the accelerator consumes 16.5mW total power with 35µW leakage providing 18.4pJ/code energy-efficiency (Fig. 8). TCAM assisted concurrent lit-len and distance tree generation, tagged register-file array based code skipping, and 3-way forward-reverse-parallel code compute circuits reduce overall decode latency down to 300 cycles enabling 2.4× higher throughput over conventional serial decode approach (Fig. 9). Absence of custom circuits enables an all-digital fully synthesizable design occupying 10,790µm² with robust ultra-low voltage operation down to 210mV where the design operates at 1.8MHz achieving 1.25M codes/s throughput with 6µW power consumption. Peak energy-efficiency of 3.1pJ/code (5.9× higher than nominal) is measured at near-threshold voltage operation of 280mV at 15MHz operation providing 10.4M codes/s throughput at 32µW total and 6µw leakage power consumption.

VII. CONCLUSION

An all-digital DEFLATE compliant dynamic Huffman decoder targeted towards accelerated processing of compressed packet header for IoT payloads in smart home network is fabricated in 14nm Tri-gate CMOS. Novel datapath circuits featuring TCAM based concurrent literal-length and distance Huffman tree recovery, register-file tagging for opportunistic code skipping, and 3-way symbol generation scheme eliminate performance bottleneck of conventional serial decoding approach enabling 2.4× faster Huffman code generation. Absence of custom circuits result in a fully-synthesizable design that operates over a wide supply of 210mV-900mV with scalable throughput ranging from 1.2 to 1160 Million Huffman codes/s. The accelerator occupies 10,790µm² and achieves 3.1pJ/code peak energy-efficiency at near threshold voltage operation at 280mV facilitating seamless integration in energy and area constrained platforms.

REFERENCES

[1] C. Deepu et al., "A hybrid data compression scheme for power reduction in wireless sensors for IoT", IEEE Trans. on BioCAS, pp. 245-254, 2017.

[2] Y. Deguchi et al., "Flash reliability boost Huffman coding (FRBH): Co-optimization of data compression and V_{TH} distribution modulation to enhance data-retention time by over 2900x", IEEE Symposium on VLSI Technology, pp. 206-207, 2017.

[3] H. Watanabe et al., "MLC/3LC NAND flash SSD cache with asymmetric error reduction Huffman coding for tiered hierarchical storage", Proceedings of the ASSCC, pp. 157-160, 2017.

[4] K. Zhu et al., "A custom GZIP decoder for DTV application", International Symposium on Circuits and Systems (ISCAS), pp. 681-684, 2013.

[5] D. Harnik et al., "A fast implementation of Deflate", Data Compression Conference, pp. 223-232, 2014.

[6] N. Reynders et al., "A 210mV 5MHz variation-resilient near-threshold JPEG encoder in 40nm CMOS," IEEE ISSCC Dig. Tech. Papers, pp. 456-457, 2014.

[7] Y. Kim et al., "A 0.5V 54µW ultra-low-power recognition processor with 93.5% accuracy geometric vocabulary tree and 47.5% database compression," IEEE ISSCC Dig. Tech. Papers, pp. 330-331, 2015.

[8] C. Angulo et al., "Accelerating Huffman decoding of seismic data on GPUs," Symposium on Signal Processing, Images and Computer Vision, 2015.

[9] E. Sitaridi et al., "Massively-parallel lossless data compression," International Conference on Parallel Processing, pp. 242-247, 2016.

Technology	14nm CMOS	
Die area	0.58mm²	
Accelerator area	10,790µm²	
Standard	DEFLATE	
Decode Latency	300 cycles	
Supply	210-900mV	
Frequency	1.8MHz-1.28GHz	
Supply	750mV	280mV
Codes per sec.	895M	10.4M
Total Power	16.5mW	32µW
Leakage Power	35µW	6µW
Energy per code	18.4pJ	3.1pJ

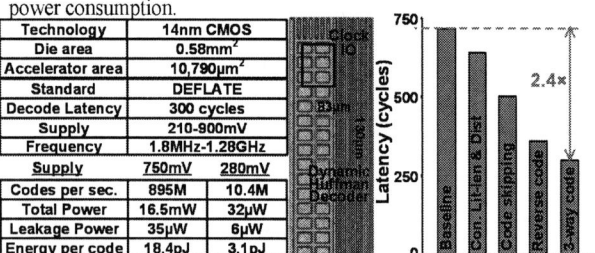

Fig.9: 14nm die-photo and improvement summary

A Wearable Auto-Patient Adaptive ECG Processor for Shockable Cardiac Arrhythmia

Syed Muhammad Abubakar[1,2], Muhammad Rizwan Khan[1], Wala Saadeh[1], and Muhammad Awais Bin Altaf[1]

Lahore University of Management Sciences (LUMS), Lahore, Pakistan[1]

Tsinghua University Beijing, China[2]

Email: awais.altaf@lums.edu.pk

Abstract— A non-machine learning patient-specific shockable cardiac arrhythmia (SCA) classification processor based on single lead electrocardiogram (ECG) is presented. The proposed SCA detection processor integrates a hardware-efficient reduced-set-of-five (RSF5) feature extraction engine to extract SCA and non-SCA, self-adaptive patient-specific threshold engine for the peak and interval detection from the ECG, and simplified decision logic to discriminate the arrhythmia in real-time. The SCAD processor consumes 0.89μJ/classification while classifying with an average sensitivity, and specificity of 98.66%, and 99.75%, respectively.

Keywords— *cardiac arrhythmia, defibrillator, ecg processor, feature extraction, FPGA, wearable*

I. INTRODUCTION

Wearable Electrocardiogram (ECG) devices are essential nowadays for personalized, preemptive, participatory, and precautionary medical practice adapted especially for the cardiovascular disorders [1]. Ventricular tachycardia (VT) and ventricular fibrillation (VF) are the leading cause of sudden cardiac deaths in the USA [2]. Timely detection of VT and VF is very crucial in delivering an electric shock therapy; continuous 24/7 monitoring with the on-sensor processing of ECG to detect shockable cardiac arrhythmias (SCA) and non-SCAs (NSCAs) along with defibrillators is very crucial [3]. [1], [4] presents single-lead on-sensor processor with high classification accuracy but targeting NSCA without defibrillation. [5], [6] presents wearable cardiac monitoring system but lacks patient adaptability whereas [3] implements a closed-loop arrhythmia diagnosis SoC but targeting implantable environment.

This paper presents a patient-adaptive ultra-low power single-channel "reduced feature set, non-machine learning" early ECG rhythm detection for SCAs targeting a wearable defibrillator with immediate response time (<20s). The system employs the reduced feature set of five (RFS5) feature extractor (FE) for both SCAs and NSCAs while utilizing non-machine learning patient adjustable threshold for proposed decision tree classifier (DTC) to achieve hardware-efficient implementation.

Fig. 1 shows the proposed SCA detection (SCAD) SoC with wearable defibrillator (WD); composed of single-lead ECG front-end, followed by an 11b SAR ADC, SCAD processor implemented on FPGA, and an external WD. For timely defibrillation, the SCAD performs auto-patient-adaptive detection to trigger WD for a short duration [biphasic shock energy: ~75J].

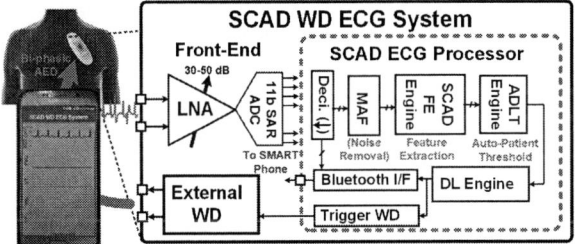

Fig. 1. Proposed wearable ECG SoC for Cardiac Arrhythmia Detection

II. SHOCKABALE CARDIAC ARRHTHMIA PROCESSOR

To integrate ECG processor on-sensor including all necessary SCA (VT, VF) and NSCA (Sinus rhythm, Premature Ventricular Contraction (PVC), Supra VT (SVT)), minimizing the hardware cost while maintaining the reliability is crucial [7][8]. The implemented SCAD ECG processor filters high frequency noise, baseline wandering and dc component while utilizing proposed moving average filter (MAF) which consumer 29% and 80% less area and power, respectively, compared to the conventional BPF. The feature extraction (FE) engine is based on proposed reduced set of five (RSF5) to extract the discriminatory features with minimal hardware requirement along with adaptive threshold value for R and T wave detection, to be used in the decision logic (DL) engine. Fig. 2 shows the implementation of SCAD FE engine; 2-sec filtered ECG data is stored in RAM which is further divided into respective P wave (PW), T wave (TW), Q wave (QW) and S wave (SW) window RAMs.

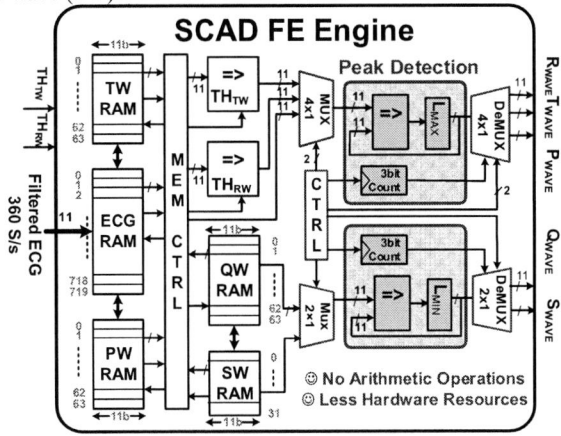

Fig. 2. Architecture of SCAD FE Engine

Peak detection is utilized to detect R-R interval based on R wave threshold which get updated automatically based on the slope change variation in the local maxima. To ensure the robustness and discard the abrupt glitch events, counter based cross check is utilized for 8-consecutive samples. Q and S wave detection will be done relative to the R-peak detection using same slope change principle. The proposed implementation achieves the comparable performance without utilization of any multiplier/divider and floating-point arithmetic blocks.

Fig. 3 shows the proposed adaptive decision logic threshold (ADTL) engine along with conventional implementation; proposed achieves an area reduction of >65% with a minimal overhead of increased processing speed and power. The RSF5 features detected by the FE update decision logic thresholds in ADTL. One feature out of RSF5 features are selected by the multiplexer (MUX) to save the value in on-chip RAM. First in first out (FIFO) based moving average of 4 values is utilized to compute/update the thresholds in a multiplexed way and decision logic thresholds are updated one by one in round-robin fashion.

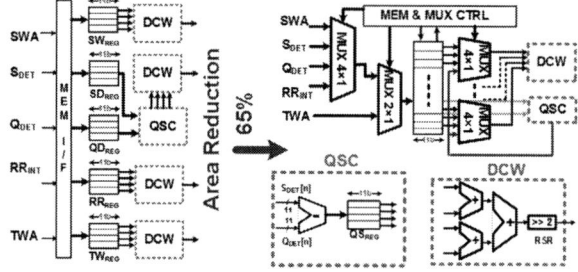

Fig. 3. Conventional vs. proposed automatic decision threshold logic.

III. MEASUREMENT RESUTLS

Fig. 4 is the FPGA measurement results with ECG-ID [9] and MIT-arrhythmia ECG database [10] that contains VF, VFL and PVC, VT cases respectively. Each ECG beat is classified using ADTL and corresponding flag is raised. Results shown are for a sinusoidal rhythm appears with amplitude equal to that of normal rhythm at a heart rate > 250 bpm, VFL rhythm is detected and VFL flag goes high. VFL rhythm may turn into VF when the amplitude of the rhythm is reduced significantly which results in shock delivery by the system and VF flag goes high.

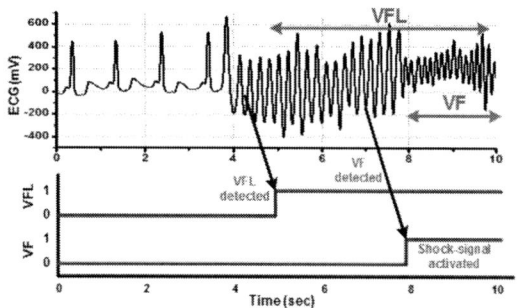

Fig. 4. Measurement results by storing the ECG data [10] on FPGA and classifying through SCAD ECG processor.

Fig. 5 shows the FPGA implementation and its performance summary based on P&R results from 180nm CMOS process. The proposed system achieves high sensitivity and specificity for both SCA and NSCA, with proposed RFS5 approach, while consuming 0.89μJ/Classification to continuously track SCA and NSCA. Table I shows the comparison of proposed with state-of-the-art works.

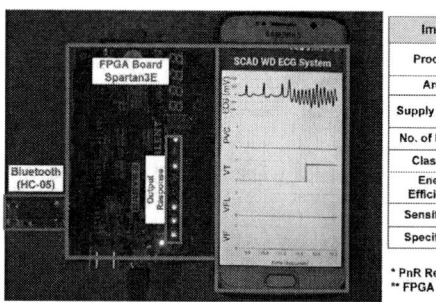

Implemented Results	
Process*	0.18μm 1P6M CMOS
Area*	0.375 x 0.375 mm
Supply Voltage	1.0V
No. of Feature	5
Classifier	Simple Threshold
Energy Efficiency*	0.89μJ/Class.
Sensitivity**	98.66%
Specificity**	99.75%

* PnR Results
** FPGA verification results

Fig. 5. FPGA implementation and performance summary.

Table 1. Comparison with state of the art works

	S. Abubakar DATE'18 [2]	D. Jeon ISSCC'14 [3]	X. Liu A-SSCC'13 [5]	S. Yin VLSI'17 [6]	This Work
Arrhythmia Detection	O	O	X	O	O
Defibrillation Shock	X	X	X	O	O
Technology	-	65 nm	180 nm	65 nm	180 nm
Supply Voltage	1.0 V	0.4 V	1.8 V	0.55V	1.0 V
Frequency	2 KHz	10 KHz	6 K Hz	2KHz	2 KHz
Classifier	ANN	Thresholds	N/A	ANN	ADLT
Power	5.94uW	45nW	0.435uW	1.06uW	115.8nW

REFERENCES

[1] W. Saadeh, et al., "A 1.1mW Hybrid OFDM Ground Effect-Resilient Body Coupled Communication Transceiver for Head and Body Area Network," in Proc. IEEE ASSCC, Nov. 2016, pp. 201-204.

[2] S. Abubakar, et al., "A wearable long-term single-lead ECG processor for early detection of cardiac arrhythmia," in Proc. IEEE/ACM DATE, pp. 961-966, Mar. 2018.

[3] D. Jeon, et al., "An Implantable 64nW ECG-Monitoring Mixed-Signal SoC for Arrhythmia Diagnosis," in IEEE ISSCC Dig. Tech. Papers, pp. 416-417, Feb. 2014.

[4] S. Izumi, et al., "A 14uA ECG Processor with Robust Heart Rate Monitor for a Wearable Healthcare System," in Proc. IEEE ESSCIRC, pp. 145-148, Sep. 2013.

[5] X. Lin, et al., "A 457-nW Cognitive Multi-Functional ECG Processor," in Proc. IEEE ASSCC, pp. 141-144, Nov. 2013.

[6] S. Yin, et al., "A 1.06 uW Smart ECG Processor in 65 nm CMOS for Real-Time Biometric Authentication and Personal Cardiac Monitoring," IEEE Symp.VLSI Cir. Dig. Tech. Papers, pp. 102-103, May. 2017.

[7] W. Saadeh, et al., "A > 89% efficient LED driver with 0.5V supply voltage for applications requiring low average current," in Proc. IEEE A-SSCC, Nov. 2013, pp.273-276.

[8] W. Saadeh, et al., "A High Accuracy and Low Latency Patient-Specific Wearable Fall Detection System," in Proc. IEEE Biomedical and Health Informatics (BHI), 2017, pp. 441-444.

[9] Creighton University Ventricular Tachyarrhythmia database [Online], Available: https://physionet.org/physiobank/database/cudb/

[10] MIT-BIH arrhythmia database [Online], Available: https://www.physionet.org/physiobank/database/mitdb.

A Capacitance-to-Digital Converter Integrated in a 32bit Microcontroller for 3D Gesture Sensing

Mitsuru Hiraki, Sugako Otani, Masao Ito, Takuya Mizokami, Masahiro Araki, and Hiroyuki Kondo
Renesas Electronics Corporation
Kodaira, Tokyo, Japan
Email: mitsuru.hiraki.ux@renesas.com

Abstract— This paper describes a capacitance-to-digital converter (CDC) integrated in a 32bit microcontroller. The analog part of the CDC that we propose is only composed of a voltage down converter, current mirror circuit, and current-controlled oscillator. Since its analog part is so simple that our CDC can be easily ported to various kinds of processes with which microcontrollers are fabricated. RF noise immunity of our CDC is enhanced by using the random phase shift scheme that we developed. The effectiveness of the scheme was experimentally verified with a CDC integrated in a 32bit microcontroller which was fabricated in 130nm CMOS process. Our CDC meets level 3 of IEC 61000-4-3 and IEC 61000-4-6 standards. With our CDC structure, users can select resolution under a trade-off between the resolution and measurement time. If the measurement time is increased up to 50ms, its resolution is as high as 2aF. 3D gesture sensing was realized using the 32bit microcontroller in which the proposed CDC was integrated.

Keywords—capacitive sensor; capacitance-to-digital converter; noise immunity; microcontroller; gesture sensing

I. INTRODUCTION

Capacitive sensors are widely used to monitor changes of various kinds of quantities including volume, weight, thickness, distance. In a system that includes a capacitive sensor, a microcontroller performs both system control and digital signal processing of the output from the capacitance-to-digital converter (CDC) which is the analog front end included in the capacitive sensor. Therefore, integrating the CDC into a microcontroller will reduce both system cost and mounting area compared with a system which uses a discrete capacitive sensor IC and a microcontroller because the total number of components on the system board is reduced. Since microcontrollers are fabricated using various kinds of processes, it is important to provide CDC process portability to easily integrate CDCs into microcontrollers.

We propose a CDC which is easy to implement using various fabrication processes. The basic structure of our CDC is described in Section II. Techniques for removing parasitic capacitance and those for enhancing noise immunity are also described in this section. Section III shows its implementation and measurement results. 3D gesture sensing using this microcontroller is demonstrated in Section IV. Section V shows the conclusion of this work.

II. CIRCUIT DESIGN OF PROPOSED CDC

A. Basic Structure

Fig. 1 shows the basic structure of the CDC that we propose. Typical CDCs use ADCs [1-4], which generally increase design complexity. The analog part of our CDC is only composed of a voltage down converter, current mirror circuit, and current-controlled oscillator. Since the structure of the analog part is so simple that our CDC can be easily ported into various kinds of processes with which microcontrollers are fabricated.

Our DCD converts the value of the capacitance C_S into the value of the counter. The circuit operation to do that is as follows. Driven by a clock signal *CLKS*, switches SW1 and SW2 turn on and off alternatively and repeatedly to charge and discharge the capacitance C_S to be measured. The stabilization capacitor smooths the charging current. The smoothed current I_R is proportional to the capacitance C_S and is expressed as fC_SV_R, where f and V_R are the switching frequency and the voltage regulated by the voltage down converter, respectively. The current mirror circuit mirrors the current I_R to the control current of the current-controlled oscillator. Since the oscillation frequency of the current-controlled oscillator is proportional to its control current, the oscillation frequency is proportional to the capacitance C_S. Thus, the capacitance C_S is digitized by counting the number of oscillations with a counter for a certain period of time.

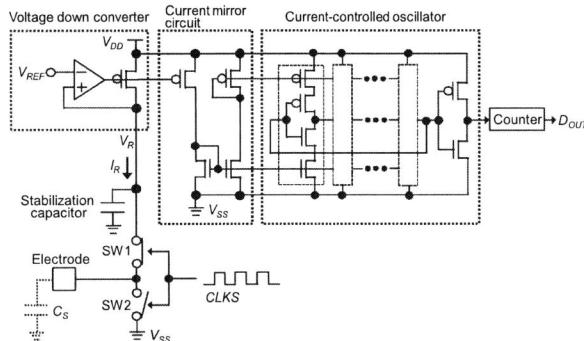

Fig. 1. Basic structure of the proposed capacitance-to-digital converter.

B. Techniques for Removing Parasitic Capacitance

Our CDC can measure not only self capacitance but also mutual capacitance. Fig. 2 shows techniques for removing parasitic capacitance in the measurement of mutual capacitance. In this figure, C_M is mutual capacitance to be measured between Electrode_A and Electrode_B. C_{PA} is parasitic capacitance at the node of Electrode_A. To exclude the parasitic capacitance in measuring the mutual capacitance, the measurement is performed twice. In the first measurement, Electrode_A is driven in phase with Electrode_B. The smoothed current I_{RA} corresponding to the first measurement is given by

$$I_{RA\_1st} = fC_{PA}V_{RA} + fC_M(V_{RA} - V_{RB}) \qquad (1)$$

where V_{RA} and V_{RB} are voltage swing of Electrode_A and Electrode_B, respectively. In the second measurement, Electrode_A is driven in opposite phase with Electrode_B. The smoothed current I_{RA} corresponding to the second measurement is expressed as follows.

$$I_{RA\_2nd} = fC_{PA}V_{RA} + fC_M(V_{RA} + V_{RB}) \qquad (2)$$

By subtracting the first measured result from the second measured result, parasitic capacitance is removed from the measured result because subtracting equation (1) from equation (2) gives equation (3) in which the term $fC_{PA}V_{RA}$ is removed.

$$I_{RA\_2nd} - I_{RA\_1st} = 2fC_M V_{RB} \qquad (3)$$

This subtraction operation between the two measured results is executed with microcontroller software.

C. Techniques for Enhancing RF Noise Immunity

Fig. 3 shows that RF noise superimposed on the sensing signal causes an error in capacitance measurement if the fundamental frequency of the noise coincides with the switching frequency. Here, we assume that the noise is capacitively-coupled with the electrode as shown in Fig. 3 (a). Fig. 3 (b) shows the timing chart for the case in which a downward phase of the noise coincides with a charging phase at every cycle. In this case, the noise causes a positive deviation from the correct current I_R. Similarly, the noise causes a negative deviation when an upward phase of the noise coincides with a charging phase at every cycle. In either case, the deviation leads to an error in capacitance measurement.

Fig. 4 shows random phase shift (RPS) scheme which we developed and used in our CDC to enhance RF noise immunity. As shown in Fig. 4 (a), phase shift function is embedded in the clock divider. The phase shift function is activated by the signal *Shift* which a random number generator (RNG) creates. This RNG is a pseudorandom number generator using a linear feedback shift register. Every time the signal *Shift* is activated, the charging phase is shifted by a half cycle as shown in Fig. 4 (b). Since the occurrence frequency of positive deviation and that of negative deviation are equal to each other, the effect of positive deviation and that of negative deviation cancel each other out. Therefore, the RPS scheme enhances the RF noise immunity of our CDC.

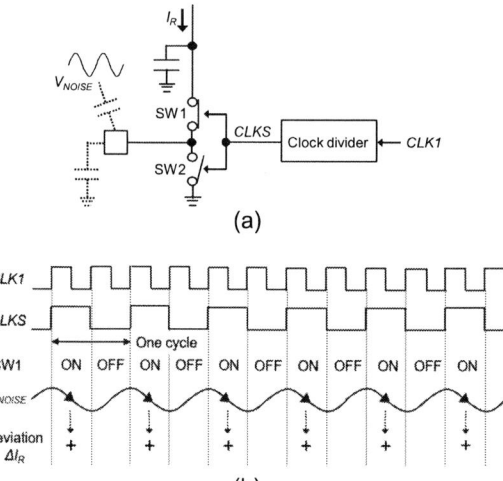

(a)

(b)

Fig. 3. Impact of noise on capacitance measurement.
(a) Block diagram. (b) Timing chart.

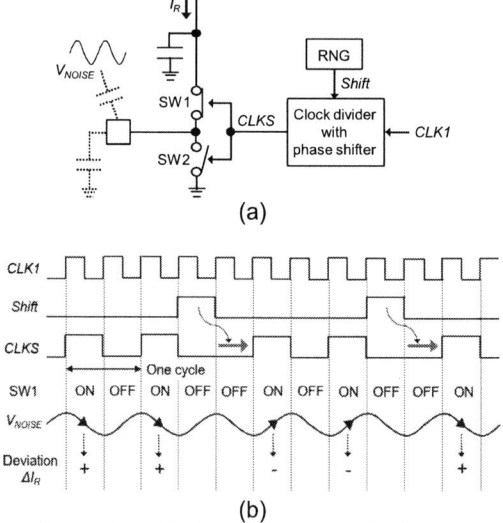

(a)

(b)

Fig. 4. Random phase shift scheme for enhancing RF noise immunity.
(a) Block diagram. (b) Timing chart.

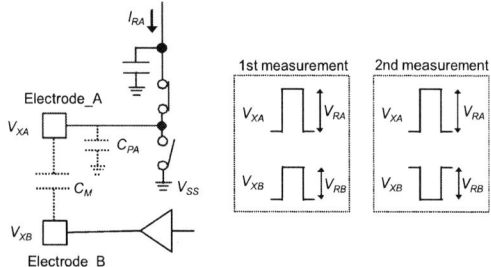

Fig. 2. Techniques for removing parasitic capacitance in mutual capacitance measurement

III. Implementation and Measurement

We integrated the proposed CDC in a 32bit microcontroller. The microcontroller chip was fabricated in 130nm CMOS process. Capacitive Sensing Unit (CSU) consists of the CDC and a control block. Table I summarizes chip features of the microcontroller. The chip micrograph is shown in Fig. 5.

To verify the effectiveness of the RPS scheme, we performed a noise immunity experiment using a touch sensing system realized with the microcontroller. The setup of this experiment is illustrated in Fig. 6. Noise frequency was varied from 0.15 MHz to 80 MHz complying with the test specification of IEC 61000-4-6. At each noise frequency, the counter output D_{OUT} was monitored and occurrence of touch detection errors were noted. As shown in Fig. 7, the deviation of counter output and the occurrence of touch detection error, which appeared remarkably near the switching frequency of 4

MHz for the system without the RPS scheme, were greatly reduced for the system with the RPS scheme. Owing to the RSP scheme, our CDC meets level 3 of IEC 61000-4-3 and IEC 61000-4-6 standards.

The resolution of our CDC improves proportionally to measurement time as shown in Fig. 8. This is because the number of oscillations which are counted by the counter increases proportionally to measurement time. Users can select resolution under a trade-off between the resolution and measurement time. Measured results show that the resolution, which is defined as the capacitance corresponding to $D_{OUT} = 1$, improves from 0.20 fF to 2.0 aF by increasing the measurement time from 0.5 ms to 50 ms. In this measurement, we selected the measured capacitance range considering use cases that we assumed. Comparison with previously reported CDCs is summarized in Table II.

IV. 3D Gesture Sensing Application

The use of human machine interfaces (HMI) are increasing in emerging IoT applications. Machines with user-friendly designs can be controlled with intuitive hand/finger movements such as swipe motions. Moreover, touch-free, 3D gesture controls are also necessary in home appliances and industry equipment because gesture recognition enables operations in conditions that are less than ideal because of water, dirt, or temperature. Other gesture recognition methods have several drawbacks. A camera-based image recognition system can be hacked, infrared sensors malfunction in sunlight and acoustic wave sensors cannot penetrate shields such as walls. To solve these problems, mutual capacitive sensors applies proximity sensor because of its shield-resistant, waterproof capability.

Table I. Features of the microcontroller.

Characteristic	Description
Technology	CMOS 130nm
Frequency	54MHz
Supply Voltage	5.0V
CPU	RXv2 32bit CISC Processor with FPU
Memory	Code Flash: 256KB, SRAM: 64KB
CSU	Capacitive Sensing Unit including CDC: 24 channels
Chip Size	3.34mm x 3.49mm

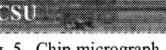

Fig. 5. Chip micrograph.

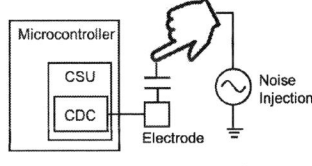

Fig. 6. Setup for experiment of RF noise immunity.

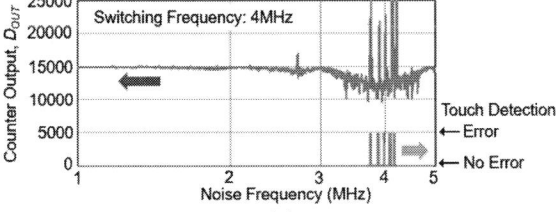

(a)

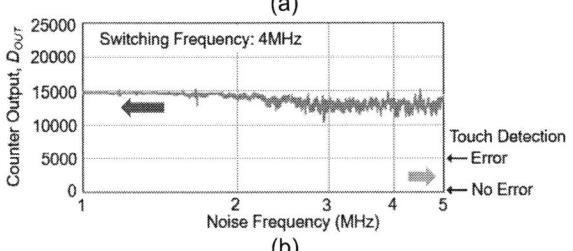

(b)

Fig. 7. Experimental results of noise immunity*.
(a) w/o RPS scheme. (b) w/ RPS scheme.
* Zooming from 1MHz to 5 MHz.

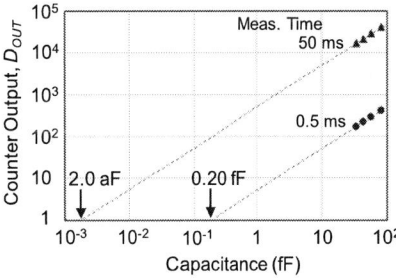

Fig. 8. Measured counter output versus capacitance.

Table II. Comparison with previously reported CDCs.

	[1]	[2]	[3]	[4]	This work	
Technology	350nm CMOS	180nm CMOS	180nm CMOS	180nm CMOS	130nm CMOS	
Method	DT- ΔΣM	SAR	SAR + ΔΣ	CT- ΔΣM	Switched Capacitor + Current-Controlled Osc.	
Meas. Time	20μs	4ms	230μs	0.125ms	0.5ms	50ms
Resolution	65aF	6.0fF	0.16fF	0.4fF	0.20fF*	2.0aF*
IEC 61000-4-3 and IEC 61000-4-6 standards	Not reported	Not reported	Not reported	Not reported	Level 3	

* Extrapolated from measured results.

A. 3D Gesture Control Using the Proposed CDC

Fig.9 shows a 3D gesture control demo system using the proposed high-precision CDC. A 3D gesture sensor consists of a main board and a sensor electrode board. The sensor electrode board has one central transmitter electrode and four receiver electrodes to measure mutual capacitance. The demo system also includes 3D position and monitoring system and a tablet which demonstrates the touch-free HMI for various applications.

The software structure consists of four layers (Fig.10 (a)). The CDC API measures capacitance using the counter output D_{OUT}. The gesture recognition library analyzes changes in the 3D position of the hand using the 3D position calculation middleware and determines the type of gesture that was made. The application program notifies results to the monitoring system and the tablet.

B. Position Calculation

Fig.11 (a) shows 3D position coordinates on the sensor electrode board. A 3D position (x, y, z) can be calculated by measuring mutual electrode capacitance of the central transmitter electrode paired with each of the four border receiver electrodes. When a hand approaches the sensor and acts as a conductor, the mutual capacitance decreases. Fig.11 (b) illustrates the measured capacitance with hand position x varying from 60mm left to 60mm right with z of 50mm and y of 0mm. The capacitance of the right electrode shown in blue decreases as the hand moves closer to the right sensor, while the capacitance of the left electrode shown in green increases. The pre-measurements of the capacitance at several positions are applied as parameters of the 3D position calculation middleware. The flow of 3D position calculation is shown in Fig.10 (b). By performing these steps at regular intervals, the position of the hand can be calculated.

The proposed CDC can realize a proximity 3D gesture control application because of its high resolution and noise immunity. The 3D gesture sensor recognizes movement in (20 x 20 x 20) cm³ space with maximum accuracy of 1mm along the z axis. We confirmed that the demo system achieved the gesture recognition rate of 100% with an experiment in which swipe gestures were repeated 1500 times at a movement speed of 60cm/s with position z of 50mm.

V. CONCLUSION

A CDC which can be easily ported to various processes due to its simple structure of the analog part was presented. The CDC was integrated in a 32bit microcontroller. Measured results show that the RPS scheme used in the CDC enhances its RF noise immunity. The CDC meets level 3 of IEC 61000-4-3 and IEC 61000-4-6 standards. Users can select CDC resolution under a trade-off between the resolution and measurement time. If the measurement time is increased up to 50ms, the resolution is as high as 2aF. 3D gesture sensing application was realized using the 32bit microcontroller in which the proposed CDC was integrated.

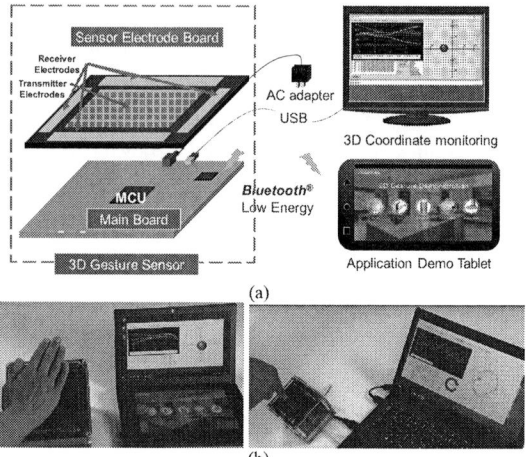

(a)

(b)

Fig. 9. 3D gesture demonstration system. (a) System configuration. (b) 3D gesutre demonstration.

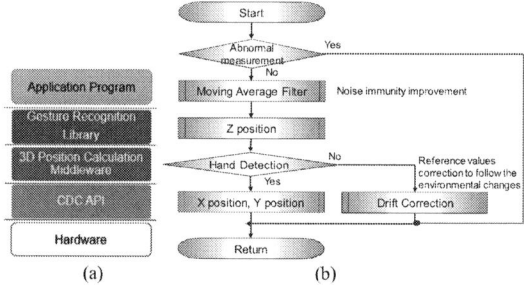

(a)   (b)

Fig. 10. (a) Software structure. (b) 3D position calculation middleware.

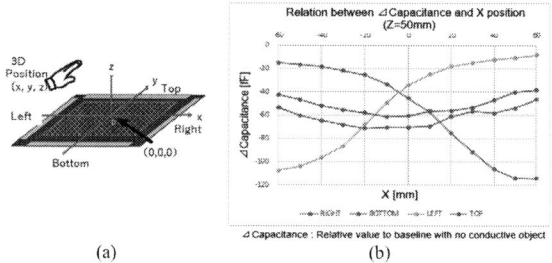

(a)   (b)

Fig. 11. (a) 3D position coordinates on a sensor electrode board. (b) Meadured result of capacitance and X position dependency.

REFERENCES

[1] S. Xia, K. Makinwa, and S. Nihtianov, "A capacitance-to-digital converter for displacement sensing with 17b resolution and 20µs conversion time," ISSCC Dig. Tech. Papers, pp. 198-199, Feb. 2012.

[2] H. Ha, D. Sylvester, D. Blaauw, and J.-Y. Sim, "A 160nW 63.9fJ/conversion-step capacitance-to-digital converter for ultra-low-power wireless sensor nodes," ISSCC Dig. Tech. Papers, pp. 220-221, Feb. 2014.

[3] S. Oh, W. Jung, K. Yang, D. Blaauw, and D. Sylvester, "15.4 b incremantal sigma-delta capacitance-to-digital converter with zoom-in 9b synchronous SAR," Symp. VLSI Circuits, pp. 222-223, June 2014.

[4] N. Narasimman, D. Nag, K. C. T. Chuan, and T. Kim, "A 1.2 V, 0.84 pJ/conv.-step ultra-low power capacitance to digital converter for microphone based auscultation," CICC, pp. 1-4, May 2017.

A 104.8TOPS/W One-Shot Time-Based Neuromorphic Chip Employing Dynamic Threshold Error Correction in 65nm

Luke R. Everson, Muqing Liu, Nakul Pande, and Chris H. Kim

Department of Electrical and Computer Engineering
University of Minnesota
Minneapolis, MN 55455 USA

Abstract- **As neural networks continue to infiltrate diverse application domains, computing will begin to move out of the cloud and onto edge devices necessitating fast, reliable, and low power solutions. To meet these requirements, we propose a time-domain core using one-shot delay measurements and a lightweight post-processing technique, Dynamic Threshold Error Correction (DTEC). This design differs from traditional digital implementations in that it uses the delay accumulated through a simple inverter chain distributed through an SRAM array to intrinsically compute resource intensive multiply-accumulate (MAC) operations. Implemented in 65nmLP CMOS we achieve, to our knowledge, the lowest reported energy efficiency for a neuromorphic processor with 52.4TSOp/s/W (104.8TOp/S/W) at 0.7V with 3b resolution for an impressive 19.1fJ/MAC.**

I. INTRODUCTION

The ever-increasing demand for higher performance and energy efficiency in machine learning (ML) applications has driven an impressive range of ASICs [1-7] aimed at meeting the needs. Digital SoCs [3-5,7] have found success by restricting the weight resolution [8], changing memory access structures, and guarding operations when the input is zero. However, all require large registers to store intermediate results and complex multiplier blocks. An emerging trend [1,2] has been to employ time-domain circuits to implement dot-products; the main kernel for ML applications. Fig. 1 details how the dot-product is computed in the time-domain and in conventional digital implementations. In time-domain the delay is modulated by the application inputs and weights to generate proportional delays.

These delays are accumulated and can be routed to a Time-to-Digital Converter (TDC) or counter to be processed for use in the deep learning application. Alternatively, the digital approach relies on many multiplier blocks and wide merging adders, typically in an array-like structure, to generate dot-products. The primary benefit of time-domain circuits is that the accumulate portion of the MAC is intrinsic to the architecture. Additionally, the processing unit can be realized as a collection of inverters making the area and active power consumption very low. Digital methods can leverage existing IP-blocks for multipliers and adders and do not require calibration, unlike the time-domain circuits. Additionally, digital circuits can handle higher bit operations more effectively due to the binary encoding. Previous time-domain neuromorphic chips have fundamental limitations. In [2], a digitally controlled oscillator was used to modulate the frequency, by switching capacitor loads representing the weights, while the number of cycles in a set sampling period was counted. While this closed loop structure has the benefit of canceling temporal noise, it must oscillate for many cycles to generate a result. Reference [1] is also a delay line based approach, but the outputs and weights are restricted to binary. More critically, their design has twice the area overhead do to the fact they utilize local reference delay lines instead of a global reference, and can limit the potential scalability of the architecture. In this work, we have addressed the shortcomings of previous designs by implementing Digitally controlled Delay Lines (DDLs) that are compared to

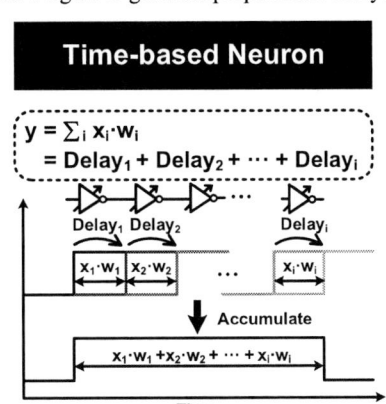

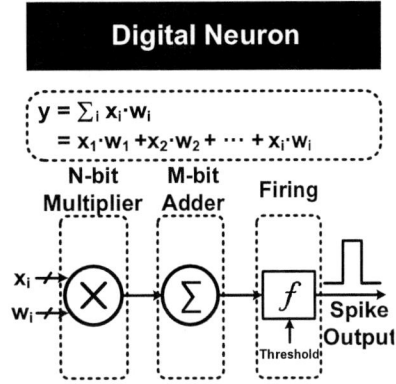

Fig. 1. Time-based neurons utilize the delay through basic circuit elements such as inverters to implement the dot-product. Digital neurons use conventional Boolean logic for arithmetic operations. Both architectures can be mapped to deep learning applications.

This research was supported in part by the National Science Foundation under award number CCF-1763761 and IGERT grant DGE-1069104.

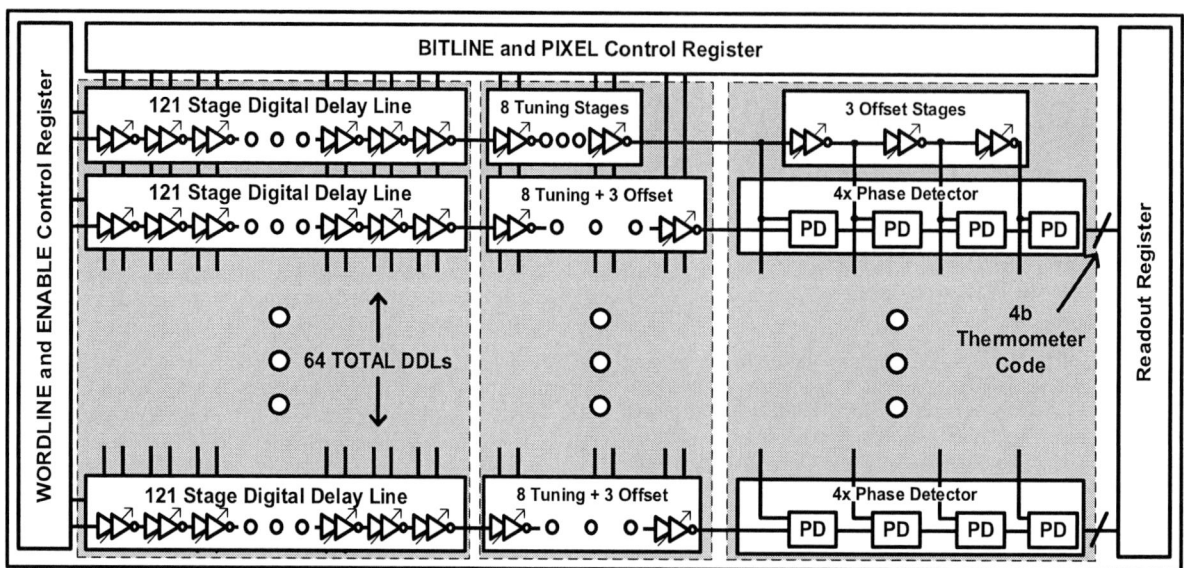

Fig. 2. Top level schematic of the time-based neuromorphic core. Layout is based on SRAM array. The core contains 64 DDLs each with 129

a shared reference delay to compute multi-bit MACs. We will explain the operating concept, novel accuracy boosting technique DTEC, and describe the chip measurement results in the following sections.

II. ONE-SHOT AND TIME-BASED NEUROMORPHIC CONCEPT

Conventionally, Boolean computations are used to realize arithmetic operations in hardware. However, time domain circuits can also be used at an advantage of lower area and power per processing unit, and reduced design complexity. The kernel of all ML algorithms can be distilled into a dot product; $y = \sum xw + b$, where x is an input vector, w a weight matrix, and b is a bias, or offset vector. Our high level architecture is shown in Fig. 2. An input pulse is presented on the left side of the core and the delay of each stage is modulated based on the application inputs. Each stage has 8 delay units (DU) with output taps which the pulse travels through as seen in Fig. 3. The number of DU enabled depends on the weight, stored locally in SRAM cells, and the input pixel, which is applied across the array on the bitlines. Each DU has two inverters to retain consistent polarity between stages. This is critical in the event that the rising and

falling propagation delays are not matched, as well as ensuring correct polarity at the TDC. The output tap is realized as a complex tristate gate and the functionality is described in Fig. 4. The first column shows the circuit schematic and corresponding connections between the different DUs. The right four columns show the activated paths, shown with black lines, depending on the values of the input and weights. DU₅ is the nominal stage delay, and is activated through the right branch of the circuit when the pixel is not present, representing "zero delay." The right table shows the mapping between the algorithm weights and the delays realized in the chip. When the input is present the

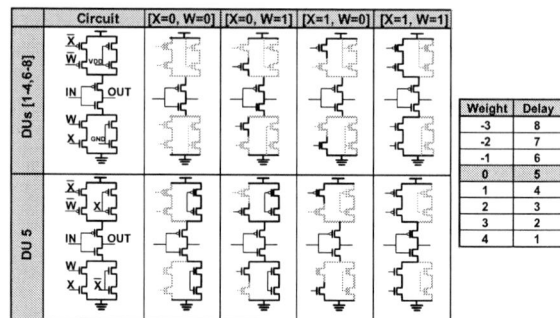

Fig. 4. Left table shows the complex tristate connections used to implement dot product based on input and weights. Right table shows 3b weight-delay mapping.

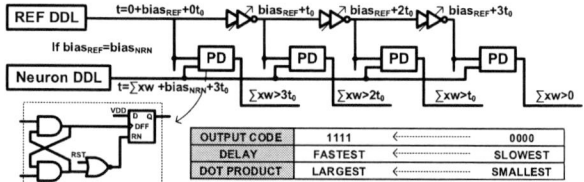

Fig. 5. Details of delay to dot product relationship.

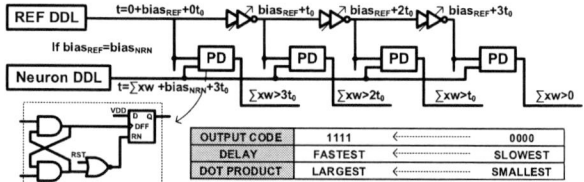

Fig. 3. (Top) Schematic of pixel stage. Complex tristates (Fig. 4)

978-1-5386-6414-8/18 $31.00 © 2018 IEEE   274

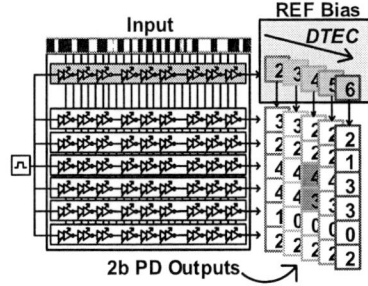

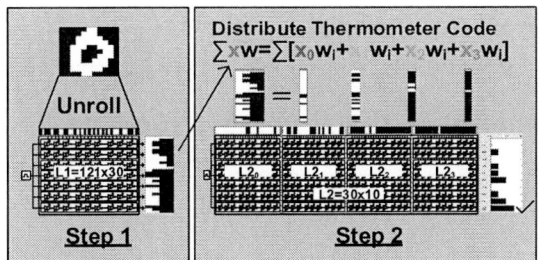

Fig. 8. Dataflow for multilayer time-based deep neural network demonstrated in this work.

Fig. 6. DTEC concept. Reference DDL bias is increased to elucidate the strongest DDL.

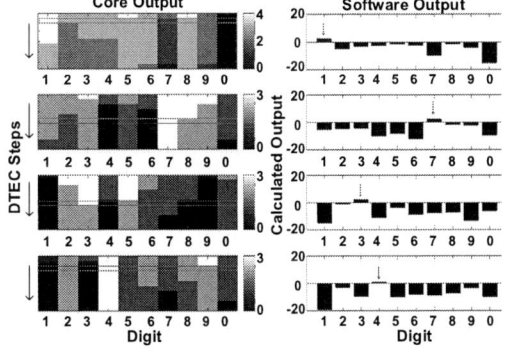

Fig. 7. Colormaps (left) show outputs from core after DTEC. Each row corresponds to results from successive values of the bias. The red rectangle highlights where DTEC has identified the dominant output. Bar plots (right) show expected results from software.

left branch is enabled in the DU corresponding the weight of the stage. The accumulation in the MAC is achieved naturally as the pulse passes sequentially through the DDL, stage by stage. The layout of each DU in the stage is pitch matched to a 6T SRAM so the layout is regular, compact, and scalable. The bias vector is applied in the same way for the last eight units. Additionally, it can be used to tune process variation, so that during evaluation those pixels are always activated. Fig. 5 shows the relationship between the time domain computation in the chip and the expected arithmetic output. The phase detector output maps roughly to the Relu transfer function. When the reference pulse beats the neuron rising edge all four thermometer bits are zero, regardless of the magnitude. The transfer function between the four bits is linear and then clips, or saturates, once the neuron pulse is faster than all the offsets.

III.   DYNAMIC THRESHOLD ERROR CORRECTION (DTEC)

In this design we opt for a 2 bit TDC due to the optimal tradeoff between small area and low power, and strong architecture performance. In networks with "winner-take-all" topologies, such as the last stage of classification networks, ambiguous predictions can occur. Unclear outputs in this work can stem from limited resolution between phase detector trip-points or activity outside of the range of the phase detector. To mitigate this issue, we propose a Dynamic Threshold Error Correction (DTEC) technique which increases the effectiveness of the 2bit TDC. As shown in Fig. 6, when two or more DDLs have the same output, DTEC works by increasing the threshold bias delay which moves the trip point of the phase detectors.

DTEC is dynamic due to the fact that the bias sweep would be terminated after the third evaluation, when the dominant DDL was identified from the phase detectors. Additionally, DTEC can be stopped after a fixed number of steps if no dominant DDL emerges to conserve power. In Fig. 7 the top row of colormaps shows ambiguous predictions from the core, while successive rows show the output as DTEC is applied. Red rectangles highlight where DTEC has successfully identified the target. Analysis of results from the 3b single layer application (section IV) show that by applying just two DTEC steps 81.64% of the correctible errors are recovered. This comes at a cost of just 0.41 additional evaluations per image. After the one-shot evaluation 73% of all images have a dominant output. The remaining 2,668 images begin DTEC and after the first step 46% are resolved and 37% after the second step leaving less than 1,000 images ambiguous. Thus, 4,108 DTEC evaluations improves the total accuracy from 69.16% to 82.14%. If three DTEC steps are applied 88.8% of errors can be recovered at an overhead of 0.51, demonstrating the dynamic scalability of the technique. DTEC is an economical and scalable approach to significantly improve application performance.

IV.   TEST CHIP MEASUREMENTS AND APPLICATION DETAILS

We evaluate the core on the MNIST benchmark [9]. Fig. 9 shows the comparison of classification accuracy on an 11x11 image for single and two layer networks between expected simulated software results, one-shot evaluation, and DTEC. To reduce the 28x28 grayscale images to 11x11 binary images, 3 pixels are sliced from all four sides of the image. Then, a fixed resizing command is applied, and finally the pixels are binary thresholded. Fig. 8 shows how the core can be used in a multi-layer deep neural net application. Each bit of the thermometer

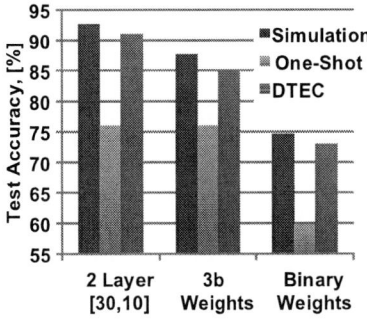

Fig. 9. Results on 11x11 MNIST for 3b two layer, 3b and binary

TABLE I. COMPARISON TABLE

	This Work		A-SSCC'16 [1]	CICC'17 [2]	ISSCC'17 [3]	ISSCC'17 [4]	ISSCC'16 [5]	ISSCC'16[6]	Science'14[7]
Chip Architecture	Time-Based		Time-Based	Time-Based	Digital	Digital	Digital	Sw. Cap	Digital
Algorithm Target	FCDNN & CNN		FCDNN & CNN	FCDNN & CNN	FCDNN & CNN	FCDNN & FFT	CNN	CNN & SGD	FCDNN & CNN
Technology [nm]	65		65	65	28 FDSOI	40	65	40	28
Chip Area [mm²]	0.644		3.61	0.24	1.87	7.1	12.25	0.012	430
Precision* [b]	[B,T,2,3]		B	3	[4-16]	[6-32]	16	3	[B,T]
On-Chip SRAM [kB]	8.06		20	3	144	270	181.5	[-]	256MB
VDD [V]	1.2 (Nom.)	0.7 (E_{MAX})	1	1.2	0.6	0.65	0.82	1	0.85
Frequency [MHz]	1700	285	23041	792	200	19.3	250	1000	0.001
Energy Efficiency** [TSop/s/W]	36.2	52.4	48.2	2.47	5.0	0.19	.18	3.86	0.04
Hardware Efficiency [GE/PE][1]	38.4		76.5	33.2	7456	18269	50637	288	6.5

*B=Binary, T=Ternary   **Synaptic Op=MAC

code is expanded as the input in the next layer. The input is divided into four segments, and the weight matrix is copied four times ($L2_0$-$L2_3$), which gives each bit equal weighting. In the example shown in Fig. 8, 30 neurons in layer 1 yield a 120 bit input to layer 2. By applying DTEC, the ambiguous results are almost completely recovered and the slight loss in accuracy is due to output differences smaller than a single tuning bit. Fig. 10 shows the tradeoff between power consumption and nominal stage delay for various supply voltages. Power is kept exceptionally low because rarely are more than two stages switching at a time in a DDL. A wide operating voltage range is enabled, due to the all-digital time-based design choices. If the design incorporated pipelining, it could achieve even greater throughput. That is, multiple pulses could be pushed into the DDL and the input could shift as well. This is ideally suited for Convolutional Nets where a weight filter slides across an image. In this case, the image could slide across the weights while input pulses are applied to the DDL. Die photo and design specs are highlighted in Fig. 11. Table I shows strong performance compared with state of the art. All comparisons

are made at the highest reported energy efficiency operating point. We report the best energy efficiency and a very competitive gate equivalent count for each processing unit at half the size of [1]. Our chip is scalable in voltage, weight resolution, and is versatile in that it is able to tackle fully connected deep networks as well as convolutional nets.

V. CONCLUSION

We described a time-based neuromorphic core based on one-shot DDLs in 65nm LP CMOS and proposed an error recovery technique, DTEC. It uses inverter delays to compute the dot product kernel, making it ideally suited for ML applications. The proposed core is validated on the MNIST dataset and achieves near simulated prediction accuracy on single and multi-layer networks after applying our error correction technique, DTEC. Maximum energy efficiency of 54.2TSOPs/s/W with 3b resolution at 0.7V makes the proposed architecture attractive for edge devices.

REFERENCES

[1] D. Miyashita, S. Kousai, T. Suzuki and J. Deguchi, "Time-domain neural network: A 48.5 TSOp/s/W neuromorphic chip optimized for deep learning and CMOS technology," *2016 IEEE Asian Solid-State Circuits Conference (A-SSCC)*, Toyama, 2016, pp. 25-28.

[2] M. Liu, L. R. Everson and C. H. Kim, "A scalable time-based integrate-and-fire neuromorphic core with brain-inspired leak and local lateral inhibition capabilities," *2017 IEEE Custom Integrated Circuits Conference (CICC)*, Austin, TX, 2017, pp. 1-4.

[3] B. Moons, R. Uytterhoeven, W. Dehaene and M. Verhelst, "14.5 Envision: A 0.26-to-10TOPS/W subword-parallel dynamic-voltage-accuracy-frequency-scalable Convolutional Neural Network processor in 28nm FDSOI," *2017 IEEE International Solid-State Circuits Conference (ISSCC)*, San Francisco, CA, 2017, pp. 246-247.

[4] S. Bang *et al.*, "14.7 A 288μW programmable deep-learning processor with 270KB on-chip weight storage using non-uniform memory hierarchy for mobile intelligence," *2017 IEEE International Solid-State Circuits Conference (ISSCC)*, San Francisco, CA, 2017, pp. 250-251.

[5] Y. H. Chen, T. Krishna, J. Emer and V. Sze, "14.5 Eyeriss: An energy-efficient reconfigurable accelerator for deep convolutional neural networks," *2016 IEEE International Solid-State Circuits Conference (ISSCC)*, San Francisco, CA, 2016, pp. 262-263.

[6] E. H. Lee and S. S. Wong, "24.2 A 2.5GHz 7.7TOPS/W switched-capacitor matrix multiplier with co-designed local memory in 40nm," *2016 IEEE International Solid-State Circuits Conference (ISSCC)*, San Francisco, CA, 2016, pp. 418-419.

[7] P.A. Merolla *et al.*, "A million spiking-neuron integrated circuit with a scalable communication network and interface," *Science*, vol. 345, no. 6197, pp. 668-673, Aug. 2014.

[8] I. Hubara, M. Courbariaux, D. Soudry, R. El-Yaniv, and Y. Bengio, "Binarized Neural Networks," 2016 NIPS, Barcelona, Spain, 2016, pp. 4107-4115, Dec. 2016.

[9] Y. LeCun, L. Bottou, Y. Bengio, and P. Haffner. "Gradient-based learning applied to document recognition." *Proceedings of the IEEE*, 86(11):2278-2324, Nov. 1998.

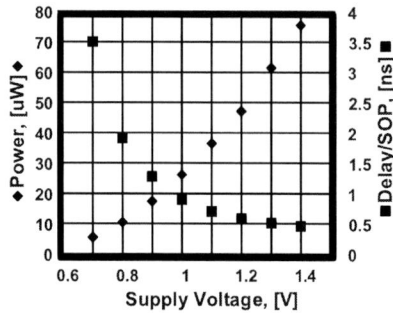

Fig. 10. Power consumption of a DDL and delay/stage versus VDD.

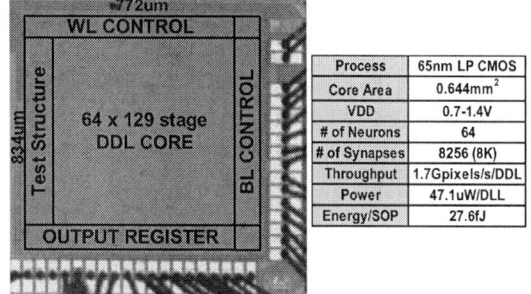

Fig. 11. Die photo and summary with reported metrics at 1.2V.

A 137-μW Area-Efficient Real-Time Gesture Recognition System for Smart Wearable Devices

Taegeun Yoo[1], Van Loi Le[1], Ju Eon Kim[2], Ngoc Le Ba[1], Kwang-Hyun Baek[2], and Tony. T. Kim[1]

[1]School of EEE, Nanyang Technological University, Singapore
[2]School of EEE, Chung-Ang University, South Korea
tgyoo@ntu.edu.sg, levanloi001@e.ntu.edu.sg

Abstract— Gesture recognition has increasingly become one of the most popular human-machine interaction techniques for smart devices. Existing gesture recognition systems suffer from either excessive power consumption or large size, limiting their applications for ultra-low power IoT and wearable devices. This paper presents an accurate, area-efficient, and ultra-low power real-time gesture recognition system for smart wearable devices. The proposed work utilizes a peak-based gesture classification engine with less memory and a low-resolution and low-power on-chip image sensor for achieving high area efficiency and low power. The feature extraction architecture removes fixed-pattern noises from the low-power on-chip image sensor for accuracy improvement and employs parallelism for recognition speed enhancement. The proposed system requires only 3.2 KB on-chip memory for processing 32×32 pixel data. Measurement results of a test chip fabricated in 65nm CMOS demonstrate that the proposed system consumes 137.0 μW at 0.8 V and 30fps while occupying only 1.78mm², which achieves the lowest power and smallest area among existing gesture recognition systems.

Keywords—gesture recognition, system on chip, feature extraction, low power processor, image sensor, wearable devices

I. INTRODUCTION

Recent demand for intelligent human-machine interactions has led to the evolution of smart wearable devices, especially for gesture recognition. One of the most popular techniques is touchscreen [1], but touch-based methods require a large-size display for applying gestures. Touch-less gesture recognition based on ultrasound [2] and infrared radiation (IR) [3] provides compact and intuitive features. However, complex analog processing for IR signals and the use of ultrasonic excitation sources result in large power consumption. This limits their applications for IoT and wearable devices where ultra-low power is required because of their limited battery sizes. Image sensors [4] - [5] have also been widely utilized for gesture recognition. However, conventional imaging-based gesture or pattern recognition systems are typically power-hungry due to sophisticated processing algorithms with large memory sizes. The vision chip in [4] can recognize 7 gestures at a high frame rate of >1000 fps by utilizing a self-organizing map neural network. However, it suffers from excessive power consumption (630 mW) and large size (82.3 mm²). The system in [5] enables real-time hand-depth sensing and hand tracking by employing a stereo-matching technique based on convolutional neural networks. Nevertheless, it also consumes high power (9.02 mW) and requires a large on-chip memory (781.5 KB).

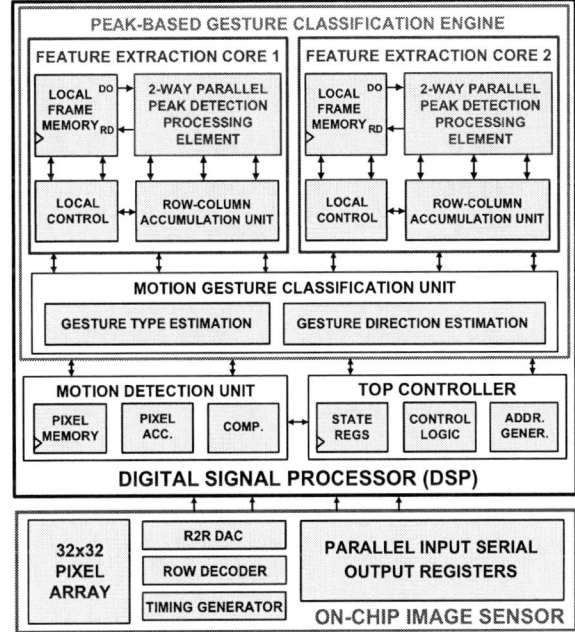

Fig. 1. Overall architecture of the proposed gesture recognition system.

In this paper, we present an accurate, area-efficient (1.78mm²), and low power (137μW) real-time gesture recognition system by utilizing an efficient memory-reduced (3.2KB) peak-based gesture classification engine. The effective classification engine allows the system to employ an on-chip low-resolution and low-power image sensor with a 32×32 pixel array and low-memory feature extraction cores while maintaining high recognition accuracy of 8 different motion hand-gestures. The recognition accuracy is further improved by eliminating fixed-pattern noises from the low-power on-chip image sensor during the feature extraction operation. Parallelism is also exploited for enhancing the recognition time.

II. PROPOSED GESTURE RECOGNITION SYSTEM

A. System Architecture

Fig. 1 shows the overall architecture of the proposed gesture recognition system that consists of an on-chip 32×32 image sensor and a low-power digital signal processor (DSP) including a peak-based gesture classification engine (PGCE), a motion detection unit, and top controller. The image sensor captures

978-1-5386-6414-8/18 $31.00 © 2018 IEEE

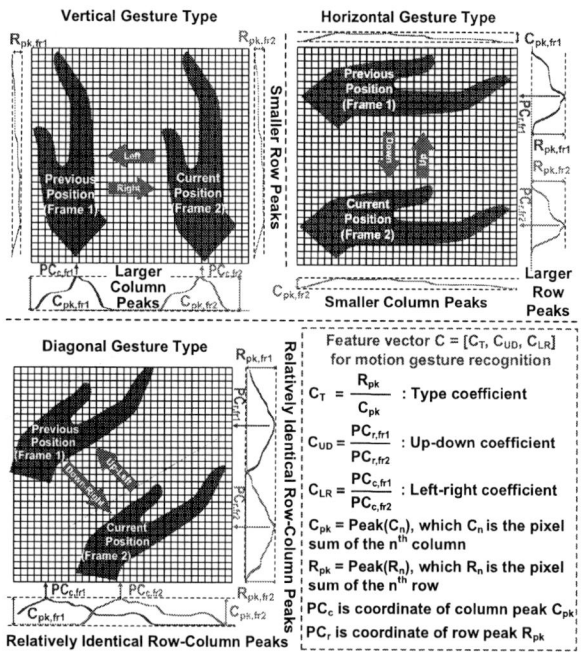

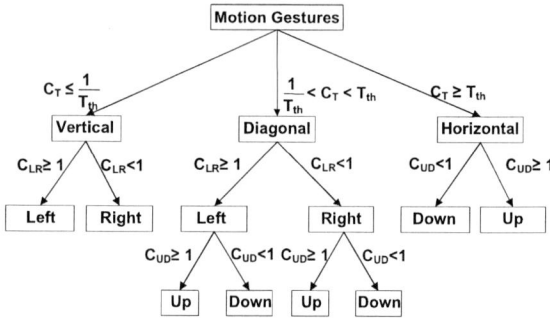

T_{th}: threshold level for gesture type selection obtained from training

(a)

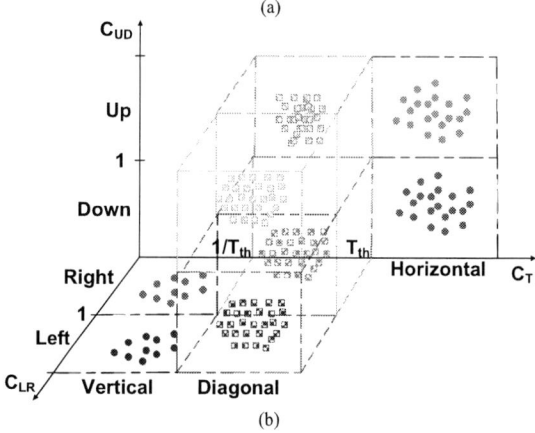

(b)

Fig. 3. The motion gesture classification algorithm including the (a) classification tree and (b) 3D feature space for gesture recognition.

Fig. 2. The fundamentals of the efficient motion gesture recognition engine based on a 3D feature vector C utilizing the row-column peaks' amplitudes and coordinates.

hand gestures while the motion detection unit detects hand motions. When a motion is detected, the PGCE saves the next two consecutive frames. Next, the PGCE's two parallel feature extraction cores calculate row and column sums and detect row-column (RC) peaks' amplitudes and positions by exploiting 2-way parallel peak detection processing elements (PDPEs). Then, a motion gesture classification unit recognizes gesture types based on the RC peaks' amplitudes while gesture directions are identified using the coordinates of the RC peaks. The detailed motion gesture recognition algorithm of the DSP is explained in the following sub-session.

B. Efficient Peak-Based Gesture Recognition Engine

Fig. 2 illustrates the fundamentals of the effective motion gesture recognition engine based on a 3D feature vector utilizing the RC peaks' amplitudes and coordinates. From extensive simulations, the behaviors of RC peaks while moving the hand gestures are observed as follows. The vertical gesture type's column peaks are larger than row peaks, whereas the horizontal type's row peaks show larger amplitudes than that of column peaks. On the other hand, the diagonal gesture type shows the relatively identical RC peaks. In terms of gesture directions, the coordinates of the RC peaks correspondingly changes with the moving hand gestures. These observations construct a 3D feature vector $C = [C_T, C_{UD}, C_{LR}]$ for motion gesture recognition exploiting the RC peaks as shown in Fig. 2.

Fig. 3 (a) and (b) explains the classification tree and the 3D feature space of the peak-based gesture recognition algorithm, respectively. The type coefficient C_T classifies 3 gesture types including vertical, diagonal, and horizontal hand-gesture types. In this work, the C_T value is estimated through off-chip training.

Once C_T tells the type of the gesture, C_{LR} and C_{UD} determine the final gesture based upon the classification tree in Fig. 3 (a). For example, if $C_T \geq T_{th}$ and $C_{UD} \geq 1$, the gesture is classified as a horizontal gesture type moving up. By utilizing the 3D feature space as shown in Fig. 3 (b), the peak-based classification engine can recognize up to 8 motion hand gestures including the 3 gesture types with 4 different directions. The efficiency of the classification engine allows the system to employ a low-memory feature extraction architecture and a low-resolution image sensor for power reduction.

C. Dual-Core Parallel Fixed-Pattern-Noise-Free (FPNF) Feature Extraction Architecture

Fig. 4 (a) and (b) show the dual-core parallel FPNF feature extraction architecture with the detailed RC accumulation and peak detection operations, respectively.

1) Fixed-Pattern-Noise Elimination in RC Accumulation

After a motion is detected, the next two frames are saved in the local frame memories and the feature extraction cores are switched to the RC accumulation stage for calculating the RC sums. To eliminate the fixed-pattern noises from the low-power low-resolution image sensor, the pixel data of the previous frame is subtracted from that of the current frame before accumulating as depicted in Fig. 4 (a). Then, the FPNF pixel data of each row or column are summed up and stored in four RC sum buffers for the peak detection in the PDPEs.

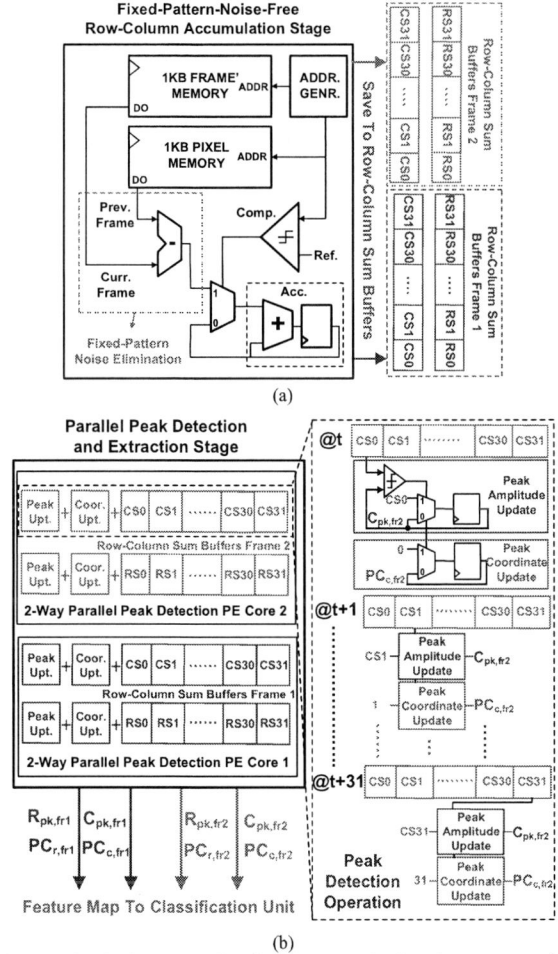

(a)

(b)

Fig. 4. The dual-core parallel fixed-pattern-noise-free feature extraction architecture with (a) the RC accumulation and (b) peak detection operations.

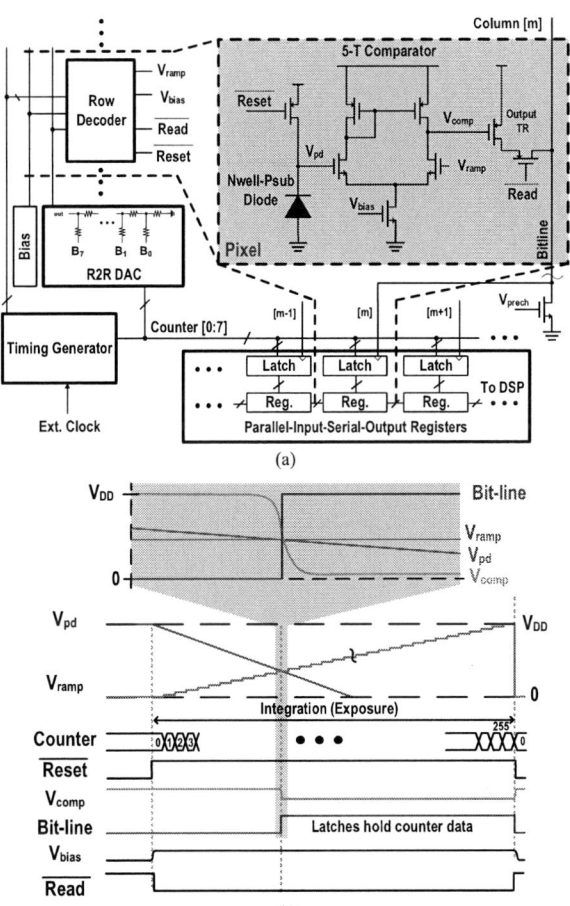

(a)

(b)

Fig. 5. Image sensor's (a) implementation and (b) timing diagram.

2) Parallel Peak Detection Processing Element (PDPE)

Fig. 4 (b) describes the parallel peak detection and extraction stage of the feature extraction cores with the detailed peak detection operations in the PDPEs. After the RC accumulation stage is finished and the RC sum data are saved in the four RC 32-sum buffers, the two 2-way parallel PDPEs start performing peak detections. The PDPEs parallelly slide digital comparators across these four buffers to detect the RC peak positions while updating the RC peaks' amplitudes and coordinates accordingly. After 32 clock cycles, the peak detection operation is done. The RC peaks' amplitudes ($R_{pk, fr1}$, $R_{pk, fr2}$, $C_{pk, fr1}$, and $C_{pk, fr2}$) and coordinates ($PC_{r, fr1}$, $PC_{r, fr2}$, $PC_{c, fr1}$, and $PC_{c, fr2}$) are detected and extracted for the motion gesture recognition. The FPNF characteristic of the feature extraction architecture improves the accuracy of the system while the parallel PDPEs reduce the feature extraction latency by 4×.

D. On-Chip Low-Resolution Low-Power Image Sensor

The on-chip low-power image sensor consists of five main blocks as shown in Fig. 5 (a): a 32×32 pixel array, an R2R DAC, a row decoder, parallel-input-serial-output (PISO) registers and

a timing generator. A pixel employs a 5-T differential-pair type comparator which is biased to achieve less than 1 nA current and gated after an integration period through V_{bias} for power reduction. To generate the reference voltage V_{ramp} for the comparator, an R2R type DAC is adopted because it does not require complex circuit implementation (only resistors) while guaranteeing rail-to-rail output swing even at extremely low supply voltages. The row decoder decodes signals from timing generator and provides control signals for a selected row. The PISO registers store parallel data from the selected row and sequentially transfer the data to the DSP.

Fig. 5 (b) shows the detailed pulse-width-modulation readout operation of the pixel. The photodiode voltage (V_{pd}) of the selected row is pre-charged to V_{DD} before the integration period. During the integration operation, V_{pd}, which gradually decreases in proportion to the intensity of light, is compared with the generated reference voltage (V_{ramp}). When V_{ramp} and V_{pd} intersect, the bit-line is pulled up to V_{DD} and the latches in each column store the counter value (Counter [0:7]). Since the intersecting time is proportional to the intensity of the light, the stored counter value at that moment represents the pixel image. After the integration period, the PISO registers start data transmission to the DSP.

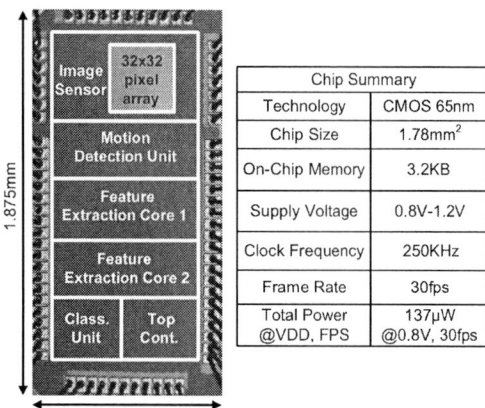

Chip Summary	
Technology	CMOS 65nm
Chip Size	1.78mm²
On-Chip Memory	3.2KB
Supply Voltage	0.8V-1.2V
Clock Frequency	250KHz
Frame Rate	30fps
Total Power @VDD, FPS	137µW @0.8V, 30fps

Fig. 6. Test chip microphotograph and performance summary.

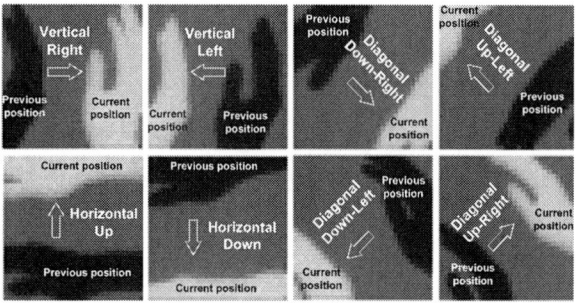

Fig. 7. Sample captured hand gestures and recognition results.

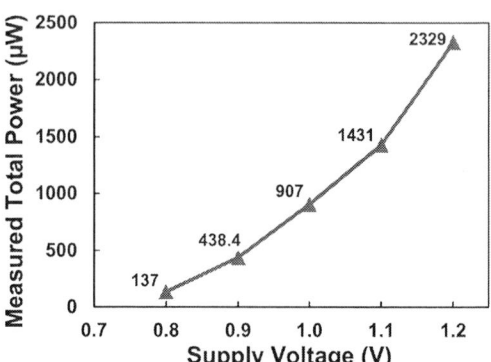

Fig. 8. Measured total power versus supply voltage at 250 KHz and 30fps.

TABLE I.
MEASUREMENT COMPARISON WITH THE EXISTING SYSTEMS

	This Work	[5] ISSCC 2018	[3] VLSIC 2016	[4] JSSC 2014	[2] ISSCC 2014	[1] ISSCC 2014
Process	65nm	65nm	65nm	180nm	180nm	130nm
Sensor (On-Chip)	Image Sensor (YES)	Stereo Camera (NO)	Infrared Sensor (NO)	Image Sensor (YES)	Ultrasonic Sensor (NO)	Sensing Electrode (NO)
Area	1.78mm²	16mm²	8.1mm²	82.3mm²	4mm²	4.2mm²
Frame Rate	30fps	33.3fps	30fps	>1000fps	30fps	240Hz
Total Power	137µW	9.02mW	260µW	630mW	401µW	19.2mW

III. MEASUREMENT RESULTS

The proposed gesture recognition system is fabricated in 65nm CMOS. The test chip microphotograph and performance summary are shown in Fig. 6. The supply voltage of the system can be scaled down to 0.8 V at 30 fps and 250 KHz. By utilizing the 1K-pixel image sensor and low memory classification engine, the on-chip memory is significantly reduced to 3.2KB. Thus, the chip area is considerably minimized to 1.78 mm².

Fig. 7 shows sample captured hand gesture images with recognition results. By exploiting the efficient peak-based gesture classification engine with the FPNF feature, the proposed system accurately identifies 8 different motion hand gestures with the recognition accuracy of 90.6 %. By utilizing parallelism for feature extraction, the system achieves real-time gesture recognition operation at 30fps with the total processing time of 4.228ms. Fig. 8 shows the measured total power of the system including the on-chip sensor when sweeping the supply voltage from 0.8 to 1.2 V at 30fps. With only 3.2KB on-chip memory, the system consumes only 137.0 µW at 0.8 V for real-time gesture recognition. Table I summarizes the measurement comparison of the proposed system and other existing systems [1-5]. The proposed system outperforms other designs in terms of total power and area, making the proposed system very well suited for IoT and wearable devices where the area and power budget are extremely critical.

IV. CONCLUSION

This paper presents an accurate, area-efficient, and ultra-low power real-time gesture recognition system for smart wearable devices. The proposed system obtains better area efficiency, low-power, and accurate gesture recognition by utilizing a peak-based gesture classification engine with less memory and an on-chip low-resolution and low-power image sensor. The feature extraction architecture eliminates fixed-pattern noises for further enhancing accuracy and also exploits parallelism for improving the recognition speed. The proposed system needs only 3.2KB on-chip memory for processing the 32×32 pixel array and can recognize 8 different hand gestures. Test chip measurement in 65nm CMOS demonstrates that the proposed system consumes 137.0µW for real-time gesture recognition at 30fps while occupying only 1.78mm², which achieves the smallest area and lowest power among the recently proposed systems [1-5]. Therefore, the proposed system is applicable to various ultra-low power applications such as IoT and wearable devices, where high area efficiency and ultra-low power consumption are topmost requirements.

REFERENCES

[1] Y. Hu et al., "3D Gesture-Sensing System for Interactive Displays Based on Extended-Range Capacitive Sensing," in ISSCC Dig. Tech. Papers, 2014, pp. 212-213.

[2] R. J. Przybyla et al., "3D Ultrasonic Gesture Recognition," in ISSCC Dig. Tech. Papers, 2014, pp. 210-211.

[3] S. Oh, et al., "A 260µW Infrared Gesture Recognition System-on-Chip for Smart Devices", in Symp. VLSIC, 2016, pp. 228-229.

[4] C. Shi et al., "A 1000 fps Vision Chip Based on a Dynamically Reconfigurable Hybrid Architecture Comprising a PE Array Processor and Self-Organizing Map Neural Network," in IEEE Journal of Solid-State Circuits, vol. 49, no. 9, pp. 2067-2082, Sept. 2014.

[5] S. Choi, J. Lee, K. Lee, and H.-J. Yoo, "A 9.02mW CNN-Stereo-Based Real-Time 3D Hand-Gesture Recognition Processor for Smart Mobile Devices," in ISSCC Dig. Tech. Papers, 2018, pp. 220-221.

A 655Mbps Successive-Cancellation Decoder for a 1024-bit Polar Code in 180nm CMOS

Hye-Yeon Yoon, Seung-Jun Hwang, and Tae-Hwan Kim

School of Electronics and Information Engineering, Korea Aerospace University

Goyang-si, Gyeonggi-do, Republic of Korea, E-mail: taehwan.kim@kau.ac.kr

Abstract—**This paper presents the implementation of a high-throughput successive-cancellation decoder for a 1024-bit polar code. The proposed decoder processes the kernels in two successive layers in a fused manner within a cycle, so as to reduce the cycle count taken to decode a codeword. In addition, the log-likelihood ratios are represented in a redundant form to realize a high-speed kernel processing. As a result, the cycle period is not lengthened even with the fused kernel processing and thus achieving a high throughput. Implemented in 180nm CMOS, the proposed decoder shows the throughput of 655 Mbps for the rate-1/2 code. In terms of the throughput efficiency, the proposed decoder is 3.51 times superior to the previous state-of-the art one.**

I. Introduction

Polar codes are the error correcting codes that can achieve the capacity asymptotically for various kinds of the practical channels. The capacity-achieving property can be realized in such a synthetic channel that is formed by simply combining and splitting the multiple realizations of the channel, which makes the encoding and decoding procedures regular and simple. Based on these good properties, polar codes have been adopted as new error-correcting schemes for various applications including the next-generation wireless communication systems [1].

The successive-cancellation (SC) decoding [2] is the standard algorithm employed to decode a polar code efficiently. However, the SC decoding has a serial flow. Furthermore, it achieves a near optimal error-rate performance only if the codeword is long. In consequence, it is very challenging to develop an efficient SC decoder that achieves a high throughput. It is noteworthy that an efficient SC decoder is also meaningful for the efficient implementation of other advanced decoding algorithms (e.g. SC-list [3]), for which the SC decoding works as the basis.

There are several previous researchers who studied efficient architectures to perform the SC decoding. Leroux *et al.* presented several efficient architectures for the SC decoders, where each processing unit (PU) is utilized in a time-multiplexed manner so as to reduce the overall complexity without diminishing the throughput severely [4], [5]. Yuan *et al.* presented the two-bit decoding architecture designed to decode out two bits within a cycle so as to reduce the cycle count taken to decode a codeword [6]. Mishra *et al.* presented the first ASIC implementation results of an SC decoder designed based on the semi-parallel architecture [7].

This study presents the implementation of a 1024-bit SC decoder. The proposed decoder is designed based on the conventional tree architecture, but several kernels involved in two successive layers are processed within a cycle in a fused manner so that the cycle count can be reduced significantly. To avoid lengthening the cycle period due to the fused kernel processing, the log-likelihood ratios (LLRs) are represented in the redundant form that has been proposed in our preliminary work [8]. The proposed decoder is designed to decode out four bits within a cycle and process two codewords simultaneously. Implemented in 180nm CMOS, the proposed decoder shows the throughput as high as 655Mbps for the rate-1/2 code. When compared to the previous state-of-the-art SC decoder, the proposed decoder shows 3.51 times higher throughput efficiency.

The rest of the paper is organized as follows. Section II explains the SC decoding briefly as the background. Section III presents the architecture of the proposed decoder. Section IV describes the implementation results of the proposed decoder. Section V draws the conclusion.

II. SC Decoding for Polar Codes

Let us consider the SC decoding for an N-bit polar code, where $N = 2^n$ for a positive integer n. The source and the codeword are denoted by N-bit vectors, $\mathbf{u}$ and $\mathbf{x}$, respectively, where the i-th bit of $\mathbf{u}$, u_i, $0 \leq i < N$, is frozen to zero if $i \in \mathcal{A}$ and $\mathcal{A}$ is given a priori. As the size of $\mathcal{A}$ is K, the code rate of K/N is realized. In the SC decoding, each of the source bit is estimated based on its LLR, which is calculated given with the channel LLRs. The LLR is calculated recursively through layers based on two kinds of the kernels, where the calculations involved in Layer l are expressed as follows:

$$
L_{l,i} = \begin{cases} f\left(L_{l-1,i}, L_{l-1,i+2^{n-l}}\right), & \text{if } i/2^{n-l} \text{ is even} \\ g\left(L_{l-1,i-2^{n-l}}, L_{l-1,i}, \hat{s}_{l,k}\right), & \text{otherwise} \end{cases} \quad (1)
$$

for $1 \leq l \leq n$ and $0 \leq k < 2^{n-1}$, $L_{l,i}$ denotes the i-th LLR in Layer l, $\hat{s}_{l,k}$ denotes the k-th partial sum in Layer l, and f and g stand for the kernels that will be delineated later. As the LLR of u_i is $L_{n,i}$, it can be calculated with the channel LLRs given by $L_{0,i} \triangleq \log\left(P\left(\mathbf{y}|x_i = 0\right)/P\left(\mathbf{y}|x_i = 1\right)\right)$, where $P(\cdot)$ means the probability, $\mathbf{y}$ is the receive codeword that has been distorted by the channel.

The two kernels in the recursive LLR calculations for the SC decoding are expressed as follows: $f(a, b) \triangleq \text{sign}(a)\text{sign}(b)\min(|a|, |b|)$ and $g(a, b, s) \triangleq (1-s)g^{(0)}(a, b)+$

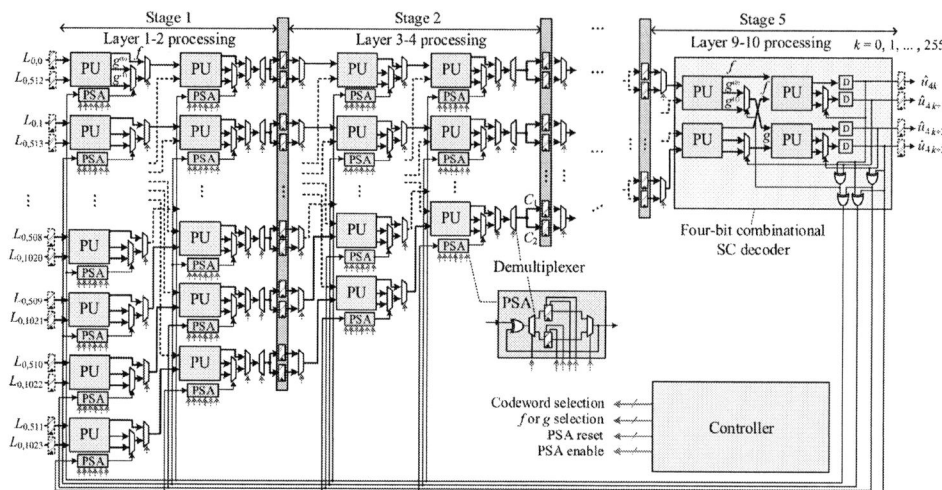

Fig. 1. Overall architecture of the proposed decoder, where PSA accumulates a decoded bit to calculate the partial sum and $\hat{u}_i$ denotes the i-th bit of the source estimate.

$sg^{(1)}(a, b)$, where a and b are the input LLRs, s is the partial sum, $g^{(0)}(a, b) \triangleq b + a$, and $g^{(1)}(a, b) \triangleq b - a$. In this study, the kernel f has been formulated based on the min-sum approximation [4] in order to achieve an efficient implementation.

III. Proposed Decoder

The proposed decoder performs the SC decoding for a 1024-bit polar code. It is designed based on the conventional tree architecture having multiple stages, but each stage is designed so as to process several kernels involved in two successive layers within a cycle in a fused manner; whereas in the conventional architecture, each stage is designed to process the kernels involved only in a single layer. As a result, the number of the stages is reduced to half of that of the conventional architecture. Fig. 1 shows the overall architecture of the proposed decoder, in which each PU processes the kernel. The fused kernel processing involved in Layer $2i - 1$ through Layer $2i$ is implemented by connecting the PUs serially in Stage i as shown in the figure. The merits owing to the fused kernel processing are two folds: first, the cycle count taken to decode a code word can be reduced by about 50%; secondly, the number of the registers for the LLR storage in between stages can also be reduced.

Though the fused kernel processing has such merits that are mentioned above, it may lengthen the cycle period because multiple kernels through layers are processed in serial within a cycle. The proposed decoder employs the redundant scheme [8] to represent the LLRs, instead of the conventional signed-magnitude or 2's complement scheme, for the purpose of realizing a fast kernel processing so as to avoid lengthening the cycle period as much as possible. In the redundant LLR representation scheme, each LLR is represented by a negating bit together with a signed value, where the negating bit is an additional bit to indicate if the value needs to be negated to get

the real one. Table I-II show how the kernels can be processed for the LLRs represented in the redundant scheme. In the tables, $x.neg$ and $x.val$ denote the negating bit and the value of an LLR x, respectively, and msb means the most significant bit. As shown in the tables, if the value of a resulting LLR needs to be negated, it is indicated by setting the negating bit, rather than negating the value explicitly. This is much faster than the kernel processing for the LLRs represented in the conventional scheme for which a resulting LLR needs to be negated explicitly in that case. As a result, the cycle period is not lengthened so much even with the fused kernel processing.

The PU is designed to process the kernel f or g for the LLRs represented in the redundant scheme. Fig. 2 shows the

TABLE I

PROCESSING OF THE KERNEL f FOR THE INPUT LLRS a AND b, WHERE $\theta = b.val + a.val$, $\gamma = b.val - a.val$, $\alpha = a.neg \oplus b.neg$, AND $\times$ MEANS A DON'T CARE TERM.

msb($a.val$)	msb($b.val$)	msb(γ)	msb(θ)	$f.neg$	$f.val$
0	0	0	$\times$	α	$a.val$
0	0	1	$\times$	α	$b.val$
0	1	$\times$	0	α	$b.val$
0	1	$\times$	1	$\overline{\alpha}$	$a.val$
1	0	$\times$	0	α	$a.val$
1	0	$\times$	1	$\overline{\alpha}$	$b.val$
1	1	0	$\times$	$\overline{\alpha}$	$b.val$
1	1	1	$\times$	$\overline{\alpha}$	$a.val$

TABLE II

PROCESSING OF THE KERNEL g FOR THE INPUTS LLRS a AND b, WHERE $\theta = b.val + a.val$, AND $\gamma = b.val - a.val$.

$a.neg$	$b.neg$	$g^{(0)}.neg$	$g^{(0)}.val$	$g^{(1)}.neg$	$g^{(1)}.val$
0	0	0	θ	0	γ
0	1	1	γ	1	θ
1	0	0	γ	0	θ
1	1	1	θ	1	γ

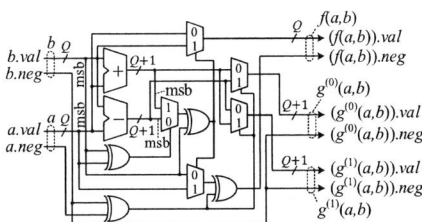

Fig. 2. Microarchitecture of the PU, where a and b are the input LLRs and their values are represented by Q bits.

microarchitecture of the PU, where each LLR is represented by a Q-bit value together with a negating bit. The proposed PU is composed of a Q-bit adder, a Q-bit subtractor, and other minor combinational logic. The proposed PU produces the resulting LLRs without negating their values explicitly, where the values are calculated by adding (subtracting) the values of the input LLRs; hence, the critical-path delay is not so longer than the logic delay of a single adder (subtractor). This is much shorter than the critical-path delay of the PU designed based on the conventional LLR representation schemes [5]–[7], which includes the logic delay taken by the explicit negation.

To reduce the cycle count further, the proposed decoder is designed to decode out four bits within a cycle. For this purpose, the last stage is designed as a four-bit combinational SC decoder as shown in Fig. 1. In the four-bit combinational SC decoder, the processing result of the kernel g is obtained by selecting between $g^{(0)}$ and $g^{(1)}$, which are pre-calculated in parallel with decoding a bit based on the processing result of the kernel f. In fact, its logic delay is not so longer than that of the other stages, and thus does not affect the cycle period.

The proposed decoder is designed to perform the SC decoding for two codewords simultaneously. To support such a simultaneous dual-codeword decoding, the proposed decoder has additional storages for the LLRs and partial sums as shown in Fig. 1. The number of the PUs is not increased; rather, the same number of the PUs are controlled efficiently to increase the utilization rate. Table III shows the scheduling for the first ten cycles of the decoding flow. In the table, f-f, f-g, g-f, and g-g stand for the types of the fused kernels involved in two successive layers, fg-fg stands for the four-bit combinational SC decoding, and $\hat{\mathbf{u}}_{i_1}^{i_2}$ represents $[\hat{u}_{i_1}, \cdots, \hat{u}_{i_2}]$. As shown in the table, each stage processes the kernels involved in two successive layers, and the two codewords, which are denoted by C_1 and C_2, are decoded simultaneously. As the cycle count taken to decode a codeword is 596 and two 1024-bit codewords can be decoded simultaneously, the overall throughput is as high as $3.43R$ bits/cycle for the rate-R code.

IV. Implementation Results

The proposed decoder has been developed as a softcore by describing its architecture in a synthesizable hardware-description language. Importing a cell library developed for a 180nm CMOS technology, the proposed decoder has been

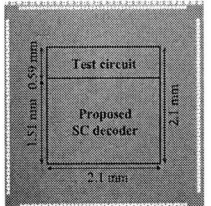

Fig. 3. Die photo of the proposed decoder.

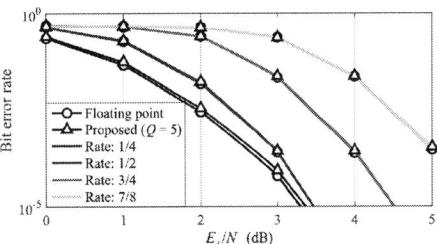

Fig. 4. Error-rate performance, where E_b and N_o mean the energy per bit and the noise variance, respectively.

synthesized into an ASIC and its die photo is shown in Fig. 3. The implementation results of the proposed decoder are summarized in Table IV along with those of the previous decoders. Each decoder in the table has been designed to decode a 1024-bit polar code based on the same decoding algorithm. The throughput of each decoder is not affected by the frozen-bit locations. The error-rate performance of the proposed decoder is close to that achieved by the floating-point simulations, as illustrated in Fig. 4. In Table IV, the throughput has been normalized by considering the technology difference and the complexity has been measured by counting the equivalent gates.

The proposed decoder shows much higher throughput than the previous decoders. When the latency is measured in terms of the cycle count taken to decode a codeword, the latency of the proposed decoder is much lower than that of the previous ones, as shown in Table IV. Despite such a low latency owing to the fused kernel processing as well as the four-bit decoding, the operating frequency is not so low. As a result, the throughput of the proposed decoder is as high as 655Mbps for the rate-1/2 code, which is much higher than that of the previous decoders. In addition, the complexity of the proposed decoder is not so high even though it supports the simultaneous dual-codeword decoding as the number of the stages is halved. In terms of the throughput efficiency, which is defined as the throughput per complexity, the proposed decoder is 3.51 times superior to the previous state-of-the-art decoder.

The functionality of the proposed decoder has been verified by probing the IO signals under the environment demonstrated in Fig. 5a. The test circuit inside the ASIC feeds the channel LLRs and frozen-bit locations to the decoder, and the decoder produces the decoded bits after several cycles. Fig. 5b shows

TABLE III
OVERALL SCHEDULING OF THE PROPOSED DECODER FOR THE FIRST TEN CYCLES.

Cycle	1	2	3	4	5	6	7	8	9	10
Stage 1	$C_1(f\text{-}f)$	$C_2(f\text{-}f)$								
Stage 2		$C_1(f\text{-}f)$	$C_2(f\text{-}f)$							
Stage 3			$C_1(f\text{-}f)$	$C_2(f\text{-}f)$						...
Stage 4				$C_1(f\text{-}f)$	$C_2(f\text{-}f)$	$C_1(f\text{-}g)$	$C_2(f\text{-}g)$	$C_1(g\text{-}f)$	$C_2(g\text{-}f)$	$C_1(g\text{-}g)$
Stage 5					$C_1(fg\text{-}fg)$	$C_2(fg\text{-}fg)$	$C_1(fg\text{-}fg)$	$C_2(fg\text{-}fg)$	$C_1(fg\text{-}fg)$	$C_2(fg\text{-}fg)$
Output					$C_1(\hat{\mathbf{u}}_0^3)$	$C_2(\hat{\mathbf{u}}_0^3)$	$C_1(\hat{\mathbf{u}}_4^7)$	$C_2(\hat{\mathbf{u}}_4^7)$	$C_1(\hat{\mathbf{u}}_8^{11})$	$C_2(\hat{\mathbf{u}}_8^{11})$

TABLE IV
IMPLEMENTATION RESULTS OF THE SC DECODERS FOR A 1024-BIT POLAR CODE.

Architecture	Semi-parallel [7]	Semi-parallel [5]	Tree [6]	Tree [8]	Proposed
LLR representation	Signed-magnitude	Signed-magnitude	Signed-magnitude	Redundant	Redundant
Simultaneous multi-codeword decoding	No	No	No	No	Dual
Latency (cycles)	1568	2080	767	1534	596
CMOS technology (nm)	180	65	45	180	180
Area (mm^2)	1.71	3.09	N.A.	N.A.	3.17
Complexity (KGE) [a]	183.64	214.37	338.50	256.34	382.04
Operating frequency (MHz)	150	500	750	377	382 [b]
Normalized throughput (Mbps) [c]	48.98	44.44	125.16	125.83	655.22
Throughput efficiency (Mbps/KGE) [d]	0.27	0.21	0.37	0.49	1.72

[a] The smallest 2-input NAND has been counted as 1 GE.
[b] Measured under the conditions of the 1.8 V supply and the 25°C.
[c] Calculated by (throughput)·(feature size) / 180 for the rate-1/2 code.
[d] Calculated by (normalized throughput) / (equivalent gate count).

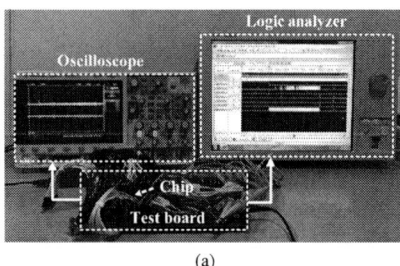

(a)

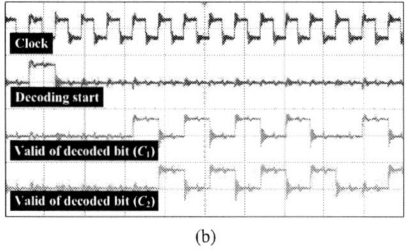

(b)

Fig. 5. Measurement (a) environment and (b) results.

some of the measurement results, where the decoded bits for two codewords are being produced out after a few cycles from the start of the decoding.

V. CONCLUSION

This paper presented the ASIC implementation of a high-throughput SC decoder for a 1024-bit polar code. The proposed decoder is designed based on the fused kernel processing as well as the four-bit decoding, in order to reduce the cycle count taken to decode a codeword. The redundant LLR representation scheme is employed to realize a fast kernel processing. The throughput efficiency of the proposed decoder is 3.51 times higher than the state-of-the-art result.

ACKNOWLEDGMENT

This work was supported by Institute for Information & Communications Technology Promotion (IITP) grant funded by the Korea government (MSIT) [2017-0-00528, The Basic Research Lab for Intelligent Semiconductor Working for the Multi-Band Smart Radar]. IDEC supported EDA tools.

REFERENCES

[1] T. R. W. M. #88, "Final report of 3GPP TSG RAN WG1 #87 v1.0.0," 3GPP, Reno, USA, Tech. Rep. R1-1701552, Nov. 2016.

[2] E. Arıkan, "Channel polarization: A method for constructing capacity-achieving codes for symmetric binary-input memoryless channels," *IEEE Trans. Information Theory*, vol. 55, no. 7, pp. 3051–3073, Jul. 2009.

[3] I. Tal and A. Vardy, "List decoding of polar codes," *IEEE Trans. Information Theory*, vol. 61, no. 5, pp. 2213–2226, May 2015.

[4] C. Leroux, I. Tal, A. Vardy, and W. J. Gross, "Hardware architectures for successive cancellation decoding of polar codes," in *Proc. Int'l Conf. Acoustics, Speech & Signal Processing*. IEEE, May 2011, pp. 1665–1668.

[5] C. Leroux, A. J. Raymond, G. Sarkis, and W. J. Gross, "A semi-parallel successive-cancellation decoder for polar codes," *IEEE Trans. Signal Processing*, vol. 61, no. 2, pp. 289–299, Jan. 2013.

[6] B. Yuan and K. K. Parhi, "Low-latency successive-cancellation polar decoder architectures using 2-bit decoding," *IEEE Trans. Circuits & Systems I: Regular Papers*, vol. 61, no. 4, pp. 1241–1254, Apr. 2014.

[7] A. Mishra, A. Raymond, L. Amaru, G. Sarkis, C. Leroux, P. Meinerzhagen, A. Burg, and W. Gross, "A successive cancellation decoder ASIC for a 1024-bit polar code in 180nm CMOS," in *Proc. Asian Solid State Circuits Conf.* IEEE, Nov. 2012, pp. 205–208.

[8] H.-Y. Yoon and T.-H. Kim, "Efficient successive-cancellation polar decoder based on redundant LLR representation," *IEEE Trans. Circuits & Systems II: Express Briefs*, To be published.

A Generated Multirate Signal Analysis RISC-V SoC in 16nm FinFET

Stevo Bailey*, Jaeduk Han*, Paul Rigge*, Richard Lin*, Eric Chang*, Howard Mao*, Zhongkai Wang*,
Chick Markley*, Adam Izraelevitz*, Angie Wang*, Nathan Narevsky*, Woorham Bae*, Steve Shauck[†],
Sergio Montano[†], Justin Norsworthy[†], Munir Razzaque[†], Wen Hau Ma[†], Akalu Lentiro[†], Matthew Doerflein[†],
Darin Heckendorn[‡], Jim McGrath[‡], Franco DeSeta[‡], Ronen Shoham[‡], Mike Stellfox[‡], Mark Snowden[‡],
Joseph Cole[‡], Dan Fuhrman[‡], Brian Richards*, Jonathan Bachrach*, Elad Alon*, and Borivoje Nikolić*

*EECS, University of California, Berkeley   [†]Northrop Grumman Corporation   [‡]Cadence Design Systems, Inc

Abstract—This paper demonstrates a signal analysis SoC consisting of a general-purpose RISC-V core with vector extensions and a fixed-function signal-processing accelerator. Both the core and the accelerators are instances produced by novel generators that allow for a wide range of parameter configurations and rapid design space exploration. The signal processing chain consists of generated instances of a time-interleaved ADC followed by a digital tuner, FIR filter, polyphase filter, and FFT all connected to the processor via an AXI4 bus. The 5 mm×5 mm chip is implemented in a 16 nm FinFET process and operates at 410 MHz at 750 mV drawing 600 mW. Presented applications show coupled functionality of the processor and accelerator performing spectrometry and radar receive processing, and a comparison with other state-of-the-art ASICs prove that generators can produce competitive designs.

I. INTRODUCTION

There is a continuing need for energy efficiency improvements through hardware specialization, yet custom IC development costs have become prohibitively large. A limiting factor in hardware design efforts is the lack of design reuse, currently relegated to (sometimes hardened) IP targeting the chosen application. Proposed solutions include raising the level of design (e.g., HLS [1]), embedding the design process into a reusable generator [2], [3], or having one expert engineer replicate a simple design [4]. However, there are no complete examples of performance-competitive, mixed-signal SoC designs reported in open literature that demonstrate a comprehensive use of both digital and mixed-signal generators for both design and verification.

This work demonstrates a complex SoC containing a signal-processing accelerator and a general-purpose processor from a set of novel, open-source digital and analog generators [5] written in Chisel [6] and BAG [7]. The general-purpose processor is a generated RISC-V core with new ISA accelerators, and the signal processing accelerator comes from a new generator of streaming signal processing functions. Section II describes the architecture produced by the generator. Testing and measurement results are given in Section III. The presented instance is general enough to apply in a variety of signal processing contexts, and Section IV demonstrates several of these applications.

II. SoC ARCHITECTURE

Figure 1 shows the system architecture. The SoC is divided into a general-purpose processor and a custom signal processing accelerator. Communication between the processor and the accelerator is handled through a memory-mapped IO manager.

A. General-Purpose Processor

The chip includes a general-purpose processor that connects a host with the chip, programs the signal-processing accelerator, moves data, and computes what other on-chip accelerators cannot. The 64-bit single-issue in-order RISC-V Rocket CPU includes a single-/double-precision (SP/DP) floating-point unit (FPU). A direct memory access (DMA) accelerator offloads memory movement between the processors from the CPU. The 4-lane high-performance Hwacha vector accelerator implements a decoupled vector-fetch architecture and can perform compute-intensive parallel workloads not handled by the signal processing accelerator. A serial adapter tethers the CPU to the FPGA host, which writes programs to the on-chip 8 MB main

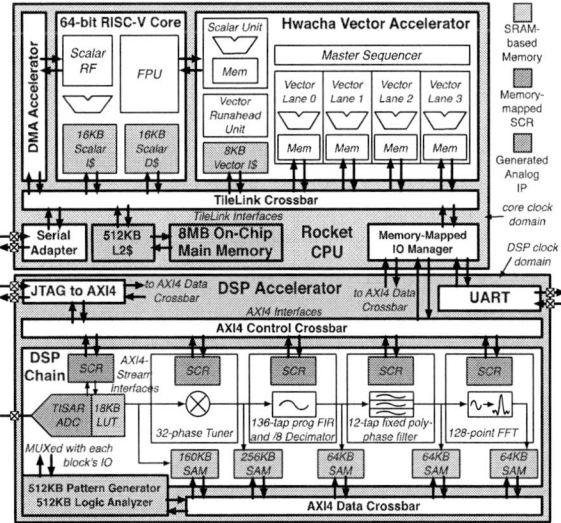

Fig. 1. Block diagram of the SoC architecture.

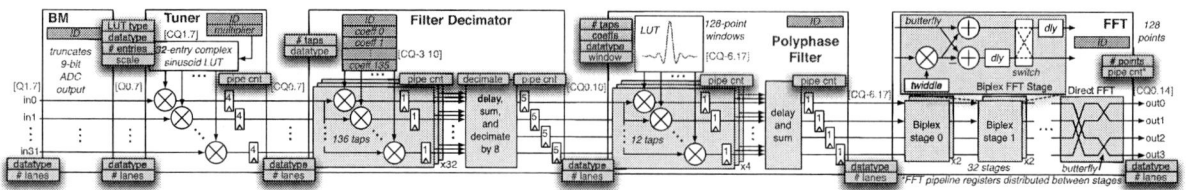

Fig. 2. Detailed diagram of the processing elements in the DSP accelerator. Red boxes indicate memory-mapped IO SCRs. Green overlays show generator parameters. Blue text gives fixed-point data type parameters chosen. CQ = complex fixed-point number.

memory before booting the core. A custom clock receiver supplies the core clock, and an asynchronous FIFO permits low-speed communication between the host and Rocket CPU.

B. Digital Signal-Processing Accelerator

The digital signal-processing accelerator architecture is reminiscent of SDF [8], [9], with actors (here processing elements) communicating through streaming interfaces rather than FIFO tokens. The Chisel generator contains memory-mapped IO registers taking commands from either the CPU or a JTAG debug port. Asynchronous FIFOs buffer data between the core clock, DSP clock, and JTAG clock domains. Separate AXI4 crossbars access status and control registers (SCRs) and data buffers (SAMs). For testing, a 512 KB pattern generator and 512 KB logic analyzer allow direct access to inputs and outputs of individual processing elements (PEs). A chain of PEs receives data from a BAG-generated 8-bit time-interleaved successive approximation (TISAR) ADC with lookup table (LUT)-based static calibration, and it also provides a clock to the PEs and AXI4 crossbars. The UART provides a backup interface into the RISC-V core, and the JTAG provides a backup interface into the accelerator.

C. Processing Elements

The selected PEs implement a signal analysis accelerator targeting spectral analysis or radar receive chain processing. Figure 2 shows the parameterization of PEs in green atop the final implementation diagram. The ADC LUT outputs 9 bits for testing, so a custom bit manipulator (BM) PE truncates this to 8 bits. The next two PEs comprise a digital down-converter (DDC). A 32-entry LUT-based digital tuner mixes the input signal with a complex sinusoid, and a fully-programmable 136-tap complex FIR filter shapes the signal and decimates it by 8. A 12-tap fixed-function polyphase filter multiplies the time-series data by a sinc function to window each FFT bin and reduce frequency-domain spectral leakage. A 128-point radix-2 FFT, comprised of 32-point biplex pipelined FFTs and a 4-point direct-form FFT, produces the complex spectrum output. The chain generator supports arbitrary ordering and duplication of PEs, so the chosen arrangement of PEs represents just one possible DSP accelerator configuration.

D. Instance Verification

Verifying and testing the generated instances is aided by coupling the design with design-for-test (DFT) structures.

Generated test benches adjust to chosen design parameters, and unit tests are run both in simulation and on the fabricated chip. Unit test vectors are generated in Python, which are then passed to a generated Unified Verification Methodology (UVM) testbench in Cadence's Verification Workbench (VWB) to verify the instance. A pattern generator and logic analyzer connect to each PE's IO and perform the same unit-level verification on the chip after fabrication. A JTAG debug module provides access to the signal-processing accelerator as a backup to the CPU. The CPU and signal-processing accelerator pass all ISA and unit-level tests, as well as comprehensive benchmarks (e.g. dhrystone) and kernels (e.g. matrix multiply).

III. TESTING RESULTS AND MEASUREMENTS

The chip is implemented in TSMC's 16nm FinFET technology and signed off at 300 MHz for both the core and DSP clock domains at 0.72 V and 125°C. Figure 3 shows the 5 mm by 5 mm annotated layout, die photo, and chip summary. The 8 MB main memory, Hwacha vector accelerator, and various other memories comprise most of the area. Also visible is the 136-tap fully programmable FIR filter, composed of

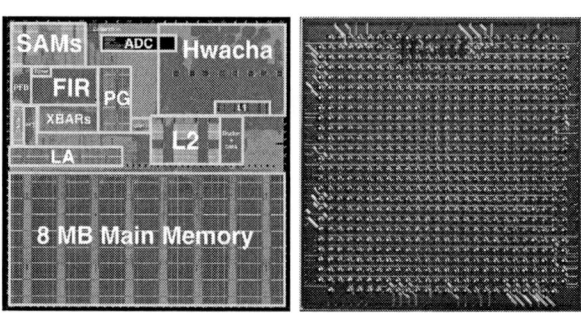

	Annotated Layout	Die Photo

Technology	16nm FinFET	
Die Area	5mm x 5mm (25mm²)	
	General-Purpose Processor	**Signal-Analysis Processor**
Area	1.1 mm² (gates) 9.2 mm² (SRAM)	1.5 mm² (gates) 0.8mm² (SRAM)
Total SRAM Size	71 Mbits	14 Mbits
Voltage	0.56 V - 0.98 V	0.56 V - 0.98 V
Max Frequency	410 MHz	417 MHz
Power	349 mW @ 0.75V, 410 MHz	210 mW @ 0.75 V, 417 MHz
Max Throughput (Mspectra/s)	0.46 (vector) 0.004 (scalar) @ 410 MHz	13 @ 417 MHz
Efficiency	23.4 GFLOPS/W (0.56V, 191MHz) (DGEMM on vector accelerator)	19.2 TOPS/W (0.56V, 192MHz) (1 op = 8-bit add ~ 17-bit mul)

Fig. 3. Chip layout, die photo, and summary

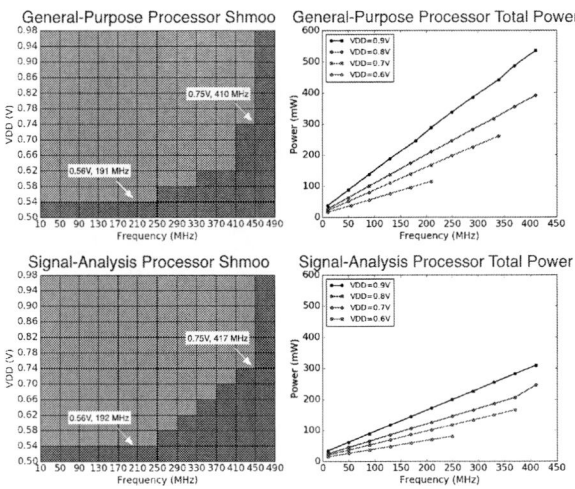

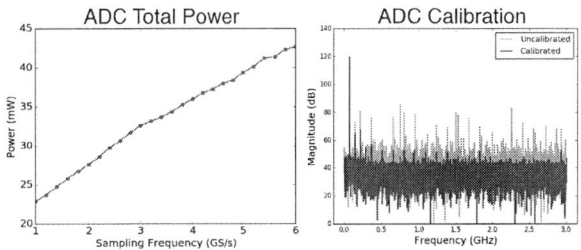

Fig. 4. Both processors function under similar operating condition ranges. The general-purpose processor consumes more power because of the 8 MB main memory.

Fig. 5. Typical ADC power consumption is less than 50 mW at 0.9V. Calibration of the ADC reduces noise and spurs.

many complex multipliers and adders. By using the Hwacha vector accelerator, at these conditions the general-purpose processor achieves 23.4 GFLOPS/W running double-precision matrix multiply on 256×256 matrices. For the general-purpose processor, throughput is measured by moving FFT output data from the SAM to the CPU memory and accumulating spectra using either the vector co-processor or the scalar ALU and DMA. Throughput is measured for the signal-analysis processor at the maximum operating frequency (see Figure 7 for more on spectral rates). Efficiency for the signal-analysis processor accounts for all PEs, and one operation is anything from a real 8-bit add to a 17-bit multiply, with complex adds and multiplies broken into their real operations.

Figure 4 shows the shmoo and power plots for two processors. Both function down to 0.56 V (the nominal supply is 0.8 V) and up to 410 MHz. A success on the shmoo plot requires the general-purpose processor to pass all ISA unit tests and the signal-analysis accelerator to pass all PE unit tests. Annotated on the shmoo plots are the corner values, i.e. the minimum voltage at the maximum frequency and the maximum frequency at the minimum supply voltage. Power measurements are averaged over the same unit tests as the shmoo. The chip consumes less than 1 W total for all modes.

Figure 5 shows the power and a typical calibration result for the 8 slice time-interleaved SAR ADC, which features a fractional radix to produce 8 real output bits from 9 total bits. The ADC reaches a max of 6.6 ENOB per slice and 6.4 ENOB total at 6 GS/s under a 0.9V supply. Analog supplies are independent from digital supplies.

IV. SIGNAL ANALYSIS APPLICATIONS

In this section we present two applications running on the SoC. Using the pattern generator to produce arbitrary waveforms and replace the ADC as the input source hastened

the development and debugging of writing these programs by decoupling program errors from ADC miscalibration, input noise, or instrument errors. The applications utilize many SoC features, including the complete signal processing chain, DMA, vector accelerator, and general-purpose processor. We compare the results with similar, fixed-function processors.

A. Spectrometry

Atmospheric spectrometers monitor molecule emissions to determine the composition of gases. Given the low SNR of these emissions, spectrometers with a wide bandwidth and long accumulation time are desired. A 512-point FFT is formed by sweeping the tuner frequency to allow the signal processor to analyze up to four frequency bands with a 128-pt FFT engine (the filter decimates the data rate by eight, but half the bands are symmetric because the input is real-valued). Figure 6 shows the signal processing path of a spectrometer input dataset. The low-pass filter, with constrained equiripple coefficients designed in MATLAB, passes the lower eighth of the spectrum to avoid aliasing when down converting. Its stopband is at 40 dB below the passband, resulting in visible aliasing above the noise floor in the combined spectrum. For each band, the tuner is set and the FFT outputs are stored in the SAM before being moved into the general-purpose processor's memory by the DMA. The power is calculated and accumulated using the vector accelerator. The use of these accelerators boosts the data processing rate for this application by over 10x, as seen in Figure 7. While not

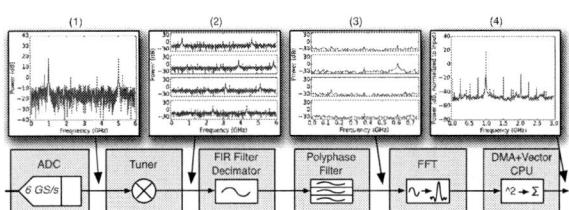

Fig. 6. Spectrometer signal processing example. Snapshots captured in the SAMs. (1) A real-valued signal is sampled through the calibrated ADC, producing a symmetric spectrum. (2) Four tuner LO frequencies are mixed with the input, producing four frequency-shifted spectra. (3) These spectra are low-pass filtered and down-converted by 8, resulting in four separate frequency bands. (4) The bands are Fourier transformed, accumulated, and combined on the CPU. This figure shows 100 accumulated spectra.

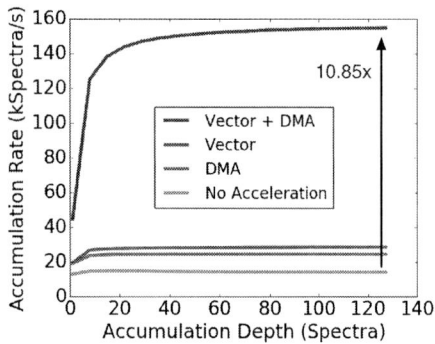

Fig. 7. Using the vector and DMA accelerators speeds up spectral accumulation by over 10x. This plot includes the overhead of sweeping the tuner frequency to monitor four frequency bands, so each spectrum is 512 channels.

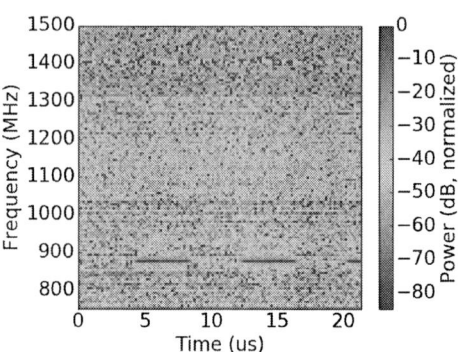

Fig. 8. Measured spectrogram of a 4 us pulse at 876 MHz.

designed specifically for spectrometry, this work is competitive with other published ASIC spectrometers, as seen in Table I.

B. Radar

Unlike atmospheric spectrometry, radar operates on fixed or variable short pulses or frequency-modulated continuous wave (FMCW) signals. These higher SNR signals require less accumulation, but processing speed limits the detectable range resolution. Figure 8 shows an example measured spectrogram of 4 µs fixed-frequency pulses, repeating every 8 µs. The tuner is set to view frequency bands containing the expected signal, and the FFT outputs a spectrum every 171 ns when the ADC operates at 6 GS/s. For unmodulated pulsed signals as in Figure 8, the minimum pulse repetition frequency (PRF) this SoC can resolve is 2.9 MHz (pulses per second), leading to a minimum range resolution of

$$\frac{c}{2 \times PRF} = 51.7\,\mathrm{m}. \tag{1}$$

It is possible to increase resolution by implementing pulse compression. The vector accelerator has sufficient throughput to convolve the received signal with the expected signal, and the FFT may be reused to perform an IFFT to recover the compressed signal. At 6 GS/s and by using a single frequency band with a 750 MHz wide linear frequency-modulated (LFM) chirp, this system has a minimum range resolution of 0.2 m.

V. CONCLUSION

This work demonstrates an ASIC, designed by using parameterized digital and analog generators, that achieves a peak efficiency of over 19 TOPS/W and 23 GFLOPS/W in 16 nm CMOS. The implemented RISC-V signal analysis SoC, generated from Chisel and BAG frameworks, performs spectrometry and radar signal processing with performance comparable to the state of the art. On-chip DFT facilitates quick bring-up and validation of the design instance. Generators used in this work are open source and may be easily adjusted and reused for a variety of applications [5].

ACKNOWLEDGMENT

This work was funded in part by the DARPA CRAFT program (HR0011-16-C-0052), BWRC, and ADEPT (Intel iSTC on Agile Design).

REFERENCES

[1] D. L. Rosenband and Arvind, "Hardware synthesis from guarded atomic actions with performance specifications," in *ICCAD*, November 2005.

[2] O. Shacham, S. Galal, S. Sankaranarayanan *et al.*, "Avoiding game over: Bringing design to the next level," in *DAC*, June 2012.

[3] B. Nikolic, "Simpler, more efficient design," in *ESSCIRC*, Sept 2015.

[4] A. Olofsson, "Epiphany-v: A 1024 processor 64-bit RISC system-on-chip," *CoRR*, vol. abs/1610.01832, 2016. [Online]. Available: http://arxiv.org/abs/1610.01832

[5] https://github.com/ucb-art/craft2-chip.

[6] J. Bachrach, H. Vo, B. Richards *et al.*, "Chisel: Constructing hardware in a scala embedded language," in *DAC*, Jun. 2012.

[7] E. Chang, J. Han, W. Bae *et al.*, "Bag2: A process-portable framework for generator-based ams circuit design," in *CICC*, Apr. 2018.

[8] E. A. Lee and D. G. Messerschmitt, "Synchronous data flow," *Proceedings of the IEEE*, vol. 75, no. 9, Sept 1987.

[9] L. Li, T. Fanni, T. Viitanen *et al.*, "Low power design methodology for signal processing systems using lightweight dataflow techniques," in *DASIP*, Oct 2016.

[10] B. Richards, N. Nicolici, H. Chen *et al.*, "A 1.5GS/s 4096-point digital spectrum analyzer for space-borne applications," in *CICC*, Sept 2009.

[11] F. Hsiao, A. Tang, Y. Kim *et al.*, "A 2.2GS/s 188mW spectrometer processor in 65nm CMOS for supporting low-power THz planetary instruments," in *CICC*, Sept 2015.

[12] S. Bailey, J. Wright, N. Mehta *et al.*, "A 28nm fdsoi 8192-point digital asic spectrometer from a chisel generator," in *CICC*, Apr. 2018.

TABLE I
COMPARISON OF STATE-OF-THE-ART ASIC SPECTROMETERS

	CICC'09 [10]	CICC'15 [11]	CICC'18 [12]	This Work
Technology	90nm CMOS	65nm CMOS	28nm FDSOI	16nm FinFET
Bandwidth	0.75 GHz	1.1 GHz	**8.5 GHz**	3.0 GHz
FFT Size	**8192 pts**	512 pts	**8192 pts**	128~512 pts
Integrated ADC	No	**Yes**	No	**Yes**
Power	1500 mW*	**188 mW**	5200 mW	586 mW
ADC Output	**8 bits**	7 bits	3 bits	**8 bits**
Can post-process	No	No	No	**Yes**
On-chip Accum. Depth	16M Spectra	1024 Spectra	65520 Spectra	**Infinite**

*excludes ADC power

A Compact High Efficiency and High Power Front-end Module for GSM/EDGE/TD-SCDMA/TD-LTE Applications in 0.13um CMOS

Shihai He, Fengxiong Peng, Linjian Xu, Hao Meng, Yongxue Qian

Beijing Huntersun Electronic Co.,Ltd. Beijing, China
Email: shihai.he@huntersun.com.cn

Abstract—This work presents a multimode multiband transmitter front-end module (TXM) fully integrated on a new 4-layer laminate substrate process supporting GSM/EDGE/TD-SCDMA/TD-LTE. In the TXM, a new compact power amplifier (PA) matching network including transformer and harmonic filter is proposed on this process with low insertion loss. An adaptive bias circuit is also proposed to enhance the PA linearity. The TXM including the 0.13um CMOS PA, SP16T antenna switch and all passive components is fully integrated in a flip-chip LGA package. Compared with other works, the measurement results of this work show 30dBm high linear power with 19% high power-added efficiency(PAE) in 915MHz EDGE mode. It also shows high linear power of 27dBm and 24dBm in TD-SCDMA mode and TD-LTE mode, respectively. The proposed TXM meets the class E2 power requirement with enough output power margin and high efficiency.

Keywords—*CMOS; power amplifier; flip-chip; front-end module; transformer; EDGE; wireless communication*

I. INTRODUCTION

For higher data rate, the wireless market has shown a remarkable development and growth from 2G to 4G communication technology and from constant envelope system GSM to unconstant envelope system EDGE, WCDMA and LTE. The high linearity, high efficiency, fully integrated PA module becomes more important to reduce the mobile phone area and cost. CMOS technology is a good candidate for low cost. However, for CMOS PA design, it needs to overcome the CMOS drawbacks like low breakdown voltage, lossy substrates and lower power capability. In the past decades, various design techniques [1-2] are used to improve the performance of CMOS PA. With techniques of distributed active-transformer [1] and adaptive bias [2], CMOS PA shows large output power and high efficiency. It also successfully demonstrates its application in commercial production for cellular GSM.

Recently, the CMOS PA TXM, which can support both linear/nonlinear modulation EDGE and GSM are explosively growing. The EDGE/GSM CMOS PA was reported for Class-E2 application [3]. A fully integrated TXM was also presented for GSM/EDGE/TD-SCDMA/TD-LTE application [4]. However, in both of the two designs, the matching circuits are implemented in integrated passive device (IPD) technique with bond-wire. This kind of TXM solution usually shows larger die area, lower PA efficiency and higher cost. In this work, a multimode multiband TXM fully integrated on a new 4-layer laminate substrate process is presented. It can support GSM/EDGE/TD-SCDMA/TD-LTE with smaller die size, lower cost and better performance.

The main blocks in this design are the multimode multiband CMOS PAs with a proposed compact matching network for Low Band (LB)/High Band (HB) frequency. The PA active parts are implemented by three-stage pseudo-differential cascade amplifiers to improve the reliability under high battery voltage. An adaptive bias circuit is proposed to improve common source transistor gm distortion. In additional, capacitive neutralization technique is implemented to decrease the AM-PM nonlinearity caused by miller effect. The proposed compact output matching network consisted of a high efficiency transformer and harmonic filter with a few surface mounted devices (SMD) components improves the TXM efficiency and output power. This work presents a completed solution for multimode multiband TXM with very low cost, high output power and high efficiency.

II. POWER AMPLIFIER CIRCUIT DESGIN

Fig. 1 shows the architecture of the proposed TXM. It is consisted of CMOS driver amplifiers and CMOS PA with on-chip inter-stage matching circuits. The output matching circuits consisted of transformer and harmonic filter are implemented on the laminate substrate. An SOI SP16T switch is used to share the antenna for both transmitter and receiver. After the SP16T switch, a coupler and RX matching circuit are designed for monitor the PA output power and RX input matching.

In the TXM, there are two RF paths for the low band and high band frequency. Each path includes a three-stage pseudo-differential cascade PA with a thin oxide MOS transistor (0.13μm) and two thick oxide MOS transistors (0.35μm) for reliability. The PA bias voltage and current in each mode can be optimized by the interface of MIPI controller circuit separately. In GMSK V_{ramp} control mode, the power control is implemented by closed-loop current control of each stage. For GMSK PA, it can be saturated with -5dBm input power when the switch in the second drive stage is off. In linear GMSK mode and V_{ramp} control EDGE/TD-SCDMA/TD-LTE mode, adaptive bias circuit is turned on for optimizing the PA efficiency while closed-loop current controller is turned off. More details for each part design are introduced in the following parts.

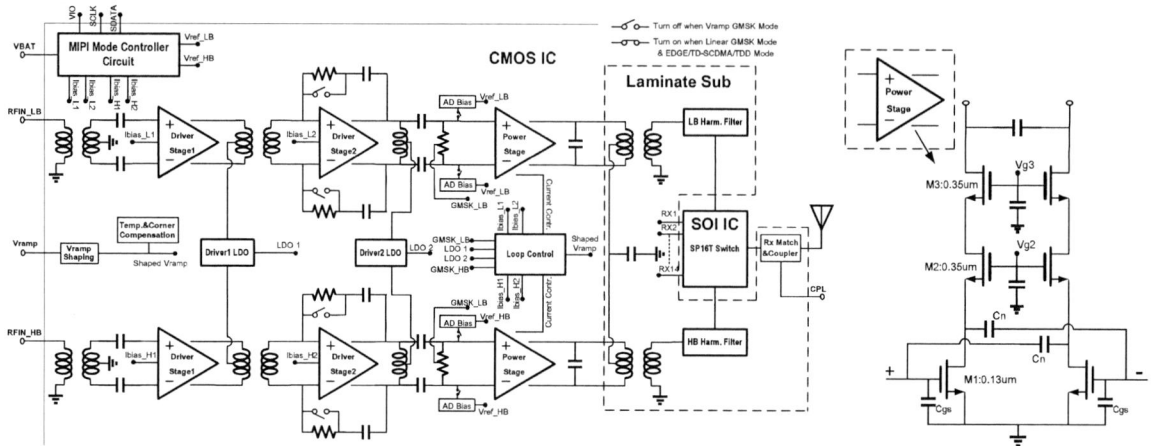

Fig. 1. Architecture diagram of the proposed front-end module (TXM).

A. Adaptive Bias Circuit

Adaptive bias technology was proved effectively to suppress the AM-AM distortion [5-6]. The proposed diode-connected bias circuit for the common source device of the power stage is shown in Fig. 2(a).The dc bias voltage (V_{GS}) is set by transistor M1, resistors R1, R2 and reference voltage V_{ref}. R4 resistance is optimized to adjust the RF input power sensitivity. As the input signal increases, the input signal starts to be clipped through the diode-connected NMOS M1 and bypass capacitor C1. Therefore, the bias voltage of M1 slightly increases. By optimizing the size of M1 and the value of C1, R1, R2, R3 and R4, the dc bias voltage rising rate to the input signal can be carefully optimized to improve output 1dB compression point (OP_{1dB}). The impedance of the bias circuit is much higher than the impedance from input capacitor C_{gs} of the PA input stage. So the gain and efficiency of power stage will not be affected. Fig. 2(b) shows simulated V_{GS} variation with sweeping the input power.

B. Capacitor Neutralizing Technology

When the cascode transistor in the output stage enters the triode region while the common source transistor is still in saturation, it generates the main source of nonlinearity in the input capacitance [7]. This causes the miller gain across C_{gd} of

the common source transistor to expand, resulting in a higher input capacitance, thus degrading the PA performance. In this work, because the PA is differential structure, cross-coupled capacitors (C_n) can be used to neutralize this effect. It also shows merits of gain enhancement and impedance matching.

C. Output Matching Circuit

Conventionally, for TXM PA the output matching networks are implemented in IPD technology [3-4]. Usually, IPD technology provides top metal with a 10um thickness copper and high resistance substrate. Although the insertion loss of transformer and matching circuit is better than that of bulk-Si substrate, it needs large area, thus increases the chip cost and size. In this work, the transformer is realized on the four-layer laminate substrate, followed by harmonic filter circuits which include spiral inductors and SMD capacitor components. The transformers occupy 1.8mm*1mm in LB and 1.5mm*1.25mm in HB. The primary winding and secondary winding are 250um width with 20um thickness. Its insertion loss is smaller than the transformer implemented on the IPD because of the metal thickness in this process. In HFSS simulation, the loss of transformers in LB and HB is 0.4dB and 0.3dB, respectively.

The differential structures of PA can suppress common mode harmonics in both LB and HB [8]. However, the suppression is not enough for the spurious emission mask. A compact harmonic suppression circuit, which is also part of the PA matching network is proposed as shown in Fig 3.

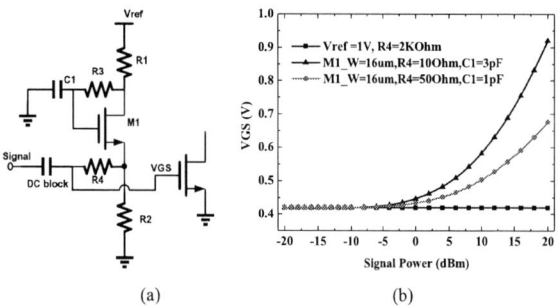

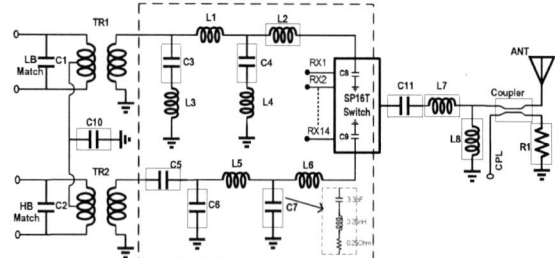

Fig. 2. Proposed adaptive bias circuit. (a) Schematic and (b)Simulation results.

Fig. 3. Proposed output matching network configuration.

In the LB, L3 and C3 provide a trap for 3rd harmonics, while L4 and C4 generate a trap for 2nd harmonics. In the HB, C6 (5.6pF) and C7 (3.3pF) with its instinct parasitic inductance are used to suppress 2nd and 3rd harmonics respectively. For high order harmonics in LB and HB, the transformer and low pass filter circuits are also effective to suppress them. The harmonic filter circuits contribute 0.8dB loss in LB and 0.5dB loss in HB by simulation. The measured loss of the SP16T switch is 0.6dB and 0.8dB in LB and HB respectively, thus the total loss from the PA output to the ANT port is about -1.8/-1.6dB for LB/HB, respectively. This output matching circuit provides optimum load impedance for a wide operating frequency rang while it deeply suppresses harmonics with a few SMD components and very low insertion loss.

D. Reliability

For handsets, the most common battery voltage is 3.8V and the operating range is typically between 3V to 4.5V. The signal swing in the drain of the PA can be two to three times of the supply voltage, especially in high efficiency PA [9]. By carefully optimizing the gate voltage swing of M2 and M3, the three stage cascode PA achieves 14.5V AC breakdown voltage. The transistor size of PA in both bands is optimized to sustain about two times of maximum operation current in each mode.

III. EXPERIMENTAL RESULTS

Fig. 4 shows the photography of the PA module in a 38-pin LGA package. The CMOS PA chip, output transformers, SMD components and an SP16T switch are mounted on a 4-layer laminate substrate. The CMOS PA was fabricated in a 0.13μm one-poly five-metal (1P5M) process with chip area of 1.2mm by 1.8mm. This chip area includes all the RF blocks, analog circuits and MIPI circuits. To verify the TXM, an evaluation board is fabricated using FR-4 printed circuit board (PCB). The input and output for this TXM are connected to 50Ohm transmission line without any off-chip matching components. The transmission line loss on PCB and all the other loss in TXM due to transformer, matching circuit and switch are included in the experiment results. In the measurement, V_{bat} is

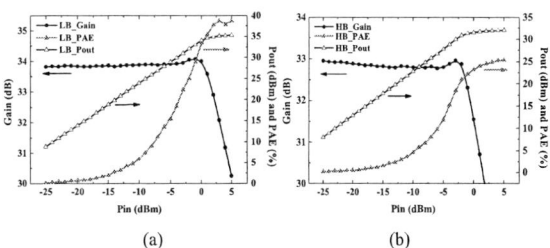

Fig. 5. Measured gain, output power and PAE vs. pin for EDGE mode. (a) at 915MHz and (b) at 1910MHz

set to 3.5V while the TXM can also work well with supply between 3.0V to 4.5V.

(1) One Tone Test: A single tone with the frequency 915MHz is applied in LB EDGE mode, and 1.91GHz is applied for HB EDGE mode to measure the gain, OP_{1dB}, saturated output power (P_{sat}) and PAE. The measurement results are plotted in Fig. 5. It achieves about 34dB gain, 34.5dBm OP_{1dB} for LB EDGE mode and about 33dB gain, 31.5dBm OP_{1dB} for HB EDGE mode, respectively. The measurement OP_{1dB} is high enough to support linear GMSK mode. It also achieves 30dB gain, 31dBm OP_{1dB} for TD-SCDMA and TD-LTE mode. In V_{ramp} control GMSK mode, it shows 34.5dBm P_{sat} with 35% PAE and 32.2dBm P_{sat} with 30% PAE in LB and HB, respectively.

(2) Envelop Test: Fig. 6(a) shows the TXM achieves 30dBm P_{out} with -36/-61dBc ACLR1/2 and 2.5% EVM at 915MHz in EDGE mode. It is almost 2dB higher than the specification and the PAE at 28dBm is 15%. If optimization is target for better PAE by tuning the bias voltage for each stage inside PA, the maximum linear output power is 28.5dBm with the PAE up to 21%. Fig. 6(b) shows 27dBm linear P_{out}(-35/-61dBc ACLR1/2 and 2.7% EVM) at 1910MHz with 16% PAE in EDGE mode. With modulated signals of TD-SCDMA at B34/B39, this TXM can support 27dBm P_{out}(-40/-63dBc ACLR1/2 and 1.6% EVM), while PAE at 24dBm is 10%. By optimizing the bias voltage, the maximum linear output power is 24.5dBm, while PAE at 24dBm is up to 13%. In TD-LTE mode, at 24dBm P_{out}, the PA achieves -36dBc EUTRA and 2.3% EVM with 10MHz 50 resource block (RB) 16QAM signals. Table I gives a summary of measured RF performance of this work. Compared with other works, even with higher loss of SP16T switch, this work shows obviously higher liner output power and better efficiency.

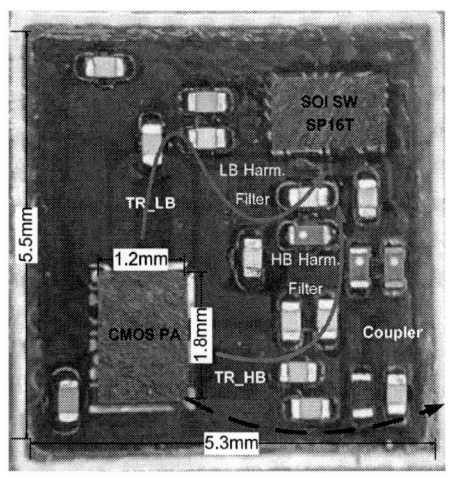
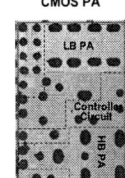

Fig. 4. Photograph of fully integrated TXM

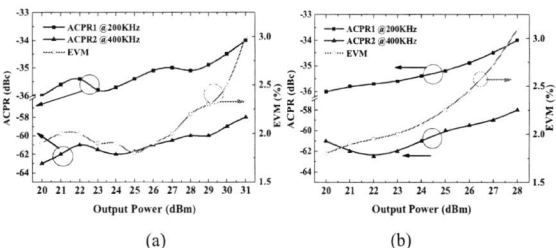

Fig. 6. Measured ACPR, and EVM vs. output power for EDGE mode. (a) at 915MHz and (b) at 1910MHz

TABLE I.   COMPARISON OF THE PROPOSED PA WITH RECENTLY REPORTED PA

			[3]	[4]		This work	
PA Process			180nm 1P6M CMOS	153nm 1P6M CMOS		130nm 1P5M CMOS	
PA Die Area (mm^2)			1.1*2.8	1.38*2.1		1.2*1.8	
Need external Filter			YES	NO		NO	
Need external matching			NO	NO		NO	
Switch			NO Switch	SP10T		SP16T	
Applications			GSM/EDGE	GSM/EDGE TD-SCDMA/TD-LTE		GSM/EDGE TD-SCDMA/TD-LTE	
Reference output			PA output	PA output	TXM output	PA output*	TXM output
GMSK	GSM/EGSM	Psat(dBm)	34.5	35.6	34	36.3	34.5
		PAE Psat(%)	55	54	36.8	53	35
	DCS/PCS	Psat(dBm)	32.5	32.8	31.5	33.8	32.2
		PAE Psat(%)	48	43	31	43.3	30
EDGE	GSM/EGSM	Pout(dBm)	28.5	29.6	28	31.8	30
		ACPR1/2	-/-57	-/-61	-/-61	-36/-61	-36/-61
	DCS/PCS	Pout(dBm)	27.5	28.3	27	28.6	27
		ACPR1/2	-/-56	-/-62	-/-62	-35/-61	-35/-61
TD-SCDMA		Psat/OP$_{1dB}$(dBm)		31.5/30.3	30.2/29	33.8/32.6	32.2/31
		Pout(dBm)		25.6	24.3	28.6 (26.1[#])	27 (24.5[#])
		ACPR1/2		-44/-65	-44/-65	-40/-61 (-47/-65)	-40/-61 (-47/-65)
TD-LTE		Pout(dBm)		22.3	21	25.6	24
		EUTRA ACLR (dBc)		-32.8	-32.8	-36	-36
Harmonics Suppression (dBm)			-	-		<-40	
Spurious Emission (dBm)			-	<-33 over VSWR 3:1		<-36 over VSWR 6:1	
Unconditional Stability and Reliability			-	-		over VSWR 20:1 @ Vbat =4.5V	

*: Estimated results from ANT port to PA matching output by test key measurement
#: Back-off output power to compare TD-SCDMA linearity performance with [4]

(3) Reliability: Since the PA may be damaged in a high VSWR under high V_{bat}. Reliability was measured with 20:1 of VSWR and open/short condition at output terminal under 4.5V V_{bat} in each mode. No damage or power degeneration or other abnormal phenomenon was observed after more than 24 hours test. Also, this TXM passed the aging test which needs the PA works well for 96 hours in an extreme environment at 125 degrees with double atmosphere pressure and 85% humidity.

IV. CONCLUSION

In this paper, a multimode multiband transmitter front-end module is proposed and fully integrated on a new 4-layer laminate substrate process. The compact PA matching circuit is proposed in the new process to minimize the insertion loss. With a new adaptive bias circuit, the TXM can support GSM/EDGE/TD-SCDMA/TD-LTE application for multiple bands with higher output power and better efficiency. It achieves 30dBm at 915 MHz, 27dBm at 1910 MHz with -35/-61dBc ACLR1/2 in EDGE mode. It also supports 27dBm P_{out}(-40/-61dBc ACLR1/2) in TD-SCDMA mode and 24dBm P_{out}(-36dBc EUTRA) in TD-LTE mode. Output power of harmonics for each mode are suppressed smaller than -40dBm. This work presents a low cost, small size solution for TXM for latest multimode multiband cellular application.

REFERENCES

[1] Aoki, Ichiro, et al. "A Fully-Integrated Quad-Band GSM/GPRS CMOS Power Amplifier." IEEE Journal of Solid-State Circuits 43.12(2008)

[2] Scott, Baker, G. Maxim, and S. Franck. "Output stage of a power amplifier having a switched-bulk biasing and adaptive biasing." US, US 8624678 B2. 2014.

[3] Kim, Woonyun, et al. "An EDGE/GSM quad-band CMOS power amplifier." Solid-State Circuits Conference Digest of Technical Papers IEEE, 2014:430-432.

[4] Tsai, Ming Da, et al. "A fully integrated multimode front-end module for GSM/EDGE/TD-SCDMA/TD-LTE applications using a Class-F CMOS power amplifier." Solid-State Circuits Conference IEEE, 2017:216-217.

[5] He, Shihai, et al. "5.25 GHz linear CMOS power amplifier with a diode-connected NMOS bias circuit." International Conference on Microwave and Millimeter Wave Technology IEEE, 2012:1-4.

[6] Jin, Sangsu, et al. "Linearization of CMOS Cascode Power Amplifiers Through Adaptive Bias Control." IEEE Transactions on Microwave Theory & Techniques 61.12(2013):4534-4543.

[7] Wongkomet, Naratip, and P. R. Gray. "Efficiency Enhancement Techniques for CMOS RF Power Amplifiers." (2007).

[8] Wang, Hua, C. Sideris, and A. Hajimiri. "A CMOS Broadband Power Amplifier With a Transformer-Based High-Order Output Matching Network." IEEE Journal of Solid-State Circuits 45.12(2010):2709-2722

[9] Mazzanti, A., et al. "Analysis of reliability and power efficiency in cascode class-E PAs." IEEE Journal of Solid-State Circuits 41.5 (2006):1222-1229

A 2.4-GHz Single-Pin Antenna Interface RF Front-End with a Function-Reuse Single-MOS VCO-PA and a Push-Pull LNA

Kai Xu [1,2], Jun Yin [1], Pui-In Mak [1], Robert Bogdan Staszewski [2], Rui P. Martins [1,3]

1 - State-Key Laboratory of Analog and Mixed-Signal VLSI, University of Macau, Macau, China {E-mail: junyin@umac.mo}
2 - University College Dublin, Dublin, Ireland {E-mail: robert.staszewski@ucd.ie}
3 – On leave from Instituto Superior Técnico, Universidade de Lisboa, Portugal

Abstract—**We propose a power-efficient sub-1V RF front-end (RFE) for 2.4GHz transceivers. It introduces the following innovations: 1) function-reuse single-MOS VCO-PA with full V_{DD} utilization while improving antenna-to-VCO isolation for better resilience to jammers; 2) a non-inverting transformer with a zero-shifting capacitor that suppresses the 2nd harmonic emission of the VCO-PA, and allows a single-pin antenna interface for both TX and RX modes; and 3) a push-pull LNA with passive gain boosting that reduces power consumption. Fabricated in 65nm CMOS, the RFE occupies merely 0.17mm². By scaling the supply voltage, the standalone VCO-PA exhibits a 20.8% (10.2%) power efficiency when delivering 0dBm (-10dBm) output. The LNA shows 11dB gain and 6.8dB NF while consuming 174µW.**

I. INTRODUCTION

Ultra-low-voltage (ULV) operation of IoT devices has been of great interest recently thanks to its feasibility of being self-powered by harvesting the ambient energy [1]-[2]. It is also well suited for Medical Body Area Network (MBAN) where low RF output power (e.g. −10dBm) is necessary. For the ubiquitous deployment sake, an ULV transmitter (TX) must be compatible with the mainstream IoT standards, such as Bluetooth low energy (BLE), while exhibiting good efficiency during the deep power back-off. It appears that the separate arrangement of the oscillator (e.g. VCO) and PA in conventional TXs cannot make that goal fully realizable. The recently introduced current-reuse PA-VCO [3] is incompatible with the ULV operation and has a limited output power due to the reduced voltage headroom. To overcome this, a six-port transformer of a recently introduced class-F DCO-PA [4] merges the DCO resonant tank with the PA matching network [Fig. 1(a)]. Yet, such a DCO-PA becomes extremely vulnerable to jammers appearing at the antenna. Besides, it requires off-chip filtering of HD_3 due to the 2nd impedance peak at the 3rd harmonic in class-F operation. In this paper, we propose a single-MOS VCO-PA [Fig. 1(b)] which improves jammer resilience and harmonic rejection. It utilizes the output matching transformer as a single-pin antenna interface for both TX and receiver (RX) modes, while offering passive gain boosting to the LNA, thus saving additional power. Fig. 2 depicts a complete diagram of the RFE.

II. FUNCTION-REUSE SINGLE-MOS VCO-PA

To enhance the antenna-to-VCO isolation of the function-reuse VCO-PA architecture, we replace the drain-to-gate (D-to-G) feedback oscillation in [4] by an S(source)-to-G feedback. It is desirable here to maximize the coupling factor k_1 and L_g/L_s ratio to obtain a large loop gain and gate voltage swing, thereby lowering the power consumption and phase noise [5]. The swing at the source node must be lowered to attain better efficiency if the transistor is simultaneously reused as a PA, but that requires small L_s and k_1 values. Transformer T_1 comprises a series-wound 3 turns on top-metal (M9) as L_g (=2nH) and a parallel stack of 2 innermost turns of AP and M9 as L_s (=0.2nH). As such, we secure a moderate $k_1 = 0.5$ without much degradation of the tank quality factor. $V_{GB} + v_g$ biasing determines the conduction angle of the M_1 transistor and the use of a small $V_{GB} < 0.3V$ stimulates

M_1 to deliver a narrow current pulse to the output, further improving the power efficiency.

We chose a step-down (in TX mode) transformer T_2 with a small turns ratio of $n = 2{:}1$ for impedance enhancement at low output power (P_{out}) and high matching network efficiency [6]. A zero-shifting capacitor C_z between the two coils (L_o, L_d) of the non-inverting T_2 rejects the HD_2 of the single-MOS M_1. The transmission zero $f_z \propto \frac{\sqrt{n}}{\sqrt{L_o \cdot C_0}}$ for a certain k_2, wherein C_0 represents the inter-winding capacitance. Then, the proposed C_z artificially tunes C_0 shifting the zero notch down to $f_z = 2f_0$ where f_0 is ~2.4GHz. We verify the effectiveness of C_z with a simplified model (Fig. 3). $|V_{out}/V_d|$ shows a 26dB rejection at 4.8GHz at a cost of a sub-1dB passband loss. Even with a ±10% capacitance variation, this technique exhibits sufficient HD_2 rejection capability. Further, a high output impedance at $3f_0$ results from L_s and C_2 obstructing the 3rd-harmonic current.

III. PASSIVE-GAIN-BOOSTING PUSH-PULL LNA

In the TX mode, M_2 and M_3 are in cut-off and it is the parasitic capacitance at the drain of M_1 that can be reflected and absorbed into the output switched-capacitor bank C_o (3 bits). Similarly, when switching to the RX mode, M_1 in off state will present the main parasitic capacitance to the LNA, which can be compensated by altering C_o to a lower code for T_2 to retain a good frequency selectivity at 2.4GHz. The reverse signal flow in the RX mode allows the push-pull LNA to benefit from the passive gain of T_2. The input impedance matching of the LNA is aided by the supply and ground bondwire for inductive degeneration showing an input resistance: $R_{in} = \frac{g_m L_{bond}}{C_{gs}}$. To save die area, we place all active components underneath the passive transformers.

IV. MEASUREMENT RESULTS

The RFE, fabricated in 65nm CMOS, occupies a compact active area of just 0.17mm² [Fig. 4(a)]. The HD_2 emission is −44.4dBm [Fig. 4(b)], which confirms the effectiveness of the zero-shifting C_z. The VCO-PA shows a 20.8% power efficiency when delivering P_{out}=0dBm at a 0.7V supply [Fig. 5(a)]. Phase noise is −126dBc/Hz at 2.5MHz offset [Fig. 5(b)]. Under a ULV supply of 0.3V, the VCO-PA sustains a relatively high efficiency >10% at -10dBm P_{out}. We verify the jammer resilience of the proposed VCO-PA to be far superior than in [4] through a −30dBm 5MHz-offset jammer [Fig. 5(c)]. The image spur detected at the output is −41dBm, which is at least 25dB better than in [4]. When the VCO-PA output pin switches to the LNA input, it exhibits 11dB gain with a noise figure (NF) of 6.8dB [Fig. 5(d)], while consuming 174µW at 0.5V.

V. CONCLUSIONS

This paper reports a 2.4GHz single-pin antenna-interface front-end featuring a single-MOS function-reuse VCO-PA and a push-pull LNA. The VCO-PA achieves high power efficiency even at a large power back-off while offering strong HD_2

rejection by means of self-driven source-to-gate oscillation and zero-shifting capacitor, C_z. Passive gain boosting and push-pull operation contribute significantly to the low power drain of LNA.

ACKNOWLEDGMENT

The authors acknowledge the support from: Macau FDCT SKL Fund, Univ. of Macau - MYRG2017-00185-AMSV, and Science Foundation Ireland 14/RP/I2921.

REFERENCES

[1] W. Yu et al., "A 0.18V 382μW Bluetooth Low-Energy (BLE) Receiver with 1.33nW Sleep Power for Energy-Harvesting Applications in 28nm CMOS," ISSCC, pp. 414-415, Feb. 2017.

[2] M. Yuan et al., "A 0.45V Sub-mW All-Digital PLL in 16nm FinFet for Bluetooth Low-Energy (BLE) Modulation and Instantaneous Channel Hopping Using 32.768kHz Reference," ISSCC, pp. 448-450, Feb. 2018.

[3] C. Li et al., "Class-C PA-VCO Cell for FSK and GFSK Transmitters," IEEE JSSC, vol. 51, pp. 1537-1546, Jul. 2016.

[4] X. Peng et al., "A 2.4-GHz ZigBee Transmitter Using a Function-Reuse Class-F DCO-PA and an ADPLL Achieving 22.6% (14.5%) System Efficiency at 6-dBm (0-dBm) P_{out}" IEEE JSSC, vol. 53, pp. 1495-1508, Jun. 2017.

[5] A. Ng et al., "A 1-V 24-GHz 17.5-mW Phase-Locked Loop in a 0.18-μm CMOS Process," IEEE JSSC, vol. 41, pp. 1236-1244, Jun. 2006.

[6] M. Babaie et al., "A Fully Integrated Bluetooth Low-Energy Transmitter in 28 nm CMOS With 36% System Efficiency at 3 dBm," IEEE JSSC, vol. 51, pp. 1547-1565, Jul. 2016.

[7] Y. Liu et al., "A 1.9 nJ/b 2.4 GHz multistandard (Bluetooth Low Energy/Zigbee/IEEE802.15.6) transceiver for personal/body-area networks," ISSCC, pp. 446-447, Feb. 2013.

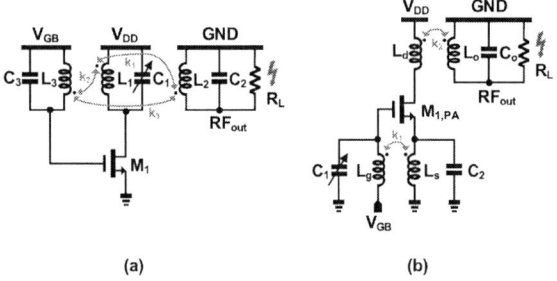

(a)   (b)

Fig. 1: Function-reuse DCO-PA: (a) in [4], and (b) in this work.

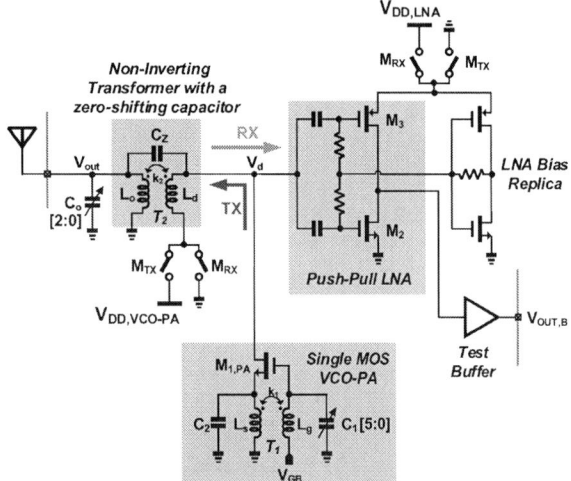

Fig. 2: Proposed single-pin antenna-interface RF front-end using a single-MOS VCO-PA and push-pull LNA.

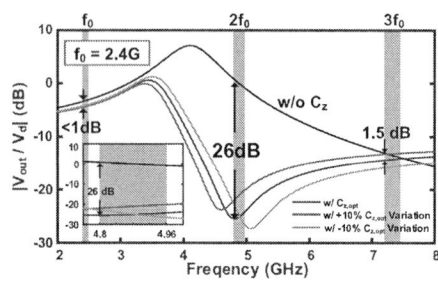

Fig. 3: Simulated HD_2 rejection with and without zero-shifting C_z.

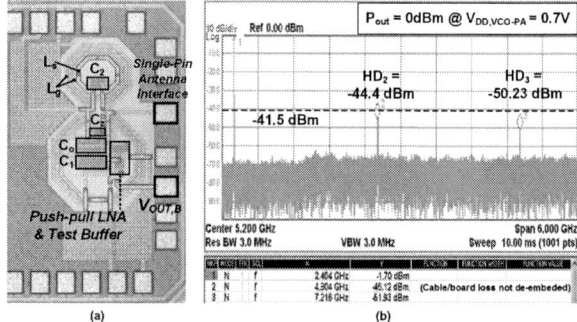

(a)   (b)

Fig. 4: (a) Die photo of the RFE. (b) Measured single-tone output spectrum.

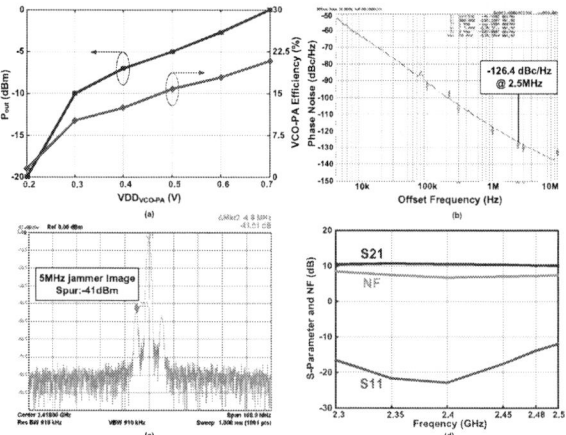

Fig. 5: Measured (a) VCO-PA P_{out} and power efficiency versus supply voltage; (b) Phase noise under P_{out} = 0dBm (c) VCO-PA under -30dBm jammer, and (d) LNA key performance metrics.

TABLE I. COMPARISON WITH STATE-OF-THE-ART VCO-PAS.

	ISSCC'13 [7]	JSSC'16 [3]		JSSC'16 [6]	JSSC'17 [4]	This work	
Supply Voltage (V)	1.2	1.2		0.5 + 1	0.3 to 0.7	0.3 to 0.7	
RF band (GHz)	2.4	2.4		2.4	2.4	2.4	
RF Power (dBm)	-10	-5	-1	-5	0	-5*	0	-10	-5	0
Power Consumption (mW) (VCO+Buffer+PA+Driver)	3.7	2.58	4.46	2.36	3.6	2.04	4.4	0.98	2	4.8
Efficiency (%)	2.7	12.2	17.9	13.4	28	15.5	22.6	10.2	15.8	20.8
Active Area (mm²)	0.71	0.2		0.85	0.39	0.17	
HD2/HD3 @ P_{out} (dBm)	N/A	-50 / -47 @ 0dBm		<-40 /-40 @ - 1dBm	-43.2 /-47.6 @ 0dBm	-44.4 /-50.2 @0dBm	
VCO Phase Noise @ 2.5 MHz (dBc/Hz) / RF power	N/A	-129 @ -1dBm		-124	-125 @ 6dBm	-126 @ 0dBm	
5MHz jammer image Spur @-30dBm interferer (dBm)	N/A	N/A		N/A	-13	-41	
Max P_{out} Variation under VSWR = 1.5 :1 @ P_{out} (dBm)	N/A	N/A		N/A	2.1dB @ 6dBm	1.1dB @ 0dBm	
VCO-PA efficiency under VSWR = 1.5 :1 @ P_{out} (%)	N/A	N/A		N/A	14 to 27.5 @ 6dBm	16 to 24.5 @ 0dBm	
Fully Integration	Yes	No		Yes	No	Yes	

A 6-8GHz 200MHz Bandwidth 9-Channel VWB Transceiver with 8 Frequency-Hopping Subbands

Haixin Song, Dang Liu, Woogeun Rhee and Zhihua Wang

Institute of Microelectronics, Tsinghua University, Beijing, China

Abstract—This paper presents a pulse-based OOK transceiver system which enables up to 9-channel transmission over 6-8GHz frequency band for the first time. Having eight frequency-hopping subbands, a very-wideband (VWB) transceiver achieves a 200MHz bandwidth with steep spectral roll-off and complies with ultra-wideband (UWB) emission mask. The VWB transmission with 1-64ns pulse duration offers a flexible communication range and a relaxed pulse generation circuit design in the transmitter. In the receiver, an asynchronous energy detection (AED) method which does not require multiphase clock generation is employed. The proposed VWB transceiver implemented in 65nm CMOS consumes 0.81mW with 20ns pulse duration at 1Mb/s data rate.

I. INTRODUCTION

The impulse radio ultra-wideband (IR-UWB) technology offers not only energy-efficient transmission but also high precision localization, which is considered a promising feature for Internet of Things (IoT) applications. For example, a communication-and-localization transceiver has been recently reported for secure mobile authentication [1]. However, the IR-UWB transceiver system has difficulty in accommodating multiple channels in the frequency domain regardless of data rate since a single pulse occupies >500MHz bandwidth with poor spectral shaping. Pulse shaping such as a Gaussian pulse can be utilized to improve the spectral efficiency with lower sidelobe leakage, but it requires a complicated pulse generator circuit design [2]. Moreover, the conventional IR-UWB transceiver suffers from a short communication range due to limited energy per bit [3]. On the receiver side, symbol level synchronization is required to search the position of pulse in the whole symbol duration [4]. Several methods are proposed to overcome the synchronization issue with additional hardware complexity or degraded energy efficiency [5]-[6]. The UWB transceiver based on multiband orthogonal frequency division multiplexing (MB-OFDM) is not competitive for low power systems.

In this work, a frequency-hopping very-wideband (VWB) technology is adopted to achieve up to nine channels with 200MHz signal bandwidth over 6-8GHz frequency band. By utilizing several subbands with frequency hopping, the VWB transceiver achieves good bandwidth efficiency as well as energy efficiency [7]. The VWB transmission also enables longer communication distance with lower data rate and significantly relaxes the pulse generation circuit design. For demodulation, a noncoherent receiver based on asynchronous OOK demodulation is designed to overcome the synchronization problem without requiring multiphase clock generation.

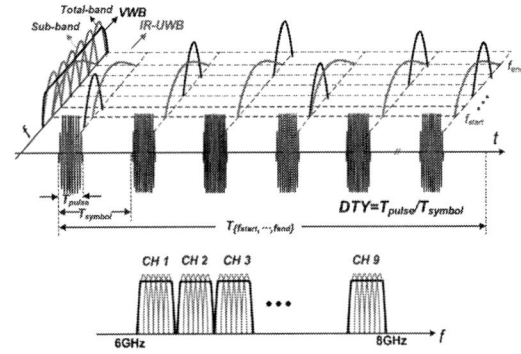

Fig. 1. VWB transmission with multiple channels.

II. SYSTEM CONSIDERATION

A. Multichannel VWB Transmission

The VWB signal in comparison with the IR-UWB signal in both time and frequency domains is illustrated in Fig. 1. Compared to the conventional IR-UWB, a single VWB pulse has longer pulse duration with the same amplitude, which results in narrower bandwidth and higher power spectral density [7]. The VWB technology has following advantages over the conventional UWB. Firstly, by having a flexible pulse duration, the number of subbands, and the center frequency for each subband, the VWB signal can easily occupy any designated frequency band with scalable bandwidth and power spectral density, which is useful for multichannel transmission. Secondly, as the number of subbands increases with narrower subband bandwidth, the overall output spectrum exhibits steeper spectral roll-off, enabling high spectrum utilization efficiency. Thirdly, with longer pulse duration, the VWB transmission provides longer communication distance with larger energy per bit than the conventional UWB.

Based on those advantages, a 9-channel VWB transmission with 200MHz bandwidth is considered over 6-8GHz frequency band as depicted in Fig. 1. With steep spectral roll-off for each channel and a 50MHz guard band between two adjacent channels, the proposed VWB transmission can offer better channel selection than the IR-UWB transmission. With such a high bandwidth efficiency, noncontiguous multiband transmission can also be considered for cognitive radio systems in the congested frequency spectrum.

B. VWB Signal Generation

Fig. 2 illustrates how the VWB signal is generated. Fig. 2(a) shows the spectrum of a single subband without frequency

978-1-5386-6414-8/18 $31.00 © 2018 IEEE

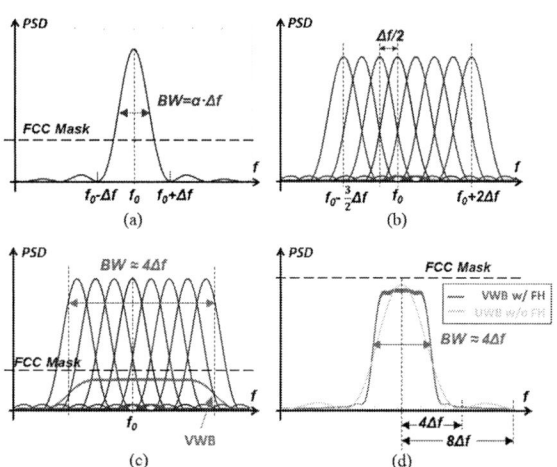

Fig. 2. (a) Spectrum of single subband, (b) spectra of 8 subbands, (c) spectrum of VWB signal, (d) comparison between VWB and UWB.

TABLE I. LINK MARGIN ANALYSIS

	VWB	Conventional UWB
Data rate (R_b)	1Mb/s	
Center frequency (f_c)	7GHz	
TX Peak power (P_{TX})	0dBm	
Pulse width (t)	20ns	5ns
Sub bandwidth (BW_{sub})	50MHz	200MHz
Number of channel (k)	8	1
Total bandwidth (BW_{tot})	200MHz	
TX average Power ($P_{TX,avr}= P_{TX} \cdot 10*log(1/ R_b /t)$)	-17dBm	-23dBm
TX/RX antenna gain (G_{TX}, G_{RX})	0dBi	
Distance (d)	6m	3m
Free space path loss ($L=20*log(4\pi d/\lambda)$)	65dB	59dB
RX average power ($P_{RX,avr}= P_{TX,avr} + G_{TX} + G_{RX} - L$)	-82dBm	
Noise power ($P_N = -174dBm/Hz+10log(BW_{sub}/1Hz)$)	-97dBm	-91dBm
RX noise figure (NF)	8dB	
Implementation loss (I)	1dB	
Required Eb/No for BER=10^{-3} (S)	20dB	
Link margin	$LM = P_{RX,avr} - P_N - NF - I - S + 10log(BW_{sub}/R_b) =$ 3dB	

hopping. For a pulse with pulse duration t_{pulse}, symbol period T_{symbol}, and carrier frequency f_0, the peak value S_f of the power spectral density (PSD) can be calculated by:

$$S_f = \frac{P_{TX}}{BW} \cdot DTY = \frac{P_{TX}}{BW} \cdot \frac{t_{pulse}}{T_{symbol}} \qquad (1)$$

where P_{TX} is the peak transmitted power, DTY is the duty cycle of the pulse, and BW is the bandwidth which is given by:

$$BW = \frac{\alpha}{t_{pulse}} = \alpha \cdot \Delta f \qquad (2)$$

Here α is a factor determined by the pulse envelope, and Δf is half of the null-to-null bandwidth. Considering a pulse with a triangle envelope, α is set to 1. Therefore, for t_{pulse} of 20ns, we can annotate the bandwidth as $BW = \Delta f = 50MHz$. With such a narrow bandwidth, the peak PSD S_f is prone to exceed the UWB indoor spectrum mask of -41.3dBm/MHz.

In order to reduce the peak PSD, we adopt the concept of the VWB by utilizing eight subbands with frequency hopping as shown in Fig. 2(b). To further enhance the spectrum efficiency,

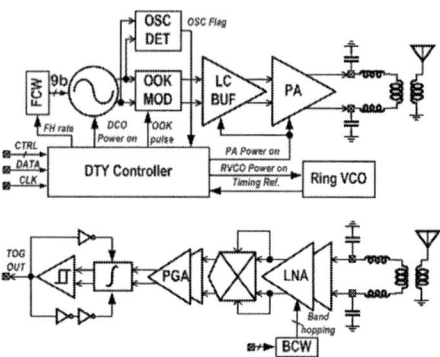

Fig. 3. Transceiver architecture.

adjacent subbands are overlapped by 50% with frequency interval of $0.5\Delta f$. Therefore, the total bandwidth BW_{total} becomes about 200MHz (= 4 x BW) as shown in Fig. 2(c). With fixed P_{TX} and DTY, the peak PSD S_f is reduced by four times compared to the condition of the single subband case according to (1). Spectrum comparison between the VWB and the UWB signals is shown in Fig. 2(d). Comparing to VWB, the pulse duration of UWB reduces to form the same bandwidth, and P_{TX} of UWB should increase by 4 times to maintain the same PSD level. VWB shows steeper spectral roll-off and lower sidelobe leakage.

Table I shows the link margin comparison of the VWB and UWB transceiver systems. With the same data rate and limited peak transmitted power, the VWB transceiver extends the communication distance by two times. For example, the communication distance of 6m is achieved with 3dB link margin for a 1Mb/s 7GHz VWB transceiver when system parameters shown in Table I are assumed.

III. IMPLEMENTATION

A. Transceiver Architecture

Fig. 3 shows the proposed transceiver architecture. Consisting of a ring voltage-controlled oscillator (RVCO), a digital-controlled oscillator (DCO), an oscillation detector and a power amplifier (PA), the transmitter generates VWB pulses with a center frequency range of 6-8GHz and a pulse duration range of 1-64ns. For OOK modulation, when data '1' is sent, the DCO and the RVCO are powered on first. When the oscillation detector detects the DCO oscillation, the duty cycle controller generates a trigger signal for the PA. Frequency hopping is realized by controlling the 9-bit control word of the DCO with a tunable hopping rate. The RVCO provides timing reference for generating the OOK envelope. When the transmission of one pulse is completed, the DCO and the RVCO are turned off for power saving. The receiver consists of a two-stage differential low noise amplifier (LNA), a Gilbert cell based squarer, and a programmable gain amplifier (PGA). The LNA performs band hopping to keep the single-channel noise bandwidth and to obtain an optimum gain for different center frequencies. The bi-direction integrator followed by the THTC performs asynchronous demodulation by generating a toggling signal when an OOK pulse is received.

B. Asynchronous OOK demodulation

In the conventional noncoherent energy detection based UWB receiver, a training sequence must be sent prior to data reception,

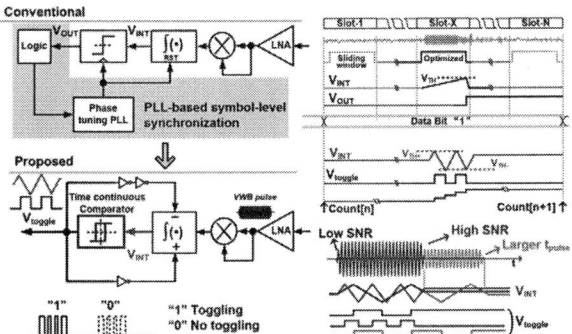

Fig. 4. AED function diagram.

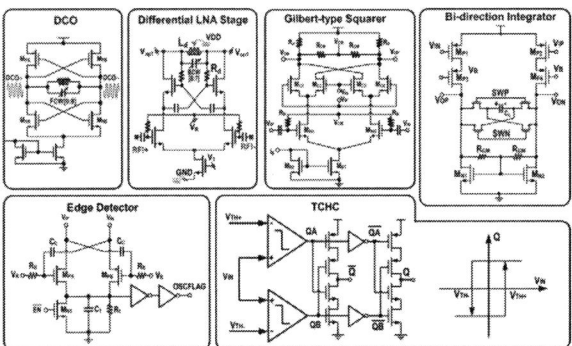

Fig. 5. Circuit schematics.

Fig. 6. Chip micrograph.

while the symbol level synchronization is applied to detect the optimized time slot in which the maximum pulse energy can be obtained. A phase tuning PLL is needed to adjust the position of the sliding window, and digital algorithm is used to determine the right position. When the pulse duration of 2ns is assumed over the symbol period of 1μs, the duty cycle of only 0.2% makes it difficult to search the optimum window, resulting in substantial amount of time for synchronization.

To avoid the hardware complexity and extra time consumption for synchronization, an asynchronous energy detection (AED) method is employed [7]. A functional block diagram is shown in Fig. 4. A time-continuous hysteresis comparator (TCHC) is designed to control the integration polarity of the bi-direction integrator. When the integrator output V_{INT} exceeds threshold V_{TH+} or V_{TH-}, the output of the TCHC toggles, which feeds back to the integrator and alternate its integration polarity. Hence, the multiphase clock generation and the time window searching can be avoided in the baseband. The AED performance comparison with high SNR and low SNR cases is shown in the right side of Fig. 4. With the high SNR, we get more toggling outputs with higher frequency. With the VWB technique, the pulse duration can be increased to have larger toggling numbers in the low SNR case.

C. Building Blocks

Schematics of critical building blocks are shown in Fig. 5. In the transmitter, the DCO covers a wide tuning range of 6-8GHz with a 9-bit binary-weighed capacitor array and performs the frequency hopping operation. For the oscillation detector, when large oscillation signal is detected at input, the two P-type transistors are turned on in rotation, which charges the capacitor C_1. The voltage on C_1 eventually exceeds $V_{dd}/2$, having the output $OSCFLAG$ set to 1.

A differential LNA in the receiver achieves a programmable band by tuning the value of the capacitor array. A transconductance enhancement method is adopted by applying a pair of cross-coupled capacitors. The Gilbert-cell based squarer converts the RF input to a baseband pulse envelope. As for the bi-direction integrator, the integration polarity depends on two switches SWP and SWN. When SWP is 1 and SWN is 0, the load capacitor C_L charges. When SWP is 0 and SWP is 1, C_L discharges. The TCHC consists of two comparators with thresholds of V_{TH+} and V_{TH-}. The output Q of the TCHC toggles only if the outputs of the two comparators Q_A and Q_B become both 0 or both 1, that is, the input voltage V_{IN} is smaller than V_{TH-} or larger than V_{TH+}.

IV. MEASUREMENT RESULTS

The prototype transceiver is fabricated in 65nm CMOS. A chip micrograph is shown in Fig. 6. The die area including PADs is 1.60mm².

Fig. 7 shows the measured transmitted spectra of multiple channels over the 6-8GHz band. As shown, 9 channels can be realized in total with approximately 200MHz channel bandwidth and a tunable guard band between two adjacent channels. Fig. 7(a) shows channel 1-3 at center frequencies of 6.0GHz, 6.25GHz and 6.5GHz respectively with the guard band of about 50MHz based on the -10dB bandwidth. Suppression of more than 10dB is achieved between adjacent channels, which can be further improved by increasing the guard band at the cost of reduced number of channels. Fig. 7(b) and Fig. 7(c) show channel 4-6 at mid frequency band and channel 7-9 at high frequency band respectively.

Fig. 8(a) shows the measured transmitted spectrum of one channel with the center frequency of 7.9GHz and the bandwidth of 200MHz. The peak PSD level is about -44.5dBm/MHz, complying with the FCC musk. Compared to the spectrum of a single subband with the bandwidth of 50MHz, the peak PSD of the VWB channel reduces by 6dB, i.e. four times. Fig. 8(b) shows that the frequency hopping VWB has steeper spectra roll-off and lower sidelobe leakage than conventional UWB with same bandwidth. With the fixed transmitted power, the data rate of the UWB is four times that of the VWB to obtain the same peak PSD level.

Fig. 9 shows the measured waveforms of the transmitted data, the receiver RFA output pulses and the AED output. The output signal of the transmitter is modulated by $TXDATA$ with the data rate of 1Mb/s. When data '1' is transmitted, the peak- to-peak pulse amplitude of the RFA output is 50mV, and toggling can be observed at the AED output with OOK demodulation. Sensitivity of -72dBm is measured at 1Mb/s, having a maximum

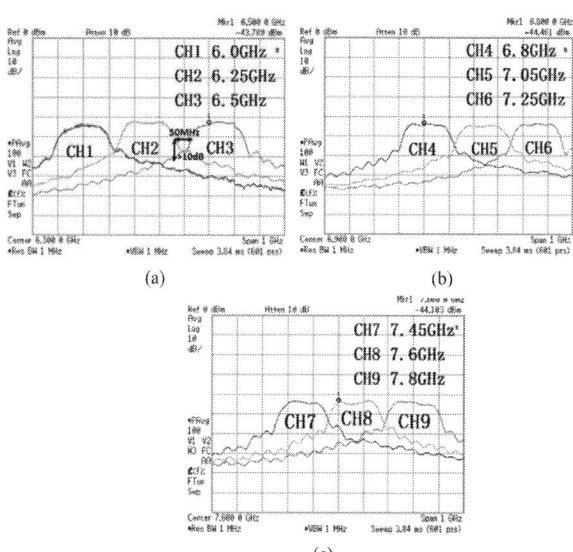

(a)

(b)

(c)

Fig. 7. Measured transmitted spectra: (a) 3 channels at lower frequency, (b) 3 channels at medium frequency, (c) 3 channels at higher frequency.

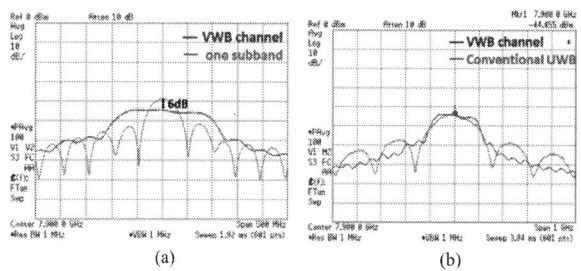

(a)

(b)

Fig. 8. (a) Measured spectra: (a) one VWB channel and one subband, (b) VWB channel and conventional UWB.

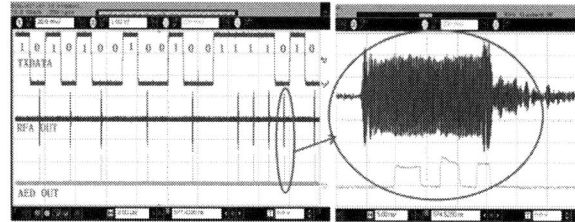

Fig. 9. Measured waveforms of TX transmitted data, RX RFA output and AED output.

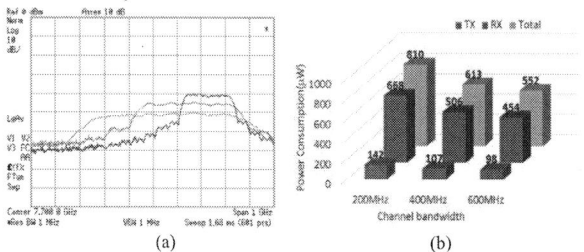

(a)

(b)

Fig. 10. (a) Measured channel bandwidth of 200MHz, 400MHz and 600MHz, (b) measured power consumption vs. channel bandwidth.

TABLE II. PERFORMANCE COMPARISON

Reference	[2]	[3]	[4]	[5]	[7]	This work
Technology	65 nm	130 nm	130nm	90nm	180nm	65 nm
Supply Votalge(V)	1.0	1.35	1.0	1.0	1.8	1.0
Area(mm²)	4.6	2.25	2.0	0.6	5.6	1.60
Architecture	Coh. UWB	Noncoh. IR-UWB	Coh. IR-UWB	Noncoh. IR-UWB	VWB	VWB
Modulation	BPSK	OOK	BPSK	S-OOK	OOK	OOK
RF Band(GHz)	3.1-10.6	7.25-8.5	6-9	3.6-4.3	3-4	6-8
Number of Channels	3*	1	1	1	1	9
Data Rate (Mb/s)	1000	5	0.85	1	1	1
Power(mW)	102.3	5.15	3.51	2.44	0.42	0.81
Efficiency(nJ/b)	0.102	1.03	4.12	2.44	0.42	0.81
Sensitivity (dBm)	-74	-70	-88	-66**	-74	-72

* Utilizing three parallel TRX signal chains   ** With high-gain reception mode

communication range of 2m according to the link margin. The communication range along with the receiver sensitivity was worse than expected. It is mainly because the bandwidth of the LNA and PGA is larger than the bandwidth of subband in the current design, significantly degrading the signal-to-noise ratio performance of the receiver.

Fig.10(a) shows scalable channel bandwidth of 200MHz, 400MHz and 600MHz by controlling the pulse duration. The transceiver consumes 0.81mW from a 1V supply at the data rate of 1Mb/s and the channel bandwidth of 200MHz. With the decreased pulse duration, the power consumption of transceiver can be further reduced to 0.61mW and 0.55mW with channel bandwidths of 400MHz and 600MHz respectively.

V. CONCLUSION

A 6-8GHz 9-channel VWB transceiver is implemented in 65nm. The nine channels are realized with bandwidth of 200MHz for each channel and guard band between two adjacent channels. Having eight frequency-hopping subbands, the VWB transceiver achieves a 200MHz bandwidth with steep spectral roll-off and complies with UWB emission mask. The transceiver consumes 0.81mW at 1Mb/s data rate.

ACKNOWLEDGMENT

This work was supported in part by NSFC under contract #61774092.

REFERENCES

[1] H. Song et al., "A secure TOF-based transceiver with low latency and sub-cm ranging for mobile authentication applications," in Proc. IEEE RFIC, June 2018, pp. 1-4.

[2] N. Kim and J. Rabaey, "A high data-rate energy-efficient triple-channel UWB-based cognitive radio", IEEE JSSC, pp. 809-820, Apr. 2016

[3] J. Maxwell et al., "A 5 Mb/s UWB-IR transceiver front-end for wireless sensor networks in 0.13 μm CMOS," IEEE JSSC, pp. 1636–1647, Jul. 2011.

[4] X. Wang et al., "A meter-range UWB transceiver chipset for around the-head audio streaming," in Proc. ISSCC, Feb. 2012, pp. 450–451.

[5] M. Crepaldi et al., "An ultra-wideband impulse-radio transceiver chipset using synchronized-OOK modulation," IEEE JSSC, pp. 2284-2299, Oct. 2011

[6] B. Vigraham and P. Kinget, "A self-duty-cycled and synchronized UWB receiver SoC consuming 375pJ/b for −76.5dBm sensitivity at 2Mb/s," in Proc. IEEE ISSCC., Feb. 2013, pp. 444–445.

[7] D. Liu et al., "A 0.42-mW 1-Mb/s 3- to 4-GHz transceiver in 0.18-μm CMOS with flexible efficiency, bandwidth, and distance control for IoT applications," IEEE JSSC, pp. 1479-1494, June 2017.

21-4 (8087)

A 0.46-2.1GHz Spurious and Oscillator-Pulling Free LO Generator for Cellular NB-IoT Transmitter with 23 dBm Integrated PAs in 28nm CMOS

Jaewon Choi[1], Nam-Seog Kim, Juyoung Han, and Thomas B. Cho

Samsung Electronics Co., Ltd, Hwaseong-si Gyeonggi-do, 18448 Korea

[1]jaewonc.choi@samsung.com

Abstract—This paper presents a spurious and oscillator pulling free LO generator (LOG) using reconfigurable of integer divider-by-2, 4 and 6 (Div2, Div4 and Div6) and mixed-mode fractional divider-by-2.5 (Div2.5) to mitigate the DCO pulling from the integrated power amplifier for a 0.46-2.1GHz cellular narrowband Internet of Things (NB-IoT) applications. The divider ratio in the designed LOG is determined by TX output power level to achieve both the mitigation of pulling and effective power consumption. An adaptive voltage bias (ADB) circuit is added to compensate PVT variation for the fractional divider. A proposed reconfigurable LOG is implemented in 28nm CMOS process and it consumes 40mA including ADPLL, DCO, and dividers. The measured phase noise of the 2GHz carrier LOG with Div2.5 is -146.3dBc/Hz at 10MHz offset. Fractional spurs are below than -90 dBc without pulling for the 1.6~2.1GHz output frequency.

Keywords—Reconfigurable of LOG, Mixed-mode Fractional Divider, Adatpvie bias circuit, Cellular NB-IoT, Integrated PA, Pulling Mitigation, Spurious Free.

I. INTRODUCTION

The market needs of narrowband Internet of Things (NB-IoT) is strongly growing because NB-IoT is a good solution for low power wide area network. For low-cost IoT applications, an integrated power amplifier (PA) for transmitter (TX) is preferred but it can be a high self-interference source which leads injection pulling of the VCO in PLL [1]. To mitigate this effect, many approaches have been reported in [1]-[8]. The integer division methods in [1], [2] utilize a simple circuit topology and can remove the direct PA pulling without spurs but the pulling due to the harmonics of PA still exits [7]. Also, its performance highly depends on layout optimization and chip floor plan. A single sideband (SSB) mixing technique has been introduced in [4]-[6] to offset the VCO frequency for less pulling but it requires external surface acoustic wave (SAW) filter to filter out spurs closed to the fundamental signal. As introduced in [7], [8], a fractional divider is an effective way to reduce direct and harmonics from PA to DCO coupling without additional SAW filters. However, transmitters of NB-IoT require stringent low out of band emission for the guard-band or in-band cases (i.e. 3.75 kHz in 180 kHz BW) as shown in Fig. 1. Even though a digital fractional divider (DFD) in [7], [8] are good methods to reduce the pulling effects for high power TX, many high spurs inherently are generated in digital fractional dividers due

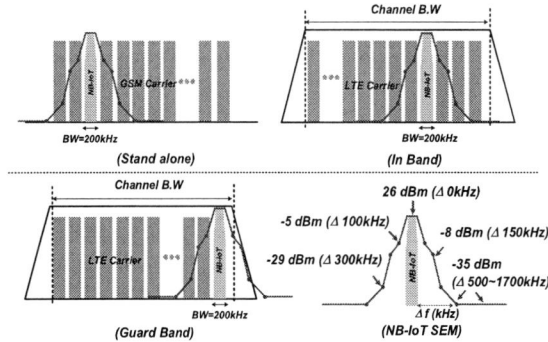

Fig. 1. Cellular NB-IoT spectrum emission

to phase mismatch between their delay cells. As a result, it requires additional digital calibration to lower spur levels and thus a different type of LO generator is required for "narrow" band IoT (NB-IoT) application.

In this paper, spurious free with a pulling mitigating reconfigurable LO generator (LOG) using a mixed-mode fractional frequency divider-by 2.5 is proposed to reduce direct and harmonic pulling effect from TX PA with no spur level without the digital calibration for a cellular NB-IoT. For a proposed fractional divider includes a duty cycle corrector (DCC) to calibrate 50% duty cycle after the frequency division and delayed locked loop (DLL) is combined to provide quadrature outputs for the TX mixers.

II. PROPOSED ARCHITECTURE AND CIRCUITS

A. LO Generator Architecture

The conceptual block diagram of the proposed LO generator is shown in Fig.2. The LOG utilizes a single DCO system (3000MHz~5600MHz) which is assigned to the different dividers based on the frequency plan for the low power requirement. The LOG is distributed to low-band (LB) and mid-band (MB) TX. Both dividers of LB and MB can be reconfigurable according to the required frequency target. For LB, Div4 and Div6 can be selected by the control logic to cover f_{tx}=460MHz~960MHz frequency range and its DCO has a relatively less pulling effect due to the large division ratio. For MB, Div2 and Div2.5 are assigned for 1.6GHz~2.1GHz. As mentioned above, the PA harmonic pulling effect cannot be ignored for the small integer division ratio (e.g. Div2).

978-1-5386-6414-8/18 $31.00 © 2018 IEEE

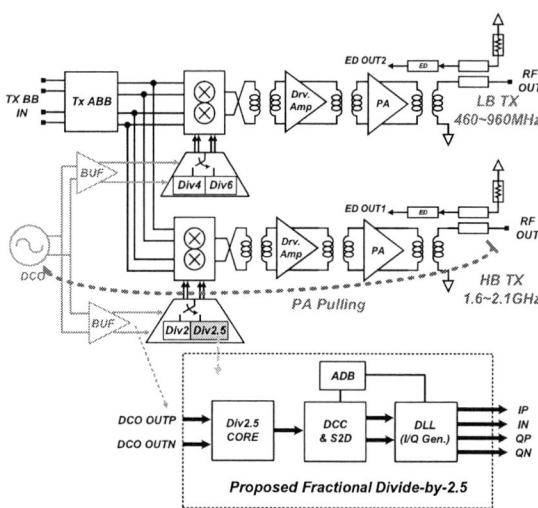

Fig. 2. Block diagram of the proposed LO generation based on single DCO with the transmitter front ends.

Thus, a MB LO divider should be selected to mitigate pulling effect according to TX output power level that determines power consumption. A designed Div2 is commonly used a latch type which has better phase noise (PN) performance, smaller area with lower power consumption than that of a fractional Div2.5. Thus, it can be assigned to low power TX with less pulling. On the other case, a fractional Div2.5, which is proposed in this paper has better pulling mitigation, can be selected for high power mode even though it consumes large power. This is acceptable because TX PA power is a dominant factor and the portion of Div2.5 power is relatively small.

B. Proposed Fractional Divider

As mentioned above, NB-IoT requires very narrow band and spectral purity for LO. A proposed divider 2.5 is mainly composed of three key blocks such as divider core (DIV_CORE), DCC with single to differential circuit (S2D), and DLL for quadrature generation as shown in Fig. 2. An input clock (CK) is from DCO output and goes into each D-latch as CK and CLKB. There are two loops upper side (QAs) and bottom side (QBs) as shown in Fig. 3 (a). Alternatively, CK and CKB are applied to each latch and make different time delays. Two pairs of latches are triggered by an input clock (CLK) and complementary clock (CLKB) to make the different clock shift between upper loop and lower loop. Upper and bottom loops have the half clock difference and NOR operation makes Div2.5 output. Detail timing diagram is shown in Fig. 3 (b). This topology utilizes the continuous sine wave from DCO and so spurious free can be achieved without any calibration scheme. Frequency division is done by the combination of latches but the duty cycle of the output at DIV_CORE is 40% and the output is also a single ended.

For TX LO, the differential signal with 50% duty cycle is required for the mixer operation and thus DCC and S2D are added. Fig. 4 shows a block diagram of DCC. Basically, DCC is an analog delay lock loop (DLL) with feedback operation. It calibrates the bias voltage of delay cells (M1 & M2) by the

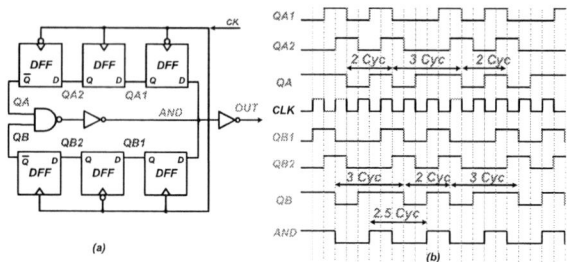

Fig. 3. (a) Block diagram and (b) timing diagram of Div2.5 Core.

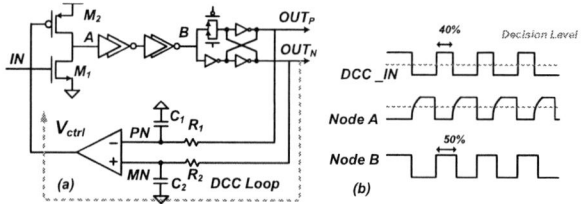

Fig. 4. (a) Schematic and (b) timing diagram of duty cycle corrector (DCC).

output voltage (V_{ctrl}) of the operational amplifier. An op-amp has R-C filter to calculate the average value of OUTP, OUTN and compares the average value at PN and MN node. When PN and MN value is equal, loop operation is stopped, and the constant voltage value is made. The basic timing diagram for operation is shown in Fig. 4 (b).

The 3rd main block of Div2.5 is a quadrature output generation circuit. DCC has only differential I-path signal, but TX requires quadrature signals. The block diagram of I/Q Gen. is shown in Fig. 5. It consists of an input buffer, delay block, op-amp, NAND gates, and a differential bias block. Like a DCC, it compares the average values of I and Q path signal with NAND outputs (IP, IN, QP, QN) with 25% duty comparison. R-C filter of op-amp can calculate the average value of the I-Q path and it is applied to the input of the op-amp. The output voltage of op-amp converges as the duty cycle of I/Q path has the same average DC value. When the average values are equal (quadrature output), delay control voltage is constant and delay operation is locked. The delay cell in Fig. 5 has NMOS/PMOS control transistor which can adjust the delay time based on the output frequency.

C. Adaptive Bias Circuit

The delay cells in the DCC and the DLL are very sensitive to PVT variation. For the robust design, delay blocks should have the same amount of the delay time regardless of different PVT condition. In this work, process and temperature tracking circuit is applied to compensate PVT variation. Basically, the amount of the delay can be controlled by its supply voltage. As shown in Fig. 5, An internal LDO reference voltage is generated by the series connection of variable resistors ladder (R_{var}) and NMOS/PMOS diodes by PTAT current as (1).

$$V_{REF} = I_{PTAT} \cdot R_{ref} + 2 \cdot (\Delta + V_{th}) \tag{1}$$

where I_{PTAT} is PTAT current, Δ is overdrive voltages and V_{th} is the threshold voltage. Thus, LDO output voltage is adaptively changed and delay time is compensated according

978-1-5386-6414-8/18 $31.00 © 2018 IEEE

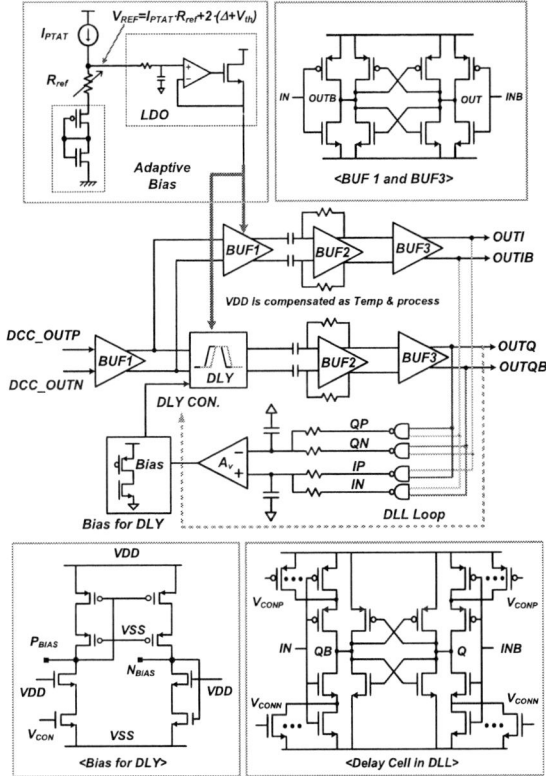

Fig. 5. Schematic of delay locked loop for I/Q Generation.

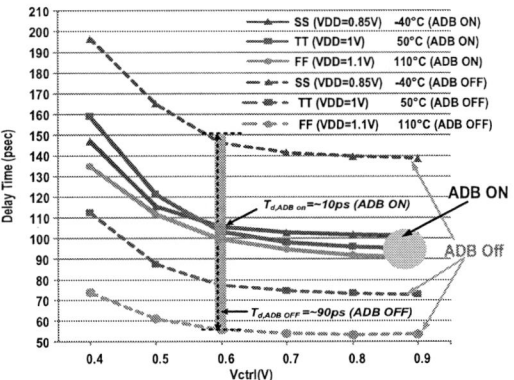

Fig. 6. Simulated results of delay time w/ADB (Solid), wo/ADB (Dot).

to PVT variation. The compensation voltage reference circuit is called an adaptive bias circuit (ADB). Fig. 6 shows the simulated delay time for PVT variations. The x-axis is the control voltage of DLL and the y-axis is the delay time for quadrature signals. The dotted line is the delay time without the ADB circuit and a solid line shows the delay with ADB circuit. It clearly shows that delay variation is significantly reduced from 90ps to less than 10ps. Thus, the proposed adaptive bias circuit is integrated into DLL for I/Q generation and it maintains the same delay as PVT variation.

III. MEASUREMENT RESULTS

The proposed LOG has been implemented in 28nm CMOS process. Flip-Chip Fine Ball Grid Array (FC-FBGA) package is used and Fig. 10 shows a die micrograph and package with balls of the transceiver. Fractional divider core size is 175um x 200um. The spectrum of the standalone divider output of the LOG is shown in Fig. 7(a). This work achieves spurious free clear LO signal (<-90dBc) without a digital calibration circuit. The external interference signal of 100 kHz offset from f_{DCO} is applied to the transceiver to observe pulling effects for each divider. As shown in Fig. 7(b), the DCO with Div2 shows the symmetrical large spurs near fundamental LO because TX 2nd harmonics is directly coupled to f_{DCO}. On other hands, Div2.5 case is no spurs compared to Div2 because f_{DCO} is affected by TX 5th harmonic whose power level is relatively small. Thus, Div2.5 mitigates DCO pulling compared to using Div2. However, Div2 is more simple circuit, and DCO frequency is

relatively lower than that of Div2.5. As results, the power consumption and the phase noise of Div2 LOG is smaller than those of Div2.5. To check this effect, the pulling power and the current ratio as (2) are measured while TX power sweeps.

$$I_{Div\_TX-PA}=I_{Div}/(I_{Div}+I_{PA}) \qquad (2)$$

where I_{Div_TX-PA} is the current of PA, and I_{Div} is for Divider. The pulling power is measured as the TX output power increases. TX-PA power is over 450mW for the maximum TX power (~23dBm). As shown in Fig. 8, there is no different pulling effect between Div2 and Div2.5 for low TX power (<10dBm). The pulling power increases with Div2, but the same for Div2.5 above 10dBm. In addition, the measured current ratio between the current of divider and PAs are shown in Fig. 8. For low TX power level, there is a large difference between the power ratio of Div2 and Div2.5. As TX power increases, PA power consumption is the dominant and both ratio of Div2 and Div2.5 are closed to each other. Thus, divider ratio can be selected by PA power level to achieve both the pulling mitigation and power efficiency. The measured phase noise is shown in Fig. 9 at 2GHz carrier frequency with Div2.5 is locked. LO phase noise at 10MHz offset is -146.3dBc/Hz (Specification at NB-IoT; -141dBc/Hz) and IPN is -43dBc (100Hz~10MHz).

IV. CONCLUSION

This paper demonstrates a spurious and oscillator-pulling free reconfigurable LO generator for cellular NB-IoT applications. The fractional frequency divider-by-2.5 with a DCC and a DLL is proposed to mitigate pulling effect and low spurs level without an additional digital calibration. The reconfigurable LOG can select the divider ratio according to TX output power level achieving both pulling mitigation and low power consumption. An adaptive voltage bias (ADB) circuit using LDO and PTAT is applied to track process, voltage and temperature variations. Therefore, ADB compensates PVT variation for the supply voltage of delay cells in LOG. A proposed LO generator is implemented in 28nm CMOS process. Compared to the previous fractional divider method for mitigating of pulling, the proposed mixed mode fractional approach provides a simple spurious free solution, and it can be a good carrier source for NB-IoT applications that require very narrow bandwidth.

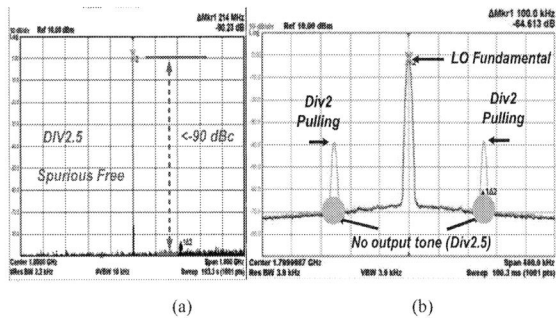

(a)   (b)

Fig. 7. (a) Measured LO output spectrum. (b) pulling power for Div2&2.5

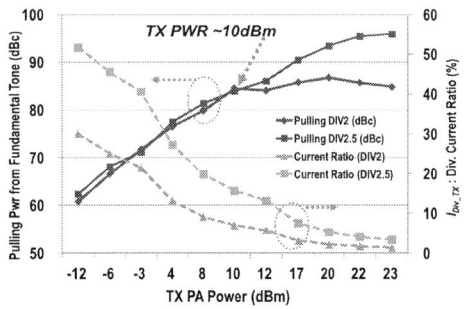

Fig. 8. Measured pulling power and current ratio for Div2 and Div2.5

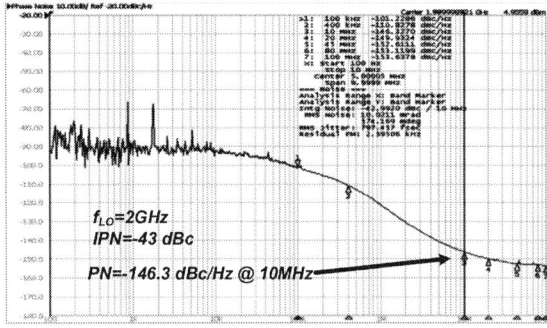

Fig. 9. Measured phase noise of the proposed LOG with Div2.5

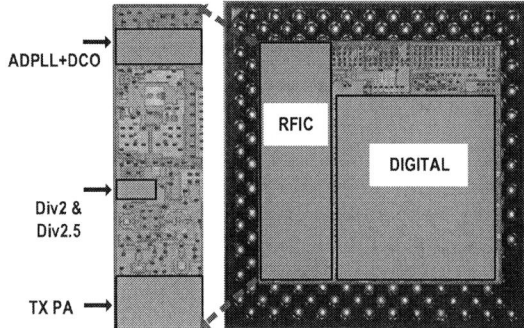

Fig. 10. Chip microphotograph and photo of the package with balls.

TABLE I.   PERFORMANCE COMPARISON

		This Work	ISSCC07 [1]	RFIC10 [3]	JSSC09 [7]	ISSCC12 [8]
Technology		28 nm	130 nm	45nm	45 nm	32 nm
Output Frequency (GHz)		0.46~2.1	2.4, 5	2.3~2.7	2.5~3.8	2.5, 3.5, 5.5
Pulling Mitigation		Mixed-mode Frac. Div.	Integer-N Div.	SSB Mixing	Digital Frac. Div. w/Cal.	Digital Frac. Div. w/Cal.
Divider Ratio		4,6,2,2.5	2	2/5,2/7,5/8	1.25	0.75,1.25 ,1.75
Pout (dBm)		**23**	**-2.5**	**N/A**	**N/A**	**N/A**
Spur Level (dBc)		**Free (2GHz)**	**< -65 (2.4GHz)**	**- 55 (2.3GHz)**	**-59 (2.5GHz)**	**-60 (2.5GHz)**
IPN (dBc)		-43 (2GHz)	N/A	-65 (2.3GHz)	N/A	N/A
Phase Noise (dBc/Hz) @10MHz		-146.3 (2GHz)	*-140 (2.4GHz)	NA	N/A	N/A
LO Gen.	Component	ADPLL + DCO+Div.	PLL + VCO+Div.	VCO+Div	Divider w/cal.	Divider w/cal.
	Power (mW)	40 (2GHz)	N/A	54.6 (2.3GHz)	18.7 (2.5GHz)	6.2 (2.5GHz)
Area (mm²)		0.87	N/A	0.65	0.99 (w/pad)	0.4

* *Normalized from 1MHz to 10MHz*

ACKNOWLEDGEMENT

Author thanks layout team in Samsung for the layout assistance and LO members in mixed signal group for the academic advise during this project.

REFERENCES

[1] M. Simon, P. Laaser, V. Filimon, H. Geltinger, D. Friedrich, Y. Raman, and R. Weigel, "An 802.11a/b/g RF transceiver in an SoC," in Proc. *IEEE ISSCC Dig. Tech. Papers*, Feb. 2007, pp. 562–622.

[2] Ishikuro, Hiroki, *et al.* A single-chip CMOS Bluetooth transceiver with 1.5 MHz IF and direct modulation transmitter. *ISSCC Dig. Tech. Papers*, 2003, 68-69.

[3] R. Sadhwani, et. al., "Multi-band multi-standard local oscillator generation for direct up/down conversion transceiver architectures supporting WiFi and WiMax bands in standard 45nm CMOS process," *IEEE Radio Frequency Integrated Circuits Symp.*, pp.149-152, 23-25 May 2010.

[4] H. Darabi *et al.*, "A 2.4-GHz CMOS transceiver for Bluetooth," *IEEE J. Solid-State Circuits*, vol. 36, no. 12, pp. 2016–2024, Dec. 2001.

[5] O. Degani *et al.*, "A 1 x 2 MIMO multi-band CMOS transceiver with an integrated front-end in 90 nm CMOS for 802.11a/g/n WLAN applications," in *Proc. IEEE ISSCC Dig. Tech. Papers*, Feb. 2008, pp.356–357.

[6] S. S. Mehta et al., "An 802.11g WLAN SoC," *IEEE J. Solid-State Circuits*,vol. 40, no. 12, pp. 2483–2491, Dec. 2005.

[7] P.Stefano *et al.*, "A 4.75-GHz fractional frequency divider-by-1.25 with TDC-based all-digital spur calibration in 45-nm CMOS." *IEEE J. Solid-State Circuits*, vol. 44, no. 12, pp. 3422–3433, Dec. 2009.

[8] C.Kailash *et al.*, "A 32nm CMOS all Digital Reconfigurable Fractional Frequency Divider for LO Generation in Multistandard Soc Radios with on-the-fly Interference Management ." *IEEE ISSCC Dig.Tech. Papers*, Feb. 2012,pp.352-354.

21-5 (8210)

A 152μW -99dBm BPSK/16-QAM OFDM Receiver for LPWAN Applications

Avish Kosari, Milad Moosavifar, David D. Wentzloff
University of Michigan, Ann Arbor, USA

Abstract—This paper presents a 152μW BPSK/16-QAM OFDM receiver operating in the 151MHz Multi-Use Radio Service (MURS) frequency band for low power and long-range IoT applications. Sub-harmonic passive mixers and an injection locked ring oscillator are used, together with a dual-IF architecture for power efficiency and blocker rejection. As a solution for LPWAN applications, the receiver is designed to operate in two modes of single-carrier and multi-carrier transmission schemes, and is implemented in a 40nm CMOS process consuming 152μW of power while achieving a sensitivity of -99dBm for BPSK modulation at 5kb/s and -77dBm for 16-QAM OFDM modulation at 384kb/s. Under only 0.9V of supply, it achieves a phase noise of -128dBc/Hz at 1MHz offset, a blocker rejection of 63dB, and is fully integrated except for the reference crystal and the balun.

Index Terms— Injection locking, IoT, long-range, low power radios, LPWAN, MURS band, sub-harmonic mixing.

I. INTRODUCTION

The rapid growth of Internet of Things (IoT) applications requiring long-range data transmission is leading to the prosperous development in long distance wireless communication systems. Recently, different standards are being proposed in sub-GHz bands which target narrowband, low data-rates, and long-range connectivity for low-power IoT applications [1].

The FCC compliant Multi-Use Radio Service (MURS) band at 151-154MHz, exhibits lower path-loss and building penetration loss compared to other higher frequency bands used for low-power wide area network (LPWAN) applications. To this end, recently a low-power MURS band transmitter was developed addressing the long-range transmission requirements of LPWAN [2]. In this work, we present a 152μW MURS band receiver addressing the low active power, high sensitivity, and interference rejection challenges and requirements of LPWAN radios. The BPSK/16QAM OFDM receiver (RX) is data rate agile and suitable for remote IoT connectivity in multipath rich environments. The RX achieves a sensitivity of -99dBm at 5kb/s, capable of 50km line-of-sight communication. This is enabled by a power efficient RX architecture (Fig. 1) and several low-power design techniques, including a dual intermediate frequency (IF) RX architecture using passive multipliers, an edge combiner sub-harmonic mixer first architecture which enables the operation of the frequency synthesizer at a much lower frequency, injection locked

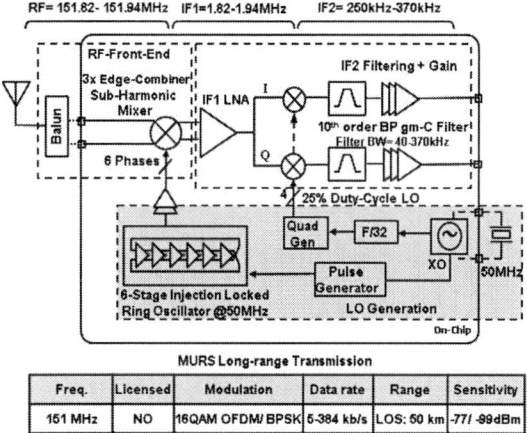

Freq.	Licensed	Modulation	Data rate	Range	Sensitivity
151 MHz	NO	16QAM OFDM/ BPSK	5-384 kb/s	LOS: 50 km	-77/ -99dBm

Fig. 1. Block Diagram of the Receiver.

local oscillators (LOs) which enable more energy efficient frequency generation compared to PLLs, and efficiently distributing the gain and noise performance requirements in the RF and IF blocks. The RX achieves a -128dBc/Hz phase noise (PN) at 1MHz offset for the LO and realizes a blocking rejection of 48dB and 63dB at 2MHz and 10MHz offsets, respectively.

II. SYSTEM ARCHITECTURE

The RX operates in two modes of single-carrier and multi-carrier transmission and is designed with the goal of achieving a similar coverage range to LPWAN technologies at a much lower power consumption. The MURS band single-carrier transmission scheme is based on 5kb/s BPSK modulation with 10kHz channel bandwidth (BW), and 60kHz channel spacing. Based on our link budget analysis, for the RX sensitivity of -99dBm and an uplink power of 10dBm, we can achieve a communication range of 50km with a path loss exponent of 2. For higher data-rate applications in denser environments, an OFDM modulated multi-carrier transmission scheme is utilized, which uses 16-QAM on 16 subcarriers with 10kHz subcarrier spacing. The multicarrier modulation symbols have a symbol duration of 125μs with a data rate of 384kb/s.

To achieve the low power consumption requirements of IoT transceivers along with the specifications for the proposed transmission scheme, a mixer-first injection locked dual IF RX architecture is employed, as shown in Fig. 1. Achieving maximum sensitivity at low current levels is a major challenge. This is enabled by two major

978-1-5386-6414-8/18 $31.00 © 2018 IEEE

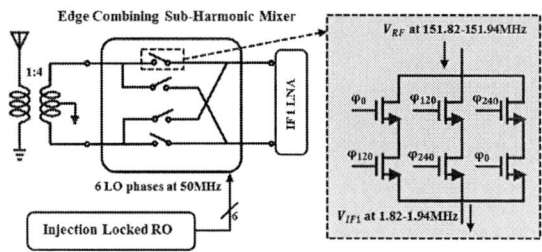

Fig. 2. RX RF front-end.

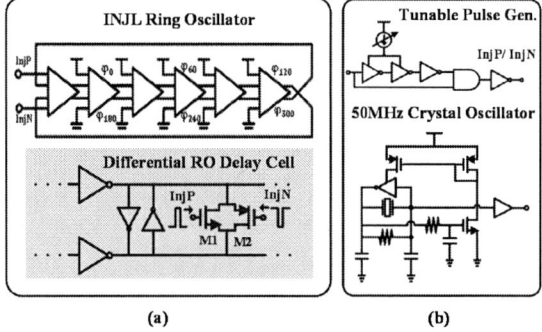

Fig. 3. Injection locked LO generation blocks (a) 6-stage RO and RO delay cell, (b) Pulse generator and 50 MHz XO.

considerations in the system-level design of the RX architecture. 1) The dual IF, sub-harmonic mixer-first architecture is chosen in order to transfer the high dynamic range and hence gain requirements of the RX to the baseband blocks as well as shifting the dual I/Q paths to the lower IF frequency where the power consumption is minimized. In addition, due to the very low IF in the second IF stage (250–370 kHz), the dynamic range and complexity of the required analog to digital converter (ADC) is highly relaxed. 2) The LO frequency generation, which is often the most power hungry block in a RX, is implemented by an injection locked frequency synthesizer at 50MHz, and therefore down-conversion is performed by a 3x sub-harmonic edge-combiner passive mixer. This also enables us to achieve the required PN at a much reduced power.

As shown in Fig. 1, the edge combiner at the input of the receiver is driven by three-phase differential LOs from the injection-locked ring oscillator (ILRO) and is followed by a low-noise amplifier (LNA) at the first IF frequency (IF1), 1.82–1.94MHz, providing 25dB of gain in the first down-conversion step. This is loaded by a quadrature passive mixer driven by 25% duty cycled LOs that are directly driven from the divide-by-32 divider from the on-chip crystal oscillator operating at 50MHz. The second mixers are then followed by very low IF (IF2) 10th order bandpass filters providing 20–54 dB of gain and a tunable bandwidth (BW) of 40–370 kHz. All the blocks are AC coupled to reject DC off-sets.

III. DESIGN IMPLEMENTATION

A. RF Front-End

Fig. 2 represents the RF front-end of the receiver. We have used an on-chip differential double balanced edge-combining sub-harmonic mixer at the input of the RX to save power and reject LO feedthrough due to the presence of random mismatches in the LO phases in the IF signal. An off-chip balun is used for impedance matching to the 50Ω source and single to differential conversion. The six differential three-phase LO signals driving the edge-combiner are fed by the ILRO operating at 50MHz and are phase shifted by 120°. As shown in Fig. 2, the voltage mode edge-combiner is realized by NMOS transistor switches to perform an AND operation on the LO waveforms in each branch and an OR operation to select one of the non-overlapping down-converted signals in each of the three main branches. Overall, the network of 24 NMOS switches, resembles a passive double balanced mixer comprising 4 combined switches operating at 3xLO frequency or 150MHz. The effect of the systematic phase error of injection locking on the output of the sub-harmonic mixer is carefully simulated and discussed in the next section.

B. Injection-Locked Frequency Synthesizer

As shown in Fig. 3, the ILRO at 50MHz is implemented by a six stage differential ring oscillator directly locked to an on-chip reference crystal oscillator [Fig. 3(a)], providing the six output phases from three differential outputs. The schematic of the crystal oscillator (XO) is shown in Fig. 3(b), where a feedback path is used to starve the primary amplifier used with the off-chip crystal, and has a measured power consumption of 29μW. The tunable pulse generator shown in Fig. 3(b) is implemented by an AND operation on the reference signal and tunable delayed versions of the reference. It outputs programmable pulse width signals, Inj_P and Inj_N, to tune the locking range and ensure that the pulse width of the injected signals is smaller than half of the crystal oscillator period. The injection differential pulses drive M_1 and M_2 transistors and short the differential output of the RO [Fig. 3(a)]. Even though the reference pulse is only injected in the first stage of the RO, all the stages are identical to increase matching. Since the injection often only happens in the first stage of the RO, ILRO systems have a systematic phase mismatch between the stages which leads to spurs occurring at LO frequency in the frequency multiplied signals in the IF stage of the receiver. The phase error between the adjacent stages in the ILRO is proportional to the frequency difference between the free running RO (f_{RO}) and the injected signal (f_{inj}) and can be written as: $\theta_{error} = \frac{\pi}{3}(f_{RO} - f_{inj})/f_{RO}$. This systematic phase error was simulated with the edge combiner mixer to ensure the LO feedthrough is small enough and can be filtered by the baseband filters. This eliminates the use of multiphase injection locking techniques and therefore,

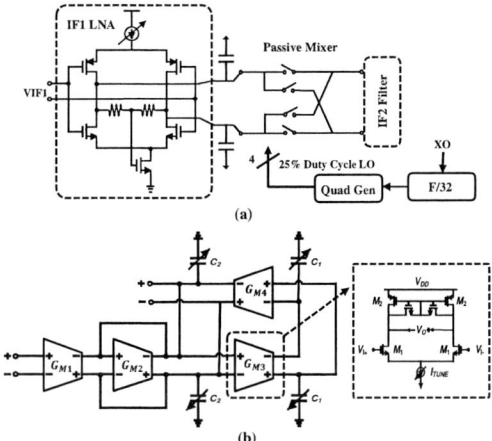

(a)

(b)

Fig 4. (a) IF1 stage and schematic of IF1 LNA, (b) Block diagram of a single stage of IF2 10th order filter.

saves power. An 8 bit current DAC is used to tune the RO frequency and minimize the effect of the phase mismatches on the receiver performance.

C. First Intermediate frequency (IF1) stage

As shown in Fig. 4(a), an inverter based topology is used for the IF1 LNA for gain and noise efficiency. Common mode feedback is provided by both the pseudo resistors and the bottom NMOS transistor, guaranteeing the output common mode stay at half V_{DD}. Since the input impedance of the LNA directly affects the receiver's input impedance and therefore noise performance through the passive mixers, the LNA's input devices as well as the tail currents are adjusted to match the desired noise level. The outputs of the LNA are directly loaded by a quadrature passive mixer driven by 25% duty cycled LO signals. The LO signals, are generated by a divide-by-32 divider driven by the XO. This further down-converts the signal for an optimum power consumption as well as a good blocker rejection performance.

D. Second Intermediate frequency (IF2) Stage

In order to enable the targeted specifications for MURS communication, a 10th-Order Chebyshev-I bandpass g_m-C filter is designed. The filter is implemented using five cascaded biquad stages. Each of the biquad stages provide a 1st-order high-pass and low-pass response with programmable pole frequency, quality factor, and gain, as shown in Fig. 4(b). In order to achieve a broad tuning range, both g_m and capacitors of each stage are designed to be tunable. The g_m variation of the G_m-cells, often causes a significant change in DC operating points. Therefore, in order to prevent DC operating point variations, and maximize the tuning range simultaneously, self-biased differential G_m-cells are used in the biquad stages. The programmability feature enables the filter to create a Chebyshev response or Bessel with different band widths,

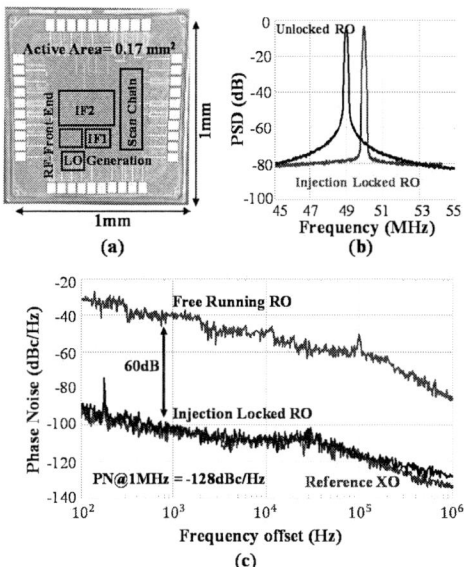

(a)   **(b)**

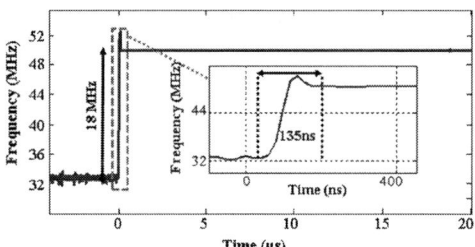

(c)

Fig 5. (a) Die micrograph of the RX, (b) Unlocked and locked RO PSD, (c) ILRO, XO and unlocked RO PN.

Fig 6. Frequency transient showing the settling time for ILRO.

making the receiver reconfigurable for different modes of operation. The filter has two modes of operation, one providing higher measured gain of 20–54dB and a BW of 40kHz for BPSK operation mode. The other mode provides a lower gain of 20–32 dB, but higher BW from 170–370 kHz for OFDM mode of operation with higher data rates. The devices in the IF2 stages are optimally dimensioned to reduce the flicker noise. The differential filtered IF2 signals are fed to an off-chip 5MS/s ADC for digitization and baseband processing.

IV. MEASUREMENTS

The RX was fabricated in a 40nm CMOS process and packaged in a 6×6mm QFN40 package. The RX core blocks, not including the I/O pads, occupy an area of 0.17mm². Fig. 5(a) shows the die photo of the receiver. The power spectral density of the locked and unlocked RO and the phase noise of the free-running RO, XO and ILRO are shown in Fig. 5(b) and 5(c), respectively. The measured PN at 1MHz offset is -128dBc/Hz that minimizes reciprocal mixing of LO PN. An off-chip SMD crystal (AMB8) is used with the on-chip XO. The power dissipation of the XO

TABLE I
PERFORMANCE SUMMARY AND COMPARISON WITH THE STATE-OF-THE-ART NARROWBAND RECEIVERS

Reference	This Work		[3] ISSCC17	[4] RFIC15	[5] ISSCC16
Technology (nm)	40		65	130	180
Architecture	Dual-IF w/ Sub-Harmonic Mixer and ILRO		Low-IF	Low-IF	Low-IF
Carrier Freq. (MHz)	151.82-151.94		850-920	433	160/960
Supply Voltage (V)	0.9		3.3	1.2/0.5	2.2/3.6
Modulation	BPSK	16QAM OFDM	UNB DBPSK	2-FSK	2-GFSK
Data-rate (kb/s)	5	384	0.100	1	2.4/37.5
Active Power (µW)	152		14500	378	57000
Sensitivity (dBm)	-99*	-77*	-136	-102.5	-122
Blocking, 3dBsens.loss (dB)	48 @2MHz/ 63 @10MHz		90 @10MHz	14 @ 0.2MHz	93 @ 2MHz
Oscillator PN (dBc/Hz)	-128 @ 1MHz		-106 @ 1MHz	N/A	-138 @ 2MHz

* Sensitivity is reported at BER of 10^{-3}

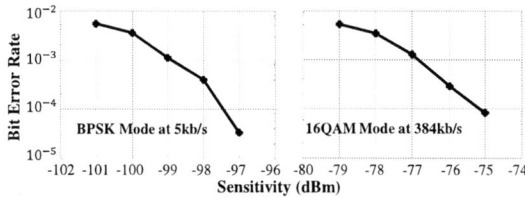

Fig. 7. BER vs. sensitivity for BPSK and 16QAM OFDM modes

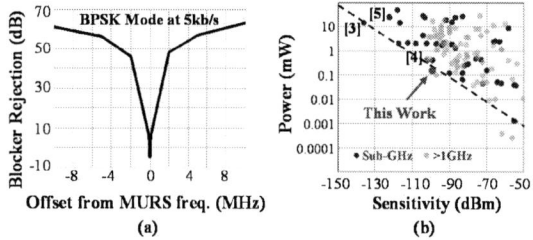

Fig. 8. (a) Blocker rejection for the BPSK mode. (b) Low-power radio survey from 2005 to present. The data of the radio survey is from top conferences and commercial RX chips [6].

and RO (and RO buffers) is 29µW and 21µW, respectively, and the pulse generator and LO dividers consume 17µW total. The maximum locking range for the ILRO is 18MHz and Fig. 6 shows the ILRO lock time of shorter than 135ns, which is measured by enabling the injection signal when the free running RO is operating at 17.6MHz offset from f_{inj}.

The measured sensitivity at BER of 10^{-3} is -99dBm for the BPSK mode at 5kb/s and is -77dBm for the 16-QAM OFDM mode at 384kb/s. Fig. 7 presents the BER performance vs. the input RF power for the two modes of operation. The blocker rejection ratio for out of band blockers is shown in Fig. 8(a). The blocker rejection ratio for 2MHz and 10MHz offsets are 48dB and 63dB, respectively, for 3dB sensitivity loss. The receiver consumes 152µW from a 0.9V power supply, and the IF1 LNA and IF2 filters consume 59µW and 26µW, respectively. The overall gain of the receiver is 79dB at the maximum gain mode, with 25dB in the RF front-end and IF1 stages and 54dB in the IF2 stage. Table I summarizes

the measured performance of the receiver and presents a comparison between this work and other state-of-the-art narrow-band receivers. According to the low power radio survey shown in Fig. 8(b) [6], this receiver sits below the power vs. sensitivity trend-line, demonstrating the best reported power amongst sub-GHz receivers with sensitivity values below -90dBm.

V. CONCLUSION

A low power MURS band receiver with an efficient system-level design and several low power design techniques is proposed for LPWAN applications. The receiver utilizes an edge-combiner mixer first with two step down-conversion architecture, and uses an injection locked ring oscillator for LO. It achieves -99dBm sensitivity with 152µW of power at 5kb/s and a 63dB blocker rejection at 10MHz offset. The ILRO, which is locked to a stable 50MHz crystal reference, achieves a PN of -128dBc/Hz at 1MHz offset. In terms of sensitivity versus power consumption tradeoff, this radio shows the lowest power consumption compared to the state-of-the-art receivers with better than -90dBm sensitivity values.

ACKNOWLEDGMENT

This work is supported by NSF under award number CNS 1422175.

REFERENCES

[1] M. Centenaro et al., "Long-range communications in unlicensed bands: the rising stars in the IoT and smart city scenarios," in *IEEE Wireless Commun.*, vol. 23, no. 5, pp. 60-67, October 2016.

[2] A. Kosari et al., "A MURS Band Digital Quadrature Transmitter with Class-B I/Q Cell Sharing for Long Range IoT Applications," in *IEEE Trans. on Circuits and Systems II:Express Briefs*, Jan. 2018.

[3] D. Lachartre et al., "A TCXO-less 100Hz-minimum-bandwidth transceiver for ultra-narrow-band sub-GHz IoT cellular networks," *IEEE Int. Solid-State Circuits Conf.*, San Francisco, CA, 2017.

[4] N. E. Roberts et al., "A 380µW Rx, 2.6mW Tx 433MHz FSK transceiver with a 102dB link budget and bit-level duty cycling," *IEEE Radio Frequency Integrated Circuits Symposium*, 2015.

[5] N. Kearney et al., "26.4 A 160-to-960MHz ETSI class-1-compliant IoE transceiver with 100dB blocker rejection, 70dB ACR and 800pA standby current," *IEEE Int. Solid-State Circuits Conf.*, 2016.

[6] David D. Wentzloff, "Low Power Radio Survey," [Online]. www.eecs.umich.edu/wics/low_power_radio_survey.html

Author Index

A

Abdo, Ibrahim	7-1 (8072)	77
Abouzeid, Fady	15-5 (8006)	197
Abubakar, Syed Muhammad	19-4 (8110)	267
Agarwal, Amit	13-2 (8055)	149
	19-3 (8024)	263
Alon, Elad	20-5 (8201)	285
Altaf, Muhammad Awais Bin	19-4 (8110)	267
Amakawa, Shuhei	17-2 (8203)	219
Ameri, Ali	17-3 (8177)	223
Anand, Chandani	3-2 (8080)	21
Anders, Mark	13-2 (8055)	149
	19-3 (8024)	263
Angevare, Jan	4-5 (8091)	43
Anwar, Mekhail	17-3 (8177)	223
Araki, Masahiro	20-1 (8032)	269
Au, Wing Lok	5-4 (8025)	59
Aurangozeb, Aurangozeb	16-2 (8192)	205
Autran, Jean-Luc	15-5 (8006)	197
Awano, Hiromitsu	11-5 (8043)	123
	13-3 (8149)	153

B

Ba, Ngoc Le	20-3 (8026)	277
Bachrach, Jonathan	20-5 (8201)	285
Bae, Woorham	20-5 (8201)	285
Baek, Kwang-Hyun	20-3 (8026)	277

Bai, Rui	6-2 (8221)	67
Bailey, Stevo	20-5 (8201)	285
Ban, Koichiro	2-1 (8063)	1
Bang, Jun-Suk	10-1 (8105)	91
Bol, David	11-1 (8189)	107
Bradley, David	2-2 (8095)	5
Bury, Erik	13-4 (8103)	157

C

Cai, Yaxin	6-2 (8221)	67
Chan, Chi-Hang	10-3 (8010)	95
	18-1 (8200)	235
Chang, Che-We	18-3 (8106)	243
Chang, Chi	2-4 (8069)	13
Chang, Eric	20-5 (8201)	285
Chang, Ik-Joon	12-3 (8228)	133
Chang, Kwuang-Han	18-5 (8067)	249
Chang, Liu	6-2 (8221)	67
Chang, Meng-Fan	12-1 (8115)	127
Chang, Tsung-Yung	2-4 (8069)	13
Chang, You-Gang	3-1 (8161)	17
Chen, Hong	5-2 (8064)	51
Chen, Hsiang-Lin	3-3 (8147)	25
Chen, Hsin-Shu	10-6 (8031)	103
	18-6 (8030)	253
Chen, Ji	2-2 (8095)	5
Chen, Ke-Horng	14-3 (8112)	171
Chen, Li-De	19-2 (8180)	261
Chen, Meng-hung	2-2 (8095)	5

Chen, Po-Hung	14-1 (8065)	165
Chen, Wei-Chi	2-4 (8069)	13
Chen, Wei-Chih	19-2 (8180)	261
Chen, Weijia	5-2 (8064)	51
Chen, Wei-Zen	16-4 (8114)	213
Chen, Xuefeng	6-2 (8221)	67
Chen, Yi-Ju	12-1 (8115)	127
Chen, Yi-Lun	14-1 (8065)	165
Chen, Yu-Sin	14-4 (8186)	175
Chen, Zhengyu	19-1 (8100)	257
Chen, Zipeng	7-2 (8157)	79
	7-3 (8184)	83
Cheng, Hao-Chung	14-1 (8065)	165
Cheng, Jhih-Siou	3-1 (8161)	17
Chi, Baoyong	7-2 (8157)	79
	7-3 (8184)	83
Chia, Nicole Shuang Yu	5-4 (8025)	59
Chiang, Patrickyin	6-2 (8221)	67
Chih, Yu-Der	2-4 (8069)	13
Chio, U-Fat	4-1 (8144)	31
Chiu, Mao-Ling	14-2 (8138)	169
Chiu, Po-Wei	15-2 (8053)	187
Cho, Sung-Yong	6-4 (8012)	73
Cho, Thomas B.	21-4 (8087)	299
Cho, Yong-Ho	12-5 (8035)	139
Choi, Hundai	12-5 (8035)	139
Choi, Jaesuk	10-1 (8105)	91
Choi, Jaewon	21-4 (8087)	299
Choi, Jung-Hwan	12-5 (8035)	139
Choi, Minsoo	16-3 (8022)	209
Chonan, Yasunori	11-3 (8060)	115

Choo, Min-Seong	6-4 (8012)	73
Chou, Chih-Hsien	3-1 (8161)	17
Chou, Chung-Cheng	2-4 (8069)	13
Chu, Fang-Chih	14-3 (8112)	171
Chu, Yuan-Hua	4-3 (8141)	37
	14-3 (8112)	171
Chuang, Kai-Hsin	13-4 (8103)	157
Chung, Haeyoung	12-5 (8035)	139
Chung, Steve S.	12-4 (8049)	135
Chung, Yung-Hui	18-3 (8106)	243
Cole, Joseph	20-5 (8201)	285

D

Dang, Le Dinh Trang	12-3 (8228)	133
Degraeve, Robin	13-4 (8103)	157
Deguchi, Jun	5-1 (8128)	47
DeSeta, Franco	20-5 (8201)	285
Doerflein, Matthew	20-5 (8201)	285
Dong, Ruibing	17-2 (8203)	219
Dorigo, Daniel De	11-4 (8182)	119
Dou, Run-Jiang	7-4 (8139)	87
Duwe, Matt	2-2 (8095)	5

E

Eberlein, Matthias	10-5 (8117)	99
Everson, Luke	20-2 (8013)	273

F

Fang, Tong	7-4 (8139)	87
Feng, Peng	17-4 (8133)	227
Flandre, Denis	11-1 (8189)	107
Fleig, David Ingvar	11-4 (8182)	119
Fuhrman, Dan	20-5 (8201)	285
Fujishima, Minoru	17-2 (8203)	219
Fujita, Mayuko	6-1 (8019)	63
Fukuda, Kouichi	6-1 (8019)	63
Fukuyama, Hiroyuki	6-3 (8130)	69
	18-2 (8126)	239
Funabashi, Masami	6-1 (8019)	63

G

Gadde, Vinod	11-5 (8043)	123
Gal-Katziri, Matan	17-5 (8020)	231
Gao, Yuan	11-2 (8136)	111
Gilio, Thierry Di	15-5 (8006)	197
Gopal, Vinodh	19-3 (8024)	263
Govindrajan, Deepar	2-2 (8095)	5
Gowder, Sujatha	2-2 (8095)	5
Gu, Jie	14-5 (8008)	179
	19-1 (8100)	257
Guilford, James	19-3 (8024)	263

H

Hajimiri, Ali	17-5 (8020)	231
Han, Jaeduk	20-5 (8201)	285
Han, Jin-Woo	12-3 (8228)	133
Han, Juyoung	21-4 (8087)	299
Hashimoto, Masanori	5-3 (8213)	55
Hatooka, Kazuya	6-1 (8019)	63
He, Anping	5-2 (8064)	51
He, Shihai	21-1 (8021)	289
Heckendorn, Darin	20-5 (8201)	285
Heng, Chun-Huat	5-4 (8025)	59
	17-1 (8015)	215
Hiao, Yu-Ling	19-2 (8180)	261
Hiraki, Mitsuru	20-1 (8032)	269
Hiratsuka, Akitaka	6-3 (8130)	69
Hong, Soonyoung	10-1 (8105)	91
Hossain, Masum	16-2 (8192)	205
Hsieh, Chih-Cheng	3-3 (8147)	25
	18-5 (8067)	249
Hsieh, Sung-En	3-3 (8147)	25
Hsu, Steven	13-2 (8055)	149
	19-3 (8024)	263
Hsu, Tzu-Hsiang	3-3 (8147)	25
Hu, Chia-Yi	18-3 (8106)	243
Hu, Kai-Yu	14-4 (8186)	175
Hu, Shang	6-2 (8221)	67

Hu, Yao-Sheng	10-6 (8031)	103
	18-6 (8030)	253
Huang, Chao-Jen	14-3 (8112)	171
Huang, Chao-Tsung	19-2 (8180)	261
Huang, Hong-Wen	14-3 (8112)	171
Huang, Kai-Jie	4-3 (8141)	37
Huang, Li-Yu	10-6 (8031)	103
Huang, Xing-Wei	3-1 (8161)	17
Huang, Yu-Ping	16-4 (8114)	213
Hung, Chia-Tse	16-4 (8114)	213
Hung, Kuo-Chih	14-3 (8112)	171
Hwang, Seung-Jun	20-4 (8066)	281
Hwang, Young-Ha	18-4 (8220)	247
Hyun, Seok-Hun	12-5 (8035)	139

I

Igarashi, Mitsuhiko	15-4 (8040)	195
Ikeda, Makoto	11-5 (8043)	123
	13-3 (8149)	153
Ishii, Yuichiro	2-3 (8028)	9
Ishikawa, Jiro	2-3 (8028)	9
Ito, Masao	20-1 (8032)	269
Iyer, Sita	2-2 (8095)	5
Izraelevitz, Adam	20-5 (8201)	285

J

Jang, Doojin	10-1 (8105)	91
Jang, Seong-Jin	12-5 (8035)	139
Je, Minkyu	10-1 (8105)	91
Jeloka, Supreet	15-1 (8073)	183
Jeon, Hyuntak	10-1 (8105)	91
Jeon, Yeseul	10-1 (8105)	91
Jeong, Deog-Kyoon	6-4 (8012)	73
	18-4 (8220)	247
Jia, Tianyu	14-5 (8008)	179
Jia, Wenyan	2-2 (8095)	5
Jiang, Wenning	18-1 (8200)	235
Jing Ming'e	12-4 (8049)	135
Jung, Yoontae	10-1 (8105)	91

K

Kaczer, Ben	13-4 (8103)	157
Kang, Seok-Yong	12-5 (8035)	139
Kang, Yi	9-2 (8236)	XVII
Kato, Shuji	6-1 (8019)	63
Katsuragi, Makihiko	7-1 (8072)	77
Kaul, Himanshu	13-2 (8055)	149
	19-3 (8024)	263
Kawai, Seitarou	7-1 (8072)	77
Khan, Muhammad Rizwan	19-4 (8110)	267
Khan, Usman	4-2 (8096)	33
Kim, Byungsub	16-3 (8022)	209

Kim, Chris H.	15-2 (8053)	187	
	20-2 (8013)	273	
Kim, Jaeha	16-1 (8027)	201	
Kim, Jinsang	12-3 (8228)	133	
Kim, Ju Eon	20-3 (8026)	277	
Kim, Jun-Bae	12-5 (8035)	139	
Kim, Kiho	12-5 (8035)	139	
Kim, Nam Sung	9-1 (8235)	XIII	
Kim, Nam-Seog	21-4 (8087)	299	
Kim, Sang-Woo	4-2 (8096)	33	
Kim, Suhwan	14-3 (8112)	171	
Kim, Tae-Hwan	20-4 (8066)	281	
Kim, Tony Tae-Hyoung	4-2 (8096)	33	
	12-6 (8166)	143	
	20-3 (8026)	277	
Kimura, Kento	7-1 (8072)	77	
Klauk, Hagen	11-4 (8182)	119	
Knight, Graham	15-1 (8073)	183	
Ko, Han-Gon	6-4 (8012)	73	
Koh, Karen Mui Ling	5-4 (8025)	59	
Koh, Seok-Tae	10-1 (8105)	91	
Komatsu, Yoshihide	6-1 (8019)	63	
Komiyama, Takao	11-3 (8060)	115	
Kondo, Hiroyuki	20-1 (8032)	269	
Kosari, Avish	21-5 (8210)	303	
Kotani, Koji	11-3 (8060)	115	
Krishnamurthy, Ram	13-2 (8055)	149	
	19-3 (8024)	263	
Kuhl, Matthias	11-4 (8182)	119	
Kuo, Chun-Chieh	14-3 (8112)	171	
Kuo, Nai-Chung	17-3 (8177)	223	

L

Lai, Chien-An	2-4 (8069)	13
Lallement, Guénolé	15-5 (8006)	197
Lam, Chi-Seng	4-1 (8144)	31
Le, Van Loi	20-3 (8026)	277
Lee, Dong-Hun	12-5 (8035)	139
Lee, Eric	2-2 (8095)	5
Lee, Kwangho	6-4 (8012)	73
Lee, Myungguk	16-3 (8022)	209
Lee, Sangyeop	17-2 (8203)	219
Lee, Taeju	10-1 (8105)	91
Lee, Yu	4-3 (8141)	37
Lentiro, Akalu	20-5 (8201)	285
Li, Jing	12-4 (8049)	135
Li, Lily	2-2 (8095)	5
Li, Shaolan	10-2 (8074)	93
Li, Shenggao	2-2 (8095)	5
Li, Yi-An	17-3 (8177)	223
Li, Zheng	7-1 (8072)	77
Liao, Pei-Chun	14-1 (8065)	165
Liao, Yu-Te	4-3 (8141)	37
Lin, Ching-Ju	14-3 (8112)	171
Lin, Ding-Guo	18-6 (8030)	253
Lin, Hsing-Hung	4-4 (8036)	39
Lin, Jhao-Huei	18-6 (8030)	253
Lin, Kai-Yue	18-6 (8030)	253
Lin, Richard	20-5 (8201)	285
Lin, Tsung-Hsien	14-2 (8138)	169
Lin, Yen-Ting	14-3 (8112)	171

Lin, Ying-Hsi	14-3 (8112)	171		Luo, Ye-Sing	4-4 (8036)	39
Lin, Yu-Cheng	2-4 (8069)	13		Lv, Hangbing	12-4 (8049)	135
Lin, Zheng-Jun	2-4 (8069)	13				
Linten, Dimitri	13-4 (8103)	157				
Liu, Bangan	7-1 (8072)	77		**M**		
Liu, Chia-Hung	4-3 (8141)	37				
Liu, Dang	21-3 (8195)	295		Ma, Jianxu	6-2 (8221)	67
Liu, Hanli	7-1 (8072)	77		Ma, Taikun	7-3 (8184)	83
Liu, Hsiao-Jung	14-3 (8112)	171		Ma, Wen Hau	20-5 (8201)	285
Liu, Jian	7-4 (8139)	87		Ma, Yao-Sheng	14-3 (8112)	171
	17-4 (8133)	227		Mak, Pui-In	21-2 (8093)	293
Liu, Kunyang	13-5 (8011)	161		Maki, Asuka	5-1 (8128)	47
Liu, Leibo	15-3 (8102)	191		Maki, Shotaro	7-1 (8072)	77
Liu, Liyuan	17-4 (8133)	227		Makinwa, Kofi	4-5 (8091)	43
Liu, Li-Yuan	7-4 (8139)	87		Maloberti, Franco	4-1 (8144)	31
Liu, Ming	12-4 (8049)	135		Manoli, Yiannos	11-4 (8182)	119
Liu, Muqing	15-2 (8053)	187		Mao, Howard	20-5 (8201)	285
	20-2 (8013)	273		Markley, Chick	20-5 (8201)	285
Liu, Shen-Iuan	4-4 (8036)	39		Martins, Rui Paulo	4-1 (8144)	31
Liu, Wenjun	12-4 (8049)	135			10-3 (8010)	95
Liu, Yibo	7-2 (8157)	79			18-1 (8200)	235
Lo, Chih-Lun	14-1 (8065)	165			21-2 (8093)	293
Loi, Le Van	12-6 (8166)	143		Mathew, Sanu	13-2 (8055)	149
Lu, Lu	12-6 (8166)	143			19-3 (8024)	263
Lu, Chih-Wen	3-1 (8161)	17		McGrath, Jim	20-5 (8201)	285
Lu, Milton	6-2 (8221)	67		Meng, Hao	21-1 (8021)	289
Lu, Minyi	13-1 (8041)	145		Mikos, Val	5-4 (8025)	59
Lu, Yan	4-1 (8144)	31		Min, Yue	13-5 (8011)	161
Lu, Yu-Ta	19-2 (8180)	261		Mitsunari, Koichi	5-3 (8213)	55
Luo, Hongrui	10-4 (8162)	97		Miura, Naoki	6-3 (8130)	69
Luo, Xueting	7-1 (8072)	77			18-2 (8126)	239

Miura, Tomohiro	2-3 (8028)	9
Miyashita, Daisuke	5-1 (8128)	47
Mizokami, Takuya	20-1 (8032)	269
Montano, Sergio	20-5 (8201)	285
Moon, Joung-Wook	12-5 (8035)	139
Moosavifar, Milad	21-5 (8210)	303
Murmann, Boris	18-1 (8200)	235
Myers, James	15-1 (8073)	183

N

Nagata, Shunya	2-3 (8028)	9
Nakamura, Daisuke	2-3 (8028)	9
Nakata, Kengo	5-1 (8128)	47
Nan, Qi	6-2 (8221)	67
Narevsky, Nathan	20-5 (8201)	285
Nguyen, Nhat	16-2 (8192)	205
Nicholson, Roan	2-2 (8095)	5
Nii, Koji	2-3 (8028)	9
	15-4 (8040)	195
Niknejad, Ali	17-3 (8177)	223
Nikolić, Borivoje	20-5 (8201)	285
Norsworthy, Justin	20-5 (8201)	285
Nosaka, Hideyuki	6-3 (8130)	69
	18-2 (8126)	239

O

Okada, Kenichi	7-1 (8072)	77

Ong, Tong-Chern	2-4 (8069)	13
Onizuka, Kohei	2-1 (8063)	1
Onodera, Hidetoshi	6-3 (8130)	69
Onoe, Seizo	1-2 (8234)	IX
Otani, Sugako	20-1 (8032)	269
Ou, I-Che	4-3 (8141)	37
Ouchi, Yukari	2-3 (8028)	9

P

Panagopoulos, Georgios	10-5 (8117)	99
Pande, Nakul	20-2 (8013)	273
Pang, Jian	7-1 (8072)	77
Park, Il-Won	12-5 (8035)	139
Park, Jun-Eun	18-4 (8220)	247
Park, Kwang-Il	12-5 (8035)	139
Pasquarella, Marcus	2-2 (8095)	5
Peng, Fengxiong	21-1 (8021)	289
Peng, Guiqiang	15-3 (8102)	191
Prabhat, Pranay	15-1 (8073)	183
Pretl, Harald	10-5 (8117)	99
Priyadarshini, Neha	3-2 (8080)	21

Q

Qi, Nan	7-2 (8157)	79
	7-3 (8184)	83
	17-4 (8133)	227
Qian, Yongxue	21-1 (8021)	289

R

Ranjandish, Reza	3-4 (8205)	27
Ravichandran, Krishnan	14-3 (8112)	171
Rawy, Karim	4-2 (8096)	33
Razzaque, Munir	20-5 (8201)	285
Rhee, Woogeun	21-3 (8195)	295
Richards, Brian	20-5 (8201)	285
Rigge, Paul	20-5 (8201)	285
Roche, Philippe	15-5 (8006)	197
Ryu, Sigang	16-1 (8027)	201

S

Saadeh, Wala	19-4 (8110)	267
Sai, Akihide	2-1 (8063)	1
Sanyal, Arindam	10-2 (8074)	93
Sarkar, Mukul	3-2 (8080)	21
Satpathy, Sudhir	13-2 (8055)	149
	19-3 (8024)	263
Schmid, Alexandre	3-4 (8205)	27
Seifaei, Masoud	11-4 (8182)	119
Seo, Dongkyu	12-3 (8228)	133
Shan, Weiwei	13-1 (8041)	145
Shang, Xinchao	13-1 (8041)	145
Sharma, Ruchi	4-2 (8096)	33
Shauck, Steve	20-5 (8201)	285
Shen, Linxiao	10-2 (8074)	93
Shi, Longxing	13-1 (8041)	145

Shibutani, Koji	15-4 (8040)	195
Shinmyo, Akinori	6-1 (8019)	63
Shinohara, Hirofumi	13-5 (8011)	161
Shirane, Atsushi	7-1 (8072)	77
Shoham, Ronen	20-5 (8201)	285
Sin, Sai-Weng	4-1 (8144)	31
Snowden, Mark	20-5 (8201)	285
Sohn, Young-Soo	12-5 (8035)	139
Son, Seuk	16-1 (8027)	201
Song, Haixin	21-3 (8195)	295
Song, Indal	12-5 (8035)	139
Song, Ki-Jae	12-5 (8035)	139
Song, Rui	2-2 (8095)	5
Song, Yoonho	18-4 (8220)	247
Spagna, Fulvio	2-2 (8095)	5
Staszewski, Robert Bogdan	21-2 (8093)	293
Stellfox, Mike	20-5 (8201)	285
Su, Chin-I	2-4 (8069)	13
Sugiyama, Shotaro	13-3 (8149)	153
Sun, Hanfeng	13-5 (8011)	161
Sun, Nan	10-2 (8074)	93
Sun, Quan	10-4 (8162)	97
Suresh, Vikram	13-2 (8055)	149
	19-3 (8024)	263
Suzuki, Kenta	12-2 (8148)	129
Suzuki, Tomoya	5-1 (8128)	47

T

Tachibana, Fumihiko	5-1 (8128)	47

Takano, Kyoya	17-2 (8203)	219
Takazawa, Yoshio	15-4 (8040)	195
Takeuchi, Ken	12-2 (8148)	129
Tan, Dawn May Leng	5-4 (8025)	59
Tanaka, Kenji	6-1 (8019)	63
	6-3 (8130)	69
Tang, Qianying	15-2 (8053)	187
Tang, Xiyuan	10-2 (8074)	93
Tao, Jingcheng	17-1 (8015)	215
Tay, Arthur	5-4 (8025)	59
Tokgoz, Korkut	7-1 (8072)	77
Tong, Luke	2-2 (8095)	5
Toyama, Yosuke	2-1 (8063)	1
Tran, Amanda	2-2 (8095)	5
Tsai, Chien-Hung	14-4 (8186)	175
Tsai, Kun-Ju	4-3 (8141)	37
Tseng, Pei-Ling	2-4 (8069)	13
Tseng, Po-Yu	3-1 (8161)	17
Tsuchiya, Akira	6-3 (8130)	69
Tsukamoto, Yasumasa	15-4 (8040)	195
Tsurumi, Kota	12-2 (8148)	129

U

U, Seng-Pan	18-1 (8200)	235
Uchida, Yuuki	15-4 (8040)	195
Usami, Ren	11-3 (8060)	115

V

Verbauwhede, Ingrid	13-4 (8103)	157
Verdico, Frank	2-2 (8095)	5
Vita, Michael De	2-2 (8095)	5

W

Wang, Angie	20-5 (8201)	285
Wang, Chien-Fan	2-4 (8069)	13
Wang, Chih-Chen	2-4 (8069)	13
Wang, Juncheng	6-2 (8221)	67
Wang, Mingyu	12-4 (8049)	135
Wang, Shufu	7-2 (8157)	79
	7-3 (8184)	83
Wang, Xiaoqing	2-2 (8095)	5
Wang, Xin	6-2 (8221)	67
Wang, Yao	15-3 (8102)	191
Wang, Zhihua	21-3 (8195)	295
Wang, Zhongkai	20-5 (8201)	285
Wei, Qiushi	15-3 (8102)	191
Wei, Shaojun	5-2 (8064)	51
	15-3 (8102)	191
Wen, Kuo-Chih	4-1 (8144)	31
Weng, Chi-Hsiang	2-4 (8069)	13
Wentzloff, David	21-5 (8210)	303
Wigton, Michelle	2-2 (8095)	5
Wu, Hui	5-2 (8064)	51
Wu, Jianxi	7-2 (8157)	79

	7-3 (8184)	83
Wu, Nan-Jian	7-4 (8139)	87
	17-4 (8133)	227
Wu, Siliang	10-2 (8074)	93

X

Xia, Tao	6-2 (8221)	67
Xiang, Yiming	13-1 (8041)	145
Xu, Jiaming	13-1 (8041)	145
Xu, Kai	21-2 (8093)	293
Xu, Linjian	21-1 (8021)	289
Xu, Pengcheng	11-1 (8189)	107
Xu, Xiaoxin	12-4 (8049)	135
Xuan, Jiangao	6-2 (8221)	67
Xue, Cheng-Xin	12-1 (8115)	127
Xue, Xiaoyong	12-4 (8049)	135

Y

Yabuuchi, Makoto	2-3 (8028)	9
Yamaguchi, Hiroyuki	11-3 (8060)	115
Yamauchi, Hiroyuki	12-1 (8115)	127
Yan, Hao	6-2 (8221)	67
Yang, Bo-Hsiang	19-2 (8180)	261
Yang, Jianguo	12-4 (8049)	135
Yang, Jia-Ping	4-3 (8141)	37
Yang, Jincheng	17-4 (8133)	227
Yang, Jun	13-1 (8041)	145

Yang, Sheng	15-1 (8073)	183
Yang, Tzu-Hsien	12-1 (8115)	127
Yang, Tzu-Hsuan	14-2 (8138)	169
Yang, Wen-Hau	14-3 (8112)	171
Yang, Xuan	13-5 (8011)	161
Yen, Shih-Cheng	5-4 (8025)	59
Yeo, Hwanseok	16-1 (8027)	201
Yi, Il-Min	18-2 (8126)	239
Yin, Bozhi	6-2 (8221)	67
Yin, Jun	21-2 (8093)	293
Yin, Shouyi	15-3 (8102)	191
Yokoyama, Yoshisato	2-3 (8028)	9
Yoo, Hye-Sung	12-5 (8035)	139
Yoo, Taegeun	12-6 (8166)	143
	20-3 (8026)	277
Yoon, Hong-Joon	4-2 (8096)	33
Yoon, Hye-Yeon	20-4 (8066)	281
Yoshida, Takashi	17-2 (8203)	219
Yoshioka, Kentaro	2-1 (8063)	1
Yu, Jaehoon	5-3 (8213)	55
Yu, Pei-Shan	14-3 (8112)	171

Z

Zeng, Xiaoyang	12-4 (8049)	135
Zhang, Hong	10-4 (8162)	97
Zhang, Kevin	1-1 (8233)	V
Zhang, Luya	17-3 (8177)	223
Zhang, Ruizhi	10-4 (8162)	97
Zhang, Yuanxi	6-2 (8221)	67

Zhang, Zhao	17-4 (8133)	227
Zhao, Jianming	11-2 (8136)	111
Zhao, Wei-Cheng	12-1 (8115)	127
Zheng, Wei	7-2 (8157)	79
	7-3 (8184)	83
Zhong, Yi	10-2 (8074)	93
Zhou, Keji	12-4 (8049)	135
Zhu, Yan	10-3 (8010)	95
	18-1 (8200)	235
Zhuo, Shenglong	6-2 (8221)	67
Zschieschang, Ute	11-4 (8182)	119

Panel Discussion

Panel Discussion

The Circuits and Systems for Mobile AI

Date	November 6, 2018 (Tuesday)
Time	15:30 - 17:30
Room	Far Eastern Grand Ballroom B, B2F
Organizer	Tsung-Hsien Lin, National Taiwan University, Taiwan
Co-organizer	Zhihua Wang, Tsinghua University, China
Moderator	Meng-Fan (Marvin) Chang, National Tsing Hua University, Taiwan
Panelists	Robert Chen-Hao Chang, National Chung Hsing University, Taiwan
	Jae-Yoon Sim, POSTECH, Korea
	Ken Takeuchi, Chuo University, Japan
	Shigeki Tomishima, Intel Corporation, USA
	Kyomin Sohn, Samsung Electronics, Korea
	Masato Motomura, Hokkaido University, Japan

Abstract:

Recently, many new technology, circuits and systems are proposed to realize AI on silicon. Analog circuits look promising for design of the neuromorphic chips but digital implementations are still the main stream technology. Moreover, PiM with non-volatile memory is researched as the next technology for the AI realization. Especially, for the mobile AI intelligence, we need ultra-low power circuits and systems.

- Do you think Digital AI solutions will win over analog? Why?
- Do you prefer Analog circuits and systems for AI chip?
- Which type of memories are good to realize AI solutions?
- Will PiM be the mainstream AI system?

**Organizer
Tsung-Hsien Lin**

Tsung-Hsien Lin received the B.S. degree in electronics engineering from National Chiao-Tung University, Taiwan. He received his MS and Ph.D. degrees in electrical engineering from UCLA, in 1997 and 2001, respectively. In 2000, he joined Broadcom Corporation, Irvine, CA, where he was a Senior Staff Scientist, during which time he involved in wireless transceiver developments. In 2004, he joined the Department of Electrical Engineering, National Taiwan University, Taiwan, where he is currently a Professor. His research interests are the design of wireless transceivers, clock and frequency generation systems, delta-sigma modulators, and transducer interface circuits.

Dr. Lin was the recipient of the Best Presentation Award for his paper presented at the 2007 IEEE VLSI-DAT Symposium, and the co-recipient of the Best Paper Award at the same Symposium in 2015. He served on the IEEE Asian Solid-State Circuit Conference (A-SSCC) Technical Program Committee (TPC) from 2005 to 2011 and was the TPC Vice-Chair for 2011 A-SSCC. He was a Guest Editor for IEEE Journal of Solid-State Circuits (JSSC) in 2012 and was an Associate Editor for the same journal from 2013 to 2015. He served on the ISSCC International Technical Program Committee from 2010 to 2016, and was the Far-East Regional Committee Chair in 2016 ISSCC. He was the TPC Chair of 2017 A-SSCC.

**Co-organizer
Zhihua Wang**

Zhihua Wang (M'99-SM'04-F'17) received the B.S., M.S., and Ph.D. degrees in Electronic Engineering in 1983, 1985 and 1990, respectively, from Tsinghua University, Beijing, China, where he has served as full professor and Deputy Director of the Institute of Microelectronics since 1997 and 2000. He was a visiting scholar at CMU (1992-1993) and KU Leuven (1993-1994), and was a visiting professor at HKUST (2014.9-2015.3). His current research mainly focuses on CMOS RFIC and biomedical applications, involving RFID, PLL, low-power wireless transceivers, and smart clinic equipment combined with leading edge RFIC and digital image processing techniques. He has co-authored 12 books/chapters, over 183 (480) papers in international journals (conferences), over 244 (29) papers in Chinese journals (conferences) and holds 123 Chinese and 8 US patents.

Prof. Wang has served as the chairman of IEEE SSCS Beijing Chapter (1999-2009), an AdCom Member of the IEEE SSCS (2016-2019), a technology program committee member of the IEEE ISSCC (2005-2011), a steering committee member of the IEEE A-SSCC (2005-), the technical program chair for A-SSCC 2013, a guest editor for IEEE JSSC Special Issues (2006.12, 2009.12 and 2014.11), an associate editor of IEEE Trans on CAS-I, II and IEEE Trans on BioCAS, and other administrative/expert committee positions in China's national science and technology projects.

**Moderator
Meng-Fan (Marvin)
Chang**

Dr. Meng-Fan (Marvin) Chang is a full Professor in National Tsing Hua University (NTHU), Taiwan. Before joining NTHU in 2006, he had spent more than 10 years working on memory circuit designs in industry, including Mentor Graphics (New Jersey, US), TSMC (Taiwan), and the IPLib (Taiwan).

Since 2010, Dr. Chang has authored more than 40+ top conference papers (including 14 ISSCC, 15 VLSI Symposia, 9 IEDM, and 5 DAC) as well as 40+ IEEE journal papers and 40+ US patents. He has been serving on technical program committees for ISSCC, IEDM (Executive committee, Chair of MT sub-committee), DAC (sub-committee chair), A-SSCC, ISCAS, and numerous conferences. He is the recipient of several prestigious national-level awards in Taiwan, including the Outstanding Research Award of MOST-Taiwan, Outstanding Electrical Engineering Professor Award, Ta-You Wu Memorial Award, Academia Sinica Junior Research Investigators Award. He currently is the Program Director of the Microelectronics Program at the Ministry of Science and Technology (MOST) in Taiwan.

His research interests include circuit design for volatile/nonvolatile/3D memory, nonvolatile and spintronics logics, circuit-device-interactions in non-CMOS devices, memristor circuits, computing-in-memory and neuromorphic circuits for deep learning and artificial intelligent (AI) chips.

**Panelist
Robert Chen-Hao
Chang**

Robert Chen-Hao Chang received B.S. and M.S. degrees in electrical engineering from National Taiwan University, Taipei, Taiwan, and the Ph.D. from the University of Southern California. He is currently a Distinguished Professor at National Chung Hsing University, Taichung, Taiwan. He served as the Chairman of the Electrical Engineering Department from 2006 to 2008, the Deputy Director General of the National Chip Implementation Center in Hsinchu, Taiwan from 2011/3 to 2014/1, and the Dean of College of Science of Technology, National Chi Nan University, Nantou, Taiwan from 2014 to 2017. From 2018, he becomes the Program Director of Semiconductor Manufacturing and Design for AI Edge Project, Ministry of Science and Technology, Taiwan. His research interests include SoC & signal processing systems and mixed-signal IC design. He served as IEEE CASS Taipei Chapter Chair from 2011 to 2012, the IEEE CASS Distinguished Lecturer of in 2013 and 2014, and IEEE A-SSCC TPC DCS Sub-Committee Chair from 2015.

**Panelist
Jae-Yoon Sim**

Jae-Yoon Sim received the B.S., M.S., and Ph.D. degrees in electrical engineering from Pohang University of Science and Technology (POSTECH), Korea, in 1993, 1995, and 1999, respectively. From 1999 to 2005, he was a Senior Engineer in the Samsung Electronics, Korea. From 2003 to 2005, he was a Postdoctoral Researcher at the University of Southern California, USA. From 2011 to 2012, he was a Visiting Scholar at the University of Michigan, Ann Arbor, MI, USA. In 2005, he joined POSTECH, where he is currently a Professor. His research interests include clock generation, serial and parallel links, data converters, neuromorphic circuits and sensor interface circuits. He has served in the Technical Program Committees of the IEEE International Solid-State Circuits Conference, Symposium on VLSI Circuits, and Asian Solid-State Circuits Conference. He is a Distinguished Professor nominated by Korea Institute of Science and Technology. He is an IEEE Distinguished Lecturer from 2018. He was a recipient of the Takuo Sugano Award and Special Author-Recognition Award at ISSCC 2001 and 2013, respectively.

**Panelist
Ken Takeuchi**

Ken Takeuchi is currently a Professor at the Department of Electrical, Electronic, and Communication Engineering of Chuo University. He is now working on the database storage system for big-data application, VLSI circuit design, signal processing and device such as the emerging non-volatile memories, 3D-integrated SSDs, low-power 3D-LSI circuits and ultra low-voltage SRAMs. Before joining Chuo University, he was an Associate Professor at the University of Tokyo from 2007 till 2012. From 1993 till 2007, he has been leading Toshiba's NAND flash memory circuit design team and commercialized six world's highest density flash memory products. He holds 228 patents worldwide. He won the Takuo Sugano Award for Outstanding Paper at ISSCC 2007. He is currently serving as the technical program chair of Symposium on VLSI Circuits. He has also served on the program committee member of International Solid-State Circuits Conference (ISSCC), Custom Integrated Circuits Conference (CICC), Asian Solid-State Circuits Conference (A-SSCC), International Memory Workshop (IMW), International Conference on Solid State Devices and Materials (SSDM) and Non-Volatile Memory Technology Symposium (NVMTS).

**Panelist
Shigeki Tomishima**

Shigeki Tomishima is a senior researcher at Intel Labs/Memory Architecture Lab, USA since 2014 to work on the advanced memory architecture research including DRAMs and other emerging memories for future computing systems.

He received his B.S. and M.S. in Solid-State Physics and Ph.D. degree in Electric Engineering from Osaka University, Japan in 1988, 1990 and 2002, respectively.

From 1990 to 2002, he worked at Mitsubishi Electric Corporation, Itami Hyogo Japan for advanced DRAM circuit and chip design.

From 2003 to 2011, he worked at Micron Technology, Boise Idaho USA for advanced DRAM array circuit design.

From 2011 to 2014, he worked Intel Corporation/Logic Technology Development, Hillsboro Oregon USA for the embedded DRAM project.

Dr. Tomishima has published 15 international conference papers, 7 invited talks and 9 journal papers, and holds 121 issued U.S. patents and more foreign patents. He received Best Paper Reviewer award from IEEE CAS in 2016 and serves as TPC on A-SSCC, TPC Chair on VLSI-DAT, and the external paper reviewer on JSSC, TVLSI, MICRO and ISCAS.

**Panelist
Kyomin Sohn**

Kyomin Sohn received the B.S. and M.S. degrees in Electrical Engineering in 1994 and 1996, respectively, from Yonsei University, Seoul. From 1996 to 2003, he was with Samsung Electronics, Korea, involved in SRAM Design Team. He designed various kinds of high-speed SRAM devices.

He received the Ph.D. degree in EECS in 2007 from KAIST, Daejeon, Korea. He rejoined Samsung Electronics in 2007, where he has been involved in DRAM Design Team. He is a Master (Technical VP) in Samsung and he is responsible for development of HBM DRAM and Future Technology.

His interests include the next generation 3D-DRAM, robust memory design, and processing-in-memory for AI applications. Since 2012, he has currently served as a Technical Program Committee member of Symposium on VLSI Circuits.

**Panelist
Masato Motomura**

Masato Motomura received B.S. and M.S. in 1985 and 1987, respectively, and Ph.D. of Electrical Engineering in 1996, all from Kyoto University. He joined NEC central research laboratories in 1987, where he worked on various hardware architectures including string search engines, multi-threaded on-chip parallel processors, embedded DRAM-FPGA hybrid systems, memory-based processors, and reconfigurable systems. From 2001 to 2008, at NEC Electronics, he led research and business development of dynamically reconfigurable processor (DRP) that he invented. He was also a visiting researcher at MIT Laboratory for Computer Science from 1991 to 1992. Now being a professor at Hokkaido University since 2011, his current research interests include reconfigurable and parallel architectures for deep neural networks, machine learning, annealing machines, and intelligent computing in general. He won the IEEE JSSC Annual Best Paper Award in 1992, IPSJ Annual Best Paper Award in 1999, and IEICE Achievement Award in 2011, ISSCC Silkroad Award as the last author in 2018, respectively. He is a member of IEEE, IEICE, IPSJ, and EAJ.

Committees

Steering Committee

Chair	Tadahiro Kuroda (Keio University, Japan)
Members	Tzi-Dar Chiueh (National Taiwan University, Taiwan)
	Nicky Lu (Etron Technology, Inc., Taiwan)
	Makoto Ikeda (The University of Tokyo, Japan)
	Yoshio Masubuchi (Toshiba Memory Corporation, Japan)
	Deog-Kyoon Jeong (Seoul National University, Korea)
	Stefan Rusu (TSMC North America, USA)
	Yong-Hyun Jun (Samsung Electronics, Korea)
	Toru Shimizu (Toyo University, Japan)
	Lawrence Loh (Mediatek, Taiwan)
	Jack Sun (TSMC, Taiwan)
	Yong Ping Xu (National University of Singapore, Singapore)
	Hoi-Jun Yoo (KAIST, Korea)
	Zhihua Wang (Tsinghua University, China)
Advisor	Takayasu Sakurai (The University of Tokyo, Japan)
	Chorng-kuang Wang (National Taiwan University, Taiwan)
Liaison	Anantha Chandrakasan (Massachusetts Institute of Technology, USA)
	Willy Sansen (KU Leuven, Belgium)

Organizing Committee

Conference Chair	Jack Sun (TSMC, Taiwan)
Organizing Committee Chair	Liang-Hung Lu (National Taiwan University, Taiwan)
Secretary	Chia-Hsiang Yang (National Taiwan University, Taiwan)
Treasurer	Tsung-Te Liu (National Taiwan University, Taiwan)
Publicity	Chih-Cheng Hsieh (National Tsing Hua University, Taiwan)
Publication	Po-Hung Chen (National Chiao Tung University, Taiwan)
Local Arrangement	Soon-Jyh Chang (National Cheng Kung University, Taiwan)
TPC Liaison	Chien-Nan Kuo (National Chiao Tung University, Taiwan)

Technical Program Committee (TPC)

TPC Chair	Hong June Park (POSTECH)
TPC Co-Chair	Mototsugu Hamada (Keio University)
TPC Vice Chair	Jae-Yoon Sim (POSTECH)
TPC Vice Co-Chair	Jun Deguchi (Toshiba Memory Corporation)
ACS Sub-Committee Chair	Hidetoshi Onodera (Kyoto University)
DC Sub-Committee Chair	Seng-Pan (Ben) U (University of Macau)
DCS Sub-Committee Chair	Robert Chen-Hao Chang (National Chung Hsing University)
SOC Sub-Committee Chair	Kazutami Arimoto (Okayama Prefectural University)
RF Sub-Committee Chair	Satoshi Tanaka (Murata Manufacturing Co., Ltd.)
WLN Sub-Committee Chair	Chulwoo Kim (Korea University)
ETA Sub-Committee Chair	Woogeun Rhee (Tsinghua University)
MEM Sub-Committee Chair	Junghwan Choi (Samsung Electronics)
Industry Program Chair	Stefan Rusu (TSMC North America)
Invited Program Chair	Hoi-Jun Yoo (KAIST)
Educational Program Co-Chair	Hoi-Jun Yoo (KAIST)
Educational Program Co-Chair	Byeong-Gyu Nam (Chungnam National University)
Student Design Contest Co-Chair	Baoyong Chi (Tsinghua University)
Student Design Contest Co-Chair	Jung-Hoon Chun (Sungkyunkwan University)
FPGA Ad-Hoc Co-Chair	Shigeki Tomishima (Intel Corporation)

Analog Circuits and Systems (ACS)

Chair	Hidetoshi Onodera (Kyoto University)
Members	Hao Yu (Southern University of Science and Technology)
	Sai-Weng Sin (University of Macau)
	Li Geng (Xi'an Jiaotong University)
	Tetsuya Hirose (Kobe University)
	Po-Chiun Huang (National Tsing Hua University)
	Takeshi Ueno (Toshiba Corporation)
	Hyun-Sik Kim (Dankook University)
	Po-Hung Chen (National Chiao Tung University)
	Ji-Yong Um (Hannam University)
	Tzu-Ming Wang (Global Mixed-mode Technology)

Data Converters (DC)

Chair	Seng-Pan (Ben) U (University of Macau)
Members	Seung-Tak Ryu (KAIST)
	Jong-Woo Lee (Samsung Electronics)
	Yan (Julia) Zhu (University of Macau)
	Liyuan Liu (Chinese Academy of Sciences)
	Yung-Yu Lin (Mediatek)
	Tsung-Heng Tsai (National Chung Cheng University)
	Jintae Kim (Konkuk University)
	Sanroku Tsukamoto (Fujitsu Laboratories Ltd.)
	Chih-Cheng Hsieh (National Tsing Hua University)
	Zule Xu (University of Tokyo)

Digital Circuits and Systems (DCS)

Chair	Robert Chen-Hao Chang (National Chung Hsing University)
Members	Tay-Jyi Lin (National Chung Cheng University)
	Byungsub Kim (POSTECH)
	Leibo Liu (Tsinghua University)
	Keiichi Kushida (Toshiba Memory Corporation)
	Jun Zhou (University of Electronic Science and Technology of China)
	Xiaoyang Zeng (Fudan University)
	Mototsugu Hamada (Keio University)
	Yoonmyung Lee (Sungkyunkwan University)
	Massimo Alioto (National University of Singapore)
	Koyo Nitta (NTT)
	Shouyi Yin (Tsinghua University)

SoC and Signal Processing (SOC)

Chair	Kazutami Arimoto (Okayama Prefectural University)
Members	Byeong-Gyu Nam (Chungnam National University)
	Satoshi Shigematsu (NTT)
	Yong Hei (Chinese Academy of Sciences)
	Chun Zhang (Tsinghua University)
	Pei-Yun Tsai (National Central University)
	Tsung-Te Liu (National Taiwan University)
	Shigeki Tomishima (Intel Corporation)
	Ji-Hoon Kim (Ewha Womans University)
	Daisuke Mizoguchi (Renesas)
	Hsi-Pin Ma (National Tsing Hua University)

Radio Frequency (RF)

Chair	Satoshi Tanaka (Murata Manufacturing Co., Ltd.)
Members	Minjae Lee (Gwangju Institute of Science and Technology)
	Huei Wang (National Taiwan University)
	Minoru Fujishima (Hiroshima University)
	Chien-Nan Kuo (National Chiao Tung University)
	Baoyong Chi (Tsinghua University)
	Davide Guermandi (IMEC)
	Tae Wook Kim (Yonsei University)
	Kai Kang (University of Electronic Science and Technology of China)
	Taizo Yamawaki (Hitachi)
	Bo Zhao (Zhejiang University)

Wireline (WLN)

Chair	Chulwoo Kim (Korea University)
Members	Wei-Zen Chen (National Chiao Tung University)
	Jun Terada (NTT)
	Jung-Hoon Chun (Sungkyunkwan University)
	Yasufumi Sakai (Fujitsu Laboratories Ltd.)
	Ching-Yuan Yang (National Chung Hsing University)
	Ziqiang Wang (Tsinghua University)
	Patrick Yin Chiang (Fudan University)
	Young-Chan Jang (Kumoh National Institute of Technology)
	Koichi Yamaguchi (Intel Corporation)
	Masum Hossain (University of Alberta)

Emerging Technologies and Applications (ETA)

Chair	Woogeun Rhee (Tsinghua University)
Members	Jerald Yoo (National University of Singapore)
	Jun Deguchi (Toshiba Memory Corporation)
	Minkyu Je (KAIST)
	Youngcheol Chae (Yonsei University)
	Nan Sun (University of Texas at Austin)
	Shuenn-Yuh Lee (National Cheng Kung University)
	Zhichao Tan (ADI)
	Chao Wang (Singapore University of Technology and Design)
	Shiro Dosho (Tokyo Institute of Technology)
	Noriyuki Miura (Kobe University)
	Ping-Hsuan Hsieh (National Tsing Hua University)

Memory (MEM)

Chair	Junghwan Choi (Samsung Electronics)
Members	Hung Jen Liao (TSMC)
	Atsushi Kawasumi (Toshiba Memory Corporation)
	Kazutaka Miyano (Micron)
	Chun Shiah (Etron)
	Meng-Fan Chang (National Tsing Hua University)
	Ken Takeuchi (Chuo University)
	Hwang Hur (SK Hynix)
	Tony T. Kim (Nanyang Technological University)
	Ik Joon Chang (Kyunghee University)
	Jun YANG (Southeast University)
	Shyh-Shyuan Sheu (Yuan Ze University)

Industry Program (IP)

Chair	Stefan Rusu (TSMC North America)
Members	Toru Shimizu (Toyo University)
	Kazuko Nishimura (Panasonic Corporation)
	Daisaburo Takashima (Toshiba Memory Corporation)
	ShaoJun Wei (Tsinghua University)
	Ting Wu (Norel Systems)
	Fan Yung Ma (Infineon)
	Surhud Khare (Intel Technology India Pvt. Ltd.)
	Masaitsu Nakajima (Socionext)

IEEE
445 Hoes Lane
Piscataway, NJ 08854-4141

ISBN 978-1-5386-6414-8